Mathematical and Computational Methods in Photonics and Phononics

Mathematical
Surveys
and
Monographs

Volume 235

Mathematical and Computational Methods in Photonics and Phononics

Habib Ammari
Brian Fitzpatrick
Hyeonbae Kang
Matias Ruiz
Sanghyeon Yu
Hai Zhang

AMERICAN MATHEMATICAL SOCIETY
Providence, Rhode Island

2010 *Mathematics Subject Classification*. Primary 47A55, 47A75, 31A10, 34A55, 35R30, 35B34, 45Q05, 30E25.

For additional information and updates on this book, visit
www.ams.org/bookpages/surv-235

Library of Congress Cataloging-in-Publication Data

Names: Ammari, Habib, author.
Title: Mathematical and computational methods in photonics and phononics / Habib Ammari [and five others].
Description: Providence, Rhode Island : American Mathematical Society, [2018] | Series: Mathematical surveys and monographs ; volume 235 | Includes bibliographical references and index.
Identifiers: LCCN 2018024210 | ISBN 9781470448004 (alk. paper)
Subjects: LCSH: Photonics–Mathematics. | Sound–Mathematics. | AMS: Operator theory – General theory of linear operators – Eigenvalue problems. msc | Potential theory – Two-dimensional theory – Integral representations, integral operators, integral equations methods. msc | Ordinary differential equations – General theory – Inverse problems. msc | Partial differential equations – Miscellaneous topics – Inverse problems. msc | Partial differential equations – Qualitative properties of solutions – Resonances. msc | Integral equations – Inverse problems – Inverse problems. msc | Functions of a complex variable – Miscellaneous topics of analysis in the complex domain – Boundary value problems. msc
Classification: LCC TA1520 .M375 2018 | DDC 621.36/50151–dc23
LC record available at https://lccn.loc.gov/2018024210

∞ The paper used in this book is acid-free and falls within the guidelines
established to ensure permanence and durability.
Visit the AMS home page at `https://www.ams.org/`

10 9 8 7 6 5 4 3 2 1 23 22 21 20 19 18

Contents

Introduction

The aim of this book is to give a self-contained presentation of recent mathematical and computational advances in photonics and phononics. The fields of photonics and phononics encompass the fundamental science of light and sound propagation and interactions in complex structures, and its technological applications.

The recent advances in nanoscience present great challenges for the applied and computational mathematics community. In nanophotonics, the aim is to control, manipulate, reshape, guide, and focus electromagnetic waves at nanometer length scales, beyond the resolution limit. In particular, one wants to break the resolution limit by reducing the focal spot and confine light to length scales that are significantly smaller than half the wavelength. Nanostructures also present exciting opportunities for designing acoustic material properties for specific applications.

Interactions between the fields of photonics and phononics and mathematics have led to the emergence of a multitude of new and unique solutions in which today's conventional technologies are approaching their limits in terms of speed, capacity and accuracy. Light and sound can be used for detection and measurement in a fast, sensitive and accurate manner, and thus photonics and phononics are uniquely positioned to revolutionize healthcare. Light-and sound-based technologies can be used effectively for the very early detection of diseases, with non-invasive imaging techniques or point-of-care applications. They are also instrumental in the analysis of processes at the molecular level, giving a greater understanding of the origin of diseases, and hence allowing prevention along with new treatments.

The main objective in this book is to report on the use of sophisticated mathematics in diffractive optics, plasmonics, super-resolution, photonic and phononic crystals, Helmholtz resonators and Minnaert bubbles, and metamaterials for electromagnetic and elastic invisibility and cloaking. The book merges highly nontrivial multi-mathematics in order to make a breakthrough in the field of mathematical modelling, imaging, and design of optical and acoustic nanodevices and nanostructures capable of light or sound enhancement, and of the focusing and guiding of light or sound at a subwavelength scale. We demonstrate the power of layer potential techniques in solving challenging problems in photonics and phononics, when they are combined with asymptotic analysis and the elegant theory of Gohberg and Sigal on meromorphic operator-valued functions.

In this book we shall consider both analytical and computational matters in photonics and phononics. The issues we consider lead to the investigation of fundamental problems in various branches of mathematics. These include asymptotic analysis, spectral analysis, mathematical imaging, optimal design, stochastic modelling, and analysis of wave propagation phenomena. Deriving mathematical foundations, and new and efficient computational frameworks and tools in photonics and phononics, requires a deep understanding of the different scales in the wave propagation problem, an accurate mathematical modelling of the nanodevices, and fine analysis of complex wave propagation phenomena.

The book is divided into five parts. In Part 1, we introduce our main mathematical and computational tools for analyzing photonic and phononic problems. Chapter 1 is devoted to the theory of characteristic values of operator-valued functions developed by Gohberg and Sigal. Chapter 2 offers a comprehensive treatment of layer potential techniques. In Chapter 3, we provide a general and unified integral equation approach based on the Gohberg-Sigal theory for deriving asymptotic expansions of eigenvalue perturbations. The main idea is to reduce the eigenvalue problems to the study of characteristic values of systems of certain integral operators.

In Part 2, we consider scattering problems by periodic structures. Chapter 4 is devoted to the analysis of scattering from diffractive gratings. In Chapters 5 and 6, we investigate the bandgap structure of the spectrum for waves in a high contrast, two-component periodic medium. We provide tools for studying photonic and phononic crystals based on layer potentials.

In Part 3, we consider direct and inverse scattering from sub-wavelength resonators. We show that plasmon resonant nanoparticles provide a possible means of achieving super-resolved imaging. We study the resonant property of high-contrast particles for different particle geometries and environments, and use them to achieve super-resolution. In Chapter 7, we introduce a layer potential formulation for plasmonic resonances and derive asymptotic formulas for the plasmonic resonances and the scattered fields in terms of the particle size. We derive an effective medium theory for suspensions of plasmonic nanoparticles. We prove that at or near plasmonic resonances, the effective medium possesses a high contrast. We also consider field enhancement for a system of plasmonic spheres and investigate plasmonic resonances for domains with corners. In Chapter 8, we discuss direct imaging of small particles. We make use of small-volume expansions for the scattered fields to design efficient non-iterative imaging algorithms. We show that the resolution limit, defined as the minimum distance required between two small particles to distinguish between them, is of the order of half the operating wavelength. In Chapter 9, we introduce super-resolution techniques in order to differentiate point sources or small particles which are located less than half the wavelength apart. We present the mathematical foundation for realizing super-resolution by using high-contrast or resonant media. The key idea is to design the surrounding medium so that the imaginary part of the corresponding Green's function is tailored in such a way as to allow us to resolve sub-wavelength details.

In Part 4, we consider the cloaking of waves and metasurfaces. In Chapter 10, we focus on interior cloaking and describe effective near-cloaking structures. In Chapter 11, the focus is placed on exterior cloaking. We link the exterior cloaking issue to the existence of anomalous localized resonance, which is associated to an infinite dimensional kernel of a non-elliptic operator. In Chapter 12, we describe the anomalous reflection properties of a thin layer of periodic plasmonic nanoparticles.

In Part 5, we consider sub-wavelength acoustic resonators. We show that Helmholtz resonators and bubbles have the potential to be basic building blocks not only for sub-wavelength acoustic imaging but also for acoustic metamaterials. In Chapter 13, we prove that the spectrum of the Helmholtz resonator essentially coincides with the spectrum of the Laplacian with Neumann boundary condition in the closed cavity, but there is an additional resonant frequency which is a subwavelength resonance, which is the key to super-resolution for acoustic waves in systems of Helmholtz resonators. In Chapter 14, we consider acoustic wave propagation in bubbly media. A distinctive feature of bubbles in fluid is the high contrast between the air density inside and outside of the bubble. This results in a quasi-static acoustic resonance, called the Minnaert resonance. Using Minnaert resonances, we prove that super-focusing properties, bandgap openings at the subwavelength scale, and double-negative refractive index phenomena can be obtained for systems of bubbles.

The bibliography provides a list of relevant references. It is by no means comprehensive. However, it should provide the reader with some useful guidance in searching for further details on the main ideas and approaches discussed in this book.

The material in this book is taught as a graduate course in applied mathematics at ETH Zürich. The Matlab computer codes used in both the book and the course can be downloaded at the official supplementary materials website for the book which is located at `www.ams.org/bookpages/surv-235`.

Some of the material in this book is from our wonderful collaborations with Toufic Abboud, Gang Bao, Giulio Ciraolo, Josselin Garnier, David Gontier, Vincent Jugnon, Hyundae Lee, Mikyoung Lim, Pierre Millien, Graeme Milton, Jean-Claude Nédélec, Fadil Santosa, Michael Vogelius, and Darko Volkov. We feel indebted to all of them.

Part 1

Mathematical and Computational Tools

CHAPTER 1

Generalized Argument Principle and Rouché's Theorem

1.1. Introduction

In this chapter we review the results of Gohberg and Sigal in [**266**] concerning the generalization to operator-valued functions of two classical results in complex analysis, the *argument principle* and *Rouché's theorem*. An efficient and reliable method, referred to as Muller's method, for finding a zero of a function defined on the complex plane is presented. This numerical method can be used for computing poles of integral operators, in particular for the computation of resonant cavities, band gap structures, and plasmonic resonances in nanoparticles. The results described in this chapter will be applied to the mathematical theory of cavities, plasmonic nanoparticles, and photonic and phononic crystals.

1.2. Argument Principle and Rouché's Theorem

To state the argument principle, we first observe that if f is holomorphic and has a zero of order n at z_0, we can write $f(z) = (z-z_0)^n g(z)$, where g is holomorphic and nowhere vanishing in a neighborhood of z_0, and therefore

$$\frac{f'(z)}{f(z)} = \frac{n}{z - z_0} + \frac{g'(z)}{g(z)}.$$

Then the function f'/f has a simple pole with residue n at z_0. A similar fact also holds if f has a pole of order n at z_0, that is, if $f(z) = (z - z_0)^{-n} h(z)$, where h is holomorphic and nowhere vanishing in a neighborhood of z_0. Then

$$\frac{f'(z)}{f(z)} = -\frac{n}{z - z_0} + \frac{h'(z)}{h(z)}.$$

Therefore, if f is meromorphic, the function f'/f will have simple poles at the zeros and poles of f, and the residue is simply the order of the zero of f or the negative of the order of the pole of f.

The argument principle results from an application of the residue formula. It asserts the following.

THEOREM 1.1 (Argument principle). *Let $V \subset \mathbb{C}$ be a bounded domain with smooth boundary ∂V positively oriented and let $f(z)$ be a meromorphic function in a neighborhood of $\overline{V}$. Let P and N be the numbers of poles and zeros of f in V, counted with their orders. If f has no poles or zeros on ∂V, then*

$$\frac{1}{2\pi\sqrt{-1}} \int_{\partial V} \frac{f'(z)}{f(z)}\, dz = N - P. \tag{1.1}$$

Rouché's theorem is a consequence of the argument principle [**490**]. It is in some sense a continuity statement. It says that a holomorphic function can be perturbed slightly without changing the number of zeros. It reads as follows.

THEOREM 1.2 (Rouché's theorem). *With V as above, suppose that $f(z)$ and $g(z)$ are holomorphic in a neighborhood of $\overline{V}$. If $|f(z)| > |g(z)|$ for all $z \in \partial V$, then $f+g$ and f have the same number of zeros in V.*

In order to explain the main results of Gohberg and Sigal in [**266**], we begin with the finite-dimensional case which was first considered by Keldyš in [**329**]; see also [**395**]. We proceed to generalize formula (1.1) in this case as follows. If a matrix-valued function $A(z)$ is holomorphic in a neighborhood of $\overline{V}$ and is invertible in $\overline{V}$ except possibly at a point $z_0 \in V$, then by Gaussian eliminations we can write

$$A(z) = E(z)D(z)F(z) \quad \text{in } V, \tag{1.2}$$

where $E(z), F(z)$ are holomorphic and invertible in V and $D(z)$ is given by

$$D(z) = \begin{pmatrix} (z-z_0)^{k_1} & & 0 \\ & \ddots & \\ 0 & & (z-z_0)^{k_n} \end{pmatrix}.$$

Moreover, the powers $k_1, k_2, \ldots, k_n$ are uniquely determined up to a permutation.

Let tr denote the trace. By virtue of the factorization (1.2), it is easy to produce the following identity:

$$\begin{aligned}
&\frac{1}{2\pi\sqrt{-1}} \operatorname{tr} \int_{\partial V} A(z)^{-1} \frac{d}{dz} A(z)\, dz \\
&\quad = \frac{1}{2\pi\sqrt{-1}} \operatorname{tr} \int_{\partial V} \left(E(z)^{-1} \frac{d}{dz} E(z) + D(z)^{-1} \frac{d}{dz} D(z) + F(z)^{-1} \frac{d}{dz} F(z) \right) dz \\
&\quad = \frac{1}{2\pi\sqrt{-1}} \operatorname{tr} \int_{\partial V} D(z)^{-1} \frac{d}{dz} D(z)\, dz \\
&\quad = \sum_{j=1}^{n} k_j,
\end{aligned}$$

which generalizes (1.1).

In the following sections, we will extend the above identity as well as the factorization (1.2) to infinite-dimensional spaces under some natural conditions.

1.3. Definitions and Preliminaries

In this section we introduce the notation which will be used in the text, gather a few definitions, and present some basic results, which are useful for the statement of the generalized Rouché theorem.

1.3.1. Compact Operators. If $\mathcal{B}$ and $\mathcal{B}'$ are two Banach spaces, we denote by $\mathcal{L}(\mathcal{B}, \mathcal{B}')$ the space of bounded linear operators from $\mathcal{B}$ into $\mathcal{B}'$. An operator $K \in \mathcal{L}(\mathcal{B}, \mathcal{B}')$ is said to be compact provided K takes any bounded subset of $\mathcal{B}$ to a relatively compact subset of $\mathcal{B}'$, that is, a set with compact closure.

The operator K is said to be of finite rank if $\operatorname{Im}(K)$, the range of K, is finite-dimensional. Clearly every operator of finite rank is compact.

The next result is called the Fredholm alternative. See, for example, [**352**].

PROPOSITION 1.3 (Fredholm alternative). *Let K be a compact operator on the Banach space $\mathcal{B}$. For $\lambda \in \mathbb{C}, \lambda \neq 0$, $\lambda I - K$ is surjective if and only if it is injective.*

1.3.2. Fredholm Operators. An operator $A \in \mathcal{L}(\mathcal{B}, \mathcal{B}')$ is said to be *Fredholm* provided the subspace $\operatorname{Ker} A$ is finite-dimensional and the subspace $\operatorname{Im} A$ is closed in $\mathcal{B}'$ and of finite codimension. Let $\operatorname{Fred}(\mathcal{B}, \mathcal{B}')$ denote the collection of all Fredholm operators from $\mathcal{B}$ into $\mathcal{B}'$.

Next, we define the index of $A \in \operatorname{Fred}(\mathcal{B}, \mathcal{B}')$ to be

$$\operatorname{ind} A = \dim \operatorname{Ker} A - \operatorname{codim} \operatorname{Im} A.$$

The following proposition shows that the index is stable under compact perturbations [**352**].

PROPOSITION 1.4. *If $A : \mathcal{B} \to \mathcal{B}'$ is Fredholm and $K : \mathcal{B} \to \mathcal{B}'$ is compact, then their sum $A + K$ is Fredholm, and*

$$\operatorname{ind}(A + K) = \operatorname{ind} A.$$

Proposition 1.4 is a consequence of the following fundamental result about the index of Fredholm operators.

PROPOSITION 1.5. *The mapping $A \mapsto \operatorname{ind} A$ is continuous in $\operatorname{Fred}(\mathcal{B}, \mathcal{B}')$; i.e., ind is constant on each connected component of $\operatorname{Fred}(\mathcal{B}, \mathcal{B}')$.*

1.3.3. Characteristic Value and Multiplicity. We now introduce the notions of characteristic values and root functions of holomorphic operator-valued functions, with which the readers might not be familiar. We refer, for instance, to the book by Markus [**383**] for the details.

Let $\mathfrak{U}(z_0)$ be the set of all operator-valued functions with values in $\mathcal{L}(\mathcal{B}, \mathcal{B}')$ which are holomorphic in some neighborhood of z_0, except possibly at z_0.

The point z_0 is called a *characteristic value* of $A \in \mathfrak{U}(z_0)$ if there exists a vector-valued function $\phi(z)$ with values in $\mathcal{B}$ such that

(i) $\phi(z)$ is holomorphic at z_0 and $\phi(z_0) \neq 0$,
(ii) $A(z)\phi(z)$ is holomorphic at z_0 and vanishes at this point.

Here, $\phi(z)$ is called a *root* function of $A(z)$ associated with the characteristic value z_0. The vector $\phi_0 = \phi(z_0)$ is called an eigenvector. The closure of the linear set of eigenvectors corresponding to z_0 is denoted by $\operatorname{Ker} A(z_0)$.

Suppose that z_0 is a characteristic value of $A \in \mathfrak{U}(z_0)$ and ϕ is an associated root function. Then there exists a number $m(\phi) \geq 1$ and a vector-valued function ψ with values in $\mathcal{B}'$, holomorphic at z_0, such that

$$A(z)\phi(z) = (z - z_0)^{m(\phi)}\psi(z), \quad \psi(z_0) \neq 0.$$

The number $m(\phi)$ is called the *multiplicity* of the root function $\phi(z)$.

For $\phi_0 \in \operatorname{Ker} A(z_0)$, we define the rank of ϕ_0, denoted by $\operatorname{rank}(\phi_0)$, to be the maximum of the multiplicities of all root functions $\phi(z)$ with $\phi(z_0) = \phi_0$.

Suppose that $n = \dim \operatorname{Ker} A(z_0) < +\infty$ and that the ranks of all vectors in $\operatorname{Ker} A(z_0)$ are finite. A system of eigenvectors ϕ_0^j, $j = 1, \ldots, n$, is called a *canonical system of eigenvectors* of A associated to z_0 if their ranks possess the following property: for $j = 1, \ldots, n$, $\operatorname{rank}(\phi_0^j)$ is the maximum of the ranks of all eigenvectors in the direct complement in $\operatorname{Ker} A(z_0)$ of the linear span of the vectors $\phi_0^1, \ldots, \phi_0^{j-1}$.

We call

$$N(A(z_0)) := \sum_{j=1}^{n} \mathrm{rank}(\phi_0^j)$$

the *null multiplicity* of the characteristic value z_0 of A. If z_0 is not a characteristic value of A, we put $N(A(z_0)) = 0$.

Suppose that $A(z)^{-1}$ exists and is holomorphic in some neighborhood of z_0, except possibly at z_0. Then the number

$$M(A(z_0)) = N(A(z_0)) - N(A(z_0)^{-1}) \tag{1.3}$$

is called the *multiplicity* of z_0. If z_0 is a characteristic value and not a pole of $A(z)$, then $M(A(z_0)) = N(A(z_0))$ while $M(A(z_0)) = -N(A(z_0)^{-1})$ if z_0 is a pole and not a characteristic value of $A(z)$.

1.3.4. Normal Points. Suppose that z_0 is a pole of $A(z)$ and the Laurent series expansion of $A(z)$ at z_0 is given by

$$A(z) = \sum_{j \geq -s} (z - z_0)^j A_j. \tag{1.4}$$

If in (1.4) the operators A_{-j}, $j = 1, \ldots, s$, have finite-dimensional ranges, then $A(z)$ is called *finitely meromorphic* at z_0.

The operator-valued function $A(z)$ is said to be of *Fredholm type* (of index zero) at the point z_0 if the operator A_0 in (1.4) is Fredholm (of index zero).

If $A(z)$ is holomorphic and invertible at z_0, then z_0 is called a *regular point* of $A(z)$. The point z_0 is called a *normal point* of $A(z)$ if $A(z)$ is finitely meromorphic, of Fredholm type at z_0, and regular in a neighborhood of z_0 except at z_0 itself.

1.3.5. Trace. Let A be a finite-rank operator acting from $\mathcal{B}$ into itself. Then there exists a finite-dimensional invariant subspace $\mathcal{C}$ of A such that A annihilates some direct complement of $\mathcal{C}$ in $\mathcal{B}$. We define the trace of A to be that of $A|_{\mathcal{C}}$, which is given in the usual way. It is desirable to recall some results about the trace operator.

PROPOSITION 1.6. *The following results hold:*

(i) $\mathrm{tr}\, A$ *is independent of the choice of* $\mathcal{C}$, *so that it is well-defined.*
(ii) tr *is linear.*
(iii) *If* B *is a finite-rank operator from* $\mathcal{B}$ *to itself, then*

$$\mathrm{tr}\, AB = \mathrm{tr}\, BA.$$

(iv) *If* M *is a finite-rank operator from* $\mathcal{B} \times \mathcal{B}'$ *to itself, given by*

$$M = \begin{pmatrix} A & B \\ C & D \end{pmatrix},$$

then $\mathrm{tr}\, M = \mathrm{tr}\, A + \mathrm{tr}\, D$.

Recall that if an operator-valued function $C(z)$ is finitely meromorphic in the neighborhood $\overline{V}$ of z_0, which contains no poles of $C(z)$ except possibly z_0, then $\int_{\partial V} C(z)\, dz$ is a finite-rank operator. The following identity will also be used frequently.

PROPOSITION 1.7. *Let $A(z)$ and $B(z)$ be two operator-valued functions finitely meromorphic in the neighborhood $\overline{V}$ of z_0, which contains no poles of $A(z)$ and $B(z)$ other than z_0. Then we have*

$$\operatorname{tr} \int_{\partial V} A(z)B(z)\, dz = \operatorname{tr} \int_{\partial V} B(z)A(z)\, dz. \tag{1.5}$$

1.4. Factorization of Operators

We say that $A \in \mathfrak{U}(z_0)$ admits a factorization at z_0 if $A(z)$ can be written as

$$A(z) = E(z)D(z)F(z), \tag{1.6}$$

where $E(z), F(z)$ are regular at z_0 and

$$D(z) = P_0 + \sum_{j=1}^{n} (z - z_0)^{k_j} P_j. \tag{1.7}$$

Here, k_j are integers, P_j's are mutually disjoint projections, $P_1, \ldots, P_n$ are rank-one operators, and $I - \sum_{j=0}^{n} P_j$ is a finite-rank operator.

THEOREM 1.8. *$A \in \mathfrak{U}(z_0)$ admits a factorization at z_0 if and only if A is finitely meromorphic and of Fredholm type of index zero at z_0.*

PROOF. Suppose that A is finitely meromorphic and of Fredholm type of index zero at z_0. We shall construct E, F, and D such that (1.6) holds. Write the Laurent series expansion of $A(z)$ as follows:

$$A(z) = \sum_{j=-\nu}^{+\infty} (z - z_0)^j A_j$$

in some neighborhood U of z_0. Since $\operatorname{ind} A_0 = 0$, $B_0 := A_0 + K_0$ is invertible for some finite-rank operator K_0 by the Fredholm alternative. Consequently,

$$B(z) := K_0 + \sum_{j=0}^{+\infty} (z - z_0)^j A_j$$

is invertible in some neighborhood U_1 of z_0 and

$$A(z) = C(z) + B(z) = B(z)[I + B(z)^{-1}C(z)], \tag{1.8}$$

where

$$C(z) = \sum_{j=-\nu}^{-1} (z - z_0)^j A_j - K_0.$$

Since $K(z) := B(z)^{-1}C(z)$ is finitely meromorphic, we can write $K(z)$ in the form

$$K(z) = \sum_{j=1}^{\nu} (z - z_0)^{-j} K_j + T_1(z),$$

where K_j, $j = 1, \ldots, \nu$, are of finite-rank and T_1 is holomorphic.

Since the operators A_j and K_j are of finite-rank, there exists a subspace $\mathfrak{N}$ of $\mathcal{B}$ of finite codimension such that

$$\begin{cases} \mathfrak{N} \subset \operatorname{Ker} A_j, \ j = -\nu, \dots, -1, \\ \mathfrak{N} \subset \operatorname{Ker} K_j, \ j = 0, \dots, \nu, \\ \mathfrak{N} \bigcap \operatorname{Im} K_j = \{0\}, \ j = 1, \dots, \nu. \end{cases}$$

Let $\mathfrak{C}$ be a direct finite-dimensional complement of $\mathfrak{N}$ in $\mathcal{B}$ and let P be the projection onto $\mathfrak{C}$ satisfying $P(I - P) = 0$. Set $P_0 := I - P$. We have

$$\begin{aligned} I + K(z) &= I + PK(z)P + P_0K(z)P \\ &= I + PK(z)P + P_0T_1(z)P, \end{aligned}$$

and therefore,

$$I + K(z) = (I + PK(z)P)(I + P_0T_1(z)P). \tag{1.9}$$

Since $P(I + K(z))P$ can be viewed as an operator from $\mathfrak{C}$ into itself and $\mathfrak{C}$ is finite-dimensional, it follows from Gaussian elimination that

$$P(I + K(z))P = E_1(z)D_1(z)F_1(z),$$

where $D_1(z)$ is diagonal and $E_1(z)$ and $F_1(z)$ are holomorphic and invertible. In view of (1.9), this implies that

$$\begin{aligned} A(z) &= B(z)(P_0 + P(I + K(z))P)(I + P_0T_1(z)P) \\ &= B(z)(P_0 + E_1(z)D_1(z)F_1(z))(I + P_0T_1(z)P) \\ &= B(z)(P_0 + E_1(z))(P_0 + D_1(z))(P_0 + F_1(z))(I + P_0T_1(z)P). \end{aligned}$$

Here $I + P_0T_1(z)P$ is holomorphic and invertible with inverse $I - P_0T_1(z)P$. Thus, taking

$$E(z) := B(z)(P_0 + E_1(z)), \qquad F(z) := (P_0 + F_1(z))(I + P_0T_1(z)P)$$

yields the desired factorization for A since $E(z)$ and $F(z)$, given by the above formulas, are holomorphic and invertible at z_0.

The converse result, that $A(z) = E(z)D(z)F(z)$ with $E(z), F(z)$ regular at z_0 and $D(z)$ satisfying (1.7) is finitely meromorphic and of Fredholm type of index zero at z_0, is easy. □

COROLLARY 1.9. *$A(z)$ is normal at z_0 if and only if $A(z)$ admits a factorization at z_0 such that $I = \sum_{j=0}^{n} P_j$ in (1.7). Moreover, we have*

$$M(A(z_0)) = k_1 + \cdots + k_n$$

for $k_1, \dots, k_n$, given by (1.7).

Suppose that $A(z)$ admits a factorization at z_0. Then we have (1.6) and (1.7). Since

$$A(z)^{-1} = F(z)^{-1}D(z)^{-1}E(z)^{-1}$$

with

$$D(z)^{-1} = P_0 + \sum_{j=1}^{n} (z - z_0)^{-k_j} P_j,$$

the following result holds.

COROLLARY 1.10. *Every normal point of $A(z)$ is a normal point of $A(z)^{-1}$.*

1.5. Main Results of the Gohberg and Sigal Theory

We now tackle our main goal of this chapter, which is to generalize the argument principle and Rouché's theorem to operator-valued functions.

1.5.1. Argument Principle. Let V be a simply connected bounded domain with rectifiable boundary ∂V. An operator-valued function $A(z)$ which is finitely meromorphic and of Fredholm type in V and continuous on ∂V is called *normal* with respect to ∂V if the operator $A(z)$ is invertible for $z \in \overline{V}$, except for a finite number of points of V which are normal points of $A(z)$.

LEMMA 1.11. *An operator-valued function $A(z)$ is normal with respect to ∂V if it is finitely meromorphic and of Fredholm type in V, continuous on ∂V, and invertible for all $z \in \partial V$.*

PROOF. To prove that A is normal with respect to ∂V, it suffices to prove that $A(z)$ is invertible except at a finite number of points in V. Choose a connected open set U with $\overline{U} \subset V$ so that $A(z)$ is invertible in $V \setminus U$. Then, for each $\xi \in U$, there exists a neighborhood U_ξ of ξ in which the factorization (1.6) holds. In U_ξ, the kernel of $A(z)$ has a constant dimension except at ξ. Since $\overline{U}$ is compact, we can find a finite covering of $\overline{U}$, *i.e.*,

$$\overline{U} \subset U_{\xi_1} \cup \cdots \cup U_{\xi_k},$$

for some points $\xi_1, \ldots, \xi_k \in U$. Therefore, $\dim \operatorname{Ker} A(z)$ is constant in $V \backslash \{\xi_1, \ldots, \xi_k\}$, and so $A(z)$ is invertible in $\overline{V} \setminus \{\xi_1, \ldots, \xi_k\}$. $\square$

Now, if $A(z)$ is normal with respect to the contour ∂V and z_i, $i = 1, \ldots, \sigma$, are all its characteristic values and poles lying in V, the full multiplicity $\mathcal{M}(A; \partial V)$ of $A(z)$ for $z \in V$ is the number of characteristic values of $A(z)$ for $z \in V$, counted with their multiplicities, minus the number of poles of $A(z)$ in V, counted with their multiplicities, namely,

$$\mathcal{M}(A; \partial V) = \sum_{i=1}^{\sigma} M(A(z_i)), \tag{1.10}$$

where $M(A(z_i))$ is defined in (1.3).

THEOREM 1.12 (Generalized argument principle). *Suppose that the operator-valued function $A(z)$ is normal with respect to ∂V. Then we have*

$$\frac{1}{2\pi\sqrt{-1}} \operatorname{tr} \int_{\partial V} A(z)^{-1} \frac{d}{dz} A(z) dz = \mathcal{M}(A; \partial V). \tag{1.11}$$

PROOF. Let z_j, $j = 1, \ldots, \sigma$, denote all the characteristic values and all the poles of A lying in V. The key of the proof lies in using the factorization (1.6) in each of the neighborhoods of the points z_j. We have

$$\frac{1}{2\pi\sqrt{-1}} \operatorname{tr} \int_{\partial V} A(z)^{-1} \frac{d}{dz} A(z) dz = \sum_{j=1}^{\sigma} \frac{1}{2\pi\sqrt{-1}} \operatorname{tr} \int_{\partial V_j} A(z)^{-1} \frac{d}{dz} A(z) dz, \tag{1.12}$$

where, for each j, V_j is a neighborhood of z_j. Moreover, in each V_j, the following factorization of A holds:

$$A(z) = E^{(j)}(z)D^{(j)}(z)F^{(j)}(z), \qquad D^{(j)}(z) = P_0^{(j)} + \sum_{i=1}^{n_j}(z - z_j)^{k_{ij}} P_i^{(j)},$$

where the powers $k_{1j}, k_{2j}, \ldots, k_{n_j j}$ are uniquely determined up to a permutation.

As for the matrix-valued case at the beginning of this chapter, it is readily verified that

$$\begin{aligned} \frac{1}{2\pi\sqrt{-1}} \operatorname{tr} \int_{\partial V_j} A(z)^{-1} \frac{d}{dz} A(z) dz &= \frac{1}{2\pi\sqrt{-1}} \operatorname{tr} \int_{\partial V_j} (D^{(j)}(z))^{-1} \frac{d}{dz} D^{(j)}(z) dz \\ &= \sum_{i=1}^{n_j} k_{ij} = M(A(z_j)). \end{aligned}$$

Now, (1.11) follows by using (1.12). □

The following is an immediate consequence of Lemma 1.11 and identities (1.5) and (1.11).

COROLLARY 1.13. *If the operator-valued functions $A(z)$ and $B(z)$ are normal with respect to ∂V, then $C(z) := A(z)B(z)$ is also normal with respect to ∂V, and*

$$\mathcal{M}(C; \partial V) = \mathcal{M}(A; \partial V) + \mathcal{M}(B; \partial V).$$

The following general form of the argument principle will be useful. It can be proven by the same argument as the one in the proof of Theorem 1.12.

THEOREM 1.14. *Suppose that $A(z)$ is an operator-valued function which is normal with respect to ∂V. Let $f(z)$ be a scalar function which is holomorphic in V and continuous in $\overline{V}$. Then*

$$\frac{1}{2\pi\sqrt{-1}} \operatorname{tr} \int_{\partial V} f(z) A(z)^{-1} \frac{d}{dz} A(z) dz = \sum_{j=1}^{\sigma} M(A(z_j)) f(z_j),$$

where z_j, $j = 1, \ldots, \sigma$, are all the points in V which are either poles or characteristic values of $A(z)$.

1.5.2. Generalization of Rouché's Theorem. A generalization of Rouché's theorem to operator-valued functions is stated below.

THEOREM 1.15 (Generalized Rouché's theorem). *Let $A(z)$ be an operator-valued function which is normal with respect to ∂V. If an operator-valued function $S(z)$ which is finitely meromorphic in V and continuous on ∂V satisfies the condition*

$$\|A(z)^{-1}S(z)\|_{\mathcal{L}(\mathcal{B},\mathcal{B})} < 1, \quad z \in \partial V,$$

then $A + S$ is also normal with respect to ∂V and

$$\mathcal{M}(A; \partial V) = \mathcal{M}(A + S; \partial V).$$

PROOF. Let $C(z) := A(z)^{-1}S(z)$. By Corollary 1.10, $C(z)$ is finitely meromorphic in V. Suppose that $z_1, z_2, \ldots, z_n$, are all of the poles of $C(z)$ in V and that $C(z)$ has the following Laurent series expansion in some neighborhood of each z_j:

$$C(z) = \sum_{k=-\nu_j}^{+\infty} (z - z_j)^k C_k^{(j)}.$$

Let $\mathfrak{N}$ be the intersection of the kernels $\operatorname{Ker} C_k^{(j)}$ for $j = 1, \ldots, n$ and $k = 1, \ldots, \nu_j$. Then, $\dim \mathcal{B}/\mathfrak{N} < +\infty$ and the restriction $C(z)|_{\mathfrak{N}}$ of $C(z)$ to $\mathfrak{N}$ is holomorphic in V.

Let $q := \max_{z\in\partial V} \|C(z)\|$, which by assumption is less than 1. Since

$$\Delta_z \|C(z)|_{\mathfrak{N}}\|^2 = 4\|\frac{\partial}{\partial z} C(z)|_{\mathfrak{N}}\|^2,$$

i.e., $\|C(z)|_{\mathfrak{N}}\|$ is subharmonic in V, we have from the maximum principle

$$\max_{z\in V} \|C(z)|_{\mathfrak{N}}\| \leq q.$$

It then follows that

$$\|(I + C(z))x\| \geq (1 - q)\|x\|, \qquad x \in \mathfrak{N},\ z \in V.$$

This implies that $(I + C(z))|_{\mathfrak{N}}$ has a closed range and $\operatorname{Ker}(I + C(z))|_{\mathfrak{N}} = 0$. Therefore, $I + C(z)$ has a closed range and a kernel of finite dimension for $z \in V \setminus \{z_1, \ldots, z_n\}$. By a slight extension of Proposition 1.5 [**494**], $\mathcal{I}(z)$ defined by

$$\mathcal{I}(z) = \dim \operatorname{Ker}(I + C(z)) - \operatorname{codim}\ \operatorname{Im}(I + C(z))$$

is continuous for $z \in \overline{V} \setminus \{z_1, \ldots, z_n\}$. Thus,

$$\operatorname{ind}(I + C(z)) = 0 \quad \text{for } z \in \overline{V} \setminus \{z_1, \ldots, z_n\}.$$

Moreover, since the Laurent series expansion of $(I + C(z))|_{\mathfrak{N}}$ in a neighborhood of z_j is given by

$$(I + C(z))|_{\mathfrak{N}} = I|_{\mathfrak{N}} + \sum_{k=0}^{+\infty} (z - z_j)^k C_k^{(j)}|_{\mathfrak{N}}, \tag{1.13}$$

it follows that $(I + C_0^{(j)})|_{\mathfrak{N}}$ has a closed range and a trivial kernel. Using Propositions 1.4 and 1.5, we have

$$\operatorname{ind}(I + C_0^{(j)}) = \operatorname{ind}(I + \sum_{k=0}^{+\infty} (z - z_j)^k C_k^{(j)}) = \operatorname{ind}(I + C(z)) = 0.$$

Thus, $(I + C_0^{(j)})$ is Fredholm. By Lemma 1.11, we deduce that $I + C(z)$ is normal with respect to ∂V.

Now we claim that $\mathcal{M}(I + C(z); \partial V) = 0$. To see this, we note that $I + tC(z)$ is normal with respect to ∂V for $0 \leq t \leq 1$. Let

$$f(t) := \mathcal{M}(I + tC(z); \partial V).$$

Then $f(t)$ attains integers as its values. On the other hand, since

$$f(t) = \frac{1}{2\pi\sqrt{-1}} \operatorname{tr} \int_{\partial V} t(I + tC(z))^{-1} \frac{d}{dz} C(z)\, dz \tag{1.14}$$

and $(I + tC(z))^{-1}$ is continuous in $[0, 1]$ in operator norm uniformly in $z \in \partial V$, $f(t)$ is continuous in $[0, 1]$. Thus, $f(1) = f(0) = 0$.

Finally, with the help of Corollary 1.13, we can conclude that the theorem holds. □

1.5.3. Generalization of Steinberg's Theorem. Steinberg's theorem asserts that if $K(z)$ is a compact operator on a Banach space, which is holomorphic in z for z in a region V in the complex plane, then $(I+K(z))^{-1}$ is meromorphic in V. See [**491**]. A generalization of this theorem to finitely meromorphic operators was first given by Gohberg and Sigal in [**266**]. The following important result holds.

THEOREM 1.16 (Generalized Steinberg's theorem). *Suppose that $A(z)$ is an operator-valued function which is finitely meromorphic and of Fredholm type in the domain V. If the operator $A(z)$ is invertible at one point of V, then $A(z)$ has a bounded inverse for all $z \in V$, except possibly for certain isolated points.*

1.6. Muller's Method

Muller's method is an efficient and reliable interpolation method for finding a zero of a function defined on the complex plane and, in particular, for determining a simple or multiple root of a polynomial. Compared to Newton's method, it has the advantage that the derivatives of the function need not be computed. Moreover, the convergence rate of Muller's method is slightly smaller than the convergence rate of Newton's method [**492**].

Muller's method can be viewed as a generalization of the secant method. The secant method is based on taking two points on the graph of a function f, and then finding an approximate root by determining the root of a linear function that passes through these two points. Muller's method, on the other hand, is based on taking three points on the graph of a function f, and then finding an approximate root by determining the root of a quadratic function that passes through these three points.

Denote by $Q_f(z)$ the quadratic interpolating polynomial for the function f that passes through the points $(z_0, f(z_0))$, $(z_1, f(z_1))$, and $(z_2, f(z_2))$, *i.e.*,

$$Q_f(z) = a(z-z_2)^2 + b(z-z_2) + c,$$

with

$$\begin{aligned} f(z_0) &= a(z_0-z_2)^2 + b(z_0-z_2) + c, \\ f(z_1) &= a(z_1-z_2)^2 + b(z_1-z_2) + c, \\ f(z_2) &= a(z_2-z_2)^2 + b(z_2-z_2) + c. \end{aligned}$$

Solving for a, b, and c we obtain

$$\begin{aligned} a &= \frac{(z_1-z_2)(f(z_0)-f(z_2)) - (z_0-z_2)(f(z_1)-f(z_2))}{(z_0-z_1)(z_0-z_2)(z_1-z_2)}, \\ b &= \frac{(z_0-z_2)^2(f(z_1)-f(z_2)) - (z_1-z_2)^2(f(z_0)-f(z_2))}{(z_0-z_1)(z_0-z_2)(z_1-z_2)}, \\ c &= f(z_2). \end{aligned}$$

To determine the root $z = z_3$ of $Q(z)$, let $\tilde{z} = z_3 - z_2$, and then

$$Q_f(z_3) = a\tilde{z}^2 + b\tilde{z} + c = 0$$

can be solved using the quadratic formula. For numerical stability we use the following version of the quadratic formula:

$$\tilde{z} = \frac{-2c}{b \pm \sqrt{b^2-4ac}},$$

where the sign of the square root is chosen so as to maximize the absolute value of the denominator. This means that the root z_3 of $Q_f(z)$, which is the next approximation of an actual root of $f(z)$, is given by

$$z_3 = z_2 + \frac{2c}{b \pm \sqrt{b^2 - 4ac}}.$$

Once z_3 has been found we set $z_i = z_{1+1}$, for $i = 0, 1, 2$. We then repeat this procedure, which results in a sequence of approximate roots, until specific termination criteria are reached; we terminate the procedure when $f(z_3) < \tau_f$ and $|z_3 - z_2| < \tau_z$, where τ_f and τ_z are some given tolerances for the value of f at the root z_3, and the distance between the roots on successive iterations, respectively.

It can be shown that the errors $\delta_i = (z_i - \xi)$ of Muller's method in the proximity of a single zero ξ of $f(z) = 0$ satisfy

$$\delta_{i+1} = \delta_i \delta_{i-1} \delta_{i-2} \bigg(-\frac{f^{(3)}(\xi)}{6f'(\xi)} + O(\delta) \bigg),$$

where $\delta = \max(|\delta_i|, |\delta_{i-1}|, |\delta_{i-2}|)$. It can also be shown that Muller's method is at least of order the largest root q of the equation $\zeta^3 - \zeta^2 - \zeta - 1 = 0$, which is approximately 1.84. A Matlab demonstration of Muller's Method can be downloaded at `www.ams.org/bookpages/surv-235`.

As an illustration, we consider the complex valued function

$$f(z) = \sin(z) + 5 + \sqrt{-1},$$

whose exact roots are given by $z_\alpha = 2\pi n - \sin^{-1}(5 + \sqrt{-1})$ or $z_\beta = 2\pi n + \pi + \sin^{-1}(5 + \sqrt{-1})$ for $n \in \mathbb{Z}$. We can obtain the roots of this function numerically using the code referenced above. For instance, if we take $n = 0$ then the exact root (to eight decimal places) is $z_\alpha = -1.36960125 - 2.31322094\sqrt{-1}$. Choosing appropriate initial guesses, say, $z_0 = 0.5$, $z_1 = 1 + 3\sqrt{-1}$, and $z_2 = -1 - 2\sqrt{-1}$, our numerical result for this root is also $-1.36960125 - 2.31322094\sqrt{-1}$.

1.7. Concluding Remarks

In this chapter, we have reviewed the main results in the theory of Gohberg and Sigal on meromorphic operator-valued functions. These results concern the generalization of the argument principle and the Rouché theorem to meromorphic operator-valued functions. Some of these results have been extended to very general operator-valued functions in [**142**, **378**] and with other types of spectrum than isolated eigenvalues in [**382**]. The theory of Gohberg and Sigal will be applied to perturbation theory of eigenvalues in Chapter 3. Other interesting applications include the investigation of scattering resonances and scattering poles [**157**, **277**]. Finally, we have described Muller's method for finding complex roots of scalar equations.

CHAPTER 2

Layer Potentials

2.1. Introduction

The mathematical and numerical framework for analyzing photonic and phononic problems described in this book relies on layer potential techniques.

In this chapter we lay the groundwork by reviewing a number of basic facts and preliminary results regarding the layer potentials associated with the Laplacian, the Helmholtz equation, the Maxwell equations, and the operator of elasticity. The most important results in this chapter are on the one hand the characterization of eigenvalues as characteristic values of layer potentials and on the other hand, the spectral properties of Neumann–Poincaré operators. Due to the vectorial aspect of the Maxwell equations and the equations of elasticity, the analysis for electromagnetism and elasticity is more delicate than for the scalar case. We also note that when dealing with exterior problems for the Helmholtz equation, Maxwell equations or time-harmonic elasticity, one should introduce a radiation condition to select the physical solution to the problem. Together with reciprocity properties satisfied by fundamental solutions to the acoustic, electromagnetic, or elastic wave propagation problems, radiation conditions yield Helmholtz-Kirchhoff identities, which play a key role in the analysis of resolution in wave imaging. We state the optical theorem, which establishes a fundamental relation between the imaginary part of the scattering amplitude and the total (or extinction) cross-section. We also investigate quasi-periodic Green's functions and associated layer potentials for the Helmholtz equation and the Lamé system. We provide spectral and spatial representations of Green's functions in periodic domains and describe analytical techniques for transforming them from slowly convergent representations into forms more suitable for computation. In particular, we discuss in some detail Ewald's method, which consists in splitting quasi-periodic Green's function into a spectral part and a spatial part to achieve exponential convergence.

2.2. Sobolev Spaces

For ease of notation we will sometimes use ∂^2 to denote the Hessian.

Let Ω be a smooth domain. We define the Hilbert space $H^1(\Omega)$ by

$$H^1(\Omega) = \Big\{ u \in L^2(\Omega) : \nabla u \in L^2(\Omega) \Big\},$$

where ∇u is interpreted as a distribution and $L^2(\Omega)$ is defined in the usual way, with

$$||u||_{L^2(\Omega)} = \left(\int_\Omega |u|^2 \right)^{1/2}.$$

The space $H^1(\Omega)$ is equipped with the norm

$$||u||_{H^1(\Omega)} = \left(\int_\Omega |u|^2 + \int_\Omega |\nabla u|^2 \right)^{1/2}.$$

If Ω is bounded, another Banach space $H^1_0(\Omega)$ arises by taking the closure of $\mathcal{C}^\infty_0(\Omega)$, the set of infinitely differentiable functions with compact support in Ω, in $H^1(\Omega)$. We will also need the subspace $H^1_{\text{loc}}(\mathbb{R}^d \setminus \overline{\Omega})$ of functions $u \in L^2_{\text{loc}}(\mathbb{R}^d \setminus \overline{\Omega})$, the set of locally square summable functions in $\mathbb{R}^d \setminus \overline{\Omega}$, such that

$$hu \in H^1(\mathbb{R}^d \setminus \overline{\Omega}), \forall\, h \in \mathcal{C}^\infty_0(\mathbb{R}^d \setminus \overline{\Omega}).$$

Furthermore, we define $H^2(\Omega)$ as the space of functions $u \in H^1(\Omega)$ such that $\partial^2 u \in L^2(\Omega)$ and the space $H^{3/2}(\Omega)$ as the interpolation space $[H^1(\Omega), H^2(\Omega)]_{1/2}$ (see, for example, the book by Bergh and Löfström [**149**]). We also define the Banach space $W^{1,\infty}(\Omega)$ by

$$W^{1,\infty}(\Omega) = \Big\{ u \in L^\infty(\Omega) : \nabla u \in L^\infty(\Omega) \Big\}, \tag{2.1}$$

where $L^\infty(\Omega)$ is defined in the usual way, with

$$||u||_{L^\infty(\Omega)} = \inf \Big\{ C \geq 0 : |u(x)| \leq C \quad \text{a.e. } x \in \Omega \Big\}.$$

The trace theorem states that the trace operator $u \mapsto u|_{\partial\Omega}$ is a bounded linear surjective operator from $H^1(\Omega)$ into $H^{1/2}(\partial\Omega)$. Here, $f \in H^{1/2}(\partial\Omega)$ if and only if $f \in L^2(\partial\Omega)$ and

$$\int_{\partial\Omega} \int_{\partial\Omega} \frac{|f(x) - f(y)|^2}{|x-y|^d} \, d\sigma(x)\, d\sigma(y) < +\infty.$$

We set $H^{-1/2}(\partial\Omega) = (H^{1/2}(\partial\Omega))^*$ and let $\langle\ ,\ \rangle_{1/2,-1/2}$ denote the duality pair between these dual spaces.

Let $T_1, \ldots, T_{d-1}$ be an orthonormal basis for the tangent plane to $\partial\Omega$ at x and let

$$\partial/\partial T = \sum_{p=1}^{d-1} (\partial/\partial T_p)\, T_p$$

denote the tangential derivative on $\partial\Omega$. We say that $f \in H^1(\partial\Omega)$ if $f \in L^2(\partial\Omega)$ and $\partial f/\partial T \in L^2(\partial\Omega)$. Furthermore, we define $H^{-1}(\partial\Omega)$ as the dual of $H^1(\partial\Omega)$ and the space $H^s(\partial\Omega)$, for $0 \leq s \leq 1$, as the interpolation space $[L^2(\partial\Omega), H^1(\partial\Omega)]_s$; see again [**149**].

Finally, we introduce Sobolev spaces of quasi-periodic functions. Let $\Lambda = (\Lambda_1, \ldots, \Lambda_n, 0, \ldots, 0) \in \mathbb{R}^d$ with $\Lambda_j > 0$ for $j = 1, \ldots, n$ and $n \leq d$. Let $\alpha = (\alpha_1, \ldots, \alpha_n, 0, \ldots, 0) \in \mathbb{R}^d$. Let $\mathcal{C}^\infty_\alpha(\mathbb{R}^d)$ be the set of functions $u \in \mathcal{C}^\infty(\mathbb{R}^d)$ satisfying:

(i) u has a compact support in $x_{n+1}, \ldots, x_d$;
(ii) $u(x + \Lambda) = e^{\sqrt{-1}\alpha\cdot\Lambda} u(x)$ for all $x \in \mathbb{R}^d$.

Recall that every function $u \in \mathcal{C}^\infty_\alpha(\mathbb{R}^d)$ can be expanded in an absolutely convergent and termwise infinitely differentiable Fourier series:

$$u(x) = \sum_{l \in \mathbb{Z}^n} u_l(x_{n+1}, \ldots, x_d) e^{\sqrt{-1}\alpha_l \cdot x},$$

where

$$\alpha_l := \alpha + 2\pi(\frac{l_1}{\Lambda_1}, \ldots, \frac{l_n}{\Lambda_n}, 0, \ldots, 0).$$

For an open set $\Omega \subset \mathbb{R}^d$, $\mathcal{C}^\infty_\alpha(\Omega)$ is the space of restrictions to Ω of functions of $\mathcal{C}^\infty_\alpha(\mathbb{R}^d)$. This enables us to consider the quasi-periodic Sobolev space given by the closure of $\mathcal{C}^\infty_\alpha(\Omega)$ in $H^1(\Omega)$, *i.e.*,

$$H^1_\alpha(\Omega) := \overline{\mathcal{C}^\infty_\alpha(\Omega)}^{H^1(\Omega)},$$

which, equipped with the $H^1(\Omega)$-norm, becomes a Hilbert space.

2.3. Layer Potentials for the Laplace Equation

A fundamental solution to the Laplacian is given by

$$\Gamma_0(x) = \begin{cases} \dfrac{1}{2\pi} \ln |x| \,, & d = 2, \\ \dfrac{1}{(2-d)\omega_d} |x|^{2-d} \,, & d \geq 3, \end{cases} \tag{2.2}$$

where ω_d denotes the area of the unit sphere in $\mathbb{R}^d$.

Given a bounded Lipschitz domain Ω in $\mathbb{R}^d, d \geq 2$, we denote, respectively, the single- and double-layer potentials of a function $\varphi \in L^2(\partial\Omega)$ as $\mathcal{S}^0_\Omega[\varphi]$ and $\mathcal{D}^0_\Omega[\varphi]$, where

$$\mathcal{S}^0_\Omega[\varphi](x) := \int_{\partial\Omega} \Gamma_0(x-y)\varphi(y)\, d\sigma(y), \quad x \in \mathbb{R}^d, \tag{2.3}$$

$$\mathcal{D}^0_\Omega[\varphi](x) := \int_{\partial\Omega} \frac{\partial}{\partial\nu(y)} \Gamma_0(x-y)\varphi(y)\, d\sigma(y)\,, \quad x \in \mathbb{R}^d \setminus \partial\Omega, \tag{2.4}$$

where $\nu(y)$ is the outward unit normal to $\partial\Omega$ at y.

Define the operator $\mathcal{K}^0_\Omega : L^2(\partial\Omega) \to L^2(\partial\Omega)$ by

$$\mathcal{K}^0_\Omega[\varphi](x) := \frac{1}{\omega_d} \text{p.v.} \int_{\partial\Omega} \frac{\langle y-x, \nu(y)\rangle}{|x-y|^d} \varphi(y)\, d\sigma(y), \tag{2.5}$$

where p.v. stands for the Cauchy principal value, and let $(\mathcal{K}^0_\Omega)^*$ be the L^2-adjoint of $\mathcal{K}^0_\Omega$. Hence, the operator $(\mathcal{K}^0_\Omega)^*$ is given by

$$(\mathcal{K}^0_\Omega)^*[\varphi](x) = \frac{1}{\omega_d} \text{p.v.} \int_{\partial\Omega} \frac{\langle x-y, \nu(x)\rangle}{|x-y|^d} \varphi(y)\, d\sigma(y), \quad \varphi \in L^2(\partial\Omega). \tag{2.6}$$

The singular integral operators $\mathcal{K}^0_\Omega$ and $(\mathcal{K}^0_\Omega)^*$ are known to be bounded on $L^2(\partial\Omega)$ [**202**]. If $\partial\Omega$ is of class $\mathcal{C}^{1,\eta}$ for some $\eta > 0$, then the operators $\mathcal{K}^0_\Omega$ and $(\mathcal{K}^0_\Omega)^*$ are compact in $L^2(\partial\Omega)$. Indeed, $\mathcal{K}^0_\Omega : L^2(\partial\Omega) \to H^s(\partial\Omega)$ is bounded for any $0 \leq s < \eta$. See, for example, [**495**].

For convenience we introduce the following notation. For a function u defined on $\mathbb{R}^d \setminus \partial\Omega$, we denote

$$u|_\pm(x) := \lim_{t\to 0^+} u(x \pm t\nu(x)), \quad x \in \partial\Omega,$$

and

$$\frac{\partial u}{\partial\nu(x)}\bigg|_\pm (x) := \lim_{t\to 0^+} \langle \nabla u(x \pm t\nu(x)), \nu(x)\rangle\,, \quad x \in \partial\Omega,$$

if the limits exist. Here $\nu(x)$ is the outward unit normal to $\partial\Omega$ at x, and $\langle\ ,\ \rangle$ denotes the scalar product in $\mathbb{R}^d$. For ease of notation we will sometimes use the dot for the scalar product in $\mathbb{R}^d$.

We relate in the next lemma the traces of the double-layer potential and the normal derivative of the single-layer potential to the operators $\mathcal{K}_\Omega^0$ and $(\mathcal{K}_\Omega^0)^*$ defined by (2.5) and (2.6).

LEMMA 2.1 (Jump relations). *If Ω is a bounded Lipschitz domain, then, for $\varphi \in L^2(\partial\Omega)$,*

$$(\mathcal{D}_\Omega^0[\varphi])\big|_{\pm}(x) = \left(\mp\frac{1}{2}I + \mathcal{K}_\Omega^0\right)[\varphi](x) \quad a.e.\ x \in \partial\Omega, \tag{2.7}$$

$$\frac{\partial}{\partial\nu}\mathcal{S}_\Omega^0[\varphi]\bigg|_{\pm}(x) = \left(\pm\frac{1}{2}I + (\mathcal{K}_\Omega^0)^*\right)[\varphi](x) \quad a.e.\ x \in \partial\Omega, \tag{2.8}$$

and

$$\frac{\partial}{\partial T}\mathcal{S}_\Omega^0[\varphi]\bigg|_{+}(x) = \frac{\partial}{\partial T}\mathcal{S}_\Omega^0[\varphi]\bigg|_{-}(x) \quad a.e.\ x \in \partial\Omega. \tag{2.9}$$

Moreover, for $\varphi \in H^{1/2}(\partial\Omega)$,

$$\frac{\partial}{\partial\nu}\mathcal{D}_\Omega^0[\varphi]\bigg|_{+} = \frac{\partial}{\partial\nu}\mathcal{D}_\Omega^0[\varphi]\bigg|_{-} \quad in\ H^{-1/2}(\partial\Omega). \tag{2.10}$$

Note that (2.8) yields the following jump relation:

$$\frac{\partial}{\partial\nu}\mathcal{S}_\Omega^0[\varphi]\bigg|_{+} - \frac{\partial}{\partial\nu}\mathcal{S}_\Omega^0[\varphi]\bigg|_{-} = \varphi \quad \text{on } \partial\Omega. \tag{2.11}$$

Note also that if Ω is of class $\mathcal{C}^{1,\eta}$ for some $0\eta > 0$, then for any $\varphi \in L^2(\partial\Omega)$, $\partial\mathcal{D}_\Omega^0[\varphi]/\partial\nu$ exists (in $H^{-1}(\partial\Omega)$) and has no jump across $\partial\Omega$. Indeed, if

$$\mathcal{N} : L^2(\partial\Omega) \to H^{-1}(\partial\Omega)$$

is the Dirichlet-to-Neumann operator defined by

$$\mathcal{N}[\varphi] = \frac{\partial u}{\partial\nu}\bigg|_{\partial\Omega},$$

where u is the solution to

$$\begin{cases} \Delta u = 0 & \text{in } \Omega, \\ u = \varphi & \text{on } \partial\Omega, \end{cases}$$

then the following formula holds:

$$\frac{\partial}{\partial\nu}\mathcal{D}_\Omega^0[\varphi]\bigg|_{\pm} = (\frac{1}{2} + (\mathcal{K}_\Omega^0)^*)\mathcal{N}[\varphi].$$

See [**495**] for the details.

We now recall the concept of capacity. Suppose $d = 2$ and let

$$(\varphi_e, a) \in L^2(\partial\Omega) \times \mathbb{R}$$

denote the unique solution of the system

$$\begin{cases} \displaystyle\frac{1}{2\pi}\int_{\partial\Omega} \ln|x-y|\varphi_e(y)d\sigma(y) + a = 0 & \text{on } \partial\Omega, \\ \displaystyle\int_{\partial\Omega} \varphi_e(y)d\sigma(y) = 1. \end{cases} \tag{2.12}$$

The logarithmic capacity of $\partial\Omega$ is defined by

$$\operatorname{cap}(\partial\Omega) := e^{2\pi a}, \tag{2.13}$$

where a is given by (2.12).

If $d=3$, there exists a unique $\varphi_e \in L^2(\partial\Omega)$ such that

$$\begin{cases} \displaystyle\int_{\partial\Omega} \frac{\varphi_e(y)}{|x-y|} d\sigma(y) = \text{constant} \quad \text{on } \partial\Omega, \\ \displaystyle\int_{\partial\Omega} \varphi_e(y) d\sigma(y) = 1. \end{cases} \tag{2.14}$$

The capacity of $\partial\Omega$ in three dimensions is defined to be

$$\frac{1}{\operatorname{cap}(\partial\Omega)} := \frac{1}{4\pi} \int_{\partial\Omega} \frac{1}{|x-y|} \varphi_e(y) d\sigma(y). \tag{2.15}$$

If u is the solution of the exterior Dirichlet problem

$$\begin{cases} \Delta u = 0 \quad \text{in } \mathbb{R}^3 \setminus \overline{\Omega}, \\ u = 1 \quad \text{on } \partial\Omega, \\ u(x) \to 0 \quad \text{as } |x| \to +\infty, \end{cases}$$

then the capacity is given by

$$\operatorname{cap}(\partial\Omega) = -\int_{\partial\Omega} \frac{\partial u}{\partial \nu}\bigg|_+ (x)\, d\sigma(x) \quad \left(= \int_{\mathbb{R}^3\setminus\overline{\Omega}} |\nabla u|^2 \, dx \right).$$

Hence, the solution u behaves like the point source $-\operatorname{cap}(\partial\Omega)\Gamma_0(x)$ as $|x| \to +\infty$.

It is clear that the capacity of the unit disk is 1 and the capacity of the unit sphere is 4π. Further interesting properties of the capacity can be found in the books by Hille [**296**], Landkof [**349**], and Armitage and Gardiner [**106**].

2.4. Neumann–Poincaré Operator

As will be seen later, the plasmonic resonances of nanoparticles are related to the spectra of the non-self-adjoint Neumann–Poincaré type operators associated with the particle shapes. We will show that plasmon resonances in nanoparticles can be treated as an eigenvalue problem for the Neumann–Poincaré operator, which leads to direct calculation of resonance values of permittivity and optimal design of nanoparticles that resonate at specified frequencies. The analysis of Neumann–Poincaré-type operators will also be the key to fathoming the blow-up of the gradient of solutions to conductivity problems as well as to cloaking by anomalous resonances. In the next subsection, by choosing a proper inner product, we prove that the non-self-adjoint operator Neumann–Poincaré $(\mathcal{K}_\Omega^0)^*$ can be symmetrized, and its spectrum is discrete and accumulates at zero, provided that $\partial\Omega$ is smooth.

2.4.1. Symmetrization of $(\mathcal{K}_\Omega^0)^*$.

Let

$$L_0^2(\partial\Omega) := \left\{ \varphi \in L^2(\partial\Omega) : \int_{\partial\Omega} \varphi \, d\sigma = 0 \right\}.$$

The following lemma holds.

LEMMA 2.2. *Assume that Ω is a bounded Lipschitz domain in $\mathbb{R}^d, d \geq 2$. The spectrum of $(\mathcal{K}_\Omega^0)^* : L^2(\partial\Omega) \to L^2(\partial\Omega)$ lies in the interval $(-1/2, 1/2]$ and therefore, the operator $(1/2)\, I + \mathcal{K}_\Omega^0$ is invertible on $L^2(\partial\Omega)$. Moreover, the operator $-(1/2)\, I + \mathcal{K}_\Omega^0$ is invertible on $L_0^2(\partial\Omega)$.*

PROOF. The argument is by contradiction. Let $\lambda \in (-\infty, -1/2] \cup (1/2, +\infty)$, and assume that $\varphi \in L^2(\partial\Omega)$ satisfies $(\lambda I - (\mathcal{K}_\Omega^0)^*)[\varphi] = 0$ and φ is not identically zero. Since $\mathcal{K}_\Omega^0[1] = 1/2$ by Green's formula, we have

$$0 = \int_{\partial\Omega} (\lambda I - (\mathcal{K}_\Omega^0)^*)[\varphi]\, d\sigma = \int_{\partial\Omega} \varphi(\lambda - \frac{1}{2})\, d\sigma,$$

and thus $\int_{\partial\Omega} \varphi d\sigma = 0$. Hence $\mathcal{S}_\Omega^0[\varphi](x) = O(|x|^{1-d})$ and $\nabla \mathcal{S}_\Omega^0[\varphi](x) = O(|x|^{-d})$ as $|x| \to +\infty$ for $d \geq 2$. Since φ is not identically zero, both of the following numbers cannot be zero:

$$A = \int_\Omega |\nabla \mathcal{S}_\Omega^0[\varphi]|^2\, dx \quad \text{and} \quad B = \int_{\mathbb{R}^d \setminus \overline{\Omega}} |\nabla \mathcal{S}_\Omega^0[\varphi]|^2\, dx.$$

In fact, if both of them are zero, then $\mathcal{S}_\Omega^0[\varphi] =$ constant in Ω and in $\mathbb{R}^d \setminus \overline{\Omega}$. Hence $\varphi = 0$ by (2.11) which is a contradiction.

On the other hand, using the divergence theorem and (2.8), we have

$$A = \int_{\partial\Omega} (-\frac{1}{2} I + (\mathcal{K}_\Omega^0)^*)[\varphi]\, \mathcal{S}_\Omega^0[\varphi]\, d\sigma \quad \text{and} \quad B = -\int_{\partial\Omega} (\frac{1}{2} I + (\mathcal{K}_\Omega^0)^*)[\varphi]\, \mathcal{S}_\Omega^0[\varphi]\, d\sigma.$$

Since $(\lambda I - (\mathcal{K}_\Omega^0)^*)[\varphi] = 0$, it follows that

$$\lambda = \frac{1}{2} \frac{B - A}{B + A}.$$

Thus, $|\lambda| < 1/2$, which is a contradiction and so, for $\lambda \in (-\infty, -\frac{1}{2}] \cup (\frac{1}{2}, +\infty)$, $\lambda I - (\mathcal{K}_\Omega^0)^*$ is one to one on $L^2(\partial\Omega)$.

If $\lambda = 1/2$, then $A = 0$ and hence $\mathcal{S}_\Omega^0[\varphi] =$ constant in Ω. Thus $\mathcal{S}_\Omega^0[\varphi]$ is harmonic in $\mathbb{R}^d \setminus \partial\Omega$, behaves like $O(|x|^{1-d})$ as $|x| \to +\infty$ (since $\varphi \in L_0^2(\partial\Omega)$), and is constant on $\partial\Omega$. By (2.8), we have $(\mathcal{K}_\Omega^0)^*[\varphi] = (1/2)\,\varphi$, and hence

$$B = -\int_{\partial\Omega} \varphi\, \mathcal{S}_\Omega^0[\varphi]\, d\sigma = C \int_{\partial\Omega} \varphi\, d\sigma = 0,$$

which forces us to conclude that $\varphi = 0$. This proves that $(1/2)\, I - (\mathcal{K}_\Omega^0)^*$ is one to one on $L_0^2(\partial\Omega)$. □

Assume that Ω is simply connected and $\partial\Omega$ is of class $\mathcal{C}^{1,\eta}$ for some $\eta > 0$. In this subsection, we symmetrize the non-self-adjoint operator $(\mathcal{K}_\Omega^0)^*$ and prove that it can be realized as a self-adjoint operator on $H^{-1/2}(\partial\Omega)$ by introducing a new inner product.

We first state the following lemma.

LEMMA 2.3. *Let $d \geq 2$. The following results hold:*

(i) *The operator $\mathcal{S}_\Omega^0$ on $H^{-1/2}(\partial\Omega)$ is self-adjoint and $-\mathcal{S}_\Omega^0 \geq 0$ on $H^{-1/2}(\partial\Omega)$.*
(ii) *The operator $(\mathcal{K}_\Omega^0)^* : H^{-1/2}(\partial\Omega) \to H^{-1/2}(\partial\Omega)$ is compact.*

By Lemma 2.3, there exists a unique square root of $-\mathcal{S}_\Omega^0$ which we denote by $\sqrt{-\mathcal{S}_\Omega^0}$; furthermore, $\sqrt{-\mathcal{S}_\Omega^0}$ is self-adjoint and $\sqrt{-\mathcal{S}_\Omega^0} \geq 0$.

Next we look into the kernel of $\mathcal{S}_\Omega^0$. If $d \geq 3$, then it is known that $\mathcal{S}_\Omega^0 : H^{-1/2}(\partial\Omega) \to H^{1/2}(\partial\Omega)$ has a bounded inverse. Suppose now that $d = 2$. If $\phi_0 \in \operatorname{Ker}(\mathcal{S}_\Omega^0)$, then the function u defined by

$$u(x) := \mathcal{S}_\Omega^0[\phi_0](x), \quad x \in \mathbb{R}^2$$

satisfies $u = 0$ on $\partial\Omega$. Therefore, $u(x) = 0$ for all $x \in \Omega$. It then follows from (2.8) that

$$(\mathcal{K}_\Omega^0)^*[\phi_0] = \frac{1}{2}\phi_0 \quad \text{on } \partial\Omega\,. \tag{2.16}$$

If $\langle \chi(\partial\Omega), \phi_0\rangle_{1/2,-1/2} = 0$ with $\chi(\partial\Omega)$ being the characteristic function of $\partial\Omega$, then $u(x) \to 0$ as $|x| \to \infty$, and hence $u(x) = 0$ for $x \in \mathbb{R}^2 \setminus \Omega$ as well. Thus $\phi_0 = 0$. The eigenfunctions of (2.16) make a one dimensional subspace of $H^{-1/2}(\partial\Omega)$, which means that $\mathrm{Ker}(\mathcal{S}_\Omega^0)$ is of at most one dimension.

Let $(\phi_e, a) \in H^{-1/2}(\partial\Omega) \times \mathbb{R}$ denote the solution of the system (2.12), then it can be shown that $\mathcal{S}_\Omega^0 : H^{-1/2}(\partial\Omega) \to H^{1/2}(\partial\Omega)$ has a bounded inverse if and only if $a \neq 0$.

The following result is well-known.

LEMMA 2.4. *The following Plemelj's symmetrization principle identity (also known as Calderón) holds:*

$$\mathcal{S}_\Omega^0(\mathcal{K}_\Omega^0)^* = \mathcal{K}_\Omega^0\mathcal{S}_\Omega^0 \quad \textit{on } H^{-1/2}(\partial\Omega)\,. \tag{2.17}$$

Consider the three-dimensional case and let $\mathcal{H}^*(\partial\Omega)$ be the space $H^{-1/2}(\partial\Omega)$ with the inner product

$$\langle u, v\rangle_{\mathcal{H}^*} = -\langle \mathcal{S}_\Omega^0[v], u\rangle_{\frac{1}{2},-\frac{1}{2}}\,, \tag{2.18}$$

which is equivalent to the original one (on $H^{-1/2}(\partial\Omega)$).

Since the operator $(\mathcal{K}_\Omega^0)^*$ can be symmetrized using Calderón identity (2.17) and hence becomes self-adjoint [**331**], it is then possible to write its spectral decomposition.

THEOREM 2.5. *For $d = 3$, the following results hold:*

(i) *The operator $(\mathcal{K}_\Omega^0)^*$ is self-adjoint in the Hilbert space $\mathcal{H}^*(\partial\Omega)$;*
(ii) *Let (λ_j, φ_j), $j = 0, 1, 2, \ldots$ be the eigenvalue and normalized eigenfunction pair of $(\mathcal{K}_\Omega^0)^*$ in $\mathcal{H}^*(\partial\Omega)$ with $\lambda_0 = 1/2$. Then, $\lambda_j \in (-\frac{1}{2}, \frac{1}{2})$ for $j \geq 1$ with $|\lambda_1| \geq |\lambda_2| \geq \ldots \to 0$ as $j \to \infty$;*
(iii) *The following spectral representation formula holds: for any $\psi \in H^{-1/2}(\partial\Omega)$,*

$$(\mathcal{K}_\Omega^0)^*[\psi] = \sum_{j=0}^{\infty} \lambda_j \langle \varphi_j, \psi\rangle_{\mathcal{H}^*} \varphi_j\,. \tag{2.19}$$

Moreover, it is clear that the following result holds.

LEMMA 2.6. *Let $d = 3$. Let $\mathcal{H}(\partial\Omega)$ be the space $H^{1/2}(\partial\Omega)$ equipped with the following equivalent inner product*

$$\langle u, v\rangle_{\mathcal{H}} = \langle v, (-\mathcal{S}_\Omega^0)^{-1}[u]\rangle_{\frac{1}{2},-\frac{1}{2}}\,. \tag{2.20}$$

Then, $\mathcal{S}_\Omega^0$ is an isometry between $\mathcal{H}^*(\partial\Omega)$ and $\mathcal{H}(\partial\Omega)$.

Furthermore, we list other useful observations and basic results in three dimensions.

LEMMA 2.7. *Let $d = 3$. The following results hold:*

(i) *We have $(-\frac{1}{2}I + (\mathcal{K}_\Omega^0)^*)(\mathcal{S}_\Omega^0)^{-1}[\chi(\partial\Omega)] = 0$.*

(ii) *The eigenspace corresponding to* $\lambda_0 = \frac{1}{2}$ *has dimension one and is spanned by the function* $\varphi_0 = c(\mathcal{S}_\Omega^0)^{-1}[\chi(\partial\Omega)]$ *for some constant* c *such that*

$$||\varphi_0||_{\mathcal{H}^*} = 1.$$

(iii) $\mathcal{H}^*(\partial\Omega) = \mathcal{H}_0^*(\partial\Omega) \oplus \{\mu\varphi_0,\ \mu \in \mathbb{C}\}$, *where* $\mathcal{H}_0^*(\partial\Omega)$ *is the zero mean subspace of* $\mathcal{H}^*(\partial\Omega)$.

In two dimensions, again based on (2.17), we show that $(\mathcal{K}_\Omega^0)^*$ can be realized as a self-adjoint operator by introducing a new inner product, slightly different from the one introduced in the three-dimensional case.

Recall that the single-layer potential $\mathcal{S}_\Omega^0 : H^{-1/2}(\partial\Omega) \to H^{1/2}(\partial\Omega)$ is not, in general, injective. Hence, $-\langle \mathcal{S}_\Omega^0[v], u\rangle_{\frac{1}{2},-\frac{1}{2}}$ does not define an inner product and the symmetrization technique described in Theorem 2.5 is no longer valid. To overcome this difficulty, a substitute of $\mathcal{S}_\Omega^0$ can be introduced as in [**98**] by

$$\widetilde{\mathcal{S}}_\Omega[\psi] = \begin{cases} \mathcal{S}_\Omega^0[\psi] & \text{if } \langle \chi(\partial\Omega), \psi\rangle_{\frac{1}{2},-\frac{1}{2}} = 0\,, \\ -\chi(\partial\Omega) & \text{if } \psi = \varphi_0\,, \end{cases} \tag{2.21}$$

where φ_0 is the unique eigenfunction of $(\mathcal{K}_\Omega^0)^*$ associated with eigenvalue $1/2$ such that $\langle \chi(\partial\Omega), \varphi_0\rangle_{\frac{1}{2},-\frac{1}{2}} = 1$. Note that, from the jump relations of the layer potentials, $\mathcal{S}_\Omega^0[\varphi_0]$ is constant.

The operator $\widetilde{\mathcal{S}}_\Omega : H^{-1/2}(\partial\Omega) \to H^{1/2}(\partial\Omega)$ is invertible. Moreover, the following Calderón identity holds $\mathcal{K}_\Omega^0\widetilde{\mathcal{S}}_\Omega = \widetilde{\mathcal{S}}_\Omega(\mathcal{K}_\Omega^0)^*$. With this, define

$$\langle u, v\rangle_{\mathcal{H}^*} = -\langle \widetilde{\mathcal{S}}_\Omega[v], u\rangle_{\frac{1}{2},-\frac{1}{2}}\,.$$

Thanks to the invertibility and positivity of $-\widetilde{\mathcal{S}}_\Omega$, this defines an inner product for which $(\mathcal{K}_\Omega^0)^*$ is self-adjoint and $\mathcal{H}^*$ is equivalent to $H^{-1/2}(\partial\Omega)$. Then, if Ω is $\mathcal{C}^{1,\eta}$, $\eta > 0$, we have the following results.

THEOREM 2.8. *Let* $d = 2$. *Let* Ω *be a* $\mathcal{C}^{1,\eta}, \eta > 0$, *bounded simply connected domain of* $\mathbb{R}^2$ *and let* $\widetilde{\mathcal{S}}_\Omega$ *be the operator defined in (2.21). Then,*

(i) *The operator* $(\mathcal{K}_\Omega^0)^*$ *is compact self-adjoint in the Hilbert space* $\mathcal{H}^*(\partial\Omega)$ *equipped with the inner product defined by*

$$\langle u, v\rangle_{\mathcal{H}^*} = -\langle \widetilde{\mathcal{S}}_D[v], u\rangle_{\frac{1}{2},-\frac{1}{2}}\,; \tag{2.22}$$

(ii) *Let* (λ_j, φ_j), $j = 0, 1, 2, \ldots$, *be the eigenvalue and normalized eigenfunction pair of* $(\mathcal{K}_\Omega^0)^*$ *with* $\lambda_0 = \frac{1}{2}$. *Then,* $\lambda_j \in (-\frac{1}{2}, \frac{1}{2})$ *with* $|\lambda_1| \geq |\lambda_2| \geq \ldots \to 0$ *as* $j \to \infty$;

(iii) $\mathcal{H}^*(\partial\Omega) = \mathcal{H}_0^*(\partial\Omega) \oplus \{\mu\varphi_0,\ \mu \in \mathbb{C}\}$, *where* $\mathcal{H}_0^*(\partial\Omega)$ *is the zero mean subspace of* $\mathcal{H}^*(\partial\Omega)$;

(iv) *The following representation formula holds: for any* $\psi \in H^{-1/2}(\partial\Omega)$,

$$(\mathcal{K}_\Omega^0)^*[\psi] = \sum_{j=0}^{\infty} \lambda_j \langle \varphi_j, \psi\rangle_{\mathcal{H}^*}\, \varphi_j\,.$$

LEMMA 2.9. *Let* $\mathcal{H}(\partial\Omega)$ *be the space* $H^{1/2}(\partial\Omega)$ *equipped with the following equivalent inner product:*

$$\langle u, v\rangle_{\mathcal{H}} = \langle v, -\widetilde{\mathcal{S}}_\Omega^{-1}[u]\rangle_{\frac{1}{2},-\frac{1}{2}}\,. \tag{2.23}$$

Then, $\widetilde{\mathcal{S}}_\Omega$ *is an isometry between* $\mathcal{H}^*(\partial\Omega)$ *and* $\mathcal{H}(\partial\Omega)$.

Note that $\widetilde{\mathcal{S}}_\Omega^{-1}[\chi(\partial\Omega)] = \varphi_0$ and

$$-(1/2)I + (\mathcal{K}_\Omega^0)^* = (-(1/2)I + (\mathcal{K}_\Omega^0)^*)\mathcal{P}_{\mathcal{H}_0^*},$$

where $\mathcal{P}_{\mathcal{H}_0^*}$ is the orthogonal projection onto $\mathcal{H}_0^*(\partial\Omega)$. In particular, we have

$$(-\frac{1}{2}I + (\mathcal{K}_\Omega^0)^*)\widetilde{\mathcal{S}}_\Omega^{-1}[\chi(\partial\Omega)] = 0.$$

In dimension two, the twin spectrum relation for the Neumann–Poincaré operator $(\mathcal{K}_\Omega^0)^*$ holds [**386**].

LEMMA 2.10. *For any $j \geq 1$, $\pm\lambda_j$ are eigenvalues of $(\mathcal{K}_\Omega^0)^*$.*

PROOF. In order to prove the twin property, suppose that λ_j is an eigenvalue of $(\mathcal{K}_\Omega^0)^*$ with an associated eigenfunction φ_j. Then $u := \mathcal{S}_\Omega^0[\varphi_j]$ is a nontrivial solution to the transmission problem

$$\begin{cases} \nabla \cdot ((1 + (k-1)\chi(\Omega))\nabla u) = 0 & \text{in } \mathbb{R}^2, \\ u(x) = O(|x|^{-1}) & \text{as } |x| \to +\infty \end{cases} \tag{2.24}$$

with $k = (2\lambda_j + 1)/(2\lambda_j - 1)$.

Let v be the harmonic conjugate of u, which is defined such that $\nabla v = \nabla^\perp u$ where

$$\nabla^\perp u = \begin{bmatrix} -\frac{\partial u}{\partial x_2} \\ \frac{\partial u}{\partial x_1} \end{bmatrix}. \tag{2.25}$$

Then v is a nontrivial solution to

$$\begin{cases} \nabla \cdot ((1 + (\frac{1}{k}-1)\chi(\Omega))\nabla v) = 0 & \text{in } \mathbb{R}^2, \\ v(x) = O(|x|^{-1}) & \text{as } |x| \to +\infty. \end{cases} \tag{2.26}$$

Therefore, by using the integral representation $v = \mathcal{S}_\Omega^0[\psi_j]$, it can be seen that

$$-\lambda_j = \frac{1 + \frac{1}{k}}{2(\frac{1}{k} - 1)}$$

is an eigenvalue of $(\mathcal{K}_\Omega^0)^*$ as well associated to the eigenfunction ψ_j. □

On the other hand, the following relation between the eigenfunctions of $(\mathcal{K}_\Omega^0)^*$ associated with $\pm\lambda_j$ holds.

LEMMA 2.11. *Let $\partial/\partial T$ denote the tangential derivative on $\partial\Omega$ and let φ_j be an eigenfunction of $(\mathcal{K}_\Omega^0)^*$ associated with λ_j. Then*

$$\frac{\frac{\partial}{\partial T}\mathcal{S}_\Omega^0[\varphi_j]}{\|\frac{\partial}{\partial T}\mathcal{S}_\Omega^0[\varphi_j]\|_{\mathcal{H}^*}}$$

is a (normalized) eigenfunction of $(\mathcal{K}_\Omega^0)^*$ associated with $-\lambda_j$.

PROOF. Let $\nu = \begin{bmatrix} \nu_1 \\ \nu_2 \end{bmatrix}$ and let $T = \begin{bmatrix} -\nu_2 \\ \nu_1 \end{bmatrix}$. From [**504**], we have

$$(\mathcal{K}_\Omega^0)^*\frac{\partial}{\partial T} = -\frac{\partial}{\partial T}\mathcal{K}_\Omega^0. \tag{2.27}$$

From (2.27) it follows that if $\phi_j \in H^{1/2}(\partial\Omega)$ is an eigenfunction of $\mathcal{K}_\Omega^0$ associated with the eigenvalue $\lambda_j \neq 1/2$, then $\partial\phi_j/\partial T$ is an eigenfunction of $(\mathcal{K}_\Omega^0)^*$ associated with the eigenvalue $-\lambda_j$. Therefore, by using Calderón's identity (2.17), we obtain Lemma 2.11.

Identity (2.27) can be proved by noticing that for $\phi \in H^{1/2}(\partial\Omega)$, the functions $\mathcal{D}_\Omega^0[\phi]$ and $\mathcal{S}_\Omega^0[\partial\phi/\partial T]$ in Ω are the harmonic conjugate functions of each other and therefore, by the jump formulas,

$$\begin{aligned}(\mathcal{K}_\Omega^0)^*[\frac{\partial\phi}{\partial T}] &= (-\frac{1}{2}+(\mathcal{K}_\Omega^0)^*)[\frac{\partial\phi}{\partial T}]+\frac{1}{2}\frac{\partial\phi}{\partial T}\\ &= \frac{\partial\mathcal{S}_\Omega^0}{\partial\nu}[\frac{\partial\phi}{\partial T}]|_- + \frac{1}{2}\frac{\partial\phi}{\partial T}\\ &= -\frac{\partial\mathcal{D}_\Omega^0}{\partial T}[\phi]|_- + \frac{1}{2}\frac{\partial\phi}{\partial T}\\ &= -\frac{\partial}{\partial T}(\frac{1}{2}I+\mathcal{K}_\Omega^0)[\phi]+\frac{1}{2}\frac{\partial\phi}{\partial T}.\end{aligned}$$

Therefore, the proof of the lemma is complete. □

In two dimensions, we will also need the following identities from [**390**, **456**].

LEMMA 2.12. *We have*

$$\frac{\partial\mathcal{D}_\Omega^0[\phi]}{\partial\nu} = \frac{\partial\mathcal{S}_\Omega^0}{\partial T}[\frac{\partial\phi}{\partial T}] \tag{2.28}$$

and

$$\mathcal{S}_\Omega^0\frac{\partial}{\partial T}\mathcal{S}_\Omega^0[\frac{\partial\phi}{\partial T}] = (\mathcal{K}_\Omega^0)^2[\phi]-\frac{1}{4}\phi \tag{2.29}$$

for $\phi \in H^{1/2}(\partial\Omega)$.

REMARK 2.13. *With the same notation as in Lemma 2.11, notice that from (2.29) it follows that*

$$\begin{aligned}\|\frac{\partial}{\partial T}\mathcal{S}_\Omega^0[\varphi_j]\|_{\mathcal{H}^*}^2 &= -\langle\mathcal{S}_\Omega^0\frac{\partial}{\partial T}\mathcal{S}_\Omega^0[\varphi_j],\frac{\partial}{\partial T}\mathcal{S}_\Omega^0[\varphi_j]\rangle_{\frac{1}{2},-\frac{1}{2}}\\ &= \langle(\mathcal{S}_\Omega^0\frac{\partial}{\partial T})^2\mathcal{S}_\Omega^0[\varphi_j],\varphi_j\rangle_{\frac{1}{2},\frac{1}{2}},\end{aligned}$$

and therefore,

$$\|\frac{\partial}{\partial T}\mathcal{S}_\Omega^0[\varphi_j]\|_{\mathcal{H}^*}^2 = \frac{1}{4}-\lambda_j^2. \tag{2.30}$$

REMARK 2.14. *When $\partial\Omega$ is Lipschitz, $(\mathcal{K}_\Omega^0)^*$ is no longer compact. Nevertheless, since it is self-adjoint, its spectrum $\sigma((\mathcal{K}_\Omega^0)^*)$ is real, consists of point and continuous spectrum, and is a closed set; see Appendix* A. *Moreover, by the spectral resolution theorem (see* [**512**]*), there is a family of projection operators $\mathcal{E}(t)$ on $\mathcal{H}^*$ (called a resolution of identity) such that*

$$(\mathcal{K}_\Omega^0)^* = \int_{t\in\sigma((\mathcal{K}_\Omega^0)^*)} t\,d\mathcal{E}(t). \tag{2.31}$$

Let b_Ω be the spectral bound of $(\mathcal{K}_\Omega^0)^*$, namely

$$b_\Omega := \sup\{|\lambda| : \lambda\in\sigma((\mathcal{K}_\Omega^0)^*)\}.$$

From the proof of Lemma 2.2 *it follows that*

$$b_\Omega = \frac{1}{2} \sup_{\varphi \in \mathcal{H}^*} \frac{\left| \int_{\mathbb{R}^d \setminus \overline{\Omega}} |\nabla \mathcal{S}_\Omega^0[\varphi]|^2 dx - \int_\Omega |\nabla \mathcal{S}_\Omega^0[\varphi]|^2 dx \right|}{\int_{\mathbb{R}^d} |\nabla \mathcal{S}_\Omega^0[\varphi]|^2 dx} \leq \frac{1}{2}.$$

If Ω *is a two-dimensional curvilinear polygon, then* $(\mathcal{K}_\Omega^0)^*$ *has an essential spectrum, which depends only on the angles of the corners and can be characterized in terms of elliptic corner singularity functions* [**153**, **449**]. *The reader is referred to Section* 7.6 *for more details.*

2.4.2. Spectral Decomposition of the Fundamental Solution. Fix $z \in \mathbb{R}^d \setminus \overline{\Omega}$. Then $\Gamma_0(\cdot - z)$ belongs to $H^{1/2}(\partial\Omega)$, and so admits the following decomposition:

$$\Gamma_0(x - z) = \sum_{j=1}^{\infty} c_j(z) \mathcal{S}_\Omega^0[\varphi_j](x) + c_0(z), \quad x \in \partial\Omega, \tag{2.32}$$

for some constants $c_j(z)$ satisfying

$$\sum_{j=1}^{\infty} |c_j(z)|^2 < \infty.$$

Since $-\langle \mathcal{S}_\Omega^0[\varphi_j], \varphi_i \rangle_{1/2,-1/2} = \delta_{ij}$, we see that

$$c_j(z) = -\mathcal{S}_\Omega^0[\varphi_j](z), \quad j = 1, 2, \ldots.$$

We also see from $\langle \chi(\partial\Omega), \varphi_0 \rangle_{1/2,-1/2} = 0$ that $c_0(z) = \mathcal{S}_\Omega^0[\varphi_0](z)$. So, we obtain the following formula:

$$\Gamma_0(x - z) = -\sum_{j=1}^{\infty} \mathcal{S}_\Omega^0[\varphi_j](z) \mathcal{S}_\Omega^0[\varphi_j](x) + \mathcal{S}_\Omega^0[\varphi_0](z), \quad x \in \partial\Omega.$$

Observe that

$$\|\sum_{j=1}^{\infty} \mathcal{S}_\Omega^0[\varphi_j](z) \mathcal{S}_\Omega^0[\varphi_j]\|_{\mathcal{H}}^2 = \sum_{j=1}^{\infty} |\mathcal{S}_\Omega^0[\varphi_j](z)|^2 < \infty.$$

Since $\| \ \|_{\mathcal{H}}$ is equivalent to the $H^{1/2}$-norm, we find from the trace theorem that the series $\sum_{j=1}^{\infty} \mathcal{S}_\Omega^0[\varphi_j](z) \mathcal{S}_\Omega^0[\varphi_j]$ converges in $H^1(\Omega)$ and is harmonic in Ω. Therefore, the following expansion of the fundamental solution Γ_0 in terms of the eigenfunctions of the Neumann–Poincaré operator $(\mathcal{K}_\Omega^0)^*$ holds.

THEOREM 2.15. *We have*

$$\Gamma_0(x - z) = -\sum_{j=1}^{\infty} \mathcal{S}_\Omega^0[\varphi_j](z) \mathcal{S}_\Omega^0[\varphi_j](x) + \mathcal{S}_\Omega^0[\varphi_0](z), \quad x \in \overline{\Omega}, z \in \mathbb{R}^d \setminus \overline{\Omega}. \tag{2.33}$$

Formula (2.33) is a general addition formula for the fundamental solution Γ_0 to the Laplace operator. It was derived in [**98**]. Addition formulas for the fundamental solution to the Laplace operator on disks, balls, ellipses, and ellipsoids are classical and well-known. That on ellipsoids is attributed to Heine (see [**214**]). The formulas describe expansions of the fundamental solution to the Laplace operator in terms of spherical harmonics (balls) and ellipsoidal harmonics (ellipses). Formula (2.33) shows that, in the general case, the addition formula is a spectral expansion by eigenfunctions of the Neumann–Poincaré operator.

2.4.3. Spectrum of the Neumann–Poincaré Operator on Disks and Ellipses. Recall that if Ω is a disk or a ball, then we may simplify the expressions defining the operators $\mathcal{K}_\Omega$ and $(\mathcal{K}_\Omega^0)^*$. The following results hold [**321**]:

(i) Suppose that Ω is a two dimensional disk with radius r_0. Then,

$$\frac{\langle x-y, \nu(x)\rangle}{|x-y|^2} = \frac{1}{2r_0} \quad \forall\, x, y \in \partial\Omega, x \neq y\,,$$

and therefore, for any $\phi \in L^2(\partial\Omega)$,

$$(\mathcal{K}_\Omega^0)^*[\phi](x) = \mathcal{K}_\Omega^0[\phi](x) = \frac{1}{4\pi r_0}\int_{\partial\Omega} \phi(y)\, d\sigma(y)\,, \tag{2.34}$$

for all $x \in \partial\Omega$.

(ii) For $d \geq 3$, if Ω is a ball with radius r_0, then, we have

$$\frac{\langle x-y, \nu(x)\rangle}{|x-y|^d} = \frac{1}{2r_0}\frac{1}{|x-y|^{d-2}} \quad \forall\, x, y \in \partial\Omega, x \neq y\,,$$

and for any $\phi \in L^2(\partial\Omega)$ and $x \in \partial\Omega$,

$$(\mathcal{K}_\Omega^0)^*[\phi](x) = \mathcal{K}_\Omega^0[\phi](x) = \frac{(2-d)}{2r_0}\mathcal{S}_\Omega^0[\phi](x)\,. \tag{2.35}$$

Another useful formula in two dimensions is the expression of $\mathcal{K}_\Omega^0[\phi](x)$, where Ω is an ellipse whose semi-axes are on the x_1- and x_2-axes and of length a_1 and a_2, respectively. Using the parametric representation $X(t) = (a_1 \cos t, a_2 \sin t), 0 \leq t \leq 2\pi$, for the boundary $\partial\Omega$, we find that

$$\mathcal{K}_\Omega^0[\phi](x) = \frac{a_1 a_2}{2\pi(a_1^2 + a_2^2)}\int_0^{2\pi} \frac{\phi(X(t))}{1 - Q\cos(t+\theta)}\, dt\,, \tag{2.36}$$

where $x = X(\theta)$ and $Q = (a_1^2 - a_2^2)/(a_1^2 + a_2^2)$.

Using (2.34), it also follows that if Ω is a disk, then the spectrum of $(\mathcal{K}_\Omega^0)^*$ is $\{0, 1/2\}$. If D is an ellipse of semi-axes a_1 and a_2, then

$$\lambda_j = \begin{cases} \dfrac{1}{2} & j = 0, \\ \pm\dfrac{1}{2}\left(\dfrac{a_1 - a_2}{a_1 + a_2}\right)^j & j \geq 1, \end{cases} \tag{2.37}$$

are the eigenvalues of $(\mathcal{K}_\Omega^0)^*$.

In three dimensions, by using (2.35) it can be shown that the spectrum of $(\mathcal{K}_\Omega^0)^*$ in the case where Ω is a ball is $1/(2(2j+1))$, for $j = 0, 1, \ldots$. Furthermore, the eigenvalues of $(\mathcal{K}_\Omega^0)^*$ for Ω being an ellipsoid can be expressed explicitly in terms of Lamé functions [**236**]. In [**236**], it is also shown that for any number $\lambda \in (-1/2, 1/2)$ there is an ellipsoid on which λ is an eigenvalue of the associated Neumann–Poincaré operator.

In two dimensions, we also recall that if the disk Ω of radius r_0 is centered at the origin, then one can easily see that for each integer $n \neq 0$

$$\mathcal{S}^0_\Omega[e^{\sqrt{-1}n\theta}](x) = \begin{cases} -\dfrac{r_0}{2|n|}\left(\dfrac{r}{r_0}\right)^{|n|} e^{\sqrt{-1}n\theta} & \text{if } |x| = r < r_0, \\ -\dfrac{r_0}{2|n|}\left(\dfrac{r_0}{r}\right)^{|n|} e^{\sqrt{-1}n\theta} & \text{if } |x| = r > r_0, \end{cases} \tag{2.38}$$

and hence

$$\frac{\partial}{\partial r}\mathcal{S}^0_\Omega[e^{\sqrt{-1}n\theta}](x) = \begin{cases} -\dfrac{1}{2}\left(\dfrac{r}{r_0}\right)^{|n|-1} e^{\sqrt{-1}n\theta} & \text{if } |x| = r < r_0, \\ \dfrac{1}{2}\left(\dfrac{r_0}{r}\right)^{|n|+1} e^{\sqrt{-1}n\theta} & \text{if } |x| = r > r_0. \end{cases} \tag{2.39}$$

We also get, for any integer n,

$$\mathcal{D}^0_\Omega[e^{\sqrt{-1}n\theta}](x) = \begin{cases} \dfrac{1}{2}\left(\dfrac{r}{r_0}\right)^{|n|} e^{\sqrt{-1}n\theta} & \text{if } |x| = r < r_0, \\ -\dfrac{1}{2}\left(\dfrac{r_0}{r}\right)^{|n|} e^{\sqrt{-1}n\theta} & \text{if } |x| = r > r_0. \end{cases}$$

It follows from (2.34) that

$$(\mathcal{K}^0_\Omega)^*[e^{\sqrt{-1}n\theta}] = 0 \quad \forall n \neq 0. \tag{2.40}$$

As $\mathcal{K}^0_\Omega[1] = 1/2$, it follows that, when Ω is a disk, $\mathcal{K}^0_\Omega$ is a rank one operator whose only non-zero eigenvalue is $1/2$. On the other hand, from $\mathcal{K}^0_\Omega[1] = 1/2$ it also follows that

$$\mathcal{S}^0_\Omega[1](x) = \begin{cases} r_0 \ln r_0 & \text{if } |x| = r < r_0, \\ r_0 \ln |x| & \text{if } |x| = r > r_0, \end{cases} \tag{2.41}$$

and hence

$$\frac{\partial}{\partial r}\mathcal{S}^0_\Omega[1](x) = \begin{cases} 0 & \text{if } |x| = r < r_0, \\ \dfrac{r_0}{r} & \text{if } |x| = r > r_0. \end{cases} \tag{2.42}$$

Let Ω_i and Ω_e be two concentric disks in $\mathbb{R}^2$ with radii $r_i < r_e$. Define $(\mathcal{K}^0_{\Omega_e\setminus\overline{\Omega}_i})^*$ by

$$(\mathcal{K}^0_{\Omega_e\setminus\overline{\Omega}_i})^* = \begin{pmatrix} -(\mathcal{K}^0_{\Omega_i})^* & -\frac{\partial}{\partial\nu^i}\mathcal{S}^0_{\Omega_e} \\ \frac{\partial}{\partial\nu^e}\mathcal{S}^0_{\Omega_i} & (\mathcal{K}^0_{\Omega_e})^* \end{pmatrix}, \tag{2.43}$$

where ν^i and ν^e are the outward normal vectors to $\partial\Omega_i$ and $\partial\Omega_e$, respectively. Let the operator $\mathbb{S}_{\Omega_e\setminus\overline{\Omega}_i}$ be given by

$$\mathbb{S}_{\Omega_e\setminus\overline{\Omega}_i} = \begin{pmatrix} \mathcal{S}^0_{\Omega_e} & \mathcal{S}^0_{\Omega_i}\big|_{\partial\Omega_e} \\ \mathcal{S}^0_{\Omega_e}\big|_{\partial\Omega_i} & \mathcal{S}^0_{\Omega_i} \end{pmatrix}.$$

Then, following the arguments given in Subsection 2.4.1, we can prove that $(\mathcal{K}^0_{\Omega_e\setminus\overline{\Omega}_i})^*$ is compact and self-adjoint for the inner product

$$\langle\varphi, \psi\rangle_{\mathcal{H}^*} := -\langle\mathbb{S}_{\Omega_e\setminus\overline{\Omega}_i}[\psi], \varphi\rangle_{1/2,-1/2} \tag{2.44}$$

for $\varphi, \psi \in H^{-1/2}(\partial\Omega_e) \times H^{-1/2}(\partial\Omega_i)$.

The following lemma from [**35**] gives the eigenvalues and eigenfunctions of the Neumann–Poincaré operator $(\mathcal{K}^0_{\Omega_e\setminus\overline{\Omega_i}})^*$ associated with the circular shell $\Omega_e\setminus\overline{\Omega_i}$ on $\mathcal{H}^*$.

LEMMA 2.16. *The eigenvalues of* $(\mathcal{K}^0_{\Omega_e\setminus\overline{\Omega_i}})^*$ *on* $\mathcal{H}^*$ *are*

$$-\frac{1}{2}, \frac{1}{2}, -\frac{1}{2}(\frac{r_i}{r_e})^n, \frac{1}{2}(\frac{r_i}{r_e})^n, \quad n = 1, 2, \ldots,$$

and corresponding eigenvectors are

$$\begin{bmatrix} 1 \\ -\frac{1}{r_e} \end{bmatrix}, \begin{bmatrix} 0 \\ 1 \end{bmatrix}, \begin{bmatrix} e^{\pm\sqrt{-1}n\theta} \\ \frac{r_i}{r_e}e^{\pm\sqrt{-1}n\theta} \end{bmatrix}, \begin{bmatrix} e^{\pm\sqrt{-1}n\theta} \\ -\frac{r_i}{r_e}e^{\pm\sqrt{-1}n\theta} \end{bmatrix}, \quad n = 1, 2, \ldots.$$

PROOF. We first prove that $\pm 1/2$ are eigenvalues of $(\mathcal{K}^0_{\Omega_e\setminus\overline{\Omega_i}})^*$ on $\mathcal{H}^*$. From (2.42) we have

$$(\mathcal{K}^0_{\Omega_e\setminus\overline{\Omega_i}})^* \begin{bmatrix} a \\ b \end{bmatrix} = \begin{pmatrix} -\frac{1}{2} & 0 \\ \frac{1}{r_e} & \frac{1}{2} \end{pmatrix} \begin{bmatrix} a \\ b \end{bmatrix},$$

where a and b are constants. So $\pm 1/2$ are eigenvalues of $(\mathcal{K}^0_{\Omega_e\setminus\overline{\Omega_i}})^*$ on $\mathcal{H}^*$.

Now we consider $(\mathcal{K}^0_{\Omega_e\setminus\overline{\Omega_i}})^*$ on $\mathcal{H}_0^*$ defined by

$$\mathcal{H}_0^* := \{\varphi \in \mathcal{H}^* : \langle \chi(\partial\Omega_e) \times \chi(\partial\Omega_i), \varphi\rangle_{1/2,-1/2} = 0\}.$$

Because of (2.40), it follows that

$$(\mathcal{K}^0_{\Omega_e\setminus\overline{\Omega_i}})^* = \begin{pmatrix} 0 & -\frac{\partial}{\partial\nu^i}\mathcal{S}^0_{\Omega_e} \\ \frac{\partial}{\partial\nu^e}\mathcal{S}^0_{\Omega_i} & 0 \end{pmatrix}$$

on $\mathcal{H}_0^*$ and hence we have from (2.39) that

$$(\mathcal{K}^0_{\Omega_e\setminus\overline{\Omega_i}})^* \begin{bmatrix} e^{\sqrt{-1}n\theta} \\ 0 \end{bmatrix} = \frac{1}{2}(\frac{r_i}{r_e})^{|n|+1} \begin{bmatrix} 0 \\ e^{\sqrt{-1}n\theta} \end{bmatrix} \tag{2.45}$$

and

$$(\mathcal{K}^0_{\Omega_e\setminus\overline{\Omega_i}})^* \begin{bmatrix} 0 \\ e^{\sqrt{-1}n\theta} \end{bmatrix} = \frac{1}{2}(\frac{r_i}{r_e})^{|n|-1} \begin{bmatrix} e^{\sqrt{-1}n\theta} \\ 0 \end{bmatrix} \tag{2.46}$$

for all $n \neq 0$, which completes the proof of the lemma. □

REMARK 2.17. *In* [**200**], *by using elliptic coordinates, the Neumann–Poincaré operator associated with two confocal ellipses is investigated and the asymptotic behavior of its eigenvalues* λ_j *as* $j \to +\infty$ *is derived.*

In three dimensions, we can compute the spectrum of the Neumann–Poincaré operator associated with concentric balls. The following lemma is needed.

LEMMA 2.18. *Let* $\Omega = \{|x| < r_0\}$ *in* $\mathbb{R}^3$. *We have for* $j = 0, 1, \ldots$

$$(\mathcal{K}^0_\Omega)^*[Y_j^l] = \frac{1}{2(2j+1)}Y_j^l(\hat{x}), \quad |x| = r_0, l = -j, \ldots, j, \tag{2.47}$$

where $\hat{x} = x/|x|$ *and* $(Y_j^l)_{l=-j,\ldots,j}$ *are the orthonormal spherical harmonics of degree* j.

PROOF. From (2.8) and (2.35), it follows that

$$\frac{\partial}{\partial r}\mathcal{S}_\Omega^0[Y_j^l]\big|_- + \frac{1}{2r_0}\mathcal{S}_\Omega^0[Y_j^l] = -\frac{1}{2}Y_j^l(\hat{x}), \quad |x| = r_0.$$

Then since $\mathcal{S}_\Omega^0[Y_j^l]$ and $|x|^j Y_j^l(\hat{x})$ are harmonic functions in Ω, we have

$$\mathcal{S}_\Omega^0[Y_j^l](x) = -\frac{1}{2j+1}\frac{r^j}{r_0^{j-1}}Y_j^l(\hat{x}) \quad \text{for } |x| = r \le r_0, \tag{2.48}$$

and (2.47) follows from (2.35). □

Let Ω_i and Ω_e be two concentric balls in $\mathbb{R}^3$ with radii $r_i < r_e$ and let the the Neumann–Poincaré operator $(\mathcal{K}_{\Omega_e\setminus\overline{\Omega}_i}^0)^*$ associated with the spherical shell $\Omega_e \setminus \overline{\Omega}_i$ be defined, analogously to the two dimensional case, by (2.43).

By (2.48), we have

$$\frac{\partial}{\partial \nu^i}\mathcal{S}_{\Omega_e}^0[Y_j^l](x) = -\frac{j}{2j+1}(\frac{r_i}{r_e})^{j-1}Y_j^l(\hat{x}), \quad |x| = r_i.$$

Similarly, we have

$$\mathcal{S}_{\Omega_i}^0[Y_j^l](x) = -\frac{1}{2j+1}\frac{r_i^{j+2}}{r^{j+1}}Y_j^l(\hat{x}), \quad |x| = r \ge r_i,$$

and hence

$$\frac{\partial}{\partial \nu^e}\mathcal{S}_{\Omega_i}^0[Y_j^l](x) = \frac{j+1}{2j+1}(\frac{r_i}{r_e})^{j+2}Y_j^l(\hat{x}), \quad |x| = r_e.$$

We now have for constants a and b

$$(\mathcal{K}_{\Omega_e\setminus\overline{\Omega}_i}^0)^*\begin{bmatrix} aY_j^l \\ bY_j^l \end{bmatrix} = \begin{pmatrix} -\frac{1}{2(2j+1)} & \frac{j}{2j+1}(\frac{r_i}{r_e})^{j-1} \\ \frac{j+1}{2j+1}(\frac{r_i}{r_e})^{j+2} & \frac{1}{2(2j+1)} \end{pmatrix}\begin{bmatrix} aY_j^l \\ bY_j^l \end{bmatrix}.$$

Thus we have the following lemma from [**35**].

LEMMA 2.19. *The eigenvalues of* $(\mathcal{K}_{\Omega_e\setminus\overline{\Omega}_i}^0)^*$ *on* $\mathcal{H}^*$ *are*

$$\pm\frac{1}{2(2j+1)}\sqrt{1+4j(j+1)(r_i/r_e)^{2j+1}}, \quad j = 0, 1, \dots,$$

and corresponding eigenfunctions are

$$\begin{bmatrix} (\sqrt{1+4j(j+1)(r_i/r_e)^{2j+1}}-1)Y_j^l \\ 2(j+1)(r_i/r_e)^{j+2}Y_j^l \end{bmatrix}, \quad \begin{bmatrix} (-\sqrt{1+4j(j+1)(r_i/r_e)^{2j+1}}-1)Y_j^l \\ 2(j+1)(r_i/r_e)^{j+2}Y_j^l \end{bmatrix},$$

for $l = -j, \dots, j$, *respectively.*

2.4.4. Neumann Poincaré Operator for Two Separated Disks and Its Spectral Decomposition. In this subsection, we consider the spectrum of the Neumann–Poincaré operator associated with two separated disks in $\mathbb{R}^2$. Let Ω_1 and Ω_2 be two separated disks. We set the Cartesian coordinates (x_1, x_2) to be such that the x_1-axis is parallel to the line joining the centers of the two disks and let $\nu^{(i)}$ be the outward normal on $\partial\Omega_i$, $i = 1, 2$.

Define the Neumann–Poincaré operator $\mathbb{K}_{\Omega_1\cup\Omega_2}^*$ associated with Ω_1 and Ω_2 by

$$\mathbb{K}_{\Omega_1\cup\Omega_2}^* := \begin{pmatrix} (\mathcal{K}_{\Omega_1}^0)^* & \frac{\partial}{\partial\nu^{(1)}}\mathcal{S}_{\Omega_2}^0 \\ \frac{\partial}{\partial\nu^{(2)}}\mathcal{S}_{\Omega_1}^0 & (\mathcal{K}_{\Omega_2}^0)^* \end{pmatrix}, \tag{2.49}$$

and define the operator $\mathbb{S}_{\Omega_1\cup\Omega_2}$ by

$$\mathbb{S}_{\Omega_1\cup\Omega_2} = \begin{pmatrix} \mathcal{S}^0_{\Omega_1} & \mathcal{S}^0_{\Omega_2}\big|_{\partial\Omega_1} \\ \mathcal{S}^0_{\Omega_1}\big|_{\partial\Omega_2} & \mathcal{S}^0_{\Omega_2} \end{pmatrix}.$$

Then, again following the arguments given in Subsection 2.4.1, we can prove that $\mathbb{K}^*_{\Omega_1\cup\Omega_2}$ is compact and self-adjoint for the inner product

$$\langle \varphi, \psi\rangle_{\mathcal{H}_0^*} := -\langle \mathbb{S}_{\Omega_1\cup\Omega_2}[\psi], \varphi\rangle_{1/2,-1/2} \tag{2.50}$$

for $\varphi, \psi \in H_0^{-1/2}(\partial\Omega_1) \times H_0^{-1/2}(\partial\Omega_2)$.

2.4.4.1. *Bipolar Coordinates.* To compute the spectrum of $\mathbb{K}^*_{\Omega_1\cup\Omega_2}$, we make use of bipolar coordinates. The following definitions are needed.

DEFINITION 2.20. *Each point $x = (x_1, x_2)$ in the Cartesian coordinate system corresponds to $(\xi, \theta) \in \mathbb{R} \times (-\pi, \pi]$ in the bipolar coordinate system through the equations*

$$x_1 = \alpha\frac{\sinh\xi}{\cosh\xi - \cos\theta} \quad \text{and} \quad x_2 = \alpha\frac{\sin\theta}{\cosh\xi - \cos\theta} \tag{2.51}$$

with a positive number α.

Notice that the bipolar coordinates can be defined using a conformal mapping. Define a conformal map Ψ by

$$z = x_1 + \sqrt{-1}x_2 = \Psi(v) = \alpha\frac{v+1}{v-1}.$$

If we write $v = e^{\xi - \sqrt{-1}\theta}$, then we can recover (2.51).

From Definition 2.20, we can see that the coordinate curves $\{\xi = c\}$ and $\{\theta = c\}$ are, respectively, the zero-level set of the following two functions:

$$f_\xi(x_1, x_2) = \left(x_1 - \alpha\frac{\cosh c}{\sinh c}\right)^2 + x_2^2 - \left(\frac{\alpha}{\sinh c}\right)^2 \tag{2.52}$$

and

$$f_\theta(x_1, x_2) = x_1^2 + \left(x_2 - \alpha\frac{\cos c}{\sin c}\right)^2 - \left(\frac{\alpha}{\sin c}\right)^2.$$

DEFINITION 2.21. *We define orthonormal basis vectors $\{e_\xi, e_\theta\}$ as follows:*

$$e_\xi := \frac{\partial x/\partial\xi}{|\partial x/\partial\xi|} \quad \text{and} \quad e_\theta := \frac{\partial x/\partial\theta}{|\partial x/\partial\theta|}.$$

Notice that, in the bipolar coordinates, the scaling factor h is

$$h(\xi, \theta) := \frac{\cosh\xi - \cos\theta}{\alpha}.$$

The gradient of any scalar function g is given by

$$\nabla g = h(\xi, \theta)\left(\frac{\partial g}{\partial\xi}e_\xi + \frac{\partial g}{\partial\theta}e_\theta\right). \tag{2.53}$$

Moreover, the normal and tangential derivatives of a function u in bipolar coordinates are

$$\begin{cases} \left.\dfrac{\partial u}{\partial\nu}\right|_{\xi=c} = \nabla u \cdot v_{\xi=c} = -\mathrm{sgn}(c)h(c,\theta)\left.\dfrac{\partial u}{\partial\xi}\right|_{\xi=c}, \\ \left.\dfrac{\partial u}{\partial T}\right|_{\xi=c} = -\mathrm{sgn}(c)h(c,\theta)\left.\dfrac{\partial u}{\partial\theta}\right|_{\xi=c}, \end{cases} \tag{2.54}$$

and the line element $d\sigma$ on the boundary $\{\xi = \xi_0\}$ is

$$d\sigma = \frac{1}{h(\xi_0, \theta)} d\theta.$$

Furthermore, the bipolar coordinate system admits separation of variables for any harmonic function f as follows:

(2.55)

$$f(\xi, \theta) = a_0 + b_0\xi + c_0\theta + \sum_{n=1}^{\infty} \big[(a_n e^{n\xi} + b_n e^{-n\xi}) \cos n\theta + (c_n e^{n\xi} + d_n e^{-n\xi}) \sin n\theta\big],$$

where a_n, b_n, c_n and d_n are constants.

We have

(2.56)

$$\begin{aligned} x + \sqrt{-1}y &= \frac{\sinh \xi - \sqrt{-1} \sin \theta}{\cosh \xi - \cos \theta} \\ &= \operatorname{sgn}(\xi)\Big(\frac{e^{\zeta} + e^{-\zeta}}{e^{\zeta} - e^{-\zeta}} = 1 + 2\sum_{n=1}^{\infty} e^{-n\xi}(\cos n\theta - \sqrt{-1} \sin n\theta)\Big) \end{aligned}$$

with $\zeta = (\xi + \sqrt{-1}\theta)/2$.

2.4.4.2. *Spectrum of* $\mathbb{K}^*_{\Omega_1 \cup \Omega_2}$. Suppose that the two disks Ω_1 and Ω_2 have the same radius r and let ϵ be their separation distance. Set

$$\alpha = \sqrt{\epsilon(r + \frac{\epsilon}{4})} \quad \text{and} \quad \xi_0 = \sinh^{-1}\Big(\frac{\alpha}{r}\Big). \tag{2.57}$$

Note that

$$\partial\Omega_j = \{\xi = (-1)^j \xi_0\} \quad \text{for } j = 1, 2. \tag{2.58}$$

To establish the spectral decomposition of $\mathbb{K}^*_{\Omega_1 \cup \Omega_2}$, we use the following lemma from [**34**].

LEMMA 2.22. *Assume that there exists* u *a nontrivial solution to the following equation:*

$$\begin{cases} \Delta u = 0 & \text{in } \Omega_1 \cup \Omega_2 \cup \mathbb{R}^2 \setminus \overline{(\Omega_1 \cup \Omega_2)}, \\ u|_+ = u|_- & \text{on } \partial\Omega_j, j = 1, 2, \\ \dfrac{\partial u}{\partial \nu}\Big|_+ = k \dfrac{\partial u}{\partial \nu}\Big|_- & \text{on } \partial\Omega_j, j = 1, 2, \\ u(x) \to 0 & \text{as } |x| \to \infty, \end{cases} \tag{2.59}$$

where $k = -\dfrac{1 + 2\lambda}{1 - 2\lambda} < 0$. *If we set*

$$\psi_j := \frac{\partial u}{\partial \nu}\Big|_+ - \frac{\partial u}{\partial \nu}\Big|_- \quad \text{on } \partial\Omega_j \text{ for } j = 1, 2,$$

then $\psi = \begin{bmatrix} \psi_1 \\ \psi_2 \end{bmatrix}$ *is an eigenvector of* $\mathbb{K}^*_{\Omega_1 \cup \Omega_2}$ *corresponding to the eigenvalue* λ.

One can see that the following function u_n is a solution to (2.59):

(2.60)
$$u_n^\pm(\xi,\theta) = \text{constant} + \begin{cases} \mp\dfrac{1}{2|n|}(e^{|n|\xi_0} \mp e^{-|n|\xi_0})e^{|n|\xi+\sqrt{-1}n\theta} & \text{for } \xi < -\xi_0, \\ \dfrac{1}{2|n|}e^{-|n|\xi_0}(e^{|n|\xi} \mp e^{-|n|\xi})e^{\sqrt{-1}n\theta} & \text{for } -\xi_0 < \xi < \xi_0, \\ \dfrac{1}{2|n|}(e^{|n|\xi_0} \mp e^{-|n|\xi_0})e^{-|n|\xi+\sqrt{-1}n\theta} & \text{for } \xi > \xi_0. \end{cases}$$

From Lemma 2.22 and (2.60), it follows that the eigenvalues and the associated eigenvectors of $\mathbb{K}^*_{\Omega_1\cup\Omega_2}$ on $\mathcal{H}_0^*$ are given by

$$\lambda_n^\pm = \pm\frac{1}{2}e^{-2|n|\xi_0}, \tag{2.61}$$

and

$$\Phi_n^\pm(\theta) = e^{\sqrt{-1}n\theta}\begin{bmatrix} h(-\xi_0,\theta) \\ \mp h(\xi_0,\theta). \end{bmatrix} \tag{2.62}$$

Note that the above eigenvectors are not normalized and both the eigenvalues and eigenvectors depend on the separation distance between Ω_1 and Ω_2.

We now compute $-\langle \mathbb{S}_{\Omega_1\cup\Omega_2}[\Phi_n^\pm], \Phi_n^\pm\rangle_{1/2,-1/2}$. From (2.60), we obtain that

$$\mathbb{S}_{\Omega_1\cup\Omega_2}[\Phi_n^\pm] = \text{constant } + \begin{bmatrix} \mp\frac{1}{2|n|}(1 \mp e^{-2|n|\xi_0})e^{\sqrt{-1}n\theta} \\ \frac{1}{2|n|}(1 \mp e^{-2|n|\xi_0})e^{\sqrt{-1}n\theta} \end{bmatrix}.$$

Thus

$$-\langle \mathbb{S}_{\Omega_1\cup\Omega_2}[\Phi_n^\pm], \Phi_n^\pm\rangle_{1/2,-1/2} = \frac{2\pi}{|n|}(1 \mp e^{-2|n|\xi_0}).$$

Therefore, we arrive at the following result, which was first proved in [**80**].

THEOREM 2.23. *We have the following spectral decomposition of* $\mathbb{K}^*_{\Omega_1\cup\Omega_2}$ *on* $\mathcal{H}_0^*$*:*

$$\mathbb{K}^*_{\Omega_1\cup\Omega_2} = \sum_{n\neq 0}\frac{1}{2}e^{-2|n|\xi_0}\Psi_n^+\otimes\Psi_n^+ + \sum_{n\neq 0}\left(-\frac{1}{2}e^{-2|n|\xi_0}\right)\Psi_n^-\otimes\Psi_n^-, \tag{2.63}$$

where $\otimes$ *denotes the tensor product and* $\Psi_n^\pm$ *are the normalized eigenvectors defined by*

$$\Psi_n^\pm(\theta) := \frac{\sqrt{|n|}e^{\sqrt{-1}n\theta}}{\sqrt{2\pi(1\mp e^{-2|n|\xi_0})}}\begin{bmatrix} h(-\xi_0,\theta) \\ \mp h(\xi_0,\theta) \end{bmatrix}. \tag{2.64}$$

Note that

(2.65)
$$(\mathcal{S}^0_{\Omega_1}[\Psi^\pm_{n,1}] + \mathcal{S}^0_{\Omega_2}[\Psi^\pm_{n,2}])(\xi,\theta) = \text{constant } + \frac{\sqrt{|n|}}{\sqrt{2\pi(1\mp e^{-2|n|\xi_0})}} \times \begin{cases} \mp\dfrac{1}{2|n|}(e^{|n|\xi_0} \mp e^{-|n|\xi_0})e^{|n|\xi+\sqrt{-1}n\theta} & \text{for } \xi < -\xi_0, \\ \dfrac{1}{2|n|}e^{-|n|\xi_0}(e^{|n|\xi} \mp e^{-|n|\xi})e^{\sqrt{-1}n\theta} & \text{for } -\xi_0 < \xi < \xi_0, \\ \dfrac{1}{2|n|}(e^{|n|\xi_0} \mp e^{-|n|\xi_0})e^{-|n|\xi+\sqrt{-1}n\theta} & \text{for } \xi > \xi_0. \end{cases}$$

2.4.5. Numerical Implementation.

2.4.5.1. *Numerical Representation.* In order to utilize the Neumann–Poincaré operator in applications we must define an appropriate numerical representation for it. We begin by parameterizing the boundary by $x(t)$ for $t \in [0, 2\pi)$. After partitioning the interval $[0, 2\pi)$ into N pieces

$$[t_1, t_2),\ [t_2, t_3), \ldots,\ [t_N, t_{N+1}),$$

with $t_1 = 0$ and $t_{N+1} = 2\pi$, we approximate the boundary $\partial\Omega = \{x(t) \in \mathbb{R}^2 : t \in [0, 2\pi)\}$ by $x^{(i)} = x(t_i)$ for $1 \le i \le N$. We then represent the infinite-dimensional operator $(\mathcal{K}_\Omega^0)^*$ acting on the density φ by a finite dimensional matrix K acting on the coefficient vector $\overline{\varphi}_i := \varphi(x^{(i)})$ for $1 \le i \le N$. We have

$$(\mathcal{K}_\Omega^0)^*[\varphi](x) = \frac{1}{2\pi}\text{p.v.}\int_{\partial\Omega} \frac{\langle x - y, \nu(x)\rangle}{|x-y|^2}\varphi(y)\,d\sigma(y),$$

for $\psi \in L^2(\partial\Omega)$ and we represent it numerically by

$$K\tilde{\psi} = \begin{pmatrix} K_{11} & K_{12} & \ldots & K_{1N} \\ K_{21} & K_{22} & \ldots & K_{2N} \\ \vdots & & \ddots & \vdots \\ K_{N1} & \ldots & \ldots & K_{NN} \end{pmatrix} \begin{pmatrix} \overline{\varphi}_1 \\ \overline{\varphi}_2 \\ \vdots \\ \overline{\varphi}_N \end{pmatrix},$$

where

$$K_{ij} = \frac{1}{2\pi}\frac{\langle x^{(i)} - x^{(j)}, \nu(x^{(i)})\rangle}{|x^{(i)} - x^{(j)}|^2}|T(x^{(i)})|(t_{j+1} - t_j) \quad i \neq j,$$

with $T(x^{(i)})$ being the tangent vector at $x^{(i)}$.

2.4.5.2. *Handling Singularities on the Diagonal.* Complications arise in the diagonal terms of K as the expression

$$\frac{\langle x^{(i)} - x^{(j)}, \nu(x^{(i)})\rangle}{|x^{(i)} - x^{(j)}|^2},$$

is singular when $i = j$. We can handle this by explicitly calculating the integrals for the diagonal terms. Let the portion of the boundary starting at $x^{(i)}$ and ending at $x^{(i+1)}$ be parameterized by $s \in [0, \varepsilon = \frac{2\pi}{N})$, which means that $\varepsilon \to 0$ as the number of discretization points $N \to \infty$. Applying this parameterization to the diagonal terms of K we have

$$K_{ii} = \frac{1}{2\pi}\int_0^\varepsilon \frac{\langle x^{(i)} - x(s), \nu(x^{(i)})\rangle}{|x^{(i)} - x(s)|^2}|T(s)|ds.$$

Denote by

$$\begin{aligned} T^{(i)} &= T(s) = r'(s), \\ \nu^{(i)} &= \nu(s), \\ a^{(i)} &= a(s) = r''(s), \end{aligned}$$

the tangent vector, the unit normal vector, and the acceleration vector respectively. Note that $x^{(i)} = x(0)$, $T^{(i)} = x'(0)$, and $a^{(i)} = x''(0)$. Taylor expanding the numerator for small s we have

$$
\begin{aligned}
\langle x^{(i)} - x(s), \nu(x^{(i)})\rangle &= \langle x^{(i)} - (x(0) + sT^{(i)} + \frac{s^2}{2}a^{(i)} + O(s^3)), \nu^{(i)}\rangle \\
&\approx -\frac{s^2}{2}\langle a^{(i)}, \nu^{(i)}\rangle,
\end{aligned}
$$

as $\varepsilon \to 0$. Similarly, we have $|x^{(i)} - x(s)|^2 \approx s^2|T^{(i)}|^2$ and $|T(s)| \approx |T^{(i)}|$ as $\varepsilon \to 0$. Therefore we can approximate the diagonal terms of K by

$$
\begin{aligned}
K_{ii} &\approx -\frac{1}{2\pi}\int_0^{\varepsilon} \frac{\langle a^{(i)}), \nu^{(i)}\rangle}{2|T^{(i)}|^2}|T^{(i)}|ds \\
&= -\frac{\varepsilon}{4\pi}\frac{\langle a^{(i)}), \nu^{(i)}\rangle}{|T^{(i)}|} \\
&= -\frac{1}{2N}\frac{\langle a^{(i)}), \nu^{(i)}\rangle}{|T^{(i)}|}.
\end{aligned}
$$

We now present some examples that demonstrate the spectrum of the Neumann–Poincaré operator in various situations. The corresponding Matlab code can be downloaded at `www.ams.org/bookpages/surv-235`.

2.4.5.3. *Spectrum of the Neumann–Poincaré Operator for an Ellipse.* We compute the spectrum of $(\mathcal{K}_{\Omega}^0)^*$ for an ellipse with semi-axes $a_1 = 10$ and $a_2 = 1$. Table 2.1 compares the first few eigenvalues obtained numerically with the eigenvalues obtained via the formula given in (2.37).

Theoretical	Numerical
0.5000	0.5000
0.4091	0.4091
−0.4091	−0.4091
0.3347	0.3347
−0.3347	−0.3347
0.2739	0.2739
−0.2739	−0.2739
0.2241	0.2241

TABLE 2.1. Spectrum of the Neumann–Poincaré operator for an ellipse.

2.4.5.4. *Spectrum of the Neumann–Poincaré Operator for Two Disks.* We now compute the spectrum of $\mathbb{K}^*_{\Omega_1\cup\Omega_2}$ for two disks with $r = 2$ and $\epsilon = 0.3$. Table 2.2 compares the first few eigenvalues obtained numerically with the eigenvalues obtained via the formula given in (2.61).

2.5. Conductivity Problem in the Free Space

2.5.1. Far-Field Expansion. Let Ω be a Lipschitz bounded domain in $\mathbb{R}^d$ and suppose that the origin $O \in \Omega$. Let $0 < k \neq 1 < +\infty$ and denote

$$
\lambda(k) := (k+1)/(2(k-1)).
$$

Theoretical	Numerical
0.5000	0.5000
0.5000	0.5000
−0.2315	−0.2315
−0.2315	−0.2315
0.2315	0.2315
0.2315	0.2315
−0.1072	−0.1072
−0.1072	−0.1072

TABLE 2.2. Spectrum of the Neumann–Poincaré operator for two disks.

Let h be a harmonic function in $\mathbb{R}^d$, and let u be the solution to the following transmission problem in free space:

$$\begin{cases} \nabla \cdot ((1+(k-1)\chi(\Omega))\nabla u_k) = 0 & \text{in } \mathbb{R}^d, \\ u_k(x) - h(x) = O(|x|^{1-d}) & \text{as } |x| \to +\infty. \end{cases} \tag{2.66}$$

For a multi-index $\alpha = (\alpha_1, \ldots, \alpha_d) \in \mathbb{N}^d$, let

$$\partial^\alpha f = \partial_1^{\alpha_1} \ldots \partial_d^{\alpha_d} f \quad \text{and} \quad x^\alpha := x_1^{\alpha_1} \ldots x_d^{\alpha_d}.$$

We can easily prove that

$$u_k(x) = h(x) + \mathcal{S}_\Omega^0 (\lambda(k)I - (\mathcal{K}_\Omega^0)^*)^{-1}[\frac{\partial h}{\partial \nu}|_{\partial\Omega}](x) \quad \text{for } x \in \mathbb{R}^d, \tag{2.67}$$

which together with the Taylor expansion

$$\Gamma_0(x-y) = \sum_{\alpha, |\alpha|=0}^{+\infty} \frac{(-1)^{|\alpha|}}{\alpha!} \partial_x^\alpha \Gamma_0(x) y^\alpha, \quad y \text{ in a compact set}, \quad |x| \to +\infty,$$

yields the far-field expansion

$$\begin{aligned} (u_k - h)(x) &= \sum_{|\alpha|, |\beta|=1}^{+\infty} \frac{(-1)^{|\alpha|}}{\alpha! \beta!} \partial_x^\alpha \Gamma_0(x) \partial^\beta h(0) \\ &\quad \times \int_{\partial\Omega} (\lambda(k)I - (\mathcal{K}_\Omega^0)^*)^{-1}\big[\nu(x) \cdot \nabla x^\alpha\big](y) y^\beta \, d\sigma(y) \end{aligned} \tag{2.68}$$

as $|x| \to +\infty$.

DEFINITION 2.24. *For $\alpha, \beta \in \mathbb{N}^d$, we define the generalized polarization tensor $M_{\alpha\beta}$ by*

$$M_{\alpha\beta}(\lambda(k), \Omega) := \int_{\partial\Omega} y^\beta \phi_\alpha(y) \, d\sigma(y), \tag{2.69}$$

where ϕ_α is given by

$$\phi_\alpha(y) := (\lambda(k)I - (\mathcal{K}_\Omega^0)^*)^{-1}\big[\nu(x) \cdot \nabla x^\alpha\big](y), \quad y \in \partial\Omega. \tag{2.70}$$

If $|\alpha| = |\beta| = 1$, we denote $M_{\alpha\beta}$ by $(m_{pq})_{p,q=1}^d$ and call $M = (m_{pq})_{p,q=1}^d$,

$$m_{pq} := \int_{\partial\Omega} y_q (\lambda(k)I - (\mathcal{K}_\Omega^0)^*)^{-1}\big[\nu_p\big](y) \, d\sigma(y), \tag{2.71}$$

with $\nu = (\nu_1, \ldots, \nu_d)$, the polarization tensor.

Formula (2.68) shows that through the generalized polarization tensors we have complete information about the far-field expansion of u_k:

$$(u_k - h)(x) = \sum_{|\alpha|,|\beta|=1}^{+\infty} \frac{(-1)^{|\alpha|}}{\alpha!\beta!} \partial_x^\alpha \Gamma_0(x) M_{\alpha\beta}(\lambda(k), \Omega) \partial^\beta h(0)$$

as $|x| \to +\infty$.

2.5.2. Polarization Tensor. In this subsection, we derive some important properties satisfied by the polarization tensor. It is worth mentioning that the concept of polarization tensor has been widely used in various areas such as the imaging of small particles and effective medium theory (see [**51**, **52**, **54**, **74**, **397**, **409**] for these applications).

For a $\mathcal{C}^{1,\eta}$, $\eta > 0$, domain Ω in $\mathbb{R}^d$, using (2.19) we can write

$$(\lambda(k)I - (\mathcal{K}_\Omega^0)^*)^{-1}[\psi] = \sum_{j=0}^{\infty} \frac{\langle \psi, \varphi_j \rangle_{\mathcal{H}^*} \varphi_j}{\lambda(k) - \lambda_j},$$

with (λ_j, φ_j) being the eigenvalues (counted with their multiplicities) and normalized eigenvectors of $(\mathcal{K}_\Omega^0)^*$ in $\mathcal{H}^*$. Hence, the entries of the polarization tensor M can be decomposed as

$$m_{pq}(\lambda(k), \Omega) = \sum_{j=1}^{\infty} \frac{\langle \nu_p, \varphi_j \rangle_{\mathcal{H}^*} \langle x_q, \varphi_j \rangle_{\frac{1}{2},-\frac{1}{2}}}{\lambda(k) - \lambda_j}. \tag{2.72}$$

Note that $\langle \chi(\partial\Omega), \nu_p \rangle_{\frac{1}{2},-\frac{1}{2}} = 0$. So, we have $\langle \nu_p, \varphi_0 \rangle_{\mathcal{H}^*} = 0$. Moreover, since

$$\begin{aligned}
\langle x_q, \varphi_j \rangle_{\frac{1}{2},\frac{1}{2}} &= \Big\langle x_q, \big(\frac{1}{2} - \lambda_j\big)^{-1} \big(\frac{1}{2}I - (\mathcal{K}_\Omega^0)^*\big)[\varphi_j] \Big\rangle_{\frac{1}{2},-\frac{1}{2}} \\
&= \frac{-1}{1/2 - \lambda_j} \Big\langle x_q, \frac{\partial \mathcal{S}_\Omega^0[\varphi_j]}{\partial \nu}\Big|_- \Big\rangle_{\frac{1}{2},-\frac{1}{2}} \\
&= \frac{-1}{1/2 - \lambda_j} \Big[\int_{\partial\Omega} \frac{\partial x_q}{\partial \nu} \mathcal{S}_\Omega^0[\varphi_j] d\sigma - \int_\Omega \Big(\Delta x_q \mathcal{S}_\Omega^0[\varphi_j] - x_q \Delta \mathcal{S}_\Omega^0[\varphi_j] \Big) dx \Big] \\
&= \frac{\langle \nu_q, \varphi_j \rangle_{\mathcal{H}^*}}{1/2 - \lambda_j}, \qquad (2.73)
\end{aligned}$$

it follows that

$$m_{pq}(\lambda(k), \Omega) = \sum_{j=1}^{\infty} \frac{\langle \nu_p, \varphi_j \rangle_{\mathcal{H}^*} \langle \nu_q, \varphi_j \rangle_{\mathcal{H}^*}}{(1/2 - \lambda_j)(\lambda(k) - \lambda_j)} = \sum_{j=1}^{\infty} \frac{\alpha_{pq}^{(j)}}{\lambda(k) - \lambda_j}. \tag{2.74}$$

Here, we have used the fact that $\mathcal{S}_\Omega^0[\varphi_j]$ is harmonic in Ω and introduced

$$\alpha_{pq}^{(j)} := \frac{1}{1/2 - \lambda_j} \langle \nu_p, \varphi_j \rangle_{\mathcal{H}^*} \langle \nu_q, \varphi_j \rangle_{\mathcal{H}^*}. \tag{2.75}$$

Notice that $\alpha_{pp}^{(j)} \geq 0$, for all $p = 1, \ldots, d$, and $j \geq 1$.

REMARK 2.25. *If Ω is a bounded Lipschitz domain, then for any k such that $\lambda(k) \notin \sigma((\mathcal{K}_\Omega^0)^*)$, it follows from (2.31) that*

$$(2.76)\quad \begin{aligned} m_{pq}(\lambda(k),\Omega) &= \int_{t\in\sigma((\mathcal{K}_\Omega^0)^*)} \frac{1}{\lambda(k)-t}\int_{\partial\Omega} \nu_p d\mathcal{E}(t)[x_q] d\sigma(x) \\ &= \int_{t\in\sigma((\mathcal{K}_\Omega^0)^*)} \frac{1}{\lambda(k)-t}\int_{\partial\Omega} \nu_q d\mathcal{E}(t)[x_p] d\sigma(x)\,. \end{aligned}$$

From (2.74), one can see that the following properties of the polarization tensor hold.

PROPOSITION 2.26. *The polarization tensor $M(\lambda(k),\Omega)$ is symmetric and if $k>1$, then $M(\lambda(k),\Omega)$ is positive definite, and it is negative definite if $0<k<1$.*

The following sum rules are from [**78, 396**].

PROPOSITION 2.27. *For $d \geq 2$, we have*

$$(2.77)\quad \sum_{j=1}^\infty \alpha_{pq}^{(j)} = \delta_{pq}|\Omega|,$$

and

$$(2.78)\quad \sum_{j=1}^\infty \lambda_j \sum_{l=1}^d \alpha_{ll}^{(j)} = \frac{(d-2)}{2}|\Omega|.$$

PROOF. Let f be a holomorphic function defined in an open set $U \subset \mathbb{C}$ containing the spectrum, $\sigma((\mathcal{K}_\Omega^0)^*)$, of $(\mathcal{K}_\Omega^0)^*$. if $f(z) = \sum_{j=0}^\infty a_j z^j$ for $z \in U$, then

$$f((\mathcal{K}_\Omega^0)^*) := \sum_{j=0}^\infty a_j ((\mathcal{K}_\Omega^0)^*)^j,$$

where

$$((\mathcal{K}_\Omega^0)^*)^j := \underbrace{(\mathcal{K}_\Omega^0)^* \circ (\mathcal{K}_\Omega^0)^* \circ \ldots \circ (\mathcal{K}_\Omega^0)^*}_{j \text{ times}}.$$

Since $(\mathcal{K}_\Omega^0)^*$ is compact and self-adjoint, we have

$$f((\mathcal{K}_\Omega^0)^*) = \sum_{j=1}^\infty f(\lambda_j)\langle \cdot, \varphi_j\rangle_{\mathcal{H}^*}\varphi_j\,.$$

Hence

$$(2.79)\quad \int_{\partial\Omega} x_p f((\mathcal{K}_\Omega^0)^*)[\nu_q](x)\, d\sigma(x) = \sum_{j=1}^\infty f(\lambda_j)\alpha_{pq}^{(j)}\,.$$

Equation (2.79) yields the summation rule (2.77) for the entries of the polarization tensor by taking $f(z)=1$. In order to prove (2.78), we set $f(z)=z$ in (2.79) and write

$$\sum_{j=1}^{\infty} \lambda_j \sum_{l=1}^{d} \alpha_{ll}^{(j)} = \sum_{l=1}^{d} \int_{\partial\Omega} x_l (\mathcal{K}_\Omega^0)^*[\nu_l](x)\, d\sigma(x),$$

$$\int_{\partial\Omega} x_l (\mathcal{K}_\Omega^0)^*[\nu_l](x)\, d\sigma(x) = \int_{\partial\Omega} x_l \left(\frac{1}{2}\nu_l(x) + \frac{\partial \mathcal{S}_\Omega^0[\nu_l]}{\partial \nu}\Big|_-(x) \right) d\sigma(x),$$

$$= \frac{|\Omega|}{2} + \int_{\partial\Omega} x_l \frac{\partial \mathcal{S}_\Omega^0[\nu_l]}{\partial \nu}\Big|_-(x) d\sigma(x). \tag{2.80}$$

Integrating by parts we arrive at

$$\sum_{l=1}^{d} \int_{\partial\Omega} x_l \frac{\partial \mathcal{S}_\Omega^0[\nu_l]}{\partial \nu}\Big|_-(x) d\sigma(x) = -\int_\Omega \left(\int_{\partial\Omega} \nu(x') \cdot \nabla_{x'} \Gamma_0(x,x') d\sigma(x') \right) dx,$$

$$= -\int_\Omega \mathcal{D}_\Omega^0[1](x) dx,$$

$$= -\,|\Omega|. \tag{2.81}$$

Hence, summing equation (2.80) for $l = 1, \ldots, d$, we get (2.78).

□

In two dimensions, by using the twin spectrum property stated in Lemma 2.10, we can rewrite the entries of the polarization tensor in the form

$$m_{pq}(\lambda(k), \Omega) = \sum_{j, \lambda_j \geq 0}^{\infty} \left[\frac{\alpha_{pq}^{(j)}}{\lambda(k) - \lambda_j} + \frac{\widetilde{\alpha}_{pq}^{(j)}}{\lambda(k) + \lambda_j} \right], \tag{2.82}$$

where $\widetilde{\alpha}_{pq}^{(j)}$ is $\alpha_{pq}^{(k)}$ for k such that $\lambda_k = -\lambda_j$.

The following result holds.

LEMMA 2.28. *For all $j \geq 1$ such that $\lambda_j \geq 0$, we have*

$$\widetilde{\alpha}_{22}^{(j)} = \alpha_{11}^{(j)},$$
$$\widetilde{\alpha}_{12}^{(j)} = -\alpha_{12}^{(j)}.$$

PROOF. For simplicity, suppose that $(\mathcal{K}_\Omega^0)^*$ has simple eigenvalues. Let $\widetilde{\varphi}_j$ be the normalized eigenfunction associated with $-\lambda_j$. Recall from Lemma 2.11 that

$$\widetilde{\varphi}_j = \frac{\frac{\partial}{\partial T} \mathcal{S}_\Omega^0[\varphi_j]}{\|\frac{\partial}{\partial T} \mathcal{S}_\Omega^0[\varphi_j]\|_{\mathcal{H}^*}},$$

where φ_j is the (normalized) eigenfunction associated with λ_j. On the other hand, from (2.75), we have

$$\widetilde{\alpha}_{pq}^{(j)} = \frac{1}{1/2 + \lambda_j} \langle \nu_p, \widetilde{\varphi}_j \rangle_{\mathcal{H}^*} \langle \nu_q, \widetilde{\varphi}_j \rangle_{\mathcal{H}^*}. \tag{2.83}$$

Since x_2 is the harmonic conjugate of x_1, the Cauchy-Riemann equations yield

$$\nu_1 = \frac{\partial x_2}{\partial T}, \quad \nu_2 = -\frac{\partial x_1}{\partial T}.$$

Hence, it follows that

$$\begin{aligned}\langle \nu_2, \widetilde{\varphi}_j \rangle_{\mathcal{H}^*} &= -\langle \mathcal{S}_\Omega^0[\widetilde{\varphi}_j], \nu_2 \rangle_{\frac{1}{2},-\frac{1}{2}} \\ &= \langle \mathcal{S}_\Omega^0[\widetilde{\varphi}_j], \frac{\partial x_1}{\partial T} \rangle_{\frac{1}{2},-\frac{1}{2}} \\ &= -\| \frac{\partial}{\partial T} \mathcal{S}_\Omega^0[\widetilde{\varphi}_j] \|_{\mathcal{H}^*} \langle x_1, \varphi_j \rangle_{\frac{1}{2},-\frac{1}{2}}.\end{aligned}$$

We then obtain from (2.30) and (2.73) that

$$\langle \nu_2, \widetilde{\varphi}_j \rangle_{\mathcal{H}^*} = -\frac{\sqrt{\frac{1}{4} - \lambda_j^2}}{1/2 - \lambda_j} \langle \nu_1, \varphi_j \rangle_{\mathcal{H}^*}.$$

Similarly, we have

$$\langle \nu_1, \widetilde{\varphi}_j \rangle_{\mathcal{H}^*} = \frac{\sqrt{\frac{1}{4} - \lambda_j^2}}{1/2 - \lambda_j} \langle \nu_2, \varphi_j \rangle_{\mathcal{H}^*}.$$

From the definitions (2.75) and (2.83) of $\alpha_{pq}^{(j)}$ and $\widetilde{\alpha}_{pq}^{(j)}$, we obtain the desired identities. □

In view of the connection of the concept of the polarization tensor to the theory of composites (see Section 7.3), it is natural for the polarization tensor to have the following bounds, which are called the Hashin-Shtrikman bounds after the names of the scientists who found optimal bounds for the effective conductivity [**285**, **397**].

PROPOSITION 2.29. *If Ω is a smooth bounded domain in $\mathbb{R}^2$, then the polarization tensor associated with Ω and the conductivity parameter $0 < k \neq 1 < +\infty$ satisfies*

$$\frac{1}{k-1} \operatorname{tr}(M(\lambda(k), \Omega)) < (1 + \frac{1}{k})|\Omega| \tag{2.84}$$

and

$$(k-1) \operatorname{tr}(M(\lambda(k), \Omega)^{-1}) \leq \frac{(1+k)}{|\Omega|}. \tag{2.85}$$

PROOF. From Lemma 2.28, we have

$$\begin{aligned}\operatorname{tr}(M(\lambda(k), \Omega)) &= \sum_{j=1}^{+\infty} \frac{\alpha_{11}^{(j)}}{\lambda(k) - \lambda_j} + \frac{\alpha_{11}^{(j)}}{\lambda(k) + \lambda_j} \\ &= \sum_{j=1}^{+\infty} \frac{2\lambda(k)\alpha_{11}^{(j)}}{\lambda(k)^2 - \lambda_j^2}.\end{aligned}$$

Since $|\lambda_j| < 1/2$ and $\alpha_{11}^{(j)} \geq 0$ for $j \geq 1$, by using the sum rule (2.77), it follows that

$$|\operatorname{tr}(M(\lambda(k), \Omega))| < \frac{2|\lambda(k)|}{\lambda(k)^2 - 1/4} |\Omega|,$$

which yields (2.84).

Concerning $\mathrm{tr}(M(\lambda(k),\Omega)^{-1})$, by rotating the coordinate system in such a way that the orthogonal eigenbasis of $M(\lambda(k),\Omega)$ is parallel to the two coordinate axes we have that

$$\mathrm{tr}(M(\lambda(k),\Omega)^{-1}) = \frac{1}{\displaystyle\sum_{j=1}^{+\infty} \frac{\alpha_{11}^{(j)}}{\lambda(k)-\lambda_j}} + \frac{1}{\displaystyle\sum_{j=1}^{+\infty} \frac{\alpha_{22}^{(j)}}{\lambda(k)-\lambda_j}}.$$

Direct calculations show that $\mathrm{tr}(M(\lambda(k),\Omega)^{-1})$ is concave in $\lambda(k) \in [1/2,+\infty)$. Moreover, the asymptotic expansion for $\lambda(k) \to +\infty$

$$\mathrm{tr}(M(\lambda(k),\Omega)^{-1}) = \frac{2\lambda(k)}{|\Omega|} + O(\lambda(k)^{-1})$$

holds. Therefore, we have

$$\big|\,\mathrm{tr}(M(\lambda(k),\Omega)^{-1})\big| \leq \frac{2|\lambda(k)|}{|\Omega|},$$

for $\lambda(k) \geq 1/2$. Lemma 2.28 implies that

$$\mathrm{tr}(M(-\lambda(k),\Omega)^{-1}) = -\,\mathrm{tr}(M(\lambda(k),\Omega)).$$

Thus,

$$\big|\,\mathrm{tr}(M(\lambda(k),\Omega)^{-1})\big| \leq \frac{2|\lambda(k)|}{|\Omega|},$$

for $\lambda(k) \in \mathbb{R}\setminus[-1/2,1/2]$, which completes the proof of the proposition. □

The bounds (2.84) and (2.85) were obtained in [**189**, **374**] and proved to be optimal in [**28**, **189**]. The proof of Proposition 2.29 given here is from [**276**]. In view of (2.76), the bounds (2.84) and (2.85) hold true for Lipschitz bounded domains.

If Ω is an ellipse of the form $R(\Omega')$ where R is a rotation by θ and Ω' is an ellipse of the form

$$\frac{x_1^2}{a_1^2} + \frac{x_2^2}{a_2^2} \leq 1,$$

then it is known (see [**52**, pp. 81-122] for example) that its polarization tensor is given by

$$\begin{aligned} M(\lambda(k),\Omega) &= (k-1)|\Omega| R \begin{pmatrix} \dfrac{a_1+a_2}{a_1+ka_2} & 0 \\ 0 & \dfrac{a_1+a_2}{ka_1+a_2} \end{pmatrix} R^t \\ &= R \begin{pmatrix} \dfrac{|\Omega|}{\lambda(k) - \frac{1}{2}\frac{a_1-a_2}{a_1+a_2}} & 0 \\ 0 & \dfrac{|\Omega|}{\lambda(k) + \frac{1}{2}\frac{a_1-a_2}{a_1+a_2}} \end{pmatrix} R^t. \end{aligned} \tag{2.86}$$

Thus for a given polarization tensor there exists a corresponding unique ellipse whose polarization tensor is the same as the given one [**166**].

In the three-dimensional case, a domain for which analogous analytical expressions for the elements of its polarization tensor M are available is the ellipsoid.

If the coordinate axes are chosen to coincide with the principal axes of Ω whose equation then becomes

$$\frac{x_1^2}{a_1^2} + \frac{x_2^2}{a_2^2} + \frac{x_3^2}{a_3^2} \leq 1, \quad 0 < a_3 \leq a_2 \leq a_1,$$

then M takes the form

(2.87)

$$M(\lambda(k), \Omega) = (k-1)|\Omega| \begin{pmatrix} \dfrac{1}{(1-A_1)+kA_1} & 0 & 0 \\ 0 & \dfrac{1}{(1-A_2)+kA_2} & 0 \\ 0 & 0 & \dfrac{1}{(1-A_3)+kA_3} \end{pmatrix},$$

where the constants A_1, A_2, and A_3 are defined by

$$\begin{aligned} A_1 &= \frac{a_2 a_3}{a_1^2} \int_1^{+\infty} \frac{1}{t^2 \sqrt{t^2 - 1 + (\frac{a_2}{a_1})^2} \sqrt{t^2 - 1 + (\frac{a_3}{a_1})^2}} \, dt, \\ A_2 &= \frac{a_2 a_3}{a_1^2} \int_1^{+\infty} \frac{1}{(t^2 - 1 + (\frac{a_2}{a_1})^2)^{\frac{3}{2}} \sqrt{t^2 - 1 + (\frac{a_3}{a_1})^2}} \, dt, \\ A_3 &= \frac{a_2 a_3}{a_1^2} \int_1^{+\infty} \frac{1}{\sqrt{t^2 - 1 + (\frac{a_2}{a_1})^2} (t^2 - 1 + (\frac{a_3}{a_1})^2)^{\frac{3}{2}}} \, dt. \end{aligned}$$

In the special case, $a_1 = a_2 = a_3$, Ω becomes a ball and $A_1 = A_2 = A_3 = 1/3$. Hence the polarization tensor associated with the ball is given by

$$M(\lambda(k), \Omega) = (k-1)|\Omega| \begin{pmatrix} \dfrac{3}{2+k} & 0 & 0 \\ 0 & \dfrac{3}{2+k} & 0 \\ 0 & 0 & \dfrac{3}{2+k} \end{pmatrix}. \tag{2.88}$$

Derivation of the above formulas can be found in [**397**].

It is worth mentioning that the polarization tensors for ellipses (or ellipsoids) satisfy the lower Hashin-Shtrikman bound (2.85). In [**319, 320**], the converse was also proved to be true.

Formula (2.86) shows that if Ω is an ellipse, then $M(\lambda(k), \Omega)$ as a meromorphic function of $\lambda(k)$ has at most two poles (given by $\pm\frac{1}{2}\frac{a-b}{a+b}$) and therefore, in view of the fact that, in $\mathcal{H}^*(\partial\Omega)$,

$$\sigma((\mathcal{K}_\Omega^0)^*) \setminus \{1/2\} = \left\{ \pm\frac{1}{2}\left(\frac{a-b}{a+b}\right)^j, \quad j = 1, 2, \ldots \right\},$$

all $\alpha_{pq}^{(j)}$ other than those corresponding to the eigenvalues $\pm\frac{1}{2}\frac{a-b}{a+b}$ vanish.

The converse is also true. If $M(\lambda(k), \Omega)$ as a meromorphic function of $\lambda(k)$ has at most two poles, then Ω is an ellipse. In particular, if the poles are 0 (of degree two), then Ω is a disk. The proof first given in [**276**] follows from the strong

Eshelby conjecture, which can be stated as follows: If for a nontrivial (c_1, c_2) the gradient in Ω of the solution u to the transmission problem

$$\begin{cases} \Delta u = 0 \quad \text{in } \mathbb{R}^2 \setminus \partial\Omega, \\ u|_+ = u|_- \quad \text{on } \partial\Omega, \\ \dfrac{\partial u}{\partial \nu}|_+ = k\dfrac{\partial u}{\partial \nu}|_- \quad \text{on } \partial\Omega, \\ u(x) - (c_1 x_1 + c_2 x_2) \to 0 \quad \text{as } |x| \to +\infty, \end{cases} \tag{2.89}$$

is constant in Ω, then Ω is an ellipse. Here, $0 < k \neq 1 < +\infty$. The strong Eshelby conjecture was proved in [**320, 377, 476**].

PROPOSITION 2.30. *Let Ω be a bounded and simply connected smooth domain in $\mathbb{R}^2$. If the meromorphic function $\lambda \mapsto M(\lambda, \Omega)$ has at most two poles, then Ω is an ellipse.*

PROOF. Let $\pm\mu$ denote the two poles of $M(\lambda, \Omega)$. Note that $M(\lambda, \Omega)$ either has two poles if $\mu \neq 0$ or one pole when $\mu = 0$. In view of (2.75), (2.82), and Lemma 2.28, we can write

$$M(\lambda, \Omega) = \begin{pmatrix} \dfrac{r_1^2}{\lambda - \mu} + \dfrac{r_2^2}{\lambda + \mu} & \dfrac{r_1 r_2}{\lambda - \mu} - \dfrac{r_1 r_2}{\lambda + \mu} \\ \dfrac{r_1 r_2}{\lambda - \mu} - \dfrac{r_1 r_2}{\lambda + \mu} & \dfrac{r_2^2}{\lambda - \mu} + \dfrac{r_1^2}{\lambda + \mu} \end{pmatrix},$$

where r_1 and $r_2 \geq 0$. Let $c = \begin{bmatrix} c_1 \\ c_2 \end{bmatrix}$ with

$$c_j := \frac{r_j}{\sqrt{r_1^2 + r_2^2}}, \quad j = 1, 2. \tag{2.90}$$

We can easily see that, for all $\lambda \in \mathbb{C} \backslash \sigma((\mathcal{K}_\Omega^0)^*)$,

$$c \cdot M(\lambda, \Omega) c = \frac{r_1^2 + r_2^2}{\lambda - \mu} = \frac{|\Omega|}{\lambda - \mu} = \int_{\partial\Omega} c \cdot x (\lambda I - (\mathcal{K}_\Omega^0)^*)^{-1}[c \cdot \nu](x) d\sigma(x).$$

Consequently, we have

$$\frac{|\Omega|}{\lambda - \mu} = \int_{\partial\Omega} c \cdot \nu(x) (\lambda I - (\mathcal{K}_\Omega^0))^{-1}[c \cdot y](x) d\sigma(x).$$

Therefore, from (2.76) it follows that $c \cdot \nu$ and $c \cdot x$ are eigenfunctions of $(\mathcal{K}_\Omega^0)^*$ and $(\mathcal{K}_\Omega^0)$, respectively, associated with the eigenvalue μ.

Since $c \cdot x$ is harmonic, we have on $\partial\Omega$

$$\begin{aligned} \mathcal{S}_\Omega^0[c \cdot \nu](x) &= \mathcal{D}_\Omega^0[c \cdot y](x) - c \cdot x \\ &= (\frac{1}{2} I + \mathcal{K}_\Omega^0)[c \cdot y](x) - c \cdot x \\ &= (\mu - \frac{1}{2}) c \cdot x, \end{aligned}$$

and hence, by the uniqueness of a solution to the Dirichlet problem

$$\mathcal{S}_\Omega^0[c \cdot \nu](x) = (\mu - \frac{1}{2}) c \cdot x, \quad x \in \Omega.$$

Therefore, the solution u to (2.89) is given by

$$\begin{aligned} u(x) &= \mathcal{S}_\Omega^0(\lambda I - (\mathcal{K}_\Omega^0)^*)^{-1}[c\cdot\nu](x) + c\cdot x \\ &= \left(\frac{\mu - \frac{1}{2}}{\lambda - \mu} - 1\right) c\cdot x. \end{aligned}$$

Since $\nabla u(x)$ is constant in Ω, it follows from the strong Eshelby conjecture that Ω is an ellipse. □

2.5.3. Conductivity Equation with Complex Coefficients.

Suppose that $k \in \mathbb{C}$. Then formula (2.67) holds true provided that $\lambda(k) \notin \sigma((\mathcal{K}_\Omega^0)^*)$ [**316**].

THEOREM 2.31. *Let $k \in \mathbb{C}$. If $\lambda(k) \notin \sigma((\mathcal{K}_\Omega^0)^*)$, then, for any harmonic function h in $\mathbb{R}^d$, the unique solution u_k to (2.66) satisfies*

$$\|\nabla(u_k - h)\|_{L^2(\mathbb{R}^d)} \le \frac{C}{\mathrm{dist}\big(\lambda(k), \sigma((\mathcal{K}_\Omega^0)^*)\big)} \|\frac{\partial h}{\partial \nu}\|_{H^{-1/2}(\partial\Omega)} \tag{2.91}$$

for some constant C independent of k. Here, dist *denotes the distance.*

PROOF. The existence of a solution to (2.66) follows from (2.67). To prove (2.91), we note that

$$\begin{aligned} \|\nabla(u_k - h)\|^2_{L^2(\mathbb{R}^d)} &= \int_\Omega \left|\nabla \mathcal{S}_\Omega^0[\varphi_k](x)\right|^2 dx + \int_{\mathbb{R}^d\setminus\overline{\Omega}} \left|\nabla \mathcal{S}_\Omega^0[\varphi_k](x)\right|^2 dx \\ &= \int_{\partial\Omega} \frac{\partial}{\partial\nu}\mathcal{S}_\Omega^0[\varphi_k]\big|_- \mathcal{S}_\Omega^0[\varphi_k]\, d\sigma - \int_{\partial\Omega} \frac{\partial}{\partial\nu}\mathcal{S}_\Omega^0[\varphi_k]\big|_+ \mathcal{S}_\Omega^0[\varphi_k]\, d\sigma \\ &= -\int_{\partial\Omega} \varphi_k \mathcal{S}_\Omega^0[\varphi_k]\, d\sigma = \|\varphi_k\|^2_{\mathcal{H}^*}, \end{aligned}$$

where φ_k is given by

$$\varphi_k = (\lambda(k)I - (\mathcal{K}_\Omega^0)^*)^{-1}[\frac{\partial h}{\partial\nu}|_{\partial\Omega}]. \tag{2.92}$$

So (2.91) follows from (2.67).

To show the uniqueness of the solution, assume that u_k^1 and u_k^2 satisfy (2.66). Let $v = u_k^1 - u_k^2$. Then v is a solution to (2.66) with $h = 0$. So we have

$$\begin{aligned} 0 &= \int_{\mathbb{R}^d} \big(\chi(\mathbb{R}^d \setminus \overline{\Omega} + k\chi(\Omega)\big)|\nabla v|^2\, dx \\ &= \int_{\mathbb{R}^d\setminus\overline{\Omega}} |\nabla v|^2\, dx + \Re k \int_\Omega |\nabla v|^2\, dx + \sqrt{-1}\Im k \int_\Omega |\nabla v|^2\, dx. \end{aligned}$$

Hence, if $\Re k > 0$, or if $\Re k \le 0$ and $\Im k \ne 0$, then

$$\int_{\mathbb{R}^d\setminus\overline{\Omega}} |\nabla v|^2\, dx = \int_\Omega |\nabla v|^2\, dx = 0.$$

So, v is constant. Since $v \to 0$ as $|x| \to \infty$, we conclude that $v = 0$.

Uniqueness for the case $k \le 0$ (and $\lambda(k) \notin \sigma((\mathcal{K}_\Omega^0)^*)$) can be proved as a limiting case of $k + \sqrt{-1}\delta$ as $\delta \to 0$. □

The following result on Lipschitz dependency on k of the solution u_k to (2.66) holds [**316**].

THEOREM 2.32. *Let $k \in \mathbb{C}$. If $\lambda(k) \notin \sigma((\mathcal{K}_\Omega^0)^*)$, then, for $|k'-k|$ small enough, there exists a positive constant C independent of k such that*

$$\|\nabla(u_k - u_{k'})\|_{L^2(\mathbb{R}^d)} \leq \frac{C|k'-k|}{\operatorname{dist}\big(\lambda(k), \sigma((\mathcal{K}_\Omega^0)^*)\big)} \|\frac{\partial h}{\partial \nu}\|_{H^{-1/2}(\partial\Omega)} \tag{2.93}$$

for all harmonic functions h in $\mathbb{R}^d$.

PROOF. Let φ_k and $\varphi_{k'}$ be defined by (2.92). We have

$$\varphi_k - \varphi_{k'} = \sum_{j=0}^{\infty} \frac{\lambda(k') - \lambda(k)}{(\lambda(k) - \lambda_j)(\lambda(k') - \lambda_j)} \langle \frac{\partial h}{\partial \nu}, \varphi_j \rangle_{\mathcal{H}^*} \varphi_j.$$

Therefore, for $|k'-k|$ small enough, there exists a positive constant C independent of k such that

$$\|\varphi_k - \varphi_{k'}\|_{H^{-1/2}(\partial\Omega)} \leq \frac{C|k-k'|}{\operatorname{dist}(\lambda(k), \sigma((\mathcal{K}_\Omega^0)^*))} \|\frac{\partial h}{\partial \nu}\|_{H^{-1/2}(\partial\Omega)}.$$

Since $u_k - u_{k'} = \mathcal{S}_\Omega^0[\varphi_k - \varphi_{k'}]$ for $x \in \mathbb{R}^d$, we obtain (2.93). □

We now investigate the behavior of the solution u_k when $\lambda(k)$ approaches one of the eigenvalues $\lambda_l \neq 0$ of $(\mathcal{K}_\Omega^0)^*$ as $\Im k \to 0$. We show that

$$\|\nabla(u_k - h)\|_{L^2(\Omega)} \sim \frac{1}{|\Im k|} \quad \text{as } \Im k \to 0, \tag{2.94}$$

as one may expect.

We first show that

$$\|\nabla \mathcal{S}_\Omega^0[\varphi]\|_{L^2(\Omega)} \approx \|\varphi\|_{\mathcal{H}^*} \tag{2.95}$$

for all $\varphi \in \mathcal{H}_0^*$. In fact, we have

$$\begin{aligned}
\|\nabla \mathcal{S}_\Omega^0[\varphi]\|_{L^2(\Omega)}^2 &= \int_{\partial\Omega} \mathcal{S}_\Omega^0[\varphi] \overline{\frac{\partial}{\partial \nu} \mathcal{S}_\Omega^0[\varphi]\big|_-} \, d\sigma \\
&= -\langle \varphi, (-\frac{1}{2} I + (\mathcal{K}_\Omega^0)^*)[\varphi] \rangle_{\mathcal{H}^*} \\
&= \sum_{j=1}^{\infty} (\frac{1}{2} - \lambda_j) |\langle \varphi, \varphi_j \rangle_{\mathcal{H}^*}|^2.
\end{aligned}$$

Since $|\lambda_j| < 1/2$ and they accumulate to 0, we have (2.95). We now see that

$$\|\nabla(u_k - h)\|_{L^2(\Omega)}^2 = \sum_{j=1}^{\infty} \frac{|\langle \frac{\partial h}{\partial \nu}, \varphi_j \rangle_{\mathcal{H}^*}|^2}{|\lambda(k) - \lambda_j|^2} = \sum_{\lambda_j = \lambda_l} \frac{|\langle \frac{\partial h}{\partial \nu}, \varphi_j \rangle_{\mathcal{H}^*}|^2}{|\lambda(k) - \lambda_l|^2} + \sum_{\lambda_j \neq \lambda_l} \frac{|\langle \frac{\partial h}{\partial \nu}, \varphi_j \rangle_{\mathcal{H}^*}|^2}{|\lambda(k) - \lambda_l|^2}.$$

Hence, we obtain (2.94) since $|\lambda(k) - \lambda_l| \sim |\Im k|$ as $|\Im k| \to 0$.

Suppose that 0 is not an eigenvalue of $(\mathcal{K}_\Omega^0)^*$. Then, since (λ_j) converge to zero, $\{0\}$ is the essential spectrum of $(\mathcal{K}_\Omega^0)^*$. We investigate the behavior of the solution u_k when $\lambda(k)$ approaches 0 as $\Im k \to 0$. For simplicity we approximate $\lambda(k)$ by $\sqrt{-1}\Im k$ and show that

$$|\Im k| \|\nabla(u_k - h)\|_{L^2(\Omega)} \to 0 \quad \text{as } \Im k \to 0. \tag{2.96}$$

We write

$$\|\nabla(u_k - h)\|_{L^2(\Omega)}^2 \approx \sum_{|\lambda_j| \leq |\Im k|} \frac{|\langle \frac{\partial h}{\partial \nu}, \varphi_j \rangle_{\mathcal{H}^*}|^2}{|\Im k|^2 + |\lambda_j|^2} + \sum_{|\lambda_j| > |\Im k|} \frac{|\langle \frac{\partial h}{\partial \nu}, \varphi_j \rangle_{\mathcal{H}^*}|^2}{|\Im k|^2 + |\lambda_j|^2} =: S_1 + S_2.$$

Since $\sum_{j=1}^{\infty}|\langle\frac{\partial h}{\partial\nu},\varphi_j\rangle_{\mathcal{H}^*}|^2<\infty$, it follows that

$$|\Im k|^2 S_1 \le \sum_{|\lambda_j|\le|\Im k|} |\langle\frac{\partial h}{\partial\nu},\varphi_j\rangle_{\mathcal{H}^*}|^2 \to 0 \quad \text{as } \Im k\to 0.$$

To show that $|\Im k|^2 S_2\to 0$, we express S_2 as

$$S_2=\sum_{l=0}^{\infty}\sum_{2^l|\Im k|<|\lambda_j|<2^{l+1}|\Im k|}\frac{|\langle\frac{\partial h}{\partial\nu},\varphi_j\rangle_{\mathcal{H}^*}|^2}{|\Im k|^2+|\lambda_j|^2}.$$

Then we see that

$$\begin{aligned}|\Im k|^2 S_2 &\le \sum_{l=0}^{\infty}\frac{1}{1+2^{2l}}\sum_{2^l|\Im k|<|\lambda_j|<2^{l+1}|\Im k|}|\langle\frac{\partial h}{\partial\nu},\varphi_j\rangle_{\mathcal{H}^*}|^2\\ &\le \sum_{l=0}^{\infty}\frac{1}{1+2^{2l}}\sum_{|\lambda_j|<2^{l+1}|\Im k|}|\langle\frac{\partial h}{\partial\nu},\varphi_j\rangle_{\mathcal{H}^*}|^2\end{aligned}$$

and so we infer from the Lebesgue's dominated convergence theorem that $|\Im k|^2 S_2\to 0$ as $\Im k\to 0$ since for each fixed l,

$$\sum_{|\lambda_j|<2^{l+1}|\Im k|}|\langle\frac{\partial h}{\partial\nu},\varphi_j\rangle_{\mathcal{H}^*}|^2\to 0$$

as $\Im k\to 0$. This completes the proof of (2.96). □

Estimates (2.94) and (2.96) are from [**98**].

2.5.4. Field Enhancement Between Closely Spaced Disks. Let Ω_1 and Ω_2 be two disks with the same radius r and conductivity k embedded in the background with conductivity 1. Let ϵ be the distance between the two disks Ω_1 and Ω_2, that is,

$$\epsilon:=\mathrm{dist}(\Omega_1,\Omega_2).$$

Let (ξ,θ) be the bipolar coordinates defined by

$$e^{\xi-\sqrt{-1}\theta}=\frac{x_1+\sqrt{-1}x_2+\alpha}{x_1+\sqrt{-1}x_2-\alpha},$$

where α is defined by (2.57).

Let u be the solution to

$$\begin{cases}\nabla\cdot\big(1+(k-1)\chi(\Omega_1\cup\Omega_2)\big)\nabla u=0 & \text{in } \mathbb{R}^2,\\ u(x)-x_1=O(|x|^{-1}) & \text{as } |x|\to+\infty.\end{cases}\tag{2.97}$$

Then u can be represented as follows:

$$u=x_1+\mathbb{S}_{\Omega_1\cup\Omega_2}[\varphi],$$

where φ is the solution to

$$(\lambda I-\mathbb{K}^*_{\Omega_1\cup\Omega_2})[\varphi]=\begin{bmatrix}\dfrac{\partial x_1}{\partial\nu}|_{\partial\Omega_1}\\ \dfrac{\partial x_1}{\partial\nu}|_{\partial\Omega_2}\end{bmatrix}.$$

Using (2.56), we have the following harmonic expansion for the linear function x_1:

$$x_1 = \operatorname{sgn}(\xi)\alpha\left[1 + 2\sum_{n=1}^{\infty} e^{-n|\xi|}\cos n\theta\right], \tag{2.98}$$

which yields

$$x_1 = \operatorname{sgn}(\xi)\alpha \sum_{m=-\infty}^{\infty} e^{-|m||\xi|+\sqrt{-1}m\theta}. \tag{2.99}$$

From (2.99), we obtain

$$\frac{\partial x_1}{\partial \nu}|_{\xi=\pm\xi_0} = \pm h(\xi_0,\theta)\alpha \sum_{m=-\infty}^{\infty} (-|m|)e^{-|m|\xi_0+\sqrt{-1}m\theta}. \tag{2.100}$$

On the other hand,

$$\begin{bmatrix} \dfrac{\partial x_1}{\partial \nu}|_{\partial\Omega_1} \\ \dfrac{\partial x_1}{\partial \nu}|_{\partial\Omega_2} \end{bmatrix} = \sum_{m=-\infty}^{\infty} \alpha\sqrt{2\pi|m|(1-e^{-2|m|\xi_0})}e^{-|m|\xi_0}\Psi_m^+ + 0\cdot\Psi_m^-, \tag{2.101}$$

where $\Psi_m^\pm$ are defined by (2.64). Hence, by Theorem 2.23,

$$\varphi = \sum_{n\neq 0} \frac{1}{\lambda-\lambda_n^+}\left(\alpha\sqrt{2\pi|n|(1-e^{-2|m|\xi_0})}e^{-|m|\xi_0}\right)\Psi_n^+. \tag{2.102}$$

Then (2.65) yields

$$u = x_1 + \sum_{n\neq 0} \frac{\alpha e^{-2|n|\xi_0}}{\lambda-\lambda_n^+}\sinh|n|\xi\, e^{\sqrt{-1}n\theta} \quad \text{for } |\xi| < |\xi_0|. \tag{2.103}$$

Now we compute $\nabla u(\xi = 0, \theta = \pi)$ at the center of the gap between Ω_1 and Ω_2. From (2.53) we have

$$\nabla u(0,\pi) = e_1 + Ee_1, \quad E = \sum_{n=1}^{\infty} \frac{4|n|e^{-2|n|\xi_0}}{\lambda-\lambda_n^+}(-1)^n,$$

where (e_1, e_2) is the orthonormal basis in Cartesian coordinates.

Let $k_n^+ = -\coth|n|\xi_0$. Note that

$$\lambda_n^+ = \frac{k_n^+ + 1}{2(k_n^+ - 1)}.$$

Then, E can be rewritten as

$$E = \sum_{n=1}^{\infty} \frac{(1-k)(k_n^+ - 1)}{k - k_n^+} 4|n|e^{-2|n|\xi_0}(-1)^n.$$

Let us assume that k is given by

$$k = k_N^+ + \sqrt{-1}\delta$$

for some $N \in \mathbb{N}$, where $\delta > 0$ is a small parameter. For small $\delta > 0$, E can be approximated by

$$E \approx \frac{(1-k)(k_N^+ - 1)}{k - k_N^+} 4Ne^{-2N\xi_0}(-1)^N. \tag{2.104}$$

Since $k_N^+ \approx -\frac{\sqrt{r}}{N\sqrt{\epsilon}}$ for small $\epsilon > 0$, we have

$$E \approx \sqrt{-1}\frac{r}{N\delta\epsilon}4e^{-2N\xi_0}(-1)^N. \tag{2.105}$$

It then follows that

$$\nabla u(0,\pi) \approx \sqrt{-1}\frac{r}{N\delta\epsilon}4e^{-2N\xi_0}(-1)^N e_1.$$

REMARK 2.33. *Estimates of the field enhancement between two disks in the limiting case when $k \to 0$ or $k \to +\infty$ were first derived in* [**67**]. *The behavior of the electric field between two nearly touching strictly convex perfect conductors or perfect insulators with smooth boundaries is investigated in* [**33**]*; see also* [**116, 117, 317, 515**]. *In* [**513**], *the singular behavior of nearly touching spheres is fully characterized. By combining the method of image charges and transformation optics, an approximate analytical formula for the electric field for two spheres is derived. The formula is highly accurate for wide ranges of complex permittivities and gap distances.*

2.5.5. Polarization Tensor of Two Particles. The polarization tensor can be defined for multiple particles. In the case of two particles $\Omega_1 \cup \Omega_2$ with the same conductivity k, it is defined as follows. Let $\mathbb{K}^*_{\Omega_1\cup\Omega_2}$ be the Neumann–Poincaré operator associated with $\Omega_1 \cup \Omega_2$ given by (2.49) and let $\nu^{(i)}$ be the outward normal on $\partial\Omega_i$, $i = 1, 2$. The polarization tensor $M = (m_{pq})_{p,q=1}^d$ associated with $\Omega_1 \cup \Omega_2$ and k is given by

$$m_{pq}(\lambda(k), \Omega_1 \cup \Omega_2) = \int_{\partial\Omega_1} y_p \phi_q^{(1)}\, d\sigma(y) + \int_{\partial\Omega_2} y_p \phi_q^{(2)}\, d\sigma(y) \quad \text{for } p, q = 1, \ldots, d,$$

where

$$\begin{bmatrix} \phi_p^{(1)} \\ \phi_p^{(2)} \end{bmatrix} = (\lambda(k)I - \mathbb{K}^*_{\Omega_1\cup\Omega_2})^{-1} \begin{bmatrix} \nu_p^{(1)}|_{\partial\Omega_1} \\ \nu_p^{(2)}|_{\partial\Omega_2} \end{bmatrix}.$$

If Ω_1 and Ω_2 are two separated disks of radius r centered at $(-1)^j(r + \frac{\epsilon}{2}, 0)$ for $j = 1, 2$ and $\epsilon > 0$ is their separation distance, then from formula (2.61) for the eigenvalues of $\mathbb{K}^*_{\Omega_1\cup\Omega_2}$, the polarization tensor associated with $\Omega_1 \cup \Omega_2$ and the conductivity k is given by the following formula:

$$M(\lambda(k), \Omega_1 \cup \Omega_2) = 8\pi\alpha^2 \begin{pmatrix} \displaystyle\sum_{j=1}^{\infty} \frac{je^{-2j\xi_0}}{\lambda(k) - \frac{1}{2}e^{-2j\xi_0}} & 0 \\ 0 & \displaystyle\sum_{j=1}^{\infty} \frac{je^{-2j\xi_0}}{\lambda(k) + \frac{1}{2}e^{-2j\xi_0}} \end{pmatrix}, \tag{2.106}$$

where α and ξ_0 are defined by (2.57). Again, it is represented in a spectral form.

2.5.6. Representation by an Equivalent Ellipse. Consider the polarization tensor for some particle(s). It can be shown that there exists a corresponding unique ellipse $\mathcal{E}$ that has precisely the same polarization tensor. We will call $\mathcal{E}$ the equivalent ellipse. The equivalent ellipse represents the essential nature of the particle. From a given polarization tensor M, we can reconstruct the parameters for the equivalent ellipse using the following formula:

$$a_1 = ea_2, \quad a_2 = \sqrt{\frac{E}{\pi e}}, \quad E = \frac{\lambda_1(e+k)}{(e+1)(k-1)}, \quad e = \frac{\lambda_2 - k\lambda_1}{\lambda_1 - k\lambda_2}, \tag{2.107}$$

where λ_1, λ_2 are the eigenvalues of M and $[e_{11}, e_{12}]^t, [e_{21}, e_{22}]^t$ are the associated normalized eigenvectors. Let $\mathcal{E}'$ be the ellipse whose semi-axes are on the x_1- and x_2-axes and of length a_1 and a_2. Let

$$\theta = \arctan \frac{e_{21}}{e_{11}}.$$

Then the equivalent ellipse is given by

$$\mathcal{E} = \begin{pmatrix} \cos\theta & -\sin\theta \\ \sin\theta & \cos\theta \end{pmatrix} \mathcal{E}'. \tag{2.108}$$

2.5.7. Numerical Results. With a particular choice of parameters we can obtain an explicit solution to the conductivity problem (2.66). Let Ω be a disk of radius $R = 5$ located at the origin in $\mathbb{R}^2$. Let us take the conductivity in Ω to be $k = 3$ which means $\lambda = 1$. We also assume that $h(x) = x_1$. It can be shown that the explicit solution is given by

$$u(r,\theta) = \begin{cases} r\cos(\theta) - \dfrac{k-1}{k+1} R^2 r^{-1} \cos(\theta), & |r| > R, \\ \dfrac{2}{k+1} r\cos(\theta), & |r| \leq R, \end{cases} \tag{2.109}$$

where (r, θ) are the polar coordinates.

Likewise, we can obtain a numerical solution. This involves inverting the operator $\lambda I - (\mathcal{K}_\Omega^0)^*$ which is possible in this case as $\lambda = 1$. A comparison between the exact solution and the numerical solution is shown in Figure 2.1 where we have evaluated the solutions on the circle $|x| = 10$. The corresponding Matlab code can be downloaded at www.ams.org/bookpages/surv-235.

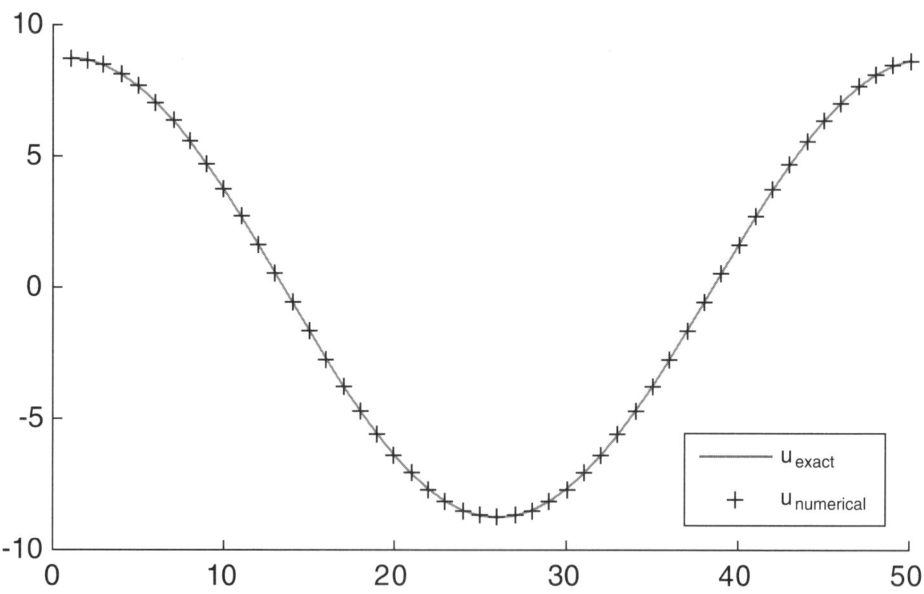

FIGURE 2.1. The exact solution and the numerical solution of the conductivity problem (2.66) evaluated on the circle $|x| = 10$.

Next, we compute the polarization tensor for an ellipse whose semi-axes are on the x_1- and x_2-axes of length $a_1 = 5$ and $a_2 = 3$. We assume $k = 3$ (or equivalently, $\lambda(k) = 1$). A comparison between the numerical values and the exact values is provided in Table 2.3. The corresponding Matlab code can be downloaded at `www.ams.org/bookpages/surv-235`.

	Theoretical	Numerical
$M(\lambda(k), \Omega)$	$\begin{pmatrix} 53.8559 & 0.0000 \\ -0.0000 & 41.8879 \end{pmatrix}$	$\begin{pmatrix} 53.8559 & 0.0000 \\ -0.0000 & 41.8879 \end{pmatrix}$

TABLE 2.3. Polarization Tensor $M(\lambda(k), \Omega)$ when Ω is an ellipse.

Finally, we consider the case of two separated disks $\Omega_1 \cup \Omega_2$ where Ω_j is a circular disk of radius $r = 1$ centered at $(-1)^j(r + \frac{\epsilon}{2}, 0)$ for $j = 1, 2$. Let the distance between the two disks be $\epsilon = 0.3$ and assume $k = 3$. A comparison between the numerical values and the exact values is provided in Table 2.4.

	Theoretical	Numerical
$M(\lambda(k), \Omega)$	$\begin{pmatrix} 6.9789 & 0.0000 \\ 0.0000 & 5.7629 \end{pmatrix}$	$\begin{pmatrix} 6.9789 & 0.0000 \\ 0.0000 & 5.7629 \end{pmatrix}$

TABLE 2.4. Polarization Tensor $M(\lambda(k), \Omega_1 \cup \Omega_2)$ when Ω_1 and Ω_2 are two disks of radius $r = 1$ separated by a distance $\epsilon = 0.3$.

From (2.107), the reconstructed parameters for the equivalent ellipse $\mathcal{E}$ defined in (2.108) turn out to be $a_1 = 1.713224, a_2 = 1.167994$, and $\theta = 0.523599$. The two disks $\Omega_1 \cup \Omega_2$ and the equivalent ellipse $\mathcal{E}$ are shown in Figure 2.2. The corresponding Matlab code can be downloaded at `www.ams.org/bookpages/surv-235`.

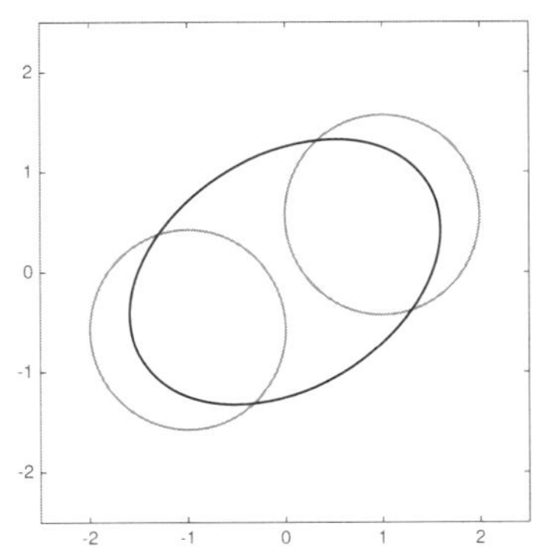

FIGURE 2.2. Two circular disks (gray) and their equivalent ellipse (black). The parameters are given as $r = 1$, $\epsilon = 0.3$, and $k = 3$.

2.6. Periodic and Quasi-Periodic Green's Functions

In this section we look into Green's functions for gratings; periodic, and quasi-periodic Green's functions; and the associated layer potentials. The results described in this section will be applied to the mathematical theory of photonic crystals, metasurfaces, and metamaterials.

2.6.1. Green's Functions for Gratings. Consider a function $G_\sharp : \mathbb{R}^2 \to \mathbb{C}$ satisfying

$$\Delta G_\sharp(x) = \sum_{n\in\mathbb{Z}} \delta_0(x + (n,0)). \tag{2.110}$$

We call $G_\sharp$ a periodic Green's function for the one-dimensional grating in $\mathbb{R}^2$.

LEMMA 2.34. *Let $x = (x_1, x_2)$. Then, $G_\sharp$, defined by*

$$G_\sharp(x) = \frac{1}{4\pi} \ln\big(\sinh^2(\pi x_2) + \sin^2(\pi x_1)\big), \tag{2.111}$$

satisfies (2.110).

PROOF. We have

$$\begin{aligned} \Delta G_\sharp(x) &= \sum_{n\in\mathbb{Z}} \delta_0(x + (n,0)) \\ &= \sum_{n\in\mathbb{Z}} \delta_0(x_2)\delta_0(x_1 + n) \\ &= \sum_{n\in\mathbb{Z}} \delta_0(x_2) e^{\sqrt{-1}2\pi n x_1}, \end{aligned} \tag{2.112}$$

where we have used the Poisson summation formula:

$$\sum_{n\in\mathbb{Z}} \delta_0(x_1 + n) = \sum_{n\in\mathbb{Z}} e^{\sqrt{-1}2\pi n x_1}.$$

On the other hand, as $G_\sharp$ is periodic in x_1 of period 1, we have

$$G_\sharp(x) = \sum_{n\in\mathbb{Z}} \beta_n(x_2) e^{\sqrt{-1}2\pi n x_1}, \tag{2.113}$$

therefore

$$\Delta G_\sharp(x) = \sum_{n\in\mathbb{Z}} (\beta_n''(x_2) + (\sqrt{-1}2\pi n)^2 \beta_n) e^{\sqrt{-1}2\pi n x_1}. \tag{2.114}$$

Comparing (2.112) and (2.114) yields

$$\beta_n''(x_2) + (\sqrt{-1}2\pi n)^2 \beta_n = \delta_0(x_2). \tag{2.115}$$

A solution to the previous equation can be found by using standard techniques for ordinary differential equations. We have

$$\begin{aligned} \beta_0 &= \frac{1}{2}|x_2| + c, \\ \beta_n &= \frac{-1}{4\pi|n|} e^{-2\pi|n||x_2|}, \quad n \neq 0, \end{aligned}$$

where c is a constant. It then follows that

$$\begin{aligned} G_\sharp(x) &= \frac{1}{2}|x_2| + c - \sum_{n\in\mathbb{Z}\setminus\{0\}} \frac{1}{4\pi|n|} e^{-2\pi|n||x_2|} e^{\sqrt{-1}2\pi n x_1} \\ &= \frac{1}{2}|x_2| + c - \sum_{n\in\mathbb{N}\setminus\{0\}} \frac{1}{2\pi n} e^{-2\pi n|x_2|} \cos(2\pi n x_1) \\ &= \frac{1}{4\pi} \ln\big(\sinh^2(\pi x_2) + \sin^2(\pi x_1)\big), \end{aligned}$$

where we have used the summation identity (see, for instance, [**280**, pp. 813-814]):

$$\sum_{n\in\mathbb{N}\setminus\{0\}} \frac{1}{2\pi n} e^{-2\pi n|x_2|} \cos(2\pi n x_1) = \frac{1}{2}|x_2| - \frac{\ln(2)}{2\pi} - \frac{1}{4\pi} \ln\big(\sinh^2(\pi x_2) + \sin^2(\pi x_1)\big),$$

and defined $c = -\dfrac{\ln(2)}{2\pi}$. □

Let us denote by $G_\sharp(x,y) := G_\sharp(x-y)$. In the following we define the one-dimensional periodic single-layer potential and the one-dimensional periodic Neumann–Poincaré operator, respectively, for a bounded domain $\Omega \Subset (-\frac{1}{2}, \frac{1}{2}) \times \mathbb{R}$ which we assume to be of class $\mathcal{C}^{1,\eta}$ for some $\eta > 0$. Let

$$\begin{aligned} \mathcal{S}_{\Omega,\sharp} : H^{-\frac{1}{2}}(\partial\Omega) &\longrightarrow H^1_{\text{loc}}(\mathbb{R}^2), H^{\frac{1}{2}}(\partial\Omega) \\ \varphi &\longmapsto \mathcal{S}_{\Omega,\sharp}[\varphi](x) = \int_{\partial\Omega} G_\sharp(x,y)\varphi(y)d\sigma(y) \end{aligned}$$

for $x \in \mathbb{R}^2$ (or $x \in \partial\Omega$) and let

$$\begin{aligned} \mathcal{K}^*_{\Omega,\sharp} : H^{-\frac{1}{2}}(\partial\Omega) &\longrightarrow H^{-\frac{1}{2}}(\partial\Omega) \\ \varphi &\longmapsto \mathcal{K}^*_{\Omega,\sharp}[\varphi](x) = \int_{\partial\Omega} \frac{\partial G_\sharp(x,y)}{\partial\nu(x)}\varphi(y)d\sigma(y) \end{aligned}$$

for $x \in \partial\Omega$. As in the previous subsections, the periodic Neumann–Poincaré operator $\mathcal{K}^*_{\Omega,\sharp}$ can be symmetrized. The following lemma holds.

LEMMA 2.35. (i) *For any $\varphi \in H^{-\frac{1}{2}}(\partial\Omega)$, $\mathcal{S}_{\Omega,\sharp}[\varphi]$ is harmonic in Ω and in $\Big(\big(-\frac{1}{2}, \frac{1}{2}\big) \times \mathbb{R}\Big) \setminus \overline{\Omega}$;*

(ii) *The following trace formula holds: for any $\varphi \in H^{-\frac{1}{2}}(\partial\Omega)$,*

$$(-\frac{1}{2}I + \mathcal{K}^*_{\Omega,\sharp})[\varphi] = \frac{\partial \mathcal{S}_{\Omega,\sharp}[\varphi]}{\partial\nu}\Big|_-;$$

(iii) *The following Calderón identity holds: $\mathcal{K}_{\Omega,\sharp}\mathcal{S}_{\Omega,\sharp} = \mathcal{S}_{\Omega,\sharp}\mathcal{K}^*_{\Omega,\sharp}$, where $\mathcal{K}_{\Omega,\sharp}$ is the L^2-adjoint of $\mathcal{K}^*_{\Omega,\sharp}$;*

(iv) *The operator $\mathcal{K}^*_{\Omega,\sharp} : H_0^{-\frac{1}{2}}(\partial\Omega) \to H_0^{-\frac{1}{2}}(\partial\Omega)$ is compact self-adjoint on $\mathcal{H}_0^*$ equipped with the following inner product:*

$$\langle u, v\rangle_{\mathcal{H}_0^*} = -\langle \mathcal{S}_{\Omega,\sharp}[v], u\rangle_{\frac{1}{2},-\frac{1}{2}}, \tag{2.116}$$

which makes $\mathcal{H}_0^*$ equivalent to $H_0^{-\frac{1}{2}}(\partial\Omega)$. Here, by E_0 we denote the zero-mean subspace of E for $E = \mathcal{H}^*$ or $H^{-\frac{1}{2}}(\partial\Omega)$.

(v) *Let (λ_j, φ_j), $j = 1, 2, \ldots$, be the eigenvalue and normalized eigenfunction pair of $\mathcal{K}_{\Omega,\sharp}^*$ in $\mathcal{H}_0^*(\partial\Omega)$, then $\lambda_j \in (-\frac{1}{2}, \frac{1}{2})$ and $\lambda_j \to 0$ as $j \to \infty$.*

PROOF. First, note that a Taylor expansion of $\sinh^2(\pi x_2) + \sin^2(\pi x_1)$ yields

$$G_\sharp(x) = \frac{\ln|x|}{2\pi} + R(x), \tag{2.117}$$

where R is a smooth function such that

$$R(x) = \frac{1}{4\pi}\ln(1 + O(x_2^2 - x_1^2)).$$

We can decompose the operators $\mathcal{S}_{\Omega,\sharp}$ and $\mathcal{K}_{\Omega,\sharp}^*$ on $\mathcal{H}_0^*(\partial\Omega)$ accordingly. Since $\mathcal{S}_{\Omega,\sharp} - \mathcal{S}_\Omega^0$ and $\mathcal{K}_{\Omega,\sharp}^* - (\mathcal{K}_\Omega^0)^*$ are smoothing operators, the proof of Lemma 2.35 follows the same arguments as those given in the previous subsections. □

2.6.2. Periodic Green's Function. In order to derive effective medium properties of sub-wavelength resonators, we shall investigate the periodic transmission problem for the Laplace operator. The results in this subsection are from [**74**].

Let $Y = (-1/2, 1/2)^d$ denote the unit cell and $\overline{D} \subset Y$. Consider the periodic transmission problem:

$$\begin{cases} \nabla \cdot \Big(1 + (k-1)\chi(D)\Big)\nabla u_p = 0 \quad \text{in } Y\,, \\ u_p - x_p \;\text{ periodic (in each direction) with period } 1\,, \\ \displaystyle\int_Y u_p\, dx = 0\,, \end{cases} \tag{2.118}$$

for $p = 1, \ldots, d$.

In order to derive a representation formula for the solution to the periodic transmission problem (2.118), we need to introduce a periodic Green's function.

Let

$$G_\sharp(x) = -\sum_{n\in\mathbb{Z}^d\setminus\{0\}} \frac{e^{\sqrt{-1}2\pi n\cdot x}}{4\pi^2|n|^2}. \tag{2.119}$$

Then we get, in the sense of distributions,

$$\Delta G_\sharp(x) = \sum_{n\in\mathbb{Z}^d\setminus\{0\}} e^{\sqrt{-1}2\pi n\cdot x} = \sum_{n\in\mathbb{Z}^d} e^{\sqrt{-1}2\pi n\cdot x} - 1\,,$$

and $G_\sharp$ has mean zero. It then follows from the Poisson summation formula:

$$\sum_{n\in\mathbb{Z}^d} e^{\sqrt{-1}2\pi n\cdot x} = \sum_{n\in\mathbb{Z}^d} \delta_0(x-n)\,, \tag{2.120}$$

that

$$\Delta G_\sharp(x) = \sum_{n\in\mathbb{Z}^d} \delta_0(x-n) - 1\,. \tag{2.121}$$

The appearance of the constant 1 in (2.121) may be somewhat peculiar. It is the volume of Y and an integration by parts shows that it should be there. In fact,

$$\int_Y \Delta G_\sharp(x)dx = \int_{\partial Y} \frac{\partial G_\sharp}{\partial \nu}\, d\sigma\,,$$

and the right-hand side is zero because of the periodicity.

The expression (2.119) for $G_\sharp$ is called a lattice sum and its asymptotic behavior has been studied extensively in many contexts in solid state physics, e.g., [**521**].

We state the next lemma in any dimension, but give in some detail a proof only for $d = 2$, leaving the proof in higher dimensions to the reader. Formulas (2.122) and (2.123) will be applied later in our study of the effective properties of systems of sub-wavelength resonators.

LEMMA 2.36. *There exists a smooth function $R_d(x)$ in the unit cell Y such that*

$$G_\sharp(x) = \begin{cases} \dfrac{1}{2\pi}\ln|x| + R_2(x), & d = 2\,, \\ \dfrac{1}{(2-d)\omega_d}\dfrac{1}{|x|^{d-2}} + R_d(x), & d \geq 3\,. \end{cases} \tag{2.122}$$

Moreover, the Taylor expansion of $R_d(x)$ at 0 for $d \geq 2$ is given by

$$R_d(x) = R_d(0) - \frac{1}{2d}(x_1^2 + \ldots + x_d^2) + O(|x|^4)\,. \tag{2.123}$$

PROOF. As mentioned above, we assume that $d = 2$. The proof we give here is not the simplest one, but has the advantage that it can be extended to other more complicated periodic Green's functions. Note that the behavior $G_\sharp(x) \sim \Gamma(x)$ as $|x| \to 0$ is to be expected since the effect of the periodic boundary conditions is negligible when x is near the origin.

We begin by writing

$$\begin{aligned} G_\sharp(x) &= -\sum_{n\in\mathbb{Z}^2\setminus\{0\}} \frac{e^{\sqrt{-1}2\pi n\cdot x}}{4\pi^2|n|^2} = -\frac{1}{4\pi^2}\sum_{n\in\mathbb{Z}^2\setminus\{0\}} \frac{\cos 2\pi n_1 x_1 \cos 2\pi n_2 x_2}{n_1^2+n_2^2} \\ &= -\frac{1}{2\pi^2}\sum_{n_1=0}^{+\infty} \cos 2\pi n_1 x_1 \sum_{n_2=1}^{+\infty} \frac{\cos 2\pi n_2 x_2}{n_1^2+n_2^2} \\ &\quad - \frac{1}{2\pi^2}\sum_{n_2=0}^{+\infty} \cos 2\pi n_2 x_2 \sum_{n_1=1}^{+\infty} \frac{\cos 2\pi n_1 x_1}{n_1^2+n_2^2} \\ &:= G_1 + G_2\,. \end{aligned}$$

After that, let us invoke three summation identities (see for instance [**203**, pp. 813-814]):

$$\sum_{n_2=1}^{+\infty} \frac{\cos 2\pi n_2 x_2}{n_1^2+n_2^2} = \begin{cases} -\dfrac{1}{2n_1^2} + \dfrac{\pi}{2n_1}\dfrac{\cosh \pi(2x_2-1)n_1}{\sinh \pi n_1} & \text{if } n_1 \neq 0\,, \\ \dfrac{\pi^2}{6} - \pi^2 x_2 + \pi^2 x_2^2 & \text{if } n_1 = 0\,, \end{cases} \tag{2.124}$$

$$\sum_{n_1=1}^{+\infty} \frac{\cos 2\pi n_1 x_1}{n_1} e^{-2\pi n_1 x_2} = \pi x_2 - \ln 2 - \frac{1}{2}\ln\Big(\sinh^2 \pi x_2 + \sin^2 \pi x_1\Big)\,. \tag{2.125}$$

We then compute

$$\begin{aligned} G_1 &= -\frac{1}{2\pi^2} \sum_{n_2=1}^{+\infty} \frac{\cos 2\pi n_2 x_2}{n_2^2} \\ &\quad - \frac{1}{2\pi^2} \sum_{n_1=1}^{+\infty} \cos 2\pi n_1 x_1 \Big(-\frac{1}{2n_1^2} + \frac{\pi}{2n_1} \frac{\cosh \pi(2x_2 - 1)n_1}{\sinh \pi n_1} \Big) \\ &= -\frac{1}{2\pi^2} \sum_{n_2=1}^{+\infty} \frac{\cos 2\pi n_2 x_2}{n_2^2} + \frac{1}{4\pi^2} \sum_{n_1=1}^{+\infty} \frac{\cos 2\pi n_1 x_1}{n_1^2} \\ &\quad - \frac{1}{4\pi} \sum_{n_1=1}^{+\infty} \frac{\cos 2\pi n_1 x_1}{n_1} \frac{\cosh \pi(2x_2 - 1)n_1}{\sinh \pi n_1} \end{aligned}$$

$$\begin{aligned} &= -\frac{1}{12} + \frac{1}{2}x_2 - \frac{1}{2}x_2^2 + \frac{1}{24} - \frac{1}{4}x_1 + \frac{1}{4}x_1^2 - \frac{1}{4\pi} \sum_{n_1=1}^{+\infty} \frac{\cos 2\pi n_1 x_1}{n_1} e^{-2\pi n_1 x_2} \\ &\quad - \frac{1}{4\pi} \sum_{n_1=1}^{+\infty} \frac{\cos 2\pi n_1 x_1}{n_1} \Big(\frac{\cosh \pi(2x_2 - 1)n_1}{\sinh \pi n_1} - e^{-2\pi n_1 x_2} \Big) \end{aligned}$$

to arrive at

$$G_1 = -\frac{1}{24} + \frac{\ln 2}{4\pi} + \frac{1}{4}(x_2 - x_1) - \frac{1}{4}(2x_2^2 - x_1^2) + \frac{1}{8\pi} \ln \Big(\sinh^2 \pi x_2 + \sin^2 \pi x_1 \Big) + r_1(x),$$

where the function $r_1(x)$ is given by

$$\begin{aligned} r_1(x) &= -\frac{1}{4\pi} \sum_{n_1=1}^{+\infty} \frac{\cos 2\pi n_1 x_1}{n_1} \Big(\frac{\cosh \pi(2x_2 - 1)n_1}{\sinh \pi n_1} - e^{-2\pi n_1 x_2} \Big) \\ &= -\frac{1}{4\pi} \sum_{n_1=1}^{+\infty} \frac{\cos 2\pi n_1 x_1}{n_1} \frac{e^{2\pi n_1 x_2} + e^{-2\pi n_1 x_2}}{e^{2\pi n_1} - 1}. \end{aligned}$$

Because of the term $e^{-\pi n_1}$, we can easily see that r_1 is a $\mathcal{C}^\infty$-function.

In the same way we can derive

$$G_2 = -\frac{1}{24} + \frac{\ln 2}{4\pi} + \frac{1}{4}(x_1 - x_2) - \frac{1}{4}(2x_1^2 - x_2^2) + \frac{1}{8\pi} \ln \Big(\sinh^2 \pi x_1 + \sin^2 \pi x_2 \Big) + r_2(x),$$

where

$$r_2(x) = -\frac{1}{4\pi} \sum_{n_1=1}^{+\infty} \frac{\cos 2\pi n_1 x_2}{n_1} \frac{e^{2\pi n_1 x_1} + e^{-2\pi n_1 x_1}}{e^{2\pi n_1} - 1}.$$

By a Taylor expansion, we readily see that

$$\begin{aligned} &\ln \Big(\sinh^2 \pi x_2 + \sin^2 \pi x_1 \Big) + \ln \Big(\sinh^2 \pi x_1 + \sin^2 \pi x_2 \Big) \\ &= 4 \ln \pi + 2 \ln(x_1^2 + x_2^2) + r_3(x), \end{aligned}$$

where $r_3(x)$ is a $\mathcal{C}^\infty$-function with $r_3(x) = O(|x|^4)$ as $|x| \to 0$. In short, we obtain

$$G_\sharp(x) = \frac{1}{2\pi} \ln |x| + R_2(x),$$

where

$$R_2(x) = C - \frac{1}{4}(x_1^2 + x_2^2) + r_1(x) + r_2(x) + r_3(x)$$

for some constant C. By a Taylor expansion again, one can see that

$$r_1(x) + r_2(x) = C + O(|x|^4) \quad \text{as } |x| \to 0,$$

for some constant C. That R_2 is harmonic follows from (2.121). This concludes the proof. □

Note that in the two-dimensional case we can expand $R_2(x)$ even further to get

$$R_2(x) = R_2(0) - \frac{1}{4}(x_1^2 + x_2^2) + \sum_{s=3}^{m} R_2^{(s)}(x) + O(|x|^{m+1}) \quad \text{as } |x| \to 0,$$

where the harmonic polynomial $R_2^{(s)}$ is homogeneous of degree s, *i.e.*,

$$R_2^{(s)}(tx) = t^s R_2^{(s)}(x)$$

for all $t \in \mathbb{R}$ and all $x \in \mathbb{R}^2$. Since

$$R_2(-x_1, x_2) = R_2(x_1, x_2) \quad \text{and } R_2(x_1, -x_2) = R_2(x_1, x_2),$$

$R_2^{(s)} \equiv 0$ if s is odd, and hence

$$R_2(x) = R_2(0) - \frac{1}{4}(x_1^2 + x_2^2) + \sum_{s=2}^{m} R_2^{(2s)}(x) + O(|x|^{m+2}) \quad \text{as } |x| \to 0. \tag{2.126}$$

We now establish a representation formula for the solution of the periodic transmission problem (2.118).

Let the periodic single-layer potential of the density function $\phi \in L^2_0(\partial\Omega)$ be defined by

$$\mathcal{S}^0_{\Omega,\sharp}[\phi](x) := \int_{\partial\Omega} G_\sharp(x-y)\phi(y)\, d\sigma(y), \quad x \in \mathbb{R}^2.$$

Lemma 2.36 shows that

$$\mathcal{S}^0_{\Omega,\sharp}[\phi](x) = \mathcal{S}^0_{\Omega}[\phi](x) + \mathcal{R}_\Omega[\phi](x), \tag{2.127}$$

where $\mathcal{R}_\Omega$ is a smoothing operator defined by

$$\mathcal{R}_\Omega[\phi](x) := \int_{\partial\Omega} R_d(x-y)\phi(y)\, d\sigma(y).$$

Thanks to (2.127), we have

$$\frac{\partial}{\partial\nu}\mathcal{S}^0_{\Omega,\sharp}[\phi]\Big|_{\pm}(x) = \frac{\partial}{\partial\nu}\mathcal{S}^0_{\Omega}[\phi]\Big|_{\pm}(x) + \frac{\partial}{\partial\nu}\mathcal{R}_\Omega[\phi](x), \quad x \in \partial\Omega.$$

Thus we can understand, with the help of Lemma 2.36, $\partial\mathcal{S}^0_{\Omega,\sharp}[\phi]/\partial\nu|_\pm$ as a compact perturbation of $\partial\mathcal{S}^0_{\Omega}[\phi]/\partial\nu|_\pm$. Based on this natural idea, we obtain the following results.

LEMMA 2.37. (i) *Let $\phi \in L^2_0(\partial\Omega)$. The following behavior holds at the boundary:*

$$\frac{\partial}{\partial\nu}\mathcal{S}^0_{\Omega,\sharp}[\phi]\Big|_{\pm}(x) = (\pm\frac{1}{2}I + (\mathcal{K}^0_{\Omega,\sharp})^*)[\phi](x) \text{ on } \partial\Omega, \tag{2.128}$$

where $(\mathcal{K}_{\Omega,\sharp}^0)^* : L^2_0(\partial\Omega) \to L^2_0(\partial\Omega)$ *is given by*

$$(\mathcal{K}_{\Omega,\sharp}^0)^*[\phi](x) = \text{p.v.} \int_{\partial\Omega} \frac{\partial}{\partial\nu(x)} G_\sharp(x-y)\phi(y)\, d\sigma(y), \quad x \in \partial D\,. \tag{2.129}$$

(ii) *If* $\phi \in L^2_0(\partial\Omega)$, *then* $\mathcal{S}_{\Omega,\sharp}^0[\phi]$ *is harmonic in* Ω *and* $Y \setminus \overline{\Omega}$.

(iii) *If* $|\lambda| \geq \frac{1}{2}$, *then the operator* $\lambda I - (\mathcal{K}_{\Omega,\sharp}^0)^*$ *is invertible on* $L^2_0(\partial\Omega)$.

PROOF. Since $(\mathcal{K}_{\Omega,\sharp}^0)^* = (\mathcal{K}_{\Omega}^0)^* + \mathcal{C}_\Omega$ where $\mathcal{C}_\Omega$ is a smoothing operator, part (i) immediately follows from (2.8). Part (ii) follows from (2.121) and the fact that $\phi \in L^2_0(\partial\Omega)$. As a consequence of parts (i) and (ii), it follows that $\lambda I - (\mathcal{K}_{\Omega,\sharp}^0)^*$ maps $L^2_0(\partial\Omega)$ into $L^2_0(\partial\Omega)$. To prove part (iii), we observe that $\mathcal{C}_\Omega$ maps $L^2(\partial\Omega)$ into $H^1(\partial\Omega)$, and hence it is a compact operator on $L^2(\partial\Omega)$. Since, by Lemma 2.2, $\lambda I - (\mathcal{K}_{\Omega}^0)^*$ is invertible on $L^2_0(\partial\Omega)$, it suffices, by applying the Fredholm alternative, to show that $\lambda I - (\mathcal{K}_{\Omega,\sharp}^0)^*$ is one-to-one on $L^2_0(\partial\Omega)$. We shall prove this fact, using the same argument as the one introduced in Lemma 2.2. Let $|\lambda| \geq 1/2$, and suppose that $\phi \in L^2_0(\partial\Omega)$ satisfies $(\lambda I - (\mathcal{K}_{\Omega,\sharp}^0)^*)[\phi] = 0$ and $\phi \neq 0$. Let

$$A := \int_\Omega |\nabla \mathcal{S}_{\Omega,\sharp}^0[\phi]|^2\, dx, \quad B := \int_{Y\setminus\overline{\Omega}} |\nabla \mathcal{S}_{\Omega,\sharp}^0[\phi]|^2\, dx\,.$$

Then $A \neq 0$. In fact, if $A = 0$, then $\mathcal{S}_{\Omega,\sharp}^0[\phi]$ is constant in Ω. Therefore $\mathcal{S}_{\Omega,\sharp}^0[\phi]$ in $Y \setminus \overline{\Omega}$ is periodic and satisfies $\mathcal{S}_{\Omega,\sharp}^0[\phi]|_{\partial\Omega} = \text{constant}$. Hence $\mathcal{S}_{\Omega,\sharp}^0[\phi] = \text{constant}$ in $Y \setminus \overline{\Omega}$. Therefore, by part (i), we get

$$\phi = \frac{\partial}{\partial\nu}\mathcal{S}_{\Omega,\sharp}^0[\phi]\Big|_+ - \frac{\partial}{\partial\nu}\mathcal{S}_{\Omega,\sharp}^0[\phi]\Big|_- = 0\,,$$

which contradicts our assumption. In a similar way, we can show that $B \neq 0$.

On the other hand, using Green's formula and periodicity, we have

$$A = \int_{\partial\Omega} (-\frac{1}{2}I + (\mathcal{K}_{\Omega,\sharp}^0)^*)[\phi]\; \mathcal{S}_{\Omega,\sharp}^0[\phi]\, d\sigma, \quad B = -\int_{\partial\Omega} (\frac{1}{2}I + (\mathcal{K}_{\Omega,\sharp}^0)^*)[\phi]\; \mathcal{S}_{\Omega,\sharp}^0[\phi]\, d\sigma\,.$$

Since $(\lambda I - (\mathcal{K}_{\Omega,\sharp}^0)^*)[\phi] = 0$, it follows that

$$\lambda = \frac{1}{2}\frac{B-A}{B+A}\,.$$

Thus, $|\lambda| < 1/2$, which is a contradiction. This completes the proof. □

The following result holds.

THEOREM 2.38. *Let* u_p *be the unique solution to the transmission problem* (2.118). *Then* u_p, $p = 1, \ldots, d$, *can be expressed as follows*

$$u_p(x) = x_p + C_p + \mathcal{S}_{\Omega,\sharp}^0(\frac{k+1}{2(k-1)}I - (\mathcal{K}_{\Omega,\sharp}^0)^*)^{-1}[\nu_p](x) \quad \text{in } Y\,, \tag{2.130}$$

where C_p *is a constant and* ν_p *is the pth component of the outward unit normal* ν *to* $\partial\Omega$.

PROOF. Observe that $u_p, p = 1, \ldots, d$, satisfies

$$\begin{cases} \Delta u_p = 0 \quad \text{in } \Omega \cup (Y \setminus \overline{\Omega}), \\ u_p|_+ - u_p|_- = 0 \quad \text{on } \partial\Omega, \\ \dfrac{\partial u_p}{\partial \nu}\Big|_+ - k \dfrac{\partial u_p}{\partial \nu}\Big|_- = 0 \quad \text{on } \partial\Omega, \\ u_p - x_p \ \text{ periodic with period } 1, \\ \displaystyle\int_Y u_p \, dx = 0. \end{cases}$$

To prove (2.130), define

$$V_p(x) = \mathcal{S}^0_{\Omega,\sharp}\big((\frac{k+1}{2(k-1)} I - (\mathcal{K}^0_{\Omega,\sharp})^*)^{-1}[\nu_p](x) \quad \text{in } Y.$$

Then routine calculations show that

$$\begin{cases} \Delta V_p = 0 \quad \text{in } \Omega \cup (Y \setminus \overline{D}), \\ V_p|_+ - V_p|_- = 0 \quad \text{on } \partial\Omega, \\ \dfrac{\partial V_p}{\partial \nu}|_+ - k \dfrac{\partial V_p}{\partial \nu}|_- = (k-1)\nu_p \quad \text{on } \partial\Omega, \\ V_p \ \text{periodic with period } 1. \end{cases} \tag{2.131}$$

Thus by choosing C_p in (2.130) so that $\int_Y u_p \, dx = 0$, we complete the proof. □

Consider a general periodic lattice in two dimensions. Suppose that the periodic lattice is given by $r_n = n_1 a^{(1)} + n_2 a^{(2)}, n = (n_1, n_2) \in \mathbb{Z}^2$. Here the vectors $a^{(1)}$ and $a^{(2)}$ determine the unit cell $Y := \{s a^{(1)} + t a^{(2)}, s, t \in (-1/2, 1/2)\}$ of the array. The reciprocal vector of r_n is given by k_n defined by $k_n \cdot a^{(i)} = n_i$, for $i = 1, 2$. The periodic Green's function of the Laplacian is defined by

$$\begin{cases} \Delta G^a_\sharp = \displaystyle\sum_{n \in \mathbb{Z}^2} \delta_0(x - r_n) - \frac{1}{|Y|}, \\ G^a_\sharp(x + r_n) = G^a_\sharp(x), \quad \forall\, n \in \mathbb{Z}^2. \end{cases}$$

Since it is possible to rotate and scale the given lattice in order to satisfy $a^{(1)} = (1, 0)$ and $a^{(2)} = (a, b)$ with $b > 0$, we can write

$$r_n = n_1(1, 0) + n_2(a, b), \quad k_n = n_1(1, -\frac{a}{b}) + n_2(0, \frac{1}{b}), \quad n = (n_1, n_2) \in \mathbb{Z}^2.$$

Analogously to (2.119), we have

$$G^a_\sharp(x) = - \sum_{n \in \mathbb{Z}^2 \setminus \{0\}} \frac{e^{\sqrt{-1} 2\pi (n_1 x_1 + (-\frac{a}{b} n_1 + \frac{1}{b} n_2) x_2}}{4\pi^2 \big(n_1^2 + (-\frac{a}{b} n_1 + \frac{1}{b} n_2)^2\big)^2}. \tag{2.132}$$

Analogously to Lemma 2.35, the Neumann–Poincaré operator $(\mathcal{K}^0_{\Omega,\sharp})^*$ can be symmetrized. The following lemma holds.

LEMMA 2.39.

(i) *The operator* $(\mathcal{K}^0_{\Omega,\sharp})^* : H_0^{-\frac{1}{2}}(\partial\Omega) \to H_0^{-\frac{1}{2}}(\partial\Omega)$ *is compact self-adjoint on the space* $\mathcal{H}^*_0(\partial\Omega)$ *equipped with the following inner product:*

$$\langle u, v \rangle_{\mathcal{H}^*_0} = -\langle \mathcal{S}^0_{\Omega,\sharp}[v], u \rangle_{\frac{1}{2}, -\frac{1}{2}}. \tag{2.133}$$

(ii) *Let* $(\lambda_{j,\sharp}, \varphi_{j,\sharp})$, $j = 1, 2, \ldots$, *be the eigenvalue and normalized eigenfunction pair of* $(\mathcal{K}^0_{\Omega,\sharp})^*$ *in* $\mathcal{H}^*_0(\partial\Omega)$, *then* $\lambda_{j,\sharp} \in (-\frac{1}{2}, \frac{1}{2})$ *and* $\lambda_{j,\sharp} \to 0$ *as* $j \to \infty$.

2.6.3. Quasi-Periodic Green's Functions. For $\alpha \in (0, 2\pi)^d$, a function u is said to be α-quasi-periodic if $e^{-\sqrt{-1}\alpha\cdot x}u$ is periodic.

Let

$$G_\alpha(x) = -\sum_{n\in\mathbb{Z}^d} \frac{e^{\sqrt{-1}(2\pi n+\alpha)\cdot x}}{|2\pi n + \alpha|^2}, \quad \alpha \in (0, 2\pi)^d. \tag{2.134}$$

We have

$$\Delta G_\alpha(x) = \sum_{n\in\mathbb{Z}^d} \delta_0(x - n)e^{\sqrt{-1}\alpha\cdot n} \quad \text{in } \mathbb{R}^d, \tag{2.135}$$

or equivalently,

$$\left(\Delta + \sqrt{-1}\alpha \cdot \nabla - |\alpha|^2\right)\left(e^{-\sqrt{-1}\alpha\cdot x}G_\alpha(x)\right) = \sum_{n\in\mathbb{Z}^d} \delta_0(x - n) \quad \text{in } \mathbb{R}^d. \tag{2.136}$$

We denote by $\mathcal{S}^0_{\Omega,\alpha}$, $\mathcal{D}^0_{\Omega,\alpha}$, and $(\mathcal{K}^0_{\Omega,\alpha})^*$ the α-quasi-periodic single- and double-layer potentials and the α-quasi-periodic Neumann–Poincaré operator associated with G_α, respectively.

Analogously to Lemma 2.35, the quasi-periodic Neumann–Poincaré operator $(\mathcal{K}^0_{\Omega,\alpha})^*$ can be symmetrized. The following lemma holds.

LEMMA 2.40.

(i) *Let* $\alpha \in (0, 2\pi)^2$. *The operator* $(\mathcal{K}^0_{\Omega,\alpha})^* : H_0^{-\frac{1}{2}}(\partial\Omega) \to H_0^{-\frac{1}{2}}(\partial\Omega)$ *is compact self-adjoint on the space* $\mathcal{H}^*_0(\partial\Omega)$ *equipped with the following inner product:*

$$\langle u, v\rangle_{\mathcal{H}^*_0} = -\langle \mathcal{S}^0_{\Omega,\alpha}[v], u\rangle_{\frac{1}{2},-\frac{1}{2}}. \tag{2.137}$$

(ii) *Let* $(\lambda_{j,\alpha}, \varphi_{j,\alpha})$, $j = 1, 2, \ldots$, *be the eigenvalue and normalized eigenfunction pair of* $(\mathcal{K}^0_{\Omega,\alpha})^*$ *in* $\mathcal{H}^*_0(\partial\Omega)$, *then* $\lambda_{j,\alpha} \in (-\frac{1}{2}, \frac{1}{2})$ *and* $\lambda_{j,\alpha} \to 0$ *as* $j \to \infty$.

2.6.4. Numerical Implementation. The periodic single-layer potential $\mathcal{S}_{\Omega,\sharp}$ can be represented numerically in the same fashion as described previously for the Neumann–Poincaré operator $(\mathcal{K}^0_\Omega)^*$ in Subsection 2.4.5. Recall that the boundary $\partial\Omega$ is parametrized by $x(t)$ for $t \in [0, 2\pi)$. After partitioning the interval $[0, 2\pi)$ into N pieces

$$[t_1, t_2),\ [t_2, t_3),\ \ldots,\ [t_N, t_{N+1}),$$

with $t_1 = 0$ and $t_{N+1} = 2\pi$, we approximate the boundary $\partial\Omega = \{x(t) \in \mathbb{R}^2 : t \in [0, 2\pi)\}$ by $x^{(i)} = x(t_i)$ for $1 \le i \le N$. We then represent the infinite dimensional operator $\mathcal{S}_{\Omega,\sharp}$ acting on the density φ by a finite dimensional matrix S acting on the coefficient vector $\overline{\varphi}_i := \varphi(x^{(i)})$ for $1 \le i \le N$. We have

$$\mathcal{S}_{\Omega,\sharp}[\varphi](x) = \int_{\partial\Omega} G_\sharp(x, y)\varphi(y)\, d\sigma(y),$$

for $\psi \in L^2(\partial\Omega)$ and we represent it numerically by

$$S\tilde{\psi} = \begin{pmatrix} S_{11} & S_{12} & \dots & S_{1N} \\ S_{21} & S_{22} & \dots & S_{2N} \\ \vdots & \vdots & \ddots & \vdots \\ S_{N1} & \dots & \dots & S_{NN} \end{pmatrix} \begin{pmatrix} \overline{\varphi}_1 \\ \overline{\varphi}_2 \\ \vdots \\ \overline{\varphi}_N \end{pmatrix},$$

where

$$S_{ij} = \frac{1}{4\pi} \ln \big(\sinh^2(\pi(x_2^{(i)} - x_2^{(j)})) + \sin^2(\pi(x_1^{(i)} - x_1^{(j)})) \big) |T(x^{(j)})| (t_{j+1} - t_j), \quad i \neq j,$$

with $T(x^{(i)})$ being the tangent vector at $x^{(i)}$. When $i = j$, we have a logarithmic singularity and therefore we must handle the diagonal terms carefully. Let us explicitly calculate the integrals for the diagonal terms. Let the portion of the boundary starting at $x^{(i)}$ and ending at $x^{(i+1)}$ be parameterized by $s \in [0, \varepsilon = \frac{2\pi}{N})$ and note that ε tends to 0 as the number of discretization points $N \to \infty$. Therefore, using the Taylor expansion (2.117) given in the proof of Lemma 2.35 the term we need to calculate for the diagonal terms is:

$$S_{ii} = \frac{1}{2\pi} \int_0^\varepsilon \ln \Big(|x^{(i)} - x(s)| \Big) |T(s)| ds.$$

Taylor expanding for small s this term becomes

$$S_{ii} = \frac{1}{2\pi} \int_0^\varepsilon \ln \Big(|x^{(i)} - (x(0) + x'(0)s + O(s^2))| \Big) |T(0) + T'(0)s + O(s^2)| ds.$$

Noting that $x^{(i)} = x(0)$ and $T^{(i)} = x'(0)$ we have

$$S_{ii} \approx \frac{|T^{(i)}|}{2\pi} \int_0^\varepsilon \ln \Big(|T^{(i)}| s \Big) ds,$$

as $\varepsilon \to 0$. As

$$\int_0^\varepsilon \ln(as) ds = \varepsilon(\ln(a\varepsilon) - 1),$$

this means that

$$\begin{aligned} S_{ii} &\approx \frac{|T^{(i)}|\varepsilon}{2\pi} \Big(\ln \Big(|T^{(i)}|\varepsilon \Big) - 1 \Big) \\ &= \frac{|T^{(i)}|}{N} \Big(\ln \Big(\frac{2\pi}{N} |T^{(i)}| \Big) - 1 \Big). \end{aligned}$$

So, we have found an explicit representation for the diagonal terms of the matrix S.

For the periodic Neumann–Poincaré operator $\mathcal{K}^*_{\Omega,\sharp}$, the terms of the corresponding discretization matrix K are given by

$$\begin{aligned} K_{ij} =& \frac{1}{2} \bigg[\frac{\nu_1^{(i)} \sin(\pi\tilde{x}_1)\cos(\pi\tilde{x}_1)}{\sinh^2(\pi\tilde{x}_2) + \sin^2(\pi\tilde{x}_1)} \\ &+ \frac{\nu_2^{(i)} \sinh(\pi\tilde{x}_2)\cosh(\pi\tilde{x}_2)}{\sinh^2(\pi\tilde{x}_2) + \sin^2(\pi\tilde{x}_1)} \bigg] |T^{(j)}|(t_{j+1} - t_j), \quad i \neq j, \end{aligned}$$

where $\tilde{x} = x^{(i)} - x^{(i+1)}$. With regard to the diagonal terms, observe that in light of (2.117) we have precisely the same singularity as the one for the non-periodic case

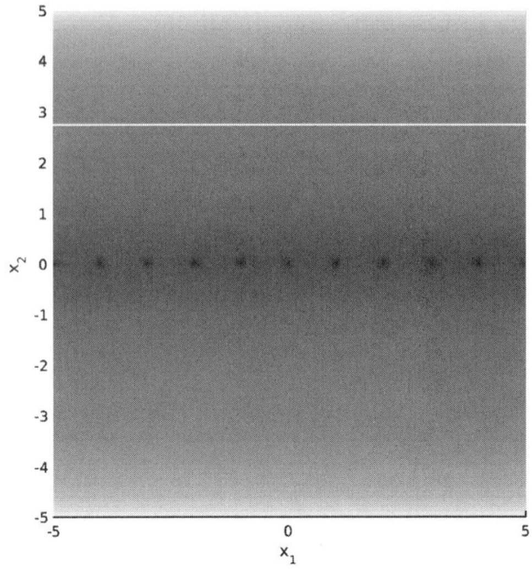

FIGURE 2.3. The periodic Green's function $G_\sharp$ for the Laplace equation.

and therefore we can use the same expression for the diagonal terms of the periodic version of the discretization matrix, that is:

$$K_{ii} \approx -\frac{1}{2N}\frac{\langle a^{(i)}, \nu^{(i)}\rangle}{|T^{(i)}|}. \tag{2.138}$$

The periodic Green's function $G_\sharp$ can be seen in Figure 2.3. The Matlab code for $G_\sharp$ and the associated layer potentials $\mathcal{S}_{\Omega,\sharp}$ and $\mathcal{K}^*_{\Omega,\sharp}$ can be downloaded at `www.ams.org/bookpages/surv-235`.

2.7. Shape Derivatives of Layer Potentials

In this section, we compute shape derivatives of layer potentials (see Appendix B.3 for the definition of the shape derivative). These calculations will be used to perform shape sensitivity analysis with respect to the eigenmodes or resonant modes of a cavity or a resonator.

Let D be a bounded domain of class $\mathcal{C}^2$ and D_ϵ be an ϵ-perturbation of D; *i.e.*, let $h \in \mathcal{C}^2(\partial D)$ and ∂D_ϵ be given by

$$\partial D_\epsilon = \left\{ \tilde{x} : \tilde{x} = x + \epsilon h(x)\nu(x),\ x \in \partial D \right\}.$$

In this section we derive full asymptotic expansions of $\mathcal{S}^\omega_{D_\epsilon}$ and $(\mathcal{K}^\omega_{D_\epsilon})^*$ in terms of ϵ.

Let $a, b \in \mathbb{R}$, with $a < b$, and let $X(t) : [a, b] \to \mathbb{R}^2$ be the arclength parametrization of ∂D; namely, X is a $\mathcal{C}^2$-function satisfying $|X'(t)| = 1$ for all $t \in [a, b]$ and

$$\partial D := \left\{ x = X(t),\ t \in [a, b] \right\}.$$

Then the outward unit normal to ∂D, $\nu(x)$, is given by $\nu(x) = R_{-\pi/2}X'(t)$, where $R_{-\pi/2}$ is the rotation by $-\pi/2$, the tangential vector at x, $T(x) = X'(t)$, and

$X'(t) \perp X''(t)$. Set the curvature $\tau(x)$ to be defined by

$$X''(t) = \tau(x)\nu(x).$$

We sometimes use $h(t)$ for $h(X(t))$ and $h'(t)$ for the tangential derivative of $h(x)$.

Then, $\tilde{X}(t) = X(t) + \epsilon h(t)\nu(x) = X(t) + \epsilon h(t) R_{-\pi/2} X'(t)$ is a parametrization of ∂D_ϵ. By $\tilde{\nu}(\tilde{x})$, we denote the outward unit normal to ∂D_ϵ at $\tilde{x}$. Then, we have

$$\begin{aligned} \tilde{\nu}(\tilde{x}) &= \frac{R_{-\pi/2}\tilde{X}'(t)}{|\tilde{X}'(t)|} \\ &= \frac{\Big(1 - \epsilon h(t)\tau(x)\Big)\nu(x) - \epsilon h'(t) X'(t)}{\sqrt{\epsilon^2 h'(t)^2 + \Big(1 - \epsilon h(t)\tau(x)\Big)^2}} \\ &= \frac{\Big(1 - \epsilon h(t)\tau(x)\Big)\nu(x) - \epsilon h'(t) T(x)}{\sqrt{\epsilon^2 h'(t)^2 + \Big(1 - \epsilon h(t)\tau(x)\Big)^2}}, \end{aligned} \tag{2.139}$$

and hence $\tilde{\nu}(\tilde{x})$ can be expanded uniformly as

$$\tilde{\nu}(\tilde{x}) = \sum_{n=0}^{+\infty} \epsilon^n \nu^{(n)}(x), \quad x \in \partial D, \tag{2.140}$$

where the vector-valued functions $\nu^{(n)}$ are bounded. In particular, the first two terms are given by

$$\nu^{(0)}(x) = \nu(x), \quad \nu^{(1)}(x) = -h'(t)T(x).$$

Likewise, we get a uniformly convergent expansion for the length element $d\sigma_\epsilon(\tilde{y})$:

$$d\sigma_\epsilon(\tilde{y}) = |\tilde{X}'(s)| ds = \sqrt{(1 - \epsilon\tau(s)h(s))^2 + \epsilon^2 h'^2(s)} ds = \sum_{n=0}^{+\infty} \epsilon^n \sigma^{(n)}(y)\, d\sigma(y), \tag{2.141}$$

where $\sigma^{(n)}$ are bounded functions such that the first few terms are given by

$$\sigma^{(0)}(y) = 1, \quad \sigma^{(1)}(y) = -\tau(y)h(y). \tag{2.142}$$

Set

$$\begin{aligned} x &= X(t),\ \tilde{x} = \tilde{X}(t) = x + \epsilon h(t) R_{-\frac{\pi}{2}} X'(t), \\ y &= X(s),\ \tilde{y} = \tilde{X}(s) = y + \epsilon h(s) R_{-\frac{\pi}{2}} X'(s). \end{aligned}$$

Since

$$\tilde{x} - \tilde{y} = x - y + \epsilon\Big(h(t)\nu(x) - h(s)\nu(y)\Big), \tag{2.143}$$

we obtain

$$|\tilde{x} - \tilde{y}|^2 = |x-y|^2 + 2\epsilon\langle x-y, h(t)\nu(x) - h(s)\nu(y)\rangle + \epsilon^2 |h(t)\nu(x) - h(s)\nu(y)|^2. \tag{2.144}$$

So, $H_0^{(1)}(\omega|\tilde{x} - \tilde{y}|)$ is equal to

$$H_0^{(1)}\bigg(\omega|x-y|(\sqrt{1 + \frac{2\epsilon\langle x-y, h(t)\nu(x) - h(s)\nu(y)\rangle + \epsilon^2|h(t)\nu(x) - h(s)\nu(y)|^2}{|x-y|^2}})\bigg).$$

Here, $H_0^{(1)}$ is the Hankel function of the first kind of order 0.

Therefore, we can write

$$H_0^{(1)}(\omega|\tilde{x}-\tilde{y}|)=\sum_{n=0}^{+\infty}\epsilon^n H_n^\omega(x,y),$$

where the series converges absolutely and uniformly. In particular,

$$H_0^\omega(x,y)=H_0^{(1)}(\omega|x-y|)$$

and

$$H_1^\omega(x,y)=\omega(H_0^{(1)})'(\omega|x-y|)\frac{\langle x-y,h(t)\nu(x)-h(s)\nu(y)\rangle}{|x-y|}.$$

Introduce a sequence of integral operators $(\mathcal{S}_{D,\omega}^{(n)})_{n\in\mathbb{N}}$, defined for any $\phi\in L^2(\partial D)$ by

$$\mathcal{S}_{D,\omega}^{(n)}[\phi](x)=-\frac{\sqrt{-1}}{4}\sum_{m=0}^{n}\int_{\partial D}H_m^\omega(x,y)\sigma^{(n-m)}(y)\,\phi(y)\,d\sigma(y)\quad\text{for } n\geq 0.$$

Let Ψ_ϵ be the diffeomorphism from ∂D onto ∂D_ϵ given by

$$\Psi_\epsilon(x)=x+\epsilon h(t)\nu(x),$$

where $x=X(t)$.

The following lemma holds.

LEMMA 2.41. *Let $N\in\mathbb{N}$. There exists C depending only on N, $\|X\|_{\mathcal{C}^2}$, and $\|h\|_{\mathcal{C}^2}$ such that for any $\tilde{\phi}\in L^2(\partial D_\epsilon)$,*

$$\left\|(\mathcal{S}_{D_\epsilon}^\omega[\tilde{\phi}])\circ\Psi_\epsilon-\mathcal{S}_D^\omega[\phi]-\sum_{n=1}^{N}\epsilon^n\mathcal{S}_{D,\omega}^{(n)}[\phi]\right\|_{L^2(\partial D)}\leq C\epsilon^{N+1}\|\phi\|_{L^2(\partial D)},\tag{2.145}$$

where $\phi:=\tilde{\phi}\circ\Psi_\epsilon$.

Turning now to the operator $(\mathcal{K}_{D_\epsilon}^\omega)^*$, we first note that

$$(\mathcal{K}_{D_\epsilon}^\omega)^*=(\mathcal{K}_{D_\epsilon}^0)^*+\mathcal{R}_{D_\epsilon},$$

where $\mathcal{R}_{D_\epsilon}$ has a smooth kernel so that we can write

$$(\mathcal{R}_{D_\epsilon}[\tilde{\phi}])\circ\Psi_\epsilon=\sum_{n=0}^{+\infty}\epsilon^n\int_{\partial D}r_n(x,y)\phi(y)\,d\sigma(y),\tag{2.146}$$

where r_n are smooth kernels and the series converges absolutely and uniformly. It suffices then to expand $(\mathcal{K}_{D_\epsilon}^0)^*$ with respect to ϵ.

From (2.144), it follows that

$$\frac{1}{|\tilde{x}-\tilde{y}|^2}=\frac{1}{|x-y|^2}\frac{1}{1+2\epsilon F(x,y)+\epsilon^2G(x,y)},\tag{2.147}$$

where

$$F(x,y)=\frac{\langle x-y,h(t)\nu(x)-h(s)\nu(y)\rangle}{|x-y|^2}$$

and

$$G(x,y)=\frac{|h(t)\nu(x)-h(s)\nu(y)|^2}{|x-y|^2}.$$

One can easily see that

$$|F(x,y)|+|G(x,y)|^{\frac{1}{2}}\leq C\|X\|_{\mathcal{C}^2}\|h\|_{\mathcal{C}^2}.$$

It follows from (2.139), (2.141), (2.143), and (2.147) that

$$\frac{\langle \tilde{x}-\tilde{y}, \tilde{\nu}(\tilde{x})\rangle}{|\tilde{x}-\tilde{y}|^2} d\sigma_\epsilon(\tilde{y}) = \Big(\frac{\langle x-y, \nu(x)\rangle}{|x-y|^2} + \epsilon\Big[\frac{\langle h(t)\nu(x) - h(s)\nu(y), \nu(x)\rangle}{|x-y|^2} - \frac{\langle x-y, \tau(x)h(t)\nu(x) + h'(t)T(x)\rangle}{|x-y|^2}\Big] - \epsilon^2 \frac{\langle h(t)\nu(x) - h(s)\nu(y), \tau(x)h(t)\nu(x)+h'(t)T(x)\rangle}{|x-y|^2}\Big) \times \frac{1}{1+2\epsilon F(x,y) + \epsilon^2 G(x,y)} \frac{\sqrt{(1-\epsilon\tau(y)h(s))^2 + \epsilon^2 h'^2(s)}}{\sqrt{(1-\epsilon\tau(x)h(t))^2 + \epsilon^2 h'^2(t)}} d\sigma(y)$$

$$:= \Big(K_0(x,y) + \epsilon K_1(x,y) + \epsilon^2 K_2(x,y)\Big) \times \frac{1}{1+2\epsilon F(x,y) + \epsilon^2 G(x,y)} \frac{\sqrt{(1-\epsilon\tau(y)h(s))^2 + \epsilon^2 h'^2(s)}}{\sqrt{(1-\epsilon\tau(x)h(t))^2 + \epsilon^2 h'^2(t)}} d\sigma(y).$$

Let

$$\frac{1}{1+2\epsilon F(x,y) + \epsilon^2 G(x,y)} \frac{\sqrt{(1-\epsilon\tau(y)h(s))^2 + \epsilon^2 h'^2(s)}}{\sqrt{(1-\epsilon\tau(x)h(t))^2 + \epsilon^2 h'^2(t)}} = \sum_{n=0}^{+\infty} \epsilon^n F_n(x,y),$$

where the series converges absolutely and uniformly. In particular, we can easily see that

$$F_0(x,y) = 1, \quad F_1(x,y) = -2F(x,y) + \tau(x)h(x) - \tau(y)h(y).$$

Then we now have

$$\frac{\langle \tilde{x}-\tilde{y}, \tilde{\nu}(\tilde{x})\rangle}{|\tilde{x}-\tilde{y}|^2} d\sigma_\epsilon(\tilde{y}) = \frac{\langle x-y, \nu(x)\rangle}{|x-y|^2} d\sigma(y) + \epsilon\Big(K_0(x,y)F_1(x,y) + K_1(x,y)\Big) d\sigma(y) + \epsilon^2 \sum_{n=0}^{+\infty} \epsilon^n \Big(F_{n+2}(x,y)K_0(x,y) + F_{n+1}(x,y)K_1(x,y) + F_n(x,y)K_2(x,y)\Big) d\sigma(y).$$

Therefore, we obtain that

$$\frac{\langle \tilde{x}-\tilde{y}, \tilde{\nu}(\tilde{x})\rangle}{|\tilde{x}-\tilde{y}|^2} d\sigma_\epsilon(\tilde{y}) = \sum_{n=0}^{+\infty} \epsilon^n \mathbb{k}_n(x,y)\, d\sigma(y),$$

where

$$\mathbb{k}_0(x,y) = \frac{\langle x-y, \nu(x)\rangle}{|x-y|^2}, \quad \mathbb{k}_1(x,y) = K_0(x,y)F_1(x,y) + K_1(x,y),$$

and for any $n \geq 2$,

$$\mathbb{k}_n(x,y) = F_n(x,y)K_0(x,y) + F_{n-1}(x,y)K_1(x,y) + F_{n-2}(x,y)K_2(x,y).$$

Introduce a sequence of integral operators $(\mathcal{K}_D^{(n)})_{n\in\mathbb{N}}$, defined for any $\phi \in L^2(\partial D)$ by

$$\mathcal{K}_D^{(n)}[\phi](x) = \int_{\partial D} \mathbb{k}_n(x,y)\phi(y)\, d\sigma(y) \quad \text{for } n \geq 0. \tag{2.148}$$

Note that $\mathcal{K}_D^{(0)} = (\mathcal{K}_D^0)^*$. Observe that the same operator with the kernel $\mathbb{k}_n(x,y)$ replaced with $K_j(x,y)$, $j = 0,1,2$, is bounded on $L^2(\partial D)$. In fact, it is an immediate consequence of the theorem of Coifman, McIntosh, and Meyer [**202**]. Therefore, each $\mathcal{K}_D^{(n)}$ is bounded on $L^2(\partial D)$.

The following lemma from [**68**] holds.

LEMMA 2.42. *Let $N \in \mathbb{N}$. There exists C depending only on N, $\|X\|_{\mathcal{C}^2}$, and $\|h\|_{\mathcal{C}^2}$ such that for any $\tilde{\phi} \in L^2(\partial D_\epsilon)$,*

$$\left\| ((\mathcal{K}_{D_\epsilon}^0)^*[\tilde{\phi}]) \circ \Psi_\epsilon - (\mathcal{K}_D^0)^*[\phi] - \sum_{n=1}^{N} \epsilon^n \mathcal{K}_D^{(n)}[\phi] \right\|_{L^2(\partial D)} \leq C \epsilon^{N+1} \|\phi\|_{L^2(\partial D)}, \tag{2.149}$$

where $\phi := \tilde{\phi} \circ \Psi_\epsilon$.

Now combining (2.146) and (2.149) immediately yields a full asymptotic expansion of $(\mathcal{K}_{D_\epsilon}^\omega)^*$ with respect to ϵ and allows us to write

$$(\mathcal{K}_{D_\epsilon}^\omega)^*[\cdot] \circ \Psi_\epsilon = (\mathcal{K}_D^\omega)^*[\cdot] + \epsilon \mathcal{K}_{D,\omega}^{(1)}[\cdot] + \epsilon^2 \mathcal{K}_{D,\omega}^{(2)}[\cdot] + \dots, \tag{2.150}$$

where each operator $\mathcal{K}_{D,\omega}^{(n)}$ is bounded on $L^2(\partial D)$.

2.8. Layer Potentials for the Helmholtz Equation

In this section we review a number of basic facts and results regarding the layer potentials associated with the Helmholtz equation. The integral equations that apply to the eigenvalue problem will be obtained from a study of these layer potentials.

2.8.1. Fundamental Solution. For $\omega > 0$, a fundamental solution $\Gamma_\omega(x)$ to the Helmholtz operator $\Delta + \omega^2$ in $\mathbb{R}^d, d = 2, 3$, is given by

$$\Gamma_\omega(x) = \begin{cases} -\dfrac{\sqrt{-1}}{4} H_0^{(1)}(\omega|x|), & d = 2, \\ -\dfrac{e^{\sqrt{-1}\omega|x|}}{4\pi|x|}, & d = 3, \end{cases} \tag{2.151}$$

for $x \neq 0$.

We recall the following behavior of the Hankel function $H_0^{(1)}$ near 0 (see, for instance, [**353**]):

$$-\frac{\sqrt{-1}}{4} H_0^{(1)}(\omega|x|) = \frac{1}{2\pi} \ln|x| + \eta_\omega + \sum_{j=1}^{+\infty} (b_j \ln(\omega|x|) + c_j)(\omega|x|)^{2j}, \tag{2.152}$$

where

$$b_j = \frac{(-1)^j}{2\pi} \frac{1}{2^{2j}(j!)^2}, \quad c_j = b_j \left(\gamma - \ln 2 - \frac{\pi\sqrt{-1}}{2} - \sum_{l=1}^{j} \frac{1}{l} \right),$$

and the constant

$$\eta_\omega = (1/2\pi)(\ln \omega + \gamma - \ln 2) - \sqrt{-1}/4, \tag{2.153}$$

γ being the Euler constant. It is known (see again [**353**]) that for large values of t we have

$$\begin{aligned} H_0^{(1)}(t) &= \sqrt{\frac{2}{\pi t}} e^{\sqrt{-1}(t - \frac{\pi}{4})} \left[1 + O\left(\frac{1}{t}\right) \right], \\ \frac{d}{dt} H_0^{(1)}(t) &= \sqrt{\frac{2}{\pi t}} e^{\sqrt{-1}(t + \frac{\pi}{4})} \left[1 + O\left(\frac{1}{t}\right) \right], \end{aligned} \quad \text{as } t \to +\infty. \tag{2.154}$$

Using (2.154) in two dimensions and the explicit form of Γ_ω in three dimensions, one can see that

$$\frac{x}{|x|} \cdot \nabla\Gamma_\omega(x) - \sqrt{-1}\omega\Gamma_\omega(x) = \begin{cases} O(|x|^{-3/2}), & d=2, \\ O(|x|^{-2}), & d=3, \end{cases} \tag{2.155}$$

as $|x| \to +\infty$.

This is exactly the Sommerfeld radiation condition one should impose in order to select the physical solution. The Sommerfeld radiation condition is also called the outgoing radiation condition and Γ_ω the outgoing fundamental solution to the Helmholtz equation.

2.8.2. Single- and Double-Layer Potentials. For a bounded Lipschitz domain Ω in $\mathbb{R}^d$ and $\omega > 0$, let $\mathcal{S}_\Omega^\omega$ and $\mathcal{D}_\Omega^\omega$ be the single- and double-layer potentials defined by Γ_ω; that is,

$$\mathcal{S}_\Omega^\omega[\varphi](x) = \int_{\partial\Omega} \Gamma_\omega(x-y)\varphi(y)\, d\sigma(y), \quad x \in \mathbb{R}^d, \tag{2.156}$$

$$\mathcal{D}_\Omega^\omega[\varphi](x) = \int_{\partial\Omega} \frac{\partial\Gamma_\omega(x-y)}{\partial\nu(y)}\varphi(y)\, d\sigma(y)\ , \quad x \in \mathbb{R}^d \setminus \partial\Omega, \tag{2.157}$$

for $\varphi \in L^2(\partial\Omega)$. Then $\mathcal{S}_\Omega^\omega[\varphi]$ and $\mathcal{D}_\Omega^\omega[\varphi]$ satisfy the Helmholtz equation

$$(\Delta + \omega^2)u = 0 \quad \text{in } \Omega \text{ and } \mathbb{R}^d \setminus \overline{\Omega}.$$

Moreover, in view of (2.155), both of them satisfy the Sommerfeld radiation condition, namely,

$$\left|\frac{\partial u}{\partial r} - \sqrt{-1}\omega u\right| = O\Big(r^{-(d+1)/2}\Big) \quad \text{as } r = |x| \to +\infty \quad \text{uniformly in } \frac{x}{|x|}. \tag{2.158}$$

Let us make note of a Green's formula to be used later. If $(\Delta + \omega^2)u = 0$ in Ω and $\partial u/\partial\nu \in L^2(\partial\Omega)$, then

$$-\mathcal{S}_\Omega^\omega\left[\frac{\partial u}{\partial\nu}\Big|_-\right](x) + \mathcal{D}_\Omega^\omega[u](x) = \begin{cases} u(x), & x \in \Omega, \\ 0, & x \in \mathbb{R}^d \setminus \overline{\Omega}. \end{cases} \tag{2.159}$$

A formula similar to (2.159) holds for the solution to the Helmholtz equation in $\mathbb{R}^d \setminus \overline{\Omega}$ subject to the Sommerfeld radiation condition (2.158).

Analogously to (2.7) and (2.8), the following formulas give the jump relations obeyed by the double-layer potential and by the normal derivative of the single-layer potential on general Lipschitz domains:

$$\frac{\partial(\mathcal{S}_\Omega^\omega[\varphi])}{\partial\nu}\bigg|_\pm (x) = \Big(\pm\frac{1}{2}I + (\mathcal{K}_\Omega^\omega)^*\Big)[\varphi](x) \quad \text{a.e. } x \in \partial\Omega, \tag{2.160}$$

$$(\mathcal{D}_\Omega^\omega[\varphi])\Big|_\pm (x) = \Big(\mp\frac{1}{2}I + \mathcal{K}_\Omega^\omega\Big)[\varphi](x) \quad \text{a.e. } x \in \partial\Omega, \tag{2.161}$$

for $\varphi \in L^2(\partial\Omega)$, where $\mathcal{K}_\Omega^\omega$ is the singular integral operator defined by

$$\mathcal{K}_\Omega^\omega[\varphi](x) = \text{p.v.} \int_{\partial\Omega} \frac{\partial\Gamma_\omega(x-y)}{\partial\nu(y)}\varphi(y)\, d\sigma(y)$$

and $(\mathcal{K}_\Omega^\omega)^*$ is the L^2-adjoint of $\mathcal{K}_\Omega^{-\omega}$, that is,

$$(\mathcal{K}_\Omega^\omega)^*[\varphi](x) = \text{p.v.} \int_{\partial\Omega} \frac{\partial\Gamma_\omega(x-y)}{\partial\nu(x)}\varphi(y) d\sigma(y).$$

Moreover, analogously to (2.10), for $\varphi \in H^{\frac{1}{2}}(\partial\Omega)$,

$$\frac{\partial}{\partial \nu}\mathcal{D}_\Omega^\omega[\varphi]\Big|_-(x) = \frac{\partial}{\partial \nu}\mathcal{D}_\Omega^\omega[\varphi]\Big|_+(x) \quad \text{in } H^{-\frac{1}{2}}(\partial\Omega). \tag{2.162}$$

The singular integral operators $\mathcal{K}_\Omega^\omega$ and $(\mathcal{K}_\Omega^\omega)^*$ are bounded on $L^2(\partial\Omega)$. Since $\Gamma_\omega(x) - \Gamma_0(x) = C + O(|x|)$ as $|x| \to 0$ where C is constant, we deduce that $\mathcal{K}_\Omega^\omega - \mathcal{K}_\Omega^0$ is bounded from $L^2(\partial\Omega)$ into $H^1(\partial\Omega)$ and hence is compact on $L^2(\partial\Omega)$. If Ω is $\mathcal{C}^{1,\eta}, \eta > 0$, then $\mathcal{K}_\Omega^0$ itself is compact on $L^2(\partial\Omega)$ and so is $\mathcal{K}_\Omega^\omega$.

2.8.3. Low-Frequency Asymptotic Expansions of Layer Potentials. We recall some basic asymptotic expansion for the layer potentials in three and two dimensions.

2.8.3.1. *Expansions in Three Dimensions.* We first consider the single-layer potential. We have the following asymptotic expansion as $\omega \to 0$:

$$\mathcal{S}_\Omega^\omega = \mathcal{S}_\Omega^0 + \sum_{j=1}^{\infty} \omega^j \mathcal{S}_{\Omega,j}, \tag{2.163}$$

where

$$\mathcal{S}_{\Omega,j}[\psi](x) = -\frac{\sqrt{-1}}{4\pi}\int_{\partial\Omega} \frac{(\sqrt{-1}|x-y|)^{j-1}}{j!}\psi(y)d\sigma(y).$$

In particular, we have

$$\mathcal{S}_{\Omega,1}[\psi](x) = -\frac{\sqrt{-1}}{4\pi}\int_{\partial\Omega} \psi(y)d\sigma(y), \tag{2.164}$$

$$\mathcal{S}_{\Omega,2}[\psi](x) = -\frac{1}{8\pi}\int_{\partial\Omega} |x-y|\psi(y)d\sigma(y). \tag{2.165}$$

LEMMA 2.43. *The norm $\|\mathcal{S}_{\Omega,j}\|_{\mathcal{L}(L^2(\partial\Omega),H^1(\partial\Omega))}$ is uniformly bounded with respect to j. Moreover, the series in (2.163) is convergent in $\mathcal{L}(L^2(\partial\Omega), H^1(\partial\Omega))$.*

We now consider the boundary integral operator $(\mathcal{K}_\Omega^\omega)^*$. We have

$$(\mathcal{K}_\Omega^\omega)^* = (\mathcal{K}_\Omega^0)^* + \omega\mathcal{K}_{\Omega,1} + \omega^2\mathcal{K}_{\Omega,2} + \ldots, \tag{2.166}$$

where

$$\begin{aligned}\mathcal{K}_{\Omega,j}[\psi](x) &= -\frac{\sqrt{-1}}{4\pi}\int_{\partial\Omega} \frac{\partial(\sqrt{-1}|x-y|)^{j-1}}{j!\partial\nu(x)}\psi(y)d\sigma(y)\\ &= -\frac{(\sqrt{-1})^j(j-1)}{4\pi j!}\int_{\partial\Omega} |x-y|^{j-3}(x-y)\cdot\nu(x)\psi(y)d\sigma(y).\end{aligned}$$

In particular, we have

$$\begin{aligned}\mathcal{K}_{\Omega,1} &= 0,\\ \mathcal{K}_{\Omega,2}[\psi](x) &= \frac{1}{8\pi}\int_{\partial\Omega} \frac{(x-y)\cdot\nu(x)}{|x-y|}\psi(y)d\sigma(y),\\ \mathcal{K}_{\Omega,3}[\psi](x) &= \frac{\sqrt{-1}}{12\pi}\int_{\partial D} (x-y)\cdot\nu(x)\psi(y)d\sigma(y).\end{aligned}$$

LEMMA 2.44. *The norm $\|\mathcal{K}_{\Omega,j}\|_{\mathcal{L}(L^2(\partial\Omega))}$ is uniformly bounded for $j \geq 1$. Moreover, the series in (2.166) is convergent in $\mathcal{L}(L^2(\partial\Omega))$.*

LEMMA 2.45. *Let $\mathcal{K}_{\Omega,j}^*$ be the adjoint of $\mathcal{K}_{\Omega,j}$ in $L^2(\Omega)$. The following identities hold:*

(i)

$$\mathcal{K}_{\Omega,2}^*[1](x) = \frac{1}{8\pi}\int_{\partial\Omega}\frac{(y-x)\cdot\nu(y)}{|y-x|}d\sigma(y) = \frac{1}{8\pi}\int_{\Omega}\nabla\cdot\frac{y-x}{|y-x|}dy = \frac{1}{4\pi}\int_{\Omega}\frac{1}{|y-x|}dy.$$

(ii)

$$\mathcal{K}_{\Omega,3}^*[1](x) = \frac{\sqrt{-1}}{12\pi}\int_{\partial\Omega}(y-x)\cdot\nu(y)d\sigma(y) = \frac{\sqrt{-1}}{12\pi}\int_{\Omega}\nabla\cdot(y-x)dy = \frac{\sqrt{-1}}{4\pi}|\Omega|.$$

2.8.3.2. *Expansions in Two Dimensions.* From (2.152), it follows that the single-layer potential for the Helmholtz equation in two dimensions has the following expansion as $\omega \to 0$:

$$\mathcal{S}_\Omega^\omega = \widehat{\mathcal{S}}_\Omega^\omega + \sum_{j=1}^{\infty}\left(\omega^{2j}\ln\omega\right)\mathcal{S}_{\Omega,j}^{(1)} + \sum_{j=1}^{\infty}\omega^{2j}\mathcal{S}_{\Omega,j}^{(2)}, \tag{2.167}$$

where

$$\widehat{\mathcal{S}}_\Omega^\omega[\psi](x) = \mathcal{S}_\Omega^0[\psi](x) + \eta_\omega\int_{\partial\Omega}\psi\, d\sigma, \tag{2.168}$$

and

$$\mathcal{S}_{\Omega,j}^{(1)}[\psi](x) = \int_{\partial\Omega} b_j|x-y|^{2j}\psi(y)d\sigma(y), \tag{2.169}$$

$$\mathcal{S}_{\Omega,j}^{(2)}[\psi](x) = \int_{\partial\Omega} |x-y|^{2j}(b_j\ln|x-y| + c_j)\psi(y)d\sigma(y). \tag{2.170}$$

We next consider the boundary integral operator $(\mathcal{K}_\Omega^\omega)^*$. We have

$$(\mathcal{K}_\Omega^\omega)^* = (\mathcal{K}_\Omega^0)^* + \sum_{j=1}^{\infty}\left(\omega^{2j}\ln\omega\right)\mathcal{K}_{\Omega,j}^{(1)} + \sum_{j=1}^{\infty}\omega^{2j}\mathcal{K}_{\Omega,j}^{(2)}, \tag{2.171}$$

where

$$\mathcal{K}_{\Omega,j}^{(1)}[\psi](x) = \int_{\partial\Omega} b_j\frac{\partial|x-y|^{2j}}{\partial\nu(x)}\psi(y)d\sigma(y),$$

$$\mathcal{K}_{\Omega,j}^{(2)}[\psi](x) = \int_{\partial\Omega}\frac{\partial\left(|x-y|^{2j}(b_j\ln|x-y|+c_j)\right)}{\nu(x)}\psi(y)d\sigma(y).$$

LEMMA 2.46. *The following expansions:*

$$\begin{aligned}\mathcal{S}_\Omega^\omega &= \widehat{\mathcal{S}}_\Omega^\omega + \omega^2\ln\omega\,\mathcal{S}_{\Omega,1}^{(1)} + \omega^2\mathcal{S}_{\Omega,1}^{(2)} + O(\omega^4\ln\omega);\\ (\mathcal{K}_\Omega^\omega)^* &= (\mathcal{K}_\Omega^0)^* + \omega^2\ln\omega\,\mathcal{K}_{\Omega,1}^{(1)} + \omega^2\mathcal{K}_{\Omega,1}^{(2)} + O(\omega^4\ln\omega),\end{aligned}$$

hold in $\mathcal{L}(L^2(\partial\Omega), H^1(\partial\Omega))$ and $\mathcal{L}(L^2(\partial\Omega), L^2(\partial\Omega))$, respectively.

LEMMA 2.47. *The following identities hold:*

(i)

$$(\mathcal{K}_{\Omega,1}^{(1)})^*[1](x) = 4\bar{b}_1|\Omega|;$$

(ii)

$$(\mathcal{K}_{\Omega,1}^{(2)})^*[1](x) = (2\bar{b}_1 + 4\bar{c}_1)|\Omega| + 4\bar{b}_1\int_\Omega \ln|x-y|dy,$$

where $\bar{b}_1$ and $\bar{c}_1$ are the complex conjugates of b_1 and c_1. Here, b_1 and c_1 are numbers appearing in (2.169) and (2.170).

PROOF. First, we have

$$\begin{aligned}(\mathcal{K}_{\Omega,1}^{(1)})^*[1](x) &= \bar{b}_1 \int_{\partial\Omega} 2(y-x)\cdot\nu(y)d\sigma(y)\\ &= \bar{b}_1 \int_{\partial\Omega} \frac{\partial |y-x|^2}{\partial\nu(y)}d\sigma(y)\\ &= \bar{b}_1 \int_{\Omega} \Delta_y |y-x|^2 dy\\ &= 4\bar{b}_1|\Omega|.\end{aligned}$$

We now prove the second identity. We have

$$\begin{aligned}(\mathcal{K}_{\Omega,1}^{(2)})^*[1](x) &= \int_{\partial\Omega} \frac{\partial\left[|y-x|^2(\bar{b}_1\ln|x-y|+\bar{c}_1)\right]}{\partial\nu(y)}d\sigma(y)\\ &= \int_{\Omega} \Delta_y[|y-x|^2(\bar{b}_1\ln|x-y|+\bar{c}_1)]dy\\ &= 4\bar{c}_1|\Omega| + \bar{b}_1\int_{\Omega}\Delta_y[|y-x|^2\ln|x-y|]dy\\ &= 4\bar{c}_1|\Omega| + \bar{b}_1\int_{\Omega} 4\ln|x-y|]dy\\ &\quad +\bar{b}_1\int_{\Omega}2dy + \bar{b}_1\int_{\Omega}|y-x|^2\Delta\ln|y-x|dy\\ &= (2\bar{b}_1+4\bar{c}_1)|\Omega| + 4\bar{b}_1\int_{\Omega}\ln|x-y|dy,\end{aligned}$$

where we have used the fact that

$$\int_{\Omega}|y-x|^2\Delta_y\ln|y-x|dy = 0 \qquad \text{for } x\in\partial\Omega.$$

This completes the proof of the Lemma. □

2.8.4. Uniqueness Results. In this subsection we consider important uniqueness results for the Helmholtz equation.

We will need the following key result from the theory of the Helmholtz equation. It will help us prove uniqueness for exterior Helmholtz problems. For its proof we refer to [**204**, Lemma 2.11] or [**390**, Lemma 9.8].

LEMMA 2.48 (Rellich's lemma). *Let $\omega\in\mathbb{R},\omega\neq 0$. Let $R_0>0$ and $B_R = \{|x|<R\}$. Let u satisfy the Helmholtz equation $\Delta u+\omega^2 u = 0$ for $|x|>R_0$. Assume, furthermore, that*

$$\lim_{R\to+\infty}\int_{\partial B_R}|u(x)|^2\,d\sigma(x) = 0.$$

Then, $u\equiv 0$ for $|x|>R_0$.

Note that the assertion of this lemma does not hold if ω is imaginary or $\omega=0$.

2.8.4.1. *Exterior Helmholtz Problems.* Now, using Lemma 2.48, we can establish the following uniqueness result for the exterior Helmholtz problem.

LEMMA 2.49. *Suppose* $d = 2$ *or* 3. *Let* $\omega \in \mathbb{R}, \omega \neq 0$. *Let* Ω *be a bounded Lipschitz domain in* $\mathbb{R}^d$. *Let* $u \in H^1_{\mathrm{loc}}(\mathbb{R}^d \setminus \overline{\Omega})$ *satisfy*

$$\begin{cases} \Delta u + \omega^2 u = 0 \quad \text{in } \mathbb{R}^d \setminus \overline{\Omega}, \\ \left| \dfrac{\partial u}{\partial r} - \sqrt{-1}\omega u \right| = O\left(r^{-(d+1)/2} \right) \quad \text{as } r = |x| \to +\infty \quad \text{uniformly in } \dfrac{x}{|x|}, \\ u = 0 \text{ or } \dfrac{\partial u}{\partial \nu} = 0 \quad \text{on } \partial\Omega. \end{cases}$$

Then, $u \equiv 0$ *in* $\mathbb{R}^d \setminus \overline{\Omega}$.

PROOF. Let $B_R = \{|x| < R\}$. For R large enough, $\Omega \subset B_R$. Notice first that by multiplying $\Delta u + \omega^2 u = 0$ by $\overline{u}$ and integrating by parts over $B_R \setminus \overline{\Omega}$, we arrive at

$$\Im \int_{\partial B_R} \overline{u} \frac{\partial u}{\partial \nu} \, d\sigma = 0.$$

Thus, we have

$$\Im \int_{\partial B_R} \overline{u} \left(\frac{\partial u}{\partial \nu} - \sqrt{-1}\omega u \right) d\sigma = -\omega \int_{\partial B_R} |u|^2.$$

Applying the Cauchy–Schwarz inequality,

$$\begin{aligned} &\left| \Im \int_{\partial B_R} \overline{u} \left(\frac{\partial u}{\partial \nu} - \sqrt{-1}\omega u \right) d\sigma \right| \\ &\qquad \leq \left(\int_{\partial B_R} |u|^2 \right)^{1/2} \left(\int_{\partial B_R} \left| \frac{\partial u}{\partial \nu} - \sqrt{-1}\omega u \right|^2 d\sigma \right)^{1/2}, \end{aligned}$$

and using the radiation condition (2.158), we get

$$\left| \Im \int_{\partial B_R} \overline{u} \left(\frac{\partial u}{\partial \nu} - \sqrt{-1}\omega u \right) d\sigma \right| \leq \frac{C}{R} \left(\int_{\partial B_R} |u|^2 \right)^{1/2},$$

for some positive constant C independent of R. Consequently, we obtain that

$$\left(\int_{\partial B_R} |u|^2 \right)^{1/2} \leq \frac{C}{R},$$

which indicates by Rellich's lemma that $u \equiv 0$ in $\mathbb{R}^d \setminus \overline{B_R}$. Hence, by the unique continuation property for $\Delta + \omega^2$, we can conclude that $u \equiv 0$ up to the boundary $\partial\Omega$. This finishes the proof. □

2.8.4.2. *Transmission Problem for the Helmholtz equation.* Let D be a bounded smooth domain in $\mathbb{R}^d$. Let μ and ε be two piecewise constant functions such that $\mu(x) = \mu_m$ and $\varepsilon(x) = \varepsilon_m$ for $x \in \mathbb{R}^d \setminus \overline{D}$ and $\mu(x) = \mu_c$ and $\varepsilon(x) = \varepsilon_c$ for $x \in D$. Suppose that $\mu_m, \varepsilon_m, \mu_c$, and ε_c are positive and let $k_m = \omega\sqrt{\varepsilon_m \mu_m}$ and $k_c = \omega\sqrt{\varepsilon_c \mu_c}$.

We consider the following transmission problem for the Helmholtz equation:

$$\begin{cases} \nabla \cdot \dfrac{1}{\mu} \nabla u + \omega^2 \varepsilon u = 0 \quad \text{in } \mathbb{R}^d, \\ u^s := u - u^i \text{ satisfies the Sommerfeld radiation condition,} \end{cases} \tag{2.172}$$

where u^i is an incident wave. Here, the Sommerfeld radiation condition reads:

$$\left|\frac{\partial u^s}{\partial r} - \sqrt{-1}k_m u^s\right| = O\left(r^{-(d+1)/2}\right) \quad \text{as } r = |x| \to +\infty \quad \text{uniformly in } \frac{x}{|x|}. \tag{2.173}$$

Notice that (2.172) can be rewritten as

$$\begin{cases} (\Delta + k_m^2)u = 0 & \text{in } \mathbb{R}^d \setminus \overline{D}, \\ (\Delta + k_c^2)u = 0 & \text{in } D, \\ u|_+ = u|_- & \text{on } \partial D, \\ \frac{1}{\mu_m}\frac{\partial u}{\partial \nu}|_+ = \frac{1}{\mu_c}\frac{\partial u}{\partial \nu}|_- & \text{on } \partial D, \\ u^s := u - u^i \text{ satisfies the Sommerfeld radiation condition.} \end{cases} \tag{2.174}$$

By using Rellich's lemma, we can prove that the following uniqueness result holds.

LEMMA 2.50. *If u satisfies (2.172) with $u^i = 0$, then $u \equiv 0$ in $\mathbb{R}^d$.*

PROOF. Using the fact that

$$\int_{\partial D} \frac{\partial u}{\partial \nu}\bigg|_+ \bar{u}\, d\sigma = \frac{\mu_m}{\mu_c}\int_{\partial D} \frac{\partial u}{\partial \nu}\bigg|_- \bar{u}\, d\sigma = \frac{\mu_m}{\mu_c}\int_D (|\nabla u|^2 - k_c^2|u|^2)\, dx\,,$$

we find that

$$\Im \int_{\partial D} \frac{\partial u}{\partial \nu}\bigg|_+ \bar{u}\, d\sigma = 0\,,$$

which gives, by applying Lemma 2.49, that $u \equiv 0$ in $\mathbb{R}^d \setminus D$. Now u satisfies $(\Delta + k_c^2)u = 0$ in D and $u = \partial u/\partial \nu = 0$ on ∂D. By the unique continuation property of $\Delta + k_c^2$, we readily get $u \equiv 0$ in D, and hence in $\mathbb{R}^d$. □

The following result from [**53**] is of importance to us for establishing a representation formula for the solution u to (2.172).

PROPOSITION 2.51. *Suppose that k_m^2 is not a Dirichlet eigenvalue for $-\Delta$ on D. For each $(F, G) \in H^1(\partial D) \times L^2(\partial D)$, there exists a unique solution $(f, g) \in L^2(\partial D) \times L^2(\partial D)$ to the system of integral equations*

$$\begin{cases} \mathcal{S}_D^{k_c}[f] - \mathcal{S}_D^{k_m}[g] = F \\ \dfrac{1}{\mu_c}\dfrac{\partial(\mathcal{S}_D^{k_c}[f])}{\partial \nu}\bigg|_- - \dfrac{1}{\mu_m}\dfrac{\partial(\mathcal{S}_D^{k_m}[g])}{\partial \nu}\bigg|_+ = G \end{cases} \quad \text{on } \partial D. \tag{2.175}$$

Furthermore, there exists a constant C independent of F and G such that

$$\|f\|_{L^2(\partial D)} + \|g\|_{L^2(\partial D)} \le C\left(\|F\|_{H^1(\partial D)} + \|G\|_{L^2(\partial D)}\right), \tag{2.176}$$

where in the three-dimensional case the constant C can be chosen independently of k_m and k_c if both k_m and k_c are sufficiently small.

PROOF. We only give the proof for $d = 3$ and $\mu_m \neq \mu_c$ leaving the general case to the reader. Let $X := L^2(\partial D) \times L^2(\partial D)$ and $Y := H^1(\partial D) \times L^2(\partial D)$, and define the operator $T : X \to Y$ by

$$T(f, g) := \left(\mathcal{S}_D^{k_c}[f] - \mathcal{S}_D^{k_m}[g], \frac{1}{\mu_c}\frac{\partial(\mathcal{S}_D^{k_c}[f])}{\partial \nu}\bigg|_- - \frac{1}{\mu_m}\frac{\partial(\mathcal{S}_D^{k_m}[g])}{\partial \nu}\bigg|_+\right).$$

We also define T_0 by

$$T_0(f,g) := \left(\mathcal{S}_D^0[f] - \mathcal{S}_D^0[g], \frac{1}{\mu} \frac{\partial(\mathcal{S}_D^0[f])}{\partial \nu}\bigg|_- - \frac{1}{\mu_m} \frac{\partial(\mathcal{S}_D^0[g])}{\partial \nu}\bigg|_+ \right).$$

We can easily see that $\mathcal{S}_D^{k_0} - \mathcal{S}_D^0 : L^2(\partial D) \to H^1(\partial D)$ is a compact operator, and so is $\frac{\partial}{\partial \nu}\mathcal{S}_D^{k_m}|_\pm - \frac{\partial}{\partial \nu}\mathcal{S}_D^0|_\pm : L^2(\partial D) \to L^2(\partial D)$. Therefore, $T - T_0$ is a compact operator from X into Y. It can be proved that $T_0 : X \to Y$ is invertible. In fact, a solution (f,g) of the equation $T_0(f,g) = (F,G)$ is given by

$$\begin{aligned} f &= g + (\mathcal{S}_D^0)^{-1}(F) \\ g &= \frac{\mu_m \mu_c}{\mu_m - \mu_c}(\lambda I + (\mathcal{K}_D^0)^*)^{-1} \left(G + \frac{1}{\mu_c}(\frac{1}{2}I - (\mathcal{K}_D^0)^*)((\mathcal{S}_D^0)^{-1}[F]) \right), \end{aligned}$$

where $\lambda = (\mu_c + \mu_m)/(2(\mu_c - \mu_m))$. From the invertibility of $\mathcal{S}_D^0$ and $\lambda I + (\mathcal{K}_D^0)^*$ we can see, by the Fredholm alternative, that it is enough to prove that T is injective.

Suppose that $T(f,g) = 0$. Then the function u defined by

$$u(x) := \begin{cases} \mathcal{S}_D^{k_m}[g](x) & \text{if } x \in \mathbb{R}^d \setminus \overline{D}, \\ \mathcal{S}_D^{k_c}[f](x) & \text{if } x \in D, \end{cases}$$

satisfies the transmission problem (2.172) with $u^i = 0$ and hence, by Lemma 2.50, $u \equiv 0$ in $\mathbb{R}^d$. In particular, $\mathcal{S}_D^{k_m}[g] = 0$ on ∂D. Since $(\Delta + k_m^2)\mathcal{S}_D^{k_m}[g] = 0$ in D and k_m^2 is not a Dirichlet eigenvalue for $-\Delta$ on D, we have $\mathcal{S}_D^{k_m}[g] = 0$ in D, and hence in $\mathbb{R}^d$. It then follows from the jump relation (2.160) that

$$g = \frac{\partial(\mathcal{S}_D^{k_m}[g])}{\partial \nu}\bigg|_+ - \frac{\partial(\mathcal{S}_D^{k_m}[g])}{\partial \nu}\bigg|_- = 0 \quad \text{on } \partial D.$$

On the other hand, $\mathcal{S}_D^{k_c} f$ satisfies $(\Delta + k_c^2)\mathcal{S}_D^{k_c}[f] = 0$ in $\mathbb{R}^d \setminus \overline{D}$ and $\mathcal{S}_D^{k_c}[f] = 0$ on ∂D. It then follows from Lemma 2.49 that $\mathcal{S}_D^{k_c}[f] = 0$. Then, in the same way as above, we can conclude that $f = 0$. This finishes the proof of solvability of (2.175). The estimate (2.176) is an easy consequence of solvability and the closed graph theorem. Finally, it can be easily proved in the three-dimensional case that if both k_m and k_c are sufficiently small, then the constant C in (2.176) can be chosen independently of k_m and k_c. We leave the details to the reader. □

By using Proposition 2.51, the following representation formula holds.

THEOREM 2.52. *Suppose that k_0^2 is not a Dirichlet eigenvalue for $-\Delta$ on D. Let u be the solution of (2.172). Then u can be represented using the single-layer potentials $\mathcal{S}_D^{k_m}$ and $\mathcal{S}_D^{k_c}$ as follows:*

$$u(x) = \begin{cases} u^i(x) + \mathcal{S}_D^{k_m}[\psi](x), & x \in \mathbb{R}^2 \setminus \overline{D}, \\ \mathcal{S}_D^{k_c}[\varphi](x), & x \in D, \end{cases} \tag{2.177}$$

where the pair $(\varphi,\psi) \in L^2(\partial D) \times L^2(\partial D)$ is the unique solution to

$$\begin{cases} \mathcal{S}_D^{k_c}[\varphi] - \mathcal{S}_D^{k_m}[\psi] = u^i \\ \dfrac{1}{\mu_c}\dfrac{\partial(\mathcal{S}_D^{k_c}[\varphi])}{\partial \nu}\bigg|_- - \dfrac{1}{\mu_m}\dfrac{\partial(\mathcal{S}_D^{k_m}[\psi])}{\partial \nu}\bigg|_+ = \dfrac{1}{\mu_m}\dfrac{\partial u^i}{\partial \nu} \end{cases} \quad \text{on } \partial D. \tag{2.178}$$

Moreover, there exists $C > 0$ independent of u^i such that

$$\|\varphi\|_{L^2(\partial D)} + \|\psi\|_{L^2(\partial D)} \leq C\Big(\|u^i\|_{L^2(\partial D)} + \|\nabla u^i\|_{L^2(\partial D)} \Big). \tag{2.179}$$

REMARK 2.53. *For a special case of the domain D, we can obtain an explicit solution to the transmission problem. Let D be a disk of radius R located at the origin in $\mathbb{R}^2$. Let $H_n^{(1)}$ be the Hankel function of the first kind of order n and let J_n be the Bessel function of order n. We assume that the incident wave is given by $u^i(x) = J_n(k_m r)e^{\sqrt{-1}n\theta}$. Then it can be shown that the explicit solution is given by*

$$u(r,\theta) = \begin{cases} J_n(k_m r)e^{\sqrt{-1}n\theta} + a_n H_n^{(1)}(k_m r)e^{\sqrt{-1}n\theta}, & |r| > R, \\ b_n J_n(k_c r)e^{\sqrt{-1}n\theta}, & |r| \leq R, \end{cases} \tag{2.180}$$

where (r,θ) are the polar coordinates and the constants a_n and b_n are given by

$$a_n = \frac{\frac{k_m}{\mu_m} J_n(k_c R)J_n'(k_m R) - \frac{k_c}{\mu_c} J_n(k_m R)J_n'(k_c R)}{\frac{k_c}{\mu_c} H_n^{(1)}(k_m R)J_n'(k_c R) - \frac{k_m}{\mu_m} J_n(k_c R)H_n'(k_m R)}, \tag{2.181}$$

$$b_n = \frac{J_n(k_m R) + a_n H_n^{(1)}(k_m R)}{J_n(k_c R)}. \tag{2.182}$$

2.9. Laplace Eigenvalues

In this section we transform eigenvalue problems of $-\Delta$ on an open bounded connected domain Ω with either Neumann, Dirichlet, Robin or mixed boundary conditions into the problem of determining the characteristic values of certain integral operator-valued functions in the complex plane. This results in a considerable advantage as it allows us to reduce the dimension of the eigenvalue problem. After discretization of the kernels of the integral operators, the problem can be turned into a complex root finding process for a scalar function; see for instance [**197**]. Many tools are available for finding complex roots of scalar functions. Muller's method described in Section 1.6 is both efficient and robust.

Moreover, with the help of the generalized argument principle, the integral formulations can also be used to study perturbations of the eigenvalues with respect to deformations of $\partial\Omega$, as we will see in Subsection 3.2.2. Furthermore, the splitting problem in the evolution of multiple eigenvalues can be easily handled. In Subsection 2.9.6, we present a method for carrying out sensitivity analysis of multiple eigenvalues with respect to deformations of $\partial\Omega$, which relies on finding a polynomial of degree equal to the geometric multiplicity of the eigenvalue such that its zeros are precisely the perturbations.

2.9.1. Eigenvalue Characterization.

We first restrict our attention to the three-dimensional case. We note that because of the holomorphic dependence of Γ_ω on ω as given in (2.151), $\mathcal{K}_\Omega^\omega$ is an operator-valued holomorphic function in $\mathbb{C}$. Indeed, the following result holds. See, for example, [**494**].

PROPOSITION 2.54 (Neumann Eigenvalue characterization). *Let $d = 3$. Suppose that Ω is of class $\mathcal{C}^{1,\eta}$ for some $\eta > 0$. Let $\omega > 0$. Then ω^2 is an eigenvalue of $-\Delta$ on Ω with Neumann boundary condition if and only if ω is a positive real characteristic value of the operator $-(1/2)I + \mathcal{K}_\Omega^\omega$.*

PROOF. Suppose that ω^2 is an eigenvalue of

$$\begin{cases} \Delta u + \omega^2 u = 0 & \text{in } \Omega, \\ \dfrac{\partial u}{\partial \nu} = 0 & \text{on } \partial\Omega. \end{cases} \tag{2.183}$$

By Green's formula (2.159), we have

$$u(x) = \mathcal{D}_\Omega^\omega[u|_{\partial\Omega}](x), \quad x \in \Omega.$$

It then follows from (2.161) that $(-I/2+\mathcal{K}_\Omega^\omega)[u|_{\partial\Omega}] = 0$ and $u|_{\partial\Omega} \neq 0$ since otherwise the unique continuation property for $\Delta + \omega^2$ would imply that $u \equiv 0$ in Ω. Thus ω is a characteristic value of $-(1/2)\, I + \mathcal{K}_\Omega^\omega$.

Suppose now that ω is a characteristic value of $-(1/2)\, I + \mathcal{K}_\Omega^\omega$; *i.e.*, there is a nonzero $\psi \in L^2(\partial\Omega)$ such that

$$\left(-\frac{1}{2} I + \mathcal{K}_\Omega^\omega\right)[\psi] = 0.$$

Then $u = \mathcal{D}_\Omega^\omega[\psi]$ on $\mathbb{R}^d \backslash \overline{\Omega}$ is a solution to the Helmholtz equation with the boundary condition $u|_+ = 0$ on $\partial\Omega$ and satisfies the radiation condition (2.158). The uniqueness result in Lemma 2.49 implies that $\mathcal{D}_\Omega^\omega[\psi] = 0$ in $\mathbb{R}^d \setminus \overline{\Omega}$. Since $\partial \mathcal{D}_\Omega^\omega[\psi]/\partial\nu$ exists and has no jump across $\partial\Omega$, we get

$$\frac{\partial \mathcal{D}_\Omega^\omega[\psi]}{\partial \nu}\bigg|_- = \frac{\partial \mathcal{D}_\Omega^\omega[\psi]}{\partial \nu}\bigg|_+ = 0 \quad \text{on } \partial\Omega.$$

Hence, we deduce that $\mathcal{D}_\Omega^\omega[\psi]$ is a solution of (2.183). Note that $\mathcal{D}_\Omega^\omega[\psi] \neq 0$ in Ω, since otherwise

$$\psi = \mathcal{D}_\Omega^\omega[\psi]\big|_- - \mathcal{D}_\Omega^\omega[\psi]\big|_+ = 0.$$

Thus ω^2 is an eigenvalue of $-\Delta$ on Ω with Neumann condition, and so the proposition is proved. $\square$

Proposition 2.54 asserts that $-(1/2)\, I + \mathcal{K}_\Omega^\omega$ is invertible on $L^2(\partial\Omega)$ for all positive ω except for a discrete set. The following result, whose proof can be found in [**494**, Proposition 7.3], shows that $(-(1/2)\, I + \mathcal{K}_\Omega^\omega)^{-1}$ has a continuation as an operator-valued meromorphic function in $\mathbb{C}$. This follows from the fact that the Neumann eigenvalues are of finite multiplicities.

PROPOSITION 2.55. *$-(1/2)\, I + \mathcal{K}_\Omega^\omega$ is invertible on $L^2(\partial\Omega)$ for all $\omega \in \mathbb{C}$ except for a discrete set, and $(-(1/2)\, I + \mathcal{K}_\Omega^\omega)^{-1}$ is an operator-valued meromorphic function on $\mathbb{C}$.*

In the two-dimensional case, Proposition 2.54 also holds true. Moreover, due to the logarithmic behavior of the Hankel function as shown by (2.152), $(-(1/2)\, I + \mathcal{K}_\Omega^\omega)^{-1}$ has a continuation to an operator-valued meromorphic function on only $\mathbb{C} \setminus \sqrt{-1}\mathbb{R}^-$.

Similarly, the eigenvalues of $-\Delta$ on Ω with Dirichlet boundary condition can be characterized as follows.

PROPOSITION 2.56 (Dirichlet Eigenvalue characterization). *Suppose that Ω is of class $\mathcal{C}^{1,\eta}$ for some $\eta > 0$. Let $\omega > 0$. Then ω^2 is an eigenvalue of $-\Delta$ on Ω with Dirichlet boundary condition if and only if ω is a positive real characteristic value of the operator $(1/2)\, I + (\mathcal{K}_\Omega^\omega)^*$.*

The problem of finding the eigenvalues of $-\Delta$ on Ω with the Robin boundary condition,

$$\frac{\partial u}{\partial \nu} + \lambda u = 0 \quad \text{on } \partial\Omega, \tag{2.184}$$

can be also transformed into the determination of a certain integral operator-valued function in the complex plane.

PROPOSITION 2.57 (Robin Eigenvalue characterization). *Suppose that Ω is of class $\mathcal{C}^{1,\eta}$ for some $\eta > 0$. Let $\omega > 0$ and $\lambda \leq 0$. Then ω^2 is an eigenvalue of $-\Delta$ on Ω with the Robin boundary condition (2.184) if and only if ω is a positive real characteristic value of the operator $-(1/2)\,I + \mathcal{K}_\Omega^\omega - \lambda \mathcal{S}_\Omega^\omega$.*

Finally, we consider the mixed boundary value problem and state the following result.

PROPOSITION 2.58 (Zaremba eigenvalue characterization). *Suppose that Ω is of class $\mathcal{C}^{1,\eta}$ for some $\eta > 0$. Let Γ_D be a subset of $\partial\Omega$ and let $\Gamma_N = \partial\Omega \setminus \overline{\Gamma_D}$. Let $\omega > 0$. Then ω^2 is an eigenvalue of $-\Delta$ on Ω with the mixed boundary conditions (also called a Zaremba eigenvalue),*

$$\begin{cases} \Delta u + \omega^2 u = 0 & \text{in } \Omega, \\ u = 0 & \text{on } \Gamma_D, \\ \dfrac{\partial u}{\partial \nu} = 0 & \text{on } \Gamma_N, \end{cases} \tag{2.185}$$

if and only if ω is a positive real characteristic value of the operator

$$\omega \mapsto \begin{bmatrix} (1/2)\,I + (\mathcal{K}_{\Gamma_D}^\omega)^* & \frac{\partial}{\partial \nu}\mathcal{D}_{\Gamma_N}^\omega\big|_{\Gamma_D} \\ -\mathcal{S}_{\Gamma_D}^\omega\big|_{\Gamma_N} & -(1/2)\,I + \mathcal{K}_{\Gamma_N}^\omega \end{bmatrix}.$$

Here, $\mathcal{S}_{\Gamma_D}^\omega, \mathcal{K}_{\Gamma_N}^\omega, (\mathcal{K}_{\Gamma_D}^\omega)^*$, and $\mathcal{D}_{\Gamma_N}^\omega$ are defined in the same way as in Section 2.3 with $\partial\Omega$ replaced with Γ_N or Γ_D.

We refer the reader to [**9**, **50**, **351**] for recent studies on the Zaremba eigenvalue problem.

2.9.2. Neumann Function. Let $0 = \mu_1 < \mu_2 \leq \mu_3 \leq \ldots$ be the eigenvalues of $-\Delta$ on Ω with Neumann conditions on $\partial\Omega$. Let u_j denote the normalized eigenfunction associated with μ_j; that is, it satisfies $\|u_j\|_{L^2(\Omega)} = 1$. Let $\omega \notin \{\sqrt{\mu_j}\}_{j\geq 1}$. Introduce $N_\Omega^\omega(x,z)$ as the Neumann function for $\Delta + \omega^2$ in Ω corresponding to a Dirac mass at z. That is, N_Ω^ω is the unique solution to

$$\begin{cases} (\Delta_x + \omega^2) N_\Omega^\omega(x,z) = -\delta_z & \text{in } \Omega, \\ \dfrac{\partial N_\Omega^\omega}{\partial \nu}\bigg|_{\partial\Omega} = 0 & \text{on } \partial\Omega. \end{cases} \tag{2.186}$$

We derive two useful facts on the Neumann function. First, we establish the following proposition, providing a purely formal proof. We refer the reader to [**463**, Theorem 9.8] for a more rigorous one where even the case $\omega = 0$ is treated.

PROPOSITION 2.59 (Spectral decomposition). *The following spectral decomposition holds pointwise:*

$$N_\Omega^\omega(x,z) = \sum_{j=1}^{+\infty} \frac{u_j(x)u_j(z)}{\mu_j - \omega^2}, \quad x \neq z \in \Omega. \tag{2.187}$$

PROOF. Consider the function

$$f(x) := \sum_{j=1}^{+\infty} a_j u_j(x), \quad x \in \Omega.$$

If $(\Delta_x + \omega^2) f(x) = -\delta_z(x)$, then we have

$$\sum_{j=1}^{+\infty} a_j (\omega^2 - \mu_j) u_j(x) = -\delta_z(x).$$

Integrating both sides of the above identity against u_k over Ω gives

$$a_k(\omega^2 - \mu_k) = -u_k(z),$$

and hence (2.187) follows. Here note that we used the orthogonality relation

$$\int_\Omega u_j\, u_k = \delta_{jk}$$

satisfied by the eigenfunctions, where δ_{jk} denotes the Kronecker symbol. □

Next, we provide an important relation between the fundamental solution Γ_ω and the Neumann function N_Ω^ω. Note that the Neumann function N_Ω^ω yields a solution operator for the Neumann problem for the Helmholtz equation. In fact, the function u defined by

$$u(z) = \int_{\partial\Omega} N_\Omega^\omega(x, z) g(x)\, d\sigma(x), \quad z \in \Omega,$$

is the unique solution to the Helmholtz equation:

$$\begin{cases} \Delta u + \omega^2 u = 0 & \text{in } \Omega, \\ \dfrac{\partial u}{\partial \nu} = g & \text{on } \partial\Omega, \end{cases} \tag{2.188}$$

provided that ω^2 is not an eigenvalue of $-\Delta$ on Ω with Neumann boundary condition. On the other hand, under this assumption, $-(1/2)\, I + \mathcal{K}_\Omega^\omega : L^2(\partial\Omega) \to L^2(\partial\Omega)$ is invertible, and so we can readily see that the solution to (2.188) can be represented as

$$u(z) = \mathcal{S}_\Omega^\omega \left(-\frac{1}{2} I + (\mathcal{K}_\Omega^\omega)^*\right)^{-1} [g](z), \quad z \in \Omega. \tag{2.189}$$

Therefore, we obtain

$$\int_{\partial\Omega} \Gamma_\omega(x - z) \left(-\frac{1}{2} I + (\mathcal{K}_\Omega^\omega)^*\right)^{-1} [g](x) d\sigma(x) = \int_{\partial\Omega} N_\Omega^\omega(x, z) g(x) d\sigma(x),$$

and hence

$$\int_{\partial\Omega} \left(-\frac{1}{2} I + \mathcal{K}_\Omega^\omega\right)^{-1} \big[\Gamma_\omega(\cdot - z)\big](x) g(x) d\sigma(x) = \int_{\partial\Omega} N_\Omega^\omega(x, z) g(x) d\sigma(x).$$

We then have the following proposition.

PROPOSITION 2.60. *The following identity relating the fundamental solution Γ_ω to the Neumann function N_Ω^ω holds:*

$$-\left(\frac{1}{2} I - \mathcal{K}_\Omega^\omega\right)^{-1} \Big[\Gamma_\omega(\cdot - z)\Big](x) = N_\Omega^\omega(x, z), \quad x \in \partial\Omega,\ z \in \Omega. \tag{2.190}$$

Finally, we recall that the Neumann function N_Ω^ω has a logarithmic singularity in two dimensions [**52**].

LEMMA 2.61. *The Neumann function N_Ω^ω has the form*

$$N_\Omega^\omega(x,z) = -\frac{1}{2\pi}\ln|x-z| + R_\Omega^\omega(x,z) \quad \textit{for } x \neq z \in \Omega, \tag{2.191}$$

where $R_\Omega^\omega(\cdot, z)$ belongs to $H^{3/2}(\Omega)$ for any $z \in \Omega$.

In dimension $d \geq 3$, the following lemma holds.

LEMMA 2.62. *Suppose that $d > 2$. The Neumann function N_Ω^ω has the form*

$$N_\Omega^\omega(x,z) = \frac{1}{(d-2)\omega_d}|x-z|^{2-d} + R_\Omega^{\omega,d}(x,z) \quad \textit{for } x \neq z \in \Omega, \tag{2.192}$$

where $R_\Omega^{\omega,d}(\cdot, z)$ belongs to $H^{3/2}(\Omega)$ for any $z \in \Omega$.

2.9.3. Dirichlet Function. Let $0 < \tau_1 < \tau_2 \leq \tau_3 \leq \dots$ be the eigenvalues of $-\Delta$ on Ω with Dirichlet conditions on $\partial\Omega$. Let v_j denote the normalized eigenfunction associated with τ_j; that is, it satisfies $\|v_j\|_{L^2(\Omega)} = 1$. Let $\omega \notin \{\sqrt{\tau_j}\}_{j\geq 1}$. The Dirichlet function $G_\Omega^\omega(x,z)$ is defined by

$$\begin{cases} (\Delta_x + \omega^2) G_\Omega^\omega(x,z) = -\delta_z & \text{in } \Omega, \\ G_\Omega^\omega = 0 & \text{on } \partial\Omega. \end{cases} \tag{2.193}$$

The following useful facts on the Dirichlet function hold.

PROPOSITION 2.63. *We have*

(i) *The following spectral decomposition holds pointwise:*

$$G_\Omega^\omega(x,z) = \sum_{j=1}^{+\infty} \frac{v_j(x)v_j(z)}{\tau_j - \omega^2}, \quad x \neq z \in \Omega. \tag{2.194}$$

(ii) *If $d > 2$, then the Dirichlet function G_Ω^ω has the form*

$$G_\Omega^\omega(x,z) = \frac{1}{(d-2)\omega_d}|x-z|^{2-d} + \widetilde{R}_\Omega^{\omega,d}(x,z) \quad \textit{for } x \neq z \in \Omega, \tag{2.195}$$

where $\widetilde{R}_\Omega^{\omega,d}(\cdot, z)$ belongs to $H^{3/2}(\Omega)$ for any $z \in \Omega$.

2.9.4. Eigenvalues in Circular Domains. Let κ_{nm} be the positive zeros of $J_n(z)$ (Dirichlet), $J_n'(z)$ (Neumann), and $J_n'(z) + \lambda J_n(z)$ (Robin). The index $n = 0, 1, 2, \dots$ counts the order of Bessel functions of first kind J_n while $m = 1, 2, \dots$ counts their positive zeros. The rotational symmetry of a disk $\Omega = \{x : |x| < R\}$ of radius R leads to an explicit representation of the eigenfunctions in polar coordinates:

$$u_{nml}(r,\theta) = J_n(\frac{\kappa_{nm} r}{R}) \times \begin{cases} \cos(n\theta), & l = 1, \\ \sin(n\theta), & l = 2 \quad (n \neq 0). \end{cases} \tag{2.196}$$

The eigenvalues of $-\Delta$ on Ω are given by κ_{nm}^2/R^2. They are independent of the index l. They are simple for $n = 0$ and of multiplicity two for $n > 0$. In the latter case, an eigenfunction is any nontrivial linear combination of u_{nm1} and u_{nm2}.

Notice that when the index n is fixed while m increases, the Bessel functions $J_n(\frac{\kappa_{nm} r}{R})$ rapidly oscillate, the amplitude of oscillations decreasing toward the boundary and the eigenfunctions u_{nml} given by (2.196) are mainly localized at the origin, yielding focusing modes. In turn, when the index m is fixed while n increases, the Bessel functions $J_n(\frac{\kappa_{nm}r}{R})$ become strongly attenuated near the origin 0 and essentially localized near the boundary. This yields the so-called whispering gallery eigenmodes. Estimates of localization are derived in [**271**].

In [**351**], the eigenfunctions of the Zaremba problem are represented in terms of infinite sums of Bessel functions in the unit circle on the complex plane and the corersponding eigenvalues are identified.

2.9.5. Shape Derivative of Laplace Eigenvalues. In this subsection, we compute shape derivatives of Laplace eigenvalues with the Neumann boundary condition by using the generalized argument principle. Let Ω be a bounded domain of class $\mathcal{C}^2$. We consider Neumann eigenvalues in the two-dimensional case and let Ω_ϵ be given by

$$\partial\Omega_\epsilon = \left\{ \tilde{x} : \tilde{x} = x + \epsilon h(x)\nu(x),\ x \in \partial\Omega \right\},$$

where $h \in \mathcal{C}^2(\partial\Omega)$ and $0 < \epsilon \ll 1$.

To fix ideas, we set μ_j for $j > 1$ to be a Neumann eigenvalue of $-\Delta$ on Ω and consider the integral operator-valued function

$$\omega \mapsto \mathcal{A}_\epsilon(\omega) := -\frac{1}{2}I + \mathcal{K}^\omega_{\Omega_\epsilon}, \tag{2.197}$$

when ω is in a small complex neighborhood of $\sqrt{\mu_j}$.

By using the compactness of $\mathcal{K}^\omega_{\Omega_\epsilon}$ and the analyticity of $H_0^{(1)}$ in $\mathbb{C} \setminus \sqrt{-1}\mathbb{R}^-$, the following results hold.

LEMMA 2.64. *The operator-valued function $\mathcal{A}_\epsilon(\omega)$ is a holomorphic Fredholm operator with index 0 in $\mathbb{C}\setminus\sqrt{-1}\mathbb{R}^-$ and $\mathcal{A}_\epsilon(\omega)^{-1}$ is a meromorphic function of ω. If ω is a real characteristic value of the operator-valued function $\mathcal{A}_\epsilon$ (or equivalently, a real pole of $(\mathcal{A}_\epsilon)^{-1}(\omega)$), then there exists j such that $\omega = \sqrt{\mu_j^\epsilon}$ (Neumann eigenvalue in Ω_ϵ).*

LEMMA 2.65. *Any $\sqrt{\mu_j}$ is a simple pole of the operator-valued function $(\mathcal{A}_0)^{-1}(\omega)$.*

LEMMA 2.66. *Let $\omega_0 = \sqrt{\mu_j}$ and suppose that μ_j is simple. Then there exists a positive constant δ_0 such that the operator-valued function $\omega \mapsto \mathcal{A}_\epsilon(\omega)$ has exactly one characteristic value in $\overline{V_{\delta_0}}(\omega_0)$, where $V_{\delta_0}(\omega_0)$ is a disk of center ω_0 and radius $\delta_0 > 0$. This characteristic value is holomorphic with respect to ϵ in $(-\epsilon_0, \epsilon_0)$ for some $\epsilon_0 > 0$. Moreover, the following assertions hold:*

(i) $\mathcal{M}(\mathcal{A}_\epsilon(\omega); \partial V_{\delta_0}) = 1$,
(ii) $\mathcal{A}_\epsilon(\omega)^{-1} = (\omega - \omega_\epsilon)^{-1}\mathcal{L}_\epsilon + \mathcal{R}_\epsilon(\omega)$,
(iii) $\mathcal{L}_\epsilon : \mathrm{Ker}((\mathcal{A}_\epsilon(\omega_\epsilon))^*) \to \mathrm{Ker}(\mathcal{A}_\epsilon(\omega_\epsilon))$ *is a finite-dimensional operator,*
(iv) $\mathcal{R}_\epsilon(\omega)$ *is a holomorphic function with respect to* $(\epsilon, \omega) \in (-\epsilon_0, \epsilon_0) \times V_{\delta_0}(\omega_0)$.

PROOF. Note that the kernel of $\mathcal{K}^\omega_{\Omega_\epsilon}$ is jointly holomorphic with respect to ϵ in $(-\epsilon_0, \epsilon_0)$ and $\omega \in V_{\delta_0}$ for ϵ_0 and δ_0 small enough; see [**176**]. Since μ_j is simple, it is clear that $\mathcal{M}(\mathcal{A}_\epsilon(\omega); \partial V_{\delta_0}) = 1$. Furthermore, from Lemmas 2.64 and 2.65, it follows that

$$\mathcal{A}_\epsilon(\omega)^{-1} = (\omega - \omega_\epsilon)^{-1}\mathcal{L}_\epsilon + \mathcal{R}_\epsilon(\omega),$$

where

$$\mathcal{L}_\epsilon : \mathrm{Ker}((\mathcal{A}_\epsilon(\omega_\epsilon))^*) \to \mathrm{Ker}(\mathcal{A}_\epsilon(\omega_\epsilon))$$

is a finite-dimensional operator and $\mathcal{R}_\epsilon(\omega)$ is a holomorphic function with respect to (ϵ, ω). □

Let $\omega_0 = \sqrt{\mu_j}$ and suppose that μ_j is simple. Then, from Theorem 1.14 it follows that $\omega_\epsilon = \sqrt{\mu_j^\epsilon}$ is given by

$$\omega_\epsilon - \omega_0 = \frac{1}{2\sqrt{-1}\pi} \operatorname{tr} \int_{\partial V_{\delta_0}} (\omega - \omega_0)\mathcal{A}_\epsilon(\omega)^{-1} \frac{\partial}{\partial \omega} \mathcal{A}_\epsilon(\omega) d\omega. \tag{2.198}$$

With the same notation as in Section 2.7, let the operator $\mathcal{K}_\Omega^{(1)}$ be defined by

$$\mathcal{K}_\Omega^{(1)}[\varphi] = \int_{\partial \Omega} k_1(x,y)\varphi(y) d\sigma(y) \tag{2.199}$$

with

$$k_1(x,y) = \frac{\sqrt{-1}\omega}{4}(L_0 M_0 N_1 + (L_0 M_1 + L_1 M_0)N_0)(x,y)$$

and the functions L_0, L_1, M_0, M_1, N_0 and N_1 being defined by

$$\begin{aligned}
L_0(x,y) &= H_1^{(1)}(\omega|x-y|), \quad M_0(x,y) = |x-y|, \quad N_0(x,y) = \tfrac{\langle y-x, \nu_y\rangle}{|x-y|^2}, \\
L_1(x,y) &= (H_1^{(1)})'(\omega|x-y|)\frac{\langle x-y, h(x)\nu(x) - h(y)\nu(y)\rangle}{|x-y|}, \\
M_1(x,y) &= \frac{\langle x-y, h(x)\nu(x) - h(y)\nu(y)\rangle}{|x-y|}, \\
N_1(x,y) &= N_0(x,y)\widetilde{F}(x,y) + K_1(x,y) \\
K_1(x,y) &= \frac{\langle h(y)\nu(y) - h(x)\nu(x), \nu(y)\rangle}{|x-y|^2} - \frac{\langle y-x, \tau(y)h(y)\nu(y) + h'(y)T(y)\rangle}{|x-y|^2}, \\
\widetilde{F}(x,y) &= -2M_1(x,y) + \tau(x)h(x) - \tau(y)h(y).
\end{aligned}$$

Here, $\tau(x)$ represents the curvature at the point $x \in \partial\Omega$.

Substituting

$$\mathcal{K}_{\Omega_\epsilon}^\omega[\tilde{\phi}] \circ \Psi_\epsilon = \mathcal{K}_\Omega^\omega[\phi] + \epsilon \mathcal{K}_\Omega^{(1)}[\phi] + O(\epsilon^2) \tag{2.200}$$

into (2.198), we obtain the following shape derivative of the Neumann eigenvalues.

THEOREM 2.67 (Shape derivative of Neumann eigenvalues). *The following asymptotic expansion holds:*

$$\sqrt{\mu_j^\epsilon} - \sqrt{\mu_j} = \frac{\epsilon}{2\sqrt{-1}\pi} \operatorname{tr} \int_{\partial V_{\delta_0}} \mathcal{A}_0(\omega)^{-1}\mathcal{K}_\Omega^{(1)}(\omega) d\omega + O(\epsilon^2), \tag{2.201}$$

where V_{δ_0} is a disk of center $\sqrt{\mu_j}$ and radius δ_0 small enough, $\mathcal{A}_0(\omega) = -(1/2)I + \mathcal{K}_\Omega^\omega$ and $\mathcal{K}_\Omega^{(1)}(\omega)$ is given by (2.199).

PROOF. If ϵ is small enough, then the following expansion is uniform with respect to ω in ∂V_{δ_0}:

$$\mathcal{A}_\epsilon(\omega)^{-1} = \mathcal{A}_0(\omega)^{-1} - \epsilon \mathcal{A}_0(\omega)^{-1}\mathcal{K}_\Omega^{(1)}(\omega)\mathcal{A}_0(\omega)^{-1} + O(\epsilon^2).$$

It then follows from Theorem 1.14 that

$$\begin{aligned}\omega_\epsilon - \omega_0 &= \frac{1}{2\sqrt{-1}\pi}\operatorname{tr}\int_{\partial V_{\delta_0}}(\omega-\omega_0)\Big[\mathcal{A}_0(\omega)^{-1}\frac{d}{d\omega}\mathcal{A}_0(\omega)\\ &\quad -\epsilon\mathcal{A}_0(\omega)^{-1}\mathcal{K}_\Omega^{(1)}(\omega)\mathcal{A}_0(\omega)^{-1}\frac{d}{d\omega}\mathcal{A}_0(\omega)+\epsilon\mathcal{A}_0(\omega)^{-1}\frac{d}{d\omega}\mathcal{K}_\Omega^{(1)}(\omega)\Big]d\omega\\ &\quad +O(\epsilon^2).\end{aligned}$$

Because of Lemma 2.65, ω_0 is a simple pole of $\mathcal{A}_0(\omega)^{-1}$ and $\mathcal{A}_0(\omega)$ is holomorphic in ω, and hence we get

$$\int_{\partial V_{\delta_0}}(\omega-\omega_0)\mathcal{A}_0(\omega)^{-1}\frac{d}{d\omega}\mathcal{A}_0(\omega)d\omega = 0. \tag{2.202}$$

Moreover, by using the property (1.5) of the trace together with the identity

$$\frac{d}{d\omega}\mathcal{A}_0(\omega)^{-1} = -\mathcal{A}_0(\omega)^{-1}\frac{d\mathcal{A}_0}{d\omega}(\omega)\mathcal{A}_0(\omega)^{-1}, \tag{2.203}$$

we arrive at

$$\omega_\epsilon - \omega_0 = -\frac{\epsilon}{2\sqrt{-1}\pi}\operatorname{tr}\int_{\partial V_{\delta_0}}(\omega-\omega_0)\frac{\partial}{\partial\omega}\left[\mathcal{A}_0^{-1}(\omega)\mathcal{K}_\Omega^{(1)}(\omega)\right]d\omega + O(\epsilon^2).$$

Now, a simple integration by parts yields the desired result. □

2.9.6. Splitting of Multiple Eigenvalues. The main difficulty in deriving asymptotic expansions of perturbations in multiple eigenvalues of the unperturbed configuration relates to their continuation. Multiple eigenvalues may evolve, under perturbations, as separated, distinct eigenvalues, and the splitting may only become apparent at high orders in their Taylor expansions with respect to the perturbation parameter [**176, 326, 462, 482**].

In this subsection, as an example, we address the splitting problem in the evaluation of the perturbations of the Neumann eigenvalues due to shape deformations. Our approach applies to the other eigenvalue perturbation problems as well.

Let ω_0^2 denote an eigenvalue of the Neumann problem for $-\Delta$ on Ω with geometric multiplicity m. We call the ω_0-group the totality of the perturbed eigenvalues $(\omega_\epsilon^i)^2$ of $-\Delta$ on Ω_ϵ for $\epsilon > 0$ that are generated by splitting from ω_0^2.

In exactly the same way as Lemma 2.66 we can show that the eigenvalues are exactly the characteristic values of $\mathcal{A}_\epsilon$ defined by (2.197). We then proceed from the generalized argument principle to investigate the splitting problem.

LEMMA 2.68. *Let $\omega_0 = \sqrt{\mu_j}$ and suppose that μ_j is a multiple Neumann eigenvalue of $-\Delta$ on Ω with geometric multiplicity m. Then there exists a positive constant δ_0 such that the operator-valued function $\omega \mapsto \mathcal{A}_\epsilon(\omega)$ defined by (2.197) has exactly m characteristic values (counted according to their multiplicity) in $\overline{V_{\delta_0}}(\omega_0)$, where $V_{\delta_0}(\omega_0)$ is a disk of center ω_0 and radius $\delta_0 > 0$. These characteristic values form the ω_0-group associated to the perturbed eigenvalue problem (3.2) and are holomorphic with respect to ϵ in $(-\epsilon_0, \epsilon_0)$ for some $\epsilon_0 > 0$. They satisfy $\omega_\epsilon^i|_{\epsilon=0} = \omega_0$ for $i = 1, \ldots, m$. Moreover, if $(\omega_\epsilon^i)_{i=1}^n$ denotes the set of distinct values of $(\omega_\epsilon^i)_{i=1}^m$, then the following assertions hold:*

(i) $\mathcal{M}(\mathcal{A}_\epsilon(\omega); \partial V_{\delta_0}) = \sum_{i=1}^n \mathcal{M}(\mathcal{A}_\epsilon(\omega_\epsilon^i); \partial V_{\delta_0}) = m,$

(ii) $\mathcal{A}_\epsilon(\omega)^{-1} = \sum_{i=1}^{n} (\omega - \omega_\epsilon^i)^{-1} \mathcal{L}_\epsilon^i + \mathcal{R}_\epsilon(\omega)$,

(iii) $\mathcal{L}_\epsilon^i : \mathrm{Ker}((\mathcal{A}_\epsilon(\omega_\epsilon^i))^*) \to \mathrm{Ker}(\mathcal{A}_\epsilon(\omega_\epsilon^i))$,

where $\mathcal{R}_\epsilon(\omega)$ is a holomorphic function with respect to $\omega \in V_{\delta_0}(\omega_0)$ and $\mathcal{L}_\epsilon^i$ for $i = 1, \ldots, n$ is a finite-dimensional operator. Here $\mathcal{M}(\mathcal{A}_\epsilon(\omega_\epsilon^i); \partial V_{\delta_0})$ is defined by (1.10).

Let, for $l \in \mathbb{N}$, $a_l(\epsilon)$ denote

$$a_l(\epsilon) = \frac{1}{2\pi\sqrt{-1}} \operatorname{tr} \int_{\partial V_{\delta_0}} (\omega - \omega_0)^l \mathcal{A}_\epsilon(\omega)^{-1} \frac{\partial}{\partial \omega} \mathcal{A}_\epsilon(\omega) d\omega.$$

By the generalized argument principle, we find

$$a_l(\epsilon) = \sum_{i=1}^{m} (\omega_\epsilon^i - \omega_0)^l \quad \text{for } l \in \mathbb{N}.$$

The following theorems from [**89**] hold.

THEOREM 2.69. *The following asymptotic expansion for $a_l(\epsilon)$ as $\epsilon \to 0$ holds:*

$$a_l(\epsilon) = \frac{\epsilon}{2\sqrt{-1}\pi} \operatorname{tr} \int_{\partial V_{\delta_0}} l(\omega - \omega_0)^{l-1} \mathcal{A}_0(\omega)^{-1} \mathcal{K}_\Omega^{(1)}(\omega) d\omega + O(\epsilon^2), \tag{2.204}$$

where V_{δ_0} is a disk of center $\omega_0 = \sqrt{\mu_j}$ and radius δ_0 small enough, $\mathcal{A}_0(\omega) = -(1/2)I + \mathcal{K}_\Omega^\omega$ and $\mathcal{K}_\Omega^{(1)}(\omega)$ is given by (2.199).

THEOREM 2.70 (Splitting of a multiple eigenvalue). *There exists a polynomial $\omega \mapsto Q_\epsilon(\omega)$ of degree m and of the form*

$$Q_\epsilon(\omega) = \omega^m + c_1(\epsilon)\omega^{m-1} + \ldots + c_i(\epsilon)\omega^{m-i} + \ldots + c_m(\epsilon)$$

such that the perturbations $\omega_\epsilon^i - \omega_0$ are precisely its zeros. The coefficients $(c_i)_{i=1}^m$ are given by the recurrence relations

$$a_{l+m} + c_1 a_{l+m-1} + \ldots + c_m a_l = 0 \quad \text{for } l = 0, 1, \ldots, m-1.$$

Based on Theorems 2.69 and 2.70, our strategy for deriving asymptotic expansions of the perturbations $\omega_\epsilon^i - \omega_0$ relies on finding a polynomial $Q_\epsilon(\omega)$ such that its zeros are precisely the perturbations $\omega_\epsilon^i - \omega_0$. We then obtain complete asymptotic expansions of the perturbations in the eigenvalues by computing the Taylor series of the polynomial coefficients.

Notice that in the cases where the multiplicity $m \in \{2, 3, 4\}$, there is no need to use Theorem 2.70, because we can explicitly have the expressions of the perturbed eigenvalues as functions of $(a_l)_{l=1}^m$. For example, if $m = 2$ which is the case when Ω is a disk, we can easily see when the splitting occurs. It suffices that one of the terms in the expansion of $2a_2(\epsilon) - a_1^2(\epsilon)$ in terms of ϵ does not vanish. Necessarily the order of splitting is even (because of the analyticity of the eigenvalues). Let $a_j(\epsilon) = \sum_n a_{j,n}\epsilon^n$ and write

$$2a_2(\epsilon) - a_1^2(\epsilon) = \sum_{n \geq 2} \alpha_n \epsilon^n, \quad \alpha_n = 2a_{2,n} - \sum_{p=1}^{n} a_{1,p} a_{1,n-p}.$$

Suppose that the splitting order is $2s$, then we obtain

$$\omega_\epsilon^j = \omega_0 + \sum_{i \geq 1} \lambda_i^{(j)} \epsilon^i, \quad j = 1, 2$$

with

$$\lambda_i^{(1)} = \lambda_i^{(2)} \quad \text{for } i \leq 2s-1,$$
$$\lambda_{2s}^{(1)} = \tfrac{a_{1,2s}}{2} - \sqrt{\alpha_{2s}}, \quad \lambda_{2s}^{(2)} = \tfrac{a_{1,2s}}{2} + \sqrt{\alpha_{2s}}.$$

Explicit formulas for $\lambda_i^{(j)}$ for $j = 1, 2$, can be obtained; see [**89**].

2.9.7. Numerical Implementation.

2.9.7.1. *Discretization of the Operator* $\mathcal{K}_\Omega^\omega$. Similarly to the case of the Neumann–Poincaré operator $(\mathcal{K}_\Omega^0)^*$ in Subsection 2.4.5 we must now define an appropriate numerical representation for the operator $\mathcal{K}_\Omega^\omega$. Suppose that the boundary $\partial\Omega$ is parametrized by $x(t)$ for $t \in [0, 2\pi)$. We first partition the interval $[0, 2\pi)$ into N pieces

$$[t_1, t_2),\ [t_2, t_3),\ \ldots,\ [t_N, t_{N+1}),$$

with $t_1 = 0$ and $t_{N+1} = 2\pi$, and then approximate the boundary $\partial\Omega = \{x(t) \in \mathbb{R}^2 : t \in [0, 2\pi)\}$ by $x^{(i)} = x(t_i)$ for $1 \leq i \leq N$.

We represent the infinite dimensional operator $\mathcal{K}_\Omega^\omega$ by a finite dimensional matrix K and the density function φ by $\overline{\varphi}_i := \varphi(x^{(i)})$ for $1 \leq i \leq N$. Then

$$\begin{aligned} \mathcal{K}_\Omega^\omega[\varphi](x) &= \int_{\partial\Omega} \frac{\partial\Gamma_\omega}{\partial\nu_y}(x, y)\varphi(y)\, d\sigma(y) \\ &= \int_{\partial\Omega} \frac{\sqrt{-1}}{4} H_1^{(1)}(\omega|x-y|)\omega \frac{\langle y - x, \nu_y\rangle}{|y-x|}\varphi(y) d\sigma(y), \end{aligned}$$

for $\psi \in L^2(\partial\Omega)$ has the numeric representation

$$K\tilde{\psi} = \begin{pmatrix} K_{11} & K_{12} & \ldots & K_{1N} \\ K_{21} & K_{22} & \ldots & K_{2N} \\ \vdots & & \ddots & \vdots \\ K_{N1} & \ldots & \ldots & K_{NN} \end{pmatrix} \begin{pmatrix} \overline{\varphi}_1 \\ \overline{\varphi}_2 \\ \vdots \\ \overline{\varphi}_N \end{pmatrix},$$

where

$$K_{ij} = \frac{\sqrt{-1}}{4} H_1^{(1)}(\omega|x^{(i)} - x^{(j)}|)\omega \frac{\langle x^{(j)} - x^{(i)}, \nu_y\rangle}{|x^{(j)} - x^{(i)}|} |T(x^{(j)})|(t_{j+1} - t_j) \quad i \neq j,$$

with $T(x^{(i)})$ being the tangent vector at $x^{(i)}$.

As in the previous section, we have singularities in the diagonal terms of the discretization matrix. Recall that $\Gamma_0(x) = \frac{1}{2\pi}\ln|x|$ and in Subsection 2.4.5.2 we showed how to compute the diagonal elements in the case of the Neumann–Poincaré operator $(\mathcal{K}_\Omega^0)^*$. In view of (2.152), the kernel $\partial\Gamma_\omega/\partial\nu_y(x, y)$ has the same singularity as that of the Neumann–Poincaré operator. Therefore we can approximate the diagonal elements of K by

$$K_{ii} \approx \frac{1}{2N} \frac{\langle a^{(i)}), \nu^{(i)}\rangle}{|T^{(i)}|}. \tag{2.205}$$

2.9.7.2. *Finding the Eigenvalues by Muller's Method.* We now describe the computation of Laplace eigenvalues with the Neumann boundary condition (or the characteristic values of $\mathcal{A}_0$) using Muller's method. Let us define a function $f : \mathbb{C} \to \mathbb{C}$ such that $f(\omega)$ is the smallest eigenvalue of $\mathcal{A}_0(\omega)$. This means that $f(\omega) = 0$ whenever ω is a characteristic value of $\mathcal{A}_0$. By applying Muller's method to the equation $f(\omega) = 0$, we can compute the characteristic value ω of $\mathcal{A}_0$.

Consider the following numerical example. Assume that Ω is a unit disk. We discretize the boundary $\partial\Omega$ with $N = 500$ points. As discussed previously, characteristic values are zeros of $J_n'(z) = 0$. The first zero is approximately 1.8412. We can also compute characteristic values numerically using Muller's Method. Upon computing a characteristic value near 1.8 we find that there is a good agreement with the exact value, as can be seen in Table 2.5. The corresponding Matlab code can be downloaded at `www.ams.org/bookpages/surv-235`.

Theoretical	Numerical
$1.8412 + 0.0000\sqrt{-1}$	$1.8421 - 0.0026\sqrt{-1}$

TABLE 2.5. Characteristic value of $\mathcal{A}$ near 1.8.

Next, we present a numerical example for computing perturbed eigenvalues using the shape derivative. We assume that Ω is a unit disk. We use polar coordinates (r, θ) to parametrize the boundary $\partial\Omega$. For the boundary perturbation, we set $\epsilon = 0.01$ and $h(\theta) = \cos(2\theta)$. We discretize the boundary $\partial\Omega_\epsilon$ with $N = 100$ points. We compute the perturbed characteristic values near $\omega_0 = 0.8412$ using Muller's method. Then we compute their approximation by using the shape derivative. A comparison between the perturbed eigenvalues obtained via Muller's Method and the approximation given by the shape derivative is provided in Table 2.6. The corresponding Matlab code can be downloaded at `www.ams.org/bookpages/surv-235`.

Muller's method	Shape derivative
$1.8623 - 0.0126\sqrt{-1}$	$1.8619 + 0.0008\sqrt{-1}$
$1.8288 - 0.0126\sqrt{-1}$	$1.8204 - 0.0007\sqrt{-1}$

TABLE 2.6. Perturbed characteristic values of the operator $\mathcal{A}_\epsilon$.

2.10. Helmholtz-Kirchhoff Identity, Scattering Amplitude and Optical Theorem

In this section we derive the Helmholtz-Kirchhoff identity, which plays a key role in understanding the resolution limit in imaging with waves, and outline the optical theorem. The optical theorem establishes a fundamental relation between the imaginary part of the scattering amplitude and the total cross-section. The scattering amplitude (or the far-field pattern) is the amplitude of the outgoing spherical or cylindrical wave scattered by a particle, relative to a plane wave. It is a function of the incidence and observation directions. The total cross-section (also called the extinction cross-section) is the sum of the scattering and absorption cross-sections, which are respectively defined as the ratio of the total radiant power scattered and absorbed by a particle in all directions, to the radiant power incident on the particle.

2.10.1. Reciprocity. An important property satisfied by the outgoing fundamental solution of the Helmholtz equation is the reciprocity property.

Let μ and ε be two piecewise smooth functions such that $\mu(x) = \mu_m$ and $\varepsilon(x) = \varepsilon_m$ for $|x| \geq R_0$ for some positive R_0. Let $k_m = \omega\sqrt{\varepsilon_m \mu_m}$. For $y \in \mathbb{R}^d$, introduce the fundamental solution $\Phi_{k_m}(x, y)$ to be the solution to

$$(\nabla_x \cdot \frac{1}{\mu(x)}\nabla_x + \omega^2 \varepsilon(x))\Phi_{k_m}(x, y) = \frac{1}{\mu_m}\delta_y(x), \tag{2.206}$$

subject to the Sommerfeld radiation condition:

$$\left|\frac{\partial \Phi_{k_m}}{\partial r} - \sqrt{-1}k_m \Phi_{k_m}\right| = O\left(r^{-(d+1)/2}\right) \quad \text{as } r = |x| \to +\infty \quad \text{uniformly in } \frac{x}{|x|}. \tag{2.207}$$

The following holds.

LEMMA 2.71. *We have, for* $x \neq y$,

$$\Phi_{k_m}(x, y) = \Phi_{k_m}(y, x)\,. \tag{2.208}$$

Identity (2.208) means that the wave recorded at x when there is a time-harmonic source at y is equal to the wave recorded at y when there is a time-harmonic source at x.

PROOF. We consider the equations satisfied by the fundamental solution with the source at y_2 and with the source at y_1 (with $y_1 \neq y_2$):

$$(\nabla_x \cdot \frac{1}{\mu}\nabla_x + \omega^2 \varepsilon)\Phi_{k_m}(x, y_2) = \frac{1}{\mu_m}\delta_{y_2}\,,$$
$$(\nabla_x \cdot \frac{1}{\mu}\nabla_x + \omega^2 \varepsilon)\Phi_{k_m}(x, y_1) = \frac{1}{\mu_m}\delta_{y_1}\,.$$

We multiply the first equation by $\Phi_{k_m}(x, y_1)$ and subtract the second equation multiplied by $\Phi_{k_m}(x, y_2)$:

$$\nabla_x \cdot \frac{\mu_m}{\mu}\Big[\Phi_{k_m}(x, y_1)\nabla_x \Phi_{k_m}(x, y_2) - \Phi_{k_m}(x, y_2)\nabla_x \Phi_{k_m}(x, y_1)\Big]$$
$$= -\Phi_{k_m}(x, y_2)\delta_{y_1} + \Phi_{k_m}(x, y_1)\delta_{y_2}\,.$$

We next integrate over the ball B_R of center 0 and radius $R > R_0$ which contains both y_1 and y_2 and use the divergence theorem:

$$\int_{\partial B_R} \nu \cdot \Big[\Phi_{k_m}(x, y_1)\nabla_x \Phi_{k_m}(x, y_2) - \Phi_{k_m}(x, y_2)\nabla_x \Phi_{k_m}(x, y_1)\Big]\, d\sigma(x)$$
$$= -\Phi_{k_m}(y_1, y_2) + \Phi_{k_m}(y_2, y_1)\,,$$

where ν is the unit outward normal to the ball B_R, which is $\nu = x/|x|$.

If $x \in \partial B_R$ and $R \to \infty$, then we have by the Sommerfeld radiation condition:

$$\nu \cdot \nabla_x \Phi_{k_m}(x, y) = \sqrt{-1}k_m \Phi_{k_m}(x, y) + O\Big(\frac{1}{R^{(d+1)/2}}\Big)\,.$$

Therefore, as $R \to \infty$,

$$-\Phi_{k_m}(y_1, y_2) + \Phi_{k_m}(y_2, y_1)$$
$$= \sqrt{-1}k_m \int_{\partial B_R} \Big[\Phi_{k_m}(x, y_1)\Phi_{k_m}(x, y_2) - \Phi_{k_m}(x, y_2)\Phi_{k_m}(x, y_1)\Big]\, d\sigma(x)$$
$$= 0\,,$$

which is the desired result. □

2.10.2. Lippmann-Schwinger Representation Formula. The following Lippmann-Schwinger representation formula for Φ_{k_m} holds.

LEMMA 2.72. *For any $x \neq y$, we have*

$$\begin{aligned} \Phi_{k_m}(x,y) &= \Gamma_{k_m}(x,y) + \int_{\mathbb{R}^d} (\frac{\mu_m}{\mu(z)} - 1)\nabla\Phi_{k_m}(z,x)\cdot\nabla\Gamma_{k_m}(z,y)\,dz \\ &\quad + k_m^2 \int_{\mathbb{R}^d} (1 - \frac{\varepsilon(z)}{\varepsilon_m})\Phi_{k_m}(z,x)\Gamma_{k_m}(z,y)\,dz\,. \end{aligned} \tag{2.209}$$

PROOF. We multiply (2.206) by Γ_{k_m} and subtract the equation satisfied by Γ_{k_m},

$$(\Delta_z + k_m^2)\Gamma_{k_m}(z,y) = \delta_y(z), \tag{2.210}$$

multiplied by $\frac{1}{\mu_m}\Phi_{k_m}$:

$$\begin{aligned} &\nabla_z \cdot \left[\tfrac{1}{\mu(z)}\Gamma_{k_m}(z,y)\nabla_z\Phi_{k_m}(z,x) - \tfrac{1}{\mu_m}\Phi_{k_m}(z,x)\nabla_z\Gamma_{k_m}(z,y)\right] \\ &= (\frac{1}{\mu(z)} - \frac{1}{\mu_m})\nabla_z\Phi_{k_m}(z,x)\cdot\nabla_z\Gamma_{k_m}(z,y) \\ &\quad + \omega^2\varepsilon_m(1 - \frac{\varepsilon(z)}{\varepsilon_m})\Phi_{k_m}(z,x)\Gamma_{k_m}(z,y) \\ &\quad + \frac{1}{\mu_m}(\Gamma_{k_m}(x,y)\delta_x(z) - \Phi_{k_m}(x,y)\delta_y(z))\,. \end{aligned}$$

We integrate over B_R (with $R > R_0$), apply the divergence theorem as in the proof of Lemma 2.71. Then sending R to infinity and using the Sommerfeld radiation condition we obtain the desired result. □

The Lippmann-Schwinger representation formula (2.209) is used as a basis for expanding the fundamental solution Φ_{k_m} when $\mu \approx \mu_m$ and $\varepsilon \approx \varepsilon_m$. If Φ_{k_m} in the right-hand side of (2.209) is replaced by Γ_{k_m}, then we obtain:

$$\begin{aligned} \Phi_{k_m}(x,y) &\approx \Gamma_{k_m}(x,y) + \int_{\mathbb{R}^d} (\frac{\mu_m}{\mu(z)} - 1)\nabla\Gamma_{k_m}(x,z)\cdot\nabla\Gamma_{k_m}(z,y)\,dz \\ &\quad + k_m^2 \int_{\mathbb{R}^d} (1 - \frac{\varepsilon(z)}{\varepsilon_m})\Gamma_{k_m}(x,z)\Gamma_{k_m}(z,y)\,dz\,, \end{aligned} \tag{2.211}$$

which is the (first-order) Born approximation for Φ_{k_m}.

2.10.3. The Helmholtz-Kirchhoff Theorem. The Helmholtz-Kirchhoff theorem plays a key role in understanding the resolution limit in imaging with waves. The following holds.

LEMMA 2.73. *Let ∂B_R be the sphere of radius R and center 0. We have*

$$\int_{\partial B_R} \left(\frac{\partial\overline{\Gamma_{k_m}}}{\partial\nu}(x,y)\Gamma_{k_m}(z,y) - \overline{\Gamma_{k_m}}(x,y)\frac{\partial\Gamma_{k_m}}{\partial\nu}(z,y)\right)d\sigma(y) = 2\sqrt{-1}\,\Im\Gamma_{k_m}(x,z)\,, \tag{2.212}$$

which yields

$$\lim_{R\to+\infty}\int_{\partial B_R} \overline{\Gamma_{k_m}}(x,y)\Gamma_{k_m}(z,y)\,d\sigma(y) = -\frac{1}{k_m}\Im\Gamma_{k_m}(x,z)\,, \tag{2.213}$$

by using the Sommerfeld radiation condition.

Identity (2.212) follows from multiplying by $\overline{\Gamma_{k_m}}$ the equation (2.210) satisfied by Γ_{k_m} and integrating by parts. Identity (2.213) can be deduced from (2.212) by using the Sommerfeld radiation condition.

Notice that identity (2.213) is valid even in inhomogeneous media. The following identity holds, which, as we will see, shows that the sharper the imaginary part of the fundamental solution Φ_{k_m} around the source is, the higher is the resolution.

THEOREM 2.74. *Let Φ_{k_m} be the fundamental solution defined in (2.206). We have*

$$\lim_{R\to+\infty}\int_{|y|=R}\overline{\Phi_{k_m}}(x,y)\Phi_{k_m}(z,y)\,d\sigma(y) = -\frac{1}{k_m}\Im\,\Phi_{k_m}(x,z)\,. \tag{2.214}$$

PROOF. As in Lemma 2.71, the proof is based essentially on the second Green's identity and the Sommerfeld radiation condition. Let us consider

$$\begin{aligned}(\nabla_y\cdot\tfrac{1}{\mu}\nabla_y+\omega^2\varepsilon)\Phi_{k_m}(y,x_2) &= \frac{1}{\mu_m}\delta_{x_2}\,,\\ (\nabla_y\cdot\tfrac{1}{\mu}\nabla_y+\omega^2\varepsilon)\Phi_{k_m}(y,x_1) &= \frac{1}{\mu_m}\delta_{x_1}\,.\end{aligned}$$

We multiply the first equation by $\overline{\Phi_{k_m}}(y,x_1)$ and we subtract the second equation multiplied by $\Phi_{k_m}(y,x_2)$:

$$\begin{aligned}\nabla_y\frac{\mu_m}{\mu}\cdot\Big[\overline{\Phi_{k_m}}(y,x_1)\nabla_y\Phi_{k_m}(y,x_2)-\Phi_{k_m}(y,x_2)\nabla_y\overline{\Phi_{k_m}}(y,x_1)\Big]&\\ = -\Phi_{k_m}(y,x_2)\delta_{x_1}+\overline{\Phi_{k_m}}(y,x_1)\delta_{x_2}\,.&\end{aligned}$$

We integrate over the ball B_R and use the divergence theorem to obtain

$$\begin{aligned}&\int_{\partial B_R}\nu\cdot\Big[\overline{\Phi_{k_m}}(y,x_1)\nabla_y\Phi_{k_m}(y,x_2)-\Phi_{k_m}(y,x_2)\nabla_y\overline{\Phi_{k_m}}(y,x_1)\Big]d\sigma(y)\\ &= -\Phi_{k_m}(x_1,x_2)+\overline{\Phi_{k_m}}(x_1,x_2)\,,\end{aligned}$$

with the help of the reciprocity property $\Phi_{k_m}(x_1,x_2)=\Phi_{k_m}(x_2,x_1)$.

This equality can be viewed as an application of the second Green's identity. The fundamental solution also satisfies the Sommerfeld radiation condition

$$\lim_{|y|\to\infty}|y|\Big(\frac{y}{|y|}\cdot\nabla_y-\sqrt{-1}k_m\Big)\Phi_{k_m}(y,x_1)=0\,,$$

uniformly in all directions $y/|y|$. Using this property, we substitute $\sqrt{-1}k_m\Phi_{k_m}(y,x_2)$ for $\nu\cdot\nabla_y\Phi_{k_m}(y,x_2)$ in the surface integral over ∂B_R, and $-\sqrt{-1}k_m\overline{\Phi_{k_m}}(y,x_1)$ for $\nu\cdot\nabla_y\overline{\Phi_{k_m}}(y,x_1)$, and we obtain the desired result. □

2.10.4. Scattering Amplitude and the Optical Theorem.

2.10.4.1. *Scattering Coefficients.* We first define the scattering coefficients of a particle D in two dimensions. Assume that k_m^2 is not a Dirichlet eigenvalue for $-\Delta$ on D. Then, the solution u to (2.172) (for $d=2$) can be represented using the single-layer potentials $\mathcal{S}_D^{k_m}$ and $\mathcal{S}_D^{k_c}$ by (2.177) where the pair $(\varphi,\psi)\in L^2(\partial D)\times L^2(\partial D)$ is the unique solution to (2.178). Moreover, by using Proposition 2.51 it follows that there exists a constant $C=C(k_c,k_m,D)$ such that

$$\|\varphi\|_{L^2(\partial D)}+\|\psi\|_{L^2(\partial D)}\le C(\|u^i\|_{L^2(\partial D)}+\|\nabla u^i\|_{L^2(\partial D)})\,. \tag{2.215}$$

Furthermore, the constant C can be chosen to be scale independent. Namely, there exists δ_0 such that if one denotes by $(\varphi_\delta, \psi_\delta)$ the solution of (2.178) with k_c and k_m respectively replaced by δk_c and δk_m, then

$$\|\varphi_\delta\|_{L^2(\partial D)} + \|\psi_\delta\|_{L^2(\partial D)} \leq C(\|u^i\|_{L^2(\partial D)} + \|\nabla u^i\|_{L^2(\partial D)}) . \tag{2.216}$$

Recall the Graf's addition formula:

$$H_0^{(1)}(k|x-y|) = \sum_{l \in \mathbb{Z}} H_l^{(1)}(k|x|)e^{\sqrt{-1}l\theta_x} J_l(k|y|)e^{-\sqrt{-1}l\theta_y} \quad \text{for } |x| > |y|, \tag{2.217}$$

where $x = (|x|, \theta_x)$ and $y = (|y|, \theta_y)$ in polar coordinates.

From (2.177) and (2.217), the following asymptotic formula holds as $|x| \to \infty$:

$$u(x) - u^i(x) = -\frac{\sqrt{-1}}{4} \sum_{l \in \mathbb{Z}} H_l^{(1)}(k_m|x|)e^{\sqrt{-1}l\theta_x} \int_{\partial D} J_l(k_m|y|)e^{-\sqrt{-1}l\theta_y} \psi(y) d\sigma(y) . \tag{2.218}$$

Let $(\varphi_{l'}, \psi_{l'})$ be the solution to (2.178) with $J_{l'}(k_m|x|)e^{\sqrt{-1}l'\theta_x}$ in place of $u^i(x)$. We define the *scattering coefficient* as follows.

DEFINITION 2.75. *The scattering coefficients $W_{ll'}$, $l, l' \in \mathbb{Z}$, associated with the permittivity and permeability distributions ε, μ and the frequency ω (or k_c, k_m, D) are defined by*

$$W_{ll'} = W_{ll'}[\varepsilon, \mu, \omega] := \int_{\partial D} J_l(k_m|y|)e^{-\sqrt{-1}l\theta_y} \psi_{l'}(y) d\sigma(y) . \tag{2.219}$$

We derive the exponential decay of the scattering coefficients. We have the following lemma for the size of $|W_{ll'}|$.

LEMMA 2.76. *There is a constant C depending on $(\varepsilon, \mu, \omega)$ such that*

$$|W_{ll'}[\varepsilon, \mu, \omega]| \leq \frac{C^{|l|+|l'|}}{|l|^{|l|}|l'|^{|l'|}} \quad \text{for all } l, l' \in \mathbb{Z} \setminus \{0\} . \tag{2.220}$$

Moreover, there exists δ_0 such that, for all $\delta \leq \delta_0$,

$$|W_{ll'}[\varepsilon, \mu, \delta\omega]| \leq \frac{C^{|l|+|l'|}}{|l|^{|l|}|l'|^{|l'|}} \delta^{|l|+|l'|} \quad \text{for all } l, l' \in \mathbb{Z} \setminus \{0\} , \tag{2.221}$$

where the constant C depends on $(\varepsilon, \mu, \omega)$ but is independent of δ.

PROOF. Let $u^i(x) = J_{l'}(k_m|x|)e^{\sqrt{-1}l'\theta_x}$ and $(\varphi_{l'}, \psi_{l'})$ be the solution to (2.178). Since

$$J_{l'}(t) \sim \frac{(-1)^{l'}}{\sqrt{2\pi|l'|}} \Big(\frac{et}{2|l'|} \Big)^{|l'|} \tag{2.222}$$

as $l' \to \infty$, we have

$$\|u^i\|_{L^2(\partial D)} + \|\nabla u^i\|_{L^2(\partial D)} \leq \frac{C^{|l'|}}{|l'|^{|l'|}}$$

for some constant C. Thus it follows from (2.215) that

$$\|\psi_{l'}\|_{L^2(\partial D)} \leq \frac{C^{|l'|}}{|l'|^{|l'|}} \tag{2.223}$$

for another constant C. So we get (2.220) from (2.219).

On the other hand, one can see from (2.216) that (2.223) still holds for some C independent of δ as long as $\delta \leq \delta_0$ for some δ_0. Note that

$$W_{ll'}[\varepsilon, \mu, \delta\omega] = \int_{\partial D} J_l(\delta k_m|y|)e^{-\sqrt{-1}l\theta_y}\psi_{l',\delta}(y)d\sigma(y)\,, \tag{2.224}$$

where $(\varphi_{l',\delta}, \psi_{l',\delta})$ is the solution to (2.178) with k_c and k_m respectively replaced by δk_c and δk_m and $J_{l'}(\delta k_m|x|)e^{\sqrt{-1}l'\theta_x}$ in the place of $u^i(x)$. So one can use (2.222) to obtain (2.221). This completes the proof. □

Recall that the family of cylindrical waves $\{J_n(k_m|y|)e^{-\sqrt{-1}n\theta_y}\}_n$ form a complete set. We have the completeness relation

$$\frac{\delta_0(r-r_0)\delta_0(\theta-\theta_0)}{r} = \sum_{l'\in\mathbb{Z}} \frac{1}{2\pi}\int_0^{+\infty} tJ_{l'}(tr)J_{l'}(tr_0)\, dt\, e^{\sqrt{-1}l'(\theta-\theta_0)}. \tag{2.225}$$

If u^i is given as

$$u^i(x) = \sum_{l'\in\mathbb{Z}} a_{l'}(u^i)J_{l'}(k_m|x|)e^{\sqrt{-1}l'\theta_x}\,, \tag{2.226}$$

where $a_{l'}(u^i)$ are constants, it follows from the principle of superposition that the solution (φ, ψ) to (2.178) is given by

$$\psi = \sum_{l'\in\mathbb{Z}} a_{l'}(u^i)\psi_{l'}\,.$$

Then one can see from (2.218) that the solution u to (2.172) can be represented as

$$u(x) - u^i(x) = -\frac{\sqrt{-1}}{4}\sum_{l\in\mathbb{Z}} H_l^{(1)}(k_m|x|)e^{\sqrt{-1}l\theta_x}\sum_{l'\in\mathbb{Z}} W_{ll'}a_{l'}(u^i) \quad \text{as } |x|\to\infty\,. \tag{2.227}$$

In particular, if u^i is given by a plane wave $e^{\sqrt{-1}k_m\xi\cdot x}$ with ξ being on the unit circle, then

$$u(x) - e^{\sqrt{-1}k_m\xi\cdot x} = -\frac{\sqrt{-1}}{4}\sum_{l\in\mathbb{Z}} H_l^{(1)}(k_m|x|)e^{\sqrt{-1}l\theta_x}\sum_{l'\in\mathbb{Z}} W_{ll'}e^{\sqrt{-1}l'(\frac{\pi}{2}-\theta_\xi)} \tag{2.228}$$

as $|x|\to\infty$, where $\xi = (\cos\theta_\xi, \sin\theta_\xi)$ and $x = (|x|, \theta_x)$. In fact, from the Jacobi-Anger expansion of plane waves

$$e^{\sqrt{-1}\xi\cdot x} = \sum_{l\in\mathbb{Z}} e^{\sqrt{-1}l(\frac{\pi}{2}-\theta_\xi)}J_l(|\xi||x|)e^{\sqrt{-1}l\theta_x}, \tag{2.229}$$

where $x = (|x|, \theta_x)$ and $\xi = (|\xi|, \theta_\xi)$ in the polar coordinates, it follows that

$$e^{\sqrt{-1}k_m\xi\cdot x} = \sum_{l'\in\mathbb{Z}} e^{\sqrt{-1}l'(\frac{\pi}{2}-\theta_\xi)}J_{l'}(k_m|x|)e^{\sqrt{-1}l'\theta_x}, \tag{2.230}$$

and

$$\psi = \sum_{l'\in\mathbb{Z}} e^{\sqrt{-1}l'(\frac{\pi}{2}-\theta_\xi)}\psi_{l'}. \tag{2.231}$$

Thus (2.228) holds. It is worth emphasizing that the expansion formulas (2.227) and (2.228) uniquely determine the scattering coefficients $W_{ll'}$, for $l, l' \in \mathbb{Z}$.

REMARK 2.77. *Let D be a disk of radius R located at the origin in $\mathbb{R}^2$. Remark 2.53 yields an explicit expression for the scattering coefficients. In fact, we have*

$$W_{ll'} = 0, \quad l \neq l',$$
$$W_{ll} = 4\sqrt{-1}a_l, \quad l \in \mathbb{Z},$$

where a_l is given by (2.181) with n replaced with l.

REMARK 2.78. *In* [**30**], *the concept of scattering coefficients is extended to heterogeneous media. The exponential decay of the heterogeneous scattering coefficients is shown and the relationship between the scattering coefficients and the scattering amplitude is established.*

2.10.4.2. *Scattering Amplitude.* Let D be a bounded domain in $\mathbb{R}^2$ with smooth boundary ∂D, and let (ε_m, μ_m) be the pair of electromagnetic parameters (permittivity and permeability) of $\mathbb{R}^2 \setminus \overline{D}$ and (ε_c, μ_c) be that of D. Then the permittivity and permeability distributions are given by

$$\varepsilon = \varepsilon_m \chi(\mathbb{R}^2 \setminus \overline{D}) + \varepsilon_c \chi(D) \quad \text{and} \quad \mu = \mu_m \chi(\mathbb{R}^2 \setminus \overline{D}) + \mu_c \chi(D)\,. \tag{2.232}$$

Given a frequency ω, set $k_c = \omega\sqrt{\varepsilon_c\mu_c}$ and $k_m = \omega\sqrt{\varepsilon_m\mu_m}$. For a function u^i satisfying $(\Delta + k_m^2)u^i = 0$ in $\mathbb{R}^2$, we consider the total wave u, *i.e.*, the solution to (2.172).

Suppose that u^i is given by a plane wave $e^{\sqrt{-1}k_m\xi\cdot x}$ with ξ being on the unit circle, then (2.228) yields

$$u(x) - e^{\sqrt{-1}k_m\xi\cdot x} = -\frac{\sqrt{-1}}{4}\sum_{l\in\mathbb{Z}} H_l^{(1)}(k_m|x|)e^{\sqrt{-1}l\theta_x}\sum_{l'\in\mathbb{Z}} W_{ll'}e^{\sqrt{-1}l'(\frac{\pi}{2}-\theta_\xi)} \tag{2.233}$$

as $|x| \to \infty$, where $W_{ll'}$, given by (2.219), are the scattering coefficients,

$$\xi = (\cos\theta_\xi, \sin\theta_\xi), \quad \text{and } x = (|x|, \theta_x).$$

The far-field pattern $A_\infty[\varepsilon, \mu, \omega]$, when the incident field is given by $e^{\sqrt{-1}k_m\xi\cdot x}$, is defined to be

$$u(x) - e^{\sqrt{-1}k_m\xi\cdot x} = -\sqrt{-1}e^{-\frac{\pi\sqrt{-1}}{4}}\frac{e^{\sqrt{-1}k_m|x|}}{\sqrt{|x|}}A_\infty[\varepsilon, \mu, \omega](\theta_\xi, \theta_x) + o(|x|^{-\frac{1}{2}}) \tag{2.234}$$

as $|x| \to \infty$.

Recall that

$$H_0^{(1)}(t) \sim \sqrt{\frac{2}{\pi t}}e^{\sqrt{-1}(t-\frac{\pi}{4})} \quad \text{as } t \to \infty\,, \tag{2.235}$$

where $\sim$ indicates that the difference between the right-hand and left-hand side is $O(t^{-1})$. If $|x|$ is large while $|y|$ is bounded, then we have

$$|x - y| = |x| - |y|\cos(\theta_x - \theta_y) + O(\frac{1}{|x|}),$$

and hence

$$H_0^{(1)}(k_m|x - y|) \sim e^{-\frac{\pi\sqrt{-1}}{4}}\sqrt{\frac{2}{\pi k_m|x|}}e^{\sqrt{-1}k_m(|x|-|y|\cos(\theta_x-\theta_y))} \quad \text{as } |x| \to \infty\,.$$

Thus, from (2.177), we get

(2.236)
$$u(x) - e^{\sqrt{-1}k_m \xi\cdot x} \sim -\sqrt{-1}e^{-\frac{\pi\sqrt{-1}}{4}} \frac{e^{\sqrt{-1}k_m|x|}}{\sqrt{8\pi k_m|x|}} \int_{\partial D} e^{-\sqrt{-1}k_m|y|\cos(\theta_x-\theta_y)}\psi(y)\, d\sigma(y)$$

as $|x| \to \infty$ and infer that the far-field pattern is given by

$$A_\infty[\varepsilon,\mu,\omega](\theta_\xi,\theta_x) = \frac{1}{\sqrt{8\pi k_m}} \int_{\partial D} e^{-\sqrt{-1}k_m|y|\cos(\theta_x-\theta_y)}\psi(y)\, d\sigma(y)\,, \tag{2.237}$$

where ψ is given by (2.231).

We now show that the scattering coefficients are basically the Fourier coefficients of the far-field pattern (the scattering amplitude) which is a 2π-periodic function in two dimensions.

Let
$$A_\infty[\varepsilon,\mu,\omega](\theta_\xi,\theta_x) = \sum_{l\in\mathbb{Z}} b_l(\theta_\xi)e^{\sqrt{-1}l\theta_x}$$

be the Fourier series of $A_\infty[\varepsilon,\mu,\omega](\theta_\xi,\cdot)$. From (2.237) it follows that

$$\begin{aligned} b_l(\theta_\xi) &= \frac{1}{2\pi}\int_0^{2\pi}\int_{\partial D} e^{-\sqrt{-1}k_m|y|\cos(\theta_x-\theta_y)}\psi(y)\, d\sigma(y)\, e^{-\sqrt{-1}l\theta_x}\, d\theta_x \\ &= \frac{1}{2\pi}\int_{\partial D}\int_0^{2\pi} e^{-\sqrt{-1}k_m|y|\cos(\theta_x-\theta_y)}e^{-\sqrt{-1}l\theta_x}\, d\theta_x\, \psi(y)\, d\sigma(\theta_y)\,. \end{aligned}$$

Since
$$\frac{1}{2\pi}\int_0^{2\pi} e^{-\sqrt{-1}k_m|y|\cos(\theta_x-\theta_y)}e^{-\sqrt{-1}l\theta_x}\, d\theta_x = J_l(k_m|y|)e^{-\sqrt{-1}l(\theta_y+\frac{\pi}{2})}\,,$$

we deduce that
$$b_l(\theta_\xi) = \int_{\partial D} J_l(k_m|y|)e^{-\sqrt{-1}l(\theta_y+\frac{\pi}{2})}\psi(y)\, d\sigma(\theta_y)\,.$$

Using (2.231) we now arrive at the following theorem.

THEOREM 2.79. *Let θ and θ' be respectively the incident and scattered direction. Then we have*

$$A_\infty[\varepsilon,\mu,\omega](\theta,\theta') = \sum_{l,l'\in\mathbb{Z}} (\sqrt{-1})^{(l'-l)}e^{\sqrt{-1}l\theta'}W_{ll'}[\varepsilon,\mu,\omega]e^{-\sqrt{-1}l'\theta}\,, \tag{2.238}$$

where the scattering coefficients $W_{ll'}$ are defined by (2.219).

We emphasize that the series in (2.238) is well-defined provided that k_m^2 is not a Dirichlet eigenvalue for $-\Delta$ on D. Moreover, it converges uniformly in θ and θ' thanks to (2.220). Furthermore, there exists $\delta_0 > 0$ such that for any $\delta \le \delta_0$ the series expansion of $A_\infty[\varepsilon,\mu,\delta\omega](\theta,\theta')$ is well-defined and its convergence is uniform in δ. This is the key point of our construction of near-cloaking structures. We also note that if u^i is given by (2.226) then the scattering amplitude, which we denote by $A_\infty[\varepsilon,\mu,\omega](u^i,\theta')$, is given by

$$A_\infty[\varepsilon,\mu,\omega](u^i,\theta') = \sum_{l\in\mathbb{Z}}(\sqrt{-1})^{-l}e^{\sqrt{-1}l\theta'}\sum_{l'\in\mathbb{Z}} W_{ll'}a_{l'}(u^i)\,. \tag{2.239}$$

The conversion of the far-field to the near field is achieved via formula (2.233).

2.10.4.3. *Optical Theorem.* Let $d = 3$. For the wave given by $\Re\left[u(x)e^{-\sqrt{-1}k_m t}\right]$, the averaged value of the energy flux vector, taken over an interval which is long compared to the period of the oscillations, is defined by

$$F(x) = -\sqrt{-1}k_m\left[\overline{u}(x)\nabla u(x) - u(x)\nabla\overline{u}(x)\right].$$

Consider the outward flow of energy through the sphere ∂B_R of radius R and center the origin:

$$\mathcal{W} = \int_{\partial B_R} F(x)\cdot\nu(x)d\sigma(x),$$

where $\nu(x)$ is the outward normal at $x \in \partial B_R$.

As the total field can be written as $u = u^i + u^s$, the flow can be decomposed into three parts:

$$\mathcal{W} = \mathcal{W}^i + \mathcal{W}^s + \mathcal{W}',$$

where

$$\begin{aligned}
\mathcal{W}^i &= -\sqrt{-1}\beta\int_{\partial B_R}\left[\overline{u^i}(x)\nabla u^i(x) - u^i(x)\nabla\overline{u^i}(x)\right]\cdot\nu(x)\,d\sigma(x),\\
\mathcal{W}^s &= -\sqrt{-1}\beta\int_{\partial B_R}\left[\overline{u^s}(x)\nabla u^s(x) - u^s(x)\nabla\overline{u^s}(x)\right]\cdot\nu(x)\,d\sigma(x),\\
\mathcal{W}' &= -\sqrt{-1}\beta\int_{\partial B_R}\Big[\overline{u^i}(x)\nabla u^s(x) - u^s(x)\nabla\overline{u^i}(x)\\
&\qquad - u^i(x)\nabla\overline{u^s}(x) + \overline{u^s}(x)\nabla u^i(x)\Big]\cdot\nu(x)\,d\sigma(x),
\end{aligned}$$

where β is a positive constant.

In the case where $u^i(x) = e^{\sqrt{-1}k_m\xi\cdot x}$ is a plane wave, we can see that $\mathcal{W}^i = 0$:

$$\begin{aligned}
\mathcal{W}^i &= -\sqrt{-1}\beta\int_{\partial B_R}\left[\overline{u^i}(x)\nabla u^i(x) - u^i(x)\nabla\overline{u^i}(x)\right]d\sigma(x),\\
&= -\sqrt{-1}\beta\int_{\partial B_R}\Big[e^{-\sqrt{-1}k_m\xi\cdot x}\sqrt{-1}k_m\xi e^{\sqrt{-1}k_m\xi\cdot x}\\
&\qquad + e^{\sqrt{-1}k_m\xi\cdot x}k_m d e^{-\sqrt{-1}k_m\xi\cdot x}\Big]\cdot\nu(x)\,d\sigma(x),\\
&= 2\beta k_m\xi\cdot\int_{\partial B_R}\nu(x)\,d\sigma(x),\\
&= 0.
\end{aligned}$$

In a non-absorbing medium with non-absorbing scatterers, $\mathcal{W}$ is equal to zero because the electromagnetic energy would be conserved by the scattering process. However, if there is an absorbing scatterer inside the medium, the conservation of energy gives the rate of absorption as

$$\mathcal{W}^a = -\mathcal{W}.$$

Therefore, we have

$$\mathcal{W}^a + \mathcal{W}^s = -\mathcal{W}'.$$

Here, $\mathcal{W}'$ is called the extinction rate. It is the rate at which the energy is removed by the scatterer from the illuminating plane wave, and it is the sum of the rate of absorption and the rate at which energy is scattered.

Denote by V the quantity

$$V(x) = \beta \left| \overline{u^i}(x) \nabla u^i(x) - u^i(x) \nabla \overline{u^i}(x) \right| .$$

In the case of a plane wave illumination, $V(x)$ is independent of x and is given by $V = 2\beta k_m$.

DEFINITION 2.80. *The scattering cross-section Q^s, the absorption cross-section Q^a and the extinction cross-section are defined by*

$$Q^s = \frac{\mathcal{W}^s}{V}, \quad Q^a = \frac{\mathcal{W}^a}{V}, \quad Q^{\text{ext}} = \frac{-\mathcal{W}'}{V} .$$

Note that these quantities are independent of x in the case of a plane wave illumination.

THEOREM 2.81 (Optical theorem). *Let $d = 3$. If $u^i(x) = e^{\sqrt{-1}k_m \xi \cdot x}$, where ξ is a unit direction of incidence, then*

(2.240)
$$Q^{\text{ext}}[\varepsilon, \mu, \omega](\xi) = Q^s[\varepsilon, \mu, \omega](\xi) + Q^a[\varepsilon, \mu, \omega](\xi) = \frac{4\pi}{k_m} \Im \left[A_\infty[\varepsilon, \mu, \omega](\xi, \xi) \right] ,$$

$$Q^s[\varepsilon, \mu, \omega](\xi) = \int_{|\hat{x}|=1} |A_\infty[\varepsilon, \mu, \omega](\xi, \hat{x})|^2 d\sigma(\hat{x}) \tag{2.241}$$

with A_∞ being the scattering amplitude defined by

$$(u - u^i)(x) = \frac{e^{\sqrt{-1}k_m |x|}}{|x|} A_\infty[\varepsilon, \mu, \omega] \left(\xi, \frac{x}{|x|} \right) + O\left(\frac{1}{|x|^2} \right) . \tag{2.242}$$

PROOF. The Sommerfeld radiation condition gives, for any $x \in \partial B_R$,

$$\nabla u^s(x) \cdot \nu(x) \sim \sqrt{-1}k_m u^s(x) . \tag{2.243}$$

Hence, from (2.242) we get

$$u^s(x) \nabla \overline{u^s}(x) \cdot \nu(x) - \overline{u^s}(x) \nabla u^s(x) \cdot \nu(x) \sim \frac{-2\sqrt{-1}k_m}{|x|^2} \left| A_\infty[\varepsilon, \mu, \omega] \left(\xi, \frac{x}{|x|} \right) \right|^2 ,$$

which yields (2.241).

We now compute the extinction rate. We have

$$\nabla u^i(x) \cdot \nu(x) = \sqrt{-1}k_m \xi \cdot \nu(x) e^{\sqrt{-1}k_m \xi \cdot x} . \tag{2.244}$$

Therefore, from (2.243) and (2.244), it follows that

$$\begin{aligned}
&\overline{u^i}(x) \nabla u^s(x) \cdot \nu(x) - u^s(x) \nabla \overline{u^i}(x) \cdot \nu(x) \\
&\sim \left[\sqrt{-1}k_m \frac{e^{\sqrt{-1}k_m(|x| - \xi \cdot x)}}{|x|} \xi \cdot \nu + \sqrt{-1}k_m \frac{e^{\sqrt{-1}k_m(|x| - \xi \cdot x)}}{|x|} \right] A_\infty[\varepsilon, \mu, \omega] \left(\xi, \frac{x}{|x|} \right) \\
&\sim \frac{\sqrt{-1}k_m e^{\sqrt{-1}k_m |x| - \xi \cdot \nu(x)}}{|x|} \left(\xi \cdot \nu(x) + 1 \right) A_\infty[\varepsilon, \mu, \omega] \left(\xi, \frac{x}{|x|} \right) .
\end{aligned}$$

For $x \in \partial B_R$, we can write

$$\begin{aligned}
&\overline{u^i}(x) \nabla u^s(x) \cdot \nu(x) - u^s(x) \nabla \overline{u^i}(x) \cdot \nu(x) \\
&\qquad \sim \frac{\sqrt{-1}k_m e^{-\sqrt{-1}k_m R \nu(x) \cdot (\xi - \nu(x))}}{R} \left(\xi \cdot \nu(x) + 1 \right) A_\infty[\varepsilon, \mu, \omega] \left(\xi, \frac{x}{|x|} \right) .
\end{aligned}$$

We now use Jones' lemma

$$\frac{1}{R}\int_{\partial B_R} \mathcal{G}(\nu(x)) e^{-\sqrt{-1}k_m \xi\cdot\nu(x)} d\sigma(x) \sim \frac{2\pi\sqrt{-1}}{k_m}\Big(\mathcal{G}(\xi)e^{-\sqrt{-1}k_m R} - \mathcal{G}(-\xi)e^{\sqrt{-1}k_m R}\Big)$$

as $R \to \infty$ for any smooth function $\mathcal{G}$, to get

$$\int_{\partial B_R} \Big[\overline{u^i}(x)\nabla u^s(x) - u^s(x)\nabla \overline{u^i}(x)\Big]\cdot \nu(x) \sim -4\pi A_\infty[\varepsilon,\mu,\omega](\xi,\xi) \quad \text{as } R\to\infty\,.$$

Therefore,

$$\mathcal{W}' = \sqrt{-1}4\pi\beta\left[A_\infty[\varepsilon,\mu,\omega](\xi,\xi) - \overline{A_\infty}[\varepsilon,\mu,\omega](\xi,\xi)\right] = -8\pi\beta\Im\left[A_\infty[\varepsilon,\mu,\omega](\xi,\xi)\right].$$

Since

$$\beta\Big|\overline{u^i}(x)\nabla u^i(x) - u^i(x)\nabla\overline{u^i}(x)\Big| = 2\beta k_m\,,$$

we obtain (2.240). □

In two dimensions, the scattering cross-section $Q^s[\varepsilon,\mu,\omega]$ satisfies

$$Q^s[\varepsilon,\mu,\omega](\theta') = \int_0^{2\pi}\Big|A_\infty[\varepsilon,\mu,\omega](\theta,\theta')\Big|^2\, d\theta\,. \tag{2.245}$$

As an immediate consequence of Theorem 2.79 we obtain

$$Q^s[\varepsilon,\mu,\omega](\theta') = 2\pi\sum_{m\in\mathbb{Z}}\Big|\sum_{l\in\mathbb{Z}}(\sqrt{-1})^{-l}W_{ll'}[\varepsilon,\mu,\omega]e^{\sqrt{-1}l\theta'}\Big|^2\,. \tag{2.246}$$

Analogously to Theorem 2.81, we can prove that

$$\Im A_\infty[\varepsilon,\mu,\omega](\theta',\theta') = -\sqrt{\frac{k_m}{8\pi}}Q^{\rm ext}[\varepsilon,\mu,\omega](\theta'),\quad \forall\,\theta'\in[0,2\pi]\,. \tag{2.247}$$

Therefore, for non-absorbing scatterers, *i.e.*, $Q^a = 0$, the above optical theorem leads to a natural constraint on $W_{ll'}$. From (2.246) and (2.247), we obtain

$$\begin{aligned}&\Im\sum_{l,l'\in\mathbb{Z}}(\sqrt{-1})^{l'-l}e^{\sqrt{-1}(l-l')\theta'}W_{ll'}[\varepsilon,\mu,\omega]\\ &\quad = -\sqrt{\frac{\pi k_m}{2}}\sum_{l'\in\mathbb{Z}}\Big|\sum_{l\in\mathbb{Z}}(\sqrt{-1})^{-l}W_{ll'}[\varepsilon,\mu,\omega]e^{\sqrt{-1}l\theta'}\Big|^2,\end{aligned} \tag{2.248}$$

$\forall\,\theta'\in[0,2\pi]$.

Since $\omega\mapsto A_\infty[\varepsilon,\mu,\omega]$ is holomorphic in the upper half-plane $\mathbb{C}^+$, A_∞ decays sufficiently rapidly as $\omega\to+\infty$, and $A_\infty[\varepsilon,\mu,-\omega] = \overline{A_\infty[\varepsilon,\mu,\omega]}$ for real values of ω, the real and imaginary parts of the scattering amplitude are connected by the Kramers-Kronig relations

$$\Re A_\infty[\varepsilon,\mu,\omega](\xi,\xi) = c_d\,\text{p.v.}\int_0^{+\infty}\frac{(\omega')^{(d+1)/2}Q^{\rm ext}[\varepsilon,\mu,\omega'](\xi)}{(\omega')^2-\omega^2}\,d\omega'\,, \tag{2.249}$$

and

$$Q^{\rm ext}[\varepsilon,\mu,\omega](\xi) = -\frac{2}{\pi\sqrt{\varepsilon_m\mu_m}}\,\text{p.v.}\int_0^{+\infty}\frac{\Re A_\infty[\varepsilon,\mu,\omega'](\xi,\xi)}{(\omega')^2-\omega^2}\,d\omega'\,, \tag{2.250}$$

for $\xi \in \mathbb{R}^d, |\xi| = 1, d = 2, 3$, where $c_3 = \sqrt{\varepsilon_m \mu_m}/(2\pi^2)$ and $c_2 = -\sqrt{\sqrt{\varepsilon_m \mu_m}/(2\pi^3)}$. Moreover, we obtain by respectively taking the limits of (2.249) and (2.250) as $\omega \to 0$ the following sum rules:

$$\Re A_\infty[\varepsilon, \mu, 0](\xi, \xi) = c_d \,\text{p.v.} \int_0^{+\infty} (\omega')^{(d-3)/2} Q^{\text{ext}}[\varepsilon, \mu, \omega'](\xi) \, d\omega' , \tag{2.251}$$

and

$$Q^{\text{ext}}[\varepsilon, \mu, 0](\xi) = -\frac{2}{\pi \sqrt{\varepsilon_m \mu_m}} \,\text{p.v.} \int_0^{+\infty} \frac{\Re A_\infty[\varepsilon, \mu, \omega'](\xi, \xi) - \Re A_\infty[\varepsilon, \mu, 0](\xi, \xi)}{(\omega')^2} \, d\omega' . \tag{2.252}$$

2.10.5. Numerical Implementation. We now discuss a numerical approach to the transmission problem (2.174) for the Helmholtz equation. This involves solving the boundary integral equation (2.178).

We begin by performing the usual boundary discretization procedure as in Subsection 2.4.5. Suppose that the boundary $\partial\Omega$ is parametrized by $x(t)$ for $t \in [0, 2\pi)$. We partition the interval $[0, 2\pi)$ into N pieces

$$[t_1, t_2), [t_2, t_3), \ldots, [t_N, t_{N+1}),$$

with $t_1 = 0$ and $t_{N+1} = 2\pi$. We then approximate the boundary $\partial\Omega = \{x(t) \in \mathbb{R}^2 : t \in [0, 2\pi)\}$ by $x^{(i)} = x(t_i)$ for $1 \leq i \leq N$. We approximate the density functions φ and ψ with $\overline{\varphi}_i := \varphi(x^{(i)})$ and $\overline{\psi}_i := \psi(x^{(i)})$ for $1 \leq i \leq N$. We also discretize the Dirichlet data $u^i|_{\partial D}$ and Neumann data $\partial u^i/\partial\nu|_{\partial D}$ of the incident wave u_i by setting $u_d = u^i(x^{(j)})$ and $u_n = \partial u^i/\partial\nu(x^{(j)})$ for $1 \leq i \leq N$. Then the integral equation (2.178) is represented numerically as

$$\begin{pmatrix} S_- & -S_+ \\ \frac{1}{\mu_c} S'_- & -\frac{1}{\mu_m} S'_+ \end{pmatrix} \begin{pmatrix} \overline{\varphi} \\ \overline{\psi} \end{pmatrix} = \begin{pmatrix} u_d \\ u_n \end{pmatrix},$$

where $S_\pm$ and $S'_\pm$ are the $N \times N$ matrices given by

$$(S_-)_{ij} = \Gamma^{k_m}(x^{(i)} - x^{(j)}) |T(x^{(j)})| (t_{j+1} - t_j), \tag{2.253}$$

$$(S_+)_{ij} = \Gamma^{k_c}(x^{(i)} - x^{(j)}) |T(x^{(j)})| (t_{j+1} - t_j), \tag{2.254}$$

$$(S'_-)_{ij} = -\frac{1}{2}\delta_{ij} + \frac{\partial \Gamma^{k_c}}{\partial \nu_x}(x^{(i)} - x^{(j)}) |T(x^{(j)})| (t_{j+1} - t_j), \tag{2.255}$$

$$(S'_+)_{ij} = \frac{1}{2}\delta_{ij} + \frac{\partial \Gamma^{k_m}}{\partial \nu_x}(x^{(i)} - x^{(j)}) |T(x^{(j)})| (t_{j+1} - t_j), \tag{2.256}$$

for $i \neq j$ and $i, j = 1, 2, \ldots, N$. The singularity for $i = j$ can be treated as explained in Subsection 2.4.5. By solving the above linear system, we obtain approximations for the density functions φ and ψ. Then we calculate the numerical solution for u using (2.177). We can also obtain the scattering coefficients $W_{ll'}$ numerically by using the definition (2.224).

Let B be a disk of radius $R = 1$ located at the origin in $\mathbb{R}^2$. Set $\omega = 2, \varepsilon_m = 1, \varepsilon_c = 1, \mu_m = 1$ and $\mu_c = 5$. Let us assume that $u^i(x) = J_3(k_m r) e^{\sqrt{-1}3\theta}$. In Figure 2.4 we compare the numerical solution to (2.178) with the exact solution by evaluating the solutions on the circle $|x| = 2$.

Next we compare the scattering coefficients W_{ll} obtained numerically for $l = 1, 2, \ldots, 7$, with theoretical results. The comparison is shown in Table 2.7 and the decay property of the scattering coefficients is clearly present.

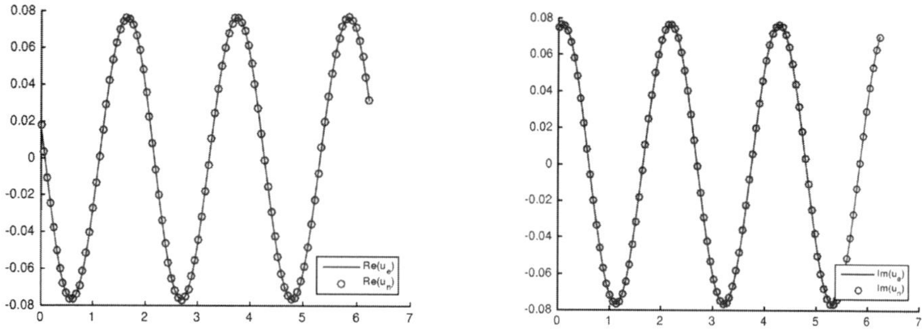

FIGURE 2.4. The exact solution u_e and the numerical solution u_n of the transmission problem (2.174) for Helmholtz equation. The inclusion D is a circular disk with radius 1. The parameters are $\omega = 2$, $\varepsilon_m = 1$, $\varepsilon_c = 1$, $\mu_m = 1$, and $\mu_c = 5$. We assume that $u^i(x) = J_3(k_m r)e^{\sqrt{-1}3\theta}$. The solutions are evaluated on the circle $|x| = 2$.

n	Theoretical	Numerical
1	$1.7866 - 1.1036\sqrt{-1}$	$1.7866 - 1.1011\sqrt{-1}$
2	$-0.9673 - 3.7540\sqrt{-1}$	$-0.9601 - 3.7545\sqrt{-1}$
3	$-0.6487 - 0.1081\sqrt{-1}$	$-0.6487 - 0.1081\sqrt{-1}$
4	$-0.0462 - 0.0005\sqrt{-1}$	$-0.0462 - 0.0005\sqrt{-1}$
5	$-0.0023 - 0.0000\sqrt{-1}$	$-0.0023 - 0.0000\sqrt{-1}$
6	$-0.0001 - 0.0000\sqrt{-1}$	$-0.0001 - 0.0000\sqrt{-1}$
7	$-0.0000 - 0.0000\sqrt{-1}$	$-0.0000 - 0.0000\sqrt{-1}$

TABLE 2.7. Scattering coefficients W_{ll} for $l = 1, 2, \ldots, 7$ when D is a unit circular disk. The parameters are $\omega = 2, \varepsilon_m = 1, \varepsilon_c = 1, \mu_m = 1$ and $\mu_c = 5$.

2.11. Scalar Wave Scattering by Small Particles

With the same notation as in Subsection 2.8.4.2, we suppose that the particle D is of the form $D = \delta B + z$, and let u be the solution of (2.172).

In this section we derive an asymptotic expansion of u as $\delta \to 0$. For simplicity, although the asymptotic expansions are valid in the two-dimensional case we only consider $d = 3$ in what follows.

We first derive an estimate of the form (2.179) with a constant C independent of δ.

PROPOSITION 2.82. *Let $D = \delta B + z$ and $(\varphi, \psi) \in L^2(\partial D) \times L^2(\partial D)$ be the unique solution of (2.178). There exists $\delta_0 > 0$ such that for all $\delta \le \delta_0$, there exists a constant C independent of δ such that*

$$\|\varphi\|_{L^2(\partial D)} + \|\psi\|_{L^2(\partial D)} \le C\Big(\delta^{-1}\|u^i\|_{L^2(\partial D)} + \|\nabla u^i\|_{L^2(\partial D)}\Big). \tag{2.257}$$

PROOF. After the scaling $x = z + \delta y$, (2.178) takes the form

$$\begin{cases} \mathcal{S}_B^{k_c\delta}[\varphi_\delta] - \mathcal{S}_B^{k_m\delta}[\psi_\delta] = \frac{1}{\delta} u_\delta^i \\ \frac{1}{\mu_c} \frac{\partial(\mathcal{S}_B^{k_c\delta}[\varphi_\delta])}{\partial \nu}\Big|_- - \frac{1}{\mu_m} \frac{\partial(\mathcal{S}_B^{k_m\delta}[\psi_\delta])}{\partial \nu}\Big|_+ = \frac{1}{\delta\mu_m} \frac{\partial u_\delta^i}{\partial \nu} \end{cases} \quad \text{on } \partial B,$$

where $\varphi_\delta(y) = \varphi(z + \delta y)$, $y \in \partial B$, *etc*, and the single-layer potentials $\mathcal{S}_B^{k_c\delta}$ and $\mathcal{S}_B^{k_m\delta}$ are defined by the fundamental solutions $\Gamma_{k_c\delta}$ and $\Gamma_{k_m\delta}$, respectively. It then follows from Theorem 2.51 that for δ small enough the following estimate holds:

$$\|\varphi_\delta\|_{L^2(\partial B)} + \|\psi_\delta\|_{L^2(\partial B)} \leq C\delta^{-1} \|u_\delta^i\|_{H^1(\partial B)},$$

for some constant C independent of δ. By scaling back, we obtain (2.257). □

Fix $n \in \mathbb{N}$, define

$$u_n^i(x) = \sum_{|l|=0}^{n} \frac{\partial^l u^i(z)}{l!} (x - z)^l,$$

and let (φ_n, ψ_n) be the unique solution of

$$\begin{cases} \mathcal{S}_D^{k_c}[\varphi_n] - \mathcal{S}_D^{k_m}[\psi_n] = u_{n+1}^i \\ \frac{1}{\mu_c} \frac{\partial(\mathcal{S}_D^{k_c}[\varphi_n])}{\partial \nu}\Big|_- - \frac{1}{\mu_m} \frac{\partial(\mathcal{S}_D^{k_m}[\psi_n])}{\partial \nu}\Big|_+ = \frac{1}{\mu_m} \frac{\partial u_{n+1}^i}{\partial \nu} \end{cases} \quad \text{on } \partial D. \tag{2.258}$$

Then $(\varphi - \varphi_n, \psi - \psi_n)$ is the unique solution of (2.258) with the right-hand sides defined by $u^i - u_{n+1}^i$. Therefore, by (2.257), we obtain

$$\begin{aligned} &\|\varphi - \varphi_n\|_{L^2(\partial D)} + \|\psi - \psi_n\|_{L^2(\partial D)} \\ &\leq C\Big(\delta^{-1}\|u^i - u_{n+1}^i\|_{L^2(\partial D)} + \|\nabla(u^i - u_{n+1}^i)\|_{L^2(\partial D)}\Big). \end{aligned} \tag{2.259}$$

By the definition of u_{n+1}^i, we have

$$\begin{aligned} \|u^i - u_{n+1}^i\|_{L^2(\partial D)} &\leq C|\partial D|^{1/2} \|u^i - u_{n+1}^i\|_{L^\infty(\partial D)} \\ &\leq C|\partial D|^{1/2} (\delta k_m)^{n+2}, \end{aligned}$$

and

$$\|\nabla(u^i - u_{n+1}^i)\|_{L^2(\partial D)} \leq C|\partial D|^{1/2} (\delta k_m)^{n+1}.$$

It then follows from (2.259) that

$$\|\varphi - \varphi_n\|_{L^2(\partial D)} + \|\psi - \psi_n\|_{L^2(\partial D)} \leq C(k_m)|\partial D|^{1/2} \delta^{n+1}. \tag{2.260}$$

By (2.178), we obtain

$$u(x) = u^i(x) + \mathcal{S}_D^{k_m}[\psi_n](x) + \mathcal{S}_D^{k_m}[\psi - \psi_n](x), \quad x \in K, \tag{2.261}$$

where $K \Subset \mathbb{R}^d \setminus \overline{D}$. Since $\operatorname{dist}(D, K) \geq c_0$, we have

$$\sup_{x \in K,\ y \in \partial D} \Big|\Gamma_{k_m}(x - y)\Big| \leq C$$

for some constant C. Hence, for each $x \in K$, we have from (2.260)

$$\begin{aligned}\left|\mathcal{S}_D^{k_m}[\psi - \psi_n](x)\right| &\leq \left[\int_{\partial D} |\Gamma_{k_m}(x-y)|^2 \, d\sigma(y)\right]^{1/2} \|\psi - \psi_n\|_{L^2(\partial D)} \\ &\leq C|\partial D|^{1/2}|\partial D|^{1/2}\delta^{n+1} \leq C'\delta^{n+d},\end{aligned}$$

where C and C' are independent of $x \in K$ and δ. Thus we conclude that

$$u(x) = u^i(x) + \mathcal{S}_D^{k_m}[\psi_n](x) + O(\delta^{n+d}), \quad \text{uniformly in } x \in K. \tag{2.262}$$

For each multi-index l, define (φ_l, ψ_l) to be the unique solution to

$$\begin{cases} \mathcal{S}_B^{k_c\delta}[\varphi_l] - \mathcal{S}_B^{k_m\delta}[\psi_l] = x^l \\ \dfrac{1}{\mu_c}\dfrac{\partial(\mathcal{S}_B^{k_c\delta}[\varphi_l])}{\partial\nu}\Big|_- - \dfrac{1}{\mu_m}\dfrac{\partial(\mathcal{S}_B^{k_m\delta}[\psi_l])}{\partial\nu}\Big|_+ = \dfrac{1}{\mu_m}\dfrac{\partial x^l}{\partial\nu} \end{cases} \quad \text{on } \partial B. \tag{2.263}$$

Then, we claim that

$$\varphi_n(x) = \sum_{|l|=0}^{n+1} \delta^{|l|-1}\frac{\partial^l u^i(z)}{l!}\varphi_l(\delta^{-1}(x-z)),$$

$$\psi_n(x) = \sum_{|l|=0}^{n+1} \delta^{|l|-1}\frac{\partial^l u^i(z)}{l!}\psi_l(\delta^{-1}(x-z)).$$

In fact, the expansions follow from the uniqueness of the solution to the integral equation (2.175) and the relation

$$\begin{aligned}&\mathcal{S}_D^{k_m}\left[\sum_{|l|=0}^{n+1} \delta^{|l|-1}\frac{\partial^l u^i(z)}{l!}\varphi_l(\delta^{-1}(\cdot - z))\right](x) \\ &= \sum_{|l|=0}^{n+1} \delta^{|l|}\frac{\partial^l u^i(z)}{l!}(\mathcal{S}_B^{k_m\delta}[\varphi_l])(\delta^{-1}(x-z)),\end{aligned}$$

for $x \in \partial D$. It then follows from (2.262) that

$$u(x) = u^i(x) + \sum_{|l|=0}^{n+1} \delta^{|l|-1}\frac{\partial^l u^i(z)}{l!}\mathcal{S}_D^{k_m}[\psi_l(\delta^{-1}(\cdot - z))](x) + O(\delta^{n+d}), \tag{2.264}$$

uniformly in $x \in K$. Note that

$$\begin{aligned}\mathcal{S}_D^{k_m}[\psi_l(\delta^{-1}(\cdot - z))](x) &= \int_{\partial D} \Gamma_{k_m}(x-y)\psi_l(\delta^{-1}(y-z)) \, d\sigma(y) \\ &= \delta^{d-1}\int_{\partial B} \Gamma_{k_m}(x-(\delta w + z))\psi_l(w) \, d\sigma(w).\end{aligned}$$

Moreover, for $x \in K$, $z \in D$, $w \in \partial B$, and sufficiently small δ, we have

$$\Gamma_{k_m}(x-(\delta w+z)) = \sum_{|l'|=0}^{\infty} \frac{\delta^{|l'|}}{l'!}\partial_z^{l'}\Gamma_{k_m}(x-z)w^{l'}.$$

Therefore, we get

$$\mathcal{S}_D^{k_m}[\psi_l(\delta^{-1}(\cdot - z))](x) = \sum_{|l'|=0}^{\infty} \frac{\delta^{|l'|+d-1}}{l'!}\partial_z^{l'}\Gamma_{k_m}(x-z)\int_{\partial B} w^{l'}\psi_l(w) \, d\sigma(w).$$

Define, for multi-indices l and l' in $\mathbb{N}^d$, the scattering tensors

$$\widetilde{W}_{ll'} := \int_{\partial B} w^{l'} \psi_l(w)\, d\sigma(w)\,. \tag{2.265}$$

Then we obtain the following theorem from (2.264).

THEOREM 2.83. *The following pointwise multipolar asymptotic expansion in $K \Subset \mathbb{R}^d \setminus \overline{D}$ holds:*

$$\begin{aligned} u(x) &= u^i(x) + \delta^{d-2} \sum_{|l'|=0}^{n+1} \sum_{|l|=0}^{n-|l'|+1} \frac{\delta^{|l|+|l'|}}{l!l'!} \partial^l u^i(z) \partial_z^{l'} \Gamma_{k_m}(x-z) \widetilde{W}_{ll'} \\ &\quad + O(\delta^{n+d})\,, \end{aligned} \tag{2.266}$$

where the remainder $O(\delta^{d+n})$ is dominated by $C\delta^{d+n}$ for some C independent of $x \in K$.

REMARK 2.84. *In view of (2.177), we obtain the following expansion:*

$$\begin{aligned} \mathcal{S}_D^{k_m}[\psi](x) &= \delta^{d-2} \sum_{|l'|=0}^{n+1} \sum_{|l|=0}^{n-|l'|+1} \frac{\delta^{|l|+|l'|}}{l!l'!} \partial^l u^i(z) \partial_z^{l'} \Gamma_{k_m}(x-z) \widetilde{W}_{ll'} \\ &\quad + O(\delta^{n+d})\,. \end{aligned} \tag{2.267}$$

Observe that ψ_l, and hence, $\widetilde{W}_{ll'}$ depends on δ, and so does u^i. Thus the formula (2.266) is not a genuine asymptotic formula. However, since it is useful for solving the inverse problem for the Helmholtz equation, we made a record of it in this remark.

REMARK 2.85. *Note that the scattering tensors defined in (2.265) are the basic building blocks for the full asymptotic expansion of the scattering coefficients given by (2.219) as $\delta \to 0$. In fact, in the three-dimensional case, after scaling (2.178) Taylor expansions yield*

$$W_{pq} = \frac{1}{k_m} \sum_{l,l' \in \mathbb{N}^3} \widetilde{W}_{ll'} \frac{(\delta k_m)^{|l|+|l'|+1}}{l!l'!} \partial^l \left[J_p(y) e^{\sqrt{-1}p\theta_y}\right]\big|_{y=0} \partial^{l'} \left[J_q(y) e^{\sqrt{-1}q\theta_y}\right]\big|_{y=0} \tag{2.268}$$

for $p, q \in \mathbb{N}$.

Now, observe that by the definition (2.263) of ψ_l, $\|\psi_l\|_{L^2(\partial B)}$ is bounded, and hence

$$|\widetilde{W}_{ll'}| \le C_{ll'}\,, \quad \forall\, l, l'\,,$$

where the constant $C_{ll'}$ is independent of δ. Since δ is small, we can derive an asymptotic expansion of (φ_l, ψ_l) using their definition (2.263). Let us briefly explain this. Let

$$T_\delta \begin{bmatrix} f \\ g \end{bmatrix} := \begin{bmatrix} \mathcal{S}_B^{k_c\delta}[f] - \mathcal{S}_B^{k_m\delta}[g] \\ \dfrac{1}{\mu_c} \dfrac{\partial(\mathcal{S}_B^{k_c\delta}[f])}{\partial\nu}\bigg|_- - \dfrac{1}{\mu_m} \dfrac{\partial(\mathcal{S}_B^{k_m\delta}[g])}{\partial\nu}\bigg|_+ \end{bmatrix} \quad \text{on } \partial B\,,$$

and let T_0 be the operator when $\delta = 0$. Then the solution (φ_l, ψ_l) of the integral equation (2.263) is given by

$$\begin{bmatrix} \varphi_l \\ \psi_l \end{bmatrix} = \Big[I + T_0^{-1}(T_\delta - T_0) \Big]^{-1} T_0^{-1} \begin{bmatrix} x^l \\ \dfrac{1}{\mu_m} \dfrac{\partial x^l}{\partial \nu} \end{bmatrix}. \tag{2.269}$$

By expanding $T_\delta - T_0$ in a power series with respect to δ, we can derive the expansions of ψ_l and $\widetilde{W}_{ll'}$. Let, for $l, l' \in \mathbb{N}^d$, $(\widehat{\varphi}_l, \widehat{\psi}_l)$ be the leading-order term in the expansion of (φ_l, ψ_l). Then $(\widehat{\varphi}_l, \widehat{\psi}_l)$ is the solution of the system of the integral equations

$$\begin{cases} \mathcal{S}_B^0[\widehat{\varphi}_l] - \mathcal{S}_B^0[\widehat{\psi}_l] = x^l \\ \dfrac{1}{\mu_c} \dfrac{\partial(\mathcal{S}_B^0[\widehat{\varphi}_l])}{\partial \nu}\Big|_- - \dfrac{1}{\mu_m} \dfrac{\partial(\mathcal{S}_B^0[\widehat{\psi}_l])}{\partial \nu}\Big|_+ = \dfrac{1}{\mu_0} \dfrac{\partial x^l}{\partial \nu} \end{cases} \quad \text{on } \partial B. \tag{2.270}$$

With regards to the simplest case, we can take $n = 1$ in (2.266) to find the leading-order term in the asymptotic expansion of $u - u^i$ as $\delta \to 0$. We first investigate the dependence of $\widetilde{W}_{ll'}$ on δ for $|l| \leq 1$ and $|l'| \leq 1$. If $|l| \leq 1$, then both sides of the first equation in (2.270) are harmonic in B, and hence

$$\mathcal{S}_B^0[\widehat{\varphi}_l] - \mathcal{S}_B^0[\widehat{\psi}_l] = x^l \quad \text{in } B.$$

Therefore we get

$$\frac{\partial(\mathcal{S}_B^0[\widehat{\varphi}_l])}{\partial \nu}\Big|_- - \frac{\partial(\mathcal{S}_B^0[\widehat{\psi}_l])}{\partial \nu}\Big|_- = \frac{\partial x^l}{\partial \nu} \quad \text{on } \partial B.$$

This identity together with the second equation in (2.270) yields

$$\frac{\mu_c}{\mu_m} \frac{\partial(\mathcal{S}_B^0[\widehat{\psi}_l])}{\partial \nu}\Big|_+ - \frac{\partial(\mathcal{S}_B^0[\widehat{\psi}_l])}{\partial \nu}\Big|_- = \Big(1 - \frac{\mu_c}{\mu_m}\Big) \frac{\partial x^l}{\partial \nu}.$$

In view of the relation (2.160), we have

$$\frac{\mu_c}{\mu_m} \Big(\frac{1}{2} I + (\mathcal{K}_B^0)^* \Big)[\widehat{\psi}_l] - \Big(-\frac{1}{2} I + (\mathcal{K}_B^0)^* \Big)[\widehat{\psi}_l] = \Big(1 - \frac{\mu_c}{\mu_m}\Big) \frac{\partial x^l}{\partial \nu},$$

where $(\mathcal{K}_B^0)^*$ is the Neumann–Poincaré operator defined in (2.6). Therefore, we have

$$\widehat{\psi}_l = (\lambda I - (\mathcal{K}_B^0)^*)^{-1} \Big(\frac{\partial x^l}{\partial \nu}\Big|_{\partial B} \Big), \tag{2.271}$$

where

$$\lambda := \frac{\frac{\mu_c}{\mu_m} + 1}{2(1 - \frac{\mu_c}{\mu_m})} = \frac{\frac{\mu_m}{\mu_c} + 1}{2(\frac{\mu_m}{\mu_c} - 1)}. \tag{2.272}$$

Observe that if $|l| = 0$, then $\widehat{\psi}_l = 0$ and $\mathcal{S}_B^0[\widehat{\varphi}_l] = 1$. Hence we obtain $\psi_l = O(\delta)$ and $\mathcal{S}_B^{k_c\delta}[\varphi_l] = 1 + O(\delta)$. Moreover, since $\mathcal{S}_B^{k_c\delta}[\varphi_l]$ depends on δ analytically and $(\Delta + k_c^2\delta^2)\mathcal{S}_B^{k_c\delta}[\varphi_l] = 0$ in B, we conclude that

$$\psi_l = O(\delta) \quad \text{and} \quad \mathcal{S}_B^{k_c\delta}[\varphi_l] = 1 + O(\delta^2), \quad |l| = 0. \tag{2.273}$$

It also follows from (2.271) that if $|l| = |l'| = 1$, then

$$\widetilde{W}_{ll'} = \int_{\partial B} x^{l'} (\lambda I - (\mathcal{K}_B^0)^*)^{-1} \Big(\frac{\partial y^l}{\partial \nu}\Big|_{\partial B} \Big)(x)\, d\sigma(x) + O(\delta). \tag{2.274}$$

The first quantity in the right-hand side of (2.274) is the polarization tensor M as defined in (2.71). In summary, we obtained that

$$\widetilde{W}_{ll'} = M + O(\delta)\,, \quad |l| = |l'| = 1\,. \tag{2.275}$$

By (2.160) and (2.263), we have

$$\begin{aligned}\psi_l &= \frac{\partial(\mathcal{S}_B^{k_m\delta}[\psi_l])}{\partial\nu}\bigg|_+ - \frac{\partial(\mathcal{S}_B^{k_m\delta}[\psi_l])}{\partial\nu}\bigg|_- \\ &= \frac{\mu_m}{\mu_c}\frac{\partial(\mathcal{S}_B^{k_c\delta}[\varphi_l])}{\partial\nu}\bigg|_- - \frac{\partial x^l}{\partial\nu} - \frac{\partial(\mathcal{S}_B^{k_m\delta}[\psi_l])}{\partial\nu}\bigg|_-\,.\end{aligned} \tag{2.276}$$

Suppose that either $l = 0$ or $l' = 0$. It then follows from the divergence theorem that

$$\begin{aligned}\int_{\partial B} x^{l'}\psi_l\,d\sigma = &-k_c^2\delta^2\frac{\mu_m}{\mu_c}\int_B x^{l'}\mathcal{S}_B^{k_c\delta}[\varphi_l]\,dx + k_m^2\delta^2\int_B x^{l'}\mathcal{S}_B^{k_m\delta}[\psi_l]\,dx \\ &+ \frac{\mu_m}{\mu_c}\int_{\partial B}\frac{\partial x^{l'}}{\partial\nu}\mathcal{S}_B^{k_c\delta}[\varphi_l]\,d\sigma - \int_{\partial B}\frac{\partial x^{l'}}{\partial\nu}\mathcal{S}_B^{k_m\delta}[\psi_l]\,d\sigma\,.\end{aligned} \tag{2.277}$$

From (2.277), we can observe the following.

$$\widetilde{W}_{ll'} = -k_c^2\delta^2\frac{\mu_m}{\mu_c}|B| + O(\delta^3) = -\delta^2\omega^2\varepsilon_c\mu_m|B| + O(\delta^3)\,, \quad |l| = |l'| = 0\,, \tag{2.278}$$

$$\widetilde{W}_{ll'} = O(\delta^2)\,, \quad |l| = 1\,,\ |l'| = 0\,, \tag{2.279}$$

$$\widetilde{W}_{ll'} = O(\delta^2)\,, \quad |l| = 0\,,\ |l'| = 1\,. \tag{2.280}$$

In fact, (2.278) and (2.280) follow from (2.273) and (2.277), and (2.279) immediately follows from (2.277). As a consequence of (2.267), (2.278), (2.279), (2.280), we obtain

$$\mathcal{S}_D^{k_m}[\psi](x) = O(\delta^d)\,, \quad \text{uniformly on } x \in K\,,$$

where $K \Subset \mathbb{R}^d \setminus \overline{D}$.

We now consider the case $|l| = 2$ and $|l'| = 0$. In this case, one can show using (2.276) that

$$\int_{\partial B}\psi_l\,d\sigma = -\int_B \Delta x^l\,dx + O(\delta^2)\,.$$

Therefore, if $|l'| = 0$, then

$$\sum_{|l|=2}\frac{1}{l!l'!}\partial^l u^i(z)\widetilde{W}_{ll'} = -\Delta u^i(z)|B| + O(\delta^2) = k_m^2 u^i(z)|B| + O(\delta^2)\,. \tag{2.281}$$

So (2.266) together with (2.275)-(2.281) yields the following dipolar expansion formula.

THEOREM 2.86. *For any $x \in K$,*

$$\begin{aligned}u(x) &= u^i(x) \\ &+ \delta^d\bigg(\nabla u^i(z)M\nabla_z\Gamma_{k_m}(x-z) + k_m^2(\frac{\varepsilon_c}{\varepsilon_m} - 1)|B|u^i(z)\Gamma_{k_m}(x-z)\bigg) \\ &\quad + O(\delta^{d+1})\,,\end{aligned} \tag{2.282}$$

where M is the polarization tensor defined in (2.71) with λ given by (2.272).

Let us now consider the case when there are several well separated particles. Let $D := \cup_{s=1}^m (\delta B_s + z_s)$. The magnetic permeability and electric permittivity of the particle $\delta B_s + z_s$ are $\mu_c^{(s)}$ and $\varepsilon_c^{(s)}$, $s = 1, \ldots, m$. By iterating formula (2.266) we can derive the following theorem.

THEOREM 2.87. *Suppose that there exists a positive constant C such that $|z_s - z_{s'}| \geq C$ for $s \neq s'$. Then the following pointwise asymptotic expansion in K holds uniformly:*

$$\begin{aligned} u(x) &= u^i(x) \\ &+ \delta^{d-2} \sum_{s=1}^m \sum_{|l'|=0}^{n+1} \sum_{|l|=0}^{n+1-|l'|} \frac{\delta^{|l|+|l'|}}{l!l'!} \partial^l u^i(z_s) \partial_z^{l'} \Gamma_{k_m}(x - z_s) W_{ll'}^{(s)} + O(\delta^{n+d}) . \end{aligned} \tag{2.283}$$

Here the scattering tensor $W_{ll'}^{(s)}$ is defined by (2.265) with B, μ_c, ε_c replaced by $B_s, \mu_c^{(s)}, \varepsilon_c^{(s)}$.

A first-order asymptotic expansion similar to (2.282) can be obtained for closely spaced particles. Let $D := \cup_{s=1}^m (\delta B_s + z)$ and let $\nu^{(s)}$ be the outward normal to ∂B_s. As before, the magnetic permeability and electric permittivity of the particle $\delta B_s + z$ are $\mu_c^{(s)}$ and $\varepsilon_c^{(s)}$, $s = 1, \ldots, m$.

Let the overall polarization tensor $M = (m_{pq})_{p,q=1}^d$ be defined by

$$m_{pq} := \sum_{s=1}^m \int_{\partial B_s} x_p \phi_q^{(s)}(x) \, d\sigma(x), \tag{2.284}$$

where the densities $\phi_q^{(s)}$ satisfy

$$(\lambda_s I - (\mathcal{K}_{B_s}^0)^*)[\phi_q^{(s)}] - \sum_{s' \neq s} \frac{\partial \mathcal{S}_{B_{s'}}^0 [\phi_q^{(s')}]}{\partial \nu^{(s)}} = \nu_q^{(s)} \quad \text{on } \partial B_s$$

with λ_s being given by (2.272) with μ_c replaced by $\mu_c^{(s)}$.

THEOREM 2.88. *Let $D := \cup_{s=1}^m (\delta B_s + z)$. For any $x \in K$, as $\delta \to 0$,*

$$\begin{aligned} u(x) &= u^i(x) \\ &+ \delta^d \Big(\nabla u^i(z) M \nabla_z \Gamma_{k_m}(x - z) + k_m^2 \big(\sum_{s=1}^m (\frac{\varepsilon_c^{(s)}}{\varepsilon_m} - 1) |B_s| \big) u^i(z) \Gamma_{k_m}(x - z) \Big) \\ &+ O(\delta^{d+1}) , \end{aligned} \tag{2.285}$$

where M is the overall polarization tensor associated with the particles B_s defined in (2.284).

REMARK 2.89. *Note that the polarization tensor of multiple particles has the same properties such as symmetry and positivity as the ones associated with a single particle.*

2.12. Quasi-Periodic Layer Potentials for the Helmholtz Equation

In this section we collect some notation and well-known results regarding quasi-periodic layer potentials for the Helmholtz equation. We refer to [**215, 373, 391, 413, 503**] for the details. This results will be used for the analysis of photonic and phononic bandgap structures. In Section 2.13 numerical schemes for the calculation of periodic Green's functions will be described. These techniques rely either on a Fourier series to compute the governing Green's function or a lattice sum representation of the Green's function via the method of images.

We denote by α the quasi-momentum variable in the Brillouin zone $B = [0, 2\pi)^2$. We introduce the two-dimensional quasi-periodic Green's function (or fundamental solution) $G^{\alpha,\omega}$, which satisfies

$$(\Delta + \omega^2) G^{\alpha,\omega}(x, y) = \sum_{n \in \mathbb{Z}^2} \delta_0(x - y - n) e^{\sqrt{-1} n \cdot \alpha}. \tag{2.286}$$

If $\omega \neq |2\pi n + \alpha|, \forall\, n \in \mathbb{Z}^2$, then by using Poisson's summation formula

$$\sum_{n \in \mathbb{Z}^2} e^{\sqrt{-1}(2\pi n + \alpha) \cdot x} = \sum_{n \in \mathbb{Z}^2} \delta_0(x - n) e^{\sqrt{-1} n \cdot \alpha}, \tag{2.287}$$

the quasi-periodic fundamental solution $G^{\alpha,\omega}$ can be represented as a sum of augmented plane waves over the reciprocal lattice:

$$G^{\alpha,\omega}(x, y) = \sum_{n \in \mathbb{Z}^2} \frac{e^{\sqrt{-1}(2\pi n + \alpha) \cdot (x - y)}}{\omega^2 - |2\pi n + \alpha|^2}. \tag{2.288}$$

Moreover, it can also be shown that $G^{\alpha,\omega}$ can be alternatively represented as a sum of images:

$$G^{\alpha,\omega}(x, y) = -\frac{\sqrt{-1}}{4} \sum_{n \in \mathbb{Z}^2} H_0^{(1)}(\omega |x - n - y|) e^{\sqrt{-1} n \cdot \alpha}, \tag{2.289}$$

where $H_0^{(1)}$ is the Hankel function of the first kind of order 0. The series in the spatial representation (2.289) of the Green's function converges uniformly for x, y in compact sets of $\mathbb{R}^2$ and $\omega \neq |2\pi n + \alpha|$ for all $n \in \mathbb{Z}^2$. From (2.289) and the well-known fact that $H_0^{(1)}(z) = (2\sqrt{-1}/\pi) \ln z + O(1)$ as $z \to 0$ (see (2.152)), it follows that $G^{\alpha,\omega}(x, y) - (1/2\pi) \ln |x - y|$ is smooth for all $x, y \in Y$. A disadvantage of the form (2.288), which is often referred to as the spectral representation of the Green's function, is that the singularity as $|x - y| \to 0$ is not explicit.

In all the sequel, we assume that $\omega \neq |2\pi n + \alpha|$ for all $n \in \mathbb{Z}^2$. Let D be a bounded domain in $\mathbb{R}^2$, with a connected Lipschitz boundary ∂D. Let ν denote the unit outward normal to ∂D. For $\omega > 0$ let $\mathcal{S}^{\alpha,\omega}$ and $\mathcal{D}^{\alpha,\omega}$ be the quasi-periodic single- and double-layer potentials[1] associated with $G^{\alpha,\omega}$ on D; that is, for a given density $\varphi \in L^2(\partial D)$,

$$\mathcal{S}^{\alpha,\omega}[\varphi](x) = \int_{\partial D} G^{\alpha,\omega}(x, y) \varphi(y)\, d\sigma(y), \quad x \in \mathbb{R}^2,$$

$$\mathcal{D}^{\alpha,\omega}[\varphi](x) = \int_{\partial D} \frac{\partial G^{\alpha,\omega}(x, y)}{\partial \nu(y)} \varphi(y)\, d\sigma(y), \quad x \in \mathbb{R}^2 \setminus \partial D.$$

[1] From now on we use $\mathcal{S}^{\alpha,\omega}$ and $\mathcal{D}^{\alpha,\omega}$ instead of $\mathcal{S}_D^{\alpha,\omega}$ and $\mathcal{D}_D^{\alpha,\omega}$ for layer potentials on D. This is to keep the notation simple.

Then, $\mathcal{S}^{\alpha,\omega}[\varphi]$ and $\mathcal{D}^{\alpha,\omega}[\varphi]$ satisfy $(\Delta+\omega^2)\mathcal{S}^{\alpha,\omega}[\varphi]=(\Delta+\omega^2)\mathcal{D}^{\alpha,\omega}[\varphi]=0$ in D and $Y\setminus\overline{D}$ where Y is the periodic cell $[0,1)^2$, and they are α-quasi-periodic. Here we assume $\overline{D}\subset Y$.

The next formulas give the jump relations obeyed by the double-layer potential and by the normal derivative of the single-layer potential on general Lipschitz domains:

$$\frac{\partial(\mathcal{S}^{\alpha,\omega}[\varphi])}{\partial\nu}\bigg|_{\pm}(x)=\Big(\pm\frac{1}{2}I+(\mathcal{K}^{-\alpha,\omega})^*\Big)[\varphi](x)\quad\text{a.e. } x\in\partial D, \tag{2.290}$$

$$(\mathcal{D}^{\alpha,\omega}[\varphi])\bigg|_{\pm}(x)=\Big(\mp\frac{1}{2}I+\mathcal{K}^{\alpha,\omega}\Big)[\varphi](x)\quad\text{a.e. } x\in\partial D, \tag{2.291}$$

for $\varphi\in L^2(\partial D)$, where $\mathcal{K}^{\alpha,\omega}$ is the operator on $L^2(\partial D)$ defined by

$$\mathcal{K}^{\alpha,\omega}[\varphi](x)=\text{p.v.}\int_{\partial D}\frac{\partial G^{\alpha,\omega}(x,y)}{\partial\nu(y)}\varphi(y)\,d\sigma(y) \tag{2.292}$$

and $(\mathcal{K}^{-\alpha,\omega})^*$ is the L^2-adjoint operator of $\mathcal{K}^{-\alpha,\omega}$, which is given by

$$(\mathcal{K}^{-\alpha,\omega})^*[\varphi](x)=\text{p.v.}\int_{\partial D}\frac{\partial G^{\alpha,\omega}(x,y)}{\partial\nu(x)}\varphi(y)\,d\sigma(y). \tag{2.293}$$

The singular integral operators $\mathcal{K}^{\alpha,\omega}$ and $(\mathcal{K}^{-\alpha,\omega})^*$ are bounded on $L^2(\partial D)$ as an immediate consequence of the fact that $G^{\alpha,\omega}(x,y)-(1/2\pi)\ln|x-y|$ is smooth for all x,y.

The following lemma is of use to us.

LEMMA 2.90. *Suppose that $\alpha\neq 0$ and ω^2 is neither an eigenvalue of $-\Delta$ in D with the Dirichlet boundary condition on ∂D nor in $Y\setminus\overline{D}$ with the Dirichlet boundary condition on ∂D and the α-quasi-periodic condition on ∂Y. Then $\mathcal{S}^{\alpha,\omega}: L^2(\partial D)\to H^1(\partial D)$ is invertible.*

PROOF. Suppose that $\phi\in L^2(\partial D)$ satisfies $\mathcal{S}^{\alpha,\omega}[\phi]=0$ on ∂D. Then $u=\mathcal{S}^{\alpha,\omega}[\phi]$ satisfies $(\Delta+\omega^2)u=0$ in D and in $Y\setminus\overline{D}$. Therefore, since ω^2 is neither an eigenvalue of $-\Delta$ in D with the Dirichlet boundary condition nor in $Y\setminus\overline{D}$ with the Dirichlet boundary condition on ∂D and the quasi-periodic condition on ∂Y, it follows that $u=0$ in Y and thus, $\phi=\partial u/\partial\nu|_+-\partial u/\partial\nu|_-=0$, as desired. □

Define

$$G^{\alpha,0}(x,y):=G_\alpha(x-y)=-\sum_{n\in\mathbb{Z}^2}\frac{e^{\sqrt{-1}(2\pi n+\alpha)\cdot(x-y)}}{|2\pi n+\alpha|^2}\quad\text{for } \alpha\neq 0,$$

where G_α is given by (2.134) for $d=2$ and

$$G^{0,0}(x,y):=G_\sharp(x-y)=-\sum_{n\in\mathbb{Z}^2\setminus\{0\}}\frac{e^{\sqrt{-1}2\pi n\cdot(x-y)}}{4\pi^2|n|^2},$$

where $G_\sharp$ is given by (2.119) for $d=2$. Note that $G^{\alpha,0}(x,y)$ for $\alpha\neq 0$ is a fundamental solution of the quasi-periodic Laplacian (2.135) in Y, while, in view of (2.121), $G^{0,0}(x,y)$ satisfies

$$\Delta_x G^{0,0}(x,y)=\delta_y-1\quad\text{in } Y \tag{2.294}$$

with periodic Dirichlet boundary conditions on ∂Y. See [**52, 74**]. The following lemma is easy to prove. It gives a complete low-frequency asymptotic expansion of $G^{\alpha,\omega}$.

LEMMA 2.91. *As $\omega \to 0$, $G^{\alpha,\omega}$ can be decomposed as*

$$G^{\alpha,\omega}(x,y) = G^{\alpha,0}(x,y) - \sum_{l=1}^{+\infty} \underbrace{\omega^{2l} \sum_{n\in\mathbb{Z}^2} \frac{e^{\sqrt{-1}(2\pi n+\alpha)\cdot(x-y)}}{|2\pi n+\alpha|^{2(l+1)}}}_{:=-G_l^{\alpha,\omega}(x,y)}, \tag{2.295}$$

for $\alpha \neq 0$, while for $\alpha = 0$, the following decomposition holds:

$$G^{0,\omega}(x,y) = \frac{1}{\omega^2} + G^{0,0}(x,y) - \sum_{l=1}^{+\infty} \underbrace{\omega^{2l} \sum_{n\in\mathbb{Z}^2\setminus\{0\}} \frac{e^{\sqrt{-1}2\pi n\cdot(x-y)}}{(4\pi^2)^{l+1}|n|^{2(l+1)}}}_{:=-G_l^{0,\omega}(x,y)}. \tag{2.296}$$

Denote by $\mathcal{S}_l^{\alpha,\omega}$ and $(\mathcal{K}_l^{-\alpha,\omega})^*$, for $l \geq 0$ and $\alpha \in [0,2\pi)^2$, the layer potentials associated with the kernel $G_l^{\alpha,\omega}(x,y)$ so that

$$\mathcal{S}^{\alpha,\omega} = \mathcal{S}^{\alpha,0} + \sum_{l=1}^{+\infty} \mathcal{S}_l^{\alpha,\omega} \quad \text{and} \quad (\mathcal{K}^{\alpha,\omega})^* = (\mathcal{K}^{\alpha,0})^* + \sum_{l=1}^{+\infty} (\mathcal{K}_l^{-\alpha,\omega})^*. \tag{2.297}$$

LEMMA 2.92. *The operator $(1/2)\,I + (\mathcal{K}^{-\alpha,0})^* : L^2(\partial D) \to L^2(\partial D)$ is invertible.*

Before proving Lemma 2.92, let us make a note of the following simple fact. If u and v are α-quasi-periodic smooth functions, then

$$\int_{\partial Y} \frac{\partial u}{\partial \nu} \overline{v}\, d\sigma = 0. \tag{2.298}$$

To prove this, it is enough to see that

$$\int_{\partial Y} \frac{\partial u}{\partial \nu} \overline{v} = \int_{\partial Y} \left[\frac{\partial (u e^{-\sqrt{-1}\alpha\cdot x})}{\partial \nu} + \sqrt{-1}\alpha\cdot\nu u e^{-\sqrt{-1}\alpha\cdot x}\right] \overline{e^{-\sqrt{-1}\alpha\cdot x} v}.$$

PROOF OF LEMMA 2.92. Let $\phi \in L^2(\partial D)$ satisfy $((1/2)\,I + (\mathcal{K}^{-\alpha,0})^*)[\phi] = 0$ on ∂D. Observe that if $\alpha = 0$, then $\int_{\partial D} \phi = 0$. In fact, by (2.294) and (2.298), we have for $x \in D$

$$\mathcal{D}^{0,0}[1](x) = -\int_{Y\setminus\overline{D}} \Delta_y G^{0,0}(x,y)dy = |Y \setminus \overline{D}|,$$

where $|\ |$ denotes the volume, and hence

$$(\frac{1}{2}I + \mathcal{K}^{0,0})[1] = |Y \setminus \overline{D}| \quad \text{on } \partial D. \tag{2.299}$$

Therefore, we have

$$|Y \setminus \overline{D}| \int_{\partial D} \phi\, d\sigma = \int_{\partial D} (\frac{1}{2}I + \mathcal{K}^{0,0})[1]\, \phi\, d\sigma = \int_{\partial D} (\frac{1}{2}I + (\mathcal{K}^{0,0})^*)[\phi]\, d\sigma = 0.$$

Consequently, for any $\alpha \in [0,2\pi)^2$, $u = \mathcal{S}^{\alpha,0}[\phi]$ is α-quasi-periodic and satisfies $\Delta u = 0$ in $Y \setminus \overline{D}$ with

$$\frac{\partial u}{\partial \nu}\bigg|_+ = (\frac{1}{2}I + (\mathcal{K}^{-\alpha,0})^*)[\phi] = 0 \quad \text{on } \partial D.$$

Therefore, it follows from (2.298) that

$$\int_{Y\setminus\overline{D}} |\nabla u|^2 = \int_{\partial Y} \frac{\partial u}{\partial \nu}\overline{u} - \int_{\partial D} \frac{\partial u}{\partial \nu}\bigg|_+ \overline{u} = 0.$$

Thus, u is constant in $Y \setminus \overline{D}$ and hence in D. This implies that

$$\phi = \frac{\partial u}{\partial \nu}\bigg|_+ - \frac{\partial u}{\partial \nu}\bigg|_- = 0,$$

as desired. □

2.13. Computations of Periodic Green's Functions

In this section, we briefly describe analytical techniques for transforming the Green's functions for the Helmholtz equation in periodic domains from the slowly convergent representations as a series of images or plane waves into forms more suitable for computation. In particular, methods derived from Kummer's transformation, lattice sums, and the use of Ewald's method are discussed. The main ideas of these techniques apply to Maxwell's equations and the Lamé system as well. We refer the reader for instance to [**158**] for a complete treatment of the subject.

2.13.1. Kummer's Transformation. The convergence of the series (2.288) and (2.289) can be improved if we use Kummer's transformation, namely, if we convert the slowly convergent series into two series which converge faster by subtracting and adding back a series which has the same asymptotic behavior as the troublesome series and which can be summed analytically [**314**]. We can accelerate the series in (2.288) by writing

$$\begin{aligned} G^{\alpha,\omega}(x,y) &= \frac{e^{\sqrt{-1}\alpha\cdot(x-y)}}{\omega^2 - |\alpha|^2} - e^{\sqrt{-1}\alpha\cdot(x-y)} \sum_{n\in\mathbb{Z}^2, n\neq 0} \frac{e^{\sqrt{-1}2\pi n\cdot(x-y)}}{4\pi^2|n|^2} \\ &\quad + \sum_{n\in\mathbb{Z}^2, n\neq 0} e^{\sqrt{-1}(2\pi n+\alpha)\cdot(x-y)} \left(\frac{1}{\omega^2 - |2\pi n+\alpha|^2} + \frac{1}{4\pi^2|n|^2} \right). \end{aligned}$$

The terms in the first summation are $O(|n|^{-3})$ as $|n| \to +\infty$ and

$$\begin{aligned} \sum_{n\in\mathbb{Z}^2, n=(n_1,n_2)\neq 0} \frac{e^{\sqrt{-1}2\pi n\cdot x}}{4\pi^2|n|^2} &= \frac{1}{2\pi^2} \sum_{n_1=0}^{+\infty} \cos 2\pi n_1 x_1 \sum_{n_2=1}^{+\infty} \frac{\cos 2\pi n_2 x_2}{n_1^2 + n_2^2} \\ &\quad + \frac{1}{2\pi^2} \sum_{n_2=0}^{+\infty} \cos 2\pi n_2 x_2 \sum_{n_1=1}^{+\infty} \frac{\cos 2\pi n_1 x_1}{n_1^2 + n_2^2} \\ &:= A_1 + A_2. \end{aligned}$$

We have (see, for example, [**52**, pp. 54–55])

$$\begin{aligned} A_1 &= \frac{1}{24} - \frac{\ln 2}{4\pi} - \frac{1}{4}(x_2 - x_1) + \frac{1}{4}(2x_2^2 - x_1^2) - \frac{1}{8\pi} \ln\left(\sinh^2 \pi x_2 + \sin^2 \pi x_1 \right) \\ &\quad + \frac{1}{4\pi} \sum_{n_1=1}^{+\infty} \frac{\cos 2\pi n_1 x_1}{n_1} \frac{e^{2\pi n_1 x_2} + e^{-2\pi n_1 x_2}}{e^{2\pi n_1} - 1}, \end{aligned}$$

and

$$A_2 = \frac{1}{24} - \frac{\ln 2}{4\pi} - \frac{1}{4}(x_1 - x_2) + \frac{1}{4}(2x_1^2 - x_2^2) - \frac{1}{8\pi}\ln\Big(\sinh^2 \pi x_1 + \sin^2 \pi x_2\Big)$$
$$+\frac{1}{4\pi}\sum_{n_1=1}^{+\infty} \frac{\cos 2\pi n_1 x_2}{n_1}\frac{e^{2\pi n_1 x_1} + e^{-2\pi n_1 x_1}}{e^{2\pi n_1} - 1},$$

where the series in A_1 and A_2 are exponentially convergent.

This acceleration process can be continued further to speed up the convergence of the series

$$\sum_{n\in\mathbb{Z}^2, n\neq 0} e^{\sqrt{-1}2\pi n\cdot(x-y)}\left(\frac{1}{\omega^2 - |2\pi n + \alpha|^2} + \frac{1}{4\pi^2|n|^2}\right)$$

if we retain more terms in the expansion

$$\frac{1}{\omega^2 - |2\pi n + \alpha|^2} = -\frac{1}{4\pi^2|n|^2} + \frac{\alpha\cdot n}{4\pi^3|n|^4} + \ldots \quad \text{as } |n| \to +\infty.$$

2.13.2. Lattice Sums. The lattice sum representation of the Green's function is an immediate consequence of a separation of variables result for $H_0^{(1)}$. For $l \in \mathbb{Z}$, let $H_l^{(1)}$ denote the Hankel function of the first kind of order l and let J_l be the Bessel function of the first kind of order l. Recall that $J_l(x) = \Re H_l^{(1)}(x)$ and $H_{-l}^{(1)}(x) = (-1)^l H_l^{(1)}(x)$ for all $x \in \mathbb{R}$. By Graf's formula, we have for $n \neq 0$:

$$H_0^{(1)}(\omega|x - n - y|) = \sum_{l\in\mathbb{Z}} J_l(\omega|x-y|)e^{\sqrt{-1}l\theta_{x-y}} H_l^{(1)}(\omega|n|)e^{\sqrt{-1}l\theta_n}, \tag{2.300}$$

where θ_n and θ_{x-y} are given by

$$n_1 + \sqrt{-1}n_2 = |n|e^{\sqrt{-1}\theta_n}, \quad n = (n_1, n_2),$$
$$(x_1 - y_1) + \sqrt{-1}(x_2 - y_2) = |x-y|e^{\sqrt{-1}\theta_{x-y}}, \quad x = (x_1, x_2),\ y = (y_1, y_2).$$

Define the lattice sums S_l^α by

$$S_l^\alpha = \sum_{n\in\mathbb{Z}^2, n\neq 0} e^{\sqrt{-1}n\cdot\alpha} H_l^{(1)}(\omega|n|)e^{\sqrt{-1}l\theta_n}. \tag{2.301}$$

Note that for $\alpha = 0$, the four-fold symmetry of the square lattice implies that $S_l^0 = 0$ for l not divisible by four. Moreover,

$$S_{-l}^\alpha = (-1)^l S_l^{-\alpha}.$$

The Green's function $G^{\alpha,\omega}$ can then be expressed as

$$G^{\alpha,\omega}(x,y) = -\frac{\sqrt{-1}}{4}H_0^{(1)}(\omega|x-y|) - \frac{\sqrt{-1}}{4}\sum_{l\in\mathbb{Z}} S_l^\alpha J_l(\omega|x-y|)e^{\sqrt{-1}l\theta_{x-y}}.$$

Rearranging terms, we write

$$\begin{aligned} G^{\alpha,\omega}(x,y) &= -\frac{\sqrt{-1}}{4}\Bigg[H_0^{(1)}(\omega|x-y|) + S_0^\alpha J_0(\omega|x-y|) \\ &\quad + \sum_{l=1}^{+\infty}\Big(S_l^\alpha e^{\sqrt{-1}l\theta_{x-y}} + S_l^{-\alpha}e^{-\sqrt{-1}l\theta_{x-y}}\Big)J_l(\omega|x-y|)\Bigg]. \end{aligned} \tag{2.302}$$

In practice, the summation (2.302) is truncated for $l < L$, leading to an evaluation procedure whose cost is proportional to L times the number of evaluation points. This cost is significantly smaller than that necessary to obtain converged values of (2.289) [**373**].

It is worth emphasizing that the lattice sums (2.301) only have to be evaluated once as they do not depend on the position at which $G^{\alpha,\omega}$ is computed. However, the computation of S_l^α must be performed with care since S_l^α becomes very large and J_l very small as l increases. The convergence of the series in S_l^α can be improved if we use Kummer's transformation together with the asymptotic expansion

$$H_l^{(1)}(z) \approx \sqrt{\frac{2}{\pi z}} e^{\sqrt{-1}(z - l\pi/2 - \pi/4)} \qquad \text{as } z \to +\infty,$$

for $l \geq 0$.

2.13.3. Ewald's Method. Ewald's method was originally developed to treat long range electrostatic interactions in periodic structures [**103**]. The key idea behind Ewald's method is to split the periodic Green's function into spectral and spatial parts that, after some careful manipulation, both converge rapidly [**103**, **190**]. So our goal in this section is to determine representations for $G_{\text{spec}}^{\alpha,\omega}$ and $G_{\text{spat}}^{\alpha,\omega}$ such that the periodic Green's function

$$G^{\alpha,\omega}(x,y) = G_{\text{spec}}^{\alpha,\omega}(x,y) + G_{\text{spat}}^{\alpha,\omega}(x,y)$$

is exponentially convergent. We begin by determining an integral representation for the Hankel function of the first kind of order zero that is often used in the literature as the starting point for a derivation of the Ewald method applied to a specific spatial and array configuration [**269**].

LEMMA 2.93. *The Hankel function of the first kind of order zero can be represented as*

$$H_0^{(1)}(\omega r) = \frac{2}{\sqrt{-1}\pi} \int_\gamma t^{-1} \exp\left(- r^2 t^2 + \frac{\omega^2}{4t^2} \right) dt, \tag{2.303}$$

where γ is an integration path in the complex plane, shown in Figure 2.5, that begins at the origin with direction $e^{-\sqrt{-1}\pi/4}$, and goes to infinity in some direction $e^{\sqrt{-1}\phi}$, with $\phi \in (-\pi/4, \pi/4)$.

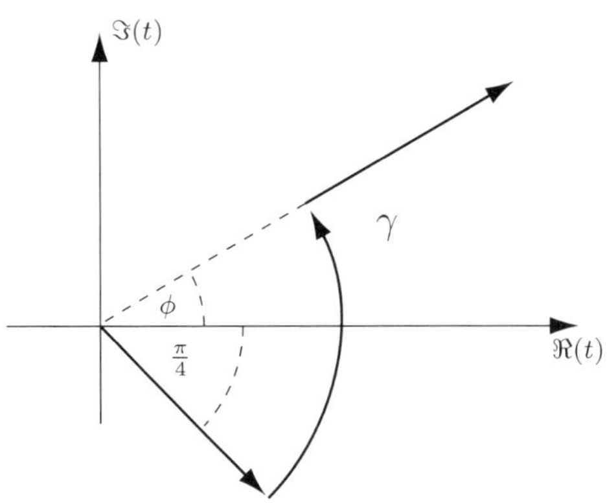

FIGURE 2.5. The integration path γ, in (2.303), that begins at the origin with direction $e^{-\sqrt{-1}\pi/4}$, and goes to infinity in some direction $e^{\sqrt{-1}\phi}$, with $\phi \in (-\pi/4, \pi/4)$.

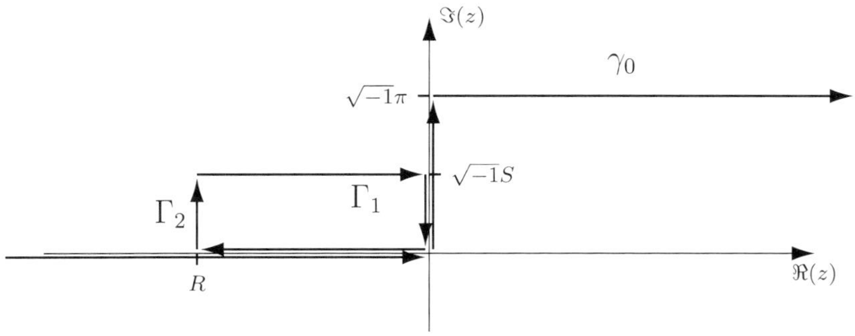

FIGURE 2.6. γ_0 is the path of integration taken in (2.304). Γ_1 and Γ_2 are specific paths of integration for which the integrals in (2.305) go towards zero as $R \to \infty$, and which cancel against γ_0.

PROOF. We have the following representation for the Hankel function of the first kind of order zero [**3**]:

$$H_0^{(1)}(x) = \frac{1}{\sqrt{-1}\pi} \int_{\gamma_0} e^{x \sinh z} dz, \qquad |\arg(x)| < \frac{\pi}{2}, \tag{2.304}$$

where the path of integration γ_0, which is shown in Figure 2.6, is given by

$$\gamma_0 = \{t : -\infty < t \leq 0\} \cup \{\sqrt{-1}t : 0 < t \leq \pi\} \cup \{t + \sqrt{-1}\pi : 0 < t < \infty\}.$$

We now define a separate path of integration for the same integrand. Let $R, S > 0$ and denote by

$$\begin{aligned} \Gamma_1 :=& \{-t : 0 \leq t \leq R\} \cup \{\sqrt{-1}t : 0 < t < S\} \cup \{-t + \sqrt{-1}S : 0 \leq t \leq R\}, \\ \Gamma_2 :=& \{-R + \sqrt{-1}t : 0 < t < S\}. \end{aligned}$$

These paths share the same starting point and end point, and as the integrand is holomorphic in z, by Cauchy's integral theorem, the integral over the contour is path independent. Therefore

$$\begin{aligned} \int_{\Gamma_1} e^{x \sinh z} dz &= \int_{\Gamma_2} e^{x \sinh z} dz \\ &= \int_0^S e^{x \sinh(-R+it)} dt. \end{aligned}$$

Suppose that $0 < \arg(x) < \pi/2$, $0 < S < \pi/2$, $t \in (0, S)$. Then the integral goes to 0 as R gets large because

$$\Re(x \sinh(-R + \sqrt{-1}t)) = -\Re(x) \cos(t) \sinh(R) - \Im(x) \sin(t) \cosh(R) < 0.$$

We have

$$\lim_{R\to\infty} \int_{\Gamma_1} e^{x \sinh z} dz = \lim_{R\to\infty} \int_{\Gamma_2} e^{x \sinh z} dz = 0. \tag{2.305}$$

We can combine the integrals on the paths Γ_1 and Γ_2 with the integral in (2.304) without changing its value, *i.e.*,

$$H_0^{(1)}(x) = \frac{1}{\sqrt{-1}\pi} \int_{\gamma_0} e^{x \sinh z} dz + \int_{\Gamma_1} e^{x \sinh z} dz + \int_{\Gamma_2} e^{x \sinh z} dz \quad \text{as } R \to \infty.$$

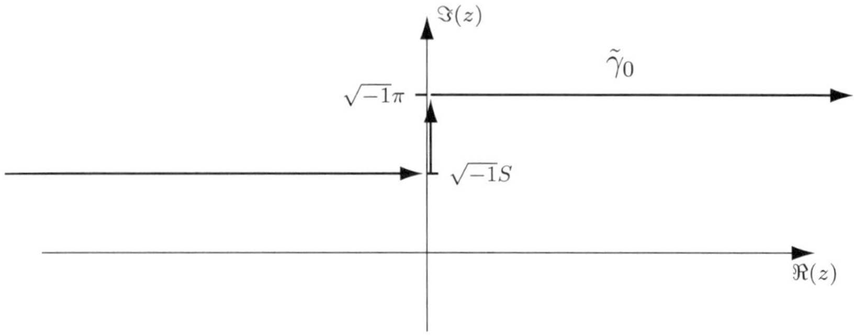

FIGURE 2.7. The integration path $\tilde{\gamma}_0$ in (2.306).

Choosing $S = \pi/2 - \arg(x)$ for $0 < \arg(x) < \pi/2$, noting that cancellation occurs due to the way the contours have been defined, and letting $R \to \infty$, we obtain the representation:

$$H_0^{(1)}(x) = \frac{1}{\sqrt{-1}\pi} \int_{\tilde{\gamma}_0} e^{x \sinh z} dz, \qquad \arg(x) < \frac{\pi}{2}, \tag{2.306}$$

where the integration path

$$\tilde{\gamma}_0 = \{t + \sqrt{-1}S : -\infty < t < 0\} \cup \{\sqrt{-1}t : S \leq t \leq \pi\} \cup \{t + \sqrt{-1}\pi : 0 < t < \infty\},$$

is shown in Figure 2.7. Rewriting this as

$$H_0^{(1)}(x) = \frac{1}{\sqrt{-1}\pi} \int_{\tilde{\gamma}_0} \exp\left(\frac{x}{2}(e^z - e^{-z})\right) dz,$$

and making the substitution $s = e^z$ gives

$$H_0^{(1)}(x) = \frac{1}{\sqrt{-1}\pi} \int_{\gamma_1} s^{-1} \exp\left(\frac{x}{2}\left(s - \frac{1}{s}\right)\right) ds,$$

where γ_1, shown in Figure 2.8, is a contour that begins at the origin with direction $e^{\sqrt{-1}(\pi/2 - \arg(x))}$, and sweeps around the origin to the point $s = -1$ before tending to minus infinity on the negative real axis.

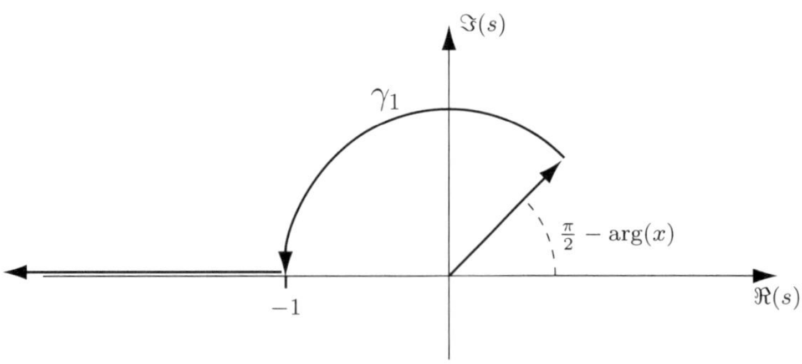

FIGURE 2.8. The integration path γ_1 in (2.306).

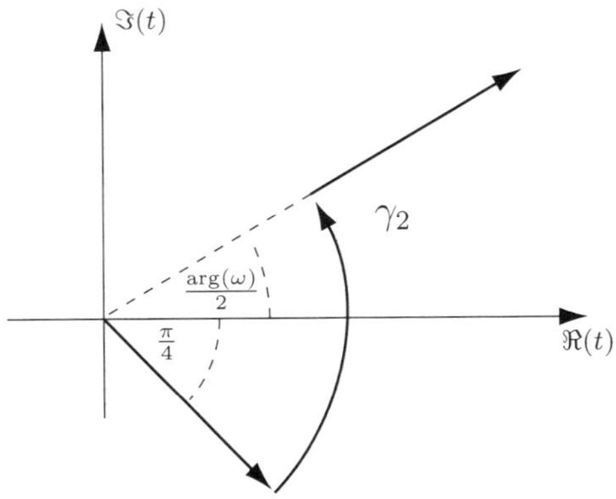

FIGURE 2.9. The integration path γ_2 in (2.307).

Setting $x = \omega r$ with $r > 0$, we obtain

$$H_0^{(1)}(\omega r) = \frac{1}{\sqrt{-1}\pi} \int_{\gamma_1} s^{-1} \exp\left(\frac{\omega r}{2}\left(s - \frac{1}{s}\right)\right) ds.$$

Making another substitution, this time with $s = -2rt^2/\omega$, we arrive at

$$H_0^{(1)}(\omega r) = \frac{2}{\sqrt{-1}\pi} \int_{\gamma_2} t^{-1} \exp\left(- r^2 t^2 + \frac{\omega^2}{4t^2}\right) dt, \tag{2.307}$$

where γ_2, shown in Figure 2.9, is an integration path in the complex plane that begins at the origin with direction $e^{-\sqrt{-1}\pi/4}$, follows the arc $|t| = \sqrt{|\omega|/(2r)}$ until the point $\sqrt{\omega/(2r)}e^{\sqrt{-1}(\arg(\omega))/2}$, and finally goes to infinity in the direction $e^{\sqrt{-1}(\arg(\omega))/2}$. The path of integration can be altered as long as (i) it begins at the origin with direction $e^{-\sqrt{-1}\pi/4}$, which ensures convergence as $|t| \to 0$, and (ii) it goes to infinity in the direction $e^{\sqrt{-1}\phi}$, with $\phi \in (-\pi/4, \pi/4)$, which ensures convergence as $|t| \to \infty$. So we have (2.303) with $\gamma = \gamma_2$.

□

By Lemma 2.93 we have

$$-\frac{\sqrt{-1}}{4} H_0^{(1)}(\omega r) = -\frac{1}{2\pi} \int_\gamma \frac{e^{-r^2 t^2 + \frac{\omega^2}{4t^2}}}{t} dt,$$

and then recalling the definition of the quasi-periodic Green's function

$$G^{\alpha,\omega}(x, y) = -\frac{\sqrt{-1}}{4} \sum_{m\in\mathbb{Z}} H_0^{(1)}(\omega|x - y - (m, 0)|)e^{\sqrt{-1}m\alpha},$$

we obtain

$$G^{\alpha,\omega}(x, y) = -\frac{1}{2\pi} \sum_{m\in\mathbb{Z}} e^{\sqrt{-1}m\alpha} \int_\gamma \frac{e^{-R_m^2 t^2 + \frac{\omega^2}{4t^2}}}{t} dt, \tag{2.308}$$

where $R_m = \sqrt{(x_2 - y_2) + (x_1 - y_1 - m)^2}$.

Let $\mathcal{E}$ be a point on the positive real axis, and let γ_1 be a contour starting from 0 following the ray $e^{-\sqrt{-1}\frac{\pi}{4}}$, then following the arc $\gamma = \mathcal{E}$ until the point $\mathcal{E}$. Let γ_2 be the contour starting at $\mathcal{E}$ following the arc $r = \mathcal{E}$ until $\mathcal{E}e^{\sqrt{-1}\phi}$, with

$\phi \in (-\frac{\pi}{4}, \frac{\pi}{4})$, and then following the ray $e^{\sqrt{-1}\phi}$ to infinity. Then the integral $\int_\gamma$ in (2.308) is equivalent to $\int_{\gamma_1} + \int_{\gamma_2}$.

LEMMA 2.94. *Consider a lossy medium such that $\Im(\omega) > 0$. Then the quasi-periodic Green's function $G^{\alpha,\omega}$ can be split into two parts such that*

$$G^{\alpha,\omega}(x,y) = G^{\alpha,\omega}_{\mathrm{spec}}(x,y) + G^{\alpha,\omega}_{\mathrm{spat}}(x,y),$$

with

$$\begin{aligned} G^{\alpha,\omega}_{\mathrm{spec}}(x,y) = &-\frac{1}{4}\sum_{p\in\mathbb{Z}} \frac{e^{-\sqrt{-1}\omega_{xp}(x_1-y_1)}}{\sqrt{-1}\omega_{yp}} \\ &\times \left[e^{\sqrt{-1}\omega_{yp}|x_2-y_2|}\mathrm{erfc}\left(\frac{\sqrt{-1}\omega_{yp}}{2\mathcal{E}} + |x_2-y_2|\mathcal{E}\right) \right. \\ &\left. + e^{-\sqrt{-1}\omega_{yp}|x_2-y_2|}\mathrm{erfc}\left(\frac{\sqrt{-1}\omega_{yp}}{2\mathcal{E}} - |x_2-y_2|\mathcal{E}\right)\right], \end{aligned}$$

$$G^{\alpha,\omega}_{\mathrm{spat}}(x,y) = -\frac{1}{4\pi}\sum_{m\in\mathbb{Z}} e^{\sqrt{-1}\alpha m}\sum_{q=0}^{\infty}\left(\frac{\omega}{2\mathcal{E}}\right)^{2q}\frac{1}{q!}E_{q+1}(R_m^2\mathcal{E}^2),$$

where $\omega_{xp} = -\alpha + \frac{2\pi p}{d}$, $\omega_{yp} = -\sqrt{\omega^2 - \omega_{xp}^2}$, $\mathrm{erfc}(z)$ is the complementary error function

$$\mathrm{erfc}(z) = \frac{2}{\sqrt{\pi}}\int_z^\infty e^{-t^2}\,dt,$$

and E_q is the qth order exponential integral which is defined as

$$E_q(z) = \int_1^\infty \frac{e^{-zt}}{t^q}\,dt.$$

PROOF. We first split (2.308) into two parts and define

$$G^{\alpha,\omega}_{\mathrm{spec}}(x,y) = -\frac{1}{2\pi}\sum_{m\in\mathbb{Z}} e^{\sqrt{-1}m\alpha}\int_{\gamma_1}\frac{e^{-R_m^2 s^2 + \frac{\omega^2}{4s^2}}}{s}\,ds, \tag{2.309}$$

and

$$G^{\alpha,\omega}_{\mathrm{spat}}(x,y) = -\frac{1}{2\pi}\sum_{m\in\mathbb{Z}} e^{\sqrt{-1}m\alpha}\int_{\gamma_2}\frac{e^{-R_m^2 s^2 + \frac{\omega^2}{4s^2}}}{s}\,ds, \tag{2.310}$$

where γ_1 and γ_2 are the complex paths of integration defined previously. Note that the convergence of $G^{\alpha,\omega}_{\mathrm{spat}}$ is already exponential as for large m it can be shown that the terms in the series behave like $e^{-n^2\mathcal{E}^2}/(n^2\mathcal{E}^2)$. The terms in $G^{\alpha,\omega}_{\mathrm{spec}}$ on the other hand decay like $1/\sqrt{m}$ due to the asymptotic behavior of $H_0^{(1)}(z)$ for large z. This term is the one we would like to accelerate.

We use the Poisson summation formula

$$\sum_{m\in\mathbb{Z}} f(m) = \frac{1}{d}\sum_{p\in\mathbb{Z}}\hat{f}(2\pi p), \tag{2.311}$$

where $\hat{f}(2\pi p)$ is the Fourier coefficient, namely,

$$\hat{f}(\beta) = \int_{-\infty}^{\infty} f(\xi)e^{-\sqrt{-1}\beta\xi}\,d\xi.$$

Let $f(m)$ be

$$f(m) = -\frac{e^{\sqrt{-1}\alpha m}}{2\pi} \int_{\gamma_1} \frac{e^{-[(x_2-y_2)^2+(x_1-y_1-m)^2]s^2+\frac{k^2}{4s^2}}}{s} ds,$$

so that

$$\hat{f}(2\pi p) = -\frac{1}{2\pi} \int_{-\infty}^{\infty} d\xi \int_{\gamma_1} ds \frac{e^{-[(x_2-y_2)^2+(x_1-y_1-\xi)^2]s^2+\frac{k^2}{4s^2}}}{s} e^{-\sqrt{-1}\omega_{xp}\xi},$$

where $\omega_{xp} = -\alpha + 2\pi p$. Noting that the s integral is convergent on the path γ_1 for $\mathcal{E} \in]-\infty, \infty[$ as $\Re(s^2) > 0$, we can switch the order of integration. Then applying the formula $\int_{-\infty}^{\infty} e^{-a\xi^2+b\xi} d\xi = \sqrt{\pi/a} e^{b^2/4a}$ results in

$$\hat{f}(2\pi p) = -\frac{e^{-\sqrt{-1}\omega_{xp}(x_1-y_1)}}{2\sqrt{\pi}} \int_{\gamma_1} \frac{e^{-(x_2-y_2)^2 s^2} e^{\omega_{yp}^2/4s^2}}{s^2} ds,$$

where $\omega_{yp} = -\sqrt{\omega^2 - \omega_{xp}^2}$ and we have taken the negative of the square root in order to ensure convergence. Making the change of variables $\tilde{s} = 1/s$ we have

$$\hat{f}(2\pi p) = -\frac{e^{-\sqrt{-1}\omega_{xp}(x_1-y_1)}}{2\sqrt{\pi}} \int_{\tilde{\gamma}_1} e^{-(x_2-y_2)^2/\tilde{s}^2} e^{(\omega_{yp}^2 \tilde{s}^2)/4} d\tilde{s},$$

and the path of integration is mapped from γ_1 onto $\tilde{\gamma}_1$. Note that since γ_1 is of the form $te^{-\sqrt{-1}\frac{\pi}{4}}$, $\tilde{\gamma}_1$ is of the form $t^{-1}e^{\sqrt{-1}\frac{\pi}{4}}$ near ∞. That is, $\Re(\omega_{zp}^2 \tilde{s}^2) < 0$ for every $p \in \mathbb{Z}$, ensuring convergence. Finally, using the identity

$$\int e^{a^2x^2 - \frac{b^2}{x^2}} dx = -\frac{\sqrt{\pi}}{4a} \left[e^{2ab} \operatorname{erfc}(ax + \frac{b}{x}) + e^{-2ab} \operatorname{erfc}(ax - \frac{b}{x}) \right] + \text{const},$$

we obtain

$$\hat{f}(2\pi p) = -\frac{e^{-\sqrt{-1}\omega_{xp}(x_1-y_1)}}{4\sqrt{-1}\omega_{yp}} \tag{2.312}$$

$$\times \Bigg[e^{\sqrt{-1}\omega_{yp}|x_2-y_2|} \operatorname{erfc}\left(\frac{\sqrt{-1}\omega_{yp}}{2\mathcal{E}} + |x_2 - y_2|\mathcal{E} \right) \tag{2.313}$$

$$+ e^{-\sqrt{-1}\omega_{yp}|x_2-y_2|} \operatorname{erfc}\left(\frac{\sqrt{-1}\omega_{yp}}{2\mathcal{E}} - |x_2 - y_2|\mathcal{E} \right) \Bigg]. \tag{2.314}$$

Inserting this into (2.311) gives us $G_{\text{spec}}^{\alpha,\omega}$.

Now we turn to $G_{\text{spat}}^{\alpha,\omega}$. Although this function is already exponentially convergent we will transform it into a form more suitable for computation. Consider the integral I in (2.310), with ϕ set to 0:

$$I = \int_{\mathcal{E}}^{\infty} \frac{e^{-R_m^2 t^2 + \frac{\omega^2}{4t^2}}}{s} ds.$$

The contours γ_1 and γ_2 when $\phi = 0$ are shown in Figure 2.10. It can be shown that after changing variables with $u = s^2$, applying the Taylor expansion $e^{\frac{\omega^2}{4u}} = \sum_{q=0}^{\infty} (\frac{\omega}{2})^{2q}/(q! u^q)$, and then changing variables again with $t = u/\mathcal{E}^2$ we have

$$I = \frac{1}{2} \sum_{q=0}^{\infty} \left(\frac{\omega}{2\mathcal{E}} \right)^{2q} \frac{1}{q!} E_{q+1}(R_m^2 \mathcal{E}^2).$$

Using this representation of I in (2.310) gives us the desired form of $G_{\text{spat}}^{\alpha,\omega}$. □

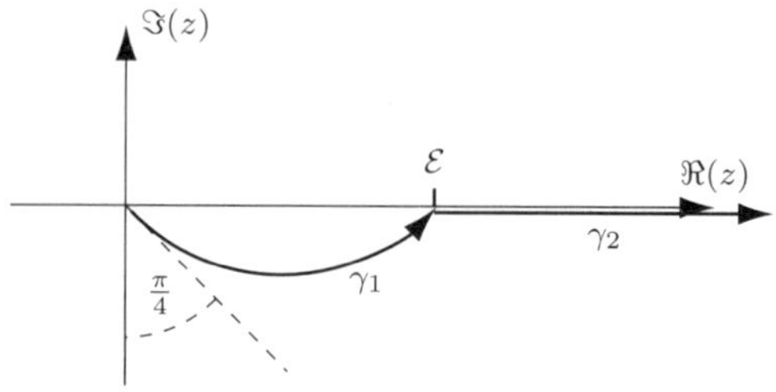

FIGURE 2.10. The complex paths of integration γ_1 and γ_2 when $\phi = 0$.

The complementary error function converges very quickly, and this is the key to the acceleration of the convergence speed of $G^{\alpha,\omega}_{\text{spec}}$. This representation of $G^{\alpha,\omega}_{\text{spec}}$ is also efficient in terms of numerical computation as only $E_1(z)$ needs to be evaluated explicitly. The higher order exponential integral terms can be computed with the recurrence relation $E_{q+1}(z) = \frac{1}{q}(e^{-z} - zE_q(z))$ for $q = 1, 2, \ldots$. Note that the optimal value of the splitting parameter $\mathcal{E}$ for wavelengths somewhat larger or smaller than the periodicity is given by $\mathcal{E} = \sqrt{\pi}/d$. It is also worth mentioning that very few terms are required in the summations in $G^{\alpha,\omega}_{\text{spec}}$ and $G^{\alpha,\omega}_{\text{spat}}$ to obtain a relative error of less than $1e{-}03$. Furthermore, although we assumed that $\Im(\omega) > 0$ in order to obtain these expressions, due to analytic continuation the expressions actually hold for all $\omega \in \mathbb{C}$.

For the quasi-periodic Neumann–Poincairé operator we need the gradient of the quasi-periodic Green's function. We note that

$$\nabla G^{\alpha,\omega}(x,y) = \nabla G^{\alpha,\omega}_{\text{spec}}(x,y) + \nabla G^{\alpha,\omega}_{\text{spat}}(x,y),$$

with

$$\begin{aligned}
\nabla G^{\alpha,\omega}_{\text{spec}}(x,y) = -\frac{1}{4}\sum_{p\in\mathbb{Z}} & \frac{e^{-\sqrt{-1}\omega_{xp}(x_1-y_1)}}{\sqrt{-1}\omega_{yp}} \\
& \Big\{[-\sqrt{-1}\hat{x}\omega_{xp} - \sqrt{-1}\hat{y}\omega_{yp}\operatorname{sgn}(x_2-y_2)] \\
& \times e^{-\sqrt{-1}\omega_{yp}|x_2-y_2|}\operatorname{erfc}\left(\frac{\sqrt{-1}\omega_{yp}}{2\mathcal{E}} - |x_2-y_2|\mathcal{E}\right) \\
& \times [-\sqrt{-1}\hat{x}\omega_{xp} + \sqrt{-1}\hat{y}\omega_{yp}\operatorname{sgn}(x_2-y_2)] \\
& \times e^{\sqrt{-1}\omega_{yp}|x_2-y_2|}\operatorname{erfc}\left(\frac{\sqrt{-1}\omega_{yp}}{2\mathcal{E}} + |x_2-y_2|\mathcal{E}\right) \\
& - \hat{z}\operatorname{sgn}(x_2-y_2)\mathcal{E}e^{-\sqrt{-1}\omega_{yp}|x_2-y_2|} \\
& \times \operatorname{erfc}'\left(\frac{\sqrt{-1}\omega_{yp}}{2\mathcal{E}} - |x_2-y_2|\mathcal{E}\right) \\
& + \hat{z}\operatorname{sgn}(x_2-y_2)\mathcal{E}e^{\sqrt{-1}\omega_{yp}|x_2-y_2|} \\
& \times \operatorname{erfc}'\left(\frac{\sqrt{-1}\omega_{yp}}{2\mathcal{E}} + |x_2-y_2|\mathcal{E}\right)\Big\},
\end{aligned}$$

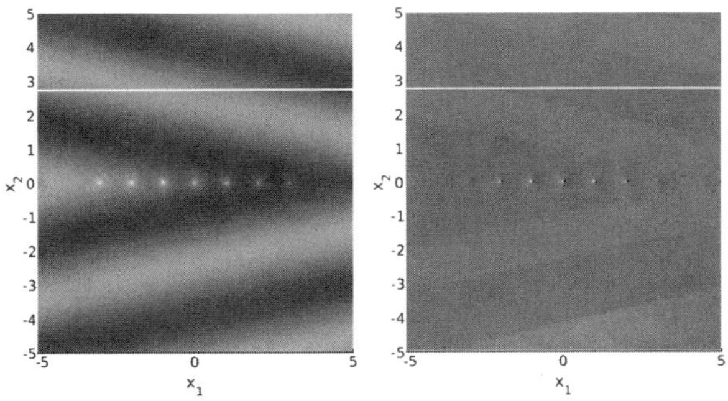

(a) $\Re(G^{\alpha,\omega})$ (b) $\Re(\frac{\partial}{\partial x_2}G^{\alpha,\omega})$

FIGURE 2.11. The quasi-periodic Green's function, and the x_2 component of its gradient, for the Helmholtz equation for a one-dimensional lattice of Dirac mass source points with periodicity 1. The quasi-momentum parameter α is set to $\pi/8$.

$$\begin{aligned}\nabla G^{\alpha,\omega}_{\text{spat}}(x,y) = \frac{\mathcal{E}^2}{2\pi}\sum_{m\in\mathbb{Z}}\Big[\hat{x}(x_1-y_1-m)+\hat{z}(x_2-y_2)\Big]e^{\sqrt{-1}\alpha m}\\ \times\sum_{q=0}^{\infty}\left(\frac{\omega}{2\mathcal{E}}\right)^{2q}\frac{1}{q!}E_q(R_m^2\mathcal{E}^2),\end{aligned}$$

where $\hat{x}$ and $\hat{y}$ are unit vectors along the x and y axes, respectively, and

$$\operatorname{erfc}(z)' = -\frac{2}{\sqrt{\pi}}e^{-z^2}.$$

Figure 2.11 shows the quasi-periodic Green's function obtained by using Ewald's method. The corresponding Matlab code can be downloaded at `www.ams.org/bookpages/surv-235`.

2.13.4. Numerical Implementation of the Operators $\mathcal{S}^{\alpha,\omega}$ and $(\mathcal{K}^{-\alpha,\omega})^*$.

In this section we discuss the numerical implementation of $\mathcal{S}^{\alpha,\omega}$ and $(\mathcal{K}^{-\alpha,\omega})^*$ assuming we are in a low frequency regime. After performing the usual boundary discretization procedure, as described in Subsection 2.4.5, we represent the infinite dimensional operator $\mathcal{S}^{\alpha,\omega}$ acting on the density φ by a finite dimensional matrix S acting on the coefficient vector $\overline{\varphi}_i := \varphi(x^{(i)})$ for $1 \leq i \leq N$. That is

$$\mathcal{S}^{\alpha,\omega}[\varphi](x) \quad = \int_{\partial\Omega} G^{\alpha,\omega}(x,y)\varphi(y)\,d\sigma(y),$$

for $\psi \in L^2(\partial\Omega)$, is represented numerically as

$$S\tilde{\psi} = \begin{pmatrix} S_{11} & S_{12} & \dots & S_{1N}\\ S_{21} & S_{22} & \dots & S_{2N}\\ \vdots & & \ddots & \vdots\\ S_{N1} & \dots & \dots & S_{NN}\end{pmatrix}\begin{pmatrix}\overline{\varphi}_1\\ \overline{\varphi}_2\\ \vdots\\ \overline{\varphi}_N\end{pmatrix},$$

where

$$S_{ij} = G^{\alpha,\omega}(x^{(i)} - x^{(j)})|T(x^{(j)})|(t_{j+1}-t_j), \quad i \neq j,$$

and $G^{\alpha,\omega}(x^{(i)}-x^{(j)})$ refers to the Ewald representation of the Green's function. This discrerization matrix S features singularities in the diagonal terms and therefore we must approximate these terms by explicit calculation. Let the portion of the boundary starting at $x^{(i)}$ and ending at $x^{(i+1)}$ be parameterized by $s \in [0, \varepsilon = \frac{2\pi}{N})$ and note that $\varepsilon \to 0$ as the number of discretization points $N \to \infty$. Observe that for $G^{\alpha,\omega} = G^{\alpha,\omega}_{\text{spec}} + G^{\alpha,\omega}_{\text{spat}}$ the singularity appears in the $G^{\alpha,\omega}_{\text{spat}}$ term precisely when $x = y$ and $m = 0$. Therefore

$$S_{ii} = \int_0^\varepsilon G^{\alpha,\omega}(x^{(i)} - x(s))|T(s)|ds \approx \int_0^\varepsilon G^{\alpha,\omega}_{\text{spat}}(x^{(i)} - x(s))|T(s)|ds,$$

as $\varepsilon \to 0$. Now retaining only the $m = 0$ term in $G^{\alpha,\omega}_{\text{spat}}$ we have

$$G^{\alpha,\omega}_{\text{spat}} \approx -\frac{1}{4\pi}\sum_{q=0}^{\infty}\left(\frac{\omega}{2\mathcal{E}}\right)^{2q}\frac{1}{q!}E_{q+1}(R_0^2\mathcal{E}^2),$$

where $R_0 = \sqrt{(x_1^{(i)} - x_1(s))^2 + (x_2^{(i)} - x_2(s))^2}$. Noting that the behavior of the exponential integrals E_{q+1} for small argument is $E_{q+1}(z) = -(-z)^q(\ln z)/q!$ gives

$$\begin{aligned}
G^{\alpha,\omega}_{\text{spat}} &\approx -\frac{1}{4\pi}\sum_{q=0}^{\infty}\left(\frac{\omega}{2\mathcal{E}}\right)^{2q}\frac{1}{q!}\left(-\frac{(-R_0^2\mathcal{E}^2)^q}{q!}\ln(R_0^2\mathcal{E}^2)\right)\\
&\approx \frac{1}{2\pi}\ln(R_0\mathcal{E}),\\
&\approx \frac{1}{2\pi}\ln(R_0),
\end{aligned}$$

where only the $q = 0$ term has been retained as $R_0 \ll 1$. Therefore,

$$\begin{aligned}
S_{ii} &\approx \frac{1}{2\pi}\int_0^\varepsilon \ln(|x^{(i)} - x(s)|)|T(s)|ds\\
&= \frac{|T(0)|\varepsilon}{2\pi}\Big(\ln(|T(0)|\varepsilon) - 1\Big)\\
&= \frac{|T(0)|}{N}\left(\ln\left(\frac{2\pi}{N}|T(0)|\right) - 1\right).
\end{aligned}$$

The discretization matrix K for the quasi-periodic Neumann–Poincaré operator $(\mathcal{K}^{-\alpha,\omega})^*$ requires no special treatment since, similarly to Subsection 2.6.4 it is clear that it features the same singularity as the non-periodic Neumann–Poincaré operator and thus the usual expression (2.138) holds for the diagonal terms of its corresponding discretized matrix. We remark that the approximations used for the diagonal terms of S and K are appropriate for low frequencies but are not stable when the frequency is high. For instance, the $q \neq 0$ terms provide a non-negligible contribution to $G^{\alpha,\omega}_{\text{spat}}$ when ω is high and cannot be ignored. Matlab code for computing the Ewald's method representations of $\mathcal{S}^{\alpha,\omega}$ and $(\mathcal{K}^{-\alpha,\omega})^*$ in low frequency regimes can be downloaded at `www.ams.org/bookpages/surv-235`.

2.13.5. Ewald's Representation of the Quasi-Biperiodic Green's Function for the Helmholtz Equation. The quasi-biperiodic Green's function, which was defined in Section 2.12, satisfies

$$(\Delta + k^2)G^{\alpha,\omega}(x,y) = \sum_{m\in\mathbb{Z}^2} \delta_0(x-y-m)e^{\sqrt{-1}m\cdot\alpha}. \tag{2.315}$$

This Green's function has the representation

$$G^{\alpha,\omega}(x,y) = -\frac{\sqrt{-1}}{4}\sum_{m\in\mathbb{Z}^2} H_0^{(1)}(\omega R_m)e^{\sqrt{-1}m\cdot\alpha}, \tag{2.316}$$

where $R_m = \sqrt{(x_1-y_1-m_1)^2+(x_2-y_2-m_2)^2}$. Through an analogous procedure to the one used in Section 2.13.3 for the quasi-periodic Green's function, it can be shown that there exists a rapidly converging Ewald representation [**440**] of the quasi-biperiodic Green's function such that

$$G^{\alpha,\omega}(x,y) = G^{\alpha,\omega}_{\text{spec}}(x,y) + G^{\alpha,\omega}_{\text{spat}}(x,y),$$

with

$$G^{\alpha,\omega}_{\text{spec}}(x,y) = -\sum_{p,q\in\mathbb{Z}} \frac{1}{\gamma_{pq}^2}e^{-\gamma_{pq}^2/4\mathcal{E}}e^{-\sqrt{-1}\omega_{pq}\cdot(x-y)},$$

and

$$G^{\alpha,\omega}_{\text{spat}}(x,y) = -\frac{1}{4\pi}\sum_{m\in\mathbb{Z}^2} e^{\sqrt{-1}\alpha\cdot m}\sum_{q=0}^{\infty}\left(\frac{\omega}{2\sqrt{\mathcal{E}}}\right)^{2q}\frac{1}{q!}E_{q+1}(R_m^2\mathcal{E}),$$

where

$$\gamma_{pq} = \sqrt{|\omega_{pq}^2-\omega^2|}, \omega_{pq} = \omega_{xp}\hat{x}+\omega_{yq}\hat{y}, \omega_{xp} = -\alpha_1+2\pi p, \omega_{yq} = -\alpha_2+2\pi q.$$

Taking the gradient of $G^{\alpha,\omega}(x,y)$ gives us the representation required for the quasi-biperiodic Neumann–Poincaré operator. We have

$$\nabla G^{\alpha,\omega}(x,y) = \nabla G^{\alpha,\omega}_{\text{spec}}(x,y) + \nabla G^{\alpha,\omega}_{\text{spat}}(x,y),$$

with

$$\nabla G^{\alpha,\omega}_{\text{spec}}(x,y) = \sqrt{-1}\sum_{p,q\in\mathbb{Z}} \frac{\omega_{pq}}{\gamma_{pq}^2}e^{-\gamma_{pq}^2/4\mathcal{E}}e^{-\sqrt{-1}\omega_{pq}\cdot(x-y)},$$

and

$$\nabla G^{\alpha,\omega}_{\text{spat}}(x,y) = \frac{\mathcal{E}}{2\pi}\sum_{m\in\mathbb{Z}^2}^{\infty}(x-y-\hat{m})e^{\sqrt{-1}\alpha\cdot\hat{m}} \times \sum_{q=0}^{\infty}\left(\frac{\omega}{2\sqrt{\mathcal{E}}}\right)^{2q}\frac{1}{q!}E_q(R_m^2\mathcal{E}).$$

The quasi-biperiodic Green's function for the Helmholtz equation is shown in Figure 2.12. The corresponding Matlab code can be downloaded at `www.ams.org/bookpages/surv-235`.

REMARK 2.95. *The quasi-biperiodic Green's function $G^{\alpha,0}$ for the Laplace equation features infinite series that are very slow to converge. In order to utilize Ewald's method and accelerate the convergence we make use of Lemma* 2.91. *We already have a Ewald representation corresponding to $G^{\alpha,0}$ for any α in the Brillouin zone $[0,2\pi)^2$ and the infinite series in* (2.295) *and* (2.296) *are relatively quick to converge. Therefore this representation of these Green's functions is appropriate for efficient numerical implementation.*

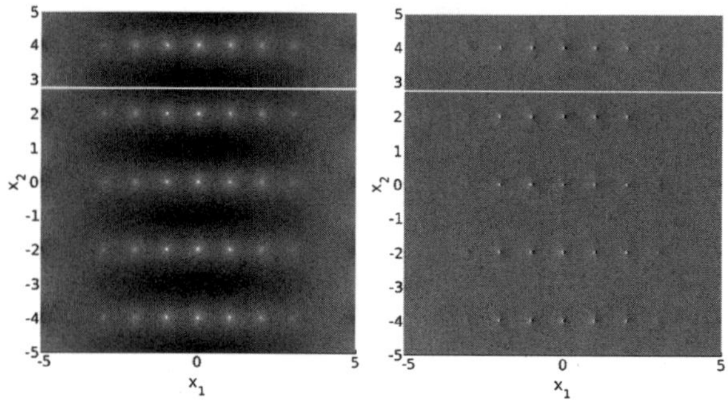

(a) $\Re(G^{\alpha,\omega})$ (b) $\Re(\frac{\partial}{\partial x_1}G^{\alpha,\omega})$

FIGURE 2.12. The quasi-biperiodic Green's function, and the x_1 component of its gradient, for the Helmholtz equation for a two-dimensional lattice of Dirac mass source points with periodicity 1 in the x_1-direction and 2 in the x_2-direction. The quasi-momentum parameter α is set to $(\pi/8, 0)$.

2.14. Integral Representation of Solutions to the Full Maxwell Equations

In this section, a few fundamental results related to electromagnetic scattering, which will be essential in what follows, are recalled.

2.14.1. Layer Potentials. Assume that D is bounded, simply connected, and of class $\mathcal{C}^{1,\eta}$ for $\eta > 0$ and let

$$H_T^s(\partial D) = \left\{\varphi \in \left(H^s(\partial D)\right)^3,\ \nu \cdot \varphi = 0\right\}$$

for $s = \pm 1/2$.

We introduce the surface gradient, surface divergence and Laplace-Beltrami operator and denote them by $\nabla_{\partial D}$, $\nabla_{\partial D}\cdot$ and $\Delta_{\partial D}$, respectively. We define the vectorial and scalar surface curl by $\vec{\operatorname{curl}}_{\partial D}\varphi = -\nu \times \nabla_{\partial D}\varphi$ for $\varphi \in H^{\frac{1}{2}}(\partial D)$ and $\operatorname{curl}_{\partial D}\varphi = -\nabla_{\partial D} \cdot (\nu \times \varphi)$ for $\varphi \in H_T^{\frac{1}{2}}(\partial D)$, respectively. We recall that

$$\begin{aligned}
\nabla_{\partial D} \cdot \nabla_{\partial D} &= \Delta_{\partial D}, \\
\operatorname{curl}_{\partial D}\vec{\operatorname{curl}}_{\partial D} &= -\Delta_{\partial D}, \\
\vec{\operatorname{curl}}_{\partial D}\operatorname{curl}_{\partial D} &= -\Delta_{\partial D} + \nabla_{\partial D}\nabla_{\partial D}\cdot, \\
\nabla_{\partial D} \cdot \vec{\operatorname{curl}}_{\partial D} &= 0, \\
\operatorname{curl}_{\partial D}\nabla_{\partial D} &= 0.
\end{aligned}$$

We introduce the following functional space:

$$H_T^{-\frac{1}{2}}(\operatorname{div}, \partial D) = \left\{\varphi \in H_T^{-\frac{1}{2}}(\partial D),\ \nabla_{\partial D} \cdot \varphi \in H^{-\frac{1}{2}}(\partial D)\right\}.$$

We define the following boundary integral operators and refer to [**52**, **411**] for their mapping properties:

$$\begin{aligned} \vec{\mathcal{S}}_D^k[\varphi] &: H_T^{-\frac{1}{2}}(\partial D) \longrightarrow H_T^{\frac{1}{2}}(\partial D) \text{ or } H^1_{\mathrm{loc}}(\mathbb{R}^3)^3 \\ \varphi &\longmapsto \vec{\mathcal{S}}_D^k[\varphi](x) = \int_{\partial D} \Gamma_k(x-y)\varphi(y)d\sigma(y); \end{aligned} \tag{2.317}$$

$$\begin{aligned} \mathcal{M}_D^k[\varphi] &: H_T^{-\frac{1}{2}}(\mathrm{div}, \partial D) \longrightarrow H_T^{-\frac{1}{2}}(\mathrm{div}, \partial D) \\ \varphi &\longmapsto \mathcal{M}_D^k[\varphi](x) = \int_{\partial D} \nu(x) \times \nabla_x \times \big(\Gamma_k(x-y)\varphi(y)\big)\, d\sigma(y); \end{aligned} \tag{2.318}$$

$$\begin{aligned} \mathcal{L}_D^k[\varphi] &: H_T^{-\frac{1}{2}}(\mathrm{div}, \partial D) \longrightarrow H_T^{-\frac{1}{2}}(\mathrm{div}, \partial D) \\ \varphi &\longmapsto \mathcal{L}_D^k[\varphi](x) = \nu(x) \times \Big(k^2 \vec{\mathcal{S}}_D^k[\varphi](x) + \nabla \mathcal{S}_D^k[\nabla_{\partial D} \cdot \varphi](x)\Big). \end{aligned} \tag{2.319}$$

The following results hold.

LEMMA 2.96. *The operator $\vec{\mathcal{S}}_D^k$ satisfies the following jump formulas on ∂D:*

$$\Big(\nu \times \nabla \times \vec{\mathcal{S}}_D^k[\varphi]\Big)\Big|_{\pm} = (\mp\frac{1}{2}I + \mathcal{M}_D^k)[\varphi],$$

and

$$\Big(\nu \times \nabla \times \nabla \times \vec{\mathcal{S}}_D^k[\varphi]\Big)\Big|_{\partial D} = \mathcal{L}_D^k[\varphi],$$

for $\varphi \in H_T^{-\frac{1}{2}}(\mathrm{div}, \partial D)$.

We will need the following lemma.

LEMMA 2.97. *The following Helmholtz decomposition holds* [**180**]:

$$H_T^{-\frac{1}{2}}(\mathrm{div}, \partial D) = \nabla_{\partial D} H^{\frac{3}{2}}(\partial D) \oplus \vec{\mathrm{curl}}_{\partial D} H^{\frac{1}{2}}(\partial D).$$

REMARK 2.98. *The Laplace-Beltrami operator $\Delta_{\partial D} : H_0^{\frac{3}{2}}(\partial D) \to H_0^{-\frac{1}{2}}(\partial D)$ is invertible. Here $H_0^{\frac{3}{2}}(\partial D)$ and $H_0^{-\frac{1}{2}}(\partial D)$ are the zero mean subspaces of $H^{\frac{3}{2}}(\partial D)$ and $H^{-\frac{1}{2}}(\partial D)$, respectively.*

The following results on the operator $\mathcal{M}_D^0$ are of great importance in the analysis of plasmonic resonances for nanoparticles. We refer to [**411**] for a proof of the following compactness property of $\mathcal{M}_D^0$.

LEMMA 2.99. *The operator $\mathcal{M}_D^0 : H_T^{-\frac{1}{2}}(\mathrm{div}, \partial D) \longrightarrow H_T^{-\frac{1}{2}}(\mathrm{div}, \partial D)$ is a compact operator.*

LEMMA 2.100. *The following identities hold* [**37**, **275**]:

$$\begin{aligned} \mathcal{M}_D^0[\vec{\mathrm{curl}}_{\partial D}\varphi] &= \vec{\mathrm{curl}}_{\partial D}\mathcal{K}_D^0[\varphi], \quad \forall \varphi \in H^{\frac{1}{2}}(\partial D), \\ \mathcal{M}_D^0[\nabla_{\partial D}\varphi] &= -\nabla_{\partial D}\Delta_{\partial D}^{-1}(\mathcal{K}_D^0)^*[\Delta_{\partial D}\varphi] + \vec{\mathrm{curl}}_{\partial D}\mathcal{R}_D[\varphi], \quad \forall \varphi \in H^{\frac{3}{2}}(\partial D), \end{aligned}$$

where $(\mathcal{K}_D^0)^*$ is the Neumann-Poincaré operator associated with the Laplace operator, and

$$\mathcal{R}_D = -\Delta_{\partial D}^{-1}\mathrm{curl}_{\partial D}\mathcal{M}_D^0\nabla_{\partial D}. \tag{2.320}$$

We now consider solving the problem

$$(\lambda I - \mathcal{M}_D^0)\,[\psi] = \varphi \tag{2.321}$$

for $(\psi, \varphi) \in \big(H_T^{-\frac{1}{2}}(\mathrm{div}, \partial D)\big)^2$ and $\lambda \notin \sigma(\mathcal{M}_D^0)$, where $\sigma(\mathcal{M}_D^0)$ is the spectrum of $\mathcal{M}_D^0$. Our motivation is to investigate plasmonic resonances for nanoparticles.

Using the Helmholtz decomposition of $H_T^{-\frac{1}{2}}(\mathrm{div}, \partial D)$ in Lemma 2.97, we can reduce (2.321) to an equivalent system of equations involving some well known operators.

DEFINITION 2.101. *For $u \in H_T^{-\frac{1}{2}}(\mathrm{div}, \partial D)$, we denote by $u^{(1)}$ and $u^{(2)}$ any two functions in $H_0^{\frac{3}{2}}(\partial D)$ and $H^{\frac{1}{2}}(\partial D)$, respectively, such that*

$$u = \nabla_{\partial D} u^{(1)} + \vec{\mathrm{curl}}_{\partial D} u^{(2)}.$$

Note that $u^{(1)}$ is uniquely defined and $u^{(2)}$ is defined up to a constant function.

LEMMA 2.102. *Assume $\lambda \neq \frac{1}{2}$, then the problem (2.321) is equivalent to*

$$(\lambda I - \widetilde{\mathcal{M}}_D) \begin{pmatrix} \psi^{(1)} \\ \psi^{(2)} \end{pmatrix} = \begin{pmatrix} \varphi^{(1)} \\ \varphi^{(2)} \end{pmatrix}, \tag{2.322}$$

where $(\varphi^{(1)}, \varphi^{(2)}) \in H_0^{\frac{3}{2}}(\partial D) \times H^{\frac{1}{2}}(\partial D)$ defined by Definition 2.101 with u replaced by φ, and

$$\widetilde{\mathcal{M}}_D = \begin{pmatrix} -\Delta_{\partial D}^{-1} (\mathcal{K}_D^0)^* \Delta_{\partial D} & 0 \\ \mathcal{R}_D & \mathcal{K}_D^0 \end{pmatrix} \tag{2.323}$$

with $\mathcal{R}_D$ being defined by (2.320).

PROOF. Let $(\psi^{(1)}, \psi^{(2)}) \in H_0^{\frac{3}{2}}(\partial D) \times H^{\frac{1}{2}}(\partial D)$ be a solution (if there is any) to (2.322) where $(\varphi^{(1)}, \varphi^{(2)}) \in H_0^{\frac{3}{2}}(\partial D) \times H^{\frac{1}{2}}(\partial D)$ satisfies

$$\varphi = \nabla_{\partial D} \varphi^{(1)} + \vec{\mathrm{curl}}_{\partial D} \varphi^{(2)}.$$

We have

$$\big(\lambda I + \Delta_{\partial D}^{-1} (\mathcal{K}_D^0)^* \Delta_{\partial D}\big)[\psi^{(1)}] = \varphi^{(1)}, \tag{2.324}$$

$$\lambda \psi^{(2)} - \mathcal{R}_D[\psi^{(1)}] - \mathcal{K}_D^0[\psi^{(2)}] = \varphi^{(2)}. \tag{2.325}$$

Taking $\nabla_{\partial D}$ in (2.324), $\vec{\mathrm{curl}}_{\partial D}$ in (2.325), adding up and using the identities of Lemma 2.100 yield

$$\big(\lambda I - \mathcal{M}_D^0\big)\,[\nabla_{\partial D} \psi^{(1)} + \vec{\mathrm{curl}}_{\partial D} \psi^{(2)}] = \nabla_{\partial D} \varphi^{(1)} + \vec{\mathrm{curl}}_{\partial D} \varphi^{(2)}.$$

Therefore

$$\psi = \nabla_{\partial D} \psi^{(1)} + \vec{\mathrm{curl}}_{\partial D} \psi^{(2)}$$

is a solution of (2.321).

Conversely, let ψ be the solution to (2.321). There exist $(\psi^{(1)}, \psi^{(2)}) \in H_0^{\frac{3}{2}}(\partial D) \times H^{\frac{1}{2}}(\partial D)$ and $(\varphi^{(1)}, \varphi^{(2)}) \in H_0^{\frac{3}{2}}(\partial D) \times H^{\frac{1}{2}}(\partial D)$ such that

$$\begin{aligned} \psi &= \nabla_{\partial D} \psi^{(1)} + \vec{\mathrm{curl}}_{\partial D} \psi^{(2)}, \\ \varphi &= \nabla_{\partial D} \varphi^{(1)} + \vec{\mathrm{curl}}_{\partial D} \varphi^{(2)}, \end{aligned}$$

and we have

$$(\lambda I - \mathcal{M}_D^0)\,[\nabla_{\partial D}\psi^{(1)} + \vec{\mathrm{curl}}_{\partial D}\psi^{(2)}] = \nabla_{\partial D}\varphi^{(1)} + \vec{\mathrm{curl}}_{\partial D}\varphi^{(2)}. \tag{2.326}$$

Taking $\nabla_{\partial D}\cdot$ in the above equation and using the identities of Lemma 2.100 yield

$$\Delta_{\partial D}\big(\lambda I + \Delta_{\partial D}^{-1}(\mathcal{K}_D^0)^*\Delta_{\partial D}\big)[\psi^{(1)}] = \Delta_{\partial D}\varphi^{(1)}.$$

Since $\psi^{(1)}, \varphi^{(1)} \in H_0^{\frac{3}{2}}(\partial D)$, we obtain

$$\big(\lambda I + \Delta_{\partial D}^{-1}(\mathcal{K}_D^0)^*\Delta_{\partial D}\big)[\psi^{(1)}] = \varphi^{(1)}.$$

Taking $\mathrm{curl}_{\partial D}$ in (2.326) and using the identities of Lemma 2.100 yield

$$\Delta_{\partial D}(\lambda\psi^{(2)} - \mathcal{R}_D[\psi^{(1)}] - \mathcal{K}_D^0[\psi^{(2)}]) = \Delta_{\partial D}\varphi^{(2)}.$$

Therefore, there exists a constant c such that

$$\lambda\psi^{(2)} - \mathcal{R}_D[\psi^{(1)}] - \mathcal{K}_D^0[\psi^{(2)}] = \varphi^{(2)} + c.$$

Since $\mathcal{K}_D^0[1] = \dfrac{1}{2}$, we have

$$\lambda\Big(\psi^{(2)} - \frac{c}{\lambda - 1/2}\Big) - \mathcal{R}_D[\psi^{(1)}] - \mathcal{K}_D^0\Big[\psi^{(2)} - \frac{c}{\lambda - 1/2}\Big] = \varphi^{(2)}.$$

Hence, $\Big(\psi^{(1)}, \psi^{(2)} - \dfrac{c}{\lambda - 1/2}\Big) \in H_0^{\frac{3}{2}}(\partial D) \times H^{\frac{1}{2}}(\partial D)$ is a solution to (2.322). □

Let us now analyze the spectral properties of $\widetilde{\mathcal{M}}_D$ defined by (2.323) in

$$H(\partial D) := H_0^{\frac{3}{2}}(\partial D) \times H^{\frac{1}{2}}(\partial D), \tag{2.327}$$

equipped with the inner product

$$\langle u, v\rangle_{H(\partial D)} = \langle \Delta_{\partial D}u^{(1)}, \Delta_{\partial D}v^{(1)}\rangle_{\mathcal{H}^*} + \langle u^{(2)}, v^{(2)}\rangle_{\mathcal{H}},$$

which is equivalent to the $H_0^{\frac{3}{2}}(\partial D) \times H^{\frac{1}{2}}(\partial D)$-norm. Recall that $\mathcal{H}^*$ and $\mathcal{H}$ are defined in (2.18) and (2.20), respectively. Note also that by abuse of notation we call $u^{(1)}$ and $u^{(2)}$ the first and second components of any $u \in H(\partial D)$.

Define

$$\begin{aligned} \sigma_1 &= \sigma(-(\mathcal{K}_D^0)^*)\backslash\Big(\sigma(\mathcal{K}_D^0) \cup \{-\frac{1}{2}\}\Big), \\ \sigma_2 &= \sigma(\mathcal{K}_D^0)\backslash\sigma(-(\mathcal{K}_D^0)^*), \\ \sigma_3 &= \sigma(-(\mathcal{K}_D^0)^*) \cap \sigma(\mathcal{K}_D^0). \end{aligned} \tag{2.328}$$

Let $\lambda_{j,1} \in \sigma_1$, $j = 1, 2\ldots$, and let $\varphi_{j,1}$ be an associated normalized eigenfunction of $(\mathcal{K}_D^0)^*$ as defined in Theorem 2.8. Note that $\varphi_{j,1} \in H_0^{-\frac{1}{2}}(\partial D)$ for $j \geq 1$. Then,

$$\psi_{j,1} := \begin{pmatrix} \Delta_{\partial D}^{-1}\varphi_{j,1} \\ (\lambda_{j,1}I - \mathcal{K}_D^0)^{-1}\mathcal{R}_D[\Delta_{\partial D}^{-1}\varphi_{j,1}] \end{pmatrix}$$

satisfies

$$\widetilde{\mathcal{M}}_D[\psi_{j,1}] = \lambda_{j,1}\psi_{j,1},$$

where $\widetilde{\mathcal{M}}_D$ is defined by (2.323).

Let $\lambda_{j,2} \in \sigma_2$ and let $\varphi_{j,2}$ be an associated normalized eigenfunction of $\mathcal{K}_D^0$. Then,

$$\psi_{j,2} = \begin{pmatrix} 0 \\ \varphi_{j,2} \end{pmatrix}$$

satisfies

$$\widetilde{\mathcal{M}}_D[\psi_{j,2}] = \lambda_{j,2}\psi_{j,2}.$$

Now, we assume for simplicity that the following condition holds.

CONDITION 2.103. *The eigenvalues of $(\mathcal{K}_D^0)^*$ are simple.*

Let $\lambda_{j,3} \in \sigma_3$, let $\varphi_{j,3}^{(1)}$ be the associated normalized eigenfunction of $(\mathcal{K}_D^0)^*$ and let $\varphi_{j,3}^{(2)}$ be the associated normalized eigenfunction of $\mathcal{K}_D^0$. Then,

$$\psi_{j,3} = \begin{pmatrix} 0 \\ \varphi_{j,3}^{(2)} \end{pmatrix}$$

satisfies

$$\widetilde{\mathcal{M}}_D[\psi_{j,3}] = \lambda_{j,3}\psi_{j,3},$$

and $\lambda_{j,3}$ has a generalized eigenfunction of rank two given by

$$\psi_{j,3,g} = \begin{pmatrix} c\Delta_{\partial D}^{-1}\varphi_{j,3}^{(1)} \\ (\lambda_{j,3}I - \mathcal{K}_D^0)^{-1}\mathcal{P}_{\mathrm{span}\{\varphi_{j,3}^{(2)}\}^\perp}\mathcal{R}_D[c\Delta_{\partial D}^{-1}\varphi_{j,3}^{(1)}] \end{pmatrix} \tag{2.329}$$

for a constant c such that $\mathcal{P}_{\mathrm{span}\{\varphi_{j,3}^{(2)}\}}\mathcal{R}_D[c\Delta_{\partial D}^{-1}\varphi_{j,3}^{(1)}] = -\varphi_{j,3}^{(2)}$. This means that

$$(\widetilde{\mathcal{M}}_D - \lambda_{j,3})^2[\psi_{j,3,g}] = 0 \quad \text{and} \quad (\widetilde{\mathcal{M}}_D - \lambda_{j,3})[\psi_{j,3,g}] \neq 0.$$

Here, $\mathrm{span}\{\varphi_{j,3}^{(2)}\}$ is the vector space spanned by $\varphi_{j,3}^{(2)}$, $\mathrm{span}\{\varphi_{j,3}^{(2)}\}^\perp$ is the orthogonal space to $\mathrm{span}\{\varphi_{j,3}^{(2)}\}$ in $\mathcal{H}(\partial D)$ (see Theorem 2.5), and $\mathcal{P}_{\mathrm{span}\{\varphi_{j,3}^{(2)}\}}$ (respectively, $\mathcal{P}_{\mathrm{span}\{\varphi_{j,3}^{(2)}\}^\perp}$) is the orthogonal (in $\mathcal{H}(\partial D)$) projection on $\mathrm{span}\{\varphi_{j,3}^{(2)}\}$ (respectively, $\mathrm{span}\{\varphi_{j,3}^{(2)}\}^\perp$).

We remark that the function $\psi_{j,3,g}$ is determined by the following equation

$$\widetilde{\mathcal{M}}_D[\psi_{j,3,g}] = \lambda_{j,3}\psi_{j,3,g} + \psi_{j,3}.$$

Consequently, the following result holds.

PROPOSITION 2.104. *The spectrum*

$$\sigma(\widetilde{\mathcal{M}}_D) = \sigma_1 \cup \sigma_2 \cup \sigma_3 = \sigma(-(\mathcal{K}_D^0)^*) \cup \sigma((\mathcal{K}_D^0)^*)\backslash\{-\frac{1}{2}\}$$

in $H(\partial D)$. Moreover, under Condition 2.103, $\widetilde{\mathcal{M}}_D$ has eigenfunctions $\psi_{j,i}$ associated to the eigenvalues $\lambda_{j,i} \in \sigma_i$ for $j = 1, 2, \ldots$ and $i = 1, 2, 3$, and generalized eigenfunctions of order one $\psi_{j,3,g}$ associated to $\lambda_{j,3} \in \sigma_3$, all of which form a non-orthogonal basis of $H(\partial D)$.

PROOF. It is clear that $\lambda I - \widetilde{\mathcal{M}}_D$ is bijective if and only if $\lambda \notin \sigma(-(\mathcal{K}_D^0)^*) \cup \sigma((\mathcal{K}_D^0)^*) \setminus \{-\frac{1}{2}\}$. Then, it is only left to show that $\psi_{j,1}, \psi_{j,2}, \psi_{j,3}, \psi_{j,3,g}$, $j = 1, 2, \ldots$, form a non-orthogonal basis of $H(\partial D)$. Indeed, let

$$\psi = \begin{pmatrix} \psi^{(1)} \\ \psi^{(2)} \end{pmatrix} \in H(\partial D).$$

Since $\psi_{j,1}^{(1)} \cup \psi_{j,3,g}^{(1)}$, $j = 1, 2, \ldots$ form an orthogonal basis of $\mathcal{H}_0^*(\partial D)$, which is equivalent to $H_0^{-\frac{1}{2}}(\partial D)$, there exist $\alpha_\kappa, \kappa \in I_1 := \{(j, 1) \cup (j, 3, g) : j = 1, 2, \ldots\}$ such that

$$\psi^{(1)} = \sum_{\kappa \in I_1} \alpha_\kappa \Delta_{\partial D}^{-1} \psi_\kappa^{(1)},$$

and

$$\sum_{\kappa \in I_1} |\alpha_\kappa|^2 \leq \infty.$$

It is clear that $\|\psi_\kappa^{(2)}\|_{H^{\frac{1}{2}}(\partial D)}$ is uniformly bounded with respect to $\kappa \in I_1$. Then

$$h := \sum_{\kappa \in I_1} \alpha_\kappa \psi_\kappa^{(2)} \in H^{\frac{1}{2}}(\partial D).$$

Since $\psi_{j,2}^{(2)} \cup \psi_{j,3}^{(2)}$, $j = 1, 2, \ldots$ form an orthogonal basis of $\mathcal{H}(\partial D)$, which is equivalent to $H^{\frac{1}{2}}(\partial D)$, there exist $\alpha_\kappa, \kappa \in I_2 := \{(j, 2) \cup (j, 3) : j = 1, 2, \ldots\}$ such that

$$\psi^{(2)} - h = \sum_{\kappa \in I_2} \alpha_\kappa \psi_\kappa^{(2)},$$

and

$$\sum_{\kappa \in I_2} |\alpha_\kappa|^2 \leq \infty.$$

Hence, there exist $\alpha_\kappa, \kappa \in I_1 \cup I_2$ such that

$$\psi = \sum_{\kappa \in I_1 \cup I_2} \alpha_\kappa \psi_\kappa,$$

and

$$\sum_{\kappa \in I_1 \cup I_2} |\alpha_\kappa|^2 \leq \infty.$$

□

To have the compactness of $\widetilde{\mathcal{M}}_D$, we need the following condition.

CONDITION 2.105. *σ_3 is finite.*

Indeed, if σ_3 is not finite, then we have

$$\widetilde{\mathcal{M}}_D(\{\psi_{j,3,g};\ j \geq 1\}) = \{\lambda_{j,3} \psi_{j,g,3} + \psi_{j,3};\ \ j \geq 1\}$$

whose closure is not compact. However, if σ_3 is finite, using Proposition 2.104 we can approximate $\widetilde{\mathcal{M}}_D$ by a sequence of finite-rank operators.

DEFINITION 2.106. *Let $\mathcal{B}$ be the basis of $H(\partial D)$ formed by the eigenfunctions and generalized eigenfunctions of $\widetilde{\mathcal{M}}_D$ as stated in Lemma 2.104. For $\psi \in H(\partial D)$, we denote by $\alpha(\psi, \psi_\kappa)$ the orthogonal projection of ψ into $\psi_\kappa \in \mathcal{B}$ such that*

$$\psi = \sum_\kappa \alpha(\psi, \psi_\kappa)\psi_\kappa.$$

The following lemma follows from the Fredholm alternative.

LEMMA 2.107. *Let*

$$\psi = \begin{pmatrix} \psi^{(1)} \\ \psi^{(2)} \end{pmatrix} \in H(\partial D).$$

Then, $\alpha(\psi, \psi_\kappa)$ is given by

$$\alpha(\psi, \psi_\kappa) = \begin{cases} \dfrac{\langle \psi, \widetilde{\psi}_\kappa \rangle_{H(\partial D)}}{\langle \psi_\kappa, \widetilde{\psi}_\kappa \rangle_{H(\partial D)}}, & \kappa = (j, i),\ i = 1, 2, \\ \dfrac{\langle \psi, \widetilde{\psi}_{\kappa'} \rangle_{H(\partial D)}}{\langle \psi_\kappa, \widetilde{\psi}_{\kappa'} \rangle_{H(\partial D)}}, & \kappa = (j, 3, g), \kappa' = (j, 3), \\ \dfrac{\langle \psi, \widetilde{\psi}_{\kappa_g} \rangle_{H(\partial D)} - \alpha(\psi, \psi_{\kappa_g})\langle \psi_{\kappa_g}, \widetilde{\psi}_{\kappa_g} \rangle_{H(\partial D)}}{\langle \psi_\kappa, \widetilde{\psi}_{\kappa_g} \rangle_{H(\partial D)}}, & \kappa = (j, 3), \kappa_g = (j, 3, g), \end{cases}$$

where

$$\widetilde{\psi}_\kappa \in \operatorname{Ker}(\overline{\lambda}_\kappa I - (\mathcal{M}_D^0)^*) \quad \textit{for } \kappa = (j, i),\ i = 1, 2, 3;$$
$$\widetilde{\psi}_\kappa \in \operatorname{Ker}(\overline{\lambda}_\kappa - (\mathcal{M}_D^0)^*)^2 \quad \textit{for } \kappa = (j, 3, g)$$

and $(\mathcal{M}_D^0)^*$ is the $H(\partial D)$-adjoint of $\mathcal{M}_D^0$.

The following remarks are in order.

REMARK 2.108. *Note that, since $\varphi_{j,1}$ and $\varphi_{j,3}^{(1)}$ form an orthogonal basis of $\mathcal{H}_0^*(\partial D)$, equivalent to $H_0^{-\frac{1}{2}}(\partial D)$, we also have*

$$\alpha(\psi, \psi_\kappa) = \begin{cases} \langle \Delta_{\partial D}\psi^{(1)}, \varphi_{j,1} \rangle_{\mathcal{H}^*}, & \kappa = (j, 1), \\ \frac{1}{c}\langle \Delta_{\partial D}\psi^{(1)}, \varphi_{j,3}^{(1)} \rangle_{\mathcal{H}^*}, & \kappa = (j, 3, g), \end{cases}$$

where c is defined in (2.329).

REMARK 2.109. *For $i = 1, 2, 3$, and $j = 1, 2, \ldots$,*

$$\begin{aligned} (\lambda I - \widetilde{\mathcal{M}}_D)^{-1}[\psi_{j,i}] &= \frac{\psi_{j,i}}{\lambda - \lambda_{j,i}}, \\ (\lambda I - \widetilde{\mathcal{M}}_D)^{-1}[\psi_{j,3,g}] &= \frac{\psi_{j,3,g}}{\lambda - \lambda_{j,3}} + \frac{\psi_{j,3}}{(\lambda - \lambda_{j,3})^2}. \end{aligned}$$

2.14.2. Layer Potential Formulation for Electromagnetic Scattering. We consider the scattering problem of a time-harmonic electromagnetic wave incident on D. The homogeneous medium is characterized by ε_m and μ_m, while D is characterized by electric permittivity ε_c and magnetic permeability μ_c, both of which depend on the frequency. Define

$$k_m = \omega\sqrt{\varepsilon_m \mu_m}, \quad k_c = \omega\sqrt{\varepsilon_c \mu_c},$$

and

$$\varepsilon_D = \varepsilon_m \chi(\mathbb{R}^3 \setminus \overline{D}) + \varepsilon_c \chi(D), \quad \mu_D = \varepsilon_m \chi(\mathbb{R}^3 \setminus \overline{D}) + \varepsilon_c \chi(D).$$

For a given incident plane wave (E^i, H^i), a solution to the Maxwell equations in free space

$$\left\{ \begin{array}{lcl} \nabla \times E^i & = & \sqrt{-1}\omega\mu_m H^i \quad \text{in } \mathbb{R}^3, \\ \nabla \times H^i & = & -\sqrt{-1}\omega\varepsilon_m E^i \quad \text{in } \mathbb{R}^3, \end{array} \right. \tag{2.330}$$

the scattering problem can be modeled by the following system of equations:

$$\left\{ \begin{array}{lcl} \nabla \times E & = & \sqrt{-1}\omega\mu_D H \quad \text{in } \mathbb{R}^3 \setminus \partial D, \\ \nabla \times H & = & -\sqrt{-1}\omega\varepsilon_D E \quad \text{in } \mathbb{R}^3 \setminus \partial D, \\ \nu \times E\big|_+ - \nu \times E\big|_- & = & \nu \times H\big|_+ - \nu \times H\big|_- = 0 \quad \text{on } \partial D, \end{array} \right. \tag{2.331}$$

subject to the Silver-Müller radiation condition:

$$\lim_{|x|\to\infty} |x| \Big(\sqrt{\mu_m}(H - H^i)(x) \times \frac{x}{|x|} - \sqrt{\varepsilon_m}(E - E^i)(x) \Big) = 0 \tag{2.332}$$

uniformly in $x/|x|$.

Using the boundary integral operators (2.317) and (2.318) and Lemma 2.96, the solution to (2.331) can be represented as

(2.333)

$$E(x) = \left\{ \begin{array}{ll} E^i(x) + \mu_m \nabla \times \vec{\mathcal{S}}_D^{k_m}[\psi](x) + \nabla \times \nabla \times \vec{\mathcal{S}}_D^{k_m}[\varphi](x), & x \in \mathbb{R}^3 \setminus \overline{D}, \\ \mu_c \nabla \times \vec{\mathcal{S}}_D^{k_c}[\psi](x) + \nabla \times \nabla \times \vec{\mathcal{S}}_D^{k_c}[\varphi](x), & x \in D, \end{array} \right.$$

and

$$H(x) = -\frac{\sqrt{-1}}{\omega\mu_D}(\nabla \times E)(x), \quad x \in \mathbb{R}^3 \setminus \partial D, \tag{2.334}$$

where the pair $(\psi, \varphi) \in \big(H_T^{-\frac{1}{2}}(\mathrm{div}, \partial D)\big)^2$ satisfies

(2.335)

$$\begin{pmatrix} \dfrac{\mu_c + \mu_m}{2} I + \mu_c \mathcal{M}_D^{k_c} - \mu_m \mathcal{M}_D^{k_m} & \mathcal{L}_D^{k_c} - \mathcal{L}_D^{k_m} \\ \mathcal{L}_D^{k_c} - \mathcal{L}_D^{k_m} & \Big(\dfrac{k_c^2}{2\mu_c} + \dfrac{k_m^2}{2\mu_m} \Big) I + \dfrac{k_c^2}{\mu_c}\mathcal{M}_D^{k_c} - \dfrac{k_m^2}{\mu_m}\mathcal{M}_D^{k_m} \end{pmatrix} \begin{bmatrix} \psi \\ \varphi \end{bmatrix} = \begin{bmatrix} \nu \times E^i \\ \sqrt{-1}\omega\nu \times H^i \end{bmatrix}\Bigg|_{\partial D}.$$

From [**497**], it follows that the system of equations (2.335) on $(H_T^{-\frac{1}{2}}(\mathrm{div}, \partial D))^2$ has a unique solution and there exists a positive constant $C = C(\varepsilon_c, \varepsilon_m, \mu_c, \mu_m, \omega)$ such that

(2.336)

$$\|\psi\|_{H_T^{-\frac{1}{2}}(\mathrm{div},\partial D)} + \|\varphi\|_{H_T^{-\frac{1}{2}}(\mathrm{div},\partial D)} \le C\big(\|E^i \times \nu\|_{H_T^{-\frac{1}{2}}(\mathrm{div},\partial D)} + \|H^i \times \nu\|_{H_T^{-\frac{1}{2}}(\mathrm{div},\partial D)}\big).$$

2.14.3. Low-Frequency Asymptotic Expansions of Layer Potentials. Low-frequency behaviors of $\mathcal{M}_D^k$ and $\mathcal{L}_D^k$ are investigated in the following lemmas.

LEMMA 2.110. *For $\varphi \in H_T^{-\frac{1}{2}}(\mathrm{div}, \partial D)$, the following asymptotic expansion as $k \to 0$ holds:*

$$\mathcal{M}_D^k[\varphi](x) = \mathcal{M}_D^0[\varphi](x) - \sum_{j=2}^{\infty} (\sqrt{-1}k)^j \mathcal{M}_D^j[\varphi](x), \quad x \in \partial D, \tag{2.337}$$

where

$$\mathcal{M}_D^j[\varphi](x) = \int_{\partial D} \frac{1}{4\pi j!}\nu(x) \times \nabla_x \times |x-y|^{j-1}\varphi(y)d\sigma(y).$$

Moreover, $\|\mathcal{M}_D^j\|_{\mathcal{L}\left(H_T^{-\frac{1}{2}}(\mathrm{div},\partial D)\right)}$ *is uniformly bounded with respect to* j*. In particular, the series converges in* $\mathcal{L}\big(H_T^{-\frac{1}{2}}(\mathrm{div},\partial D)\big)$ *and* $k \mapsto \mathcal{M}_D^k$ *is holomorphic in* k*.*

PROOF. A Taylor expansion of $\Gamma_k(x-y)$ yields

$$\Gamma_k(x-y) = -\sum_{j=0}^{\infty} \frac{(\sqrt{-1}k|x-y|)^j}{j!4\pi|x-y|} = -\frac{1}{4\pi|x-y|} + \sum_{j=1}^{\infty}\frac{(\sqrt{-1}k)^j}{4\pi j!}|x-y|^{j-1}.$$

Hence, (2.337) holds.

Note that from the regularity of $|x-y|^{j-1}$, $j \geq 2$, $\|\mathcal{M}_D^j[\varphi]\|_{H_T^{-\frac{1}{2}}(\mathrm{div},\partial D)}$ is bounded uniformly with respect to j, and therefore, $\|\mathcal{M}_D^j\|_{\mathcal{L}\left(H_T^{-\frac{1}{2}}(\mathrm{div},\partial D)\right)}$ is uniformly bounded with respect to j as well. □

LEMMA 2.111. *For* $\varphi \in H_T^{-\frac{1}{2}}(\mathrm{div},\partial D)$*, the following asymptotic expansion as* $\omega \to 0$ *holds:*

$$(\mathcal{L}_D^{k_c} - \mathcal{L}_D^{k_m})[\varphi](x) = \sum_{j=1}^{\infty}\omega^{j+1}\mathcal{L}_D^j[\varphi](x), \quad x \in \partial D,$$

where

$$\begin{aligned}\mathcal{L}_D^j[\varphi](x) &= C_j\nu(x)\times\Big(\int_{\partial D}|x-y|^{j-2}\varphi(y)d\sigma(y)\\ &\quad -\int_{\partial D}\frac{|x-y|^{j-2}(x-y)}{j+1}\nabla_{\partial D}\cdot\varphi(y)d\sigma(y)\Big),\end{aligned}$$

with

$$C_j = \frac{(\sqrt{-1})^j((\sqrt{\varepsilon_c\mu_c})^{j+1} - (\sqrt{\varepsilon_m\mu_m})^{j+1})}{4\pi(j-1)!}.$$

Moreover, $\|\mathcal{L}_D^j\|_{\mathcal{L}\left(H_T^{-\frac{1}{2}}(\mathrm{div},\partial D)\right)}$ *is uniformly bounded with respect to* j*. In particular, the series converges in* $\mathcal{L}\big(H_T^{-\frac{1}{2}}(\mathrm{div},\partial D)\big)$ *and* $k \mapsto \mathcal{L}_D^k$ *is holomorphic in* k*.*

PROOF. The proof is similar to that of Lemma 2.110. □

2.14.4. Coordinate Transformation and Invariance in Electromagnetism. It is a remarkable fact that Maxwell's equations under coordinate transformations can be written in an identical Cartesian form, after transforming the electromagnetic parameters and the electromagnetic fields. As will be shown later, this result is useful for the design of invisibility cloaks.

Suppose that we make a coordinate transformation $x \mapsto F(x)$, possibly singular. Let DF denote the Jacobian matrix. Consider the following Maxwell equations:

$$\begin{aligned}\nabla\times E &= \sqrt{-1}\omega\mu(x)H \quad \text{in } \mathbb{R}^3,\\ \nabla\times H &= -\sqrt{-1}\omega\varepsilon(x)E \quad \text{in } \mathbb{R}^3,\end{aligned}$$

subject to the Silver-Müller radiation condition (2.332), where $\varepsilon(x) = \varepsilon_m$ and $\mu(x) = \mu_m$ for $|x|$ large enough and (E^i, H^i) is an incident plane wave.

LEMMA 2.112. *Let F be a diffeomorphism of $\mathbb{R}^3$ onto $\mathbb{R}^3$ such that $F(x)$ is the identity for $|x|$ large enough. Define $\widetilde{E}$ and $\widetilde{H}$ by $\widetilde{E}(y) = E(F^{-1}(y))$ and $\widetilde{H}(y) = H(F^{-1}(y))$. Then, $(\widetilde{E}, \widetilde{H})$ satisfies*

$$\begin{aligned} \nabla_y \times \widetilde{E}(y) &= \sqrt{-1}\omega F_\star[\mu](y)\widetilde{H}(y) \quad \text{in } \mathbb{R}^3, \\ \nabla_y \times \widetilde{H}(y) &= -\sqrt{-1}\omega F_\star[\varepsilon](y)\widetilde{E}(y) \quad \text{in } \mathbb{R}^3, \end{aligned}$$

together with the Silver-Müller radiation condition (2.332) where the incident fields are $H^i(F^{-1}(y))$ and $H^i(F^{-1}(y))$, where for a function $q(x)$,

$$F_\star[q](y) = \frac{DF(x)q(x)DF^t(x)}{\det DF(x)}$$

with $x = F^{-1}(y)$ and t being the transpose.

A similar result holds for the Helmholtz equation.

LEMMA 2.113. *Let F be a diffeomorphism of $\mathbb{R}^d$ onto $\mathbb{R}^d$, for $d = 2, 3$, such that $F(x)$ is identity for $|x|$ large enough. Suppose that v is a solution to*

$$\nabla \cdot \frac{1}{\mu}\nabla v + \omega^2 \varepsilon v = 0 \quad \text{in } \mathbb{R}^d,$$

and $v - v^i$ satisfies the Sommerfeld radiation condition, where v^i is an incident plane wave. Then $\widetilde{v}$ defined by $\widetilde{v}(y) = v(F^{-1}(y))$ satisfies

$$\nabla_y \cdot F_\star[\frac{1}{\mu}](y)\nabla_y \widetilde{v}(y) + \omega^2 \frac{\varepsilon(x)}{\det DF(x)}\widetilde{v}(y) = 0 \quad \text{in } \mathbb{R}^d, \tag{2.338}$$

and $\widetilde{v}(y) - v^i(F^{-1}(y))$ satisfies the Sommerfeld radiation condition.

2.14.5. Multipole Solutions to the Maxwell Equations. For a wave number $k > 0$, $l' = -l, \ldots, l$ and $l = 1, 2, \ldots$, the function

$$v_{ll'}(k; x) = h_l^{(1)}(k|x|)Y_l^{l'}(\hat{x}) \tag{2.339}$$

satisfies the Helmholtz equation $\Delta v + k^2 v = 0$ in $\mathbb{R}^3 \setminus \{0\}$ and the Sommerfeld radiation condition:

$$\lim_{|x|\to\infty} |x|(\frac{\partial v_{ll'}}{\partial |x|}(k; x) - \sqrt{-1}k v_{ll'}(k; x)) = 0.$$

Here, $Y_l^{l'}$ is the spherical harmonic defined on the unit sphere S, $\hat{x} = x/|x|$, and $h_l^{(1)}$ is the spherical Hankel function of the first kind and order l which satisfies the Sommerfeld radiation condition in three dimensions. Similarly, we define $\tilde{v}_{ll'}(x)$ by

$$\tilde{v}_{ll'}(k; x) = j_l(k|x|)Y_l^{l'}(\hat{x}), \tag{2.340}$$

where j_l is the spherical Bessel function of the first kind. The function $\tilde{v}_{ll'}$ satisfies the Helmholtz equation in all $\mathbb{R}^3$.

In the same manner, one can construct solutions to the Maxwell system with the vector version of spherical harmonics. Define the vector spherical harmonics as

$$U_{ll'} = \frac{1}{\sqrt{l(l+1)}}\nabla_S Y_l^{l'}(\hat{x}) \quad \text{and} \quad V_{ll'} = \hat{x} \times U_{ll'}, \tag{2.341}$$

for $l' = -l, \ldots, l$ and $l = 1, 2, \ldots$. Here, $\hat{x} \in S$ and ∇_S denotes the surface gradient on the unit sphere S. The vector spherical harmonics defined in (2.341) form a complete orthogonal basis for $H_T^{-\frac{1}{2}}(\mathrm{div}, S)$.

Through multiplication of the vector spherical harmonics with the Hankel function, one can construct the so-called multipole solutions to the Maxwell system. To keep the analysis simple, one separates the solutions into transverse electric, $E \cdot x = 0$, and transverse magnetic, $H \cdot x = 0$. Define the exterior transverse electric multipoles to the Maxwell equations in free space as

$$\begin{cases} E_{ll'}^{TE}(k;x) = -\sqrt{l(l+1)}h_l^{(1)}(k|x|)V_{ll'}(\hat{x}), \\ H_{ll'}^{TE}(k;x) = -\dfrac{\sqrt{-1}}{\omega\mu}\nabla \times E_{ll'}^{TE}(k;x), \end{cases} \tag{2.342}$$

and the exterior transverse magnetic multipoles as

$$\begin{cases} E_{ll'}^{TM}(k;x) = \dfrac{\sqrt{-1}}{\omega\varepsilon}\nabla \times H_{ll'}^{TM}(k;x), \\ H_{ll'}^{TM}(k;x) = -\sqrt{l(l+1)}h_l^{(1)}(k|x|)V_{ll'}(\hat{x}). \end{cases} \tag{2.343}$$

The exterior electric and magnetic multipoles satisfy the radiation condition. In the same manner, one defines the interior multipoles $(\widetilde{E}_{ll'}^{TE}, \widetilde{H}_{ll'}^{TE})$ and $(\widetilde{E}_{ll'}^{TM}, \widetilde{H}_{ll'}^{TM})$ with $h_l^{(1)}$ replaced by j_l, *i.e.*,

$$\begin{cases} \widetilde{E}_{ll'}^{TE}(k;x) = -\sqrt{l(l+1)}j_l^{(1)}(k|x|)V_{ll'}(\hat{x}), \\ \widetilde{H}_{ll'}^{TE}(k;x) = -\dfrac{\sqrt{-1}}{\omega\mu}\nabla \times \widetilde{E}_{ll'}^{TE}(k;x), \end{cases} \tag{2.344}$$

and

$$\begin{cases} \widetilde{E}_{ll'}^{TM}(k;x) = \dfrac{\sqrt{-1}}{\omega\varepsilon}\nabla \times \widetilde{H}_{ll'}^{TM}(k;x), \\ \widetilde{H}_{ll'}^{TM}(k;x) = -\sqrt{l(l+1)}j_l^{(1)}(k|x|)V_{ll'}(\hat{x}). \end{cases} \tag{2.345}$$

Note that one has

$$\nabla \times E_{ll'}^{TE}(k;x) = \frac{\sqrt{l(l+1)}}{|x|}\mathcal{H}_l(k|x|)U_{ll'}(\hat{x}) + \frac{l(l+1)}{|x|}h_l^{(1)}(k|x|)Y_l^{l'}(\hat{x})\hat{x}, \tag{2.346}$$

$$\nabla \times \widetilde{E}_{ll'}^{TE}(k;x) = \frac{\sqrt{l(l+1)}}{|x|}\mathcal{J}_l(k|x|)U_{ll'}(\hat{x}) + \frac{l(l+1)}{|x|}j_l^{(1)}(k|x|)Y_l^{l'}(\hat{x})\hat{x}, \tag{2.347}$$

where $\mathcal{H}_l(t) = h_l^{(1)}(t) + t\left(h_l^{(1)}\right)'(t)$ and $\mathcal{J}_l(t) = j_l(t) + tj_l'(t)$.

The solutions to the Maxwell system can be represented as separated variable sums of the multipole solutions; see [**411**, Section 5.3]. With multipole solutions and the Helmholtz solutions in (2.339) and (2.340), it is also possible to expand the fundamental solution Γ_k to the Helmholtz operator.

Let p be a fixed vector in $\mathbb{R}^3$. For $|x| > |y|$, the following addition formula holds (see [**404**, Section 9.3.3]):

$$\begin{aligned}\Gamma_k(x-y)p = &-\sum_{l=1}^{\infty}\frac{\sqrt{-1}k}{l(l+1)}\frac{\varepsilon}{\mu}\sum_{l'=-l}^{l} E_{ll'}^{TM}(k;x)\overline{\widetilde{E}_{ll'}^{TM}(k;y)}\cdot p \\ &-\sum_{l=1}^{\infty}\frac{\sqrt{-1}k}{l(l+1)}\sum_{l'=-l}^{l} E_{ll'}^{TE}(k;x)\overline{\widetilde{E}_{ll'}^{TE}(k;y)}\cdot p \\ &-\frac{\sqrt{-1}}{k}\sum_{l=1}^{\infty}\sum_{l'=-l}^{l}\nabla v_{ll'}(k;x)\overline{\nabla \tilde{v}_{ll'}(k;y)}\cdot p,\end{aligned} \tag{2.348}$$

with $v_{ll'}$ and $\tilde{v}_{ll'}$ being defined by (2.339) and (2.340).

Plane wave solutions to the Maxwell equations have an expansion using the multipole solutions as well (see [**332**]). The incoming wave

$$E^i(x) = \sqrt{-1}k(q\times p)\times q e^{\sqrt{-1}kq\cdot x},$$

where $q\in S$ is the direction of propagation and the vector $p\in\mathbb{R}^3$ is the direction of polarization, is expressed as

$$E^i(x) = \sqrt{-1}k\sum_{l=1}^{\infty}\frac{4\pi(\sqrt{-1})^l}{\sqrt{l(l+1)}}\sum_{l'=-l}^{l}\left[-\sqrt{-1}(V_{ll'}(q)\cdot c)\widetilde{E}_{ll'}^{TE}(x)-\sqrt{\frac{\varepsilon}{\mu}}(U_{ll'}(q)\cdot c)\widetilde{E}_{ll'}^{TM}(x)\right], \tag{2.349}$$

where $c = (q\times p)\times q$.

2.14.6. Scattering Coefficients and their Properties. This subsection introduces the notion of scattering coefficients associated to the Maxwell equations and provides some of their properties. It extends the notions and results established in the previous section for the Helmholtz equation.

2.14.6.1. *Notion of Scattering Coefficients.* From (2.348) (with k_m in the place of k) and (2.333) it follows that, for sufficiently large $|x|$,

$$(E-E^i)(x) = \sum_{l=1}^{\infty}\frac{\sqrt{-1}k_m}{l(l+1)}\sum_{l'=-l}^{l}\Big(\alpha_{ll'}E_{ll'}^{TE}(k_m;x) + \beta_{ll'}E_{ll'}^{TM}(k_m;x)\Big), \tag{2.350}$$

where

$$\begin{aligned}\alpha_{ll'} &= -\sqrt{-1}\omega\varepsilon_m\mu_m\int_{\partial D}\overline{\widetilde{E}_{ll'}^{TM}(k_m;y)}\cdot\varphi(y) - k_m^2\int_{\partial D}\overline{\widetilde{E}_{ll'}^{TE}(k_m;y)}\cdot\psi(y),\\ \beta_{ll'} &= \sqrt{-1}\omega\varepsilon_m\mu_m\int_{\partial D}\overline{\widetilde{E}_{ll'}^{TE}(k_m;y)}\cdot\varphi(y) - \omega^2\varepsilon_m^2\int_{\partial D}\overline{\widetilde{E}_{ll'}^{TM}(k_m;y)}\cdot\psi(y).\end{aligned}$$

Here, the densities ψ and φ are defined by (2.335).

Let $(\varphi_{pp'}^{TE},\psi_{pp'}^{TE})$ be the solution to (2.335) when

$$E^i = \widetilde{E}_{pp'}^{TE}(k_m;y) \quad \text{and } H^i = \widetilde{H}_{pp'}^{TE}(k_m;y),$$

and $(\varphi_{pp'}^{TM},\psi_{pp'}^{TM})$ when

$$E^i = \widetilde{E}_{pp'}^{TM}(k_m;y) \quad \text{and } H^i = \widetilde{H}_{pp'}^{TM}(k_m;y).$$

DEFINITION 2.114 (Scattering Coefficients). *The scattering coefficients*

$$\left(W_{ll',pp'}^{TE,TE}, W_{ll',pp'}^{TE,TM}, W_{ll',pp'}^{TM,TE}, W_{ll',pp'}^{TM,TM}\right)$$

associated with the permittivity and the permeability distributions ε, μ and the frequency ω (or k_c, k_m, D) are defined to be

$$\begin{aligned}
W_{ll',pp'}^{TE,TE} &= -\sqrt{-1}\omega\varepsilon_m\mu_m \int_{\partial D} \overline{\widetilde{E}_{ll'}^{TM}}(k_m; y) \cdot \varphi_{pp'}^{TE}(y)\, d\sigma(y) \\
&\quad -k_m^2 \int_{\partial D} \overline{\widetilde{E}_{ll'}^{TE}}(k_m; y) \cdot \psi_{pp'}^{TE}(y)\, d\sigma(y), \\
W_{ll',pp'}^{TE,TM} &= -\sqrt{-1}\omega\varepsilon_m\mu_m \int_{\partial D} \overline{\widetilde{E}_{ll'}^{TM}}(k_m; y) \cdot \varphi_{pp'}^{TM}(y)\, d\sigma(y) \\
&\quad -k_m^2 \int_{\partial D} \overline{\widetilde{E}_{ll'}^{TE}}(k_m; y) \cdot \psi_{pp'}^{TM}(y)\, d\sigma(y), \\
W_{ll',pp'}^{TM,TE} &= \sqrt{-1}\omega\varepsilon_m\mu_m \int_{\partial D} \overline{\widetilde{E}_{ll'}^{TE}}(k_m; y) \cdot \varphi_{pp'}^{TE}(y)\, d\sigma(y) \\
&\quad -\omega^2\varepsilon_m^2 \textstyle\int_{\partial D} \overline{\widetilde{E}_{ll'}^{TM}}(k_m; y) \cdot \psi_{pp'}^{TE}(y)\, d\sigma(y), \\
W_{ll',pp'}^{TM,TM} &= \sqrt{-1}\omega\varepsilon_m\mu_m \int_{\partial D} \overline{\widetilde{E}_{ll'}^{TE}}(k_m; y) \cdot \varphi_{pp'}^{TM}(y)\, d\sigma(y) \\
&\quad -\omega^2\varepsilon_m^2 \textstyle\int_{\partial D} \overline{\widetilde{E}_{ll'}^{TM}}(k_m; y) \cdot \psi_{pp'}^{TM}(y)\, d\sigma(y).
\end{aligned}$$

As will be seen, the scattering coefficients appear naturally in the expansion of the scattering amplitude. One first obtains the following estimates for the decay of the scattering coefficients.

LEMMA 2.115. *There exists a constant C depending on $(\varepsilon, \mu, \omega)$ such that*

$$\left| W_{ll',pp'}^{TE,TE}[\varepsilon, \mu, \omega] \right| \leq \frac{C^{l+p}}{l^l p^p} \tag{2.351}$$

for all $l, l', p, p' \in \mathbb{N} \setminus \{0\}$. The same estimates hold for $W_{ll',pp'}^{TE,TM}$, $W_{ll',pp'}^{TM,TE}$, and $W_{ll',pp'}^{TM,TM}$.

PROOF. Let (φ, ψ) be the solution to (2.335) with $E^i(y) = \widetilde{E}_{pp'}^{TE}(k_m; y)$ and $H^i = -\frac{\sqrt{-1}}{\omega\mu_m}\nabla \times E^i$. Recall that the spherical Bessel function j_p behaves as

$$j_p(t) = \frac{t^p}{1 \times 3 \times \cdots \times (2p+1)}\left(1 + O\left(\frac{1}{p}\right)\right) \quad \text{as } p \to \infty,$$

uniformly on compact subsets of $\mathbb{R}$. Using Stirling's formula

$$p! = \sqrt{2\pi p}(p/e)^p(1 + o(1)),$$

one has

$$j_p(t) = O\left(\frac{C^p t^p}{p^p}\right) \quad \text{as } p \to \infty, \tag{2.352}$$

uniformly on compact subsets of $\mathbb{R}$ with a constant C independent of p. Thus one has

$$\|E^i\|_{H_T^{-\frac{1}{2}}(\mathrm{div}, \partial D)} + \|H^i\|_{H_T^{-\frac{1}{2}}(\mathrm{div}, \partial D)} \leq \frac{C'^p}{p^p}$$

for some constant C'. It then follows from (2.336) that

$$\|\varphi\|_{L^2(\partial D)} + \|\psi\|_{L^2(\partial D)} \leq \frac{C^p}{p^p}$$

for another constant C. So one gets (2.351) from the definition of the scattering coefficients. □

Suppose now that the incoming wave is of the form

$$E^i(x) = \sum_{p=1}^{\infty} \sum_{p'=-p}^{p} \Big(a_{pp'} \widetilde{E}_{pp'}^{TE}(k_m; x) + b_{pp'} \widetilde{E}_{pp'}^{TM}(k_m; x) \Big) \tag{2.353}$$

for some constants $a_{pp'}$ and $b_{pp'}$. Then the solution (φ, ψ) to (2.335) is given by

$$\varphi = \sum_{p=1}^{\infty} \sum_{p'=-p}^{p} \Big(a_{pp'} \varphi_{pp'}^{TE} + b_{pp'} \varphi_{pp'}^{TM} \Big),$$

$$\psi = \sum_{p=1}^{\infty} \sum_{p'=-p}^{p} \Big(a_{pp'} \psi_{pp'}^{TE} + b_{pp'} \psi_{pp'}^{TM} \Big).$$

By (2.350) and Definition 2.114, the solution E to (2.331) can be represented as (2.354)

$$(E - E^i)(x) = \sum_{l=1}^{\infty} \frac{\sqrt{-1} k_m}{l(l+1)} \sum_{l'=-l}^{l} \Big(\alpha_{ll'} E_{ll'}^{TE}(k_m; x) \ + \ \beta_{ll'} E_{ll'}^{TM}(k_m; x) \Big), \quad |x| \to \infty,$$

where

$$\begin{cases} \alpha_{ll'} = \displaystyle\sum_{p=1}^{\infty} \sum_{p'=-p}^{p} \Big(a_{pp'} W_{ll',pp'}^{TE,TE} + b_{pp'} W_{ll',pp'}^{TE,TM} \Big), \\ \beta_{ll'} = \displaystyle\sum_{p=1}^{\infty} \sum_{p'=-p}^{p} \Big(a_{pp'} W_{ll',pp'}^{TM,TE} + b_{pp'} W_{ll',pp'}^{TM,TM} \Big). \end{cases} \tag{2.355}$$

Using (2.354), (2.355), and the behavior of the spherical Bessel functions, the far-field pattern of the scattered wave $(E - E^i)$ can be estimated. We define the scattering amplitude $A_\infty[\varepsilon, \mu, \omega]$ by

$$E(x) - E^i(x) = \frac{e^{\sqrt{-1} k_m |x|}}{k_m |x|} A_\infty[\varepsilon, \mu, \omega](\hat{x}) + o(|x|^{-1}) \quad \text{as } |x| \to \infty. \tag{2.356}$$

Since the spherical Bessel function $h_l^{(1)}$ and its derivative behave like

$$\begin{cases} h_l^{(1)}(t) \sim \dfrac{1}{t} e^{\sqrt{-1} t} e^{-\sqrt{-1} \frac{l+1}{2} \pi} & \text{as } t \to \infty, \\ (h_l^{(1)})'(t) \sim \dfrac{1}{t} e^{\sqrt{-1} t} e^{-\sqrt{-1} \frac{l}{2} \pi} & \text{as } t \to \infty, \end{cases}$$

one can easily see by using (2.346) that

$$\begin{cases} E_{ll'}^{TE}(k_m; x) \sim \dfrac{e^{\sqrt{-1} k_m |x|}}{k_m |x|} e^{-\sqrt{-1} \frac{l+1}{2} \pi} \ \big(- \sqrt{l(l+1)} \big) V_{ll'}(\hat{x}) & \text{as } |x| \to \infty, \\ E_{ll'}^{TM}(k_m; x) \sim \dfrac{e^{\sqrt{-1} k_m |x|}}{k_m |x|} \sqrt{\dfrac{\mu_m}{\varepsilon_m}} e^{-\sqrt{-1} \frac{l+1}{2} \pi} \big(- \sqrt{l(l+1)} \big) U_{ll'}(\hat{x}) & \text{as } |x| \to \infty. \end{cases}$$

Therefore, the following result holds.

PROPOSITION 2.116. *If E^i is given by (2.353), then the corresponding scattering amplitude can be expanded as*

$$A_\infty[\varepsilon, \mu, \omega](\hat{x}) = \sum_{l=1}^{\infty} \frac{-(\sqrt{-1})^{-l} k_m}{\sqrt{l(l+1)}} \sum_{l'=-l}^{l} \Big(\alpha_{ll'} V_{ll'}(\hat{x}) + \beta_{ll'} \sqrt{\frac{\mu_m}{\varepsilon_m}} U_{ll'}(\hat{x}) \Big), \tag{2.357}$$

where $\alpha_{ll'}$ and $\beta_{ll'}$ are defined by (2.355).

Consider the case where the incident wave E^i is given by a plane wave $e^{\sqrt{-1}k_m d\cdot x}c$ with $d \in S$ and $d \cdot c = 0$. It follows from (2.349) that

$$\begin{aligned} e^{\sqrt{-1}k\cdot x}c &= \sum_{p=1}^{\infty} \frac{4\pi(\sqrt{-1})^p}{\sqrt{p(p+1)}} \sum_{p'=-p}^{p} \Big[-\sqrt{-1}\big(V_{pp'}(d)\cdot c\big)\widetilde{E}_{pp'}^{TE}(k_m;x) \\ &\quad - \sqrt{\frac{\varepsilon_m}{\mu_m}}\big(U_{pp'}(d)\cdot c\big)\widetilde{E}_{pp'}^{TM}(k_m;x)\Big], \end{aligned}$$

and therefore,

$$a_{pp'} = -\frac{4\pi(\sqrt{-1})^{p+1}}{\sqrt{p(p+1)}}(V_{pp'}(d)\cdot c) \quad \text{and} \quad b_{pp'} = -\frac{4\pi(\sqrt{-1})^p}{\sqrt{p(p+1)}}\sqrt{\frac{\varepsilon_m}{\mu_m}}(U_{pp'}(d)\cdot c).$$

Hence, the scattering amplitude, denoted by $A_\infty[\varepsilon,\mu,\omega](c,d;\hat{x})$, is given by (2.357) with

$$(2.358) \qquad \begin{aligned} \alpha_{ll'} &= \sum_{p=1}^{\infty}\sum_{p'=-p}^{p} \frac{4\pi(\sqrt{-1})^p}{\sqrt{p(p+1)}}\Big[-\sqrt{-1}(V_{pp'}(d)\cdot c)W_{ll',pp'}^{TE,TE} \\ &\quad - \sqrt{\frac{\varepsilon_m}{\mu_m}}(U_{pp'}(d)\cdot c)W_{ll',pp'}^{TE,TM}\Big], \\ \beta_{ll'} &= \sum_{p=1}^{\infty}\sum_{p'=-p}^{p} \frac{4\pi(\sqrt{-1})^p}{\sqrt{p(p+1)}}\Big[-\sqrt{-1}(V_{pp'}(d)\cdot c)W_{ll',pp'}^{TM,TE} \\ &\quad - \sqrt{\frac{\varepsilon_m}{\mu_m}}(U_{pp'}(d)\cdot c)W_{ll',pp'}^{TM,TM}\Big], \end{aligned}$$

which shows that the scattering coefficients appear in the expansion of the scattering amplitude.

The low-frequency behavior of the scattering coefficients is now investigated. The following result holds.

LEMMA 2.117. *There exists $\delta_0 > 0$ such that, for all $\delta \le \delta_0$,*

$$(2.359) \qquad \left| W_{ll',pp'}^{TE,TE}[\varepsilon,\mu,\delta\omega] \right| \le \frac{C^{l+p}}{l^l p^p}\delta^{l+p+1},$$

for all $l, p \in \mathbb{N}\setminus\{0\}, l' = -l,\ldots,l, p' = -p,\ldots,p$, where the constant C depends on (ε,μ,ω) but is independent of δ. The same estimate holds for $W_{ll',pp'}^{TE,TM}$, $W_{ll',pp'}^{TM,TE}$, and $W_{ll',pp'}^{TM,TM}$.

PROOF. Let (φ,ψ) be the solution to (2.335) with $E^{i,\delta}(y) = \widetilde{E}_{pp'}^{TE}(\delta k_m; y)$ and $H^{i,\delta} = -\frac{\sqrt{-1}}{\delta\omega\mu_m}\nabla\times E^i$. Then, from (2.352), it follows that

$$\|E^{i,\delta}\|_{H_T^{-\frac{1}{2}}(\mathrm{div},\partial D)} + \|H^{i,\delta}\|_{H_T^{-\frac{1}{2}}(\mathrm{div},\partial D)} \le \frac{C^p}{p^p}\delta^p,$$

where C is independent of δ, and hence

$$\|\varphi^\delta\|_{L^2(\partial D)} + \delta\|\psi^\delta\|_{L^2(\partial D)} \le \frac{C^p}{p^p}\delta^p$$

for $\delta \le \delta_0$ for some δ_0. So one obtains (2.359) from Definition 2.114 of the scattering coefficients. □

2.14.6.2. *Multi-Layer Structure and its Scattering Coefficients.* Here we consider a multi-layered structure and explain how to compute its scattering coefficients. A numerical example is also presented. The multi-layered structure is defined as follows: For positive numbers $r_1, \dots, r_{L+1}$ with $2 = r_1 > r_2 > \cdots r_{L+1} = 1$, let

$$A_j := \{x : r_{j+1} \leq |x| < r_j\}, \quad j = 1, \dots, L,$$
$$A_0 := \mathbb{R}^3 \setminus \overline{B_2}, \quad A_{L+1}(= D) := \{x : |x| < 1\},$$

where B_2 denotes the ball of center 0 and radius 2, and let

$$\Gamma_j = \{|x| = r_j\}, \quad j = 1, \dots, L+1.$$

Let (μ_j, ε_j) be the pair of permeability and permittivity parameters of A_j for $j = 1, \dots, L+1$. Set $\mu_0 = 1$ and $\varepsilon_0 = 1$. Then define the permeability and permittivity distributions of the layered structure to be

$$\mu = \sum_{j=0}^{L+1} \mu_j \chi(A_j) \quad \text{and} \quad \varepsilon = \sum_{j=0}^{L+1} \varepsilon_j \chi(A_j). \tag{2.360}$$

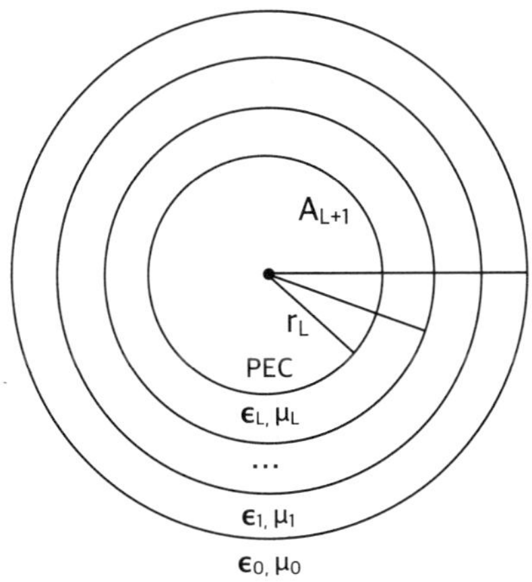

FIGURE 2.13. A multi-layered structure.

The transmission condition on each interface Γ_j is given by

$$[\hat{x} \times E] = [\hat{x} \times H] = 0. \tag{2.361}$$

Assume that the core A_{L+1} is perfectly conducting (PEC), that is,

$$E \times \nu = 0 \quad \text{on } \Gamma_{L+1} = \partial A_{L+1}. \tag{2.362}$$

The scattering coefficients $\left(W^{TE,TE}_{(n,m)(p,q)}, W^{TE,TM}_{(n,m)(p,q)}, W^{TM,TE}_{(n,m)(p,q)}, W^{TM,TM}_{(n,m)(p,q)}\right)$ are defined as before, namely, if E^i is given as in (2.353), the scattered field $E - E^i$ can be expanded as (2.354) and (2.355).

Thanks to the symmetry of the layered (radial) structure, the scattering coefficients are much simpler than the general case. In fact, if the incident field is given by $E^i = \widetilde{E}^{TE}_{n,m}$, then the solution E to (2.331) subject to (2.332) takes the form

$$E(x) = \tilde{a}_j \widetilde{E}^{TE}_{n,m}(x) + a_j E^{TE}_{n,m}(x), \quad x \in A_j, \quad j = 0, \dots, L, \tag{2.363}$$

with $\tilde{a}_0 = 1$. From (2.346) and (2.347), the interface condition (2.361) amounts to

$$\begin{bmatrix} j_n(k_j r_j) & h_n^{(1)}(k_j r_j) \\ \frac{1}{\mu_j}\mathcal{J}_n(k_j r_j) & \frac{1}{\mu_j}\mathcal{H}_n(k_j r_j) \end{bmatrix} \begin{bmatrix} \tilde{a}_j \\ a_j \end{bmatrix} = \begin{bmatrix} j_n(k_{j-1} r_j) & h_n^{(1)}(k_{j-1} r_j) \\ \frac{1}{\mu_{j-1}}\mathcal{J}_n(k_{j-1} r_j) & \frac{1}{\mu_{j-1}}\mathcal{H}_n(k_{j-1} r_j) \end{bmatrix} \begin{bmatrix} \tilde{a}_{j-1} \\ a_{j-1} \end{bmatrix}, \quad j = 1, \ldots, L, \tag{2.364}$$

where $\mathcal{H}_n(t) = h_n^{(1)}(t) + t\left(h_n^{(1)}\right)'(t)$ and $\mathcal{J}_n(t) = j_n(t) + t j_n'(t)$, and the boundary condition on the perfectly conducting surface Γ_{L+1} is

$$\begin{bmatrix} j_n(k_L) & h_n^{(1)}(k_L) \\ 0 & 0 \end{bmatrix} \begin{bmatrix} \tilde{a}_L \\ a_L \end{bmatrix} = \begin{bmatrix} 0 \\ 0 \end{bmatrix}. \tag{2.365}$$

Since the matrices appearing in (2.364) are invertible, one can see that there exist a_j and $\tilde{a}_j$, $j = 0, 1, \ldots L$, satisfying (2.364) and (2.365). Similarly, one can see that if the incident field is given by $E^i = \widetilde{E}_{n,m}^{TM}(x)$, then the solution E takes the form

$$E(x) = \tilde{b}_j \widetilde{E}_{n,m}^{TM}(x) + b_j E_{n,m}^{TM}(x), \quad x \in A_j, \quad j = 0, 1, \ldots, L \tag{2.366}$$

for some constants b_j and $\tilde{b}_j$ ($\tilde{b}_0 = 1$). One can see now from (2.363) and (2.366) that the scattering coefficients satisfy

$$\begin{aligned} W_{(n,m)(p,q)}^{TE,TM} &= W_{(n,m)(p,q)}^{TM,TE} = 0 \quad \text{for all } (m,n) \text{ and } (p,q), \\ W_{(n,m)(p,q)}^{TE,TE} &= W_{(n,m)(p,q)}^{TM,TM} = 0 \quad \text{if } (m,n) \neq (p,q), \end{aligned}$$

and, since (2.363) and (2.366) hold independently of m, one has

$$\begin{aligned} W_{(n,0)(n,0)}^{TE,TE} &= W_{(n,m)(n,m)}^{TE,TE}, \\ W_{(n,0)(n,0)}^{TM,TM} &= W_{(n,m)(n,m)}^{TM,TM} \qquad \text{for } -n \leq m \leq n. \end{aligned}$$

Moreover, if one writes

$$W_n^{TE} := W_{(n,0)(n,0)}^{TE} \quad \text{and} \quad W_n^{TM} := W_{(n,0)(n,0)}^{TM},$$

then one has

$$W_n^{TE} = -\frac{\sqrt{-1}n(n+1)}{k_0} a_0 \quad \text{and} \quad W_n^{TE} = -\frac{\sqrt{-1}n(n+1)}{k_0} b_0. \tag{2.367}$$

Suppose now that $\widetilde{E}_{n,0}^{TE}$ is the incident field and the solution E is given by

$$E(x) = \tilde{a}_j \widetilde{E}_{n,0}^{TE}(x) + a_j E_{n,0}^{TE}(x), \quad x \in A_j, \quad j = 0, \ldots, L,$$

with $\tilde{a}_0 = 1$, where the coefficients $\tilde{a}_j$'s and a_j's are determined by (2.364) and (2.365). From (2.364) it follows that

$$\begin{bmatrix} \tilde{a}_j \\ a_j \end{bmatrix} = \begin{bmatrix} j_n(k_j r_j) & h_n^{(1)}(k_j r_j) \\ \frac{1}{\mu_j}\mathcal{J}_n(k_j r_j) & \frac{1}{\mu_j}\mathcal{H}_n(k_j r_j) \end{bmatrix}^{-1} \begin{bmatrix} j_n(k_{j-1} r_j) & h_n^{(1)}(k_{j-1} r_j) \\ \frac{1}{\mu_{j-1}}\mathcal{J}_n(k_{j-1} r_j) & \frac{1}{\mu_{j-1}}\mathcal{H}_n(k_{j-1} r_j) \end{bmatrix} \begin{bmatrix} \tilde{a}_{j-1} \\ a_{j-1} \end{bmatrix},$$

for $j = 1, \ldots, L$.

Define

(2.368)

$$M_j = \begin{bmatrix} j_n(k_j r_j) & h_n^{(1)}(k_j r_j) \\ \frac{1}{\mu_j}\mathcal{J}_n(k_j r_j) & \frac{1}{\mu_j}\mathcal{H}_n(k_j r_j) \end{bmatrix}, \quad N_j = \begin{bmatrix} j_n(k_{j-1} r_j) & h_n^{(1)}(k_{j-1} r_j) \\ \frac{1}{\mu_{j-1}}\mathcal{J}_n(k_{j-1} r_j) & \frac{1}{\mu_{j-1}}\mathcal{H}_n(k_{j-1} r_j) \end{bmatrix}.$$

Substituting these relations into (2.365) yields

$$\begin{bmatrix} 0 \\ 0 \end{bmatrix} = P_n^{TE}[\varepsilon, \mu, \omega] \begin{bmatrix} \tilde{a}_0 \\ a_0 \end{bmatrix}, \tag{2.369}$$

where

(2.370)

$$\begin{aligned} P_n^{TE}[\varepsilon, \mu, \omega] &:= \begin{bmatrix} p_{n,1}^{TE} & p_{n,2}^{TE} \\ 0 & 0 \end{bmatrix} \\ &= (-\sqrt{-1}\omega)^L \left(\prod_{j=1}^{L} \mu_j^{\frac{3}{2}} \varepsilon_j^{\frac{1}{2}} r_j \right) \begin{bmatrix} j_n(k_L) & h_n^{(1)}(k_L) \\ 0 & 0 \end{bmatrix} \prod_{j=1}^{L} M_j^{-1} N_j. \end{aligned}$$

Then (2.369) yields

$$W_n^{TE} = -\frac{\sqrt{-1}n(n+1)}{k_0} a_0 = -\frac{\sqrt{-1}n(n+1)}{k_0} \frac{p_{n,1}^{TE}}{p_{n,2}^{TE}}. \tag{2.371}$$

Similarly, for W_n^{TM}, one looks for another solution E of the form

$$E(x) = \tilde{b}_j \widetilde{E}_{n,0}^{TM}(x) + b_j E_{n,0}^{TM}(x), \quad x \in A_j, \quad j = 0, \ldots, L,$$

with $\tilde{b}_0 = 1$. The transmission conditions become

$$\begin{aligned} &\begin{bmatrix} \frac{1}{\varepsilon_j}\mathcal{J}_n(k_j r_j) & \frac{1}{\varepsilon_j}\mathcal{H}_n(k_j r_j) \\ j_n(k_j r_j) & h_n^{(1)}(k_j r_j) \end{bmatrix} \begin{bmatrix} \tilde{b}_j \\ b_j \end{bmatrix} \\ &= \begin{bmatrix} \frac{1}{\varepsilon_{j-1}}\mathcal{J}_n(k_{j-1} r_j) & \frac{1}{\varepsilon_{j-1}}\mathcal{H}_n(k_{j-1} r_j) \\ j_n(k_{j-1} r_j) & h_n^{(1)}(k_{j-1} r_j) \end{bmatrix} \begin{bmatrix} \tilde{b}_{j-1} \\ b_{j-1} \end{bmatrix}, \quad j = 1, \ldots, N+1, \end{aligned} \tag{2.372}$$

and the boundary condition on the inner most layer, which is perfectly conducting, is

$$\begin{bmatrix} \mathcal{J}_n(k_L) & \mathcal{H}_n(k_L) \\ 0 & 0 \end{bmatrix} \begin{bmatrix} \tilde{b}_L \\ b_L \end{bmatrix} = \begin{bmatrix} 0 \\ 0 \end{bmatrix}. \tag{2.373}$$

Define

(2.374)

$$\widetilde{M}_j = \begin{bmatrix} \frac{1}{\varepsilon_j}\mathcal{J}_n(k_j r_j) & \frac{1}{\varepsilon_j}\mathcal{H}_n(k_j r_j) \\ j_n(k_j r_j) & h_n^{(1)}(k_j r_j) \end{bmatrix}, \quad \widetilde{N}_j = \begin{bmatrix} \frac{1}{\varepsilon_{j-1}}\mathcal{J}_n(k_{j-1} r_j) & \frac{1}{\varepsilon_{j-1}}\mathcal{H}_n(k_{j-1} r_j) \\ j_n(k_{j-1} r_j) & h_n^{(1)}(k_{j-1} r_j) \end{bmatrix}.$$

From (2.372) and (2.373), one obtains

$$\begin{bmatrix} 0 \\ 0 \end{bmatrix} = P_n^{TM}[\varepsilon, \mu, \omega] \begin{bmatrix} \tilde{b}_0 \\ b_0 \end{bmatrix}, \tag{2.375}$$

where
(2.376)

$$\begin{aligned} P_n^{TM}[\varepsilon,\mu,\omega] &:= \begin{bmatrix} p_{n,1}^{TM} & p_{n,2}^{TM} \\ 0 & 0 \end{bmatrix} \\ &= (\sqrt{-1}\omega)^L \left(\prod_{j=1}^{L} \mu_j^{\frac{1}{2}}\varepsilon_j^{\frac{3}{2}} r_j\right) \begin{bmatrix} \mathcal{J}_n(k_L) & \mathcal{H}_n(k_L) \\ 0 & 0 \end{bmatrix} \prod_{j=1}^{L} \widetilde{M}_j^{-1}\widetilde{N}_j. \end{aligned}$$

From the definition of W_n^{TM} and (2.375),

$$W_n^{TE} = -\frac{\sqrt{-1}n(n+1)}{k_0}\frac{b_0}{\tilde{b}_0} = -\frac{\sqrt{-1}n(n+1)}{k_0}\frac{p_{n,1}^{TM}}{p_{n,2}^{TM}}. \tag{2.377}$$

It is worth emphasizing that $p_{n,2}^{TE} \neq 0$ and $p_{n,2}^{TM} \neq 0$. In fact, if $p_{n,2}^{TE} = 0$, then (2.369) can be fulfilled with $\tilde{a}_0 = 0$ and $a_0 = 1$. This means that there exists (μ, ε) on $\mathbb{R}^3 \setminus \overline{D}$ such that the following problem has a solution:

$$\begin{cases} \nabla \times E = \sqrt{-1}\omega\mu H & \text{in } \mathbb{R}^3 \setminus \overline{D}, \\ \nabla \times H = -\sqrt{-1}\omega\varepsilon E & \text{in } \mathbb{R}^3 \setminus \overline{D}, \\ (x \times E)\big|_+ = 0 & \text{on } \partial D, \\ E(x) = E_{n,0}^{TE}(x) & \text{for } |x| > 2. \end{cases}$$

Applying the following Green's theorem on $\Omega = \{x :\ 1 < |x| < R\}$,

$$\begin{aligned} &\int_\Omega \big(E\cdot\Delta F + \nabla\times E\cdot\nabla\times F + \nabla\cdot E\ \nabla\cdot F\big)dx \\ &= \int_{\partial\Omega} \big(\nu\times E\cdot\nabla\times F + \nu\cdot E\ (\nabla\cdot F)\big)d\sigma(x) \end{aligned}$$

with $F = \overline{E_{n,0}^{TE}}(x)$ and the boundary condition on the perfectly conducting surface $\{|x| = 1\}$, it follows that

$$\int_{|x|=R} (\nu\times E)\cdot\overline{H}d\sigma(x) = \sqrt{-1}k_0\int_\Omega (|H|^2 - |E|^2)dx.$$

In particular, the left-hand side is real-valued. Hence,

$$\begin{aligned} \int_{|x|=R} |H\times\nu - E|^2 d\sigma(x) &= \int_{|x|=R} \big(|H\times\nu|^2 + |E|^2 - 2\Re((\nu\times E)\cdot\overline{H}\big)d\sigma(x) \\ &= \int_{|x|=R} \big(|H\times\nu|^2 + |E|^2\big)d\sigma(x). \end{aligned}$$

From the radiation condition, the left-hand side goes to zero as $R \to \infty$, which contradicts the behavior of the Hankel functions. One can show that $p_{n,2}^{TM} \neq 0$ in a similar way.

2.14.6.3. *Numerical Example.* We now demonstrate how to compute the scattering coefficients W_n^{TE} and W_n^{TM} numerically. The corresponding Matlab code can be downloaded at `www.ams.org/bookpages/surv-235`.

For simplicity, we consider only W_n^{TE}. Recall that

$$W_n^{TE} = -\frac{\sqrt{-1}n(n+1)}{k_0}a_0, \tag{2.378}$$

with the constant a_0 being determined by (2.364) and (2.365). From (2.364), we obtain

$$\begin{bmatrix}\tilde{a}_0/a_L \\ a_0/a_L\end{bmatrix} = (M_1^{-1}N_1)(M_2^{-1}N_2)\dots(M_L^{-1}N_L)\begin{bmatrix}\tilde{a}_L/a_L \\ 1\end{bmatrix}, \tag{2.379}$$

where M_j and N_j are defined by (2.368).

From (2.365), we immediately see that

$$\frac{\tilde{a}_L}{a_L} = -\frac{h_n^{(1)}(k_L r_{L+1})}{j_n(k_L r_{L+1})}.$$

Therefore, we can compute $\tilde{a}_0/a_L$ and a_0/a_L. But, since $\tilde{a}_0 = 1$, we can also compute a_0 and then W_n^{TE}.

Now we present a numerical example. We set the parameters for the structure as follows: the number of layers L is $L = 3$, the radii of the layers are $r_1 = 2, r_2 = 5/3, r_3 = 4/3, r_4 = 1$, and the material parameters are $(\varepsilon_0, \mu_0) = (1, 1), (\varepsilon_0, \mu_0) = (0.5, 0.5), (\varepsilon_0, \mu_0) = (2, 2), (\varepsilon_0, \mu_0) = (0.5, 0.5)$. The numerical result for W_n^{TE} and W_n^{TM} for $n = 1, 2, \dots, 7$ is shown in Table 2.8. The decaying behavior of W_n^{TE} and W_n^{TM} as n increases is clearly shown.

n	W_n^{TE}	W_n^{TM}
1	$-0.9991 + 0.9572\sqrt{-1}$	$-0.7473 + 1.6644\sqrt{-1}$
2	$-0.7527 + 0.0960\sqrt{-1}$	$-0.7650 + 0.0992\sqrt{-1}$
3	$-0.1642 + 0.0022\sqrt{-1}$	$-0.1643 + 0.0023\sqrt{-1}$
4	$-0.0191 + 0.0000\sqrt{-1}$	$-0.0191 + 0.0000\sqrt{-1}$
5	$-0.0013 + 0.0000\sqrt{-1}$	$-0.0013 + 0.0000\sqrt{-1}$
6	$-0.0001 + 0.0000\sqrt{-1}$	$-0.0001 + 0.0000\sqrt{-1}$
7	$-0.0000 + 0.0000\sqrt{-1}$	$-0.0000 + 0.0000\sqrt{-1}$

TABLE 2.8. Scattering coefficients for a multi-layer spherical shell.

2.14.7. The Helmholtz-Kirchhoff Theorem. Let

$$\mathbf{G}_{k_m}(x) = \varepsilon_m\left(\Gamma_{k_m}(x)I + \frac{1}{k_m^2}\nabla_x\nabla_x\Gamma_{k_m}(x)\right) \tag{2.380}$$

be the Dyadic Green (matrix valued) function for the full Maxwell equations. $\mathbf{G}_{k_m}$ satisfies

$$\nabla\times\nabla\times\mathbf{G}_{k_m} - k_m^2\mathbf{G}_{k_m} = \delta_0 I \quad \text{in } \mathbb{R}^3,$$

together with the Silver-Müller radiation condition

$$\lim_{|x|\to\infty}|x|\left(\nabla\times\mathbf{G}_{k_m}(x)\times\frac{x}{|x|} - \sqrt{-1}k_m\mathbf{G}_{k_m}(x)\right) = 0$$

uniformly in $x/|x|$.

Here, the tensor $\nabla\nabla$ is defined by

$$\nabla\nabla = (\partial_{ij}^2)_{i,j=1}^3. \tag{2.381}$$

The following Helmholtz-Kirchhoff identity holds [**464**].

PROPOSITION 2.118. *Let ∂B_R be the sphere of radius R and center 0. We have*

$$\int_{\partial B_R} \left(\frac{\partial \overline{\mathbf{G}_{k_m}}}{\partial \nu}(x-y)\mathbf{G}_{k_m}(z-y) - \overline{\mathbf{G}_{k_m}}(x-y)\frac{\partial \mathbf{G}_{k_m}}{\partial \nu}(z-y) \right) d\sigma(y) = 2\sqrt{-1}\Im\, \mathbf{G}_{k_m}(x-z)\,, \tag{2.382}$$

which yields

$$\lim_{R\to+\infty} \int_{\partial B_R} \overline{\mathbf{G}_{k_m}}(x-y)\mathbf{G}_{k_m}(z-y)\, d\sigma(y) = -\frac{1}{k_m}\Im\, \mathbf{G}_{k_m}(x-z)\,, \tag{2.383}$$

by using the Silver-Müller radiation condition.

2.14.8. The Optical Theorem. The optical cross-section theorem for the scattering of electromagnetic waves can be stated as follows [**300**].

PROPOSITION 2.119. *Assume that the incident fields are plane waves given by*

$$\begin{aligned} E^i(x) &= ce^{\sqrt{-1}k_m d\cdot x}, \\ H^i(x) &= \sqrt{\frac{\varepsilon_m}{\mu_m}}\, d \times ce^{\sqrt{-1}k_m d\cdot x}, \end{aligned}$$

where $c \in \mathbb{R}^3$ and $d \in S$ are such that $c \cdot d = 0$. Then, the extinction cross-section Q^{ext}, defined by

$$Q^{\text{ext}} := -\frac{1}{|c|^2}\sqrt{\frac{\mu_m}{\varepsilon_m}}\Re\left[\int_{\partial D} \left(\overline{E^i} \times (H - H^i) + (E - E^i) \times \overline{H^i}\right)\cdot \nu\, d\sigma\right],$$

satisfies

$$Q^{\text{ext}} = \frac{4\pi}{k_m}\Im\left[\frac{c\cdot A_\infty(c,d;d)}{|c|^2}\right],$$

where the scattering amplitude $A_\infty(c,d;\hat{x})$ is defined by (2.357) with $\alpha_{ll'}$ and $\beta_{ll'}$ are given by (2.358).

Analogously to the scalar case, the extinction cross-section Q^{ext} is defined as the ratio of the sum of the mean powers absorbed and scattered by D to the mean intensity power flow in the incident field. The latter quantity is given by

$$-\frac{1}{2}\Re\big[(E^i \times \overline{H^i})\cdot d\big]$$

which reduces to

$$\frac{1}{2}\sqrt{\frac{\varepsilon_m}{\mu_m}}|c|^2.$$

2.14.9. Electromagnetic Scattering by Small Particles. We consider the scattering problem of a time-harmonic electromagnetic wave incident on a particle D. The homogeneous medium is characterized by ε_m and μ_m, while D is characterized by electric permittivity ε_c and magnetic permeability μ_c. We assume that $\varepsilon_m, \varepsilon_c, \mu_m$, and μ_c are positive constants and define

$$k_m = \omega\sqrt{\varepsilon_m\mu_m}, \quad k_c = \omega\sqrt{\varepsilon_c\mu_c},$$

and

$$\varepsilon_D = \varepsilon_m\chi(\mathbb{R}^3 \setminus \overline{D}) + \varepsilon_c\chi(D), \quad \mu_D = \varepsilon_m\chi(\mathbb{R}^3 \setminus \overline{D}) + \varepsilon_c\chi(D).$$

For a given incident plane wave (E^i, H^i), the solution to the the scattering problem for the Maxwell equations in free space (2.330) can be modeled by the system of equations (2.331) subject to the Silver-Müller radiation condition (2.332).

Let $D = z + \delta B$ where B contains the origin and $|B| = O(1)$. The following result follows from [**76, 90, 92**]. It gives the leading-order term in the asymptotic expansion of the scattered electric field E^s far-away from the particle.

THEOREM 2.120. *For $D = z + \delta B \Subset \mathbb{R}^3$ of class $\mathcal{C}^{1,\alpha}$ for $\alpha > 0$ and $K \Subset \mathbb{R}^3 \backslash \overline{D}$, the following far-field expansion holds uniformly in K*

$$(2.384) \qquad \begin{aligned} E^s(x) &= -\frac{\sqrt{-1}\omega\mu_m}{\varepsilon_m}\nabla \times \mathbf{G}_{k_m}(x-z)M(\lambda_\mu, D)H^i(z) \\ &\quad -\omega^2\mu_m\mathbf{G}_{k_m}(x-z)M(\lambda_\varepsilon, D)E^i(z) + O(\delta^4), \end{aligned}$$

where $\mathbf{G}_{k_m}(x-z)$ is the Dyadic Green (matrix valued) function for the full Maxwell equations defined by (2.380) and $M(\lambda_\mu, D)$ and $M(\lambda_\varepsilon, D)$ are the polarization tensors associated with D and the contrasts λ_μ and λ_ε given by (2.72) with $k = \mu_m/\mu_c$ and $k = \varepsilon_c/\varepsilon_m$, respectively.

2.15. Integral Representation of Solutions to the Lamé System

Let Ω be a bounded domain in $\mathbb{R}^d$ with a connected Lipschitz boundary. Let λ and μ be the Lamé constants for Ω satisfying the strong convexity condition

$$(2.385) \qquad \mu > 0 \text{ and } d\lambda + 2\mu > 0.$$

The corresponding Lamé system is given by

$$\mathcal{L}^{\lambda,\mu}u = \mu\triangle u + (\lambda+\mu)\nabla\nabla\cdot u,$$

and the conormal derivative $\partial u/\partial\nu$ is defined by

$$(2.386) \qquad \frac{\partial u}{\partial \nu} = \lambda(\nabla\cdot u)n + \mu(\nabla u + \nabla u^t)n,$$

where the superscript t denotes the transpose and n is the unit normal to the boundary $\partial\Omega$. We introduce the symmetric gradient as

$$(2.387) \qquad \nabla^s u := \frac{1}{2}(\nabla u + \nabla u^t)$$

and define the elasticity tensor $\mathbb{C} = (C_{ijkl})_{i,j,k,l=1}^d$ by

$$(2.388) \qquad C_{ijkl} = \lambda\delta_{ij}\delta_{kl} + \mu(\delta_{ik}\delta_{jl} + \delta_{il}\delta_{jk}).$$

With this notation, we have

$$\mathcal{L}^{\lambda,\mu}u = \nabla\cdot\mathbb{C}\nabla^s u,$$

and

$$\frac{\partial u}{\partial\nu} = (\mathbb{C}\nabla^s u)n.$$

2.15.1. Fundamental Solutions.

In the three-dimensional case, the Kupradze matrix $\mathbf{\Gamma}^\omega = (\Gamma^\omega_{ij})_{i,j=1}^3$ of the fundamental solution to the operator $\mathcal{L}^{\lambda,\mu} + \omega^2$ is given by

$$(2.389) \qquad \Gamma^\omega_{ij}(x) = -\frac{\delta_{ij}}{4\pi\mu|x|}e^{\frac{\sqrt{-1}\omega|x|}{c_s}} + \frac{1}{4\pi\omega^2}\partial_i\partial_j\frac{e^{\frac{\sqrt{-1}\omega|x|}{c_p}} - e^{\frac{\sqrt{-1}\omega|x|}{c_s}}}{|x|},$$

where ∂_j denotes $\partial/\partial x_j$ and

$$c_s = \sqrt{\mu}, \quad c_p = \sqrt{\lambda + 2\mu}.$$

See [**342**, Chapter 2]. One can easily show that Γ^ω_{ij} has the series representation:

$$\Gamma^\omega_{ij}(x) = -\frac{1}{4\pi}\sum_{n=0}^{+\infty}\frac{\sqrt{-1}^n}{(n+2)n!}\Big(\frac{n+1}{c_s^{n+2}}+\frac{1}{c_p^{n+2}}\Big)\omega^n\delta_{ij}|x|^{n-1} + \frac{1}{4\pi}\sum_{n=0}^{+\infty}\frac{\sqrt{-1}^n(n-1)}{(n+2)n!}\Big(\frac{1}{c_s^{n+2}}-\frac{1}{c_p^{n+2}}\Big)\omega^n|x|^{n-3}x_ix_j. \tag{2.390}$$

If $\omega = 0$, then $\mathbf{\Gamma}^0$ is the Kelvin matrix of the fundamental solution to the Lamé system; *i.e.*,

$$\Gamma^0_{ij}(x) = -\frac{\gamma_1}{4\pi}\frac{\delta_{ij}}{|x|} - \frac{\gamma_2}{4\pi}\frac{x_ix_j}{|x|^3}, \tag{2.391}$$

where

$$\gamma_1 = \frac{1}{2}\left(\frac{1}{\mu}+\frac{1}{2\mu+\lambda}\right) \quad\text{and}\quad \gamma_2 = \frac{1}{2}\left(\frac{1}{\mu}-\frac{1}{2\mu+\lambda}\right). \tag{2.392}$$

In the two-dimensional case, the fundamental solution $\mathbf{\Gamma}^\omega = (\Gamma^\omega_{ij})^2_{i,j=1}$ to the operator $\mathcal{L}^{\lambda,\mu}+\omega^2$, $\omega\neq 0$, is given by

$$\Gamma^\omega_{ij}(x) = -\frac{\sqrt{-1}}{4\mu}\delta_{ij}H_0^{(1)}\left(\frac{\omega|x|}{c_s}\right) + \frac{\sqrt{-1}}{4\omega^2}\partial_i\partial_j\left(H_0^{(1)}\left(\frac{\omega|x|}{c_p}\right) - H_0^{(1)}\left(\frac{\omega|x|}{c_s}\right)\right). \tag{2.393}$$

See [**8**] and [**342**, Chapter 2]. For $\omega = 0$, we set $\mathbf{\Gamma}^0$ to be the Kelvin matrix of fundamental solutions to the Lamé system; *i.e.*,

$$\Gamma^0_{ij}(x) = \frac{\gamma_1}{2\pi}\delta_{ij}\ln|x| - \frac{\gamma_2}{2\pi}\frac{x_ix_j}{|x|^2}. \tag{2.394}$$

2.15.2. Single- and Double-Layer Potentials. Analogously to the Laplace operator, the single- and double-layer potentials for the operator $\mathcal{L}^{\lambda,\mu}+\omega^2$ are defined by

$$\mathcal{S}^\omega_\Omega[\varphi](x) = \int_{\partial\Omega}\mathbf{\Gamma}^\omega(x-y)\varphi(y)\,d\sigma(y), \quad x\in\mathbb{R}^d, \tag{2.395}$$

$$\mathcal{D}^\omega_\Omega[\varphi](x) = \int_{\partial\Omega}\frac{\partial}{\partial\nu(y)}\mathbf{\Gamma}^\omega(x-y)\varphi(y)\,d\sigma(y), \quad x\in\mathbb{R}^d\setminus\partial\Omega, \tag{2.396}$$

for $\varphi\in L^2(\partial\Omega)^d$. Here, the conormal derivative of $\mathbf{\Gamma}^\omega$ is defined by

$$\frac{\partial}{\partial\nu(y)}\mathbf{\Gamma}^\omega(x-y)b = \frac{\partial}{\partial\nu(y)}\big(\mathbf{\Gamma}^\omega(x-y)b\big) \tag{2.397}$$

for any constant vector b.

The following formulas give the jump relations obeyed by the double-layer potential and by the conormal derivative of the single-layer potential:

$$\frac{\partial(\mathcal{S}^\omega_\Omega[\varphi])}{\partial\nu}\bigg|_\pm(x) = \Big(\pm\frac{1}{2}I + (\mathcal{K}^\omega_\Omega)^*\Big)[\varphi](x) \text{ a.e. } x\in\partial\Omega, \tag{2.398}$$

$$(\mathcal{D}^\omega_\Omega[\varphi])\Big|_\pm(x) = \Big(\mp\frac{1}{2}I + \mathcal{K}^\omega_\Omega\Big)[\varphi](x) \text{ a.e. } x\in\partial\Omega, \tag{2.399}$$

where $\mathcal{K}^\omega_\Omega$ is the operator defined by

$$\mathcal{K}^\omega_\Omega[\varphi](x) = \text{p.v.}\int_{\partial\Omega}\frac{\partial\mathbf{\Gamma}^\omega(x-y)}{\partial\nu(y)}\varphi(y)\,d\sigma(y) \tag{2.400}$$

and $(\mathcal{K}_\Omega^\omega)^*$ is the L^2-adjoint of $\mathcal{K}_\Omega^{-\omega}$; that is,

$$(\mathcal{K}_\Omega^\omega)^*[\varphi](x) = \text{p.v.} \int_{\partial\Omega} \frac{\partial \mathbf{\Gamma}^\omega(x-y)}{\partial \nu(x)} \varphi(y)\, d\sigma(y).$$

See [**212, 342**].

Let Ψ be the vector space of all linear solutions to the equation $\mathcal{L}^{\lambda,\mu} u = 0$ and $\partial u/\partial \nu = 0$ on $\partial\Omega$, or alternatively,

$$\Psi = \Big\{ \psi : \partial_i \psi_j + \partial_j \psi_i = 0, \ 1 \le i, j \le d \Big\}. \tag{2.401}$$

Define a subspace of $L^2(\partial\Omega)^d$ by

$$L^2_\Psi(\partial\Omega) = \Big\{ f \in L^2(\partial\Omega)^d : \int_{\partial\Omega} f \cdot \psi \, d\sigma = 0 \text{ for all } \psi \in \Psi \Big\}.$$

In particular, since Ψ contains constant functions, we get

$$\int_{\partial\Omega} f \, d\sigma = 0$$

for any $f \in L^2_\Psi(\partial\Omega)$. We also know that if u is smooth and satisfies $\mathcal{L}^{\lambda,\mu} u = 0$ in Ω, then $\partial u/\partial \nu \big|_{\partial\Omega} \in L^2_\Psi(\partial\Omega)$.

We recall Green's formulas for the Lamé system, which can be obtained by integration by parts. The first formula is

$$\int_{\partial\Omega} u \cdot \frac{\partial v}{\partial \nu} \, d\sigma = \int_\Omega u \cdot \mathcal{L}^{\lambda,\mu} v + E(u, v), \tag{2.402}$$

where $u \in H^1(\Omega)^d$, $v \in H^{3/2}(\Omega)^d$, and

$$E(u, v) = \int_\Omega \lambda (\nabla \cdot u)(\overline{\nabla \cdot v}) + \frac{\mu}{2} (\nabla u + \nabla u^t) \cdot (\overline{\nabla v} + \overline{\nabla v}^t). \tag{2.403}$$

Formula (2.402) yields Green's second formula

$$\int_{\partial\Omega} \left(u \cdot \frac{\partial v}{\partial \nu} - v \cdot \frac{\partial u}{\partial \nu} \right) = \int_\Omega \left(u \cdot \mathcal{L}^{\lambda,\mu} v - v \cdot \mathcal{L}^{\lambda,\mu} u \right), \tag{2.404}$$

where $u, \ v \in H^{3/2}(\Omega)^d$.

Formula (2.404) shows that if $u \in H^{3/2}(\Omega)^d$ satisfies $\mathcal{L}^{\lambda,\mu} u = 0$ in Ω, then $\partial u/\partial \nu \big|_{\partial\Omega} \in L^2_\Psi(\partial\Omega)$.

2.15.3. Helmholtz Decompositions and Radiation Conditions. Let us formulate the *radiation conditions* for the elastic waves when $\Im \omega \ge 0$ and $\omega \ne 0$.

Any smooth solution u to the constant-coefficient equation $(\mathcal{L}^{\lambda,\mu} + \omega^2) u = 0$ can be decomposed as follows [**342**, Theorem 2.5]:

$$u = u_p + u_s, \tag{2.405}$$

where u_p and u_s are given by

$$u_p = (\kappa_s^2 - \kappa_p^2)^{-1} (\triangle + \kappa_s^2) u,$$
$$u_s = (\kappa_p^2 - \kappa_s^2)^{-1} (\triangle + \kappa_p^2) u,$$

with

$$\kappa_s = \frac{\omega}{c_s} = \frac{\omega}{\sqrt{\mu}} \quad \text{and} \quad \kappa_p = \frac{\omega}{c_p} = \frac{\omega}{\sqrt{\lambda + 2\mu}}. \tag{2.406}$$

Then u_s and u_p satisfy the equations

$$\begin{cases} (\triangle + \kappa_s^2) u_s = 0, & \nabla \times u_s = 0, \\ (\triangle + \kappa_p^2) u_p = 0, & \nabla \cdot u_p = 0. \end{cases} \tag{2.407}$$

We impose on u_s and u_p the radiation condition (2.158) for solutions of the Helmholtz equation by requiring that

$$\begin{cases} \partial_r u_s(x) - \sqrt{-1}\kappa_s u_s(x) = o(r^{-1}), \\ \partial_r u_p(x) - \sqrt{-1}\kappa_p u_p(x) = o(r^{-1}), \end{cases} \quad \text{as } r = |x| \to +\infty. \tag{2.408}$$

We say that u satisfies the Sommerfeld-Kupradze radiation condition if it can be decomposed in the form (2.405) with u_s and u_p satisfying (2.407) and (2.408). By a straightforward calculation, one can see that the single- and double-layer potentials satisfy the radiation condition. We refer to [**7**, **342**] for details.

We recall the following uniqueness results for the exterior problem [**342**].

LEMMA 2.121. *Let u be a solution to $(\mathcal{L}^{\lambda,\mu} + \omega^2)u = 0$ in $\mathbb{R}^d \setminus \overline{\Omega}$ satisfying the radiation condition. If either $u = 0$ or $\partial u / \partial \nu = 0$ on $\partial \Omega$, then u is identically zero in $\mathbb{R}^d \setminus \overline{\Omega}$.*

In dimension d, the Kupradze matrix $\mathbf{\Gamma}^\omega$ can be decomposed into shear and pressure components [**10**]:

$$\mathbf{\Gamma}^\omega(x) = \mathbf{\Gamma}_s^\omega(x) + \mathbf{\Gamma}_p^\omega(x), \quad x \in \mathbb{R}^d, \quad x \neq 0, \tag{2.409}$$

where

$$\mathbf{\Gamma}_p^\omega(x) = -\frac{1}{\mu \kappa_s^2} \nabla\nabla \Gamma_{\kappa_p}(x) \quad \text{and} \quad \mathbf{\Gamma}_s^\omega(x) = \frac{1}{\mu \kappa_s^2} (\kappa_s^2 I + \nabla\nabla)\Gamma_{\kappa_s}(x). \tag{2.410}$$

Here, the tensor $\nabla\nabla$ is defined by

$$\nabla\nabla = (\partial_{ij}^2)_{i,j=1}^d,$$

and the functions Γ_{κ_p} and Γ_{κ_s} are defined by (2.151).

2.15.4. Transmission Problem. Let $\tilde{\lambda}, \tilde{\mu}$ be another pair of Lamé parameters such that

$$(\lambda - \tilde{\lambda})(\mu - \tilde{\mu}) \geq 0, \quad (\lambda - \tilde{\lambda})^2 + (\mu - \tilde{\mu})^2 \neq 0. \tag{2.411}$$

Later in this book, we will consider the following transmission problem:

$$\begin{cases} \mathcal{L}^{\lambda,\mu} u + \omega^2 u = 0 & \text{in } \Omega \setminus \overline{D}, \\ \mathcal{L}^{\tilde{\lambda},\tilde{\mu}} u + \omega^2 u = 0 & \text{in } D, \\ \dfrac{\partial u}{\partial \nu} = g & \text{on } \partial \Omega, \\ u\big|_+ - u\big|_- = 0 & \text{on } \partial D, \\ \dfrac{\partial u}{\partial \nu}\Big|_+ - \dfrac{\partial u}{\partial \tilde{\nu}}\Big|_- = 0 & \text{on } \partial D. \end{cases} \tag{2.412}$$

Let $\tilde{\mathcal{S}}_D^\omega$ denote the single-layer potential defined by (2.395) with λ, μ replaced by $\tilde{\lambda}, \tilde{\mu}$. We also denote by $\partial u / \partial \tilde{\nu}$ the conormal derivative associated with $\tilde{\lambda}, \tilde{\mu}$. We now have the following solvability result which can be viewed as a compact perturbation result of the case $\omega = 0$.

THEOREM 2.122. *Suppose that* $(\lambda-\tilde{\lambda})(\mu-\tilde{\mu}) \geq 0$ *and* $0 < \tilde{\lambda}, \tilde{\mu} < +\infty$. *Suppose that* $\Im\omega \geq 0$ *and* ω^2 *is not a Dirichlet eigenvalue for* $-\mathcal{L}^{\lambda,\mu}$ *on* D. *For any given* $(F,G) \in H^1(\partial D)^d \times L^2(\partial D)^d$, *there exists a unique pair* $(f,g) \in L^2(\partial D)^d \times L^2(\partial D)^d$ *such that*

$$\begin{cases} \tilde{\mathcal{S}}_D^\omega[f]|_- - \mathcal{S}_D^\omega[g]|_+ = F, \\ \dfrac{\partial}{\partial \tilde{\nu}} \tilde{\mathcal{S}}_D^\omega[f]\Big|_- - \dfrac{\partial}{\partial \nu} \mathcal{S}_D^\omega[g]\Big|_+ = G. \end{cases}$$

If $\omega = 0$ *and* $G \in L^2_\Psi(\partial D)$, *then* $g \in L^2_\Psi(\partial D)$. *Moreover, if* $F \in \Psi$ *and* $G = 0$, *then* $g = 0$.

PROOF. For $\omega = 0$, the theorem is proved in [**233**]. Here, we only consider the case $\omega \neq 0$, which can be treated as a compact perturbation of the case $\omega = 0$. In fact, let us define the operators $T, T_0 : L^2(\partial D)^d \times L^2(\partial D)^d \to H^1(\partial D)^d \times L^2(\partial D)^d$ by

$$T(f,g) := \left(\tilde{\mathcal{S}}_D^\omega[f]|_- - \mathcal{S}_D^\omega[g]|_+, \frac{\partial}{\partial \tilde{\nu}} \tilde{\mathcal{S}}_D^\omega[f]\Big|_- - \frac{\partial}{\partial \nu} \mathcal{S}_D^\omega[g]\Big|_+ \right)$$

and

$$T_0(f,g) := \left(\tilde{\mathcal{S}}_D^0[f]|_- - \mathcal{S}_D^0[g]|_+, \frac{\partial}{\partial \tilde{\nu}} \tilde{\mathcal{S}}_D^0[f]\Big|_- - \frac{\partial}{\partial \nu} \mathcal{S}_D^0[g]\Big|_+ \right).$$

It is easily checked that $T - T_0$ is a compact operator. Since we know that T_0 is invertible, by the Fredholm alternative, it is enough to show that T is injective. Suppose that $T(f,g) = 0$. Then the function u given by

$$u(x) := \begin{cases} \mathcal{S}_D^\omega[g](x), & x \in \mathbb{R}^d \setminus \overline{D}, \\ \tilde{\mathcal{S}}_D^\omega[f](x), & x \in D, \end{cases}$$

is a solution to the transmission problem

$$\begin{cases} \mathcal{L}^{\lambda,\mu} u + \omega^2 u = 0 & \text{in } \mathbb{R}^d \setminus \overline{D}, \\ \mathcal{L}^{\tilde{\lambda},\tilde{\mu}} u + \omega^2 u = 0 & \text{in } D, \\ u\big|_+ - u\big|_- = 0 & \text{on } \partial D, \\ \dfrac{\partial u}{\partial \nu}\Big|_+ - \dfrac{\partial u}{\partial \tilde{\nu}}\Big|_- = 0 & \text{on } \partial D, \end{cases}$$

satisfying the radiation condition. By the uniqueness of a solution to this transmission problem, see for instance [**342**, Chapter 3], we have $u = 0$. From the assumption on ω, we conclude that $f = g = 0$. This completes the proof. □

For transmission problems such as (2.412), the following representation formula holds.

THEOREM 2.123. *Let* $\Im\omega \geq 0$. *Suppose that* ω^2 *is not a Dirichlet eigenvalue for* $-\mathcal{L}^{\lambda,\mu}$ *on* D. *Let* u *be a solution of (*2.412*) and* $f := u|_{\partial\Omega}$. *Define*

$$H(x) := \mathcal{D}_\Omega^\omega[f](x) - \mathcal{S}_\Omega^\omega[g](x), \quad x \in \mathbb{R}^d \setminus \partial\Omega. \tag{2.413}$$

Then u *can be represented as*

$$u(x) = \begin{cases} H(x) + \mathcal{S}_D^\omega[\psi](x), & x \in \Omega \setminus \overline{D}, \\ \tilde{\mathcal{S}}_D^\omega[\phi](x), & x \in D, \end{cases} \tag{2.414}$$

where the pair $(\phi, \psi) \in L^2(\partial D)^d \times L^2(\partial D)^d$ *is the unique solution of*

$$\begin{cases} \tilde{\mathcal{S}}_D^\omega[\phi] - \mathcal{S}_D^\omega[\psi] = H|_{\partial D}, \\ \dfrac{\partial}{\partial \tilde{\nu}} \tilde{\mathcal{S}}_D^\omega[\phi] - \dfrac{\partial}{\partial \nu} \mathcal{S}_D^\omega[\psi] = \dfrac{\partial H}{\partial \nu}\Big|_{\partial D}. \end{cases} \tag{2.415}$$

Moreover, we have

$$H(x) + \mathcal{S}_D^\omega[\psi](x) = 0, \quad x \in \mathbb{R}^d \setminus \overline{\Omega}. \tag{2.416}$$

PROOF. We consider the following two-phase transmission problem:

$$\begin{cases} \mathcal{L}^{\lambda,\mu} v + \omega^2 v = 0 & \text{in } (\Omega \setminus \overline{D}) \cup (\mathbb{R}^d \setminus \overline{\Omega}), \\ \mathcal{L}^{\tilde{\lambda},\tilde{\mu}} v + \omega^2 v = 0 & \text{in } D, \\ v\big|_- - v\big|_+ = f, \ \dfrac{\partial v}{\partial \nu}\Big|_- - \dfrac{\partial v}{\partial \nu}\Big|_+ = g & \text{on } \partial\Omega, \\ v\big|_- - v\big|_+ = 0, \ \dfrac{\partial v}{\partial \tilde{\nu}}\Big|_- - \dfrac{\partial v}{\partial \nu}\Big|_+ = 0 & \text{on } \partial D, \end{cases} \tag{2.417}$$

with the radiation condition. This problem has a unique solution. See [**342**, Chapter 3]. It is easily checked that both v and $\tilde{v}$ defined by

$$v(x) = \begin{cases} u(x), & x \in \Omega, \\ 0, & x \in \mathbb{R}^d \setminus \overline{\Omega}, \end{cases}$$

and

$$\tilde{v}(x) = \begin{cases} H(x) + \mathcal{S}_D^\omega[\psi](x), & x \in \mathbb{R}^d \setminus \left(\overline{D} \cup \partial\Omega\right), \\ \tilde{\mathcal{S}}_D^\omega[\phi](x), & x \in D, \end{cases}$$

are solutions to (2.417). Hence $v = \tilde{v}$, which concludes the proof of the theorem. □

2.15.5. Eigenvalue Characterization. Let κ be an eigenvalue of $-\mathcal{L}^{\lambda,\mu}$ in Ω with the Neumann condition on $\partial\Omega$ and let u denote an eigenfunction associated with κ; *i.e.*,

$$\begin{cases} \mathcal{L}^{\lambda,\mu} u + \kappa u = 0 & \text{in } \Omega, \\ \dfrac{\partial u}{\partial \nu} = 0 & \text{on } \partial\Omega. \end{cases} \tag{2.418}$$

We note that since $-\mathcal{L}^{\lambda,\mu}$ is elliptic, it has discrete eigenvalues of finite multiplicities. The following proposition from [**342**, Chapter 7] is of importance to us.

PROPOSITION 2.124 (Eigenvalue characterization). *The necessary and sufficient condition for* (2.418) *to have a nontrivial solution is that* κ *is nonnegative and* $\sqrt{\kappa}$ *coincides with one of the characteristic values of* $(1/2)\, I - \mathcal{K}_\Omega^\omega$. *If* $\kappa = \omega_0^2$ *is an eigenvalue of* (2.418) *with multiplicity* m, *then* $\big((1/2)\, I - \mathcal{K}_\Omega^{\omega_0}\big)[\phi] = 0$ *has* m *linearly independent solutions. Moreover, for every eigenvalue* $\kappa > 0$, $\sqrt{\kappa}$ *is a simple pole of the operator-valued function* $\omega \mapsto ((1/2)\, I - \mathcal{K}_\Omega^\omega)^{-1}$.

2.15.6. Neumann Function. Let $0 \le \kappa_1 \le \kappa_2 \le \dots$ be the eigenvalues of $-\mathcal{L}^{\lambda,\mu}$ in Ω with the Neumann condition on $\partial\Omega$. For $\omega \notin \{\sqrt{\kappa_j}\}_{j\ge1}$, let $\mathbf{N}_\Omega^\omega(x,z)$ be the Neumann function for $\mathcal{L}^{\lambda,\mu}+\omega^2$ in Ω corresponding to a Dirac mass at z. That is, $\mathbf{N}_\Omega^\omega$ is the solution to

$$\begin{cases} (\mathcal{L}^{\lambda,\mu}+\omega^2)\mathbf{N}_\Omega^\omega(x,z) = -\delta_z(x)I, & x \in \Omega, \\ \dfrac{\partial \mathbf{N}_\Omega^\omega}{\partial \nu}(x,z) = 0, & x \in \partial\Omega. \end{cases} \tag{2.419}$$

Then the following relation, which can be proved similarly to (2.190), holds (see [**51**]):

$$\Big(-\frac{1}{2}I + \mathcal{K}_\Omega^\omega\Big)[\mathbf{N}_\Omega^\omega(\cdot,z)](x) = \mathbf{\Gamma}^\omega(x,z), \quad x \in \partial\Omega,\ z \in \Omega. \tag{2.420}$$

Let $(u_j)_{j\ge1}$ denote the set of orthogonal eigenfunctions associated with $(\kappa_j)_{j\ge1}$, with $\|u_j\|_{L^2(\Omega)} = 1$. Then we have the following spectral decomposition:

$$\mathbf{N}_\Omega^\omega(x,z) = \sum_{j=1}^{+\infty} \frac{u_j(x)u_j(z)^t}{\kappa_j - \omega^2}. \tag{2.421}$$

Here we regard u_j as a column vector, and hence $u_j(x)\,u_j(z)^t$ is a $d\times d$ matrix-valued function. We refer the reader to [**463**] for a proof of (2.421).

2.15.7. Dirichlet Function. Now we turn to the properties of the Dirichlet function. Let $0 \le \tau_1 \le \tau_2 \le \dots$ be the eigenvalues of $-\mathcal{L}^{\lambda,\mu}$ in Ω with the Dirichlet condition on $\partial\Omega$. For $\omega \notin \{\sqrt{\tau_j}\}_{j\ge1}$, let $\mathbf{G}_\Omega^\omega(x,z)$ be the Dirichlet function for $\mathcal{L}^{\lambda,\mu}+\omega^2$ in Ω corresponding to a Dirac mass at z. That is, for $z\in\Omega$, $\mathbf{G}_\Omega^\omega(\cdot,z)$ is the matrix-valued solution to

$$\begin{cases} (\mathcal{L}^{\lambda,\mu}+\omega^2)\mathbf{G}_\Omega^\omega(x,z) = -\delta_z(x)I, & x \in \Omega, \\ \mathbf{G}_\Omega^\omega(x,z) = 0, & x \in \partial\Omega. \end{cases} \tag{2.422}$$

Then for any $x \in \partial\Omega$, and $z \in \Omega$ we can prove in the same way as (2.420) that

$$\Big(\frac{1}{2}I + (\mathcal{K}_\Omega^\omega)^*\Big)[\frac{\partial \mathbf{G}_\Omega^\omega}{\partial \nu}(\cdot,z)](x) = -\frac{\partial \mathbf{\Gamma}^\omega}{\partial \nu}(x,z). \tag{2.423}$$

Moreover, we mention the following important properties of $\mathbf{G}_\Omega^\omega$:

(i) Let $(v_j)_{j\ge1}$ denote the set of orthogonal eigenvectors associated with $(\tau_j)_{j\ge1}$, with $\|v_j\|_{L^2(\Omega)} = 1$. Then we have the following spectral decomposition:

$$\mathbf{G}_\Omega^\omega(x,z) = \sum_{j=1}^{+\infty} \frac{v_j(x)v_j(z)^t}{\tau_j - \omega^2}. \tag{2.424}$$

(ii) For $x \in \partial\Omega$, $z\in\Omega$, $y \in \partial B$, and $\epsilon \to 0$,

$$\mathbf{G}_\Omega^\omega(x,\epsilon y + z) = \sum_{|\beta|=0}^{+\infty} \frac{1}{\beta!}\epsilon^{|\beta|}\partial_z^\beta \mathbf{G}_\Omega^\omega(x,z)y^\beta. \tag{2.425}$$

2.15.8. Neumann–Poincaré Operator. The Neumann–Poincaré operator for the Lamé system $\mathcal{K}_\Omega^0$ is defined by (2.400) for $\omega = 0$. It is connected to $\mathcal{L}^{\lambda,\mu}$ in the following way. The Dirichlet boundary value problem for the Lamé system

$$\begin{cases} \mathcal{L}^{\lambda,\mu} u = 0 & \text{in } \Omega, \\ u = g & \text{on } \partial\Omega, \end{cases} \tag{2.426}$$

can be solved by using the double-layer potential $u = \mathcal{D}_\Omega^0[f]$ and finding the solution of the integral equation

$$\left(\frac{1}{2}I + \mathcal{K}_\Omega^0\right)[f] = g \quad \text{on } \partial\Omega. \tag{2.427}$$

In this subsection, we show that $\mathcal{K}_\Omega^0$ can be realized as a self-adjoint operator on $H^{1/2}(\partial\Omega)^d$ by introducing a new inner product in a way parallel to the case of the Laplace operator. But, there is a significant difference between Neumann–Poincaré operators for the Laplace operator and the Lamé operator. The Neumann–Poincaré operator for the Lamé operator is *not* compact even if the domain has a smooth boundary, which means that we cannot infer directly that the Neumann–Poincaré operator has point spectrum (eigenvalues). However, it can be shown that the elasto-static Neumann–Poincaré operator on planar domains with $\mathcal{C}^{1,\eta}$, $\eta > 0$, boundaries has only point spectrum. In fact, for $d = 2$, one can show that on such domains [**97**]

$$(\mathcal{K}_\Omega^0)^2 - \kappa_*^2 I \ \text{ is compact,} \tag{2.428}$$

where

$$\kappa_* := \frac{\mu}{2(2\mu + \lambda)}. \tag{2.429}$$

As an immediate consequence of (2.428), it follows that the spectrum of $\mathcal{K}_\Omega^0$ consists of eigenvalues which accumulate at $\pm\mu/(2(2\mu+\lambda))$. Eigenvalues of $\mathcal{K}_\Omega^0$ are computed explicitly on disks and ellipses. It turns out that $\pm\mu/(2(2\mu+\lambda))$ are eigenvalues of infinite multiplicities (there are two other eigenvalues of finite multiplicities) on disks, while on ellipses $\pm\mu/(2(2\mu+\lambda))$ are accumulation points of eigenvalues, but not eigenvalues, and the rates of convergence to $\pm\mu/(2(2\mu+\lambda))$ are exponential.

Let

$$\mathcal{H}_\Psi^*(\partial\Omega) := \left\{ f \in H^{-1/2}(\partial\Omega)^d \ : \ \langle \psi, f \rangle_{1/2,-1/2} = 0 \text{ for all } \psi \in \Psi \right\}, \tag{2.430}$$

where Ψ is defined by (2.401).

The following lemma collects some facts to be used in the sequel, proofs of which can be found in [**194, 212, 402**].

LEMMA 2.125. *We have*

(i) $\mathcal{K}_\Omega^0$ *is bounded on* $H^{1/2}(\partial\Omega)^d$, *and* $(\mathcal{K}_\Omega^0)^*$ *is on* $H^{-1/2}(\partial\Omega)^d$.
(ii) *The spectrum of* $(\mathcal{K}_\Omega^0)^*$ *on* $H^{-1/2}(\partial\Omega)^d$ *lies in* $(-1/2, 1/2]$.
(iii) $(1/2)I - (\mathcal{K}_\Omega^0)^*$ *is invertible on* $\mathcal{H}_\Psi^*(\partial\Omega)$.
(iv) $\mathcal{S}_\Omega^0$ *as an operator defined on* $\partial\Omega$ *is bounded from* $H^{-1/2}(\partial\Omega)^d$ *into* $H^{1/2}(\partial\Omega)^d$.
(v) $\mathcal{S}_\Omega^0 : H^{-1/2}(\partial\Omega)^d \to H^{1/2}(\partial\Omega)^d$ *is invertible in three dimensions.*

In two dimensions $\mathcal{S}_\Omega^0$ may not be invertible. In fact, there is a bounded domain $\partial\Omega$ on which $\mathcal{S}_\Omega^0[\varphi] = 0$ on $\partial\Omega$ for some $\varphi \neq 0$ (see Subsection 2.15.9).

LEMMA 2.126. *Ψ is the eigenspace of $\mathcal{K}_\Omega^0$ on $H^{1/2}(\partial\Omega)^d$ corresponding to $1/2$.*

PROOF. Let $f \in \Psi$. Then $f = v|_{\partial\Omega}$ where v satisfies $\mathcal{L}^{\lambda,\mu} v = 0$ in Ω and $\partial v/\partial \nu = 0$ on $\partial\Omega$. So, we have for $x \in \mathbb{R}^d \setminus \overline{\Omega}$

$$\begin{aligned}\mathcal{D}_\Omega^0[f](x) &= \int_{\partial\Omega} \frac{\partial \mathbf{\Gamma}^0}{\partial \nu(y)}(x-y) f(y) d\sigma(y) \\ &= \int_{\partial\Omega} \left[\frac{\partial \mathbf{\Gamma}^0}{\partial \nu(y)}(x-y) v(y) - \mathbf{\Gamma}^0(x-y) \frac{\partial v}{\partial \nu(y)} \right] d\sigma(y) = 0.\end{aligned}$$

So we infer from (2.399) that

$$\mathcal{K}_\Omega^0[f] = \frac{1}{2} f. \tag{2.431}$$

Conversely, if (2.431) holds, then we have from (2.399) that $\mathcal{D}_\Omega^0[f]|_- = f$ and $\mathcal{D}_\Omega^0[f](x) = 0$ for $x \in \mathbb{R}^d \setminus \overline{\Omega}$. So $\partial \mathcal{D}_\Omega^0[f]/\partial\nu|_- = \partial \mathcal{D}_\Omega^0[f]/\partial\nu|_+ = 0$. It implies that $f \in \Psi$. This completes the proof. □

Let $N_d := \frac{d(d+1)}{2}$, which is the dimension of Ψ. Let $\{f^{(j)}\}_{j=1}^{N_d}$ be a basis of Ψ such that

$$\langle f^{(i)}, f^{(j)} \rangle_{1/2,-1/2} = \delta_{ij}, \tag{2.432}$$

where δ_{ij} is the Kronecker's delta. Since $\partial \mathcal{S}_\Omega^0[f^{(j)}]/\partial\nu|_- \in \mathcal{H}_\Psi^*$ and $1/2I - (\mathcal{K}_\Omega^0)^*$ is invertible on $\mathcal{H}_\Psi^*$, there is a unique $\widetilde{\varphi}^{(j)} \in \mathcal{H}_\Psi^*$ such that

$$\left(\frac{1}{2} I - (\mathcal{K}_\Omega^0)^* \right) [\widetilde{\varphi}^{(j)}] = \frac{\partial}{\partial \nu} \mathcal{S}_\Omega^0[f^{(j)}]|_- = \left(-\frac{1}{2} I + (\mathcal{K}_\Omega^0)^* \right) [f^{(j)}].$$

Define $\varphi^{(j)} := \widetilde{\varphi}^{(j)} + f^{(j)}$. Then, we have

$$(\mathcal{K}_\Omega^0)^* [\varphi^{(j)}] = \frac{1}{2} \varphi^{(j)}. \tag{2.433}$$

Moreover, we have

$$\langle f^{(i)}, \varphi^{(j)} \rangle_{1/2,-1/2} = \langle f^{(i)}, \widetilde{\varphi}^{(j)} \rangle_{1/2,-1/2} + \langle f^{(i)}, f^{(j)} \rangle_{1/2,-1/2} = \delta_{ij}, \tag{2.434}$$

which, in particular, implies that the $\varphi^{(j)}$'s are linearly independent.

Let

$$W := \operatorname{span} \left\{ \varphi^{(1)}, \ldots, \varphi^{(N_d)} \right\}, \tag{2.435}$$

and let

$$\mathcal{H}_W := \{ f \in H^{1/2}(\partial\Omega)^d \ : \ \langle f, \varphi \rangle_{1/2,-1/2} = 0 \text{ for all } \varphi \in W \}. \tag{2.436}$$

LEMMA 2.127. *The following results hold.*

(i) *Each $\varphi \in H^{-1/2}(\partial\Omega)^d$ is uniquely decomposed as*

$$\varphi = \varphi' + \varphi'' := \varphi' + \sum_{j=1}^{N_d} \langle f^{(j)}, \varphi \rangle_{1/2,-1/2} \varphi^{(j)}, \tag{2.437}$$

and $\varphi' \in \mathcal{H}_\Psi^*$.

(ii) *Each* $f \in H^{1/2}(\partial\Omega)^d$ *is uniquely decomposed as*

$$f = f' + f'' := f' + \sum_{j=1}^{N_d} \langle f, \varphi^{(j)} \rangle_{1/2,-1/2} f^{(j)}, \tag{2.438}$$

and $f' \in \mathcal{H}_W$.

(iii) $\mathcal{S}_\Omega^0$ *maps* W *into* Ψ, *and* $\mathcal{H}_\Psi^*$ *into* $\mathcal{H}_W$.

(iv) W *is the eigenspace of* $(\mathcal{K}_\Omega^0)^*$ *corresponding to the eigenvalue* $1/2$.

PROOF. For $\varphi \in H^{-1/2}(\partial\Omega)^d$, and let φ'' be as in (2.437). Then, one can immediately see from (2.434) that $\langle f^{(j)}, \varphi' \rangle_{1/2,-1/2} = 0$ for all j, and hence $\varphi' \in \mathcal{H}_\Psi^*$. Uniqueness of the decomposition can be proved easily. (ii) can be proved similarly.

Thanks to (2.433) we have $\partial \mathcal{S}_\Omega^0[\varphi^{(j)}]/\partial\nu|_- = 0$, and so $\mathcal{S}_\Omega^0[\varphi^{(j)}]|_{\partial\Omega} \in \Psi$. If $\varphi \in \mathcal{H}_\Psi^*$, then

$$\langle \mathcal{S}_\Omega^0[\varphi], \varphi^{(j)} \rangle_{1/2,-1/2} = \langle \mathcal{S}_\Omega^0[\varphi^{(j)}], \varphi \rangle_{1/2,-1/2} = 0$$

for all j. So, $\mathcal{S}_\Omega^0$ maps $\mathcal{H}_\Psi^*$ into $\mathcal{H}_W$. This proves (iii).

Suppose that $(\mathcal{K}_\Omega^0)^*[\varphi] = 1/2\varphi$ and that φ admits the decomposition (2.437). Then $(\mathcal{K}_\Omega^0)^*[\varphi'] = 1/2\varphi'$. So we have from (iii) that $\mathcal{S}_\Omega^0[\varphi'] \in \Psi$, and hence $\langle \mathcal{S}_\Omega^0[\varphi'], \varphi' \rangle_{1/2,-1/2} = 0$. Since $\int_{\partial\Omega} \varphi' d\sigma = 0$, we have from (2.398)

$$-\langle \mathcal{S}_\Omega^0[\varphi'], \varphi' \rangle_{1/2,-1/2} = \langle \mathcal{S}_\Omega^0[\varphi'], \frac{\partial}{\partial\nu} \mathcal{S}_\Omega^0[\varphi']|_- \rangle_{1/2,-1/2} - \langle \mathcal{S}_\Omega^0[\varphi'], \frac{\partial}{\partial\nu} \mathcal{S}_\Omega^0[\varphi']|_+ \rangle_{1/2,-1/2}$$

$$= \int_\Omega \mathbb{C}\nabla^s \mathcal{S}_\Omega^0[\varphi'] : \nabla^s \mathcal{S}_\Omega^0[\varphi'] \, dx \tag{2.439}$$

$$+ \int_{\mathbb{R}^d \setminus \overline{\Omega}} \mathbb{C}\nabla^s \mathcal{S}_\Omega^0[\varphi'] : \nabla^s \mathcal{S}_\Omega^0[\varphi'] \, dx, \tag{2.440}$$

where $\mathbb{C}$ defined by (2.388) is the elasticity tensor associated with (λ, μ) and $A : B = \sum_{i,j=1}^d a_{ij} b_{ij}$ for matrices $A = (a_{ij})$ and $B = (b_{ij})$. So $\mathcal{S}_\Omega^0[\varphi'] \in \Psi$. Thus we have $\varphi' = \frac{\partial}{\partial\nu} \mathcal{S}_\Omega^0[\varphi']|_+ - \frac{\partial}{\partial\nu} \mathcal{S}_\Omega^0[\varphi']|_- = 0$, and hence $\varphi \in W$. Thus (iv) is proved. This completes the proof. □

REMARK 2.128. *In three dimensions, it was recently proved in* [**100**] *that the Neumann–Poincaré operator defined on a smooth bounded domain is polynomially compact. Its spectrum consists of three non-empty sequences of eigenvalues accumulating to* $\kappa_*, -\kappa_*$, *and* 0, *where* κ_* *is the number given in* (2.429).

2.15.9. Symmetrization of the Neumann–Poincaré Operator. In this subsection we introduce a new inner product on $H^{-1/2}(\partial\Omega)^d$ (and $H^{1/2}(\partial\Omega)^d$) which makes the Neumann–Poincaré operator operator $(\mathcal{K}_\Omega^0)^*$ self-adjoint.

In three dimensions, $\mathcal{S}_\Omega^0[\varphi](x) = O(|x|^{-1})$ as $|x| \to \infty$. Using this fact, one can show that $-\mathcal{S}_\Omega^0$ is positive-definite. In fact, similarly to (2.440) we obtain

$$\begin{aligned} -\langle \mathcal{S}_\Omega^0[\varphi], \varphi \rangle_{1/2,-1/2} &= \int_\Omega \mathbb{C}\nabla^s \mathcal{S}_\Omega^0[\varphi] : \nabla^s \mathcal{S}_\Omega^0[\varphi] \, dx + \int_{\mathbb{R}^3 \setminus \overline{\Omega}} \mathbb{C}\nabla^s \mathcal{S}_\Omega^0[\varphi] : \nabla^s \mathcal{S}_\Omega^0[\varphi] \, dx \\ &\geq 0. \end{aligned}$$

If $\langle \mathcal{S}_\Omega^0[\varphi], \varphi \rangle_{1/2,-1/2} = 0$, then $\mathcal{S}_\Omega^0[\varphi]$ belongs to Ψ. Thus we have $\varphi = \partial \mathcal{S}_\Omega^0[\varphi]/\partial\nu|_+ - \partial \mathcal{S}_\Omega^0[\varphi]/\partial\nu|_- = 0$. So, if we define

$$\langle \varphi, \psi \rangle_{\mathcal{H}^*} := -\langle \mathcal{S}_\Omega^0[\varphi], \varphi \rangle_{1/2,-1/2}, \tag{2.441}$$

it is an inner product on $H^{-1/2}(\partial\Omega)^3$.

In two dimensions, the same argument shows that $-\mathcal{S}_\Omega^0$ is positive-definite on $\mathcal{H}_\Psi^*$. In fact, if $\varphi \in \mathcal{H}_\Psi^*$, then $\mathcal{S}_\Omega^0[\varphi](x) = O(|x|^{-1})$ as $|x| \to \infty$, and hence we can apply the same argument as in three dimensions. However, $-\mathcal{S}_\Omega^0$ may fail to be positive on W: if Ω is the disk of radius r (centered at 0), then we have

$$\mathcal{S}_\Omega^0[c](x) = \left[\alpha_1 r \ln r - \frac{\alpha_2 r}{2}\right] c \quad \text{for } x \in \Omega. \tag{2.442}$$

for any constant vector $c = (c_1, c_2)^t$. It shows that $-\mathcal{S}_\Omega^0$ can be positive or negative depending on r. To see (2.442), we note that

$$\begin{aligned}\mathcal{S}_\Omega^0[c]_i(x) &= \frac{\alpha_1 c_i}{2\pi}\int_{\partial\Omega} \ln|x-y| d\sigma(y) - \frac{\alpha_2}{2\pi}\sum_{j=1}^2 c_j \int_{\partial\Omega} \frac{(x-y)_i(x-y)_j}{|x-y|^2} d\sigma(y) \\ &= \alpha_1 c_i \mathcal{S}[1](x) - \alpha_2\Big(x_i c\cdot\nabla\mathcal{S}[1](x) - c\cdot\nabla\mathcal{S}[y_i](x)\Big),\end{aligned}$$

where $\mathcal{S}$ is the electro-static single-layer potential, namely,

$$\mathcal{S}[f](x) = \frac{1}{2\pi}\int_{\partial\Omega} \ln|x-y| f(y)\, d\sigma(y), \tag{2.443}$$

which was already defined before. Here, we use a different notation to avoid confusion with $\mathcal{S}_\Omega^0$.

It is known (see [**34**]) that $\mathcal{S}[1](x) = r\ln r$ and $\mathcal{S}[y_i](x) = -\frac{r x_i}{2}$ for $x \in \Omega$. So we have (2.442).

We introduce a variance of $\mathcal{S}_\Omega^0$ in two dimensions. For $\varphi \in H^{-1/2}(\partial\Omega)^2$, define using the decomposition (2.437)

$$\widetilde{\mathcal{S}}_\Omega^0[\varphi] := \mathcal{S}_\Omega^0[\varphi'] + \sum_{j=1}^3 \langle f^{(j)}, \varphi\rangle_{1/2,-1/2} f^{(j)}. \tag{2.444}$$

We emphasize that $\widetilde{\mathcal{S}}_\Omega^0[\varphi] = \mathcal{S}_\Omega^0[\varphi]$ for all $\varphi \in \mathcal{H}_\Psi^*$ and $\widetilde{\mathcal{S}}_\Omega^0[\varphi^{(j)}] = f^{(j)}$, $j = 1, 2, 3$. In view of (2.434) and Lemma 2.127 (iii), we have

$$-\langle \widetilde{\mathcal{S}}_\Omega^0[\varphi], \varphi\rangle_{1/2,-1/2} = -\langle \mathcal{S}_\Omega^0[\varphi'], \varphi'\rangle_{1/2,-1/2} + \sum_{j=1}^3 |\langle f^{(j)}, \varphi\rangle_{1/2,-1/2}|^2. \tag{2.445}$$

So, $-\widetilde{\mathcal{S}}_\Omega^0$ is positive-definite on $H^{-1/2}(\partial\Omega)^2$. In fact, since

$$-\langle \mathcal{S}_\Omega^0[\varphi'], \varphi'\rangle_{1/2,-1/2} \geq 0,$$

we have $-\langle \widetilde{\mathcal{S}}_\Omega^0[\varphi], \varphi\rangle_{1/2,-1/2} \geq 0$. If $-\langle \widetilde{\mathcal{S}}_\Omega^0[\varphi], \varphi\rangle_{1/2,-1/2} = 0$, then

$$-\langle \mathcal{S}_\Omega^0[\varphi'], \varphi'\rangle_{1/2,-1/2} = 0$$

and

$$\sum_{j=1}^3 |\langle f^{(j)}, \varphi\rangle_{1/2,-1/2}|^2 = 0.$$

So, $\varphi' = 0$ and $\langle f^{(j)}, \varphi\rangle_{1/2,-1/2} = 0$ for all j, and hence $\varphi = 0$.

Let us also denote $\mathcal{S}_\Omega^0$ in three dimensions by $\widetilde{\mathcal{S}}_\Omega^0$ for convenience. Define

$$\langle \varphi, \psi\rangle_{\mathcal{H}^*} := -\langle \widetilde{\mathcal{S}}_\Omega^0[\psi], \varphi\rangle_{1/2,-1/2}, \quad \varphi, \psi \in H^{-1/2}(\partial\Omega)^d. \tag{2.446}$$

PROPOSITION 2.129. *$\langle\cdot,\cdot\rangle_{\mathcal{H}^*}$ is an inner product on $H^{-1/2}(\partial\Omega)^d$. The norm induced by $\langle\cdot,\cdot\rangle_{\mathcal{H}^*}$, denoted by $\|\cdot\|_*$, is equivalent to the $H^{-1/2}$-norm.*

PROOF. Positive-definiteness of $-\widetilde{\mathcal{S}}^0_\Omega$ implies that $\widetilde{\mathcal{S}}^0_\Omega : H^{-1/2}(\partial\Omega)^d \to H^{1/2}(\partial\Omega)^d$ is bijective. So, we have

$$\|\varphi\|_{-1/2} \approx \|\widetilde{\mathcal{S}}^0_\Omega[\varphi]\|_{1/2}.$$

Here and throughout this chapter $A \lesssim B$ means that there is a constant C such that $A \leq CB$, and $A \approx B$ means $A \lesssim B$ and $B \lesssim A$. It then follows from the definition (2.446) that

$$|\langle\varphi,\varphi\rangle_{\mathcal{H}^*}| \leq \|\varphi\|_{H^{-1/2}}\|\widetilde{\mathcal{S}}^0_\Omega[\varphi]\|_{1/2} \lesssim \|\varphi\|^2_{H^{-1/2}}.$$

We have from the Cauchy Schwarz inequality

$$|\langle\widetilde{\mathcal{S}}^0_\Omega[\psi],\varphi\rangle_{1/2,-1/2}| = |\langle\varphi,\psi\rangle_{\mathcal{H}^*}| \leq \|\varphi\|_{\mathcal{H}^*}\|\psi\|_{\mathcal{H}^*} \lesssim \|\varphi\|_{\mathcal{H}^*}\|\widetilde{\mathcal{S}}^0_\Omega[\psi]\|_{H^{1/2}}.$$

So we have

$$\|\varphi\|_{-1/2} = \sup_{\psi\neq 0} \frac{|\langle\widetilde{\mathcal{S}}^0_\Omega[\psi],\varphi\rangle_{1/2,-1/2}|}{\|\widetilde{\mathcal{S}}^0_\Omega[\psi]\|_{1/2}} \lesssim \|\varphi\|_*.$$

This completes the proof. □

We may define a new inner product on $H^{1/2}(\partial\Omega)^d$ by

(2.447)

$$\langle f,g\rangle_{\mathcal{H}} := \langle(\widetilde{\mathcal{S}}^0_\Omega)^{-1}[f],(\widetilde{\mathcal{S}}^0_\Omega)^{-1}[g]\rangle_{\mathcal{H}^*} = -\langle g,(\widetilde{\mathcal{S}}^0_\Omega)^{-1}[f]\rangle_{1/2,-1/2}, \quad f,g \in H^{1/2}(\partial\Omega)^d.$$

PROPOSITION 2.130. *$\langle\cdot,\cdot\rangle_{\mathcal{H}}$ is an inner product on $H^{1/2}(\partial\Omega)^d$. The norm induced by $\langle\cdot,\cdot\rangle$, denoted by $\|\cdot\|_{\mathcal{H}}$, is equivalent to $\|\cdot\|_{H^{1/2}}$. Moreover, $\widetilde{\mathcal{S}}^0_\Omega$ is an isometry between $H^{-1/2}(\partial\Omega)^d$ and $H^{1/2}(\partial\Omega)^d$.*

As for the Laplace operator, the Neumann–Poincaré operator $(\mathcal{K}^0_\Omega)^*$ associated to the Lamé system can be realized as a self-adjoint operator on $H^{-1/2}(\partial\Omega)^d$ using Plemelj's symmetrization principle which states that

$$\mathcal{S}^0_\Omega(\mathcal{K}^0_\Omega)^* = \mathcal{K}^0_\Omega\mathcal{S}^0_\Omega. \tag{2.448}$$

This relation is a consequence of the Green's formula. In fact, if $\mathcal{L}^{\lambda,\mu}u = 0$ in Ω, then we have for $x \in \mathbb{R}^d \setminus \overline{\Omega}$

$$\mathcal{S}^0_\Omega\left[\frac{\partial u}{\partial\nu}\Big|_-\right](x) - \mathcal{D}^0_\Omega[u|_-](x) = 0.$$

Substituting $u(x) = \mathcal{S}^0_\Omega[\varphi](x)$ for some $\varphi \in H^{-1/2}(\partial\Omega)^d$ into the above relation yields

$$\mathcal{S}^0_\Omega\left(-\frac{1}{2}I + (\mathcal{K}^0_\Omega)^*\right)[\varphi](x) = \mathcal{D}^0_\Omega\mathcal{S}^0_\Omega[\varphi](x), \quad x \in \mathbb{R}^d \setminus \overline{\Omega}.$$

Letting x approach $\partial\Omega$, we have from (2.399)

$$\mathcal{S}^0_\Omega\left(-\frac{1}{2}I + (\mathcal{K}^0_\Omega)^*\right)[\varphi](x) = \left(-\frac{1}{2}I + \mathcal{K}^0_\Omega\right)\mathcal{S}^0_\Omega[\varphi](x), \quad x \in \partial\Omega.$$

So we have (2.448).

The relation (2.448) holds with $\mathcal{S}^0_\Omega$ replaced by $\widetilde{\mathcal{S}}^0_\Omega$, namely,

$$\widetilde{\mathcal{S}}^0_\Omega(\mathcal{K}^0_\Omega)^* = \mathcal{K}^0_\Omega\widetilde{\mathcal{S}}^0_\Omega. \tag{2.449}$$

In fact, if $\varphi \in W$, then $(\mathcal{K}_\Omega^0)^*[\varphi] = 1/2\varphi$ and $\widetilde{\mathcal{S}}_\Omega^0[\varphi] \in \Psi$. So, we have

$$\widetilde{\mathcal{S}}_\Omega^0(\mathcal{K}_\Omega^0)^*[\varphi] = \mathcal{K}_\Omega^0\widetilde{\mathcal{S}}_\Omega^0[\varphi].$$

This proves (2.449).

PROPOSITION 2.131. *The Neumann–Poincaré operators $(\mathcal{K}_\Omega^0)^*$ and $\mathcal{K}_\Omega^0$ are self-adjoint with respect to $\langle\cdot,\cdot\rangle_{\mathcal{H}^*}$ and $\langle\cdot,\cdot\rangle_{\mathcal{H}}$, respectively.*

PROOF. According to (2.449), we have

$$\begin{aligned}\langle\varphi,(\mathcal{K}_\Omega^0)^*[\psi]\rangle_{\mathcal{H}^*} &= -\langle\widetilde{\mathcal{S}}_\Omega^0(\mathcal{K}_\Omega^0)^*[\psi],\varphi\rangle_{1/2,-1/2} = -\langle\mathcal{K}_\Omega^0\widetilde{\mathcal{S}}_\Omega^0[\psi],\varphi\rangle_{1/2,-1/2}\\ &= -\langle\widetilde{\mathcal{S}}_\Omega^0[\psi],(\mathcal{K}_\Omega^0)^*[\varphi]\rangle_{1/2,-1/2} = \langle(\mathcal{K}_\Omega^0)^*[\varphi],\psi\rangle_{\mathcal{H}^*}.\end{aligned}$$

So $(\mathcal{K}_\Omega^0)^*$ is self-adjoint. That $\mathcal{K}_\Omega^0$ is self-adjoint can be proved similarly. □

2.15.10. Spectrum of the Neumann–Poincaré Operators on Smooth Planar Domains. In this subsection we prove (2.428) when $\partial\Omega$ is $\mathcal{C}^{1,\eta}$ for some $\eta > 0$. For that purpose we look into $\mathcal{K}_\Omega^0$ in more explicit form. The definition (2.397) and straightforward computations show that

$$\frac{\partial}{\partial\nu(y)}\mathbf{\Gamma}^0(x-y) = \frac{\mu}{2\mu+\lambda}K_1(x,y) - K_2(x,y), \tag{2.450}$$

where

$$K_1(x,y) = \frac{n(y)(x-y)^t - (x-y)n(y)^t}{\omega_d|x-y|^d}, \tag{2.451}$$

$$K_2(x,y) = \frac{\mu}{2\mu+\lambda}\frac{(x-y)\cdot n(y)}{\omega_d|x-y|^d}I + \frac{2(\mu+\lambda)}{2\mu+\lambda}\frac{(x-y)\cdot n(y)}{\omega_d|x-y|^{d+2}}(x-y)(x-y)^t, \tag{2.452}$$

where ω_d is 2π if $d=2$ and 4π if $d=3$, and I is the $d\times d$ identity matrix. Let

$$T_j[\varphi](x) := \text{p.v.}\int_{\partial\Omega} K_j(x,y)\varphi(y)\,d\sigma(y), \quad x\in\partial\Omega,\ j=1,2, \tag{2.453}$$

so that

$$\mathcal{K}_\Omega^0 = \frac{\mu}{2\mu+\lambda}T_1 - T_2. \tag{2.454}$$

Note that each term of K_2 features the term $(x-y)\cdot n(y)$. Since $\partial\Omega$ is $\mathcal{C}^{1,\eta}$, we have

$$|(x-y)\cdot n(y)| \le C|x-y|^{1+\eta}$$

for some constant C due to the orthogonality of $x-y$ and $n(y)$. So we have

$$|K_2(x,y)| \le C|x-y|^{-d+1+\alpha}.$$

So T_2 is compact on $H^{1/2}(\partial\Omega)^d$ (see, for example, [**243**]), and T_1 is responsible for the non-compactness of K.

2.15.11. Compactness and Spectrum of $(\mathcal{K}_\Omega^0)^2 - \kappa_*^2$.

PROPOSITION 2.132. *Let Ω be a bounded $\mathcal{C}^{1,\eta}$ domain in $\mathbb{R}^2$ for some $\eta > 0$. Then $(\mathcal{K}_\Omega^0)^2 - \kappa_*^2 I$ is compact on $H^{1/2}(\partial\Omega)^2$.*

PROOF. In view of (2.454), it suffices to show that $T_1^2 - \frac{1}{4}I$ is compact.

In two dimensions we have

$$K_1(x,y) = \frac{1}{2\pi|x-y|^2}\begin{bmatrix} 0 & K(x,y) \\ -K(x,y) & 0 \end{bmatrix},$$

where

$$K(x,y) := -n_2(y)(x_1 - y_1) + n_1(y)(x_2 - y_2).$$

Let

$$\mathcal{R}[\varphi](x) = \frac{1}{2\pi}\text{p.v.}\int_{\partial\Omega} \frac{K(x,y)}{|x-y|^2}\varphi(y)\,d\sigma(y). \tag{2.455}$$

Then we have

$$T_1[\varphi] = \begin{bmatrix} \mathcal{R}[\varphi_2] \\ -\mathcal{R}[\varphi_1] \end{bmatrix}. \tag{2.456}$$

For $x \in \partial\Omega$, set $\Omega_\epsilon := \Omega \setminus B_\epsilon(x)$ where $B_\epsilon(x)$ is the disk of radius ϵ centered at x. For $\varphi \in H^{1/2}(\partial\Omega)$, let u be the solution to $\Delta u = 0$ in Ω with $u = \varphi$ on $\partial\Omega$. Since

$$\nabla \times \Big(\frac{x-y}{|x-y|^2}\Big) = 0, \quad x \neq y,$$

we obtain from Stokes' formula

$$\begin{aligned}\mathcal{R}[\varphi](x) &= \lim_{\epsilon\to 0}\frac{1}{2\pi}\int_{\partial\Omega_\epsilon} \frac{-(x_1-y_1)n_2(y) + (x_2-y_2)n_1(y)}{|x-y|^2}\varphi(y)\,d\sigma(y) \\ &= \lim_{\epsilon\to 0}\frac{1}{2\pi}\int_{\Omega_\epsilon} \frac{-(x_1-y_1)\partial_2 u(y) + (x_2-y_2)\partial_1 u(y)}{|x-y|^2}\,dy.\end{aligned}$$

Let v be a harmonic conjugate of u in Ω and $\psi := v|_{\partial\Omega}$ so that

$$\psi = \mathcal{T}[\varphi], \tag{2.457}$$

where $\mathcal{T}$ is the Hilbert transformation on $\partial\Omega$. Then we have from the divergence theorem

$$\begin{aligned}\mathcal{R}[\varphi](x) &= \lim_{\epsilon\to 0}\frac{1}{2\pi}\int_{\Omega_\epsilon} \frac{(x_1-y_1)\partial_1 v(y) + (x_2-y_2)\partial_2 v(y)}{|x-y|^2}\,dy \\ &= \lim_{\epsilon\to 0}\frac{1}{2\pi}\int_{\partial\Omega_\epsilon} \frac{(x-y)\cdot n(y)}{|x-y|^2}\psi(y)\,d\sigma(y).\end{aligned}$$

Observe that

$$\frac{1}{2\pi}\int_{\partial\Omega_\epsilon} \frac{(x-y)\cdot n(y)}{|x-y|^2}\psi(y)\,d\sigma(y)$$

is the electro-static double-layer potential of y, and $x \notin \Omega_\epsilon$. So by the jump formula (2.399) of the the double-layer potential, we have

$$\lim_{\epsilon\to 0}\frac{1}{2\pi}\int_{\partial\Omega_\epsilon} \frac{(x-y)\cdot n(y)}{|x-y|^2}\psi(y)\,d\sigma(y) = -\frac{1}{2}\psi(x) + \mathcal{K}_\Omega^0[\psi](x),$$

where

$$\mathcal{K}[\psi](x) := \frac{1}{2\pi}\int_{\partial\Omega} \frac{(x-y)\cdot n(y)}{|x-y|^2}\psi(y)\,d\sigma(y), \quad x \in \partial\Omega. \tag{2.458}$$

It is worth mentioning that $\mathcal{K}$ is the electro-static Neumann–Poincaré operator.

So far we have shown that

$$\mathcal{R}[\varphi] = -\frac{1}{2}\mathcal{T}[\varphi] + \mathcal{K}\mathcal{T}[\varphi]. \tag{2.459}$$

Since $\mathcal{T}$ is bounded and $\mathcal{K}$ is compact on $H^{1/2}(\partial\Omega)$, we have

$$\mathcal{R} = -\frac{1}{2}\mathcal{T} + \text{compact operator}. \tag{2.460}$$

Since $\mathcal{T}^2 = -I$, we infer that $\mathcal{R}^2 + \frac{1}{4}I$ is compact, and so is $T_1^2 - \frac{1}{4}I$. This completes the proof. □

Since $(\mathcal{K}_\Omega^0)^2 - \kappa_*^2 I$ is compact and self-adjoint, it has eigenvalues converging to 0. The proof of Proposition 2.132 shows that neither $\mathcal{K}_\Omega^0 - \kappa_* I$ nor $\mathcal{K}_\Omega^0 + \kappa_* I$ is compact, so we obtain the following theorem.

THEOREM 2.133. *Let Ω be a bounded domain in $\mathbb{R}^2$ with $\mathcal{C}^{1,\eta}$ boundary for some $\eta > 0$.*

(i) *The spectrum of $\mathcal{K}_\Omega^0$ on $H^{1/2}(\partial\Omega)^2$ consists of eigenvalues accumulating at $\pm\kappa_*$, and their multiplicities are finite if they are not equal to κ_* or $-\kappa_*$.*

(ii) *The spectrum of $(\mathcal{K}_\Omega^0)^*$ on $H^{-1/2}(\partial\Omega)^2$ is the same as that of $\mathcal{K}_\Omega^0$ on $H^{1/2}(\partial\Omega)^2$.*

(iii) *The set of linearly independent eigenfunctions of $\mathcal{K}_\Omega^0$ defines a complete orthogonal system of $H^{1/2}(\partial\Omega)^2$.*

(iv) *φ is an eigenfunction of $(\mathcal{K}_\Omega^0)^*$ on $H^{-1/2}(\partial\Omega)^2$ if and only if $\widetilde{\mathcal{S}}_\Omega^0[\varphi]$ is an eigenfunction of $\mathcal{K}_\Omega^0$ on $H^{1/2}(\partial\Omega)^2$.*

2.15.12. Spectral Expansion of the Fundamental Solution. Let $\{\psi_j\}$ be a complete orthonormal (with respect to the inner product $\langle\cdot,\cdot\rangle_{\mathcal{H}^*}$) system of $\mathcal{H}_\Psi^*$ consisting of eigenfunctions of $(\mathcal{K}_\Omega^0)^*$ on $\partial\Omega$ in two dimensions. Then they, together with $\varphi^{(j)}$, $j = 1, 2, 3$, defined in Subsection 2.15.9, define an orthonormal system of $H^{-1/2}(\partial\Omega)^d$. Then by Theorem 2.133 (iv) $\{\mathcal{S}_\Omega^0[\psi_j]\}$ together with $f^{(i)}$ is a complete orthonormal system of $H^{1/2}(\partial\Omega)^d$ with respect to the inner product $\langle\cdot,\cdot\rangle_{\mathcal{H}}$.

Let $\mathbf{\Gamma}^0(x-y)$ be the Kelvin matrix defined in (2.394). If $x \in \mathbb{R}^2 \setminus \overline{\Omega}$ and $y \in \partial\Omega$, then there are (column) vector-valued functions a_j and b_i such that

$$\mathbf{\Gamma}^0(x-y) = \sum_{j=1}^{\infty} a_j(x)\mathcal{S}_\Omega^0[\psi_j](y)^t + \sum_{i=1}^{3} b_i(x) f^{(i)}(y)^t. \tag{2.461}$$

It then follows that

$$\begin{aligned}\int_{\partial\Omega} \mathbf{\Gamma}^0(x-y)\psi_l(y)\,d\sigma(y) &= \sum_{j=1}^{\infty} a_j(x)\langle\mathcal{S}_\Omega^0[\psi_j], \psi_l\rangle_{1/2,-1/2} \\ &\quad + \sum_{i=1}^{3} b_i(x)\langle f^{(i)}, \psi_l\rangle_{1/2,-1/2} \\ &= -\sum_{j=1}^{\infty} a_j(x)\langle\psi_j, \psi_l\rangle_{\mathcal{H}^*} + \sum_{i=1}^{3} b_i(x)\langle\varphi^{(i)}, \psi_l\rangle_{\mathcal{H}^*} = -a_l(x).\end{aligned}$$

In other words, we obtain $a_l(x) = -\mathcal{S}_\Omega^0[\psi_l](x)$. Likewise, one can show that

$$b_i(x) = \widetilde{\mathcal{S}}_\Omega^0[\varphi^{(i)}](x).$$

So, we obtain

$$\mathbf{\Gamma}^0(x-y) = -\sum_{j=1}^{\infty} \mathcal{S}_\Omega^0[\psi_j](x)\mathcal{S}_\Omega^0[\psi_j](y)^t + \sum_{i=1}^{3} \widetilde{\mathcal{S}}_\Omega^0[\varphi^{(i)}](x) f^{(i)}(y)^t, \quad x \in \mathbb{R}^2 \setminus \overline{\Omega},\ y \in \partial\Omega.$$

Since both sides of the above are solutions of the Lamé equation in y for a fixed x, we obtain the following theorem from the uniqueness of the solution to the Dirichlet boundary value problem.

THEOREM 2.134 (expansion in two dimensions). *Let Ω be a bounded domain in $\mathbb{R}^2$ with $\mathcal{C}^{1,\eta}$ boundary for some $\eta > 0$ and let $\{\psi_j\}$ be a complete orthonormal system of $\mathcal{H}_\Psi^*$ consisting of eigenfunctions of $(\mathcal{K}_\Omega^0)^*$. Let $\mathbf{\Gamma}^0(x-y)$ be the Kelvin matrix of the fundamental solution to the Lamé system. It holds that*

(2.462)
$$\mathbf{\Gamma}^0(x-y) = -\sum_{j=1}^{\infty} \mathcal{S}_\Omega^0[\psi_j](x)\mathcal{S}_\Omega^0[\psi_j](y)^t + \sum_{i=1}^{3} \widetilde{\mathcal{S}}_\Omega^0[\varphi^{(i)}](x) f^{(i)}(y)^t, \quad x \in \mathbb{R}^2 \setminus \overline{\Omega},\ y \in \Omega.$$

In three dimensions one can prove the following theorem similarly.

THEOREM 2.135 (expansion in three dimensions). *Let Ω be a bounded domain in $\mathbb{R}^3$. Let $\{\psi_j\}$ be eigenfunctions of $(\mathcal{K}_\Omega^0)^*$, which form a complete orthonormal system of $H^{-1/2}(\partial\Omega)^d$. It holds that*

$$\mathbf{\Gamma}^0(x-y) = -\sum_{j=1}^{\infty} \mathcal{S}_\Omega^0[\psi_j](x)\mathcal{S}_\Omega^0[\psi_j](y)^t, \quad x \in \mathbb{R}^3 \setminus \overline{\Omega},\ y \in \Omega. \tag{2.463}$$

Theorems 2.134 and 2.135 extend formula (2.33) to the Kelvin matrix of the fundamental solution to the Lamé system. Using explicit forms of eigenfunctions to be derived in the next subsection, one can compute in two dimensions the expansion formula on disks and ellipses explicitly.

2.15.13. Spectrum of the Neumann–Poincaré Operator on Disks and Ellipses. In this subsection we write down the spectrum of the Neumann–Poincaré operator on disks and ellipses. Detailed derivation of the spectrum is presented in [**97**].

Suppose that Ω is a disk. The spectrum of $(\mathcal{K}_\Omega^0)^*$ is as follows:

Eigenvalues:

$$\frac{1}{2}, \quad -\frac{\lambda}{2(2\mu+\lambda)}, \quad \pm\kappa_*. \tag{2.464}$$

It is worth mentioning that the second eigenvalue above is less than $1/2$ in absolute value because of the strong convexity condition (2.385).

Eigenfunctions:

(i) $1/2$:

$$(1,0)^t, \quad (0,1)^t, \quad (y,-x)^t, \tag{2.465}$$

(ii) $-\frac{\lambda}{2(2\mu+\lambda)}$:

$$(x,y)^t, \tag{2.466}$$

(iii) κ_*:

$$\begin{bmatrix}\cos m\theta\\ \sin m\theta\end{bmatrix}, \quad \begin{bmatrix}-\sin m\theta\\ \cos m\theta\end{bmatrix}, \quad m=2,3,\dots, \tag{2.467}$$

(iv) $-\kappa_*$:

$$\begin{bmatrix}\cos m\theta\\ -\sin m\theta\end{bmatrix}, \quad \begin{bmatrix}\sin m\theta\\ \cos m\theta\end{bmatrix}, \quad m=1,2,\dots. \tag{2.468}$$

We emphasize that these eigenfunctions are not normalized.

We now describe eigenvalues and eigenfunctions on ellipses. Suppose that Ω is an ellipse of the form

$$\frac{x_1^2}{a^2}+\frac{x_2^2}{b^2}<1, \quad a\geq b>0. \tag{2.469}$$

Set $R:=\sqrt{a^2-b^2}$. Then the elliptic coordinates (ρ,ω) are defined by

$$x_1=R\cosh\rho\cos\omega, \quad x_2=R\sinh\rho\sin\omega, \quad \rho\geq 0,\ 0\leq\omega\leq 2\pi, \tag{2.470}$$

in which the ellipse Ω is given by $\partial\Omega=\{(\rho,\omega):\rho=\rho_0\}$, where ρ_0 is defined to be $a=R\cosh\rho_0$ and $b=R\sinh\rho_0$. Define

$$h_0(\omega):=R\sqrt{\sinh^2\rho_0+\sin^2\omega}. \tag{2.471}$$

For notational convenience we set

$$q:=(\lambda+\mu)\sinh 2\rho_0 \tag{2.472}$$

and

$$\gamma_n^{\pm}:=\sqrt{e^{4n\rho_0}\mu^2+(\lambda+\mu)(\lambda+3\mu)+nq(\pm 2e^{2n\rho_0}\mu+nq)}. \tag{2.473}$$

The spectrum of $(\mathcal{K}_\Omega^0)^*$ is as follows

Eigenvalues:

$$\frac{1}{2}, \quad k_{j,n},\ j=1,\dots,4, \tag{2.474}$$

where

$$\begin{aligned} k_{1,n}&=\frac{e^{-2n\rho_0}}{2(\lambda+2\mu)}(-qn+\gamma_n^-), \quad n\geq 1,\\ k_{2,n}&=\frac{e^{-2n\rho_0}}{2(\lambda+2\mu)}(qn+\gamma_n^+), \quad n\geq 2,\\ k_{3,n}&=\frac{e^{-2n\rho_0}}{2(\lambda+2\mu)}(-qn-\gamma_n^-), \quad n\geq 1,\\ k_{4,n}&=\frac{e^{-2n\rho_0}}{2(\lambda+2\mu)}(qn-\gamma_n^+), \quad n\geq 1. \end{aligned} \tag{2.475}$$

Eigenfunctions:

(i) $1/2$:

$$\frac{1}{h_0(\omega)}\begin{bmatrix}1\\0\end{bmatrix}, \quad \frac{1}{h_0(\omega)}\begin{bmatrix}0\\1\end{bmatrix}, \quad \frac{1}{h_0(\omega)}\begin{bmatrix}\big((\lambda+\mu)e^{-2\rho_0}-(\lambda+3\mu)\big)\sin\omega\\ \big((\lambda+\mu)e^{-2\rho_0}+(\lambda+3\mu)\big)\cos\omega\end{bmatrix}, \tag{2.476}$$

(ii) $k_{j,n}$, $j = 1, 2, 3, 4$:

(2.477)
$$\begin{aligned}
\varphi_{1,n} &= \psi_{1,n} + \frac{p_n}{\kappa_* + k_{1,n}} \psi_{3,n}, \quad n \geq 1, \\
\varphi_{2,n} &= \psi_{2,n} + \frac{p_n}{\kappa_* + k_{2,n}} \psi_{4,n}, \quad n \geq 2, \\
\varphi_{3,n} &= \frac{\kappa_* + k_{3,n}}{p_n} \psi_{1,n} + \psi_{3,n}, \quad n \geq 1, \\
\varphi_{4,n} &= \frac{\kappa_* + k_{4,n}}{p_n} \psi_{2,n} + \psi_{4,n}, \quad n \geq 1,
\end{aligned}$$

where

(2.478)
$$p_n = \left(\frac{1}{2} - (\mu/(2(2\mu + \lambda))) \right) e^{-2n\rho_0},$$

and

(2.479)
$$\begin{aligned}
\psi_{1,n}(\omega) &= \frac{1}{h_0(\omega)} \begin{bmatrix} \cos n\omega \\ \sin n\omega \end{bmatrix}, \quad \psi_{2,n}(\omega) = \frac{1}{h_0(\omega)} \begin{bmatrix} -\sin n\omega \\ \cos n\omega \end{bmatrix}, \\
\psi_{3,n}(\omega) &= \frac{1}{h_0(\omega)} \begin{bmatrix} \cos n\omega \\ -\sin n\omega \end{bmatrix}, \quad \psi_{4,n}(\omega) = \frac{1}{h_0(\omega)} \begin{bmatrix} \sin n\omega \\ \cos n\omega \end{bmatrix}.
\end{aligned}$$

A remark on $k_{2,1}$ in (2.475) is in order. It is given by

$$k_{2,1} = \frac{e^{-2\rho_0}}{2(\lambda + 2\mu)} (q + \gamma_1^+),$$

where

$$\gamma_1^+ := \sqrt{e^{4\rho_0} \mu^2 + (\lambda + \mu)(\lambda + 3\mu) + q(2e^{2\rho_0} \mu + q)}.$$

Since

$$\mu^2 e^{4\rho_0} + (\lambda + \mu)(\lambda + 3\mu) + q(2e^{2\rho_0} \mu + q) = \frac{1}{4} \left[(\lambda + 3\mu) e^{2\rho_0} + (\lambda + \mu) e^{-2\rho_0} \right]^2,$$

we have $\lambda_{2,1} = \frac{1}{2}$ and the corresponding eigenfunction is

$$\varphi_{2,1} = h_0^{-1}(\omega) \begin{bmatrix} \big((\lambda + \mu) e^{-2\rho_0} - (\lambda + 3\mu) \big) \sin \omega \\ \big((\lambda + \mu) e^{-2\rho_0} + (\lambda + 3\mu) \big) \cos \omega \end{bmatrix}.$$

So it is listed as an eigenfunction for $1/2$.

Let us now look into the asymptotic behavior of the eigenvalues as $n \to \infty$. One can easily see from the definition (2.473) that

$$\gamma_n^\pm = \mu e^{2n\rho_0} \pm qn \mp \frac{(\lambda + \mu)(\lambda + 3\mu) q}{2\mu^2} n e^{-2n\rho_0} + e^{-2n\rho_0} O(1),$$

where $O(1)$ indicates constants bounded independently of n. So one can infer from (2.475) that

(2.480)
$$\begin{aligned}
k_{1,n} &= \kappa_* - \frac{q}{\lambda + 2\mu} n e^{-2n\rho_0} + n^2 e^{-4n\rho_0} O(1), \\
k_{2,n} &= \kappa_* + \frac{q}{\lambda + 2\mu} n e^{-2n\rho_0} + n^2 e^{-4n\rho_0} O(1), \\
k_{3,n} &= -\kappa_* - \frac{(\lambda + \mu)(\lambda + 3\mu) q}{4\mu^2 (\lambda + 2\mu)} n e^{-4n\rho_0} + e^{-4n\rho_0} O(1), \\
k_{4,n} &= -\kappa_* + \frac{(\lambda + \mu)(\lambda + 3\mu) q}{4\mu^2 (\lambda + 2\mu)} n e^{-4n\rho_0} + e^{-4n\rho_0} O(1),
\end{aligned}$$

as $n \to \infty$. In particular, we see that $k_{1,n}$ and $k_{2,n}$ converge to κ_* while $k_{3,n}$ and $k_{4,n}$ to $-\kappa_*$ as $n \to \infty$. We emphasize that the convergence rates are exponential.

2.15.14. Reciprocity Property and Helmholtz-Kirchhoff Identities. An important property satisfied by the fundamental solution $\mathbf{\Gamma}^\omega$ is the reciprocity property.

If the medium is homogeneous, then one can see from (2.389) and (2.393) that $\mathbf{\Gamma}^\omega(x, y)$ is symmetric and

$$\mathbf{\Gamma}^\omega(y, x) = \mathbf{\Gamma}^\omega(x, y), \quad x \neq y. \tag{2.481}$$

If the medium is not homogeneous, then the following holds:

$$\mathbf{\Gamma}^\omega(y, x) = \left[\mathbf{\Gamma}^\omega(x, y)\right]^t, \quad x \neq y. \tag{2.482}$$

Identity (2.482) states that the nth component of the displacement at x due to a point source excitation at y in the mth direction is identical to the mth component of the displacement at y due to a point source excitation at x in the nth direction. It can be proved analogously to (2.208).

Let $\mathbf{\Gamma}^\omega_\alpha$ for $\alpha = p, s$ be respectively the pressure and shear components of the Kupradze matrix $\mathbf{\Gamma}^\omega$, defined in (2.410). The following Helmholtz-Kirchhoff identities hold [**25, 26**].

PROPOSITION 2.136 (Helmholtz-Kirchhoff Identities). *Let $\Omega \subset \mathbb{R}^d$ be a ball with radius R. Then, for all $x, z \in \Omega$, we have*

$$\lim_{R\to+\infty} \int_{\partial\Omega} \mathbf{\Gamma}^\omega_\alpha(x, y)\overline{\mathbf{\Gamma}^\omega_\alpha}(y, z) d\sigma(y) = -\frac{1}{\omega c_\alpha} \Im \left\{\mathbf{\Gamma}^\omega_\alpha(x, z)\right\}, \ \alpha = p, s, \tag{2.483}$$

and

$$\lim_{R\to+\infty} \int_{\partial\Omega} \mathbf{\Gamma}^\omega_s(x, y)\overline{\mathbf{\Gamma}^\omega_p}(y, z) d\sigma(y) = 0. \tag{2.484}$$

We emphasize that the proofs of the Helmholtz-Kirchhoff identities in Proposition 2.136 require the medium to be homogeneous, and we cannot expect these identities to be true in a heterogeneous medium because of mode conversion between pressure and shear waves.

2.15.15. The Scattering Coefficients and the Scattering Amplitude. This subsection is dedicated to defining the elastic scattering coefficients in two dimensions first introduced in [**1**]. We first recall some background material on cylindrical eigenfunctions of the Lamé equation and present the multipolar expansions of the exterior scattered elastic field.

Consider a time harmonic incident elastic field U satisfying

$$(\mathcal{L}^{\lambda,\mu} + \omega^2)U(x) = 0, \quad \forall x \in \mathbb{R}^2. \tag{2.485}$$

Then the total displacement field due to the presence of D, represented by u, satisfies the transmission problem

$$\begin{cases} (\mathcal{L}^{\lambda,\mu} + \omega^2)u(x) = 0, \quad x \in \mathbb{R}^2 \setminus \overline{D}, \\ (\mathcal{L}^{\widetilde{\lambda},\widetilde{\mu}} + \omega^2)u(x) = 0, \quad x \in D, \\ u(x)|_+ = u(x)|_-, \quad x \in \partial D, \\ \dfrac{\partial u}{\partial \nu}|_+ = \dfrac{\partial u}{\partial \widetilde{\nu}}|_-, \quad x \in \partial D, \\ (u - U)(x) \text{ satisfies Kupradze's radiation condition when } |x| \to \infty\,. \end{cases} \tag{2.486}$$

The total field u admits the integral representation

$$u(x,\omega)=\begin{cases}U(x,\omega)+\mathcal{S}_D^{\omega}[\psi](x,\omega), & x\in\mathbb{R}^2\setminus\overline{D},\\ \widetilde{\mathcal{S}}_D^{\omega}[\varphi](x,\omega), & x\in D,\end{cases} \tag{2.487}$$

where unknown densities $\varphi,\psi\in L^2(\partial D)^2$ satisfy the system of integral equations

$$\begin{pmatrix}\widetilde{\mathcal{S}}_D^{\omega} & -\mathcal{S}_D^{\omega}\\ \dfrac{\partial}{\partial\widetilde{\nu}}\widetilde{\mathcal{S}}_D^{\omega}\Big|_- & -\dfrac{\partial}{\partial\nu}\mathcal{S}_D^{\omega}\Big|_+\end{pmatrix}\begin{pmatrix}\varphi\\ \psi\end{pmatrix}=\begin{pmatrix}U\\ \dfrac{\partial U}{\partial\nu}\end{pmatrix}\Bigg|_{\partial D}. \tag{2.488}$$

2.15.15.1. *Cylindrical Elastic Waves and Multipolar Expansions.* We define $\hat{x}:=x/|x|$ for all $x\in\mathbb{R}^2\setminus\{0\}$ and write $\mathbb{S}:=\{x\in\mathbb{R}^2 : |x|^2=1\}$. The position vector $x\in\mathbb{R}^2$ can be equivalently expressed as $x=(|x|\cos\varphi_x,|x|\sin\varphi_x)$ where $\varphi_x\in[0,2\pi)$ denotes the polar angle of x. Denote by $\{e_r,e_\theta\}$ the orthonormal basis vectors for the polar coordinate system in two dimensions, that is,

$$e_r=(\cos\varphi_x,\sin\varphi_x),\quad e_\theta=-(\sin\varphi_x,\cos\varphi_x).$$

Consider the surface vector harmonics in two-dimensions

$$P_m(\hat{x})=e^{\sqrt{-1}m\varphi_x}e_r \quad\text{and}\quad S_m(\hat{x})=e^{\sqrt{-1}m\varphi_x}e_\theta \quad\text{for all } m\in\mathbb{Z}. \tag{2.489}$$

It is known, see [**406**] for instance, that these cylindrical surface vector potentials enjoy the orthogonality properties

$$\int_{\mathbb{S}}P_n(\hat{x})\cdot\overline{P_m(\hat{x})}d\sigma(\hat{x})=2\pi\delta_{nm}, \tag{2.490}$$

$$\int_{\mathbb{S}}S_n(\hat{x})\cdot\overline{S_m(\hat{x})}d\sigma(\hat{x})=2\pi\delta_{nm}, \tag{2.491}$$

$$\int_{\mathbb{S}}P_m(\hat{x})\cdot\overline{S_m(\hat{x})}d\sigma(\hat{x})=0, \tag{2.492}$$

for all $n,m\in\mathbb{Z}$, where δ_{nm} is the Kronecker's delta function and $d\sigma$ is the infinitesimal differential element on $\mathbb{S}$.

Let $H_m^{(1)}$ and J_m be the Hankel and Bessel functions of first kind of order $m\in\mathbb{Z}$, respectively. Then, for each $\kappa>0$, we construct the functions $v_m(\cdot,\kappa)$ and $w_m(\cdot,\kappa)$ by

$$v_m(x,\kappa):=H_m^{(1)}(\kappa|x|)e^{\sqrt{-1}m\varphi_x} \quad\text{and}\quad w_m(x,\kappa):=J_m(\kappa|x|)e^{\sqrt{-1}m\varphi_x}. \tag{2.493}$$

It is easy to verify that v_m are outgoing radiating solutions (*i.e.*, satisfying the Sommerfeld radiation condition) to the Helmholtz equation $\Delta v+\kappa^2 v=0$ in $\mathbb{R}^2\setminus\{0\}$ and that w_m are entire functions which satisfy $\Delta v+\kappa^2 v=0$ in $\mathbb{R}^2$.

Using surface vector harmonics P_m, S_m and functions v_m, w_m, we define

$$\begin{aligned}H_m^p(x,\kappa_p):=&\nabla v_m(x,\kappa_p)\\ =&\kappa_p\left(H_m^{(1)}(\kappa_p|x|)\right)'P_m(\hat{x})+\frac{\sqrt{-1}m}{|x|}H_m^{(1)}(\kappa_p|x|)S_m(\hat{x}),\end{aligned} \tag{2.494}$$

$$\begin{aligned}H_m^s(x,\kappa_s):=&\nabla\times(e_z v_m(x,\kappa_s))\\ =&\frac{\sqrt{-1}m}{|x|}H_m^{(1)}(\kappa_s|x|)P_m(\hat{x})-\kappa_s\left(H_m^{(1)}(\kappa_s|x|)\right)'S_m(\hat{x}),\end{aligned} \tag{2.495}$$

and

$$J_m^p(x,\kappa_p) := \nabla w_m(x,\kappa_p)$$

$$= \kappa_p \left(J_m(\kappa_p|x|)\right)' P_m(\hat{x}) + \frac{\sqrt{-1}m}{|x|} J_m(\kappa_p|x|) S_m(\hat{x}), \tag{2.496}$$

$$J_m^s(x,\kappa_s) := \nabla \times (e_z w_m(x,\kappa_s))$$

$$= \frac{\sqrt{-1}m}{|x|} J_m(\kappa_s|x|) P_m(\hat{x}) - \kappa_s \left(J_m(\kappa_s|x|)\right)' S_m(\hat{x}), \tag{2.497}$$

for all $\kappa_\alpha > 0$ and $m \in \mathbb{Z}$, where $e_z = (0,0,1)$ is a unit normal vector to the (x_1, x_2)−plane and

$$\left(H_m^{(1)}\right)'(t) := \frac{d}{dt}\left[H_m^{(1)}(t)\right] \quad \text{and} \quad (J_m)'(t) := \frac{d}{dt}\left[J_m(t)\right]. \tag{2.498}$$

For simplicity, we suppress the dependence of J_m^α and H_m^α on wavenumbers κ_α, $\alpha = p, s$, henceforth.

The functions J_m^p and J_m^s are the interior longitudinal and transverse eigenvectors of the Lamé system in $\mathbb{R}^2$. Similarly, H_m^p and H_m^s are the exterior eigenvectors of the Lamé system in $\mathbb{R}^2 \setminus \{0\}$ [**108**]. The following result on the completeness and linear independence of the interior eigenvectors (J_m^p, J_m^s) and exterior eigen-vectors (H_m^p, H_m^s) with respect to $L^2(\partial D)^2$−norm holds. The interested readers are referred to [**477**, Lemmas 1-3] for further details.

LEMMA 2.137. *Let $D \subset \mathbb{R}^2$ be a bounded simply connected domain containing origin and ∂D be smooth. Then the set $\{H_m^p, H_m^s : m \in \mathbb{Z}\}$ is complete and linearly independent in $L^2(\partial D)^2$. Moreover, if ω^2 is not a Dirichlet eigenvalue of the Lamé equation on D, then the set $\{J_m^p, J_m^s : m \in \mathbb{Z}\}$ is also complete and linearly independent in $L^2(\partial D)^2$.*

As a direct consequence of Lemma 2.137, corresponding to every incident field U satisfying (2.485), there exist constants $a_m^p, a_m^s \in \mathbb{C}$ for all $m \in \mathbb{Z}$ such that

$$U(x) = \sum_{m\in\mathbb{Z}} \left(a_m^s J_m^s(x) + a_m^p J_m^p(x)\right), \quad x \in \partial D. \tag{2.499}$$

In particular, a general plane incident wave of the form

$$\begin{aligned} U(x) &= \frac{1}{c_s^2} e^{\sqrt{-1}\kappa_s x\cdot d} d^\perp + \frac{1}{c_p^2} e^{\sqrt{-1}\kappa_p x\cdot d} d \\ &= -\left(\frac{\sqrt{-1}}{c_s^2\kappa_s} \nabla \times \left[e_z\, e^{\sqrt{-1}\kappa_s x\cdot d}\right] + \frac{\sqrt{-1}}{c_p^2\kappa_p}\left[\nabla e^{\sqrt{-1}\kappa_p x\cdot d}\right]\right) \end{aligned} \tag{2.500}$$

can be written in the form (2.499) with

$$a_m^\beta := a_m^\beta(U) = -\frac{\sqrt{-1}}{c_\beta^2 \kappa_\beta} e^{\sqrt{-1}m(\pi/2-\theta)}, \quad \beta \in \{p, s\}, \tag{2.501}$$

where $d = (\cos\theta, \sin\theta) \in S$ is the direction of incidence and $d^\perp$ is a vector perpendicular to d. In fact, this decomposition is a simple consequence of the Jacobi-Anger decomposition (2.229) of the scalar plane wave $e^{\sqrt{-1}\kappa x\cdot d}$.

Moreover, according to [**501**], for all $x, y \in \mathbb{R}^2$ such that $|x| > |y|$ and for any vector $p \in \mathbb{R}^2$ independent of x,

$$\begin{aligned}\mathbf{\Gamma}^\omega(x,y)p &= -\frac{\sqrt{-1}}{4c_s^2}\sum_{n\in\mathbb{Z}} H_n^s(x)\left[\overline{J_n^s(y)}\cdot p\right] \\ &\quad -\frac{\sqrt{-1}}{4c_p^2}\sum_{n\in\mathbb{Z}} H_n^p(x)\left[\overline{J_n^p(y)}\cdot p\right].\end{aligned} \tag{2.502}$$

2.15.15.2. *Scattering Coefficients of Elastic Particles.* Note that the multipolar expansion (2.502) of the fundamental solution $\mathbf{\Gamma}^\omega$ enables us to derive the expansion

$$\begin{aligned}\mathcal{S}_D^\omega[\psi](x) = &-\frac{\sqrt{-1}}{4c_p^2}\sum_{n\in\mathbb{Z}} H_n^p(x)\int_{\partial D}\left[\overline{J_n^p(y)}\cdot\psi(y)\right]d\sigma(y) \\ &-\frac{\sqrt{-1}}{4c_s^2}\sum_{n\in\mathbb{Z}} H_n^s(x)\int_{\partial D}\left[\overline{J_n^s(y)}\cdot\psi(y)\right]d\sigma(y)\end{aligned} \tag{2.503}$$

for all $x \in \mathbb{R}^2 \setminus \overline{D}$ sufficiently far from the boundary ∂D. Consequently, by virtue of expansion (2.503) and the integral representation (2.487), the scattered field can be expanded as

$$u(x) - U(x) = -\frac{\sqrt{-1}}{4}\sum_{n\in\mathbb{Z}}\left[\frac{b_n^s}{c_s^2}H_n^s(x) + \frac{b_n^p}{c_p^2}H_n^p(x)\right], \tag{2.504}$$

where

$$b_n^\alpha = \int_{\partial D}\left[\overline{J_n^\alpha(y)}\cdot\psi(y)\right]d\sigma(y), \quad \alpha\in\{p,s\}, \quad \forall\, n\in\mathbb{Z}. \tag{2.505}$$

DEFINITION 2.138. *Let $(\varphi_m^\beta, \psi_m^\beta)$, $m \in \mathbb{Z}$, be the solution of (2.488) corresponding to $U = J_m^\beta$. Then the elastic scattering coefficients $W_{m,n}^{\alpha,\beta}$ of $D \Subset \mathbb{R}^2$ are defined by*

$$W_{m,n}^{\alpha,\beta} = W_{m,n}^{\alpha,\beta}[D,\lambda_0,\lambda_1,\mu_0,\mu_1,\omega] := \int_{\partial D}\left[\overline{J_n^\alpha(y)}\cdot\psi_m^\beta(y)\right]d\sigma(y) \tag{2.506}$$

for $m, n \in \mathbb{Z}$, where α and β indicate wavemodes p or s.

Analogously to Lemma 2.76, the following result on the decay rate of the elastic scattering coefficients holds.

LEMMA 2.139. *There exist constants $C_{\alpha,\beta} > 0$ for each wave-mode $\alpha, \beta = p, s$ such that*

$$\left|W_{m,n}^{\alpha,\beta}[D,\lambda_0,\lambda_1,\mu_0,\mu_1,\omega]\right| \le \frac{C_{\alpha,\beta}^{|n|+|m|-2}}{|n|^{|n|-1}|m|^{|m|-1}} \tag{2.507}$$

for all $m, n \in \mathbb{Z}$ and $|m|, |n| \to +\infty$.

2.15.15.3. *Connections with Scattered Field and Far-Field Amplitudes.* Consider a general plane incident field U of the form (2.500) admitting decomposition (2.499)–(2.501). By the principle of superposition, the solution (φ, ψ) of (2.488) is given by

$$\psi(x) = \sum_{m\in\mathbb{Z}}\left[a_m^p\psi_m^p + a_m^s\psi_m^s\right] \quad \text{and} \quad \varphi(x) = \sum_{m\in\mathbb{Z}}\left[a_m^p\varphi_m^p + a_m^s\varphi_m^s\right]. \tag{2.508}$$

This, together with Definition 2.138 of the scattering coefficients and the expansion (2.504), renders the asymptotic expansion

$$u(x) - U(x) = \sum_{n\in\mathbb{Z}} \left[\gamma_n^p H_n^p(x) + \gamma_n^s H_n^s(x)\right] \tag{2.509}$$

of the scattered field for all $x \in \mathbb{R}^2 \setminus D$ sufficiently far from ∂D, where

$$\gamma_n^\beta = \sum_{m\in\mathbb{Z}} \left(d_m^p W_{m,n}^{\beta,p} + d_m^s W_{m,n}^{\beta,s}\right), \quad \beta \in \{p, s\}, \tag{2.510}$$

with

$$d_m^\beta := \frac{\sqrt{-1}a_m^\beta}{4\rho_0\omega^2}. \tag{2.511}$$

In order to substantiate the connection between the elastic scattering coefficients and far-field scattering amplitudes we recall that the cylindrical Hankel functions $H_n^{(1)}$ admit the far-field behavior

$$H_n^{(1)}(\kappa|x|) = \frac{e^{\sqrt{-1}\kappa|x|}}{\sqrt{|x|}}\sqrt{\frac{2}{\pi\kappa}}e^{-\sqrt{-1}\pi(n/2+1/4)} + O\left(|x|^{-3/2}\right), \tag{2.512}$$

$$\left(H_n^{(1)}(\kappa|x|)\right)' = \sqrt{-1}\kappa\frac{e^{\sqrt{-1}\kappa|x|}}{\sqrt{|x|}}\sqrt{\frac{2}{\pi\kappa}}e^{-\sqrt{-1}\pi(n/2+1/4)} + O\left(|x|^{-3/2}\right) \tag{2.513}$$

as $|x| \to \infty$ (see, for instance, [**423**, Formulas 10.2.5 and 10.17.11]). Consequently, the far-field behavior of the functions H_n^p and H_n^s can be predicted as

$$H_n^p(x) \sim \frac{e^{\sqrt{-1}\kappa_p|x|}}{\sqrt{|x|}}A_n^{\infty,p}P_n(\hat{x}) \quad \text{and} \quad H_n^s(x) \sim \frac{e^{\sqrt{-1}\kappa_s|x|}}{\sqrt{|x|}}A_n^{\infty,s}S_n(\hat{x}), \quad \text{as } |x| \to \infty, \tag{2.514}$$

where

$$A_n^{\infty,p} := (\sqrt{-1}+1)\kappa_p e^{-\sqrt{-1}n\pi/2}\sqrt{\frac{1}{\pi\kappa_p}} \tag{2.515}$$

and

$$A_n^{\infty,s} := -(\sqrt{-1}+1)\kappa_s e^{-\sqrt{-1}n\pi/2}\sqrt{\frac{1}{\pi\kappa_s}}. \tag{2.516}$$

Thus, for all $x \in \mathbb{R}^2 \setminus D$ such that $|x| \to \infty$ the scattered field $(u - U)$ in (2.509) admits the asymptotic expansion

$$u(x) - U(x) = \frac{e^{\sqrt{-1}\kappa_p|x|}}{\sqrt{|x|}}\sum_{n\in\mathbb{Z}} \left[\gamma_n^p A_n^{\infty,p} P_n(\hat{x})\right] + \frac{e^{\sqrt{-1}\kappa_s|x|}}{\sqrt{|x|}}\sum_{n\in\mathbb{Z}} \left[\gamma_n^s A_n^{\infty,s} S_n(\hat{x})\right]. \tag{2.517}$$

On the other hand, the Kupradze radiation condition guarantees the existence of two holomorphic functions $u_p^\infty : \mathbb{S} \to \mathbb{C}^2$ and $u_s^\infty : \mathbb{S} \to \mathbb{C}^2$ such that

$$u(x) - U(x) = \frac{e^{\sqrt{-1}\kappa_s|x|}}{\sqrt{|x|}}u_p^\infty(\hat{x}) + \frac{e^{\sqrt{-1}\kappa_p|x|}}{\sqrt{|x|}}u_s^\infty(\hat{x}) + O\left(\frac{1}{|x|^{3/2}}\right), \quad \text{as } |x| \to \infty. \tag{2.518}$$

The functions u_p^∞ and u_s^∞ are respectively known as the longitudinal and transverse far-field patterns or the scattering amplitudes. Comparing (2.517) and (2.518) the following result is readily proved, which substantiates the proposition that the far-field scattering amplitudes admit natural expansions in terms of scattering coefficients [**1**].

THEOREM 2.140. *Let U be the incident plane field given by* (2.499). *Then the corresponding longitudinal and transverse scattering amplitudes can be written as*

$$u_p^\infty[D,\lambda_0,\lambda_1,\mu_0,\mu_1,\omega](\hat{x}) = \sum_{n\in\mathbb{Z}} \gamma_n^p A_n^{\infty,p} P_n(\hat{x}) \tag{2.519}$$

$$u_s^\infty[D,\lambda_0,\lambda_1,\mu_0,\mu_1,\omega](\hat{x}) = \sum_{n\in\mathbb{Z}} \gamma_n^s A_n^{\infty,s} S_n(\hat{x}), \tag{2.520}$$

where the coefficients γ_n^p and γ_n^s are defined in (2.510).

Properties of the elastic scattering coefficients resulting from the principles of reciprocity and conservation of energy in elastic media are derived in [**1**].

2.15.16. Elastic Scattering by Small Particles. Suppose that an elastic inclusion D in $\mathbb{R}^d$ is given by $D = \delta B + z$, where B is a bounded Lipschitz domain in $\mathbb{R}^d$. Suppose that D has the pair of Lamé constants $(\widetilde{\lambda},\widetilde{\mu})$ satisfying (2.385) and (2.411) and denote by $\widetilde{\rho}$ its density. Let $y \in \mathbb{R}^d \setminus \overline{D}$ and assume that there exists $c_0 > 0$ such that $\mathrm{dist}(y, D) > c_0$. Let $u_0(x) = \mathbf{\Gamma}^\omega(x,y)\theta$ be an incident displacement field, where θ is a unit vector in $\mathbb{R}^d$.

Let u_δ be the solution to the transmission problem

$$\begin{cases} (\mathcal{L}^{\lambda,\mu} + \omega^2\rho)u_\delta = 0 \quad \text{in } \mathbb{R}^d\setminus\overline{D}, \\ (\mathcal{L}^{\widetilde{\lambda},\widetilde{\mu}} + \omega^2\widetilde{\rho})u_\delta = 0 \quad \text{in } D, \\ u_\delta\big|_- = u_\delta\big|_+ \quad \text{on } \partial D, \\ \dfrac{\partial u_\delta}{\partial\widetilde{\nu}}\bigg|_- = \dfrac{\partial u_\delta}{\partial\nu}\bigg|_+ \quad \text{on } \partial D, \\ (u_\delta - u_0)(x) \text{ satisfies the Sommerfeld-Kupradze radiation condition.} \end{cases} \tag{2.521}$$

The leading-order term in the asymptotic expansion of $u_\delta - u_0$ as $\delta \to 0$ is expressed in terms of the elastic moment tensor, a concept that extends the notion of polarization tensors to linear elasticity.

The elastic moment tensor associated with the domain B and the Lamé parameters $(\lambda,\mu;\widetilde{\lambda},\widetilde{\mu})$ is defined as follows: For $i,j = 1,\ldots,d$, let f_{ij} and g_{ij} solve

$$\begin{cases} \widetilde{\mathcal{S}}_B^0[f_{ij}]|_- - \mathcal{S}_B^0[g_{ij}]|_+ = x_i e_j|_{\partial B}, \\ \dfrac{\partial}{\partial\widetilde{\nu}}\widetilde{\mathcal{S}}_B^0[f_{ij}]\bigg|_- - \dfrac{\partial}{\partial\nu}\mathcal{S}_B^0[g_{ij}]\bigg|_+ = \dfrac{\partial(x_i e_j)}{\partial\nu}|_{\partial B}, \end{cases} \tag{2.522}$$

where $(e_1,\ldots,e_d)$ is the canonical basis of $\mathbb{R}^d$. Then the elastic moment tensor $\mathcal{M} := (m_{ijpq})_{i,j,p,q=1}^d$ is defined by

$$m_{ijpq} := \int_{\partial B} x_p e_q \cdot g_{ij}\, d\sigma. \tag{2.523}$$

The following lemma holds [**52**]. It gives an equivalent representation of $\mathcal{M}$.

LEMMA 2.141. *Suppose that $0 < \widetilde{\lambda},\widetilde{\mu} < +\infty$. For $i,j,p,q = 1,\ldots,d$,*

$$m_{ijpq} = \int_{\partial B}\left[-\frac{\partial(x_p e_q)}{\partial\nu} + \frac{\partial(x_p e_q)}{\partial\widetilde{\nu}}\right]\cdot v_{ij}\, d\sigma, \tag{2.524}$$

where v_{ij} is the unique solution of the transmission problem

$$
(2.525)\qquad \begin{cases} \mathcal{L}^{\lambda,\mu} v_{ij} = 0 & \text{in } \mathbb{R}^d \setminus \overline{B}, \\ \mathcal{L}^{\widetilde{\lambda},\widetilde{\mu}} v_{ij} = 0 & \text{in } B, \\ v_{ij}|_+ - v_{ij}|_- = 0 & \text{on } \partial B, \\ \left.\dfrac{\partial v_{ij}}{\partial \nu}\right|_+ - \left.\dfrac{\partial v_{ij}}{\partial \widetilde{\nu}}\right|_- = 0 & \text{on } \partial B, \\ v_{ij}(x) - x_i e_j = O(|x|^{1-d}) & \text{as } |x| \to +\infty. \end{cases}
$$

PROOF. Note first that v_{ij} defined by

$$
(2.526)\qquad v_{ij}(x) := \begin{cases} \mathcal{S}_B^0[g_{ij}](x) + x_i e_j, & x \in \mathbb{R}^d \setminus \overline{B}, \\ \widetilde{\mathcal{S}}_B^0[f_{ij}](x), & x \in B, \end{cases}
$$

is the solution of (2.525). Using (2.398) (for $\omega = 0$) and (2.522) we compute

$$
\begin{aligned}
m_{ijpq} &= \int_{\partial B} x_p e_q \cdot g_{ij}\, d\sigma \\
&= \int_{\partial B} x_p e_q \cdot \Big[\frac{\partial}{\partial \nu} \mathcal{S}_B^0[g_{ij}]\big|_+ - \frac{\partial}{\partial \nu} \mathcal{S}_B^0[g_{ij}]\big|_- \Big]\, d\sigma \\
&= -\int_{\partial B} x_p e_q \cdot \frac{\partial(x_i e_j)}{\partial \nu}\, d\sigma - \int_{\partial B} x_p e_q \cdot \Big[\frac{\partial}{\partial \nu} \mathcal{S}_B^0[g_{ij}]\big|_- - \frac{\partial}{\partial \widetilde{\nu}} \widetilde{\mathcal{S}}_B^0[f_{ij}]\big|_- \Big]\, d\sigma \\
&= -\int_{\partial B} \frac{\partial(x_p e_q)}{\partial \nu} \cdot x_i e_j\, d\sigma - \int_{\partial B} \Big[\frac{\partial(x_p e_q)}{\partial \nu} \cdot \mathcal{S}_B^0[g_{ij}] - \frac{\partial(x_p e_q)}{\partial \widetilde{\nu}} \cdot \widetilde{\mathcal{S}}_B^0[f_{ij}] \Big]\, d\sigma \\
&= \int_{\partial B} \Big[-\frac{\partial(x_p e_q)}{\partial \nu} + \frac{\partial(x_p e_q)}{\partial \widetilde{\nu}} \Big] \cdot \widetilde{\mathcal{S}}_B^0[f_{ij}]\, d\sigma,
\end{aligned}
$$

and hence (2.524) is established. □

The following symmetry and positive-definiteness properties of the elastic moment tensor hold [**52**, **71**].

THEOREM 2.142. *Let $\mathcal{M}$ be the elastic moment tensor associated with the domain B, and $(\widetilde{\lambda}, \widetilde{\mu})$ and (λ, μ) be the Lamé parameters of B and the background, respectively. Then,*

(i) *Symmetry: For $p, q, i, j = 1, \ldots, d$,*

$$
(2.527)\qquad m_{ijpq} = m_{ijqp}, \quad m_{ijpq} = m_{jipq}, \quad \text{and} \quad m_{ijpq} = m_{pqij}.
$$

(ii) *Positive-definiteness: Suppose that (2.411) holds. If $\widetilde{\mu} > \mu$ ($\widetilde{\mu} < \mu$, resp.), then $\mathcal{M}$ is positive (negative, resp.) definite on the space M_d^S of $d \times d$ symmetric matrices.*

The following asymptotic expansion of $u_\delta - u_0$ as $\delta \to 0$ holds [**25**, **27**].

THEOREM 2.143. *Let $K \Subset \mathbb{R}^d \setminus \overline{D}$. For $\omega\delta \ll 1$, the following asymptotic expansion holds uniformly for all $x \in K$:*

$$
(2.528)\qquad \begin{aligned} u_\delta(x) - u_0(x) = -\delta^d \Big(&\nabla u_0(z) : \mathbb{M} \nabla_z \mathbf{\Gamma}^\omega(x,z) \\ &+ \omega^2(\rho - \widetilde{\rho}) |B| \mathbf{\Gamma}^\omega(x,z) u_0(z) \Big) + O(\delta^{d+1}). \end{aligned}
$$

2.16. Quasi-Periodic Layer Potentials for the Lamé System

In this section we collect some notation and well-known results regarding quasi-periodic layer potentials for the Lamé system in $\mathbb{R}^2$. We refer to [**283**, **391**, **458**] for the details. Ewald's method described in Subsection 2.13.3 can be used to efficiently compute quasi-periodic layer potentials for the Lamé system.

As before, we assume that the unit cell $Y = [0,1)^2$ is the periodic cell and the quasi-momentum variable, denoted by α, ranges over the Brillouin zone $B = [0, 2\pi)^2$. We introduce the two-dimensional quasi-periodic fundamental solution $\mathbf{G}^{\alpha,\omega} = (G_{ij}^{\alpha,\omega})$ for $\omega \neq 0$, which satisfies

$$(\mathcal{L}^{\lambda,\mu} + \omega^2)\mathbf{G}^{\alpha,\omega}(x,y) = \sum_{n\in\mathbb{Z}^2} \delta_0(x-y-n)e^{\sqrt{-1}n\cdot\alpha} I, \tag{2.529}$$

where I denotes the identity matrix. Here we assume that $\kappa_s, \kappa_p \neq |2\pi n + \alpha|$ for all $n \in \mathbb{Z}^2$ where κ_s and κ_p are given by (2.406).

Using Poisson's summation formula (2.287), we have

$$\sum_{n\in\mathbb{Z}^2} \delta_0(x-y-n)e^{\sqrt{-1}n\cdot\alpha} I = \sum_{n\in\mathbb{Z}^2} e^{\sqrt{-1}(2\pi n+\alpha)\cdot(x-y)} I.$$

We plug this equation into (2.529) and then take the Fourier transform of both sides of (2.529) to obtain

$$\begin{aligned}\widehat{G}_{ij}^{\alpha,\omega}(\xi,y) =& (2\pi)^2 \left[\frac{\delta_{ij}}{c_s^2}\frac{1}{\kappa_s^2-\xi^2} + \frac{\xi_i\xi_j}{\omega^2}\left(\frac{1}{\kappa_p^2-\xi^2} - \frac{1}{\kappa_s^2-\xi^2}\right)\right] \\ &\times \sum_{n\in\mathbb{Z}^2} e^{-\sqrt{-1}(2\pi n+\alpha)\cdot y}\delta_0(\xi + 2\pi n + \alpha),\end{aligned}$$

where $\xi^2 = \xi\cdot\xi$ and $\widehat{\ }$ denotes the Fourier transform. Then taking the inverse Fourier transform, we can see that the quasi-periodic fundamental solution $\mathbf{G}^{\alpha,\omega}$ can be represented as a sum of augmented plane waves over the reciprocal lattice:

$$\begin{aligned}G_{ij}^{\alpha,\omega}(x,y) &= \frac{\delta_{ij}}{c_s^2}\sum_{n\in\mathbb{Z}^2} \frac{e^{\sqrt{-1}(2\pi n+\alpha)\cdot(x-y)}}{\kappa_s^2 - |2\pi n+\alpha|^2} \\ &\quad + \frac{\kappa_s^2-\kappa_p^2}{\omega^2}\sum_{n\in\mathbb{Z}^2} \frac{e^{\sqrt{-1}(2\pi n+\alpha)\cdot(x-y)}(2\pi n+\alpha)_i(2\pi n+\alpha)_j}{(\kappa_p^2-|2\pi n+\alpha|^2)(\kappa_s^2-|2\pi n+\alpha|^2)}.\end{aligned} \tag{2.530}$$

Moreover, it can also be easily shown that

$$\mathbf{G}^{\alpha,\omega}(x,y) = \sum_{n\in\mathbb{Z}^2} \mathbf{\Gamma}^{\omega}(x-n-y)e^{\sqrt{-1}n\cdot\alpha},$$

where $\mathbf{\Gamma}^{\omega}$ is the Kupradze fundamental solution defined by (2.393).

When $\omega = 0$, we define $\mathbf{G}^{\alpha,0}$ by

$$\begin{aligned}G_{ij}^{\alpha,0}(x,y) &:= \frac{1}{\mu}\sum_{n\in\mathbb{Z}^2} e^{\sqrt{-1}(2\pi n+\alpha)\cdot(x-y)}\Bigg(\frac{-\delta_{ij}}{|2\pi n+\alpha|^2} \\ &\qquad + \frac{\lambda+\mu}{\lambda+2\mu}\frac{(2\pi n+\alpha)_i(2\pi n+\alpha)_j}{|2\pi n+\alpha|^4}\Bigg)\end{aligned} \tag{2.531}$$

if $\alpha \neq 0$, while if $\alpha = 0$, we set

$$G_{ij}^{0,0}(x,y) := \frac{1}{\mu} \sum_{n \neq (0,0)} e^{\sqrt{-1}2\pi n\cdot(x-y)} \left(\frac{-\delta_{ij}}{|2\pi n|^2} + \frac{\lambda+\mu}{\lambda+2\mu} \frac{4\pi^2 n_i n_j}{|2\pi n|^4} \right). \tag{2.532}$$

Then $\mathbf{G}^{\alpha,0}$ is quasi-periodic and satisfies

$$\mathcal{L}^{\lambda,\mu}\mathbf{G}^{\alpha,0}(x,y) = \sum_{n\in\mathbb{Z}^2} \delta_0(x-y-n)I \quad \text{if } \alpha \neq 0, \tag{2.533}$$

$$\mathcal{L}^{\lambda,\mu}\mathbf{G}^{0,0}(x,y) = \sum_{n\in\mathbb{Z}^2} \delta_0(x-y-n)I - I. \tag{2.534}$$

See [**52, 66**] for a proof.

Let D be a bounded domain in $\mathbb{R}^2$ with a connected Lipschitz boundary ∂D. Let $\mathcal{S}^{\alpha,\omega}$ and $\mathcal{D}^{\alpha,\omega}$ be the quasi-periodic single- and double-layer potentials associated with $\mathbf{G}^{\alpha,\omega}$; that is, for a given density $\varphi \in L^2(\partial D)^2$,

$$\mathcal{S}^{\alpha,\omega}[\varphi](x) = \int_{\partial D} \mathbf{G}^{\alpha,\omega}(x,y)\varphi(y)\,d\sigma(y), \quad x \in \mathbb{R}^2,$$

$$\mathcal{D}^{\alpha,\omega}[\varphi](x) = \int_{\partial D} \frac{\partial \mathbf{G}^{\alpha,\omega}(x,y)}{\partial \nu(y)}\varphi(y)\,d\sigma(y), \quad x \in \mathbb{R}^2 \setminus \partial D,$$

where $\partial/\partial\nu(y)$ denotes the conormal derivative with respect to $y \in \partial D$. Then, $\mathcal{S}^{\alpha,\omega}[\varphi]$ and $\mathcal{D}^{\alpha,\omega}[\varphi]$ are solutions to

$$(\mathcal{L}^{\lambda,\mu} + \omega^2)u = 0$$

in D and $Y \setminus \overline{D}$ and they are α-quasi-periodic.

The next formulas give the jump relations obeyed by the quasi-periodic double-layer potential and by the conormal derivative of the quasi-periodic single-layer potential on general Lipschitz domains:

$$\frac{\partial(\mathcal{S}^{\alpha,\omega}[\varphi])}{\partial \nu}\Big|_{\pm}(x) = \Big(\pm \frac{1}{2}I + (\mathcal{K}^{-\alpha,\omega})^* \Big)[\varphi](x) \quad \text{a.e. } x \in \partial D, \tag{2.535}$$

$$(\mathcal{D}^{\alpha,\omega}[\varphi])\big|_{\pm}(x) = \Big(\mp \frac{1}{2}I + \mathcal{K}^{\alpha,\omega} \Big)[\varphi](x) \quad \text{a.e. } x \in \partial D, \tag{2.536}$$

for $\varphi \in L^2(\partial D)^2$, where $\mathcal{K}^{\alpha,\omega}$ is the operator on $L^2(\partial D)^2$ defined by

$$\mathcal{K}^{\alpha,\omega}[\varphi](x) = \text{p.v.} \int_{\partial D} \frac{\partial \mathbf{G}^{\alpha,\omega}(x,y)}{\partial \nu(y)}\varphi(y)\,d\sigma(y) \tag{2.537}$$

and $(\mathcal{K}^{-\alpha,\omega})^*$ is the L^2-adjoint operator of $\mathcal{K}^{-\alpha,\omega}$, which is given by

$$(\mathcal{K}^{-\alpha,\omega})^*[\varphi](x) = \text{p.v.} \int_{\partial D} \frac{\partial \mathbf{G}^{\alpha,\omega}(x,y)}{\partial \nu(x)}\varphi(y)\,d\sigma(y). \tag{2.538}$$

The formulas (2.535) and (2.536) hold because $\mathbf{G}^{\alpha,\omega}(x,y)$ has the same kind of singularity at $x = y$ as that of $\mathbf{\Gamma}^{\omega}(x-y)$.

The following lemma will be of use in the next chapter.

LEMMA 2.144. *For any constant vector* ϕ

$$(\frac{1}{2}I + \mathcal{K}^{0,0})[\phi] = |Y \setminus \overline{D}|\phi \quad \textit{on } \partial D, \tag{2.539}$$

and for any $\psi \in L^2(\partial D)^2$,

$$\int_{\partial D} \left(\frac{1}{2}I + (\mathcal{K}^{0,0})^*\right)[\psi] = |Y \setminus \overline{D}| \int_{\partial D} \psi. \tag{2.540}$$

PROOF. By Green's formula and (2.534) we have for any constant vector ϕ

$$\mathcal{D}^{0,0}[\phi](x) = \int_D \mathcal{L}^{\lambda,\mu}\mathbf{G}^{0,0}(x,y)\phi(y)\,dy = \phi(x) - \int_D \phi\,d\sigma, \quad x \in D,$$

and hence, since $((1/2)\,I + \mathcal{K}^{0,0})[\phi] = \mathcal{D}^{0,0}[\phi]|_-$ and $|Y| = 1$, we readily get (2.539).

The identity (2.540) is a consequence of (2.539). In fact, for any constant vector ϕ, we have

$$\begin{aligned}\int_{\partial D} \phi \cdot \left(\frac{1}{2}I + (\mathcal{K}^{0,0})^*\right)[\psi] &= \int_{\partial D} \left(\frac{1}{2}I + \mathcal{K}^{0,0}\right)[\phi] \cdot \psi \\ &= |Y \setminus \overline{D}| \int_{\partial D} \phi \cdot \psi,\end{aligned}$$

from which (2.540) follows. □

2.17. Concluding Remarks

In this chapter, we have reviewed layer potential techniques associated with the Laplacian, the Helmholtz equation, the Maxwell equations, and the time-harmonic elasticity system. Our main concern has been on the one hand to characterize the eigenvalues of the Laplacian and Lamé systems with Dirichlet or Neumann boundary conditions as characteristic values of certain layer potentials which are in general meromorphic operator-valued functions and on the other hand, to analyze the Neumann–Poincaré operators associated with the Laplacian and the Lamé systems. We have also introduced Helmholtz-Kirchhoff identities which play a key role in the analysis of resolution in imaging and investigated quasi-periodic layer potentials for the Helmholtz equation and the time-harmonic elasticity system. We have provided spectral and spatial representations of the Green's functions in periodic domains and described analytical techniques for transforming them from slowly convergent representations into forms more suitable for computation. These results will be useful for studying photonic and phononic crystals.

CHAPTER 3

Perturbations of Cavities and Resonators

3.1. Introduction

This chapter is devoted to investigating perturbations in optical cavities and resonators. Optical cavities and resonators have attracted much interest in photonic technologies in recent years. Several different designs for resonators have been studied experimentally and theoretically with a particular emphasis on high quality factor cavities. The detection of nanoparticles inside cavities and the design of a resonator which possesses a resonance with the largest possible quality factor have been considered as challenging problems in photonics.

Resonance is a solution to the wave equation which is spatially localized while its dependence on time is harmonic except for decay due to radiation. Finding resonance in a linear wave equation with a radiation boundary condition involves solving a nonlinear eigenvalue problem. The quality factor of a resonator can be defined as the ratio between the real and imaginary parts of a resonance.

In this chapter, the theory of Gohberg and Sigal is used to establish an asymptotic theory for perturbations in eigenvalue problems due to the presence of small particles inside a cavity.

Using integral equations we formulate the nonlinear eigenvalue problem for a resonator as a characteristic value problem for a meromorphic operator-valued function. Then, on the one hand, Muller's method can be used for computing resonances, while on the other hand, the generalized argument principle yields a sensitivity analysis of the quality factor with respect to the material properties of the resonator.

3.2. Optical Cavities

3.2.1. Eigenvalue Perturbations Due to Small Particles. In this section we provide a rigorous derivation of a full asymptotic formula for perturbations in the eigenvalues caused by the presence of a conductive particle of small diameter with conductivity different from that of the background. To fix ideas, we consider Neumann boundary conditions on the boundary of the background medium. Dirichlet, Robin or mixed boundary conditions can be treated in exactly the same way; see Section 2.9.

Let $0 = \mu_1 < \mu_2 \leq \ldots$ be the eigenvalues of $-\Delta$ in Ω with Neumann conditions, namely, the eigenvalues of the problem:

$$\begin{cases} \Delta u + \mu u = 0 & \text{in } \Omega, \\ \dfrac{\partial u}{\partial \nu} = 0 & \text{on } \partial\Omega, \end{cases} \tag{3.1}$$

arranged in an increasing order and counted according to multiplicity. Let $(u_j)_{j\geq 1}$ be an orthonormal basis of $L^2(\Omega)$ of normalized eigenvectors. Fix j and suppose that the eigenvalue μ_j is simple. It is proved in [**12**, **13**, **499**] that the eigenvalues are generically simple. As we said in the introduction, by generic, we mean the existence of arbitrary small deformations of $\partial\Omega$ such that in the deformed domain the eigenvalue is simple. However, we emphasize that our assumption is not essential in what follows and is made only for ease of exposition. In fact, we have dwelt briefly on the splitting of nonsimple eigenvalues in Section 2.9.6.

Let $D = \epsilon B + z$ be the conductivity particle of small diameter which causes the perturbation of the eigenvalue. Fix j and suppose that the unperturbed eigenvalue μ_j is simple. Then there exists a simple eigenvalue μ_j^ϵ, near μ_j, associated to the normalized eigenfunction u_j^ϵ, satisfying the following problem:

$$
(3.2) \qquad \begin{cases} \Delta u^\epsilon + \omega^2 u^\epsilon = 0 & \text{in } \Omega \setminus \overline{D}, \\ \Delta u^\epsilon + \dfrac{\omega^2}{k} u^\epsilon = 0 & \text{in } D, \\ u^\epsilon|_+ - u^\epsilon|_- = 0 & \text{on } \partial D, \\ \dfrac{\partial u^\epsilon}{\partial \nu}\Big|_+ - k\dfrac{\partial u^\epsilon}{\partial \nu}\Big|_- = 0 & \text{on } \partial D, \\ \dfrac{\partial u^\epsilon}{\partial \nu} = 0 & \text{on } \partial\Omega, \end{cases}
$$

with $\omega = \sqrt{\mu_j^\epsilon}$ and the conductivity k of the particle D is such that $0 < k \neq 1$.

From Chapter 2, we know that the solution of (3.2) can be represented as

$$
(3.3) \qquad u(x) = \begin{cases} \mathcal{D}_\Omega^\omega[\psi](x) + \mathcal{S}_D^\omega[\phi](x) & \text{in } \Omega \setminus \overline{D}, \\ \mathcal{S}_D^{\frac{\omega}{\sqrt{k}}}[\theta](x) & \text{in } D, \end{cases}
$$

where the triplet of densities $(\psi := u|_{\partial\Omega}, \phi, \theta) \in L^2(\partial\Omega) \times L^2(\partial D) \times L^2(\partial D)$ satisfies the following system of integral equations:

$$
(3.4) \qquad \begin{cases} \left(\dfrac{1}{2}I - \mathcal{K}_\Omega^\omega\right)[\psi] - \mathcal{S}_D^\omega[\phi] = 0 & \text{on } \partial\Omega, \\ \mathcal{D}_\Omega^\omega[\psi] + \mathcal{S}_D^\omega[\phi] - \mathcal{S}_D^{\frac{\omega}{\sqrt{k}}}[\theta] = 0 & \text{on } \partial D, \\ \epsilon\left[\dfrac{\partial}{\partial\nu}\left(\mathcal{D}_\Omega^\omega[\psi] + \mathcal{S}_D^\omega[\phi]\right)\Big|_+ - k\dfrac{\partial}{\partial\nu}\left(\mathcal{S}_D^{\frac{\omega}{\sqrt{k}}}[\theta]\right)\Big|_-\right] = 0 & \text{on } \partial D. \end{cases}
$$

By using the jump formula (2.160), we reduce the eigenvalue problem to the calculation of the asymptotic expressions of the characteristic values of the operator-valued function $\mathcal{A}_\epsilon(\omega)$ given by

$$
\omega \mapsto \mathcal{A}_\epsilon(\omega) := \begin{pmatrix} \dfrac{1}{2}I - \mathcal{K}_\Omega^\omega & -\mathcal{S}_D^\omega & 0 \\ \mathcal{D}_\Omega^\omega & \mathcal{S}_D^\omega & -\mathcal{S}_D^{\frac{\omega}{\sqrt{k}}} \\ \epsilon\dfrac{\partial}{\partial\nu}\mathcal{D}_\Omega^\omega & \epsilon(\dfrac{1}{2}I + (\mathcal{K}_D^\omega)^*) & -\epsilon k(-\dfrac{1}{2}I + (\mathcal{K}_D^{\frac{\omega}{\sqrt{k}}})^*) \end{pmatrix}.
$$

We shall now expand the operator-valued function $\mathcal{A}_\epsilon(\omega)$ in terms of ϵ. We first consider the three-dimensional case. We need the following lemma.

LEMMA 3.1. *Let $d = 3$. Let $\psi \in L^2(\partial\Omega)$ and let $\varphi \in L^2(\partial D)$. Define $\tilde{\varphi}(x) = \epsilon\varphi(\epsilon x + z)$ for $x \in \partial B$. Then, we have*

$$\mathcal{S}_D^\omega[\varphi](\epsilon x + z) = -\frac{1}{4\pi}\sum_{n=0}^{+\infty}\frac{1}{n!}(\sqrt{-1}\omega\epsilon)^n\int_{\partial B}|x-y|^{n-1}\tilde{\varphi}(y)\,d\sigma(y), \quad x \in \partial B,$$

$$\mathcal{D}_\Omega^\omega[\psi](\epsilon x + z) = \sum_{n=0}^{+\infty}\epsilon^n\sum_{|\alpha|=n}\frac{1}{\alpha!}\partial^\alpha\mathcal{D}_\Omega^\omega[\psi](z)x^\alpha, \quad x \in \partial B,$$

$$\mathcal{S}_D^\omega[\varphi](x) = \sum_{n=0}^{+\infty}(-1)^n\epsilon^{n+1}\sum_{|\alpha|=n}\frac{1}{\alpha!}\partial^\alpha\Gamma_\omega(x-z)\Big(\int_{\partial B}y^\alpha\tilde{\varphi}(y)\,d\sigma(y)\Big), \quad x \in \partial\Omega.$$

PROOF. For any $\tilde{x}, \tilde{y} \in \partial D$, we have

$$\mathcal{S}_D^\omega[\varphi](\tilde{x}) = \int_{\partial D}\Gamma_\omega(\tilde{x}-\tilde{y})\varphi(\tilde{y})\,d\sigma(\tilde{y}).$$

By the change of variables $\tilde{x} = \epsilon x + z$ and $\tilde{y} = \epsilon y + z$, we obtain that

$$\begin{aligned}\mathcal{S}_D^\omega[\varphi](\tilde{x}) &= \epsilon^2\int_{\partial B}\Gamma_\omega(\epsilon(x-y))\varphi(\epsilon y + z)\,d\sigma(y)\\ &= \epsilon\int_{\partial B}\Gamma_\omega(\epsilon(x-y))\tilde{\varphi}(y)\,d\sigma(y).\end{aligned}$$

The first formula immediately follows from the following Taylor expansion of $\Gamma_\omega(\epsilon x)$ as $\epsilon \to 0$:

$$\Gamma_\omega(\epsilon x) = -\frac{1}{4\pi\epsilon}\sum_{n=0}^{+\infty}\frac{1}{n!}(\sqrt{-1}\omega\epsilon)^n|x|^{n-1}.$$

Since $(\Delta + \omega^2)\mathcal{D}_\Omega^\omega[\psi] = 0$ in Ω, $\mathcal{D}_\Omega^\omega[\psi]$ is a smooth function in Ω and its Taylor expansion at z yields the second formula.

Finally, for any $x \in \partial\Omega$, it is easy to see that

$$\begin{aligned}\mathcal{S}_D^\omega[\varphi](x) &= \int_{\partial D}\Gamma_\omega(x-\tilde{y})\varphi(\tilde{y})\,d\sigma(\tilde{y})\\ &= \epsilon\int_{\partial B}\Gamma_\omega(x-z-\epsilon y)\tilde{\varphi}(y)\,d\sigma(y)\\ &= \epsilon\int_{\partial B}\sum_{n=0}^{+\infty}(-1)^n\epsilon^n\sum_{|\alpha|=n}\frac{1}{\alpha!}\partial^\alpha\Gamma_\omega(x-z)y^\alpha\tilde{\varphi}(y)\,d\sigma(y)\\ &= \sum_{n=0}^{+\infty}(-1)^n\epsilon^{n+1}\sum_{|\alpha|=n}\frac{1}{\alpha!}\partial^\alpha\Gamma_\omega(x-z)\int_{\partial B}y^\alpha\tilde{\varphi}(y)\,d\sigma(y),\end{aligned}$$

which completes the proof of the lemma. □

With Lemma 3.1 in hand, we only need to write the expansion of $\partial\mathcal{D}_\Omega^\omega/\partial\nu$ and $(\mathcal{K}_D^\omega)^*$.

On the one hand, we have, for $\psi \in L^2(\partial\Omega)$,

$$\frac{\partial}{\partial\nu}\big(\mathcal{D}_\Omega^\omega[\psi](\epsilon x + z)\big) = \sum_{n=1}^{+\infty}\epsilon^n\sum_{|\alpha|=n}\frac{1}{\alpha!}\partial^\alpha\mathcal{D}_\Omega^\omega[\psi](z)\frac{\partial x^\alpha}{\partial\nu}, \quad x \in \partial B. \tag{3.5}$$

On the other hand, using the Taylor expansion, we get

$$\begin{aligned}\frac{\partial}{\partial\nu(x)}\Gamma_\omega(\epsilon(x-y)) &= -\frac{\langle x-y,\nu(x)\rangle}{4\pi|x-y|^2}\Big[-\frac{1}{\epsilon|x-y|}\\ &\quad+\sum_{n=0}^{+\infty}(\frac{1}{n!}-\frac{1}{(n+1)!})(\sqrt{-1}\omega)^{n+1}\epsilon^n|x-y|^n\Big],\end{aligned}\tag{3.6}$$

and obtain the following expansions.

LEMMA 3.2. *Let $d=3$. Let $\varphi\in L^2(\partial D)$. Define $\tilde{\varphi}(x)=\epsilon\varphi(\epsilon x+z), x\in\partial B$. Then, for $x\in\partial B$, we have*

$$\begin{aligned}&\epsilon(\mathcal{K}_D^\omega)^*[\varphi](\epsilon x+z)=(\mathcal{K}_B^0)^*[\tilde{\varphi}](x)\\ &-\frac{1}{4\pi}\sum_{n=1}^{+\infty}(\frac{1}{n!}-\frac{1}{(n+1)!})(\sqrt{-1}\omega\epsilon)^{n+1}\int_{\partial B}\langle x-y,\nu(x)\rangle|x-y|^{n-2}\tilde{\varphi}(y)\,d\sigma(y),\end{aligned}$$

where $(\mathcal{K}_B^0)^*$ is given by (2.6).

Define $\tilde{\phi}(x)=\epsilon\phi(\epsilon x+z)$ and $\tilde{\theta}(x)=\epsilon\theta(\epsilon x+z)$, $x\in\partial B$. By Lemmas 3.1 and 3.2, the system of equations (3.4) now takes the form

$$\widetilde{\mathcal{A}}_\epsilon(\omega)\begin{pmatrix}\psi\\ \tilde{\phi}\\ \tilde{\theta}\end{pmatrix}=0,$$

where $\widetilde{\mathcal{A}}_\epsilon(\omega)$ has the expansion

$$\widetilde{\mathcal{A}}_\epsilon(\omega)=\sum_{n=0}^{+\infty}(\omega\epsilon)^n\mathcal{A}_n(\omega),\tag{3.7}$$

with

$$\mathcal{A}_0(\omega):=\begin{pmatrix}\frac{1}{2}I-\mathcal{K}_\Omega^\omega & 0 & 0\\ \mathcal{D}_\Omega^\omega[\cdot](z) & \mathcal{S}_B^0 & -\mathcal{S}_B^0\\ 0 & \frac{1}{2}I+(\mathcal{K}_B^0)^* & -k(-\frac{1}{2}I+(\mathcal{K}_B^0)^*)\end{pmatrix},\tag{3.8}$$

and, for $n\geq 1$, writing $\mathcal{A}_n(\omega)=((\mathcal{A}_n(\omega))_{ll'})_{l,l'=1,2,3}$, we have

$$\begin{aligned}&(\mathcal{A}_n(\omega))_{11}=(\mathcal{A}_n(\omega))_{13}=0,\\ &(\mathcal{A}_n(\omega))_{12}=(-1)^n\omega^{-n}\sum_{|\alpha|=n-1}\frac{1}{\alpha!}\partial^\alpha\Gamma_\omega(x-z)\Big(\int_{\partial B}y^\alpha\cdot\,d\sigma(y)\Big),\\ &(\mathcal{A}_n(\omega))_{22}=-\frac{1}{4\pi}\frac{1}{n!}\sqrt{-1}^n\int_{\partial B}|x-y|^{n-1}\cdot\,d\sigma(y),\\ &(\mathcal{A}_n(\omega))_{23}=\frac{1}{4\pi}\frac{1}{n!}(\frac{\sqrt{-1}}{\sqrt{k}})^n\int_{\partial B}|x-y|^{n-1}\cdot\,d\sigma(y),\\ &(\mathcal{A}_n(\omega))_{21}=\frac{1}{\omega^n}\sum_{|\alpha|=n}\frac{1}{\alpha!}\partial^\alpha\mathcal{D}_\Omega^\omega[\cdot](z)x^\alpha,\\ &(\mathcal{A}_n(\omega))_{31}=\frac{1}{\omega^n}\sum_{|\alpha|=n}\frac{1}{\alpha!}\partial^\alpha\mathcal{D}_\Omega^\omega[\cdot](z)\frac{\partial x^\alpha}{\partial\nu},\end{aligned}$$

$(\mathcal{A}_1(\omega))_{32} = (\mathcal{A}_1(\omega))_{33} = 0$, and

$$(\mathcal{A}_n(\omega))_{32} = -\frac{1}{4\pi}\left(\frac{1}{(n-1)!} - \frac{1}{n!}\right)\sqrt{-1}^n \int_{\partial B} \langle x-y, \nu(x)\rangle |x-y|^{n-3} \cdot d\sigma(y),$$

$$(\mathcal{A}_n(\omega))_{33} = \frac{k}{4\pi}\left(\frac{1}{(n-1)!} - \frac{1}{n!}\right)\left(\frac{\sqrt{-1}}{\sqrt{k}}\right)^n \int_{\partial B} \langle x-y, \nu(x)\rangle |x-y|^{n-3} \cdot d\sigma(y),$$

for $n \geq 2$.

In three dimensions, it can be shown that the right-lower 2×2 block of $\mathcal{A}_0(\omega)$, given by

$$\begin{pmatrix} \mathcal{S}_B^0 & -\mathcal{S}_B^0 \\ \frac{1}{2}I + (\mathcal{K}_B^0)^* & -k(-\frac{1}{2}I + (\mathcal{K}_B^0)^*) \end{pmatrix},$$

is invertible. In fact, the inverse is given by

$$\frac{1}{k-1}\begin{pmatrix} k(\lambda I - (\mathcal{K}_B^0)^*)^{-1}(\frac{1}{2}I - (\mathcal{K}_B^0)^*)(\mathcal{S}_B^0)^{-1} & (\lambda I - (\mathcal{K}_B^0)^*)^{-1} \\ -(\lambda I - (\mathcal{K}_B^0)^*)^{-1}(\frac{1}{2}I + (\mathcal{K}_B^0)^*)(\mathcal{S}_B^0)^{-1} & (\lambda I - (\mathcal{K}_B^0)^*)^{-1} \end{pmatrix},$$

where $\lambda := (k+1)/(2(k-1))$. Therefore the invertibility of $\mathcal{A}_0(\omega)$ holds for any $\omega \notin \{\sqrt{\mu_j}\}_{j\geq 1}$.

In the two-dimensional case, (3.5) holds true. Moreover, we have

(3.9)
$$\begin{aligned}\frac{\partial}{\partial \nu(x)}\Gamma_\omega(\epsilon(x-y)) &= \frac{\langle x-y, \nu(x)\rangle}{2\pi|x-y|^2}\Bigg[1 + \sum_{n=1}^{+\infty}(-1)^n \frac{(\omega\epsilon)^{2n}}{2^{2n}n!(n-1)!}|x-y|^{2n} \\ &\quad \times \left(\ln(\omega\epsilon|x-y|) + \ln\eta + \frac{1}{2n} - \sum_{j=1}^n \frac{1}{j}\right)\Bigg].\end{aligned}$$

Therefore, it follows that

(3.10)
$$\begin{aligned}\epsilon(\mathcal{K}_D^\omega)^*[\varphi](\epsilon x + z) &= (\mathcal{K}_B^0)^*[\tilde{\varphi}](x) + \sum_{n=1}^{+\infty}(-1)^n \frac{(\omega\epsilon)^{2n}}{2^{2n+1}\pi n!(n-1)!} \\ &\quad \times \int_{\partial B}\langle x-y, \nu(x)\rangle |x-y|^{2(n-1)}\left(\ln(\omega\epsilon|x-y|) + \ln\eta + \frac{1}{2n} - \sum_{j=1}^n \frac{1}{j}\right)\tilde{\varphi}(y)\, d\sigma(y).\end{aligned}$$

Similarly to (3.7), one can compute an analogous asymptotic expansion for $\mathcal{A}_\epsilon(\omega)$ in two dimensions, where the leading-order term $\mathcal{A}_0(\omega)$ is defined by (3.8). Analogously to the three-dimensional case, it can be shown that $\mathcal{A}_0(\omega)$ is invertible for any $\omega \notin \{\sqrt{\mu_j}\}_{j\geq 1}$.

Before proceeding further, using the generalized argument principle to construct the complete asymptotic expansions for μ_j^ϵ with respect to ϵ, we provide a rigorous study of the integral operator-valued function $\omega \mapsto \widetilde{\mathcal{A}}_\epsilon(\omega)$, when ω is in a small complex neighborhood of $\sqrt{\mu_j}$. The next three lemmas, analogous to Lemmas 2.64, 2.65, and 2.66, are immediate.

LEMMA 3.3. *The operator-valued function $\widetilde{\mathcal{A}}_\epsilon(\omega)$ is a holomorphic Fredholm operator with index 0 in $\mathbb{C}$ (while in the two-dimensional case in $\mathbb{C}\setminus\sqrt{-1}\mathbb{R}^-$) and $(\widetilde{\mathcal{A}}_\epsilon)(\omega)^{-1}$ is a meromorphic function of ω. If ω is a real characteristic value of the* *operator-valued function* $\widetilde{\mathcal{A}}_\epsilon$ *(or equivalently, a real pole of* $(\widetilde{\mathcal{A}}_\epsilon)(\omega)^{-1}$*), then there exists* j *such that* $\omega = \sqrt{\mu_j^\epsilon}$.

LEMMA 3.4. *Any* $\sqrt{\mu_j}$ *is a simple pole of the operator-valued function* $\omega \mapsto (\mathcal{A}_0)(\omega)^{-1}$.

LEMMA 3.5. *Let* $\omega_0 = \sqrt{\mu_j}$ *and suppose that* μ_j *is simple. Then there exists a positive constant* δ_0 *such that the operator-valued function* $\omega \mapsto \widetilde{\mathcal{A}}_\epsilon(\omega)$ *has exactly one characteristic value in* $\overline{V_{\delta_0}}(\omega_0)$*, where* $V_{\delta_0}(\omega_0)$ *is a disk of center* ω_0 *and radius* $\delta_0 > 0$*. This characteristic value is holomorphic with respect to* ϵ *in* $(-\epsilon_0, \epsilon_0)$ *for some* $\epsilon_0 > 0$*. Moreover, the following assertions hold:*

(i) $\mathcal{M}(\widetilde{\mathcal{A}}_\epsilon(\omega); \partial V_{\delta_0}) = 1$,

(ii) $(\widetilde{\mathcal{A}}_\epsilon)(\omega)^{-1} = (\omega - \omega_\epsilon)^{-1}\mathcal{L}_\epsilon + \mathcal{R}_\epsilon(\omega)$,

(iii) $\mathcal{L}_\epsilon : \mathrm{Ker}((\widetilde{\mathcal{A}}_\epsilon(\omega_\epsilon))^*) \to \mathrm{Ker}(\widetilde{\mathcal{A}}_\epsilon(\omega_\epsilon))$,

where $\mathcal{R}_\epsilon(\omega)$ *is a holomorphic function with respect to* $(\epsilon, \omega) \in (-\epsilon_0, \epsilon_0) \times V_{\delta_0}(\omega_0)$ *and* $\mathcal{L}_\epsilon$ *is a finite-dimensional operator.*

We are now ready to apply the generalized argument principle. Since we deal with simple characteristic values, the relevant formula will be as follows, which is an immediate consequence of Theorem 1.14.

LEMMA 3.6. *Let* $\omega_0 = \sqrt{\mu_j}$ *and suppose that* μ_j *is simple. Then* $\omega_\epsilon = \sqrt{\mu_j^\epsilon}$ *is given by*

$$\omega_\epsilon - \omega_0 = \frac{1}{2\pi\sqrt{-1}} \operatorname{tr} \int_{\partial V_{\delta_0}} (\omega - \omega_0)\widetilde{\mathcal{A}}_\epsilon(\omega)^{-1} \frac{\partial}{\partial \omega} \widetilde{\mathcal{A}}_\epsilon(\omega) d\omega. \tag{3.11}$$

Substituting the formula (3.7) into (3.11), we obtain the following complete asymptotic expansion with respect to ϵ for the eigenvalue perturbations in the three-dimensional case.

THEOREM 3.7 (Eigenvalue perturbations). *The following asymptotic expansion holds:*

$$\omega_\epsilon - \omega_0 = \frac{1}{2\sqrt{-1}\pi} \sum_{p=1}^{+\infty} \frac{1}{p} \sum_{n=p}^{+\infty} \epsilon^n \operatorname{tr} \int_{\partial V_{\delta_0}} B_{n,p}(\omega) d\omega, \tag{3.12}$$

where

$$B_{n,p}(\omega) = (-1)^p \sum_{\substack{n_1+\ldots+n_p=n \\ n_i \geq 1}} \mathcal{A}_0(\omega)^{-1}\mathcal{A}_{n_1}(\omega) \ldots \mathcal{A}_0(\omega)^{-1}\mathcal{A}_{n_p}(\omega)\omega^n.$$

PROOF. If ϵ is small enough, then the following Neumann series converges uniformly with respect to ω in ∂V_{δ_0}:

$$\widetilde{\mathcal{A}}_\epsilon(\omega)^{-1} = \sum_{p=0}^{+\infty} \left[\mathcal{A}_0(\omega)^{-1}(\mathcal{A}_0(\omega) - \widetilde{\mathcal{A}}_\epsilon(\omega))\right]^p \mathcal{A}_0(\omega)^{-1},$$

and hence we may deduce that $\omega_\epsilon - \omega_0$ is equal to

$$\frac{1}{2\pi\sqrt{-1}} \sum_{p=0}^{+\infty} \operatorname{tr} \int_{\partial V_{\delta_0}} (\omega - \omega_0) \left[\mathcal{A}_0(\omega)^{-1}(\mathcal{A}_0(\omega) - \widetilde{\mathcal{A}}_\epsilon(\omega))\right]^p \mathcal{A}_0(\omega)^{-1} \frac{d}{d\omega} \widetilde{\mathcal{A}}_\epsilon(\omega) d\omega.$$

As in the proof of Theorem 2.67, by using the property (1.5) of the trace together with identity (2.203) we have

$$
\begin{aligned}
&\operatorname{tr} \int_{\partial V_{\delta_0}} (\omega - \omega_0) \frac{1}{p} \frac{\partial}{\partial \omega} \left[\mathcal{A}_0(\omega)^{-1} (\mathcal{A}_0(\omega) - \widetilde{\mathcal{A}}_\epsilon(\omega)) \right]^p d\omega \\
&= \operatorname{tr} \Bigg[\int_{\partial V_{\delta_0}} (\omega - \omega_0) \Big[\mathcal{A}_0(\omega)^{-1} (\mathcal{A}_0(\omega) - \widetilde{\mathcal{A}}_\epsilon(\omega)) \Big]^{p-1} \mathcal{A}_0(\omega)^{-1} \frac{\partial}{\partial \omega} (\mathcal{A}_0(\omega) - \mathcal{A}_\epsilon(\omega)) d\omega \\
&\quad - \int_{\partial V_{\delta_0}} (\omega - \omega_0) \left[\mathcal{A}_0(\omega)^{-1} (\mathcal{A}_0(\omega) - \mathcal{A}_\epsilon(\omega)) \right]^p \mathcal{A}_0(\omega)^{-1} \frac{\partial}{\partial \omega} \mathcal{A}_0(\omega) d\omega \Bigg],
\end{aligned}
$$

and therefore,

$$
\begin{aligned}
\omega_\epsilon - \omega_0 = &- \frac{1}{2\sqrt{-1}\pi} \sum_{p=1}^{+\infty} \operatorname{tr} \int_{\partial V_{\delta_0}} (\omega - \omega_0) \frac{1}{p} \frac{\partial}{\partial \omega} \left[\mathcal{A}_0(\omega)^{-1} (\mathcal{A}_0(\omega) - \mathcal{A}_\epsilon(\omega)) \right]^p d\omega \\
&+ \frac{1}{2\pi\sqrt{-1}} \operatorname{tr} \int_{\partial V_{\delta_0}} (\omega - \omega_0) \mathcal{A}_0(\omega)^{-1} \frac{d}{d\omega} \mathcal{A}_0(\omega) d\omega.
\end{aligned}
$$

Since ω_0 is a simple pole of $\mathcal{A}_0(\omega)^{-1}$ and $\mathcal{A}_0(\omega)$ is holomorphic in ω, we readily get

$$
\int_{\partial V_{\delta_0}} (\omega - \omega_0) \mathcal{A}_0(\omega)^{-1} \frac{d}{d\omega} \mathcal{A}_0(\omega) d\omega = 0.
$$

Thus, it follows that

$$
\omega_\epsilon - \omega_0 = - \frac{1}{2\pi\sqrt{-1}} \sum_{p=1}^{+\infty} \operatorname{tr} \int_{\partial V_{\delta_0}} (\omega - \omega_0) \frac{1}{p} \frac{\partial}{\partial \omega} \left[\mathcal{A}_0(\omega)^{-1} (\mathcal{A}_0(\omega) - \mathcal{A}_\epsilon(\omega)) \right]^p d\omega.
$$

Now, a simple integration by parts yields

$$
\omega_\epsilon - \omega_0 = \frac{1}{2\pi\sqrt{-1}} \sum_{p=1}^{+\infty} \frac{1}{p} \operatorname{tr} \int_{\partial V_{\delta_0}} \Big[\mathcal{A}_0(\omega)^{-1} (\mathcal{A}_0(\omega) - \mathcal{A}_\epsilon(\omega)) \Big]^p d\omega.
$$

Notice from (3.7) that

$$
\begin{aligned}
&\Big(\mathcal{A}_0(\omega)^{-1} (\mathcal{A}_0(\omega) - \mathcal{A}_\epsilon(\omega)) \Big)^p \\
&\quad = (-1)^p \sum_{n=p}^{+\infty} \epsilon^n \sum_{\substack{n_1 + \ldots + n_p = n \\ n_i \geq 1}} \mathcal{A}_0(\omega)^{-1} \mathcal{A}_{n_1}(\omega) \ldots \mathcal{A}_0(\omega)^{-1} \mathcal{A}_{n_p}(\omega) \omega^n.
\end{aligned}
$$

Therefore, upon inserting this into the latter formula, we arrive at the desired asymptotic expansion (3.12). □

With regards to the simplest case, we can find the leading-order term in the asymptotic expansion of $\mu_j^\epsilon - \mu_j$ as $\epsilon \to 0$. By (3.12), the leading-order of the expansion of $\omega_\epsilon - \omega_0$ is the ϵ-order term and its coefficient is given by

$$
-\frac{1}{2\sqrt{-1}\pi} \operatorname{tr} \int_{\partial V_{\delta_0}} \mathcal{A}_0(\omega)^{-1} \mathcal{A}_1(\omega) \omega d\omega. \tag{3.13}
$$

We can now recover the following result from [**76**, **79**] giving the leading-order term in the asymptotic expansion of the eigenvalue perturbations.

COROLLARY 3.8. *Suppose μ_j is a simple eigenvalue associated with the normalized eigenfunction u_j. The following asymptotic expansion holds:*

$$\mu_j^\epsilon - \mu_j = \epsilon^d \nabla u_j(z) \cdot M(\lambda(k), B) \nabla u_j(z) + o(\epsilon^d), \tag{3.14}$$

where $M(\lambda(k), B)$, defined by (2.74), is the polarization tensor associated with the domain B and the conductivity k.

In view of the positivity properties of M, we can then deduce from formula (3.14) the sign of the variation of a given eigenvalue in terms of the conductivity of the particle. Furthermore, the Hashin-Shtrikman bounds (2.84) and (2.85) lead, in view of (3.14), to perturbation bounds for the eigenvalues.

Turning now to the behavior of the perturbed eigenfunction near the particle D or at the boundary Ω, we can prove from [**79**] that the following inner and outer expansions of the perturbed eigenfunction with respect to ϵ hold.

LEMMA 3.9. *Let u_j^ϵ be the normalized eigenfunction associated with μ_j^ϵ.*

(i) *The following inner expansion holds for x near z:*

$$u_j^\epsilon(x) = u_j(z) + \epsilon \sum_{l=1}^{d} \partial_l u_j(z) \psi_l \left(\frac{x-z}{\epsilon} \right) + o(\epsilon), \tag{3.15}$$

where ψ_l is defined by

$$\begin{cases} \nabla \cdot (1 + (k-1)\chi(B)) \nabla \psi_l = 0 & \text{in } \mathbb{R}^d, \\ \psi_l(x) - x_l = O(|x|^{1-d}) & \text{as } |x| \to +\infty. \end{cases} \tag{3.16}$$

(ii) *The following outer expansion holds uniformly for $x \in \partial\Omega$:*

$$(u_j^\epsilon - u_j)(x) = -\epsilon^d \nabla u_j(z) \cdot M(\lambda(k), B) \nabla \mathcal{N}_\Omega^{\omega_j}(x, z) + o(\epsilon^d), \tag{3.17}$$

where $\mathcal{N}_j$ is the solution to

$$\begin{cases} (\Delta_x + \omega_j^2) \mathcal{N}_\Omega^{\omega_j}(x, y) = -\delta_y + u_j(x) u_j(y) & \text{in } \Omega, \\ \dfrac{\partial \mathcal{N}_\Omega^{\omega_j}}{\partial \nu} = 0 \quad \text{on } \partial\Omega, \\ \displaystyle\int_\Omega \mathcal{N}_\Omega^{\omega_j} u_j = 0. \end{cases} \tag{3.18}$$

PROOF. We only outline the derivation of the asymptotic expansions (3.15) and (3.17) of u_j^ϵ leaving the details to the reader.

Note first that one can show that the polarization tensor $M = (m_{ll'})$ can be rewritten as

$$m_{ll'} = (k-1) \int_{\partial B} \psi_l \frac{\partial x_{l'}}{\partial \nu} \, d\sigma = (k-1) \int_{\partial B} \frac{\partial \psi_l}{\partial \nu} \bigg|_- x_{l'} \, d\sigma, \tag{3.19}$$

where ψ_l is the solution to (3.16).

For any $f \in L_0^2(\Omega)$, define $T^\epsilon[f] = v^\epsilon$, where v^ϵ is the solution to

$$\begin{cases} \nabla \cdot (1 + (k-1)\chi(D)) \nabla v^\epsilon = -f & \text{in } \Omega, \\ \dfrac{\partial v^\epsilon}{\partial \nu} = 0 & \text{on } \partial\Omega, \end{cases}$$

and $T[f] = v$, where v is the solution to

$$\begin{cases} \Delta v = -f & \text{in } \Omega, \\ \dfrac{\partial v}{\partial \nu} = 0 & \text{on } \partial\Omega. \end{cases}$$

By the spectral decomposition, T^ϵ and T can be expressed as follows:

$$T^\epsilon = \sum_l \frac{1}{\mu_l^\epsilon} u_l^\epsilon \otimes u_l^\epsilon, \tag{3.20}$$

and

$$T = \sum_l \frac{1}{\mu_l} u_l \otimes u_l. \tag{3.21}$$

Now let V be a disk centered at $1/\omega_j^2$, with radius so small that $1/\omega_l^2 \notin V$ if $l \neq j$. Decomposition (3.21) shows that, for any $\mu \in \partial V$,

$$(\mu - T)^{-1}[u_l] = \frac{1}{\mu - \dfrac{1}{\omega_l^2}} u_l, \quad \forall\, l.$$

On the other hand, (3.20) shows that

$$u_j^\epsilon \otimes u_j^\epsilon = \frac{1}{2\pi\sqrt{-1}} \int_{\partial V} (\mu - T^\epsilon)^{-1}\, d\mu.$$

Since

$$\begin{aligned} (\mu - T^\epsilon)^{-1}[u_j] &= (\mu - T)^{-1}[u_j] + (\mu - T)^{-1}(T^\epsilon - T)(\mu - T)^{-1}[u_j] + \text{ h.o.t.} \\ &= \frac{1}{\mu - \dfrac{1}{\omega_j^2}} \left[u_j + (\mu - T)^{-1}(T^\epsilon - T)[u_j] + \text{ h.o.t.} \right] \end{aligned}$$

we have

$$\langle u_j, u_j^\epsilon \rangle\, u_j^\epsilon = u_j + \frac{1}{2\pi\sqrt{-1}} \int_{\partial V} \frac{1}{\mu - \dfrac{1}{\omega_j^2}} (\mu - T)^{-1}(T^\epsilon - T)[u_j]\, d\mu + \text{ h.o.t.} \tag{3.22}$$

Here h.o.t. stands for higher-order term in terms of ϵ. Set $\Psi_l(x) := \psi_l(x) - x_l$. According to [**79**], the following expansion with respect to ϵ holds;

$$(T^\epsilon - T)[u_j] = \frac{\epsilon}{\omega_j^2} \sum_{l=1}^{d} \partial_l u_j(z) \Psi_l\Big(\frac{x - z}{\epsilon}\Big) + \text{ h.o.t.} \quad \text{in } \Omega,$$

and consequently,

$$(\mu - T)^{-1}(T^\epsilon - T)[u_j] = \frac{\epsilon}{\omega_j^2} \sum_{l=1}^{d} \partial_l u_j(z) (\mu - T)^{-1}\Big[\Psi_l\Big(\frac{\cdot - z}{\epsilon}\Big)\Big] + \text{ h.o.t.} \quad \text{in } \Omega.$$

From the definition of T, we can readily get

$$(\mu - T)^{-1}\Big[\Psi_l\Big(\frac{\cdot - z}{\epsilon}\Big)\Big] = \frac{1}{\mu}\left[\Psi_l\Big(\frac{\cdot - z}{\epsilon}\Big) + \frac{1}{\mu} \int_\Omega N_\Omega^{1/\sqrt{\mu}}(\cdot, y) \Psi_l\Big(\frac{y - z}{\epsilon}\Big)\, dy \right],$$

where $N_\Omega^{1/\sqrt{\mu}}$ is defined by (2.186). But

$$\Psi_l(\frac{\cdot - z}{\epsilon}) + \frac{1}{\mu}\int_\Omega N_\Omega^{1/\sqrt{\mu}}(\cdot, y)\Psi_l(\frac{y-z}{\epsilon})\,dy$$
$$= (1-k)\int_{\partial D} N_\Omega^{1/\sqrt{\mu}}(\cdot, y)\frac{\partial \psi_l}{\partial \nu}\bigg|_-(\frac{y-z}{\epsilon})\,d\sigma(y) + \text{ h.o.t.}$$

Therefore, we get from the definition (3.19) of the polarization tensor and the fact that ψ_l is harmonic in B that for $x \in \partial\Omega$,
(3.23)

$$\Psi_l(\frac{x-z}{\epsilon}) + \frac{1}{\mu}\int_\Omega N_\Omega^{1/\sqrt{\mu}}(x, y)\Psi_l(\frac{y-z}{\epsilon})\,dy = -\epsilon^{d-1}\sum_{l'=1}^d m_{ll'}\partial_{l'}N_\Omega^{1/\sqrt{\mu}}(x,z) + \text{ h.o.t.}$$

Inserting (3.23) into (3.22) and using the spectral decomposition (2.187) of $N_\Omega^{1/\sqrt{\mu}}$, we finally obtain

$$\langle \cdot, u_j^\epsilon\rangle\, u_j^\epsilon = u_j - \epsilon^d \nabla u_j(z)\cdot M(\lambda(k), B)\nabla \mathcal{N}_\Omega^{\omega_j}(x,z) + \text{ h.o.t.} \quad \text{on } \partial\Omega, \tag{3.24}$$

where $\mathcal{N}_\Omega^{\omega_j}$ is defined by (3.18) or equivalently by the following spectral representation:

$$\mathcal{N}_\Omega^{\omega_j}(x,z) = \sum_{l\neq j}\frac{u_l(x)u_l(z)}{\omega_l^2 - \omega_j^2}, \qquad x \neq z \in \Omega.$$

Since

$$\langle u_j^\epsilon, u_j\rangle^2 = 1 + \frac{\epsilon^d(1-k)}{2\pi\sqrt{-1}\omega_j^2}\sum_{l=1}^d \partial_l u_j(z)$$
$$\times \int_{\partial B}\frac{\partial \psi_l}{\partial \nu}\bigg|_-(y)\left(\int_{\partial V}\frac{1}{\mu(\mu - \frac{1}{\omega_j^2})}\int_\Omega N_\Omega^{1/\sqrt{\mu}}(x, \epsilon y + z)u_j(x)\,dx\right)d\sigma(y) + \text{ h.o.t.}$$
$$= 1 + o(\epsilon^d),$$

by using once again (2.187), the desired outer expansion follows immediately from (3.24).

The inner expansion follows in exactly the same manner as the outer expansion by observing that

$$(1-k)\int_{\partial D} N_\Omega^{1/\sqrt{\mu}}(x,y)\frac{\partial \psi_l}{\partial \nu}\bigg|_-(\frac{y-z}{\epsilon})\,d\sigma(y)$$
$$= (k-1)\int_{\partial D}\Gamma_0(x,y)\frac{\partial \psi_l}{\partial \nu}\bigg|_-(\frac{y-z}{\epsilon})\,d\sigma(y) + \text{h.o.t.}$$
$$= \Psi_l(\frac{x-z}{\epsilon}) + \text{h.o.t.},$$

for x near z, where Γ_0 is the fundamental solution of the Laplacian given by (2.2). □

Note that if we consider the eigenvalue problem

$$(3.25)\qquad \begin{cases} \Delta u_j^\epsilon + \mu_j^\epsilon u_j^\epsilon = 0 & \text{in } \Omega \setminus \overline{D}, \\ \Delta u_j^\epsilon + k\, \mu_j^\epsilon u_j^\epsilon = 0 & \text{in } D, \\ u_j^\epsilon|_+ - u_j^\epsilon|_- = 0 & \text{on } \partial D, \\ \dfrac{\partial u_j^\epsilon}{\partial \nu}\Big|_+ - \dfrac{\partial u_j^\epsilon}{\partial \nu}\Big|_- = 0 & \text{on } \partial D, \\ \dfrac{\partial u_j^\epsilon}{\partial \nu} = 0 & \text{on } \partial \Omega, \end{cases}$$

then a much simpler formula than (3.14) holds. In fact, if we suppose that μ_j is a simple eigenvalue associated with the normalized eigenfunction u_j (satisfying $\int_\Omega (1+(k-1)\chi(D))|u_j|^2 = 1$), then we have

$$(3.26)\qquad \mu_j^\epsilon - \mu_j = (1-k)|D||u_j(z)|^2 + o(\epsilon^d).$$

See, for example, [**76**] for the details and for a higher-order expansion. Moreover,

$$(u_j^\epsilon - u_j)(x) = -\mu_j(1-k)|D|u_j(z)\mathcal{N}_\Omega^{\sqrt{\mu_j}}(x,z) + o(|D|)$$

holds uniformly in $x \in \partial\Omega$.

3.2.2. Eigenvalue Perturbations Due to Shape Deformations. Let D_ϵ be an ϵ-perturbation of D; *i.e.*, let $h \in \mathcal{C}^2(\partial D)$ and ∂D_ϵ be given by

$$\partial D_\epsilon = \left\{ \tilde{x} : \tilde{x} = x + \epsilon h(x)\nu(x),\ x \in \partial D \right\}.$$

Consider the following eigenvalue problem:

$$(3.27)\qquad \begin{cases} \Delta u^\epsilon + \omega^2 u^\epsilon = 0 & \text{in } \Omega \setminus \overline{D_\epsilon}, \\ \Delta u^\epsilon + \dfrac{\omega^2}{k} u^\epsilon = 0 & \text{in } D_\epsilon, \\ u^\epsilon|_+ - u^\epsilon|_- = 0 & \text{on } \partial D_\epsilon, \\ \dfrac{\partial u^\epsilon}{\partial \nu}\Big|_+ - k\dfrac{\partial u^\epsilon}{\partial \nu}\Big|_- = 0 & \text{on } \partial D_\epsilon, \\ \dfrac{\partial u^\epsilon}{\partial \nu} = 0 & \text{on } \partial \Omega. \end{cases}$$

In exactly the same manner as in the previous section, we reduce the eigenvalue problem (3.27) to the calculation of the asymptotic expressions of the characteristic values of the operator-valued function $\mathcal{A}_\epsilon(\omega)$ given by

$$\omega \mapsto \mathcal{A}_\epsilon(\omega) := \begin{pmatrix} \frac{1}{2}I - \mathcal{K}_\Omega^\omega & -\mathcal{S}_{D_\epsilon}^\omega & 0 \\ \mathcal{D}_\Omega^\omega & \mathcal{S}_{D_\epsilon}^\omega & -\mathcal{S}_{D_\epsilon}^{\frac{\omega}{\sqrt{k}}} \\ \frac{\partial}{\partial\nu}\mathcal{D}_\Omega^\omega & \frac{1}{2}I + (\mathcal{K}_{D_\epsilon}^\omega)^* & -k(-\frac{1}{2}I + (\mathcal{K}_{D_\epsilon}^{\frac{\omega}{\sqrt{k}}})^*) \end{pmatrix}.$$

To derive a full asymptotic expansion of the perturbations in the eigenvalues, we shall expand the operator-valued function $\mathcal{A}_\epsilon(\omega)$ in terms of ϵ. For this purpose, we only need to construct high-order expansions of $\mathcal{S}_{D_\epsilon}^\omega$ and $(\mathcal{K}_{D_\epsilon}^\omega)^*$.

Suppose that ω_0 is a simple eigenvalue of (3.27) for $\epsilon = 0$. Let V_{δ_0} be a disk of center ω_0 and radius $\delta_0 > 0$ so that ω_0 is the only characteristic value of $\mathcal{A}_0(\omega)$ in $\overline{V_{\delta_0}}$.

With (2.145) and (2.150) in hand, we write

$$\mathcal{A}_\epsilon(\omega) = \sum_{n=0}^{+\infty} (\omega\epsilon)^n \mathcal{A}_n(\omega) \quad \text{for } \omega \in \overline{V_{\delta_0}}.$$

Therefore, from

$$\omega_\epsilon - \omega_0 = \frac{1}{2\pi\sqrt{-1}} \operatorname{tr} \int_{\partial V_{\delta_0}} (\omega - \omega_0)\mathcal{A}_\epsilon(\omega)^{-1} \frac{\partial}{\partial\omega}\mathcal{A}_\epsilon(\omega) d\omega,$$

we obtain the following complete asymptotic expansion for the eigenvalue perturbations due to a shape deformation of the particle.

THEOREM 3.10 (Eigenvalue perturbations). *For ϵ small enough, the following asymptotic expansion holds:*

$$\omega_\epsilon - \omega_0 = \frac{1}{2\sqrt{-1}\pi} \sum_{p=1}^{+\infty} \frac{1}{p} \sum_{n=p}^{+\infty} \epsilon^n \operatorname{tr} \int_{\partial V_{\delta_0}} B_{n,p}(\omega) d\omega, \tag{3.28}$$

where

$$B_{n,p}(\omega) = (-1)^p \sum_{\substack{n_1+\ldots+n_p=n \\ n_i \geq 1}} \mathcal{A}_0(\omega)^{-1}\mathcal{A}_{n_1}(\omega)\ldots\mathcal{A}_0(\omega)^{-1}\mathcal{A}_{n_p}(\omega)\omega^n.$$

The leading-order of the expansion of $\omega_\epsilon - \omega_0$ is the ϵ-order term and its coefficient is given by

$$-\frac{1}{2\sqrt{-1}\pi} \operatorname{tr} \int_{\partial V_{\delta_0}} \mathcal{A}_0(\omega)^{-1}\mathcal{A}_1(\omega)\omega d\omega.$$

Tedious calculations yield

$$\omega_\epsilon - \omega_0 = \epsilon\frac{k-1}{2\omega_0}\int_{\partial D} h\left[k(\frac{\partial u^0}{\partial\nu}|_-)^2 + (\frac{\partial u^0}{\partial T})^2\right] d\sigma + O(\epsilon^2), \tag{3.29}$$

where u^0 satisfying $\int_\Omega |u^0|^2 = 1$ is the normalized eigenvalue of (3.27) for $\epsilon = 0$. As will be seen later, this is exactly the shape derivative of ω_ϵ. Another way of deriving (3.29) is given in [**21**]. It is based on fine gradient estimates from [**368**] (see also [**367**]) together with Osborn's result on spectral approximation for compact operators in [**425**].

Let v be the solution to

$$\begin{cases} \Delta v + \omega_0^2 v = 0 & \text{in } \Omega \setminus \overline{D}, \\ \Delta v + \omega_0^2 v = 0 & \text{in } D, \\ v|_+ - v|_- = -h\dfrac{\partial u^0}{\partial\nu}|_- & \text{on } \partial D, \\ \dfrac{\partial v}{\partial\nu}\Big|_+ - k\dfrac{\partial v}{\partial\nu}\Big|_- = -\dfrac{\partial}{\partial T}h\dfrac{\partial u^0}{\partial T} & \text{on } \partial D, \\ \dfrac{\partial v}{\partial\nu} = 0 & \text{on } \partial\Omega, \\ \displaystyle\int_\Omega v\, u^0 = 0. \end{cases} \tag{3.30}$$

It can be shown that the asymptotic expansion of u^ϵ

$$u^\epsilon - u^0 = \epsilon(k-1)v + o(\epsilon) \tag{3.31}$$

holds uniformly on $\partial\Omega$.

Observe that if we define $\mathcal{N}^{\omega_0}_{\Omega,D}$ as the solution to

$$\begin{cases} \Big(\nabla\cdot(1+(k-1)\chi(D))\nabla+\omega_0^2\Big)\mathcal{N}^{\omega_0}_{\Omega,D}(x,y) = -\delta_y + u^0(x)u^0(y) \quad \text{in } \Omega, \\ \dfrac{\partial \mathcal{N}^{\omega_0}_{\Omega,D}}{\partial\nu} = 0 \quad \text{on } \partial\Omega, \\ \displaystyle\int_\Omega \mathcal{N}^{\omega_0}_{\Omega,D} u^0 = 0, \end{cases}$$

then v admits the following integral representation:

$$v(x) = \int_{\partial D} h(y)\left[\frac{\partial u^0}{\partial T}(y)\frac{\partial \mathcal{N}^{\omega_0}_{\Omega,D}}{\partial T}(x,y) + k\frac{\partial u^0}{\partial\nu}\bigg|_-(y)\frac{\partial \mathcal{N}^{\omega_0}_{\Omega,D}}{\partial\nu}\bigg|_-(x,y)\right]d\sigma(y).$$

In exactly the same manner as in the derivation of (3.29), if we consider the eigenvalue problem

$$\begin{cases} \Delta u^\epsilon + \omega^2 u^\epsilon = 0 & \text{in } \Omega\setminus\overline{D_\epsilon}, \\ \Delta u^\epsilon + \dfrac{\omega^2}{k}u^\epsilon = 0 & \text{in } D_\epsilon, \\ u^\epsilon|_+ - u^\epsilon|_- = 0 & \text{on } \partial D_\epsilon, \\ \dfrac{\partial u^\epsilon}{\partial\nu}\bigg|_+ - \dfrac{\partial u^\epsilon}{\partial\nu}\bigg|_- = 0 & \text{on } \partial D_\epsilon, \\ \dfrac{\partial u^\epsilon}{\partial\nu} = 0 \text{ or } u^\epsilon = 0 & \text{on } \partial\Omega, \end{cases}$$

or equivalently,

$$\begin{cases} \Delta u^\epsilon + \omega^2(\chi(\Omega\setminus\overline{D_\epsilon}) + \dfrac{1}{k}\chi(D_\epsilon))u^\epsilon = 0 & \text{in } \Omega, \\ \dfrac{\partial u^\epsilon}{\partial\nu} = 0 \text{ or } u^\epsilon = 0 & \text{on } \partial\Omega, \end{cases} \tag{3.32}$$

we can prove that the following asymptotic formula holds:

$$\omega_\epsilon - \omega_0 = \frac{\epsilon}{2}\omega_0(\frac{1}{k}-1)\int_{\partial D} h|u^0|^2\,d\sigma + O(\epsilon^2), \tag{3.33}$$

where u^0 is the eigenvalue of (3.32) for $\epsilon = 0$ satisfying the normalization

$$\int_\Omega \left(\chi(\Omega\setminus\overline{D}) + \frac{1}{k}\chi(D)\right)|u^0|^2 = 1.$$

Furthermore, the asymptotic expansion of u^ϵ

$$u^\epsilon - u^0 = \epsilon\omega_0^2(1-\frac{1}{k})w + o(\epsilon) \tag{3.34}$$

holds uniformly on $\partial\Omega$, where

$$\begin{cases} \Delta w + \omega_0^2 w = 0 & \text{in } \Omega\setminus\overline{D}, \\ \Delta w + \omega_0^2 w = 0 & \text{in } D, \\ w|_+ - w|_- = 0 & \text{on } \partial D, \\ \dfrac{\partial w}{\partial\nu}\bigg|_+ - \dfrac{\partial w}{\partial\nu}\bigg|_- = hu^0 & \text{on } \partial D, \\ \dfrac{\partial w}{\partial\nu} = 0 \quad \text{or } w = 0 & \text{on } \partial\Omega, \\ \displaystyle\int_\Omega w\,u^0 = 0. \end{cases}$$

REMARK 3.11. *In* [**76**, **91**], *asymptotic formulas for perturbations in the eigenvalues of the full Maxwell equations due to the presence of small dielectric particles are derived.*

3.3. Optical Resonators

3.3.1. Integral Formulation of Resonances. The integral formulation described below can be applied to resonance problems for Maxwell's and elasticity equations. However, for simplicity, we will limit the presentation to a scalar problem. Consider the solution u to the following problem:

$$\begin{cases} \Delta u + \omega^2 n(x) u = 0, \\ u \text{ satisfies the Sommerfeld radiation condition.} \end{cases} \tag{3.35}$$

We assume that $n-1$ is compactly supported in a bounded domain $D \subset \mathbb{R}^d$ for $d = 2, 3$, and is assumed to be known.

Keeping in mind that (2.151) holds for $\omega \in \mathbb{C}$, we can formally rewrite the solution to (3.35) in integral form as

$$u(x) + \omega^2 \int_D (n(y) - 1)\Gamma_\omega(x-y)u(y)dy = 0\,, \quad x \in \mathbb{R}^d. \tag{3.36}$$

We call a particular $\omega \in \mathbb{C}$ a resonance if it yields nontrivial solutions $u(x)$ of (3.36). Hence, resonance is a solution to the wave equation which is spatially localized while its time dependence is harmonic except for decay due to radiation. The decay rate, which is proportional to the imaginary part of the resonance value, depends on the material properties of the resonator. If ω is a resonance, then we call (nontrivial) functions u satisfying (3.35) the resonant modes.

Define the operator $\mathcal{A}_0(\omega)$ as

$$\mathcal{A}_0(\omega)[u] = u(x) + \omega^2 \int_D (n(y) - 1)\Gamma_\omega(x-y)u(y)dy.$$

Notice that the adjoint $\mathcal{A}_0(\omega)^*$ of $\mathcal{A}_0(\omega)$ is given by

$$\mathcal{A}_0(\omega)^*[v] = v(x) + (n(x) - 1)\omega^2 \int_D \Gamma_\omega(x-y)v(y)dy.$$

Now given $n(x)$, we have the nonlinear eigenvalue problem $\mathcal{A}_0(\omega)[u] = 0$, which can be solved using Muller's method. We can easily prove that ω_0 is a resonance if and only if it is a characteristic value of the meromorphic operator-valued function $\omega \mapsto \mathcal{A}_0(\omega)$.

3.3.2. Optimization of the Quality Factor. We define the quality factor Q as

$$Q = \left|\frac{\Re \omega}{\Im \omega}\right|,$$

where $\omega \in \mathbb{C}$ is a resonance. The quality factor is inversely proportional to the decay rate.

In order to compute the sensitivity of Q to changes in $n(x)$, we can make use of the generalized argument principle.

Write $n_\epsilon(x) = n(x) + \epsilon\mu(x)$, where μ is compactly supported in D and ϵ is a small parameter and let ω_0 be a characteristic value of $\mathcal{A}_0(\omega)$. Denote by $\mathcal{A}_\epsilon(\omega)$ the operator-valued function associated with n_ϵ. Then there exists a positive constant

δ_0 such that the operator-valued function $\omega \mapsto \mathcal{A}_\epsilon(\omega)$ has exactly one characteristic value in $\overline{V_{\delta_0}}(\omega_0)$, where $V_{\delta_0}(\omega_0)$ is a disk of center ω_0 and radius $\delta_0 > 0$.

Analogously to Lemma 3.6, we have

$$\omega_\epsilon - \omega_0 = \frac{1}{2\sqrt{-1}\pi} \operatorname{tr} \int_{\partial V_{\delta_0}} (\omega - \omega_0)\mathcal{A}_\epsilon(\omega)^{-1} \frac{\partial}{\partial \omega}\mathcal{A}_\epsilon(\omega) d\omega,$$

and hence, the leading-order of the expansion of $\omega_\epsilon - \omega_0$ is the ϵ-order term and its coefficient is given by

$$-\frac{1}{2\pi\sqrt{-1}} \operatorname{tr} \int_{\partial V_{\delta_0}} \mathcal{A}_0(\omega)^{-1}\mathcal{A}_1^{(\mu)}(\omega)\omega d\omega, \tag{3.37}$$

where the operator $\mathcal{A}_1^{(\mu)}$ is defined by

$$\mathcal{A}_1^{(\mu)}(\omega)[u] = \omega^2 \int_\Omega \mu(y)\Gamma_\omega(x-y)u(y)dy.$$

Formula (3.37) yields the Fréchet derivative of the quality factor Q with respect to n. Given an admissible set of functions $n(x)$, an optimal control can be used to maximize the quality factor of the resonator D.

3.4. Elastic Cavities

Let Ω be an elastic medium in $\mathbb{R}^3$ with a connected Lipschitz boundary whose Lamé constants are λ, μ. We consider the eigenvalue problem for the Lamé system of linear elasticity:

$$\mathcal{L}^{\lambda,\mu}u + \kappa u = \mu\Delta u + (\lambda+\mu)\nabla\nabla\cdot u + \kappa u = 0 \quad \text{in } \Omega, \tag{3.38}$$

with the Neumann boundary condition $\partial u/\partial \nu = 0$ on $\partial\Omega$. Here the conormal derivative $\partial u/\partial\nu$ is defined by (2.386).

Suppose that Ω contains a small particle D of the form $D = z + \epsilon B$, where B is a bounded Lipschitz domain containing the origin, ϵ is a small parameter, and z indicates the location of the particle. Due to the presence of the particle D, the eigenvalues of the domain Ω are perturbed. Our goal in this section is to find an asymptotic expansion for the perturbation of eigenvalues due to the presence of the particle. Let $\kappa_1 \le \kappa_2 \le \ldots$ be the eigenvalues of (3.38) and let $\kappa_1^\epsilon \le \kappa_2^\epsilon \le \ldots$ be the eigenvalues in the presence of the particle. The main result of this section is a complete asymptotic expansion of $\kappa_j^\epsilon - \kappa_j$ as $\epsilon \to 0$.

The main ingredients in deriving the results of this section are again the integral equations and the theory of meromorphic operator-valued functions. Using integral representations of solutions to the harmonic oscillatory linear elastic equation, we reduce this problem to the study of characteristic values of integral operators in the complex plane.

The elastic particles we deal with are soft particles. A soft particle is characterized by the transmission conditions on its boundary. We will explicitly calculate the leading-order term, which is of order of the volume of the particle, and is expressed in terms of the eigenfunctions and the elastic moment tensor. We confine our attention to the eigenvalues of the Neumann boundary value problem. The Dirichlet boundary case can be treated in a similar way with only minor modifications of the techniques presented here. We also confine our attention to the three-dimensional case. The two-dimensional case can be dealt with in an almost identical way.

We now investigate the perturbation of eigenvalues due to the presence of a small soft elastic particle. Suppose that the elastic medium Ω contains a small particle D of the form $D = z + \epsilon B$, whose Lamé constants are $\tilde{\lambda}, \tilde{\mu}$ satisfying $(\lambda - \tilde{\lambda})(\mu - \tilde{\mu}) \geq 0$ and $0 < \tilde{\lambda}, \tilde{\mu} < +\infty$. Let κ_k be an eigenvalue of $-\mathcal{L}^{\lambda,\mu}$ and let κ_k^ϵ be the perturbed eigenvalue in the presence of the particle. Then the eigenfunction u_k^ϵ corresponding to the (simple) eigenvalue κ_k^ϵ is the solution to (2.412) with $\omega^2 = \kappa_k^\epsilon$. Recall that $\tilde{\mathcal{S}}_D^\omega$ and $\tilde{\mathcal{K}}_D^\omega$ denote the single-layer potential and the boundary integral operator, respectively, defined by (2.395) and (2.400) with λ, μ replaced by $\tilde{\lambda}, \tilde{\mu}$.

Let u be the solution to

$$(3.39)\qquad \begin{cases} \mathcal{L}^{\lambda,\mu} u + \omega^2 u = 0 & \text{in } \Omega \setminus \overline{D}, \\ \mathcal{L}^{\tilde{\lambda},\tilde{\mu}} u + \omega^2 u = 0 & \text{in } D, \\ u\big|_+ - u\big|_- = 0 & \text{on } \partial D, \\ \dfrac{\partial u}{\partial \nu}\Big|_+ - \dfrac{\partial u}{\partial \tilde{\nu}}\Big|_- = 0 & \text{on } \partial D, \\ \dfrac{\partial u}{\partial \nu} = g & \text{on } \partial \Omega. \end{cases}$$

We may assume as before that ω^2 is not a Dirichlet eigenvalue for $-\mathcal{L}^{\lambda,\mu}$ on D. By Theorem 2.123, u can be represented as

$$(3.40)\qquad u = \begin{cases} \mathcal{D}_\Omega^\omega[\psi] + \mathcal{S}_D^\omega[\phi] & \text{in } \Omega \setminus \overline{D}, \\ \tilde{\mathcal{S}}_D^\omega[\theta] & \text{in } D, \end{cases}$$

where ψ, ϕ, and θ satisfy the system of integral equations

$$(3.41)\qquad \begin{cases} (\dfrac{1}{2} I - \mathcal{K}_\Omega^\omega)[\psi] - \mathcal{S}_D^\omega[\phi] = 0 & \text{on } \partial \Omega, \\ \mathcal{D}_\Omega^\omega[\psi] + \mathcal{S}_D^\omega[\phi] - \tilde{\mathcal{S}}_D^\omega[\theta] = 0 & \text{on } \partial D, \\ \dfrac{\partial (\mathcal{D}_\Omega^\omega[\psi])}{\partial \nu} + \dfrac{\partial (\mathcal{S}_D^\omega[\phi])}{\partial \nu}\Big|_+ - \dfrac{\partial (\tilde{\mathcal{S}}_D^\omega[\theta])}{\partial \tilde{\nu}}\Big|_- = 0 & \text{on } \partial D. \end{cases}$$

Conversely, $(\phi, \psi, \theta) \in L^2(\partial \Omega)^3 \times L^2(\partial D)^3 \times L^2(\partial D)^3$ satisfying (3.41) yields the solution to (3.39) via the representation formula (3.40).

In order to derive an asymptotic expansion for κ^ϵ, we begin by establishing the following, which generalizes Lemma 3.1 to the elasticity case.

LEMMA 3.12. *Let $\psi \in L^2(\partial \Omega)^3$ and $\phi \in L^2(\partial D)^3$. If $\tilde{\phi}(x) = \epsilon \phi(\epsilon x + z)$ for $x \in \partial B$, then we have*

$$(3.42)\quad \mathcal{S}_D^\omega[\phi](x) = \sum_{n=0}^{+\infty} (-1)^n \epsilon^{n+1} \sum_{|\alpha| = n} \frac{1}{\alpha!} \partial^\alpha \mathbf{\Gamma}^\omega (x - z) \int_{\partial B} y^\alpha \tilde{\phi}(y) d\sigma(y), \ x \in \partial \Omega,$$

$$(3.43)\quad \mathcal{D}_\Omega^\omega[\psi](\epsilon x + z) = \sum_{n=0}^{+\infty} \epsilon^n \sum_{|\alpha| = n} \frac{1}{\alpha!} \partial^\alpha (\mathcal{D}_\Omega^\omega \psi)(z) x^\alpha, \ x \in \partial B,$$

and for $x \in \partial B$ *and* $i = 1, 2, 3$,

(3.44)

$$\begin{aligned}
&\mathcal{S}_D^\omega[\phi]_i(\epsilon x + z) \\
&= -\frac{1}{4\pi}\sum_{n=0}^{+\infty}(\epsilon\omega)^n \frac{(\sqrt{-1})^n}{(n+2)n!}\Bigg[\Big(\frac{n+1}{c_T^{n+2}} + \frac{1}{c_L^{n+2}}\Big)\epsilon_{ij}\int_{\partial B}|x-y|^{n-1}\tilde{\phi}_j(y)\,d\sigma(y) \\
&\quad -\Big(\frac{n-1}{c_T^{n+2}} - \frac{n-1}{c_L^{n+2}}\Big)\int_{\partial B}|x-y|^{n-3}(x_i-y_i)(x_j-y_j)\tilde{\phi}_j(y)\,d\sigma(y)\Bigg],
\end{aligned}$$

where $\mathcal{S}_D^\omega[\phi]_i$ *denotes the* i*th component of* $\mathcal{S}_D^\omega[\phi]$.

PROOF. The series (3.43) is exactly a Taylor expansion of $\mathcal{D}_\Omega^\omega[\psi](\epsilon x + z)$ at z. By a change of variables, we have that, for any $x \in \partial\Omega$,

$$\mathcal{S}_D^\omega[\phi](x) = \int_{\partial D}\mathbf{\Gamma}^\omega(x-\tilde{y})\phi(\tilde{y})d\sigma(\tilde{y}) = \epsilon\int_{\partial B}\mathbf{\Gamma}^\omega(x-z-\epsilon y)\tilde{\phi}(y)d\sigma(y).$$

Using the Taylor expansion of $\mathbf{\Gamma}^\omega(x-z-\epsilon y)$ at $x-z$, we readily get (3.42). Similarly, (3.44) immediately follows from a change of variables and (2.390). This completes the proof. □

Let $\varphi(x) = \epsilon\phi(\epsilon x + z)$ and $\vartheta(x) = \epsilon\theta(\epsilon x + z)$. Then using Lemma 3.12, (3.41) can be written as follows:

$$\mathcal{A}_\epsilon^\omega\begin{pmatrix}\psi\\ \varphi\\ \vartheta\end{pmatrix} = 0,\ \mathcal{A}_\epsilon^\omega = \sum_{n=0}^{+\infty}(\omega\epsilon)^n\mathcal{A}_n^\omega,$$

where

$$\mathcal{A}_0^\omega = \begin{pmatrix}(\frac{1}{2}I - \mathcal{K}_\Omega^\omega) & 0 & 0\\ \mathcal{D}_\Omega^\omega[\cdot](z) & \mathcal{S}_B^0 & -\tilde{\mathcal{S}}_B^0\\ 0 & \frac{1}{2}I + (\mathcal{K}_B^0)^* & \frac{1}{2}I - (\tilde{\mathcal{K}}_B^0)^*\end{pmatrix},$$

and for $n = 1, 2, \ldots$, $\mathcal{A}_n^\omega$ is equal to

$$\begin{pmatrix}0 & \displaystyle\frac{(-1)^n}{\omega^n}\sum_{|\alpha|=n-1}\frac{1}{\alpha!}\partial^\alpha\mathbf{\Gamma}^\omega(x-z)\int_{\partial B}y^\alpha\cdot d\sigma(y) & 0\\ \displaystyle\frac{1}{\omega^n}\sum_{|\alpha|=n}\frac{1}{\alpha!}x^\alpha\partial^\alpha\mathcal{D}_\Omega^\omega(\cdot)(z) & S_n & -\tilde{S}_n\\ \displaystyle\frac{1}{\omega^n}\sum_{|\alpha|=n}\frac{1}{\alpha!}\frac{\partial(x^\alpha I)}{\partial\nu}\partial^\alpha\mathcal{D}_\Omega^\omega(\cdot)(z) & K_n & -\tilde{K}_n\end{pmatrix}.$$

Here S_n is the operator from $L^2(\partial B)^3$ into $H^1(\partial B)^3$ defined by

$$S_n[\varphi]_i = \sum_{j=1}^3 \big(S_n\big)_{ij}[\varphi_j],$$

with

$$\begin{aligned}
\big(S_n\big)_{ij} &= -\frac{1}{4\pi}\frac{(\sqrt{-1})^n}{(n+2)n!}\Big(\frac{n+1}{c_T^{n+2}} + \frac{1}{c_L^{n+2}}\Big)\epsilon_{ij}\int_{\partial B}|x-y|^{n-1}\cdot d\sigma(y) \\
&\quad + \frac{1}{4\pi}\frac{(\sqrt{-1})^n(n-1)}{(n+2)n!}\Big(\frac{1}{c_T^{n+2}} - \frac{1}{c_L^{n+2}}\Big)\int_{\partial B}|x-y|^{n-3}(x_i-y_i)(x_j-y_j)\cdot d\sigma(y),
\end{aligned}$$

and $K_n = \partial S_n/\partial \nu$, regarding S_n as an operator mapping functions on ∂B to functions in B. The operators $\tilde{S}_n$ and $\tilde{K}_n$ are defined in exactly the same way with c_T and c_L replaced by $\tilde{c}_T = \sqrt{\tilde{\mu}}$ and $\tilde{c}_L = \sqrt{\tilde{\lambda} + 2\tilde{\mu}}$.

With this notation, the following theorem holds.

THEOREM 3.13 (Eigenvalue perturbations). *Let κ_j be a simple Neumann eigenvalue for $-\mathcal{L}^{\lambda,\mu}$ in Ω without the particle and let κ_j^ϵ be the Neumann eigenvalue when Ω contains the particle. Let $\omega_0 := \sqrt{\kappa_j}$ and $\omega_\epsilon := \sqrt{\kappa_j^\epsilon}$. Then we have*

$$\omega_\epsilon - \omega_0 = \frac{1}{2\sqrt{-1}\pi} \sum_{p=1}^{+\infty} \frac{1}{p} \sum_{n=p}^{+\infty} \epsilon^n \operatorname{tr} \int_{\partial V_j} \mathcal{B}_{n,p}(\omega) d\omega, \tag{3.45}$$

where V_j is a neighborhood of $\sqrt{\kappa_j}$ and

$$\mathcal{B}_{n,p}(\omega) = (-1)^p \sum_{\substack{n_1+\cdots+n_p=n \\ n_i \geq 1}} (\mathcal{A}_0^\omega)^{-1} \mathcal{A}_{n_1}^\omega \ldots (\mathcal{A}_0^\omega)^{-1} \mathcal{A}_{n_p}^\omega \omega^n. \tag{3.46}$$

We now state the following theorem.

THEOREM 3.14. *Let κ_k be a simple Neumann eigenvalue for $-\mathcal{L}^{\lambda,\mu}$ in Ω without the particle, let κ_k^ϵ be the Neumann eigenvalue when Ω contains the particle, and let u_k be the corresponding eigenfunction such that $\|u_k\|_{L^2(\Omega)} = 1$. Then we have*

$$\kappa_k^\epsilon - \kappa_k = \epsilon^3 \sum_{i,j,p,q=1}^{3} m_{ijpq} \partial_i (u_k)_j(z) \partial_p (u_k)_q(z) + O(\epsilon^4), \tag{3.47}$$

where $(u_k)_j$ denotes the jth component of u_k and $M = (m_{ijpq})$ defined by (2.523) is the elastic moment tensor associated with B and the elastic parameters $\tilde{\lambda}$ and $\tilde{\mu}$.

We note that because of the symmetry of the elastic moment tensor $m_{ijpq} = m_{jipq} = m_{ijqp}$ (see (2.527)), (3.47) can be written in a more compact form using the standard notation of the contraction and the strain for tensors:

$$\kappa_k^\epsilon - \kappa_k = \epsilon^3 \nabla^s u_k(z) : M \nabla^s u_k(z) + O(\epsilon^4). \tag{3.48}$$

It is worth mentioning that if the particle is harder (softer, resp.) than the background, *i.e.*, $\tilde{\mu} > \mu$ and $\tilde{\lambda} \geq \lambda$ ($\tilde{\mu} < \mu$ and $\tilde{\lambda} \leq \lambda$, resp.), then M is positive (negative, resp.) definite (see Theorem 2.142), and hence $\kappa_k^\epsilon > \kappa_k$ ($\kappa_k^\epsilon > \kappa_k$, resp) provided that ϵ is small enough and $\nabla^s u_k(z) \neq 0$. Formula (3.48) makes it possible to deduce the sign of the variation of a given eigenvalue in terms of the elastic parameters of the particle.

PROOF OF THEOREM 3.14. We first observe from (3.45) that the ϵ-order term is given by

$$-\frac{1}{2\sqrt{-1}\pi} \operatorname{tr} \int_{\partial V_k} (\mathcal{A}_0^\omega)^{-1} \mathcal{A}_1^\omega \omega d\omega, \tag{3.49}$$

the ϵ^2-order term is given by

$$\frac{1}{2\sqrt{-1}\pi} \operatorname{tr} \int_{\partial V_k} \Big[-(\mathcal{A}_0^\omega)^{-1} \mathcal{A}_2^\omega + \frac{1}{2}\big((\mathcal{A}_0^\omega)^{-1} \mathcal{A}_1^\omega\big)^2 \Big] \omega^2 d\omega, \tag{3.50}$$

and the ϵ^3-order term is given by

$$\frac{1}{2\sqrt{-1}\pi}\operatorname{tr}\int_{\partial V_k}\Big[-(\mathcal{A}_0^\omega)^{-1}\mathcal{A}_3^\omega+(\mathcal{A}_0^\omega)^{-1}\mathcal{A}_1^\omega(\mathcal{A}_0^\omega)^{-1}\mathcal{A}_2^\omega \\ -\frac{1}{3}\big((\mathcal{A}_0^\omega)^{-1}\mathcal{A}_1^\omega\big)^3\Big]\omega^3\,d\omega. \tag{3.51}$$

Introduce

$$\begin{pmatrix} \mathcal{S}_B^0 & -\tilde{\mathcal{S}}_B^0 \\ \frac{1}{2}I+(\mathcal{K}_B^0)^* & \frac{1}{2}I-(\tilde{\mathcal{K}}_B^0)^* \end{pmatrix}^{-1} = \begin{pmatrix} A_1 & A_2 \\ A_3 & A_4 \end{pmatrix}, \tag{3.52}$$

where the invertibility is guaranteed by Theorem 2.122. As another direct consequence of this theorem, we also have that

$$A_1(f),\ A_2(g)\in L^2_\Psi(\partial B) \tag{3.53}$$

for any $f\in H^1(\partial B)^3$ and $g\in L^2_\Psi(\partial B)$ and

$$A_1(f)=0 \quad \text{for any} \quad f\in\Psi. \tag{3.54}$$

Explicit calculations show that $\big(\mathcal{A}_0^\omega\big)^{-1}$ takes the following form:

$$\big(\mathcal{A}_0^\omega\big)^{-1} = \begin{pmatrix} \big(\frac{1}{2}I-\mathcal{K}_\Omega^\omega\big)^{-1} & 0 & 0 \\ 0 & A_1 & A_2 \\ (\tilde{\mathcal{S}}_B^0)^{-1}[I]\big(\mathcal{D}_\Omega^\omega(\frac{1}{2}I-\mathcal{K}_\Omega^\omega)^{-1}[\cdot]\big)(z) & A_3 & A_4 \end{pmatrix}.$$

Since A_i, $i=1,2,3,4$, are independent of ω, we have

$$\operatorname{tr}\int_{\partial V_k}(\mathcal{A}_0^\omega)^{-1}\mathcal{A}_n^\omega\omega^n\,d\omega=0 \tag{3.55}$$

for any positive integer n.

From (3.53) and (3.54) we readily find $\big(\mathcal{A}_0^\omega\big)^{-1}\mathcal{A}_1^\omega\omega$ is equal to

$$\begin{pmatrix} 0 & T_1 & 0 \\ \sum_{|\alpha|=1}\Big(A_1(x^\alpha I)+A_2\Big(\frac{\partial(x^\alpha I)}{\partial\nu}\Big)\Big)\partial^\alpha\mathcal{D}_\Omega^\omega[\cdot](z) & 0 & 0 \\ \sum_{|\alpha|=1}\Big(A_3(x^\alpha I)+A_4\Big(\frac{\partial(x^\alpha I)}{\partial\nu}\Big)\Big)\partial^\alpha\mathcal{D}_\Omega^\omega[\cdot](z) & T_2 & -\omega A_3\tilde{S}_1 \end{pmatrix},$$

where

$$T_1=-\big(\frac{1}{2}I-\mathcal{K}_\Omega^\omega\big)^{-1}\mathbf{\Gamma}^\omega(x-z)\int_{\partial B}\cdot\,d\sigma(y),$$
$$T_2=-(\tilde{\mathcal{S}}_B^0)^{-1}\mathcal{D}_\Omega^\omega\Big[\big(\frac{1}{2}I-\mathcal{K}_\Omega^\omega\big)^{-1}\mathbf{\Gamma}^\omega(x-z)\Big](z)\int_{\partial B}\cdot\,d\sigma(y)+A_3S_1.$$

Using (3.53), we can write that

$$\int_{\partial B}A_1(x^\alpha I)+A_2\Big(\frac{\partial(x^\alpha I)}{\partial\nu}\Big)\,d\sigma(y)=0 \quad \text{if } |\alpha|=1$$

and easily check that

$$\text{tr}\int_{\partial V_k}\left((\mathcal{A}_0^\omega)^{-1}\mathcal{A}_1^\omega\right)^n\omega^n d\omega = 0, \tag{3.56}$$

for any positive integer n. Now combining (3.49)–(3.51), (3.55), and (3.56) gives

$$\omega_\epsilon - \omega_0 = \frac{\epsilon^3}{2\sqrt{-1}\pi}\,\text{tr}\int_{\partial V_k}(\mathcal{A}_0^\omega)^{-1}\mathcal{A}_1^\omega(\mathcal{A}_0^\omega)^{-1}\mathcal{A}_2^\omega\omega^3 d\omega + O(\epsilon^4). \tag{3.57}$$

Indeed, we have

$$\left(\mathcal{A}_0^\omega\right)^{-1}\mathcal{A}_2^\omega\omega^2 = \begin{pmatrix} 0 & T_3 & 0 \\ T_5 & \omega^2(A_1S_2 + A_2K_2) & -\omega^2(A_1\tilde{S}_2 + A_2\tilde{K}_2) \\ T_6 & T_4 & -\omega^2(A_3\tilde{S}_2 + A_4\tilde{K}_2) \end{pmatrix},$$

where

$$\begin{aligned}
T_3 &= \sum_{|\alpha|=1}\left(\frac{1}{2}I - \mathcal{K}_\Omega^\omega\right)^{-1}\partial^\alpha\mathbf{\Gamma}^\omega(x-z)\int_{\partial B} y^\alpha\cdot d\sigma(y), \\
T_4 &= \sum_{|\alpha|=1}(\tilde{\mathcal{S}}_B^0)^{-1}(I)\mathcal{D}_\Omega^\omega\Big[\left(\frac{1}{2}I - \mathcal{K}_\Omega^\omega\right)^{-1}\partial^\alpha\mathbf{\Gamma}^\omega(x-z)\Big](z)\int_{\partial B} y^\alpha\cdot d\sigma(y) \\
&\quad + \omega^2(A_3S_2 + A_4K_2), \\
T_5 &= \sum_{|\alpha|=2}\frac{1}{\alpha!}\Big(A_1(x^\alpha I) + A_2\Big(\frac{\partial(x^\alpha I)}{\partial\nu}\Big)\Big)\partial^\alpha\mathcal{D}_\Omega^\omega(\cdot)(z), \\
T_6 &= \sum_{|\alpha|=2}\frac{1}{\alpha!}\Big(A_3(x^\alpha I) + A_4\Big(\frac{\partial(x^\alpha I)}{\partial\nu}\Big)\Big)\partial^\alpha\mathcal{D}_\Omega^\omega(\cdot)(z).
\end{aligned}$$

Using the following identity, whose proof will be given later,

$$\frac{1}{2\sqrt{-1}\pi}\,\text{tr}\int_{\partial V_k}\Big(T_1T_5 - \omega^2T_2(A_1\tilde{S}_2 + A_2\tilde{K}_2)\Big)\,d\omega = 0, \tag{3.58}$$

it follows from (3.58) that

$$\begin{aligned}
&\frac{1}{2\sqrt{-1}\pi}\,\text{tr}\int_{\partial V_k}(\mathcal{A}_0^\omega)^{-1}\mathcal{A}_1^\omega(\mathcal{A}_0^\omega)^{-1}\mathcal{A}_2^\omega\omega^3 d\omega \\
&= \frac{1}{2\sqrt{-1}\pi}\,\text{tr}\sum_{|\alpha|=1}\int_{\partial V_k}\Big(A_1(x^\alpha I) + A_2\Big(\frac{\partial(x^\alpha I)}{\partial\nu}\Big)\Big)\partial^\alpha\mathcal{D}_\Omega^\omega[T_3(\cdot)](z)\,d\omega \\
&= \frac{1}{2\sqrt{-1}\pi}\,\text{tr}\sum_{|\alpha|=1}\Big(A_1(x^\alpha I) + A_2\Big(\frac{\partial(x^\alpha I)}{\partial\nu}\Big)\Big)\int_{\partial V_k}\partial^\alpha\mathcal{D}_\Omega^\omega[T_3(\cdot)](z)\,d\omega.
\end{aligned} \tag{3.59}$$

By (2.420) and (2.421) we have

$$\begin{aligned}
\left(\frac{1}{2}I - \mathcal{K}_\Omega^\omega\right)^{-1}\partial^\alpha\mathbf{\Gamma}^\omega(\cdot - z)(x) &= -\partial_z^\alpha\left(\frac{1}{2}I - \mathcal{K}_\Omega^\omega\right)^{-1}[\mathbf{\Gamma}^\omega(\cdot - z)](x) \\
&= \sum_{j=1}^{+\infty}\frac{1}{\kappa_j - \omega^2}u_j(x)\partial^\alpha u_j(z)^t.
\end{aligned} \tag{3.60}$$

By Green's formula, the following relation holds:

$$\mathcal{D}_\Omega^\omega[u_j](z) = u_j(z) + (\kappa_j - \omega^2)\int_\Omega\mathbf{\Gamma}^\omega(z-y)u_j(y)dy. \tag{3.61}$$

Combining (3.61) with (2.421), we obtain

$$\begin{aligned}\frac{1}{2\sqrt{-1}\pi}\int_{\partial V_j}\mathcal{D}_\Omega^\omega[\mathbf{N}_\Omega^\omega(\cdot,z)](z)d\omega &= \frac{1}{2\sqrt{-1}\pi}u_j(z)u_j(z)^t\int_{\partial V_j}\frac{1}{\kappa_j-\omega^2}d\omega\\ &= \frac{-1}{2\sqrt{\kappa_j}}u_j(z)u_j(z)^t.\end{aligned}\tag{3.62}$$

We also have from (3.61) that

$$\partial^\alpha\mathcal{D}_\Omega^\omega[u_k](z) = \partial^\alpha u_k(z) + (\kappa_k-\omega^2)\int_\Omega \partial^\alpha\mathbf{\Gamma}^\omega(z-y)u_k(y)\,dy.\tag{3.63}$$

Using (3.60) and (3.63), it follows that

$$\frac{1}{2\sqrt{-1}\pi}\int_{\partial V_k}\partial^\alpha\mathcal{D}_\Omega^\omega[T_3(\cdot)](z)\,d\omega = -\frac{1}{2\sqrt{\kappa_k}}\sum_{|\beta|=1}\partial^\alpha u_k(z)\partial^\beta u_k(z)^t\int_{\partial B}y^\beta\cdot d\sigma(y).\tag{3.64}$$

Substituting (3.64) into (3.59), we obtain

$$\begin{aligned}&\frac{1}{2\sqrt{-1}\pi}\operatorname{tr}\int_{\partial V_k}(\mathcal{A}_0^\omega)^{-1}\mathcal{A}_1^\omega(\mathcal{A}_0^\omega)^{-1}\mathcal{A}_2^\omega\omega^3d\omega\\ &= -\frac{1}{2\sqrt{\kappa_k}}\operatorname{tr}\sum_{|\alpha|=|\beta|=1}\Big(A_1(x^\alpha I)+A_2\Big(\frac{\partial(x^\alpha I)}{\partial\nu}\Big)\Big)\partial^\alpha u_k(z)\partial^\beta u_k(z)^t\int_{\partial B}y^\beta\cdot d\sigma(y)\\ &= -\frac{1}{2\sqrt{\kappa_k}}\operatorname{tr}\sum_{|\alpha|=|\beta|=1}\partial^\alpha u_k(z)\partial^\beta u_k(z)^t\int_{\partial B}y^\beta\Big(A_1(x^\alpha I)+A_2\Big(\frac{\partial(x^\alpha I)}{\partial\nu}\Big)\Big)d\sigma(y)\\ &= -\frac{1}{2\sqrt{\kappa_k}}\sum_{|\alpha|=|\beta|=1}\partial^\beta u_k(z)^t\Big[\int_{\partial B}y^\beta\Big(A_1(x^\alpha I)+A_2\Big(\frac{\partial(x^\alpha I)}{\partial\nu}\Big)\Big)d\sigma(y)\Big]\partial^\alpha u_k(z).\end{aligned}\tag{3.65}$$

But, by the definition of A_1 and A_2, the (i,j)-component of

$$\int_{\partial B}y^\beta\Big(A_1(x^\alpha I)+A_2\Big(\frac{\partial(x^\alpha I)}{\partial\nu}\Big)\Big)d\sigma(y)$$

is equal to $-m_{\alpha j\beta i}$. Now, plugging (3.65) into (3.57), we arrive as desired at the following asymptotic formula:

$$\omega_\epsilon-\omega_0 = \frac{\epsilon^3}{2\sqrt{\kappa_k}}\sum_{i,j,\alpha,\beta=1}^3 m_{\alpha j\beta i}\partial_\beta(u_k)_i(z)\partial_\alpha(u_k)_j(z)+O(\epsilon^4).$$

In order to complete the proof of the theorem, we verify identity (3.58). As before, it is easy to see that

$$\begin{aligned}&\frac{1}{2\sqrt{-1}\pi}\operatorname{tr}\int_{\partial V_k}T_1T_5d\omega\\ &= -\frac{1}{2\sqrt{\kappa_k}}\sum_{|\alpha|=2}\frac{1}{\alpha!}u_k(z)^t\Big[\int_{\partial B}\Big(A_1(x^\alpha I)+A_2\Big(\frac{\partial(x^\alpha I)}{\partial\nu}\Big)\Big)d\sigma(y)\Big]\partial^\alpha u_k(z)\\ &= -\frac{1}{2\sqrt{\kappa_k}}u_k(z)^t\int_{\partial B}A_2\Big(\frac{\partial u_k^{(2)}}{\partial\nu}\Big)\,d\sigma(y),\end{aligned}$$

where $u_k^{(2)}(x) = \sum_{|\alpha|=2} \frac{1}{\alpha!} x^\alpha \partial^\alpha u_k(z)$. Using (3.62), we also have

$$\begin{aligned}
&\frac{1}{2\sqrt{-1}\pi} \operatorname{tr} \int_{\partial V_k} \omega^2 T_2(A_1 \tilde{S}_2 + A_2 \tilde{K}_2)\, d\omega \\
&= -\frac{1}{2\sqrt{-1}\pi} \operatorname{tr} \int_{\partial V_k} \omega^2 \mathcal{D}_\Omega^\omega \Big[\big(\frac{1}{2} I - \mathcal{K}_\Omega^\omega\big)^{-1} \mathbf{\Gamma}^\omega (x - z) \Big](z) \int_{\partial B} A_2 \tilde{K}_2 (\tilde{\mathcal{S}}_B^0)^{-1} d\sigma(y)\, d\omega \\
&= -\frac{\sqrt{\kappa_k}}{2} \operatorname{tr} u_k(z) u_k(z)^t \int_{\partial B} A_2 \tilde{K}_2 (\tilde{\mathcal{S}}_B^0)^{-1}\, d\sigma(y) \\
&= -\frac{\sqrt{\kappa_k}}{2} u_k(z)^t \int_{\partial B} A_2 \tilde{K}_2 (\tilde{\mathcal{S}}_B^0)^{-1} [u_k(z)]\, d\sigma(y).
\end{aligned}$$

Inserting the Taylor expansion of u_k at z into $(\mathcal{L}^{\lambda,\mu} + \kappa_k) u_k = 0$ yields

$$\mathcal{L}^{\lambda,\mu} u_k^{(2)} + \kappa_k u_k(z) = 0. \tag{3.66}$$

Since $\tilde{\mathcal{S}}_B^\omega = \tilde{\mathcal{S}}_B^0 + \sum_{n=1}^{+\infty} \omega^n \tilde{S}_n$, we have

$$\mathcal{L}^{\tilde{\lambda},\tilde{\mu}} \tilde{S}_2 (\tilde{\mathcal{S}}_B^0)^{-1} [u_k(z)] + u_k(z) = 0. \tag{3.67}$$

By the definition of A_n and the jump relation of the single-layer potential, we have

$$A_2(f) = \frac{\partial}{\partial \nu} \mathcal{S}_B^0 A_2(f)\Big|_+ - \frac{\partial}{\partial \nu} \mathcal{S}_B^0 A_2(f)\Big|_- = f + \frac{\partial}{\partial \tilde{\nu}} \tilde{\mathcal{S}}_B^0 A_4(f)\Big|_- - \frac{\partial}{\partial \nu} \mathcal{S}_B^0 A_2(f)\Big|_-,$$

and hence

$$\int_{\partial B} A_2(f) d\sigma = \int_{\partial B} f d\sigma. \tag{3.68}$$

From (3.66), (3.67), and (3.68), we conclude that

$$\begin{aligned}
&\int_{\partial B} A_2\Big(\frac{\partial u_k^{(2)}}{\partial \nu}\Big) - \kappa_k A_2 \tilde{K}_2 (\tilde{\mathcal{S}}_B^0)^{-1} [u_k(z)]\, d\sigma(y) \\
&\quad = \int_{\partial B} \frac{\partial u_k^{(2)}}{\partial \nu} - \kappa_k \tilde{K}_2 (\tilde{\mathcal{S}}_B^0)^{-1} [u_k(z)]\, d\sigma(y) \\
&\quad = \int_B \mathcal{L}^{\lambda,\mu} u_k^{(2)} - \kappa_k \mathcal{L}^{\tilde{\lambda},\tilde{\mu}} \tilde{S}_2 (\tilde{\mathcal{S}}_B^0)^{-1} [u_k(z)]\, dy = 0,
\end{aligned}$$

which completes the proof. □

3.5. Eigenvalue Perturbations Due to Shape Deformations

Let $\Omega \Subset \mathbb{R}^2$. As in Subsection 3.2.2, we consider D_ϵ to be an ϵ-perturbation of $D \Subset \Omega$. The boundary ∂D_ϵ is then given by

$$\partial D_\epsilon = \Big\{ \tilde{x} : \tilde{x} = x + \epsilon h(x) N(x),\ x \in \partial D \Big\},$$

where $h \in \mathcal{C}^2(\partial D)$. Here N is the outward normal to ∂D, T denotes the tangential vector, and τ is the curvature of ∂D.

Consider the following eigenvalue problem:

$$
(3.69) \quad \begin{cases} \mathcal{L}^{\lambda,\mu} u^\epsilon + \omega_\epsilon^2 u^\epsilon = 0 & \text{in } \Omega \setminus \overline{D_\epsilon}, \\ \mathcal{L}^{\tilde{\lambda},\tilde{\mu}} u^\epsilon + \omega_\epsilon^2 u^\epsilon = 0 & \text{in } D_\epsilon, \\ u^\epsilon \big|_+ - u^\epsilon \big|_- = 0 & \text{on } \partial D_\epsilon, \\ \dfrac{\partial u^\epsilon}{\partial \nu}\Big|_+ - \dfrac{\partial u^\epsilon}{\partial \tilde{\nu}}\Big|_- = 0 & \text{on } \partial D_\epsilon, \\ \dfrac{\partial u^\epsilon}{\partial \nu} = 0 & \text{on } \partial \Omega, \end{cases}
$$

with the normalization $\int_\Omega |u^\epsilon|^2 = 1$.

The following theorem from [**22**] holds.

THEOREM 3.15. *The leading-order term in the perturbations of the eigenvalues due to the interface changes is given by*

$$
\omega_\epsilon^2 - \omega_0^2 = \epsilon \int_{\partial D} h(x)\, \mathcal{M}[u^0](x) : \nabla^s u^0(x)\, d\sigma(x) + o(\epsilon),
$$

where

$$
\mathcal{M}[u^0] = a \nabla \cdot u^0 I + b \nabla^s u^0 + c\Big(\frac{\partial (u^0 \cdot T)}{\partial T} + \tau u^0 \cdot \nu \Big) T \otimes T + d \frac{\partial (u^0 \cdot N)}{\partial N} N \otimes N,
$$

with

$$
\begin{cases} a &= (\widetilde{\lambda} - \lambda) \dfrac{\lambda + 2\mu}{\widetilde{\lambda} + 2\widetilde{\mu}}, \\ b &= 2(\widetilde{\mu} - \mu) \dfrac{\mu}{\widetilde{\mu}}, \\ c &= 2(\widetilde{\mu} - \mu) \Big(\dfrac{2\widetilde{\lambda} + 2\widetilde{\mu} - \lambda}{\widetilde{\lambda} + 2\widetilde{\mu}} - \dfrac{\mu}{\widetilde{\mu}} \Big), \\ d &= 2(\widetilde{\mu} - \mu) \dfrac{\widetilde{\mu}\lambda - \mu\widetilde{\lambda}}{\widetilde{\mu}(\widetilde{\lambda} + 2\widetilde{\mu})}. \end{cases}
$$

Here, I is the identity matrix, $a \otimes b := (a_i b_j)$ is the tensor product between vectors in $\mathbb{R}^2$, and T is the tangent vector to ∂D.

3.6. Concluding Remarks

In this chapter we have rigorously derived asymptotic expansions for the eigenvalues of the Laplacian and the Lamé system in singularly perturbed cavities. These asymptotics are with respect to the size of the perturbation. We have also formulated the problem of finding resonances of an optical resonator as a characteristic value problem for a meromorphic operator-valued function.

Part 2

Diffraction Gratings and Band-Gap Materials

CHAPTER 4

Diffraction Gratings

4.1. Introduction

Diffractive optics is a fundamental yet vigorously growing technology which continues to be a source of novel optical devices. Significant recent technology advances have led to the development of high precision micromachining techniques which permit the creation of gratings (periodic structures) and other diffractive structures with tiny features. Current and potential application areas include corrective lenses, antireflective interfaces, beam splitters, and sensors.

Because of the small structural features, light propagation in micro-optical structures is generally governed by diffraction. In order to accurately predict the energy distributions of an incident field in a given structure, the numerical solution of the full Maxwell equations is required. If the field configurations are built up of harmonic electromagnetic waves that are transverse, then the Maxwell equations can be reduced to two scalar Helmholtz equations. Computational models also allow the exciting possibility of obtaining completely new structures through the solution of optimal design problems.

The basic electromagnetic theory of gratings has been studied extensively since Rayleigh's time (1907). Recent advances have been greatly accelerated due to several new approaches and numerical methods including differential methods, integral methods, analytical continuation, and variational methods, to name but a few.

Diffractive optical elements, as opposed to the traditional optical lenses, have many advantages. They are light, small, and inexpensive. Often diffractive structures exhibit certain periodicity. There are two classes of grating structures:

(i) linear grating (one-dimensional gratings),
(ii) crossed gratings (biperiodic or two-dimensional gratings).

The chapter begins with basic electromagnetic theory for diffraction gratings. We introduce the basic physics and present the system of Maxwell's equations as well as the two fundamental polarizations. The well-known grating formula is also derived. Then the method of boundary variations is also discussed. This method is based on the observation that since electromagnetic fields behave analytically with respect to perturbations of a scattering surface, they can be represented by convergent power series in a perturbation parameter. The effect of small defects in a diffraction grating is addressed using an integral representation formulation.

4.2. Electromagnetic Theory of Gratings

4.2.1. Time-Harmonic Maxwell's Equations. The electromagnetic wave propagation is governed by Maxwell's equations. Throughout, we shall restrict our attention to time-harmonic electromagnetic fields with time dependence ($e^{-\sqrt{-1}\omega t}$), *i.e.*,

$$E(x,t) = E(x)e^{-\sqrt{-1}\omega t}, \tag{4.1}$$

$$H(x,t) = H(x)e^{-\sqrt{-1}\omega t} \tag{4.2}$$

for some operating frequency $\omega > 0$ with E and H being respectively the electric and magnetic field.

The time-harmonic Maxwell equations take the following form:

$$\nabla \times E = \sqrt{-1}\omega\mu H, \tag{4.3}$$

$$\nabla \times H = -\sqrt{-1}\omega\varepsilon E, \tag{4.4}$$

where μ is the magnetic permeability and ε is the electric permittivity. Note that from (4.3) and (4.4), it follows that

$$\nabla \cdot (\varepsilon E) = 0, \tag{4.5}$$

$$\nabla \cdot (\mu H) = 0. \tag{4.6}$$

The fields are further assumed to be nonmagnetic and $\mu = \mu_0$ (usually the magnetic permeability of vacuum). Then (4.6) becomes

$$\nabla \cdot H = 0.$$

It follows from (4.3-4.4) that the following jump conditions hold:

(i) the tangential components of E and H must be continuous crossing an interface,

(ii) the normal components of εE and H must be continuous crossing an interface.

In a homogeneous and isotropic medium, ε does not depend on x. By taking the curl of (4.3) we obtain that

$$-\Delta E + \nabla(\nabla \cdot E) = \sqrt{-1}\omega\mu_0 \nabla \times H.$$

Using (4.4), we have

$$-\Delta E + \nabla(\nabla \cdot E) = \omega^2 \varepsilon\mu_0 E$$

or the Helmholtz equation

$$\Delta E + k^2 E = 0 \tag{4.7}$$

with $k = \omega\sqrt{\varepsilon\mu_0}$.

Similarly, H satisfies

$$\Delta H + k^2 H = 0.$$

Note that in a dielectric medium k^2 is real and positive. The wavelength λ is given by $\lambda = (2\pi)/k$.

4.2.2. Grating Geometry and Fundamental Polarizations. Throughout, a grating is always assumed to be infinitely wide.

Figure 4.1 shows the grating geometry. We denote the period, height, and incident angle by Λ, h, and θ, respectively.

An alternative way to specify the periodicity is by means of the permittivity ε.

For a 1-D grating (linear grating):

$$\varepsilon(x_1 + n\Lambda, x_2) = \varepsilon(x_1, x_2), \quad n \in \mathbb{Z}.$$

In the case of a crossed grating with period $\Lambda = (\Lambda_1, \Lambda_2)$ we have

$$\varepsilon(x_1 + n_1\Lambda_1,\ x_2 + n_2\Lambda_2,\ x_3) = \varepsilon(x_1, x_2, x_3),\ \forall\ n_1, n_2 \in \mathbb{Z}.$$

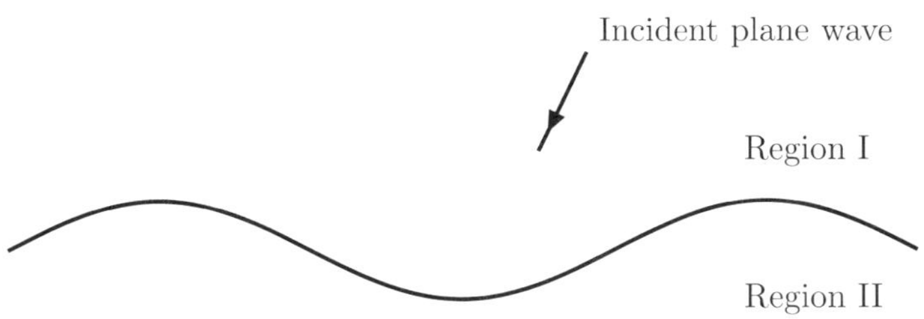

FIGURE 4.1. Grating geometry.

We assume that above the interface ε is real and positive. However, below the interface the parameter ε can be real which corresponds to a dielectric medium; complex corresponding to an absorbing or lossy medium; or perfectly conducting medium.

In the next three subsections we shall discuss two separate cases: The perfectly conducting grating and the dielectric one.

Suppose that a grating is illuminated under the incidence θ by a plane wave of unit amplitude propagating in Region I (Figure 4.1). The incident vector K_i lies in the (x_1, x_2) plane

$$K_i = k_1(\sin\theta, -\cos\theta, 0).$$

The electromagnetic fields are assumed to be independent of x_3. We consider the following two fundamental cases of polarization: TE (transverse electric) and TM (transverse magnetic).

In TE polarization, the electric field is parallel to the grooves, *i.e.*, points in the x_3-direction. So, E takes the form

$$E = u(x_1, x_2)e_3$$

where u is a scalar function and (e_1, e_2, e_3) is the standard basis of $\mathbb{R}^3$.

In TM polarization, the magnetic field is parallel to the grooves, *i.e.*,

$$H = u(x_1, x_2)e_3.$$

As we shall see, the resulting Maxwell's equations in these two polarizations are quite different.

4.2.3. Perfectly Conducting Gratings. In this subsection, the grating is assumed to be perfectly conducting. In order to treat the two fundamental polarizations simultaneously, we denote $u = E_3(x_1, x_2)$ in TE polarization; $u = H_3(x_1, x_2)$ in TM polarization, where the subscript 3 stands for the third component. Assume that the grating is expressed by $x_2 = f(x_1)$. Then $u = 0$ in Region II ($x_2 < f(x_1)$). In Region I, the field u satisfies

$$\Delta u + k^2 u = 0 \text{ if } x_2 > f(x_1). \tag{4.8}$$

We next derive the boundary condition of u on $x_2 = f(x_1)$ with f being a periodic function of period Λ. Using the continuity of the tangential component of E crossing an interface and the fact that E is zero in Region II, we have

$$\nu \times E = 0 \quad \text{on } x_2 = f(x_1), \tag{4.9}$$

where ν is the outward normal to Region II.

In TE polarization, $E = (0, 0, u)$, hence (4.9) implies that

$$u(x_1, f(x_1)) = 0, \tag{4.10}$$

i.e., a homogeneous Dirichlet boundary condition.

In TM polarization, $H = (0, 0, u)$. We obtain by using Maxwell's equations and the condition (4.9) that

$$\left.\frac{\partial u}{\partial \nu}\right|_{x_2 = f(x_1)} = 0, \tag{4.11}$$

which is a homogeneous Neumann boundary condition.

Define the scattered field as the difference between the total field u and the incident field, *i.e.*,

$$u^s = u - u^i. \tag{4.12}$$

Here, $u^i = e^{\sqrt{-1}(\alpha x_1 - \beta x_2)}$ with

$$\begin{cases} \alpha = k_1 \sin\theta, \\ \beta = k_1 \cos\theta. \end{cases} \tag{4.13}$$

Since the incident field u^i satisfies the Helmholtz equation everywhere, we can easily show that

$$\Delta u^s + k_1^2 u^s = 0 \qquad \text{for } x_2 > f(x_1). \tag{4.14}$$

From (4.10) and (4.11), u^s satisfies either one of the following boundary conditions:

TE polarization:

$$u^s = -u^i \qquad \text{on } x_2 = f(x_1). \tag{4.15}$$

TM polarization:

$$\frac{\partial u^s}{\partial \nu} = -\frac{\partial u^i}{\partial \nu} \qquad \text{on } x_2 = f(x_1). \tag{4.16}$$

Next, since the problem is posed in an unbounded domain, a radiation condition is needed. We assumed that u^s is bounded when x_2 goes to infinity and consisted of outgoing plane waves. This radiation condition is also referred to as the outgoing wave condition.

The grating problem can be stated as: find a function that satisfies the Helmholtz equation (4.14), a boundary condition on $\{x_2 = f(x_1)\}$, and the outgoing wave condition.

Motivated by uniqueness, we shall seek the so–called "quasi–periodic" solutions, *i.e.*, solutions u^s such that $u^s(x_1, x_2)e^{-\sqrt{-1}\alpha x_1}$ is a periodic function of period Λ with respect to x_1 for every x_2. In fact, if the grating problem attains a unique solution then we want to show that

$$v(x_1, x_2) = u(x_1, x_2)e^{-\sqrt{-1}\alpha x_1}$$

is a periodic function of period Λ, *i.e.*,

$$v(x_1 + \Lambda, x_2) = v(x_1, x_2)$$

or equivalently

$$u^s(x_1 + \Lambda, x_2)e^{-\sqrt{-1}\alpha\Lambda} = u^s(x_1, x_2). \tag{4.17}$$

Because of uniqueness, if $w(x_1, x_2) = u^s(x_1 + \Lambda, x_2)e^{-\sqrt{-1}\alpha\Lambda}$ is also a solution of the grating problem, then it must be identical to u^s. It is obvious that w satisfies the Helmholtz equation (4.14). The boundary conditions (4.15) and (4.16) are also satisfied by observing that u^i is a quasi-periodic function and using the boundary condition of u^s.

4.2.4. Grating Formula. Since $u^s(x_1, x_2)e^{-\sqrt{-1}\alpha x_1}$ is periodic in x_1, it follows by using a Fourier series expansion that

$$\begin{aligned} u^s(x_1, x_2) &= e^{\sqrt{-1}\alpha x_1} \sum_{n\in\mathbb{Z}} V_n(x_2) e^{\sqrt{-1}n\frac{2\pi}{\Lambda}x_1} \\ &= \sum_{n\in\mathbb{Z}} V_n(x_2) e^{\sqrt{-1}\alpha_n x_1} \end{aligned} \tag{4.18}$$

with

$$\alpha_n = \alpha + \frac{2\pi n}{\Lambda}, \tag{4.19}$$

or equivalently,

$$\alpha_n = k_1 \sin\theta + n\frac{2\pi}{\Lambda}. \tag{4.20}$$

Thus, in order to solve for u^s it suffices to determine $V_n(x_2)$.

Now in the region $\{x_2 > \max\{f(x_1)\}\}$, $u^s(x_1, x_2)$ satisfies the Helmholtz equation. Substituting (4.18) into the Helmholtz equation gives

$$\sum_{n\in\mathbb{Z}} \left[\frac{d^2V_n(x_2)}{dx_2^2} + (k_1^2 - \alpha_n^2)V_n(x_2)\right] e^{\sqrt{-1}n\frac{2\pi}{\Lambda}x_1} = 0.$$

Hence

$$\frac{d^2V_n(x_2)}{dx_2^2} + (k_1^2 - \alpha_n^2)V_n(x_2) = 0 \quad \text{for all } n \in \mathbb{Z}.$$

Define

$$\beta_n = \begin{cases} \sqrt{k_1^2 - \alpha_n^2} & \text{if } k_1^2 > \alpha_n^2, \\ \sqrt{-1}\sqrt{\alpha_n^2 - k_1^2} & \text{if } k_1^2 \le \alpha_n^2. \end{cases} \tag{4.21}$$

Then, solving the simple ordinary differential equation yields

$$V_n(x_2) = A_n e^{-\sqrt{-1}\beta_n x_2} + B_n e^{\sqrt{-1}\beta_n x_2}.$$

The radiation condition implies that $A_n = 0$. Actually if $|k_1| \ge |\alpha_n|$ then $e^{-\sqrt{-1}\beta_n x_2}$ represents incoming waves instead. If $|k_1| < |\alpha_n|$ then $e^{-\sqrt{-1}\beta_n x_2}$ is unbounded when x_2 goes to infinity. Therefore we arrive at the Rayleigh expansion of the form

$$\begin{aligned} u^s(x_1, x_2) &= \sum_{|\alpha_n|<k_1} B_n e^{\sqrt{-1}\alpha_n x_1 + \sqrt{-1}\beta_n x_2} \quad \text{“outgoing waves”} \\ &+ \sum_{|\alpha_n|\ge k_1} B_n e^{\sqrt{-1}\alpha_n x_1 + \sqrt{-1}\beta_n x_2} \quad \text{“evanescent waves”}. \end{aligned} \tag{4.22}$$

Denote

$$U = \{n, |\alpha_n| < k_1\}.$$

Each term ($n \in U$) of the outgoing waves in (4.22) represents a propagating plane wave, which is called the scattered wave in the nth order. If $|n|$ is large ($n \notin U$), then the corresponding term in (4.22) represents an evanescent wave $B_n e^{-|\beta_n|x_2} e^{\sqrt{-1}\alpha_n x_1}$

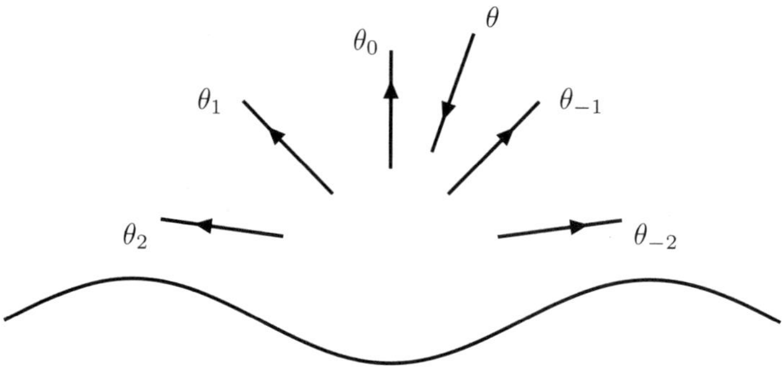

FIGURE 4.2. Geometric interpretation of the grating formula.

which propagates along the x_1-axis and is exponentially damped with respect to x_2. The scattered wave in the nth order takes the form:

$$\psi_n(x_1, x_2) = B_n e^{\sqrt{-1}\alpha_n x_1 + \sqrt{-1}\sqrt{k_1^2-\alpha_n^2}\, x_2} \quad \text{for } n \in U. \tag{4.23}$$

Since $|\alpha_n/k_1| < 1$, we denote

$$\frac{\alpha_n}{k_1} = \sin\theta_n, \; -\frac{\pi}{2} < \theta_n < \frac{\pi}{2}. \tag{4.24}$$

From (4.19), we have

$$\frac{\alpha_n}{k_1} = \sin\theta_n = \sin\theta + \frac{2\pi n}{k_1\Lambda} \tag{4.25}$$

and (4.23) becomes

$$\psi_n(x_1, x_2) = B_n e^{\sqrt{-1}k_1(x_1\sin\theta_n + x_2\cos\theta_n)}, \tag{4.26}$$

where θ_n is the angle of diffraction.

Thus we have derived the following grating formula:

$$\sin\theta_n = \sin\theta + n\frac{\lambda_1}{\Lambda} \quad \text{or} \quad k_1\sin\theta_n = k_1\sin\theta + \frac{n2\pi}{\Lambda}, \tag{4.27}$$

where λ_1 is the wavelength in Region I and

$$k_1 = \frac{2\pi}{\lambda_1}.$$

In the next theorem we state a reciprocity property.

THEOREM 4.1. *Let θ and θ_n be the angle of incidence and the angle of diffraction of the nth order, respectively. Then when the angle of incidence is $\theta' = -\theta_n$, the nth scattered order propagates in the direction defined by $\theta'_n = -\theta$.*

The grating efficiency E_n is the measurement of energy in the nth propagating order. It is defined as

$$E_n = \frac{\phi_n^s}{\phi^i}, \tag{4.28}$$

where ϕ^i and ϕ_n^s are the fluxes of the Poynting vectors associated with the incident wave and the nth scattered wave respectively through a unit rectangle in which one

FIGURE 4.3. The reciprocity theorem.

side is parallel to the x_1-axis while the other is parallel to the x_3-axis. It is easy to show that

$$E_n = |B_n|^2 \frac{\cos\theta_n}{\cos\theta}. \tag{4.29}$$

We next state a simple result which is convenient in many applications. The proof is based on integration by parts.

LEMMA 4.2. *Assume that u_1 and u_2 are two functions which satisfy the Helmholtz equation*

$$\Delta u + k^2 u = 0$$

and either a homogeneous Dirichlet or a Neumann boundary condition. Then for any $x_2 > \max f(x_1)$,

$$\int_0^\Lambda (u_1 \frac{\partial u_2}{\partial x_2} - u_2 \frac{\partial u_1}{\partial x_2}) dx_1 = 0. \tag{4.30}$$

THEOREM 4.3 (The conservation of energy).

$$\sum_{n\in U} E_n = 1. \tag{4.31}$$

This is to say, the incident energy is equal to the scattered energy.

PROOF. Let u be a solution of the Helmholtz equation with either the Dirichlet or the Neumann boundary condition. Since k_1 is real, $\overline{u}$ also satisfies the equation and the boundary condition. By applying Lemma 4.2 to u and $\overline{u}$, we get

$$\frac{1}{\Lambda}\int_0^\Lambda \left(u\frac{\partial \overline{u}}{\partial x_2} - \overline{u}\frac{\partial u}{\partial x_2}\right) dx_1 = 0 \text{ for } x_2 > \max f(x_1) \tag{4.32}$$

or

$$\frac{1}{\Lambda}\Im\left\{\int_0^\Lambda u\frac{\partial \overline{u}}{\partial x_2}\right\} = 0 \text{ for } x_2 > \max f(x_1). \tag{4.33}$$

Next,

$$\begin{aligned} u = e^{\sqrt{-1}\alpha x_1 - \sqrt{-1}\beta x_2} &+ \sum_{n\in U} B_n e^{\sqrt{-1}\alpha_n x_1 + \sqrt{-1}\beta_n x_2} \\ &+ \sum_{n\notin U} B_n e^{\sqrt{-1}\alpha_n x_1 + \sqrt{-1}\beta_n x_2} \end{aligned} \tag{4.34}$$

and

$$\overline{u} = e^{-\sqrt{-1}\alpha x_1+\sqrt{-1}\beta x_2} + \sum_{n\in U} \overline{B}_n e^{-\sqrt{-1}\alpha_n x_1-\sqrt{-1}\beta_n x_2}$$

$$+ \sum_{n\notin U} \overline{B}_n e^{-\sqrt{-1}\alpha_n x_1-\sqrt{-1}\beta_n x_2}. \tag{4.35}$$

Substituting (4.34) and (4.35) into (4.33), we find

$$\beta = \sum_{n\in U} \beta_n |B_n|^2,$$

or equivalently

$$\sum_{n\in U} E_n = 1.$$

□

4.2.5. Dielectric Gratings. Suppose that Region II is filled with a material of real permittivity ε_2.

The solution of the grating problem satisfies:

In Region I,

$$\Delta u + k_1^2 u = 0 \qquad \text{if } x_2 > f(x_1). \tag{4.36}$$

In Region II,

$$\Delta u + k_2^2 u = 0 \qquad \text{if } x_2 < f(x_1). \tag{4.37}$$

Also, outgoing wave conditions are satisfied by $u^s = u - u^i$ (for $x_2 \to +\infty$) and by u (for $x_2 \to -\infty$).

From the jump conditions across the interface and Maxwell's equations, we have that u is continuous, $\partial u/\partial\nu$ is continuous in TE polarization, and $(1/\varepsilon)\partial u/\partial\nu$ is continuous in TM polarization.

Again, the quasi-periodicity of the field follows from the uniqueness of the solution. Then for $x_2 > \max f(x_1)$

$$u(x_1, x_2) = e^{\sqrt{-1}\alpha x_1} \sum_{n\in\mathbb{Z}} V_n(x_2) e^{\sqrt{-1}n\frac{2\pi}{\Lambda}x_1}. \tag{4.38}$$

Substituting (4.38) into (4.36) and (4.37), we obtain the Rayleigh expansion above the groove

$$u(x_1, x_2) = e^{(\sqrt{-1}\alpha x_1-\sqrt{-1}\beta x_2)} + \sum R_n e^{\sqrt{-1}\alpha_n x_1+\sqrt{-1}\beta_{n1} x_2} \tag{4.39}$$

with $\alpha_n = k_1 \sin\theta + n\frac{2\pi}{\Lambda}$ and $\beta_{n1}^2 = k_1^2 - \alpha_n^2$.

If $x_2 < \min f(x_1)$

$$u(x_1, x_2) = \sum_{n\in\mathbb{Z}} T_n \, e^{\sqrt{-1}\alpha_n x_1-\sqrt{-1}\beta_{n2} x_2}$$

with

$$\beta_{n2}^2 = k_2^2 - \alpha_n^2.$$

These two expansions contain propagating and evanescent waves depending on the value of n.

For $j = 1, 2$, denote

$$U_j = \{n, \beta_{nj}^2 > 0\}.$$

Then if $n \in U_1$, $\alpha_n^2 < k_1^2$, we have

$$\alpha_n = k_1 \sin\theta + n\frac{2\pi}{\Lambda} = k_1 \sin\theta_{n1}, \qquad -\frac{\pi}{2} < \theta_{n1} < \frac{\pi}{2}, \qquad (4.40)$$
$$\beta_{n1} = k_1 \cos\theta_{n1},$$

and $R_n e^{\sqrt{-1}\alpha_n x_1 + \sqrt{-1}\beta_{n1} x_2}$ represents a plane wave propagating in the θ_{n1}-direction. Similarly, if $n \in U_2$, then

$$\alpha_n = k_2 \sin\theta + n\frac{2\pi}{\Lambda} = k_2 \sin\theta_{n2}, \qquad -\frac{\pi}{2} < \theta_{n2} < \frac{\pi}{2}, \qquad (4.41)$$
$$\beta_{n2} = k_2 \cos\theta_{n2},$$

and $T_n e^{\sqrt{-1}\alpha_n x_1 - \sqrt{-1}\beta_{n2} x_2}$ stands for a transmitted plane wave propagating in the θ_{n2}-direction.

Equations (4.40) and (4.41) are the grating formulas.

4.3. Variational Formulations

4.3.1. Model Problems: TE and TM Polarizations. Consider a time-harmonic electromagnetic plane wave incident on a slab of some optical material in $\mathbb{R}^3$, which is periodic in the x_1-direction. Throughout, the medium is assumed to be nonmagnetic and invariant in the x_3-direction. We study the diffraction problem in TM (traverse magnetic) polarization, *i.e.*, the magnetic field is transversal to the (x_1, x_2)-plane. The case when the electric field is transversal to the (x_1, x_2)-plane is called TE (transverse electric) polarization. These two polarizations are of primary importance since any other polarization may be decomposed into a simple combination of them. The differential equations derived from time harmonic Maxwell's equations are quite different for the TE and TM cases: In the TE case, $(\Delta + k^2)u = 0$, where E (the electric field vector) $= u(x_1, x_2)e_3$; In the TM case,

$$\nabla \cdot (\frac{1}{k^2}\nabla u) + u = 0,$$

where the magnetic field vector $H(x) = u(x_1, x_2)e_3$. In both cases,

$$k = \omega\sqrt{\varepsilon\mu_0} = \omega q,$$

where q is the index of refraction of the medium.

Let us first specify the problem geometry. Let S_1 and S_2 be two simple curves embedded in the strip

$$\Omega = \{(x_1, x_2) \in \mathbb{R}^2 : -b < x_2 < b\},$$

where b is some positive constant. The medium in the region Ω between S_1 and S_2 is inhomogeneous. Above the surface S_1 and below the surface S_2, the media are assumed to be homogeneous. The entire structure is taken to be periodic in the x_1-direction. Without loss of generality, we assume that S_1 and S_2 are periodic of period Λ with respect to $\mathbb{Z}$.

Let $\Omega_1 = \{x = (x_1, x_2) \in \mathbb{R}^2 : x_2 > b\}$, $\Omega_2 = \{x = (x_1, x_2) \in \mathbb{R}^2 : x_2 < -b\}$. Define the boundaries $\Gamma_1 = \{x_2 = b\}$, $\Gamma_2 = \{x_2 = -b\}$. Assume that $S_1 > S_2$ pointwise, *i.e.*, if $(x_1, x_2) \in S_1$ and $(x_1, x_2') \in S_2$, then $x_2 > x_2'$. The curves S_1 and S_2 divide Ω into three connected components. Denote the component which meets Γ_1 by Ω_1^+; the component which meets Γ_2 by Ω_2^+; and $\Omega_0 = \Omega \setminus (\overline{\Omega_1^+} \cup \overline{\Omega_2^+})$.

Suppose that the whole space is filled with material with a periodic dielectric coefficient function ε of period Λ,

$$\varepsilon(x) = \begin{cases} \varepsilon_1 & \text{in } \Omega_1^+ \cup \overline{\Omega}_1, \\ \varepsilon_0(x) & \text{in } \Omega_0, \\ \varepsilon_2 & \text{in } \Omega_2^+ \cup \overline{\Omega}_2, \end{cases}$$

where $\varepsilon_0(x) \in L^\infty$, ε_1 and ε_2 are constants, ε_1 is real and positive, and $\Re\varepsilon_2 > 0$, $\Im\varepsilon_2 \geq 0$. The case $\Im\varepsilon_2 > 0$ accounts for materials which absorb energy (see, for instance, [**156**]). For convenience, we also use the "index of refraction" $q = \sqrt{\varepsilon\mu_0}$

$$q(x) = \begin{cases} q_1 & \text{in } \Omega_1^+ \cup \overline{\Omega}_1, \\ q_0(x) & \text{in } \Omega_0, \\ q_2 & \text{in } \Omega_2^+ \cup \overline{\Omega}_2, \end{cases}$$

where ε is the dielectric coefficient given above and μ_0 is the free space magnetic permeability, which is constant.

We want to solve the Helmholtz equation derived from Maxwell's system of equations

$$\nabla \cdot (\frac{1}{q^2}\nabla u) + \omega^2 u = 0 \quad \text{in } \mathbb{R}^2, \tag{4.42}$$

when an incoming plane wave

$$u^i(x_1, x_2) = e^{\sqrt{-1}\alpha x_1 - \sqrt{-1}\beta x_2}$$

is incident on S_1 from Ω_1, where α and β are given by (4.13) with $-\pi/2 < \theta < \pi/2$ being the angle of incidence.

We are interested in "quasi-periodic" solutions u, that is, solutions $u(x_1, x_2)$ such that $u(x_1, x_2)e^{-\sqrt{-1}\alpha x_1}$ are Λ-periodic in x_1. Define

$$u_\alpha(x_1, x_2) = u(x_1, x_2)e^{-\sqrt{-1}\alpha x_1}.$$

It is easily seen that if u satisfies (4.42) then u_α satisfies

$$\nabla_\alpha \cdot (\frac{1}{q^2}\nabla_\alpha u_\alpha) + \omega^2 u_\alpha = 0 \qquad \text{in } \mathbb{R}^2, \tag{4.43}$$

where the operator ∇_α is defined by

$$\nabla_\alpha = \nabla + \sqrt{-1}(\alpha, 0).$$

We expand u_α in a Fourier series:

$$u_\alpha(x_1, x_2) = \sum_{n\in\mathbb{Z}} u_\alpha^{(n)}(x_2)e^{\sqrt{-1}\frac{2\pi n}{\Lambda}x_1}, \tag{4.44}$$

where

$$u_\alpha^{(n)}(x_2) = \frac{1}{\Lambda}\int_0^\Lambda u_\alpha(x_1, x_2)e^{-\sqrt{-1}\frac{2\pi n}{\Lambda}x_1}dx_1.$$

Introduce the sets

$$\Gamma_1' = \{x \in \mathbb{R}^2 : x_2 = b_1\}, \ \ \Gamma_2' = \{x_2 = -b_1\},$$

with $0 < b_1 < b$ being such that $\Omega_0 \subseteq \{-b_1 < x_2 < b_1\}$. Let

$$D_1 = \{x \in \mathbb{R}^2 : x_2 > b_1\} \quad \text{and} \quad D_2 = \{x \in \mathbb{R}^2 : x_2 < -b_1\}.$$

Define for $j = 1, 2$ the coefficients

$$\beta_j^n(\alpha) = e^{\sqrt{-1}\gamma_j^n/2}|k_j^2 - \alpha_n^2|^{1/2} = e^{\sqrt{-1}\gamma_j^n/2}|\omega^2 q_j^2 - \alpha_n^2|^{1/2}, \quad n \in \mathbb{Z}, \tag{4.45}$$

where α_n is defined by (4.19), $k_j = \omega q_j$, and

$$\gamma_j^n = \arg(k_j^2 - \alpha_n^2), \quad 0 \le \gamma_j^n < 2\pi. \tag{4.46}$$

We assume that

$$k_j^2 \ne \alpha_n^2 \quad \text{for all } n \in \mathbb{Z}, \text{ and } \ j = 1, 2. \tag{4.47}$$

This condition excludes "resonance" cases and ensures that a fundamental solution for (4.43) exists inside D_1 and D_2. In particular, for real k_2, we have the following equivalent form of (4.45)

$$\beta_j^n(\alpha) = \begin{cases} \sqrt{k_j^2 - \alpha_n^2}, & k_j^2 > \alpha_n^2, \\ \sqrt{-1}\sqrt{\alpha_n^2 - k_j^2}, & k_j^2 < \alpha_n^2. \end{cases}$$

Notice that if $\Im k_j > 0$, then (4.47) is certainly satisfied.

From the knowledge of the fundamental solution (see, for instance, [**195**, **220**]), it follows that inside D_1 and D_2, u_α can be expressed as a sum of plane waves:

$$u_\alpha|_{D_j} = \sum_{n\in\mathbb{Z}} a_j^n e^{\pm\sqrt{-1}\beta_j^n(\alpha)x_2 + \sqrt{-1}\frac{2\pi n}{\Lambda}x_1}, \quad j = 1, 2, \tag{4.48}$$

where the a_j^n are complex scalars.

We next impose a radiation condition on the scattering problem. Since β_j^n is real for at most finitely many n, there are only a finite number of propagating plane waves in the sum (4.48), the remaining waves are exponentially damped (so-called evanescent waves) or radiate (unbounded) as $|x_2| \to \infty$. We will insist that u_α is composed of bounded outgoing plane waves in D_1 and D_2, plus the incident incoming wave u^i in D_1.

From (4.44) and (4.48) we then have the condition that the coefficients $u_\alpha^{(n)}(x_2)$ in (4.44) are given by

(4.49)

$$u_\alpha^{(n)}(x_2) = \begin{cases} u_\alpha^{(n)}(b)e^{\sqrt{-1}\beta_1^n(\alpha)(x_2-b)} & \text{in } D_1 \quad \text{for } n \ne 0, \\ u_\alpha^{(0)}(b)e^{\sqrt{-1}\beta(x_2-b)} + e^{-\sqrt{-1}\beta x_2} - e^{\sqrt{-1}\beta(x_2-2b)} & \text{in } D_1 \quad \text{for } n = 0, \\ u_\alpha^{(n)}(-b)e^{-\sqrt{-1}\beta_2^n(\alpha)(x_2+b)} & \text{in } D_2. \end{cases}$$

From (4.49) we can then calculate the normal derivative of $u_\alpha^n(x_2)$ on $\Gamma_j, j = 1, 2$:

$$\left.\frac{\partial u_\alpha^{(n)}}{\partial \nu}\right|_{\Gamma_j} = \begin{cases} \sqrt{-1}\beta_1^n(\alpha)u_\alpha^{(n)}(b) & \text{on } \Gamma_1 \quad \text{for } n \ne 0, \\ \sqrt{-1}\beta u_\alpha^{(0)}(b) - 2\sqrt{-1}\beta e^{-\sqrt{-1}\beta b} & \text{on } \Gamma_1 \quad \text{for } n = 0, \\ \sqrt{-1}\beta_2^n(\alpha)u_\alpha^{(n)}(-b) & \text{on } \Gamma_2. \end{cases} \tag{4.50}$$

Thus from (4.48) and (4.50), it follows that

$$\left.\frac{\partial u_\alpha}{\partial \nu}\right|_{\Gamma_1} = \sum_{n\in\mathbb{Z}} \sqrt{-1}\beta_1^n(\alpha)u_\alpha^{(n)}(b)e^{\sqrt{-1}\frac{2\pi n}{\Lambda}x_1} - 2\sqrt{-1}\beta e^{-\sqrt{-1}\beta b}, \tag{4.51}$$

$$\left.\frac{\partial u_\alpha}{\partial \nu}\right|_{\Gamma_2} = \sum_{n\in\mathbb{Z}} \sqrt{-1}\beta_2^n(\alpha)u_\alpha^{(n)}(-b)e^{\sqrt{-1}\frac{2\pi n}{\Lambda}x_1}, \tag{4.52}$$

where the outward normal vector $\nu = (0, 1)$ on Γ_1 and $= (0, -1)$ on Γ_2.

In particular, the above discussion yields the following simple result.

LEMMA 4.4. *Suppose that $\alpha_n^2 > k_1^2$. Then*

$$u_\alpha^{(n)}(b) = u_\alpha^{(n)}(b_1) e^{-(b-b_1)\sqrt{\alpha_n^2 - k_1^2}}.$$

Similarly, if $\alpha_n^2 > |k_2|^2$, then

$$|u_\alpha^{(n)}(-b)| = |u_\alpha^{(n)}(-b_1)| e^{-(b-b_1)\sin(\gamma_2^n/2)\sqrt[4]{(\alpha_n^2 - \Re(k_2^2))^2 + (\Im(k_2^2))^2}}.$$

PROOF. The first identity is a simple consequence of (4.49) since k_1^2 is real. Recall that from (4.46),

$$\gamma_2^n = \arg(\Re(k_2^2) - \alpha_n^2 + \sqrt{-1}\Im(k_2^2)).$$

Using (4.49), we have

$$u_\alpha^{(n)}(-b) = u_\alpha^{(n)}(-b_1) e^{-(b-b_1)|\beta_2^n|(\sin(\gamma_2^n/2) - \sqrt{-1}\cos(\gamma_2^n/2))},$$

and hence

$$|u_\alpha^{(n)}(-b)| = |u_\alpha^{(n)}(-b_1)| e^{-(b-b_1)\sin(\gamma_2^n/2)\sqrt[4]{(\alpha_n^2 - \Re(k_2^2))^2 + (\Im(k_2^2))^2}},$$

which completes the proof. □

REMARK 4.5. *Actually, when $\alpha_n^2 \gg |k_2|^2$, the angle $\gamma_2^n/2 \leq \pi/2$ will approach $\pi/2$. Thus, there exists a fixed constant σ_0, such that*

$$\delta_0 \leq \sin(\gamma_2^n/2) \leq 1 . \tag{4.53}$$

Since the fields u_α are Λ-periodic in x_1, we can move the problem from $\mathbb{R}^2$ to the quotient $\mathbb{R}^2/(\Lambda\mathbb{Z} \times \{0\})$. In what follows, we shall identify Ω with the cylinder $\Omega/(\Lambda\mathbb{Z} \times \{0\})$, and similarly for the boundaries $\Gamma_j \equiv \Gamma_j/\Lambda\mathbb{Z}$. Thus from now on, all functions defined on Ω and Γ_j are implicitly Λ-periodic in the x_1 variable.

For functions $f \in H^{\frac{1}{2}}(\Gamma_j)$ (the Sobolev space of Λ-periodic complex valued functions), define, in the sense of distributions, the operator T_j^α by

$$T_j^\alpha[f](x_1) = \sum_{n \in \mathbb{Z}} \sqrt{-1}\beta_j^n(\alpha) f^{(n)} e^{\sqrt{-1}\frac{2\pi n}{\Lambda}x_1}, \tag{4.54}$$

where

$$f^{(n)} = \frac{1}{\Lambda}\int_0^\Lambda f(x_1) e^{-\sqrt{-1}\frac{2\pi n}{\Lambda}x_1}\, dx_1.$$

It is necessary in our study to understand the continuity properties of the above "Dirichlet-to-Neumann" maps. Fortunately, this is trivial by observing that T_j^α is a standard pseudo-differential operator (a convolution operator) of order one which can be seen from the definition of $\beta_j^n(\alpha)$. Thus the standard theory on pseudo-differential operators (see, for instance, [**494**]) applies.

LEMMA 4.6. *For $j = 1, 2$, the operator $T_j^\alpha : H^{\frac{1}{2}}(\Gamma_j) \to H^{-\frac{1}{2}}(\Gamma_j)$ is continuous.*

The scattering problem can be formulated as follows: find $u_\alpha \in H^1(\Omega)$ such that

$$\nabla_\alpha \cdot (\frac{1}{q^2}\nabla_\alpha u_\alpha) + \omega^2 u_\alpha = 0 \quad \text{in } \Omega, \tag{4.55}$$

$$T_1^\alpha[u_\alpha] - \frac{\partial u_\alpha}{\partial \nu} = 2\sqrt{-1}\beta e^{-\sqrt{-1}\beta b} \quad \text{on } \Gamma_1, \tag{4.56}$$

$$T_2^\alpha[u_\alpha] - \frac{\partial u_\alpha}{\partial \nu} = 0 \quad \text{on } \Gamma_2. \tag{4.57}$$

An equivalent form of the above system is

$$
(4.58)\qquad \begin{cases} \nabla_\alpha \cdot (\frac{1}{q^2}\nabla_\alpha \widetilde{u}_\alpha) + \omega^2 \widetilde{u}_\alpha = -f \quad \text{in } \Omega, \\ T_1^\alpha[\widetilde{u}_\alpha] - \frac{\partial \widetilde{u}_\alpha}{\partial \nu} = 0 \quad \text{on } \Gamma_1, \\ T_2^\alpha[\widetilde{u}_\alpha] - \frac{\partial \widetilde{u}_\alpha}{\partial \nu} = 0 \quad \text{on } \Gamma_2, \end{cases}
$$

where $f \in (H^1(\Omega))'$ and $\widetilde{u}_\alpha = u_\alpha - u_0$ with u_0 a fixed smooth function. In fact, u_0 may be constructed in the following way: Let u_0 be a smooth Λ-periodic function supported near the boundary Γ_1. It can be further arranged that $u_0(x_1, b) = 0$ and $-\partial_{x_2} u_0 = 2\sqrt{-1}\beta e^{-\sqrt{-1}\beta b}$ on Γ_1. Clearly, $\widetilde{u}_\alpha = u_\alpha - u_0$ solves the above equation with $f = \nabla_\alpha \cdot (\frac{1}{q^2}\nabla_\alpha u_0) + \omega^2 u_0 \in (H^1(\Omega))'$, the dual space of $H^1(\Omega)$.

For simplicity of notation, we shall denote $\widetilde{u}_\alpha$ by u_α. One can then write down a variational form equivalent to (4.58): Given $f \in (H^1(\Omega))'$, find $u_\alpha \in H^1(\Omega)$ such that

$$
(4.59)\qquad a(u_\alpha, \phi) = \langle f, \phi \rangle, \quad \forall \phi \in H^1(\Omega).
$$

Here, the sesquilinear form a is defined by

$$
\begin{aligned} a(w_1, w_2) &= \int_\Omega \frac{1}{q^2}\nabla w_1 \cdot \nabla \overline{w_2} - \int_\Omega (\omega^2 - \frac{\alpha^2}{q^2}) w_1 \overline{w_2} - \sqrt{-1}\alpha \int_\Omega \frac{1}{q^2}(\partial_{x_1} w_1)\overline{w_2} \\ &\quad + \sqrt{-1}\alpha \int_\Omega \frac{1}{q^2} w_1 \overline{\partial_{x_1} w_2} - \int_{\Gamma_1} \frac{1}{q_1^2} T_1^\alpha[w_1]\overline{w_2} - \int_{\Gamma_2} \frac{1}{q_2^2} T_2^\alpha[w_1]\overline{w_2}, \end{aligned}
$$

where $\int_{\Gamma_j}$ represents the dual pairing of $H^{-\frac{1}{2}}(\Gamma_j)$ with $H^{\frac{1}{2}}(\Gamma_j)$.

We first state the existence and uniqueness of the solution to the continuous scattering problem. The proof is from [**121**, **129**, **217**].

THEOREM 4.7. *For all but a countable set of frequencies ω_j, $|\omega_j| \to +\infty$, the diffraction problem has a unique solution $u_\alpha \in H^1(\Omega)$.*

For simplicity, from now on, we shall remove the subscript and superscript and denote u_α, T_j^α by u, T_j, respectively. In the proof of Theorem 4.7, we denote $k = q\omega$ to illustrate the explicit dependence on the frequency parameter ω.

PROOF. Since $k = q\omega$, we can write $a(w_1, w_2)$ as

$$
a(w_1, w_2) = B_1(w_1, w_2) + \omega^2 B_2(w_1, w_2),
$$

where

$$
\begin{aligned} B_1(w_1, w_2) &= \int_\Omega \frac{1}{q^2}\nabla w_1 \cdot \nabla \overline{w_2} + 2\int_\Omega \frac{\alpha^2}{q^2} w_1 \overline{w_2} - \sqrt{-1}\alpha \int_\Omega \frac{1}{q^2}(\partial_{x_1} w_1)\overline{w_2} \\ &\quad + \sqrt{-1}\alpha \int_\Omega \frac{1}{q^2} w_1 \overline{\partial_{x_1} w_2} - \int_{\Gamma_1} \frac{1}{q_1^2} T_1[w_1]\overline{w_2} - \int_{\Gamma_2} \frac{1}{q_2^2} T_2[w_1]\overline{w_2}, \\ B_2(w_1, w_2) &= -\int_\Omega (1 + \frac{\alpha^2}{k^2}) w_1 \overline{w_2}\,. \end{aligned}
$$

It follows that

$$
\begin{aligned} B_1(u, u) &= \int_\Omega \frac{1}{q^2}|\nabla u|^2 + 2\int_\Omega \frac{\alpha^2}{q^2}|u|^2 - 2\alpha \int_\Omega \frac{1}{q^2}\Im(u\,\overline{\partial_{x_1} u}) - \int_{\Gamma_1} \frac{1}{q_1^2} T_1[u]\overline{u} \\ &\quad - \int_{\Gamma_2} \frac{1}{q_2^2} T_2[u]\overline{u}. \end{aligned}
$$

Next denote $\frac{1}{q^2} = \frac{1}{\varepsilon\mu_0}$ by $\sigma' - \sqrt{-1}\sigma''$. Clearly, $\sigma' > 0$ and $\sigma'' \geq 0$. Also, denote $\frac{1}{q_2^2}$ by $\sigma_2' - \sqrt{-1}\sigma_2''$, where $\sigma_2' > 0$ and $\sigma_2'' \geq 0$. Thus

$$\begin{aligned}
\Re\{B_1(u,u)\} &= \int_\Omega \sigma'|\nabla u|^2 + 2\int_\Omega \alpha^2\sigma'|u|^2 - 2\alpha\int_\Omega \sigma'\Im(u\,\overline{\partial_{x_1} u}) \\
&\quad -\Re\{\int_{\Gamma_1}\frac{1}{q_1^2}T_1[u]\overline{u} + \int_{\Gamma_2}\frac{1}{q_2^2}T_2[u]\overline{u}\} \\
&\geq \int_\Omega \frac{\sigma'}{2}|\nabla u|^2 - \Re\{\int_{\Gamma_1}\frac{1}{q_1^2}T_1[u]\overline{u} + \int_{\Gamma_2}\frac{1}{q_2^2}T_2[u]\overline{u}\},
\end{aligned}$$

and

$$\begin{aligned}
-\Im\{B_1(u,u)\} &= \int_\Omega \sigma''|\nabla u|^2 + 2\int_\Omega \alpha^2\sigma''|u|^2 - 2\alpha\int_\Omega \sigma''\Im(u\,\overline{\partial_{x_1} u}) \\
&\quad +\Im\{\int_{\Gamma_1}\frac{1}{q_1^2}T_1[u]\overline{u} + \int_{\Gamma_2}\frac{1}{q_2^2}T_2[u]\overline{u}\} \\
&\geq \int_\Omega \frac{\sigma''}{2}|\nabla u|^2 + \Im\{\int_{\Gamma_1}\frac{1}{q_1^2}T_1[u]\overline{u} + \int_{\Gamma_2}\frac{1}{q_2^2}T_2[u]\overline{u}\}.
\end{aligned}$$

Further,

$$\begin{aligned}
-\int_{\Gamma_1}\frac{1}{q_1^2}T_1[u]\overline{u} &= -\sum \frac{1}{q_1^2}\Lambda\sqrt{-1}\beta_1^n|u^{(n)}|^2 \\
&= \sum \frac{1}{q_1^2}\Lambda(\Im\beta_1^n)\,|u^{(n)}|^2 - \sqrt{-1}\sum\frac{1}{n_1^2}\Lambda\Re\beta_1^n|u^{(n)}|^2,
\end{aligned}$$

and it is easy to see that

$$\begin{aligned}
-\int_{\Gamma_2}\frac{1}{q_2^2}T_2[u]\overline{u} &= -\sum \frac{1}{q_2^2}\sqrt{-1}\Lambda\beta_2^n|u^{(n)}(-b)|^2 \\
&= \sum_n \Lambda|\beta_2^n||u^{(n)}(-b)|^2 p_n,
\end{aligned}$$

where $p_n = p_n' - \sqrt{-1}p_n''$ with

$$p_n' = -\sigma_2''\cos(\gamma_2^n/2) + \sigma_2'\sin(\gamma_2^n/2)$$

and

$$p_n'' = \sigma_2'\cos(\gamma_2^n/2) + \sigma_2''\sin(\gamma_2^n/2).$$

Recall that

$$\gamma_2^n = \arg(\Re(k_2^2) - \alpha_n^2 + \sqrt{-1}\Im(k_2^2))$$

where $k_2 = \omega q_2$, and $0 \leq \gamma_2^n < 2\pi$. Then it follows that $p_n'' > 0$ for all n and the set $\{n : p_n' < 0\}$ is finite. It is also easy to verify that $|p_n''| > |p_n'|$ for $n \in \{n : p_n' < 0\}$. Moreover, for fixed $\omega \notin \mathcal{B}$ where $\mathcal{B}$ is defined by

$$\mathcal{B} := \{\omega : \beta_j^n(\omega) = 0, \;\; j = 1, 2\},$$

we have

$$|\beta_j^n| \geq C(1 + |n|^2)^{1/2}, \; j = 1, 2.$$

Combining the above estimates, we have

$$\begin{aligned}
|B_1(u,u)| &\geq C\Big[\int_\Omega |\nabla u|^2 + \|u\|^2_{H^{1/2}(\Gamma_1)} + \sum_{n\in\Lambda}(|p_n''| - |p_n'|)|u^{(n)}(-b)|^2 \\
&\quad + \sum_{n\notin\Lambda} |p_n''||u^{(n)}(-b)|^2\Big] \\
&\geq C[\int_\Omega |\nabla u|^2 + \|u\|^2_{H^{1/2}(\Gamma_1)} + \|u\|^2_{H^{1/2}(\Gamma_2)}] \\
&\geq C\|u\|^2_{H^1(\Omega)},
\end{aligned}$$

where the last inequality may be obtained by applying some standard elliptic estimates; see [**264**]. Therefore, we have shown that

$$|B_1(u,u)| \geq C\|u\|^2_{H^1(\Omega)}, \tag{4.60}$$

i.e., B_1 is a bounded coercive sesquilinear form over $H^1(\Omega)$. The Lax-Milgram lemma then gives the existence of a bounded invertible map $A_1 = A_1(\omega) : H^1(\Omega) \to (H^1(\Omega))'$ such that $\langle A_1 u, v\rangle = B_1(u,v)$. Moreover, A_1^{-1} is bounded. Notice that the operator $A_2 : H^1(\Omega) \to (H^1(\Omega))'$ defined by $\langle A_2 u, v\rangle = B_2(u,v)$ is compact and independent of ω.

Holding $\omega_0 \notin \mathcal{B}$ fixed, consider the operator $A(\omega_0,\omega) = A_1(\omega_0) + \omega^2 A_2$. Since A_1 is bounded invertible and A_2 is compact, we see that $A(\omega_0,\omega)^{-1}$ exists by Fredholm theory for all $\omega \notin \mathcal{E}(\omega_0)$, where $\mathcal{E}(\omega_0)$ is some discrete set. It is clear that

$$\|A_1(\omega) - A_1(\omega_0)\| \to 0, \quad \text{as } \omega \to \omega_0.$$

Thus, since $\|A(\omega,\omega) - A(\omega_0,\omega)\| = \|A_1(\omega) - A_1(\omega_0)\|$ is small for $|\omega - \omega_0|$ sufficiently small, it follows from the stability of bounded invertibility (see, for instance, Kato [**326**, Chapter 4]) that $A(\omega,\omega)^{-1}$ exists and is bounded for $|\omega - \omega_0|$ sufficiently small, $\omega \notin \mathcal{E}(\omega_0)$. Since $\omega_0 > 0$ can be an arbitrary real number, we have shown that $A(\omega,\omega)^{-1}$ exists for all but a discrete set of points. □

4.3.2. Biperiodic Structures. Consider a time-harmonic electromagnetic plane wave incident on a biperiodic structure in $\mathbb{R}^3$. The periodic structure separates two homogeneous regions. The medium inside the structure is heterogeneous. The diffraction problem is then to predict energy distributions of the reflected and transmitted waves. In this subsection, we study some mathematical aspects of the diffraction problem for biperiodic structures. We introduce a variational formulation of the diffraction problem by dielectric gratings. Our main result is concerned with the well-posedness of the model problem. It is shown that for all but possibly a discrete set of frequencies, there is a unique quasi-periodic weak solution to the diffraction problem. Our proof is based on the Hodge decomposition and a compact embedding result. An energy conservation for the weak solution is also proved. An important step of our approach is to reduce the original diffraction problem with an infinite configuration to another problem with a bounded domain. This is done by introducing a pair of transparent boundary conditions. We emphasize that the variational approach is very general. In particular, the material coefficients ε and μ are only assumed to be bounded functions. The geometry can be extremely general as well. The incident angles and grating shapes may be arbitrary. Moreover, a class of finite element methods can be formulated based on the variational approach.

4.3.3. Diffraction Problem. We first specify the geometry of the problem. Let Λ_1 and Λ_2 be two positive constants, such that the material functions ε and μ satisfy, for any n_1, $n_2 \in \mathbb{Z}$,

$$\begin{aligned} \varepsilon(x_1 + n_1\Lambda_1, x_2 + n_2\Lambda_2, x_3) &= \varepsilon(x_1, x_2, x_3), \\ \mu(x_1 + n_1\Lambda_1, x_2 + n_2\Lambda_2, x_3) &= \mu(x_1, x_2, x_3). \end{aligned}$$

In addition, it is assumed that, for some fixed positive constant b and sufficiently small $\delta > 0$,

$$\begin{aligned} \varepsilon(x) &= \varepsilon_1, \quad \mu(x) = \mu_1 \ \text{ for } x_3 > b - \delta, \\ \varepsilon(x) &= \varepsilon_2, \quad \mu(x) = \mu_2 \ \text{ for } x_3 < -b + \delta, \end{aligned}$$

where ε_1, ε_2, μ_1, and μ_2 are positive constants. All of these assumptions are physical.

We make the following general assumptions: $\varepsilon(x)$, $\mu(x)$, and $\beta(x)$ are all real valued L^∞ functions, $\varepsilon(x) \geq \varepsilon_0$ and $\mu(x) \geq \mu_0$, where ε_0 and μ_0 are positive constants.

Let $\Omega = \{x \in \mathbb{R}^3 : -b < x_3 < b\}$, $\Omega_1 = \{x \in \mathbb{R}^3 : x_3 > b\}$, $\Omega_2 = \{x \in \mathbb{R}^3 : x_3 < -b\}$.

Consider a plane wave in Ω_1

$$E^i = s e^{\sqrt{-1} q\cdot x}, \quad H^i = p e^{\sqrt{-1} q\cdot x}, \tag{4.61}$$

incident on Ω. Here $q = (\alpha_1, \alpha_2, -\beta) = \omega\sqrt{\varepsilon_1\mu_1}(\cos\theta_1\cos\theta_2, \cos\theta_1\sin\theta_2, -\sin\theta_1)$ is the incident wave vector whose direction is specified by θ_1 and θ_2, with $0 < \theta_1 < \pi$ and $0 < \theta_2 \leq 2\pi$. The vectors s and p satisfy

$$s = \frac{1}{\omega\varepsilon_1}(p \times q), \quad q\cdot q = \omega^2\varepsilon_1\mu_1, \quad p \cdot q = 0. \tag{4.62}$$

We are interested in biperiodic solutions, *i. e.*, solutions E and H such that the fields E_α, H_α defined by, for $\alpha = (\alpha_1, \alpha_2, 0)$,

$$E_\alpha = e^{-\sqrt{-1}\alpha\cdot x} E(x_1, x_2, x_3), \tag{4.63}$$

$$H_\alpha = e^{-\sqrt{-1}\alpha\cdot x} H(x_1, x_2, x_3), \tag{4.64}$$

are periodic in the x_1-direction of period Λ_1 and in the x_2-direction of period Λ_2.

Denote

$$\nabla_\alpha = \nabla + \sqrt{-1}\alpha = \nabla + \sqrt{-1}(\alpha_1, \alpha_2, 0).$$

It is easy to see from (4.3) and (4.4) that E_α and H_α satisfy

$$\nabla_\alpha \times \Big(\frac{1}{\mu}\nabla_\alpha \times E_\alpha\Big) - \omega^2\varepsilon E_\alpha = 0, \tag{4.65}$$

$$\nabla_\alpha \times E_\alpha = \sqrt{-1}\omega\mu(x)\, H_\alpha. \tag{4.66}$$

In order to solve the system of differential equations, we need boundary conditions in the x_3-direction. These conditions may be derived by the radiation condition, the periodicity of the structure, and the Green functions. To do so, we can expand E_α in a Fourier series since it is Λ periodic:

$$E_\alpha(x) = E^i_\alpha(x) + \sum_{n\in\mathbb{Z}} U^{(n)}_\alpha(x_3) e^{\sqrt{-1}(\frac{2\pi n_1}{\Lambda_1}x_1 + \frac{2\pi n_2}{\Lambda_2}x_2)}, \tag{4.67}$$

where $E^i_\alpha(x) = E^i(x)e^{-\sqrt{-1}\alpha\cdot x}$ and

$$U^{(n)}_\alpha(x_3) = \frac{1}{\Lambda_1\Lambda_2}\int_0^{\Lambda_1}\int_0^{\Lambda_2}(E_\alpha(x) - E^i_\alpha(x))e^{-\sqrt{-1}(\frac{2\pi n_1}{\Lambda_1}x_1+\frac{2\pi n_2}{\Lambda_2}x_2)}dx_1dx_2.$$

Denote

$$\Gamma_1 = \{x \in \mathbb{R}^3 : x_3 = b\} \text{ and } \Gamma_2 = \{x_3 = -b\}.$$

Define for $j = 1, 2$ the coefficients

$$\beta^{(n)}_j(\alpha) = \begin{cases} \sqrt{\omega^2\varepsilon_j\mu_j - |\alpha_n|^2}, & \omega^2\varepsilon_j\mu_j > |\alpha_n|^2, \\ \sqrt{-1}\sqrt{|\alpha_n|^2 - \omega^2\varepsilon_j\mu_j}, & \omega^2\varepsilon_j\mu_j < |\alpha_n|^2, \end{cases} \tag{4.68}$$

where

$$\alpha_n = \alpha + (2\pi n_1/\Lambda_1, 2\pi n_2/\Lambda_2, 0).$$

We assume that $\omega^2\varepsilon_j\mu_j \neq |\alpha_n|^2$ for all $n \in \mathbb{Z}^2$, $j = 1, 2$. This condition excludes "resonances".

For convenience, we also introduce the following notation:

$$\Lambda^+_j = \{n \in \mathbb{Z}^2 : \ \Im(\beta^{(n)}_j) = 0\},$$
$$\Lambda^-_j = \{n \in \mathbb{Z}^2 : \ \Im(\beta^{(n)}_j) \neq 0\}.$$

Observe that inside Ω_j $(j = 1, 2)$, $\varepsilon = \varepsilon_j$ and $\mu = \mu_j$, Maxwell's equations then become

$$(\Delta_\alpha + \omega^2\varepsilon_j\mu_j)E_\alpha = 0\,, \tag{4.69}$$

where $\Delta_\alpha = \Delta + 2\sqrt{-1}\alpha\cdot\nabla - |\alpha|^2$.

Since the medium in Ω_j $(j = 1, 2)$ is homogeneous, the method of separation of variables implies that E_α can be expressed as a sum of plane waves:

$$E_\alpha|_{\Omega_j} = E^i_\alpha(x) + \sum_{n=(n_1,n_2)\in\mathbb{Z}^2} A^{(n)}_j e^{\pm\sqrt{-1}\beta^{(n)}_j x_3 + \sqrt{-1}(\frac{2\pi n_1}{\Lambda_1}x_1+\frac{2\pi n_2}{\Lambda_2}x_2)}, \quad j = 1, 2, \tag{4.70}$$

where the $A^{(n)}_j$ are constant (complex) vectors, where $E^i_\alpha(x) = 0$ in Ω_2.

We next impose a radiation condition on the scattering problem. Due to the (infinite) periodic structure, the usual Sommerfeld or Silver-Müller radiation condition is no longer valid. Instead, the following radiation condition based on the diffraction theory is employed: Since β^n_j is real for at most finitely many n, there are only a finite number of propagating plane waves in the sum (4.70), the remaining waves are exponentially decaying (or unbounded) as $|x_3| \to \infty$. We will insist that E_α is composed of bounded outgoing plane waves in Ω_1 and Ω_2, plus the incident (incoming) wave in Ω_1.

From (4.67) and (4.68) we deduce

$$E^{(n)}_\alpha(x_3) = \begin{cases} U^{(n)}_\alpha(b)e^{\sqrt{-1}\beta^{(n)}_1(x_3-b)} & \text{in } \Omega_1, \\ U^{(n)}_\alpha(-b)e^{-\sqrt{-1}\beta^{(n)}_2(x_3+b)} & \text{in } \Omega_2. \end{cases} \tag{4.71}$$

By matching the two expansions (4.67) and (4.70), we get

$$A^{(n)}_1 = U^{(n)}_\alpha(b)e^{-\sqrt{-1}\beta^{(n)}_1 b} \text{ on } \Gamma_1, \tag{4.72}$$

$$A^{(n)}_2 = U^{(n)}_\alpha(-b)e^{-\sqrt{-1}\beta^{(n)}_2 b} \text{ on } \Gamma_2. \tag{4.73}$$

Furthermore, since in the regions $\{x:\ x_3 > b-\delta\} \cup \{x:\ x_3 < -b+\delta\}$,

$$\nabla \cdot E = 0\,,\ \ \nabla \cdot E^i = 0$$

or

$$\nabla_\alpha \cdot E_\alpha = 0\,,\ \ \nabla_\alpha \cdot E^i_\alpha = 0,$$

we have from (4.70) that

$$\alpha_n \cdot U^{(n)}_\alpha(b) + \beta^{(n)}_1 U^{(n)}_{\alpha,3}(b) = 0 \ \text{ on } \Gamma_1, \tag{4.74}$$

$$\alpha_n \cdot U^{(n)}_\alpha(-b) - \beta^{(n)}_2 U^{(n)}_{\alpha,3}(-b) = 0 \ \text{ on } \Gamma_2. \tag{4.75}$$

LEMMA 4.8. *There exist boundary pseudo-differential operators B_j $(j = 1, 2)$ of order one, such that*

$$\nu \times (\nabla_\alpha \times (E_\alpha - E^i_\alpha)) = B_1 P[E_\alpha - E^i_\alpha] \ \text{ on } \Gamma_1, \tag{4.76}$$

$$\nu \times (\nabla_\alpha \times E_\alpha) = B_2 P[E_\alpha] \ \text{ on } \Gamma_2, \tag{4.77}$$

where the operator B_j is defined by

(4.78)

$$B_j[f] = -\sqrt{-1}\sum_{n\in\mathbb{Z}^2} \frac{1}{\beta^{(n)}_j}\{(\beta^{(n)}_j)^2 (f^{(n)}_1, f^{(n)}_2, 0) + (\alpha_n \cdot f^{(n)})\alpha_n\} e^{\sqrt{-1}(\frac{2\pi n_1}{\Lambda_1}x_1 + \frac{2\pi n_2}{\Lambda_2}x_2)},$$

where P is the projection onto the plane orthogonal to ν, i.e.,

$$P[f] = -\nu \times (\nu \times f),$$

and

$$f^{(n)} = \Lambda_1^{-1}\Lambda_2^{-1}\int_0^{\Lambda_1}\int_0^{\Lambda_2} f(x) e^{-\sqrt{-1}(\frac{2\pi n_1}{\Lambda_1}x_1 + \frac{2\pi n_2}{\Lambda_2}x_2)} dx_1 dx_2.$$

Here ν is the outward normal to Ω.

The proof may be given by using the expansion (4.70) together with (4.72–4.75), and some simple calculation.

REMARK 4.9. *The Dirichlet-to-Neumann operator B carries the information on radiation condition in an explicit form. Here it is crucial to assume that $\beta^{(n)}$ is nonzero.*

We introduce the L^2 scalar product

$$\langle f,\, g\rangle = \int_A f\overline{g}\,,$$

where A is the domain.

Denote by B^*_j the adjoint of B_j, that is,

$$\langle B_j[f],\, g\rangle = \left\langle f,\, B^*_j[g]\right\rangle,$$

for any f and g in $L^2(\Gamma_j)$.

It is easily seen that the adjoint operator of B_j in the above lemma is given by

(4.79)

$$B^*_j[f] = \sqrt{-1}\sum_{n\in\mathbb{Z}^2} \frac{1}{\overline{\beta}^{(n)}_j}\{(\overline{\beta}^{(n)}_j)^2 (f^{(n)}_1, f^{(n)}_2, 0) + (\alpha_n \cdot f^{(n)})\alpha_n\} e^{\sqrt{-1}(\frac{2\pi n_1}{\Lambda_1}x_1 + \frac{2\pi n_2}{\Lambda_2}x_2)}.$$

Define

$$\Lambda = \Lambda_1\mathbb{Z} \times \Lambda_2\mathbb{Z} \times \{0\} \subset \mathbb{R}^3.$$

Since the fields E_α are Λ-periodic, we can move the problem from $\mathbb{R}^3$ to the quotient space $\mathbb{R}^3/\Lambda$. For the remainder of the section, we shall identify Ω with the cube Ω/Λ, and similarly for the boundaries $\Gamma_j \equiv \Gamma_j/\Lambda$. Thus from now on,

all functions defined on Ω and Γ_j are implicitly Λ-periodic.

Define $\nabla_\alpha\cdot$ by $\nabla_\alpha \cdot u = (\partial_{x_1} + \sqrt{-1}\alpha_1)u_1 + (\partial_{x_2} + \sqrt{-1}\alpha_2)u_2$.

Let H^m be the mth order L^2-based Sobolev spaces of complex valued functions. We denote by $H_p^m(\Omega)$ the subset of all functions in $H^m(\Omega)$ which are the restrictions to Ω of the functions in $H^m_{\mathrm{loc}}(\mathbb{R}^2 \times (-b, b))$ that are Λ-periodic. Similarly we define $H_p^m(\Omega_j)$ and $H_p^m(\Gamma_j)$. In the future, for simplicity, we shall drop the subscript p. We shall also drop the subscript α from E_α, E_α^i, ∇_α, and $\nabla_\alpha\cdot$. Finally, we introduce the space $H(\mathrm{curl}, \Omega) := \{E \in L^2(\Omega) : \nabla \times E \in L^2(\Omega)\}$.

Therefore, the diffraction problem can be reformulated as follows:

$$
\left\{
\begin{array}{l}
\nabla \times (\frac{1}{\mu}\nabla \times E) - \omega^2 \varepsilon E = 0 \text{ in } \Omega, \\
\nu \times (\nabla \times E) = B_1 P[E] - f \text{ on } \Gamma_1, \\
\nu \times (\nabla \times E) = B_2 P[E] \text{ on } \Gamma_2,
\end{array}
\right.
\tag{4.80}
$$

where

$$
f = \frac{1}{\mu_1}(B_1 P[E^i]|_{\Gamma_1} - \nu \times (\nabla \times E^i)|_{\Gamma_1}).
\tag{4.81}
$$

The weak form of the above boundary value problem is to find $E \in H(\mathrm{curl}, \Omega)$, such that for any $F \in H(\mathrm{curl}, \Omega)$

$$
\begin{aligned}
&\int_\Omega \frac{1}{\mu} \nabla \times E \cdot \overline{\nabla \times F} \\
&\quad - \int_\Omega \omega^2 \varepsilon E \cdot \overline{F} + \int_{\Gamma_1} \frac{1}{\mu_1} B_1 P[E] \cdot \overline{F} + \int_{\Gamma_2} \frac{1}{\mu_2} B_2 P[E] \cdot \overline{F} = \int_{\Gamma_1} f \cdot \overline{F}.
\end{aligned}
\tag{4.82}
$$

4.3.4. The Hodge Decomposition and a Compactness Result. We present a version of the Hodge decomposition and compactness lemma. The results are crucial in the proof of our theorem on existence and uniqueness. We also state a useful trace regularity estimate. We remark that for simplicity, no attempt is made to give the most general forms of these results.

Let us begin with a simple property of the operator B_j. From now on, we define $\nabla_{\Gamma_j}\cdot$ as the surface divergence on Γ_j.

PROPOSITION 4.10. *For $j = 1, 2$ and $q \in H^1(\Omega)$*

$$
-\Re \int_{\Gamma_j} B_j P[\nabla q] \cdot \overline{\nabla q} \geq 0\,.
$$

PROOF. Using the definitions of the operator B_j in (4.78) and $\beta_j^{(n)}$ in (4.68), we have by integration by parts on the surface

$$\begin{aligned} -\Re\int_{\Gamma_j} B_j P[\nabla q]\cdot\overline{\nabla q} &= \Re\int_{\Gamma_j} \nabla_\Gamma\cdot B_j P[\nabla q]\cdot\overline{q} \\ &= \Re\sum_{n\in\mathbb{Z}^2}\left\{\sqrt{-1}\beta_j^{(n)}|\alpha_n|^2|q^{(n)}|^2+\frac{\sqrt{-1}}{\beta_j^{(n)}}|\alpha_n|^2|q^{(n)}|^2\right\} \\ &= \sum_{n\in\Lambda_j^-}\frac{(-|\beta_j^{(n)}|^2+|\alpha_n|^2)}{|\beta_j^{(n)}|}|\alpha_n|^2|q^{(n)}|^2 \\ &= \sum_{n\in\Lambda_j^-}\varepsilon_j\mu_j\omega^2\frac{|\alpha_n|^2}{|\beta_j^{(n)}|}|q^{(n)}|^2\ge 0. \end{aligned}$$

Recall that ∇, $\nabla\cdot$, are the shorthand notations of $\nabla+\sqrt{-1}\alpha$, $\nabla_\alpha\cdot$, respectively. □

LEMMA 4.11. *For any function $f\in(H^1(\Omega))'$ which is smooth near Γ_1 and Γ_2, the boundary value problem*

$$(4.83)\qquad \begin{cases} \nabla\cdot(\varepsilon\nabla p) &= f \text{ in } \Omega, \\ \varepsilon_1\frac{\partial p}{\partial\nu} &= -\frac{1}{\mu_1}\nabla_\Gamma\cdot B_1P[\nabla p] \text{ on } \Gamma_1, \\ \varepsilon_2\frac{\partial p}{\partial\nu} &= -\frac{1}{\mu_2}\nabla_\Gamma\cdot B_2P[\nabla p] \text{ on } \Gamma_2, \end{cases}$$

has a unique solution in $H_0^1(\Omega)=\{q:\ q\in H^1(\Omega),\ \int_\Omega q=0\}$.

PROOF. We examine the weak form of the boundary value problem (4.83). For any $q\in H_0^1(\Omega)$, multiplying both sides of (4.83) by $\overline{q}$ and integrating over Ω yield

$$\int_\Omega \nabla\cdot(\varepsilon\nabla p)\cdot\overline{q}=\int_\Omega f\cdot\overline{q}.$$

After using the boundary conditions integration by parts gives that

$$(4.84)\qquad \int_\Omega\varepsilon\nabla p\cdot\overline{\nabla q}+\int_{\Gamma_1}\frac{1}{\mu_1}\nabla_\Gamma\cdot B_1P[\nabla p]\cdot\overline{q}+\int_{\Gamma_2}\frac{1}{\mu_2}\nabla_\Gamma\cdot B_2P[\nabla p]\cdot\overline{q}=-\int_\Omega f\cdot\overline{q}.$$

Denote the left-hand side of (4.84) by $b(p,q)$. Keeping in mind that p and q are periodic, from integration by parts on the boundary, we obtain

$$b(p,q)=\int_\Omega\varepsilon\nabla p\cdot\overline{\nabla q}-\int_{\Gamma_1}\frac{1}{\mu_1}B_1P[\nabla p]\cdot\overline{P[\nabla q]}-\int_{\Gamma_2}\frac{1}{\mu_2}B_2P[\nabla p]\cdot\overline{P[\nabla q]}.$$

The variational problem takes the form: to find $p\in H_0^1(\Omega)$, such that

$$b(p,q)=-\int_\Omega f\cdot\overline{q},\quad\forall q\in H_0^1(\Omega).$$

It is now obvious from Proposition 4.10 that

$$\begin{aligned} \Re\, b(p,p) &= \int_\Omega\varepsilon|\nabla p|^2-\Re\left\{\int_{\Gamma_1}\frac{1}{\mu_1}B_1P[\nabla p]\cdot\overline{P[\nabla p]}+\int_{\Gamma_2}\frac{1}{\mu_2}B_2P[\nabla p]\cdot\overline{P[\nabla p]}\right\} \\ &\ge C||\nabla p||^2_{L^2(\Omega)}. \end{aligned}$$

Therefore by a version of Poincaré's inequality ($\int_\Omega p=0$), we obtain

$$\Re\, b(p,p)\ge C\|p\|^2_{H^1(\Omega)}.$$

The proof is complete by a direct application of the Lax-Milgram lemma. □

Next, we present an embedding result. Let $W(\Omega)$ be a functional space defined by

$$\Big\{u : \ u \in H(\mathrm{curl}, \Omega), \ \ \nabla \cdot (\varepsilon u) = 0 \text{ in } \Omega, \text{ and} \tag{4.85}$$

$$\omega^2 \varepsilon_j u \cdot \nu = -\frac{1}{\mu_j} \nabla_\Gamma \cdot B_j P[u] \text{ on } \Gamma_j\,, \ \ j = 1, 2\Big\}. \tag{4.86}$$

LEMMA 4.12. *The embedding from $W(\Omega)$ to $L^2(\Omega)$ is compact.*

PROOF. Let u be a function in $W(\Omega)$. Define an extension of u by

$$\widetilde{u} = \begin{cases} u_1 \text{ in } \Omega_1, \\ u \text{ in } \Omega, \\ u_2 \text{ in } \Omega_2, \end{cases}$$

where u_j $(j = 1, 2)$ satisfies

$$\begin{aligned} &\nabla \times \nabla \times u_j - \omega^2 \varepsilon_j \mu_j u_j = 0 \text{ in } \Omega_j\,, \\ &u_j \times \nu = u \times \nu \text{ on } \Gamma_j, \\ &\text{the radiation condition at the infinity.} \end{aligned}$$

Since the medium in Ω_j is homogeneous, it may be shown that

$$\omega^2 \varepsilon_j u_j \cdot \nu = -\frac{1}{\mu_j} \nabla_\Gamma \cdot B_j P[u] \text{ on } \Gamma_j\,, \quad j = 1, 2. \tag{4.87}$$

In the following, we outline the proof of (4.87). In fact, it is easy to see that the function u_j satisfies the boundary condition

$$\nu \times \nabla \times u_j = B_j P[u_j].$$

Hence

$$\nabla_\Gamma \cdot (\nu \times \nabla \times u_j) = \nabla_\Gamma \cdot (B_j P[u_j]). \tag{4.88}$$

But

$$\nabla_\Gamma \cdot (\nu \times \nabla \times u_j) = -(\nabla \times \nabla \times u_j) \cdot \nu,$$

which together with the Maxwell equation for u_j yield that

$$-\omega^2 \varepsilon_j \mu_j u_j \cdot \nu = \nabla_\Gamma \cdot B_j P[u_j]. \tag{4.89}$$

From (4.88), (4.89), the boundary identity (4.87) follows.

Therefore from $[\widetilde{u} \times \nu] = 0$, it follows that $[\widetilde{u} \cdot \nu] = 0$ on Γ_j and then

$$\nabla \cdot (\varepsilon \widetilde{u}) = 0 \text{ in } \overline{\Omega} \cup \Omega_1 \cup \Omega_2.$$

It follows from $[\widetilde{u} \times \nu] = 0$ on Γ_j and the radiation condition that $\widetilde{u} \in H(\mathrm{curl}, D)$ for any bounded domain $D \subset \overline{\Omega} \cup \Omega_1 \cup \Omega_2$.

Now let $\{\widetilde{u}_j\}$ be a sequence of functions in W that converges weakly to zero in $W(\Omega)$. Construct a cutoff function χ with the properties: χ is supported in $\widetilde{\Omega} \Supset \Omega$ and $\chi = 1$ in Ω. Here $\widetilde{\Omega} = \{-b' \leq x_3 \leq b', \ \ 0 < x_1 < \Lambda_1, \ \ 0 < x_2 < \Lambda_2\}$ with $b' > b$.

Hence

$$\{\chi \widetilde{u}_j\} \subset \widetilde{W} = \Big\{v : \ v \in H(\mathrm{curl}, \widetilde{\Omega}), \ \ \nabla \cdot (\varepsilon v) = 0, \ \ \nu \times v = 0 \text{ on } x_3 = b', -b'\Big\}.$$

It follows from a well known result of Weber [**508**] that the embedding from $\widetilde{W}(\widetilde{\Omega})$ to $L^2(\widetilde{\Omega})$ is compact. Therefore the sequence $\{\widetilde{u}_j\}$ converges strongly to zero in $L^2(\Omega)$, which completes the proof. □

We now state a useful trace regularity result.

PROPOSITION 4.13. *Let D be a bounded domain. For any $\eta > 0$, there is a constant $C(\eta)$ such that the following estimate*

$$\|\nu \times u\|_{H^{-1/2}(\partial D)} \leq \eta \|\nabla \times u\|_{L^2(D)} + C(\eta)\|u\|_{L^2(D)}$$

holds.

PROOF. The proof is straightforward. For the sake of completeness, we sketch it here.

For any function $\phi \in H^{1/2}(\partial D)$, consider an auxiliary problem

$$\begin{cases} \nabla \times \nabla \times w + \dfrac{1}{\eta^2} w & = & 0 \text{ in } D\,, \\ -\nu \times (\nu \times w) & = & \phi \text{ on } \partial D. \end{cases}$$

The result of the proposition follows immediately from estimating $|\langle \nu \times, \phi \rangle|$. □

4.3.5. Existence and Uniqueness of a Solution. In this subsection, we investigate questions on existence and uniqueness for the model problem. Our main result is as follows.

THEOREM 4.14. *For all but possibly a countable set of frequencies ω_j, $\omega_j \to +\infty$, the variational problem (4.82) admits a unique weak solution E in $H(\text{curl}, \Omega)$.*

PROOF. The proof is based on the Lax-Milgram lemma. We first decompose the field E into two parts

$$E = u + \nabla p\,, \quad u \in H(\text{curl}, \Omega), \;\; p \in H^1(\Omega).$$

By choosing $E = u + \nabla p$, $F = v$ in (4.82), we arrive at

$$\begin{aligned} (4.90) \quad & \int_\Omega \frac{1}{\mu} \nabla \times u \cdot \overline{\nabla \times v} \\ & -\omega^2 \int_\Omega \varepsilon u \cdot \overline{v} + \int_{\Gamma_1} \frac{1}{\mu_1} B_1 P[u] \cdot \overline{v} + \int_{\Gamma_2} \frac{1}{\mu_2} B_2 P[u] \cdot \overline{v} \\ & -\omega^2 \int_\Omega \varepsilon \nabla p \cdot \overline{v} + \int_{\Gamma_1} \frac{1}{\mu_1} B_1 P[\nabla p] \cdot \overline{v} + \int_{\Gamma_2} \frac{1}{\mu_2} B_2 P[\nabla p] \cdot \overline{v} = \int_{\Gamma_1} f \cdot \overline{v}. \end{aligned}$$

Similarly by choosing $E = u + \nabla p$, $F = \nabla q$ in (4.82), we get

$$\begin{aligned} (4.91) \; & -\omega^2 \int_\Omega \varepsilon u \cdot \overline{\nabla q} + \int_{\Gamma_1} \frac{1}{\mu_1} B_1 P[u] \cdot \overline{\nabla q} + \int_{\Gamma_2} \frac{1}{\mu_2} B_2 P[u] \cdot \overline{\nabla q} \\ & -\omega^2 \int_\Omega \varepsilon \nabla p \cdot \overline{\nabla q} + \int_{\Gamma_1} \frac{1}{\mu_1} B_1 P[\nabla p] \cdot \overline{\nabla q} + \int_{\Gamma_2} \frac{1}{\mu_2} B_2 P[\nabla p] \cdot \overline{\nabla q} = \int_{\Gamma_1} f \cdot \overline{\nabla q}. \end{aligned}$$

We use the following Hodge decomposition:

$$E = u + \nabla p\,,$$

where $p \in H^1(\Omega)$ and $u \in W(\Omega)$. The functional space W consists of all functions $U \in H(\mathrm{curl}, \Omega)$ that satisfy

$$\left\{ \begin{array}{lcll} \nabla \cdot (\varepsilon u) & = & 0 & \text{in } \Omega, \\ \omega^2 \varepsilon_1 u \cdot \nu & = - & \dfrac{1}{\mu_1} \nabla_\Gamma \cdot B_1 P[u] \text{ on } \Gamma_1, \\ \omega^2 \varepsilon_2 u \cdot \nu & = - & \dfrac{1}{\mu_2} \nabla_\Gamma \cdot B_2 P[u] \text{ on } \Gamma_2. \end{array} \right. \tag{4.92}$$

The fact that this decomposition is valid follows from Lemma 4.11. Actually, it is obvious to see that for any given E, Lemma 4.11 implies that there is a function p, such that $\nabla \cdot (\varepsilon \nabla p) = \nabla \cdot (\varepsilon E)$ and the suitable boundary conditions. Therefore, $u = E - \nabla p$ solves the problem (4.92).

Moreover, according to Lemma 4.12, the embedding from $W(\Omega)$ to $L^2(\Omega)$ is compact. We point out that the embedding from $H(\mathrm{curl}, \Omega)$ to $L^2(\Omega)$ is not compact.

Denote the left-hand sides of (4.90), (4.91) by $a_1(u, v)$, $a_2(p, q)$, respectively.

After some simple calculation, we obtain for u, $v \in W$, p, $q \in H^1$ that

$$\begin{aligned} a_1(u, v) \; &= \; \int_\Omega \frac{1}{\mu} \nabla \times u \cdot \overline{\nabla \times v} - \omega^2 \int_\Omega \varepsilon u \cdot \overline{v} \\ &+ \frac{1}{\mu_1} \int_{\Gamma_1} B_1 P[u] \cdot \overline{v} + \frac{1}{\mu_2} \int_{\Gamma_2} B_2 P[u] \cdot \overline{v} \\ &- \int_{\Gamma_1} \frac{1}{\mu_1} p \, \nabla_\Gamma \cdot \overline{((B_1^* - B_1) P[v]} \\ &- \int_{\Gamma_2} \frac{1}{\mu_2} p \, \nabla_\Gamma \cdot (\overline{(B_2^* - B_2) P[v]}) \end{aligned} \tag{4.93}$$

and

$$\begin{aligned} a_2(p, q) \; &= \; -\omega^2 \int_\Omega \varepsilon \nabla p \cdot \overline{\nabla q} \\ &+ \frac{1}{\mu_1} \int_{\Gamma_1} B_1 P[\nabla p] \cdot \overline{\nabla q} + \int_{\Gamma_2} \frac{1}{\mu_2} B_2 P[\nabla p] \cdot \overline{\nabla q}. \end{aligned} \tag{4.94}$$

By taking $v = u$, $q = p$, we deduce from (4.93), (4.94) that

$$\begin{aligned} & a_1(u, u) - a_2(p, p) = \int_\Omega d |\nabla \times u|^2 - \omega^2 \int_\Omega \varepsilon |u|^2 \\ & + \frac{1}{\mu_1} \int_{\Gamma_1} B_1 P[u] \overline{u} + \frac{1}{\mu_2} \int_{\Gamma_2} B_2 P[u] \cdot \overline{u} \\ & - \int_{\Gamma_1} \frac{1}{\mu_1} p \, \nabla_\Gamma \cdot (\overline{(B_1^* - B_1) P[v]}) - \int_{\Gamma_2} \frac{1}{\mu_2} p \, \nabla_\Gamma \cdot (\overline{(B_2^* - B_2) P[v]}) \\ & + \omega^2 \int_\Omega \varepsilon |\nabla p|^2 - \frac{1}{\mu_1} \int_{\Gamma_1} B_1 P[\nabla p] \cdot \overline{\nabla p} - \int_{\Gamma_2} \frac{1}{\mu_2} B_2 P[\nabla p] \cdot \overline{\nabla p} \\ & = \int_{\Gamma_1} f \cdot (\overline{u - \nabla p}). \end{aligned} \tag{4.95}$$

Thus, we have

$$
(4.96)\qquad
\begin{aligned}
&\Re\Big\{a_1(u,u)-a_2(p,p)\Big\}\ge d_0\|\nabla\times u\|^2_{L^2(\Omega)}+\nabla\times u\cdot\overline{u})\\
&-\omega^2\int_\Omega\varepsilon|u|^2+\Re\Big\{\frac{1}{\mu_1}\int_{\Gamma_1}B_1P[u]\overline{u}+\frac{1}{\mu_2}\int_{\Gamma_2}B_2P[u]\cdot\overline{u}\Big\}\\
&-\Re\Big\{\int_{\Gamma_1}\frac{1}{\mu_1}p\,\nabla_\Gamma\cdot\overline{((B_1^*-B_1)P[v])}+\int_{\Gamma_2}\frac{1}{\mu_2}p\,\nabla_\Gamma\cdot\overline{((B_2^*-B_2)P[v])}\Big\}\\
&+\omega^2\int_\Omega\varepsilon|\nabla p|^2-\Re\Big\{\frac{1}{\mu_1}\int_{\Gamma_1}B_1P[\nabla p]\cdot\overline{\nabla p}-\int_{\Gamma_2}\frac{1}{\mu_2}B_2P[\nabla p)\cdot\overline{\nabla p}\Big\}.
\end{aligned}
$$

We now estimate the terms on the right-hand side of (4.96) one by one.

It follows from the boundary condition (4.78) that

$$
\begin{aligned}
\Re\int_{\Gamma_j}\frac{1}{\mu_j}B_jP[u]\overline{u} &= \frac{1}{\mu_j}\sum_{n\in\Lambda_j^-}\Big\{|\beta_j^{(n)}||P[u^{(n)}]|^2-\frac{1}{|\beta_j^{(n)}|}|\alpha_n\cdot P[u^{(n)}]|^2\Big\}\\
&\ge \frac{1}{\mu_j}\sum_{n\in\Lambda_j^-}\frac{1}{|\beta_j^{(n)}|}(|\beta_j^{(n)}|^2-|\alpha_n|^2)|P[u^{(n)}]|^2\\
&\ge -\omega^2\varepsilon_j||\nu\times u||^2_{H^{-1/2}(\Gamma_j)},
\end{aligned}
$$

where to get the last estimate, we have used the expression (4.68). An application of Proposition 4.13 then leads to

$$
\Re\Big\{\int_{\Gamma_1}\frac{1}{\mu_1}B_1P[u]\overline{u}+\int_{\Gamma_2}\frac{1}{\mu_2}B_2P[u]\overline{u}\Big\}\ge-\eta\|\nabla\times u\|^2_{L^2(\Omega)}-C(\eta)||u||^2_{L^2(\Omega)}\,.
$$

We next estimate the term

$$
-\Re\Big\{\int_{\Gamma_j}\frac{1}{\mu_j}p\,\nabla_\Gamma\cdot\overline{((B_j^*-B_j)P[v])}\Big\}.
$$

From (4.78) and (4.79),

$$
\begin{aligned}
&\nabla_\Gamma\cdot((B_j^*-B_j)P[v])\\
=\;&\nabla_\Gamma\cdot\sum_{n\in\mathbb{Z}^2}\Big\{(\sqrt{-1}\beta_j^{(n)}+\sqrt{-1}\overline{\beta_j^{(n)}})(v_1^{(n)},v_2^{(n)},0)+(\frac{\sqrt{-1}}{\beta_j^{(n)}})\\
&+\frac{\sqrt{-1}}{\overline{\beta_j^{(n)}}})(\alpha_n\cdot v^{(n)})\alpha_n\Big\}e^{\sqrt{-1}(\frac{2\pi n_1}{\Lambda_1}x_1+\frac{2\pi n_2}{\Lambda_2}x_2)}\\
=\;&-\sum_{n\in\Lambda_j^+}2\Big\{|\beta_j^{(n)}|\alpha_n\cdot v^{(n)}+\frac{1}{|\beta_j^{(n)}|}\alpha_n\cdot v^{(n)}|\alpha_n|^2\Big\}e^{\sqrt{-1}(\frac{2\pi n_1}{\Lambda_1}x_1+\frac{2\pi n_2}{\Lambda_2}x_2)}.
\end{aligned}
$$

Thus

$$
\begin{aligned}
&-\Re\Big\{\int_{\Gamma_j}\frac{1}{\mu_j}p\,\nabla_\Gamma\cdot\overline{((B_j^*-B_j)P[v])}\Big\}\\
&=\Re\sum_{n\in\Lambda_j^+}\frac{2}{\mu_j}p^{(n)}\Big\{|\beta_j^{(n)}|\alpha_n\cdot\overline{v}^{(n)}+\frac{1}{|\beta_j^{(n)}|}\alpha_n\cdot\overline{v}^{(n)}|\alpha_n|^2\Big\}\\
&=\Re\sum_{n\in\Lambda_j^+}\Big\{2\omega^2\varepsilon_j|\beta_j^{(n)}|^{-1}p^{(n)}\alpha_n\cdot\overline{v}^{(n)}\Big\}\\
&\le C\|p\|_{H^{1/2}(\Gamma_j)}\|\nu\times v\|_{H^{-1/2}(\Gamma_j)}.
\end{aligned}
$$

Hence Proposition 4.13 and the trace theorem may be used once again to obtain that

$$
-\sum_{j=1,2}\Re\Big\{\int_{\Gamma_j}\frac{1}{\mu_j}p\,\nabla_\Gamma\cdot\overline{((B_j^*-B_j)P[v])}\Big\}\le\eta\|p\|^2_{H^1(\Omega)}+\eta\|\nabla\times v\|_{L^2(\Omega)}+C(\eta)\|v\|_{L^2(\Omega)}.
$$

Finally by Proposition 4.10

$$
-\Re\int_{\Gamma_j}\frac{1}{\mu_j}B_jP[\nabla p]\cdot\overline{\nabla p}=\Re\int_{\Gamma_j}\frac{1}{\mu_j}\nabla_\Gamma\cdot B_jP[\nabla p]\cdot\overline{p}\ge 0.
$$

Combining the above estimates, we have shown for any $u\in W$ and $p\in H^1$ that the following Garding type estimate holds:

$$
\Re\Big\{a_1(u,u)-a_2(p,p)\Big\}\ge C_1\|u\|^2_{H(\mathrm{curl},\Omega)}+C_2\|p\|^2_{H^1(\Omega)}-C_3(\|u\|^2_{L^2(\Omega)}+\|p\|^2_{L^2(\Omega)}).
$$

Denote the left-hand side of (4.82) by $a_\omega(E,F)$. Since the embedding from W to L^2 is compact and the dependence of the bilinear form $a(,)$ on ω is holomorphic outside a discrete set Λ (the set of resonances frequencies $\omega_j^{(n)}=\frac{1}{\varepsilon_j}|\alpha_n|^2, n\in\mathbb{Z}^2, j=1,2$), the meromorphic Fredholm theorem holds. To prove the theorem it suffices then to find a frequency $\omega\in C\setminus\Lambda$ such that the bilinear form $a_\omega(,)$ is injective. Let us choose $\omega=\sqrt{-1}\lambda$, for some positive constant λ. If $E\in H(\mathrm{curl},\Omega)$ is such that $a_{i\lambda}(E,F)=0$ for any $F\in H(\mathrm{curl},\Omega)$ then define the extension of E by

$$
\widetilde{E}=\begin{cases}E_1 \text{ in } \Omega_1,\\ E \text{ in } \Omega\,,\\ E_2 \text{ in } \Omega_2,\end{cases}
$$

where E_j $(j=1,2)$ is the unique solution in $H_{\mathrm{loc}}(\mathrm{curl},\Omega_j)$ of the Maxwell equations

$$\nabla\times\nabla\times E_j-\omega^2\varepsilon_j\mu_jE_j=0 \text{ in } \Omega_j\,, \tag{4.97}$$

$$E_j\times\nu=E\times\nu \text{ on } \Gamma_j\,, \tag{4.98}$$

(4.99) the radiation condition at the infinity.

From the (transparent) boundary condition satisfied by E on Γ_j it follows that

$$
[\widetilde{E}\times\nu]=[\frac{1}{\mu}\nabla\times\widetilde{E}\times\nu]=[\varepsilon\widetilde{E}\cdot\nu]=0
$$

on Γ_j. Moreover, since ω is a pure complex number, $\widetilde{E}$ is exponentially decaying as $|x_3|\to+\infty$. It follows that $\widetilde{E}$ is a solution in $H(\mathrm{curl},\mathbb{R}^3)$ (*i.e.*, of finite energy) of the homogeneous Maxwell equations and so,

$$\int_{\mathbb{R}^3} \frac{1}{\mu} |\nabla \times \widetilde{E}|^2 = 0,$$

which implies that $\widetilde{E} = 0$ in $\mathbb{R}^3$. The uniqueness of a solution to the problem for this particular choice of frequency ω gives the claim. The proof is complete. □

4.3.6. Energy Conservation. In this subsection we study the energy distribution for our diffraction problem. In general, the energy is distributed away from the grating structure through the propagating plane waves which consist of propagating reflected modes in Ω_1 and propagating transmitted modes in Ω_2. It is measured by the coefficients of each term of the sum (4.70).

Since no energy absorption takes place, the coefficients of propagating reflected plane waves are

$$\begin{aligned} r_n &= E^{(n)}(b) e^{-\sqrt{-1}\beta_1^{(n)} b} \quad \text{for } n \neq 0, n \in \Lambda_1^+, \\ r_0 &= U^{(0)}(b) e^{-\sqrt{-1}\beta_1^{(0)} b} \quad \text{for } n = 0, \end{aligned}$$

where again $\Lambda_1^+ = \{n \in \mathbb{Z}^2 : \ \Im(\beta_1^{(n)}) = 0\}$. Hence, the energy of each reflected mode may be defined as

$$\frac{\beta_1^{(n)} |r_n|^2}{\beta}$$

and the total energy of all reflected modes is

$$\mathcal{E}_r = \frac{1}{\beta} \sum_{n \in \Lambda_1^+} \beta_1^{(n)} |r_n|^2.$$

Similarly, the coefficients of each propagating transmitted mode are

$$t_n = E^{(n)}(-b) e^{-\sqrt{-1}\beta_2^{(n)} b} \quad \text{for } n \in \Lambda_2^+$$

where $\Lambda_2^+ = \{n \in \mathbb{Z}^2 : \ \Im(\beta_2^{(n)}) = 0\}$. The energy of each transmitted mode is defined by

$$\frac{\mu_1 \beta_2^{(n)} |t_n|^2}{\mu_2 \beta}$$

and the total energy of all transmitted modes is

$$\mathcal{E}_t = \frac{\mu_1}{\mu_2 \beta} \sum_{n \in \Lambda_2^+} \beta_2^{(n)} |t_n|^2.$$

REMARK 4.15. *In optics literature, the numbers $\mathcal{E}_r$ and $\mathcal{E}_t$ are called reflected and transmitted* efficiencies, *respectively. They represent the proportion of energy distributed in each propagating mode. The sum of reflected and transmitted efficiency is referred to as the grating* efficiency [**450**].

The following result states that in the case of no energy absorption the total energy is conserved, *i.e.*, the incident energy is the same as the total energy of the propagating waves.

THEOREM 4.16. *Assume that the material coefficients $\varepsilon_0(x)$, ε_1, ε_2, $\mu(x)$, μ_1, and μ_2 are all real and positive. Then*

$$\mathcal{E}_r + \mathcal{E}_t = |s|^2 .$$

Thus the total energy that leaves the medium is the same as that of the incident field.

PROOF. Multiplying both sides of the equation (4.80) by $\overline{E}$ and integrating it over Ω, we obtain

$$\begin{aligned}(4.100)\quad &\int_\Omega d|\nabla \times E|^2 \\ &-\int_\Omega \omega^2\varepsilon|E|^2 + \int_{\Gamma_1}\frac{1}{\mu_1}B_1P[E]\cdot\overline{E} + \int_{\Gamma_2}\frac{1}{\mu_2}B_2P[E]\cdot\overline{E} = \int_{\Gamma_1} f\cdot\overline{(}E),\end{aligned}$$

where f is defined by (4.81).

Taking the imaginary part of (4.100), we get

$$\sum_{n\in\Lambda_1^+}\frac{1}{\mu_1}\beta_1^{(n)}|E^{(n)}|^2 + \sum_{n\in\Lambda_2^+}\frac{1}{\mu_2}\beta_2^{(n)}|E^{(n)}|^2 = \frac{1}{\mu_1}\Im\left(2\sqrt{-1}\beta\int_{\Gamma_1} s\cdot\overline{E}e^{-\sqrt{-1}\beta b}\,dx\right).$$

The proof is completed by noting that

$$\begin{aligned}|r_0|^2 &= |U^{(0)}e^{-\sqrt{-1}\beta b}|^2 = |(E^{(0)} - (E^i)^{(0)})e^{-\sqrt{-1}\beta b}|^2 \\ &= |E^{(0)}|^2 - 2\Re(s\cdot\overline{E^{(0)}}e^{-\sqrt{-1}\beta b}) + |s|^2.\end{aligned}$$

□

4.4. Boundary Integral Formulations

The boundary integral equation method was one of the first numerical methods used in grating theory. It has been used for the investigation of diffraction gratings of different kinds. In this section we present boundary integral formulations for scattering problems by dielectric periodic and biperiodic gratings.

4.4.1. Dielectric Periodic Gratings. In this subsection we establish an integral formulation for the diffraction problem from a one-dimensional dielectric grating. We consider (4.36) and (4.37) subject to the quasi-periodic radiation conditions on u^s derived in Subsection 4.2.5. As before, we denote the period Λ and let $\Gamma = \{x_2 = f(x_1)\}/(\Lambda\mathbb{Z}\setminus\{0\})$.

We introduce the quasi-periodic Green's function for the grating, which satisfies

$$(4.101)\qquad (\Delta + k^2)G^{\alpha,k}(x,y) = \sum_{n\in\mathbb{Z}}\delta_0(x - y - (n\Lambda, 0))e^{\sqrt{-1}n\alpha\Lambda}.$$

We have

$$(4.102)\qquad G^{\alpha,k}(x,y) = -\frac{\sqrt{-1}}{4}\sum_{n\in\mathbb{Z}}H_0^{(1)}(k|x - (n\Lambda,0) - y|)e^{\sqrt{-1}n\alpha\Lambda},$$

where $H_0^{(1)}$ is the Hankel function of the first kind of order 0.

If $k \neq |\alpha_n|, \forall\ n \in \mathbb{Z}$, where α_n is defined by (4.19), then by using Poisson's summation formula

$$(4.103)\qquad \sum_{n\in\mathbb{Z}} e^{\sqrt{-1}(\frac{2\pi n}{\Lambda}+\alpha)x_1} = \sum_{n\in\mathbb{Z}}\delta_0(x_1 - n\Lambda)e^{\sqrt{-1}n\alpha\Lambda},$$

we can equivalently represent $G^{\alpha,k}$ as

$$(4.104)\qquad G^{\alpha,k}(x,y) = \sum_{n\in\mathbb{Z}}\frac{e^{\sqrt{-1}\alpha_n(x_1-y_1)+\sqrt{-1}\beta_n(x_2-y_2)}}{k^2 - \alpha_n^2},$$

where β_n is given by

$$\beta_n = \begin{cases} \sqrt{k^2 - \alpha_n^2} & k^2 > \alpha_n^2, \\ \sqrt{-1}\sqrt{\alpha_n^2 - k^2} & k^2 < \alpha_n^2. \end{cases} \tag{4.105}$$

Let $\mathcal{S}_\Gamma^{\alpha,k}$ be the quasi-periodic single-layer potential associated with $G^{\alpha,k}$ on Γ; that is, for a given density $\varphi \in L^2(\Gamma)$,

$$\mathcal{S}_\Gamma^{\alpha,k}[\varphi](x) = \int_\Gamma G^{\alpha,k}(x,y)\varphi(y)\,d\sigma(y), \quad x \in \mathbb{R}^2.$$

Analogously to (2.172), u can be represented using the single-layer potentials $\mathcal{S}_\Gamma^{\alpha,k_1}$ and $\mathcal{S}_\Gamma^{\alpha,k_2}$ as follows:

$$u(x) = \begin{cases} u^i(x) + \mathcal{S}_\Gamma^{\alpha,k_1}[\psi](x), & x \in \{x = (x_1,x_2) : x_2 > f(x_1)\}, \\ \mathcal{S}_\Gamma^{\alpha,k_2}[\varphi](x), & x \in \{x = (x_1,x_2) : x_2 < f(x_1)\}, \end{cases} \tag{4.106}$$

where the pair $(\varphi,\psi) \in L^2(\Gamma) \times L^2(\Gamma)$ satisfies

$$\begin{cases} \mathcal{S}_\Gamma^{\alpha,k_2}[\varphi] - \mathcal{S}_\Gamma^{\alpha,k_1}[\psi] = u^i \\ \left.\dfrac{\partial(\mathcal{S}_\Gamma^{\alpha,k_2}[\varphi])}{\partial\nu}\right|_- - \left.\dfrac{\partial(\mathcal{S}_\Gamma^{\alpha,k_1}[\psi])}{\partial\nu}\right|_+ = \dfrac{\partial u^i}{\partial\nu} \end{cases} \quad \text{on } \Gamma. \tag{4.107}$$

THEOREM 4.17. *For all but possibly a countable set of frequencies ω_j, $\omega_j \to +\infty$, the system of integral equations (4.107) has a unique solution*

$$(\varphi,\psi) \in H^{-1/2}(\Gamma) \times H^{-1/2}(\Gamma).$$

PROOF. Since the Fredholm alternative applies for (4.107), it is enough to prove uniqueness. Let $(\varphi,\psi) \in H^{-1/2}(\Gamma) \times H^{-1/2}(\Gamma)$ be a solution to (4.107) and let v be given by (4.106) with $u^i = 0$. Then, v satisfies the variational problem (4.59) and Theorem 4.7 yields that for all but a discrete set of ω, $v = 0$. □

4.4.2. Dielectric Biperiodic Gratings. We consider the diffraction problem in Subsection 4.3.3. Assume that the biperiodic grating is expressed by $x_3 = f(x_1,x_2)$. We denote by $\Gamma = \{x_3 = f(x_1,x_2)\}/((\Lambda_1\mathbb{Z} \setminus \{0\}) \times \Lambda_2\mathbb{Z} \setminus \{0\}))$, where λ_j is the period of the grating in the direction x_j for $j = 1,2$. Suppose that

$$\begin{aligned} \varepsilon(x) &= \varepsilon_1, \ \mu(x) = \mu_1 \text{ for } x_3 > f(x_1,x_2), \\ \varepsilon(x) &= \varepsilon_2, \ \mu(x) = \mu_2 \text{ for } x_3 < f(x_1,x_2), \end{aligned}$$

where ε_1, ε_2, μ_1, and μ_2 are positive constants.

Analogously to (2.333), the electric field E can be represented as

$$E(x) = \begin{cases} E^i(x) + \mu_m \nabla \times \vec{\mathcal{S}}_\Gamma^{\alpha,k_1}[\varphi](x) + \nabla \times \nabla \times \vec{\mathcal{S}}_\Gamma^{\alpha,k_1}[\psi](x), \\ \qquad x \in \{x = (x_1,x_2,x_3) : x_3 > f(x_1,x_2)\}, \\ \mu_c \nabla \times \vec{\mathcal{S}}_\Gamma^{\alpha,k_2}[\varphi](x) + \nabla \times \nabla \times \vec{\mathcal{S}}_\Gamma^{\alpha,k_2}[\psi](x), \\ \qquad x \in \{x = (x_1,x_2,x_3) : x_3 < f(x_1,x_2)\}, \end{cases} \tag{4.108}$$

where the pair $(\varphi, \psi) \in \left(H_T^{-\frac{1}{2}}(\text{div}, \Gamma)\right)^2$ satisfies

(4.109)

$$\begin{pmatrix} \frac{\mu_2+\mu_1}{2} I + \mu_2 \mathcal{M}_\Gamma^{\alpha,k_2} - \mu_1 \mathcal{M}_\Gamma^{\alpha,k_1} & \mathcal{L}_\Gamma^{\alpha,k_2} - \mathcal{L}_\Gamma^{\alpha,k_1} \\ \mathcal{L}_\Gamma^{\alpha,k_2} - \mathcal{L}_\Gamma^{\alpha,k_1} & \left(\frac{k_2^2}{2\mu_2} + \frac{k_1^2}{2\mu_1}\right) I + \frac{k_m^2}{\mu_2} \mathcal{M}_\Gamma^{\alpha,k_2} - \frac{k_1^2}{\mu_1} \mathcal{M}_\Gamma^{\alpha,k_1} \end{pmatrix} \begin{bmatrix} \varphi \\ \psi \end{bmatrix}$$
$$= \begin{bmatrix} \nu \times E^i \\ \sqrt{-1}\omega \nu \times H^i \end{bmatrix} \Bigg|_\Gamma ,$$

where $\mathcal{M}_\Gamma^{\alpha,k}$ and $\mathcal{L}_\Gamma^{\alpha,k}$ are respectively defined by (2.318) and (2.319) with Γ_k replaced with $G^{\alpha,k}$ and ∂D with Γ.

The following result can be proved similarly to Theorem 4.17.

THEOREM 4.18. *For all but possibly a countable set of frequencies ω_j, $\omega_j \to +\infty$, the system of integral equations (4.109) has a unique solution*

$$(\varphi, \psi) \in \left(H_T^{-\frac{1}{2}}(\text{div}, \Gamma)\right)^2.$$

4.5. Optimal Design of Grating Profiles

For simplicity, we focus the presentation to the dielectric periodic gratings described in Subsection 4.4.1. Assume that the material above the periodic interface $S := \{x_2 = f(x_1)\}$ has refractive index k_1 and the material below the interface has index k_2. Both k_1 and k_2 are assumed to be real. Define

$$a_S(x) = \begin{cases} k_1^2 & \text{if } x \text{ is above } S, \\ k_2^2 & \text{if } x \text{ is below } S. \end{cases}$$

Let $b > \max |f(x_1)|$ and let an incoming plane wave

$$u^i = e^{\sqrt{-1}\alpha x_1 - \sqrt{-1}\beta x_3}$$

be incident on S from $x_2 > f(x_1)$ with

$$\alpha = \omega k_1 \sin\theta, \quad \beta = \omega k_1 \cos\theta,$$

and $-\pi/2 < \theta < \pi/2$ being the angle of incidence.

Then consider the scattering problem

$$(\Delta_\alpha + a_S)u = 0 \quad \text{in } \{-b < x_2 < b\}, \tag{4.110}$$

$$(T_1^\alpha - \frac{\partial}{\partial x_2})u = 2\sqrt{-1}\beta e^{-\sqrt{-1}\beta b} \quad \text{on } \{x_2 = b\}, \tag{4.111}$$

$$(T_2^\alpha - \frac{\partial}{\partial x_2})u = 0 \quad \text{on } \{x_2 = -b\}, \tag{4.112}$$

where $\Delta_\alpha = \triangle + 2\sqrt{-1}\alpha\partial_{x_1} - \alpha^2$, and periodic boundary conditions are assumed in x_1. The operators T_j^α, $j = 1, 2$, are defined by (4.54).

Suppose that the materials, the period of the structure, and the frequency of the incoming waves are fixed. There are then a fixed number of propagating modes, each of which corresponds to an index n for which the propagation constant β_j^n is real-valued. Let us define the set of indices of the reflected propagating modes

$$P_r = \{n \in \mathbb{Z} : \beta_1^n(\alpha) \in \mathbb{R}\},$$

and indices of transmitted modes

$$P_t = \{m \in \mathbb{Z} : \beta_2^m(\alpha) \in \mathbb{R}\}.$$

The coefficients of each propagating reflected mode are determined by the trace of the solution u on the artificial boundary $\{x_2 = b\}$:

$$\begin{aligned} r_n &= u_n(b)e^{-\sqrt{-1}\beta_1 b} & \text{for } n \neq 0, \quad n \text{ in } P_r, \\ r_0 &= u_0(b)e^{-\sqrt{-1}\beta b} - \text{constant} & \text{for } n = 0, \end{aligned}$$

where

$$u_n(x_2) = \frac{1}{\Lambda}\int_0^\Lambda u(x_1, x_2)e^{-\sqrt{-1}\frac{2\pi n}{\Lambda}x_1}\,dx_1.$$

Similarly, the coefficients of the propagating transmitted modes are

$$t_m = u_m(-b)e^{-\sqrt{-1}\beta_2 b} \qquad \text{for } m \text{ in } P_t.$$

Writing the reflection and transmission coefficients as vectors

$$r = (r_n)_{n\in P_r}, \quad t = (t_m)_{m\in P_t},$$

denote the pair $(r, t) = F$. The coefficients r_n and t_m, and hence F, are functions of the interface profile S. Denote this dependence by $F(S)$.

A general optimal design problem is to find a profile S such that $F(S)$ is as close as possible to some specified diffraction pattern g. Asking that $F(S)$ is close to g in a least-squares sense, one obtains the problem

$$\min_{S\in\mathcal{S}} J(S) = \|F(S) - g\|^2, \tag{4.113}$$

where $\mathcal{S}$ is some admissible class of profiles. One could of course generalize further and specify a range of incidence angles or a range of frequencies (or both).

To implement the least squares approach, we calculate the gradient of the cost functional (4.113) with respect to the interface profile S. The representation formula (4.106) is useful. From Section 2.7, the calculation of the sensitivity of u on $x_2 = b$ with respect to changes of the profile is straightforward and therefore, the Fréchet derivative of J can be easily obtained.

4.6. Numerical Implementation

In this section we use the boundary integral representation of the dielectric periodic grating described in Subsection 4.4.1 to numerically determine the electric field in the case of a periodic array of spherical particles located on the x_1-axis. Denote by Ω_1 and Ω_2 the region outside the particles and the region representing the particles, respectively. Let ε_j and μ_j represent the corresponding material parameters. Let $k_j = \omega\sqrt{\varepsilon_j\mu_j}$ $(j = 1, 2)$ be the wavenumber outside and inside the particles, respectively.

The discretization of the system is performed in precisely the same manner as described in Subsection 2.13.4 and leads to the system of equations

$$\begin{pmatrix} S_- & -S_+ \\ \frac{1}{\mu_2}S'_- & -\frac{1}{\mu_1}S'_+ \end{pmatrix}\begin{pmatrix} \overline{\varphi} \\ \overline{\psi} \end{pmatrix} = \begin{pmatrix} u_d \\ \frac{1}{\mu_1}u_n \end{pmatrix},$$

where $S_\pm$ and $S'_\pm$ are $N \times N$ matrices given by

$$(S_-)_{ij} = G^{\alpha,k_2}(x^{(i)} - x^{(j)})|T(x^{(j)})|(t_{j+1} - t_j), \tag{4.114}$$

$$(S_+)_{ij} = G^{\alpha,k_1}(x^{(i)} - x^{(j)})|T(x^{(j)})|(t_{j+1} - t_j), \tag{4.115}$$

$$(S'_-)_{ij} = -\frac{1}{2}\delta_{ij} + \frac{\partial G^{\alpha,k_2}}{\partial \nu_x}(x^{(i)} - x^{(j)})|T(x^{(j)})|(t_{j+1} - t_j), \tag{4.116}$$

$$(S'_+)_{ij} = \frac{1}{2}\delta_{ij} + \frac{\partial G^{\alpha,k_1}}{\partial \nu_x}(x^{(i)} - x^{(j)})|T(x^{(j)})|(t_{j+1} - t_j), \tag{4.117}$$

for $i \neq j$ and $i, j = 1, 2, \ldots, N$, and where $G^{\alpha,k}$ is the quasi-periodic Green's function defined by (4.101). Once we solve this system for the density functions $\overline{\varphi}$ and $\overline{\psi}$, the electric field can be calculated using

$$u(x) = \begin{cases} u_d(x) + S_+[\overline{\psi}](x), & x \in \Omega_1\,, \\ S_-[\overline{\varphi}](x), & x \in \Omega_2\,. \end{cases} \tag{4.118}$$

Since $G^{\alpha,k}$ is extremely slow to converge we must use the Ewald representation of the Green's function to accelerate the convergence; see Subsection 2.13.3. Recall that the Ewald representation of the quasi-periodic Green's function is given by

$$G^{\alpha,k}(x,y) = G^{\alpha,k}_{\text{spec}}(x,y) + G^{\alpha,k}_{\text{spat}}(x,y),$$

with

$$\begin{aligned} G^{\alpha,k}_{\text{spec}}(x,y) = &-\frac{1}{4\Lambda}\sum_{n\in\mathbb{Z}} \frac{e^{-\sqrt{-1}\alpha_n(x_1-y_1)}}{\sqrt{-1}\beta_n} \\ &\times \left[e^{\sqrt{-1}\beta_n|x_2-y_2|}\operatorname{erfc}\left(\frac{\sqrt{-1}\beta_n}{2E} + |x_2-y_2|\mathcal{E}\right)\right. \\ &\left. + e^{-\sqrt{-1}\beta_n|x_2-y_2|}\operatorname{erfc}\left(\frac{\sqrt{-1}\beta_n}{2E} - |x_2-y_2|\mathcal{E}\right)\right], \end{aligned}$$

$$G^{\alpha,k}_{\text{spat}}(x,y) = -\frac{1}{4\pi}\sum_{m\in\mathbb{Z}} e^{\sqrt{-1}\alpha m\Lambda} \sum_{q=0}^{\infty}\left(\frac{k}{2\mathcal{E}}\right)^{2q}\frac{1}{q!}E_{q+1}(R_m^2\mathcal{E}^2),$$

where $\alpha_n = -\alpha + \frac{2\pi n}{\Lambda}$, $\beta_n = -\sqrt{k^2 - \alpha_n^2}$, erfc$(z)$ is the complementary error function

$$\operatorname{erfc}(z) = \frac{2}{\sqrt{\pi}}\int_z^\infty e^{-t^2}\,dt,$$

and E_q is the qth order exponential integral which is defined as

$$E_q(z) = \int_1^\infty \frac{e^{-zt}}{t^q}\,dt.$$

We set $\Lambda = 1$ and the radius of the particles to be 0.4. We set the incident plane wave to be $u^i(x_1, x_2) = 3e^{\sqrt{-1}(\alpha x_1 - \beta x_2)}$ where $\alpha = k_1\sin(\theta)$, $\beta = k_1\cos(\theta)$ with $\theta = \pi/8$. As we are considering a non-magnetic material we set the permeability to be $\mu_1 = \mu_2 = 1$. For the permittivity we set $\varepsilon_1 = 1$ and $\varepsilon_2 = 5$. We set the operating frequency to be $\omega = 1$. The resulting incident, scattered, and total fields are shown in Figure 4.4. The corresponding Matlab code can be downloaded at `www.ams.org/bookpages/surv-235`.

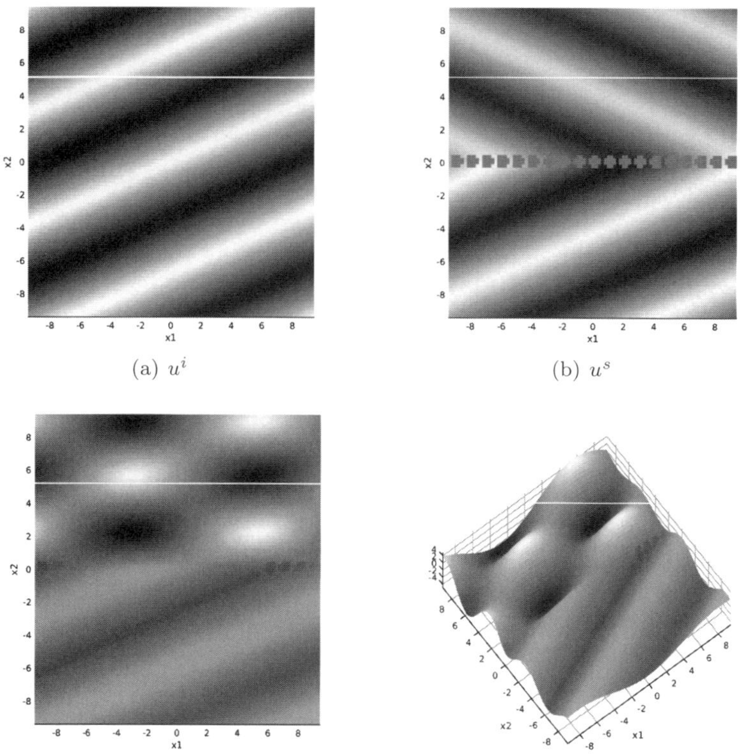

(a) u^i (b) u^s

(c) u (d) u

FIGURE 4.4. The incident electric field, scattered electric field, and total electric field for a dielectric grating consisting of a periodic array of spherical particles on the x_1-axis.

4.7. Concluding Remarks

In this chapter we have established uniqueness and existence results for solutions to electromagnetic scattering problems by gratings. We refer the reader to [**102**, **229**–**231**] for the analysis of the scattering of elastic waves by diffraction gratings. The results of Section 2.7 together with those in the previous section can be used to perform optimal design of periodic interfaces that give rise to a specified diffraction pattern [**19**, **123**, **138**, **139**]. Note that by making assumption (4.47), we have excluded Wood anomalies. In [**167**, **168**], a methodology based on use of a certain shifted Green function is introduced. It provides a solver for problems of scattering by gratings which is valid and accurate, in particular, at and around Wood anomaly frequencies, at which the quasi-periodic Green function ceases to exist. In [**177**, **178**], this new approach has been extended to the computation of two-dimensional periodic Green's functions in three dimensions. In [**265**], a different approach to improve the robustness of the integral formulation at and near Wood anomalies is combined with a fast multipole scheme for solving scattering problems by infinite periodic arrays of obstacles.

CHAPTER 5

Photonic Band Gaps

5.1. Introduction

Photonic crystals are structures constructed from electromagnetic materials arranged in a periodic array. They have attracted enormous interest in the last decade because of their unique optical and electromagnetic properties. Such structures have been found to exhibit interesting spectral properties with respect to classical wave propagation, including the appearance of band gaps [**313**, **465**, **510**].

In order to study the propagation of light in a photonic crystal, we shall use the Maxwell equations. In general the electromagnetic fields are complicated functions of time and space. If the field configurations are built up of harmonic electromagnetic waves that are transverse, we can reduce the Maxwell equations to two scalar Helmholtz equations. Throughout this chapter, we will focus on this scalar model which is also the underlying model for the acoustic analog of photonic crystals.

Our aim is to analyze the contrast and geometry dependence of the band gap of the frequency spectrum for waves in photonic crystals. We consider photonic crystals consisting of a background medium which is perforated by an array of arbitrary-shaped holes periodic along each of the two orthogonal coordinate axes in the plane. The background medium is of higher index. It has been proved that the high contrast of a photonic crystal favors spectral gaps; see [**246**, **294**, **295**, **315**, **475**, **518**, **519**].

In this chapter we adopt the high-contrast model to give a full understanding of the relationship between variations in the index ratio or in the geometry of the holes and variations in the band gap structure of the photonic crystal. We provide such a high-order sensitivity analysis using a boundary integral approach with rigorous justification based on the generalized Rouché theorem.

Carrying out a band structure calculation for a given photonic crystal involves a family of eigenvalue problems, as the quasi-momentum is varied over the first Brillouin zone. We show that these eigenvalues are the characteristic values of meromorphic operator-valued functions that are of Fredholm type of index zero. We then proceed from the generalized Rouché theorem to construct their complete asymptotic expressions as the index ratio goes to infinity. We also provide their complete expansions in terms of infinitesimal changes in the geometry of the holes.

Our integral formulation in this chapter of the photonic band gap problem offers an efficient approach to the computation of the band gap structure which is based on a combination of boundary element methods and Muller's method described in Section 1.6.

In this chapter we confine our attention to the two-dimensional case to demonstrate our approach and results. The asymptotic results for the band gap structure with respect to the index ratio and the geometry of the holes can be obtained in three dimensions with only minor modifications of the techniques presented here.

5.2. Floquet Transform

In this section, the Floquet transform, which in the periodic case plays the role of the Fourier transform, is established and the structure of spectra of periodic elliptic operators is discussed.

Let $f(x)$ be a function decaying sufficiently fast. We define the Floquet transform of f as follows:

$$\mathcal{U}[f](x,\alpha) = \sum_{n\in\mathbb{Z}^d} f(x-n)e^{\sqrt{-1}\alpha\cdot n}. \tag{5.1}$$

This transform is an analogue of the Fourier transform for the periodic case. The parameter α is called the quasi-momentum, and it is an analogue of the dual variable in the Fourier transform. If we shift x by a period $m \in \mathbb{Z}^d$, then we get the Floquet condition

$$\mathcal{U}[f](x+m,\alpha) = e^{\sqrt{-1}\alpha\cdot m}\mathcal{U}[f](x,\alpha), \tag{5.2}$$

which shows that it suffices to know the function $\mathcal{U}[f](x,\alpha)$ on the unit cell $Y := [0,1)^d$ in order to recover it completely as a function of the x-variable. Moreover, $\mathcal{U}[f](x,\alpha)$ is periodic with respect to the quasi-momentum α:

$$\mathcal{U}[f](x,\alpha+2\pi m) = \mathcal{U}[f](x,\alpha), \quad m \in \mathbb{Z}^d. \tag{5.3}$$

Therefore, α can be considered as an element of the torus $\mathbb{R}^d/(2\pi\mathbb{Z}^d)$. Another way of saying this is that all information about $\mathcal{U}[f](x,\alpha)$ is contained in its values for α in the fundamental domain B of the dual lattice $2\pi\mathbb{Z}^d$. This domain is referred to as the (first) Brillouin zone.

The following result is an analogue of the Plancherel theorem when one uses the Fourier transform. Suppose that the measures $d\alpha$ and the dual torus $\mathbb{R}^d/(2\pi\mathbb{Z}^d)$ are normalized. The following theorem holds. See [**341**] for a proof.

THEOREM 5.1 (Plancherel-type theorem). *The transform*

$$\mathcal{U} : L^2(\mathbb{R}^d) \to L^2(\mathbb{R}^d/(2\pi\mathbb{Z}^d), L^2(Y))$$

is isometric. Its inverse is given by

$$\mathcal{U}^{-1}[g](x) = \int_{\mathbb{R}^d/(2\pi\mathbb{Z}^d)} g(x,\alpha)\, d\alpha,$$

*where the function $g(x,\alpha) \in L^2(\mathbb{R}^d/(2\pi\mathbb{Z}^d), L^2(Y))$ is extended from Y to all $x \in \mathbb{R}^d$ according to the Floquet condition (*5.2*).*

The books [**324, 341, 461**] give more detailed treatments of this subject.

5.3. Structure of Spectra of Periodic Elliptic Operators

In this section we will briefly discuss spectral properties of periodic elliptic operators. See [**340, 341, 461, 484**] for details and references.

Consider a linear partial differential operator $L(x,\partial_x)$, whose coefficients are periodic with respect to $\mathbb{Z}^d, d = 2,3$. A natural question is about the type of spectrum (absolutely continuous, singular continuous, point) of L (see Appendix A). It is not hard to prove that for a periodic elliptic operator of any order, the singular continuous spectrum is empty. For any second-order periodic operator of elliptic type, it is likely that no eigenvalues can arise. Although it has been unanimously believed by physicists for a long time, proving this statement turns out to be a difficult mathematical problem. See [**340**].

Due to periodicity, the operator commutes with the Floquet transform

$$\mathcal{U}[Lf](x,\alpha) = L(x,\partial_x)\mathcal{U}[f](x,\alpha).$$

For each α, the operator $L(x,\partial_x)$ now acts on functions satisfying the corresponding Floquet condition (5.2). In other words, although the differential expression of the operator stays the same, its domain changes with α. Denoting this operator by $L(\alpha)$, we see that the Floquet transform $\mathcal{U}$ expands the periodic partial differential operator L in $L^2(\mathbb{R}^d)$ into the direct integral of operators

$$\int_{\mathbb{R}^d/(2\pi\mathbb{Z}^d)}^{\oplus} L(\alpha)\, d\alpha. \tag{5.4}$$

The key point in the direct fiber decomposition (5.4) is that the operators $L(\alpha)$ act on functions defined on a torus, while the original operator acts in $\mathbb{R}^d$.

If L is a self-adjoint operator, one can prove the main spectral statement:

$$\sigma(L) = \bigcup_{\alpha\in B} \sigma(L(\alpha)), \tag{5.5}$$

where σ denotes the spectrum.

If L is elliptic, the operators $L(\alpha)$ have compact resolvents and hence discrete spectra. If L is bounded from below, the spectrum of $L(\alpha)$ accumulates only at $+\infty$. Denote by $\mu_n(\alpha)$ the nth eigenvalue of $L(\alpha)$ (counted in increasing order with their multiplicity). The function $\alpha \mapsto \mu_n(\alpha)$ is continuous in B. It is one branch of the dispersion relations and is called a band function. We conclude that the spectrum $\sigma(L)$ consists of the closed intervals (called the spectral bands)

$$\left[\min_\alpha \mu_n(\alpha), \max_\alpha \mu_n(\alpha)\right],$$

where $\min_\alpha \mu_n(\alpha) \to +\infty$ when $n \to +\infty$. In dimension $d \geq 2$, the spectral bands normally do overlap, which makes opening gaps in the spectrum of L a mathematically hard problem. But, it is still conceivable that at some locations the bands might not overlap and hence open a gap in the spectrum. It is commonly believed that the number of gaps one can open in a periodic medium in dimension $d \geq 2$ is finite. In the case of the periodic Schrödinger operator, this constitutes the Bethe-Sommerfeld conjecture. Since the major and inspirational work by Skriganov [**483**], significant progress has been made on this problem. See, for example, [**324, 325, 340, 442–444**].

5.4. Boundary Integral Formulation

5.4.1. Problem Formulation. The photonic crystal we consider in this chapter consists of a homogeneous background medium of constant index k which is perforated by an array of arbitrary-shaped holes periodic along each of the two orthogonal coordinate axes in $\mathbb{R}^2$. These holes are assumed to be of index 1. We assume that the structure has unit periodicity and define the unit cell $Y := [0, 1]^2$.

We seek eigenfunctions u of

$$(5.6)\qquad \begin{cases} \nabla \cdot (1 + (k-1)\chi(Y \setminus \overline{D}))\nabla u + \omega^2 u = 0 & \text{in } Y, \\ e^{-\sqrt{-1}\alpha \cdot x} u \text{ is periodic in the whole space,} \end{cases}$$

where $\chi(Y \setminus \overline{D})$ is the indicator function of $Y \setminus \overline{D}$. Problem (5.6) can be rewritten as

$$(5.7)\qquad \begin{cases} k\Delta u + \omega^2 u = 0 & \text{in } Y \setminus \overline{D}, \\ \Delta u + \omega^2 u = 0 & \text{in } D, \\ u|_+ = u|_- & \text{on } \partial D, \\ k\dfrac{\partial u}{\partial \nu}\Big|_+ = \dfrac{\partial u}{\partial \nu}\Big|_- & \text{on } \partial D, \\ e^{-\sqrt{-1}\alpha \cdot x} u \text{ is periodic in the whole space.} \end{cases}$$

For each quasi-momentum variable α, let $\sigma_\alpha(D, k)$ be the (discrete) spectrum of (5.6). Then the spectral band of the photonic crystal is given by

$$\bigcup_{\alpha \in [0, 2\pi]^2} \sigma_\alpha(D, k).$$

We shall investigate the behavior of $\sigma_\alpha(D, k)$ when $k \to +\infty$ in Section 5.5 and that under perturbation of D in Section 5.7.

Note first that if D is invariant under the transformations

$$(5.8)\quad (x_1, x_2) \mapsto (-x_1, -x_2), \qquad (x_1, x_2) \mapsto (-x_1, x_2), \qquad (x_1, x_2) \mapsto (x_2, x_1),$$

then all possible eigenvalues associated with (5.7) for any $\alpha \in [0, 2\pi]^2$ must occur with α restricted to the triangular region (the reduced Brillouin zone)

$$(5.9)\qquad T := \Big\{ \alpha = (\alpha_1, \alpha_2) : 0 \le \alpha_1 \le \pi, 0 \le \alpha_2 \le \alpha_1 \Big\}.$$

Consequently, to search for band gaps associated with D with the symmetries (5.8), it suffices to take $\alpha \in T$ rather than $\alpha \in [0, 2\pi]^2$.

Note also that a change of variables $x' = sx$ and a simultaneous change of the spectral parameter $\omega' = s\omega$ reduce the problem (5.7) to the similar one with the rescaled material property $(1 + (k-1)\chi(sY \setminus \overline{sD}))$. This means that in rescaling the material property of a medium, we do not need to recompute the spectrum, since its simple rescaling would suffice. Another important scaling property deals with the values of the material property. It is straightforward to compute that if we multiply the material property by a scaling factor s, the spectral problem for the new material parameter $s(1+(k-1)\chi(sY \setminus \overline{sD}))$ can be reduced to the old one by rescaling the eigenvalues according to the formula $\omega' = \sqrt{s}\omega$. These two scaling properties mean that there is no fundamental length nor a fundamental material property value for the spectral problem (5.7) [**340**].

Suppose now that ω^2 is not an eigenvalue of $-\Delta$ in $Y \setminus \overline{D}$ with the Dirichlet boundary condition on ∂D and the quasi-periodic condition on ∂Y and ω^2/k is not an eigenvalue of $-\Delta$ in D with the Dirichlet boundary condition. Following the same argument as in (3.3), one can show that the solution u to (5.6) can be represented as

$$(5.10)\qquad u(x) = \begin{cases} \mathcal{S}^{\alpha,\omega}[\phi](x), & x \in D, \\ H(x) + \mathcal{S}^{\alpha, \frac{\omega}{\sqrt{k}}}[\psi](x), & x \in Y \setminus \overline{D}, \end{cases}$$

for some densities ϕ and ψ in $L^2(\partial D)$, where the function H is given by

$$H(x) = -\mathcal{S}_Y^{\alpha, \frac{\omega}{\sqrt{k}}}[\frac{\partial u}{\partial \nu}|_{\partial Y}] + \mathcal{D}_Y^{\alpha, \frac{\omega}{\sqrt{k}}}[u|_{\partial Y}], \quad x \in Y.$$

Here, the quasi-periodic single- and double-layer potentials are introduced in Section 2.12. In order to keep the notation simple, we use $\mathcal{S}^{\alpha,\omega}$ and $\mathcal{D}^{\alpha,\omega}$ instead of $\mathcal{S}_D^{\alpha,\omega}$ and $\mathcal{D}_D^{\alpha,\omega}$ for layer potentials on D.

Now by (2.298) we have $H \equiv 0$, and hence

$$u(x) = \begin{cases} \mathcal{S}^{\alpha,\omega}[\phi](x), & x \in D, \\ \mathcal{S}^{\alpha, \frac{\omega}{\sqrt{k}}}[\psi](x), & x \in Y \setminus \overline{D}. \end{cases} \tag{5.11}$$

A proof of the representation formula (5.11) will be given later in Section 5.8.

In view of the transmission conditions in (5.7), the pair $(\phi, \psi) \in L^2(\partial D) \times L^2(\partial D)$ satisfies the following system of integral equations:

$$\begin{cases} \mathcal{S}^{\alpha,\omega}[\phi] - \mathcal{S}^{\alpha, \frac{\omega}{\sqrt{k}}}[\psi] = 0 & \text{on } \partial D, \\ \Big(-\frac{1}{2}I + (\mathcal{K}^{-\alpha,\omega})^* \Big)[\phi] - k\Big(\frac{1}{2}I + (\mathcal{K}^{-\alpha, \frac{\omega}{\sqrt{k}}})^*\Big)[\psi] = 0 & \text{on } \partial D. \end{cases} \tag{5.12}$$

The converse is also true. If $(\phi, \psi) \in L^2(\partial D) \times L^2(\partial D)$ is a nonzero solution of (5.12), then u given by (5.11) is an eigenfunction of (5.6) associated to the eigenvalue ω^2.

Suppose $\alpha \neq 0$. Let $\mathcal{A}^{\alpha,k}(\omega)$ be the operator-valued function defined by

$$\mathcal{A}^{\alpha,k}(\omega) := \begin{pmatrix} \mathcal{S}^{\alpha,\omega} & -\mathcal{S}^{\alpha, \frac{\omega}{\sqrt{k}}} \\ \frac{1}{k}\Big(\frac{1}{2}I - (\mathcal{K}^{-\alpha,\omega})^*\Big) & \frac{1}{2}I + (\mathcal{K}^{-\alpha, \frac{\omega}{\sqrt{k}}})^* \end{pmatrix}. \tag{5.13}$$

Then, ω^2 is an eigenvalue corresponding to u with a given quasi-momentum α if and only if ω is a characteristic value of $\mathcal{A}^{\alpha,k}$.

For $\alpha = 0$, let $\widetilde{\mathcal{A}}^{0,k}$ be given by

$$\widetilde{\mathcal{A}}^{0,k}(\omega) := \begin{pmatrix} \mathcal{S}^{0,\omega} & -\frac{1}{k}\mathcal{S}^{0, \frac{\omega}{\sqrt{k}}} \\ \frac{1}{2}I - (\mathcal{K}^{0,\omega})^* & \frac{1}{2}I + (\mathcal{K}^{0, \frac{\omega}{\sqrt{k}}})^* \end{pmatrix}. \tag{5.14}$$

By a change of functions, it is easy to see that ω is a characteristic value of $\widetilde{\mathcal{A}}^{0,k}$ if and only if ω^2 is an eigenvalue of (5.6) for $\alpha = 0$.

Consequently, we have now a new way of looking at the spectrum of (5.6) by examining the characteristic values of $\mathcal{A}^{\alpha,k}$ and $\widetilde{\mathcal{A}}^{0,k}$.

The following lemma will be useful later.

LEMMA 5.2. *The operator-valued function $\mathcal{A}^{\alpha,k}$ is a holomorphic Fredholm operator with index 0 in $\mathbb{C} \setminus \sqrt{-1}\mathbb{R}^-$. Moreover, $\omega \mapsto (\mathcal{A}^{\alpha,k})^{-1}(\omega)$ is a meromorphic function and its poles are on the real axis.*

PROOF. Because of the logarithmic behavior of quasi-periodic Green's functions, we shall restrict the set on which we define the operator $\mathcal{A}^{\alpha,k}$ to $\mathbb{C} \setminus \sqrt{-1}\mathbb{R}^-$. To see that the operator-valued function $\mathcal{A}^{\alpha,k}$ is a holomorphic Fredholm operator with index 0 in $\mathbb{C} \setminus \sqrt{-1}\mathbb{R}^-$, it suffices to write

$$\mathcal{A}^{\alpha,k}(\omega) = \begin{pmatrix} \mathcal{S}^{\alpha,0} & -\mathcal{S}^{\alpha,0} \\ \frac{1}{2k}I & \frac{1}{2}I \end{pmatrix} + \begin{pmatrix} \mathcal{S}^{\alpha,\omega} - \mathcal{S}^{\alpha,0} & -\mathcal{S}^{\alpha, \frac{\omega}{\sqrt{k}}} + \mathcal{S}^{\alpha,0} \\ \frac{1}{k}(\mathcal{K}^{-\alpha,\omega})^* & (\mathcal{K}^{-\alpha,\omega})^* \end{pmatrix} := \mathcal{A}^\alpha + \mathcal{B}^\alpha(\omega).$$

Since $\mathcal{A}^\alpha$ is invertible and $\mathcal{B}^\alpha$ is compact and holomorphic in ω, it follows that $\mathcal{A}^{\alpha,k}$ is a holomorphic Fredholm operator with index 0. By the generalization of the Steinberg theorem given in Chapter 1 (Theorem 1.16), the invertibility of $\mathcal{A}^{\alpha,k}(\omega)$ at $\omega = 0$ shows that $\omega \mapsto (\mathcal{A}^{\alpha,k})^{-1}(\omega)$ is a meromorphic function. Let ω_0 be a pole of $(\mathcal{A}^{\alpha,k})^{-1}(\omega)$. Then ω_0 is a characteristic value of $\mathcal{A}^{\alpha,k}$. Set (ϕ, ψ) to be a root function associated with ω_0. Define

$$u(x) = \begin{cases} \mathcal{S}^{\alpha,\omega_0}[\phi](x), & x \in D, \\ \mathcal{S}^{\alpha,\frac{\omega_0}{\sqrt{k}}}[\psi](x), & x \in Y \setminus \overline{D}. \end{cases}$$

Then, integrating by parts, we obtain that

$$\int_Y (1 + (k-1)\chi(Y \setminus \overline{D}))|\nabla u|^2 - \omega_0^2 \int_Y |u|^2 = 0,$$

which shows that ω_0 is real. □

It can be easily seen that the same result holds for $\widetilde{\mathcal{A}}^{0,k}$.

5.4.2. Numerical Approach for Band Structure Calculations. Band structure calculations reduce then to the computation of the characteristic values of $\mathcal{A}^{\alpha,k}$ for α moving through the Brillouin zone. It is important to note that in this formulation, one is no longer seeking eigenvalues of a differential equation. Instead one is seeking nontrivial solutions to a homogeneous linear system in which the spectral parameter ω plays a nonlinear role. The advantage gained is that we avoid having to discretize the whole cell Y, but only discretize the material interfaces themselves. To find such solutions numerically, we first have to discretize all the integrals in (5.13) and (5.14).

After the integrals are discretized, we obtain a rather involved linear system which, for a fixed value of ω, we can write in the form $\mathcal{A}^{\alpha,k}(\omega)[x] = 0$. The unknown vector x represents point values of the densities ϕ and ψ on ∂D. Thus, if N points are used to discretize ∂D, there are $2N$ unknowns. Lemma 5.2 ensures that the entries of the matrix $\mathcal{A}^{\alpha,k}$ are holomorphic, nonlinear functions of ω. Finding the characteristic values corresponds to finding values of ω for which the system of equations $\mathcal{A}^{\alpha,k}(\omega)[x] = 0$ has nontrivial solutions. An efficient strategy first described in [**197**] is based on determining a new function of ω:

$$f(\omega) := \frac{1}{\langle x, \mathcal{A}^{\alpha,k}(\omega)^{-1}[y] \rangle},$$

where x and y are two fixed *random* vectors. It is straightforward to verify that the function $f(\omega)$ is a holomorphic function of its argument. Moreover, since

$$||\mathcal{A}^{\alpha,k}(\omega)^{-1}|| = +\infty$$

when ω corresponds to a characteristic value, we have that $f(\omega) = 0$. In short, the singular matrix problem has been turned into a complex root finding process for the function. Muller's method described in Section 1.6 can be used to find complex roots of $f(\omega)$. This approach is both efficient and robust [**197, 211**]. In this subsection we discuss the details of its implementation and present numerical examples.

5.4.2.1. *Empty Resonance.* The appropriate Green's function for the layer potentials used in the previous subsection is the quasi-biperiodic Green's function $G_\sharp^{\alpha,\omega}$ which satisfies

$$(5.15) \qquad (\Delta + \omega^2) G_\sharp^{\alpha,\omega}(x,y) = \sum_{m \in \mathbb{Z}^2} \delta_0(x-y-m) e^{\sqrt{-1} m \cdot \alpha}.$$

If $\omega \neq |2\pi m + \alpha|, \forall\, m \in \mathbb{Z}^2$, then $G_\sharp^{\alpha,\omega}$ has the spectral representation (2.288). In the context of the standard boundary integral approach to numerical computation, when the parameters ω and α are such that $\omega \sim |2\pi m + \alpha|$ for any $m \in \mathbb{Z}^2$, the quasi-periodic Green's function $G_\sharp^{\alpha,\omega}$ can have highly aberrant behavior that makes determining characteristic values of $\mathcal{A}^{\alpha,k}(\omega)$ impossible. This phenomenon, which is known as empty resonance, is due to the resonance of the empty unit cell Y with refractive index 1 everywhere and quasi-periodic boundary conditions.

In order to deal with this issue it is necessary to use an approach that is less susceptible to the problem, or an approach that avoids it altogether. We will briefly discuss the Barnett-Greengard method [**140**] for quasi-periodic fields which was developed specifically to tackle the problem of empty resonances. We will then present a numerical example in which the photonic crystal band structure is calculated using the multipole method and incorporates lattice sums, an approach which was found to be much less susceptible to the empty resonance problem.

5.4.2.2. *Barnett-Greengard Method.* The Barnett-Greengard method avoids the problem of empty resonances by introducing a new integral representation for the problem that doesn't use the quasi-periodic Green's function. Instead, the usual free-space Green's function is used and the quasi-periodicity is enforced through auxiliary layer potentials defined on the boundary of the unit cell.

The quasi-periodicity condition in (5.6) can equivalently be written as a set of boundary conditions on the unit cell Y. Let L represent the left wall of the unit cell and B represent the bottom wall. Define $a := e^{\sqrt{-1}k_1}$ and $b := e^{\sqrt{-1}k_2}$. Then the quasi-periodicity condition can be stated as:

$$\begin{cases} u|_{L+e_1} &= a u|_L, \\ \dfrac{\partial u}{\partial \nu}|_{L+e_1} &= a \dfrac{\partial u}{\partial \nu}|_L, \\ u|_{B+e_2} &= b u|_B, \\ \dfrac{\partial u}{\partial \nu}|_{B+e_2} &= b \dfrac{\partial u}{\partial \nu}|_B. \end{cases}$$

Recall that the usual boundary integral formulation enables the determination of characteristic values of the operator valued function $\mathcal{A}^{\alpha,k}(\omega)$ given in (5.13) by finding the values ω such that the equation

$$\mathcal{A}^{\alpha,k}(\omega)[\Psi] = 0,$$

has a non-trivial solution $\Psi \in L^2(\partial D) \times L^2(\partial D)$. We note that the elements of $\mathcal{A}^{\alpha,k}(\omega)$ are quasi-periodic layer potentials. The Barnett-Greengard method uses an analogous equation

$$\mathcal{E}^{\alpha,k}(\omega)[\Psi] = \kappa,$$

where

$$\mathcal{E}^{\alpha,k}(\omega) := \begin{pmatrix} A & B \\ C & Q \end{pmatrix}, \ \Psi = \begin{pmatrix} \eta \\ \xi \end{pmatrix}, \ \kappa = \begin{pmatrix} m \\ d \end{pmatrix},$$

and the operators A, B, C, and Q, which will be explained shortly, involve layer potentials which utilize the free-space Green's function. η represents surface potentials for the inclusion, and ξ represents auxiliary surface potentials defined on the boundary of the unit cell. m and d are called the mismatch and the discrepancy, respectively. m represents the amount by which the matching conditions at the interface fail to be satisfied and is defined as:

$$m := \begin{pmatrix} u|_+ - u|_- \\ \frac{\partial u}{\partial \nu}|_+ - \frac{\partial u}{\partial \nu}|_- \end{pmatrix}.$$

The discrepancy d represents the amount by which the quasi-periodicity conditions on the boundary of unit cell fail to be satisfied:

$$d := \begin{pmatrix} u|_L - a^{-1}u|_{L+e_1} \\ \frac{\partial u}{\partial \nu}|_L - a^{-1}\frac{\partial u}{\partial \nu}|_{L+e_1} \\ u|_B - b^{-1}u|_{B+e_2} \\ \frac{\partial u}{\partial \nu}|_B - b^{-1}\frac{\partial u}{\partial \nu}|_{B+e_2} \end{pmatrix}.$$

The aim is to find non-trivial surface potentials such that the mismatch and discrepancy are both zero. With that in mind, the characteristic values of the operator valued function $\mathcal{E}^{\alpha,k}(\omega)$ are the values ω such that the equation

$$\mathcal{E}^{\alpha,k}(\omega)\Psi = 0,$$

has a non-trivial solution $\Psi \in L^2(\partial D)^4$.

Before we discuss the operators used to construct $\mathcal{E}^{\alpha,k}(\omega)$ let us introduce the generalized layer potentials:

$$\widetilde{\mathcal{S}}_{D_1,D_2}[\varphi](x) = \int_{\partial D_2} \sum_{m,n\in\{-1,0,1\}} a^m b^n G^\omega(x, y + me_1 + ne_2)\varphi(y)\, d\sigma(y),$$

$$\widetilde{\mathcal{D}}_{D_1,D_2}[\varphi](x) = \int_{\partial D_2} \sum_{m,n\in\{-1,0,1\}} a^m b^n \frac{\partial G^\omega}{\partial \nu(y)}(x, y + me_1 + ne_2)\varphi(y)\, d\sigma(y),$$

$$\widetilde{\mathcal{D}}^*_{D_1,D_2}[\varphi](x) = \int_{\partial D_2} \sum_{m,n\in\{-1,0,1\}} a^m b^n \frac{\partial G^\omega}{\partial \nu(x)}(x, y + me_1 + ne_2)\varphi(y)\, d\sigma(y),$$

$$\widetilde{\mathcal{T}}_{D_1,D_2}[\varphi](x) = \int_{\partial D_2} \sum_{m,n\in\{-1,0,1\}} a^m b^n \frac{\partial^2 G^\omega}{\partial \nu(x)\partial \nu(y)}(x, y + me_1 + ne_2)\varphi(y)\, d\sigma(y)$$

for $x \in D_1$.

These layer potentials involve summations over the nearest 3×3 neighboring images. This direct summation over the nearest neighbors, such that their contribution will be excluded from the auxiliary quasi-periodic representation, has been found to result in much improved convergence rates in the fast multipole literature. If the curves D_1 and D_2 both represent the inclusion D we drop subscripts and use the notation $\widetilde{\mathcal{S}}^\omega$ for the generalized single-layer potential, and similarly for the other layer potentials.

Now we are in position to describe the role of the operators A, B, C, and Q. These operators are arrived at by substituting the representation formula

$$u(x) = \begin{cases} \mathcal{S}[\phi](x) + \mathcal{D}[\psi](x) & x \in D, \\ \widetilde{\mathcal{S}}[\phi](x) + \widetilde{\mathcal{D}}[\psi](x) + u_{QP}[\xi](x) & x \in Y\backslash\overline{D}, \end{cases}$$

into the expressions for m and d. u_{QP} is an auxiliary field that is represented by a set of layer potentials on the specific borders of the neighboring cells that touch the borders of the unit cell, and ξ represents the auxiliary densities, associated with u_{QP} which are defined on these borders. The operator A is similar to the $\mathcal{A}^{\alpha,k}(\omega)$ operator in the usual boundary integral formulation. It describes the effect of the inclusion densities on the mismatch and is defined as

$$A := \begin{pmatrix} I & 0 \\ 0 & I \end{pmatrix} + \begin{pmatrix} \widetilde{\mathcal{D}} - \mathcal{D} & \widetilde{\mathcal{S}} - \mathcal{S} \\ \widetilde{\mathcal{T}} - \mathcal{T} & \widetilde{\mathcal{D}}^* - \mathcal{D}^* \end{pmatrix}.$$

The operator C describes the effect of the inclusion densities on the discrepancy and is defined as:

$$C := \begin{pmatrix} \widetilde{\mathcal{D}}_{L,\partial D} - a^{-1}\widetilde{\mathcal{D}}_{L+e_1,\partial D} & -\widetilde{\mathcal{S}}_{L,\partial D} - a^{-1}\widetilde{\mathcal{S}}_{L+e_1,\partial D} \\ \widetilde{\mathcal{T}}_{L,\partial D} - a^{-1}\widetilde{\mathcal{T}}_{L+e_1,\partial D} & -\widetilde{\mathcal{D}}^*_{L,\partial D} - a^{-1}\widetilde{\mathcal{D}}^*_{L+e_1,\partial D} \\ \widetilde{\mathcal{D}}_{B,\partial D} - b^{-1}\widetilde{\mathcal{D}}_{B+e_2,\partial D} & -\widetilde{\mathcal{S}}^{\omega}_{B,\partial D} - b^{-1}\widetilde{\mathcal{S}}_{B+e_2,\partial D} \\ \widetilde{\mathcal{T}}_{B,\partial D} - b^{-1}\widetilde{\mathcal{T}}_{B+e_2,\partial D} & -\widetilde{\mathcal{D}}^*_{B,\partial D} - b^{-1}\widetilde{\mathcal{D}}^*_{B+e_2,\partial D} \end{pmatrix}.$$

Due to symmetry and translation invariance it can be shown that significant cancellation occurs when summing over the nearest neighbor terms, and therefore the operator C can be further optimized.

The operator Q describes the effect of the auxiliary densities on the discrepancy and is defined as:

$$Q := I + \begin{pmatrix} Q_{LL} & Q_{LB} \\ Q_{BL} & Q_{BB} \end{pmatrix},$$

where Q_{LL}, Q_{LB}, Q_{BL}, and Q_{BB} are respectively defined by

$$\begin{pmatrix} \sum\limits_{m\in\{-1,1\},n\in\{-1,0,1\}} ma^m b^k \mathcal{D}_{L,L+me_1+ne_2} & - \sum\limits_{m\in\{-1,1\},n\in\{-1,0,1\}} ma^m b^k \mathcal{S}_{L,L+me_1+ne_2} \\ \sum\limits_{m\in\{-1,1\},n\in\{-1,0,1\}} ma^m b^k \mathcal{T}_{L,L+me_1+ne_2} & - \sum\limits_{m\in\{-1,1\},n\in\{-1,0,1\}} ma^m b^k \mathcal{D}^*_{L,L+me_1+ne_2} \end{pmatrix},$$

$$Q_{LB} = \begin{pmatrix} Q^{11}_{LB} & Q^{12}_{LB} \\ Q^{21}_{LB} & Q^{22}_{LB} \end{pmatrix}$$

with

$$Q^{11}_{LB} = \sum_{m\in\{0,1\}} b^m (a\mathcal{D}_{L,B+e_1+me_2} - a^{-2}\mathcal{D}_{L,B-2e_1+me_2}),$$

$$Q^{12}_{LB} = \sum_{m\in\{0,1\}} b^m (-aS_{L,B+e_1+me_2} + a^{-2}S_{L,B-2e_1+me_2}),$$

$$Q^{21}_{LB} = \sum_{m\in\{0,1\}} b^m (a\mathcal{T}_{L,B+e_1+me_2} - a^{-2}\mathcal{T}_{L,B-2e_1+me_2}),$$

$$Q^{22}_{LB} = \sum_{m\in\{0,1\}} b^m (-aD^*_{L,B+e_1+me_2} + a^{-2}D^*_{L,B-2e_1+me_2}),$$

$$Q_{BL} = \begin{pmatrix} Q^{11}_{BL} & Q^{12}_{BL} \\ Q^{21}_{BL} & Q^{22}_{BL} \end{pmatrix}$$

with

$$Q^{11}_{BL} = \sum_{m\in\{0,1\}} a^m(b\mathcal{D}_{B,L+me_1+e_2} - b^{-2}\mathcal{D}_{B,L+me_1-2e_2}),$$
$$Q^{12}_{BL} = \sum_{m\in\{0,1\}} a^m(-b\mathcal{S}_{B,L+me_1+e_2} + b^{-2}\mathcal{S}_{B,L+me_1-2e_2}),$$
$$Q^{21}_{BL} = \sum_{m\in\{0,1\}} a^m(b\mathcal{T}_{B,L+me_1+e_2} - b^{-2}\mathcal{T}_{B,L+me_1-2e_2}),$$
$$Q^{22}_{BL} = \sum_{m\in\{0,1\}} a^m(-b\mathcal{D}^*_{B,L+me_1+e_2} + b^{-2}\mathcal{D}^*_{B,L+me_1-2e_2}),$$

and

$$\begin{pmatrix} \sum\limits_{m\in\{-1,1\},n\in\{-1,0,1\}} ma^m b^k \mathcal{D}_{B,B+me_1+ne_2} & -\sum\limits_{m\in\{-1,1\},n\in\{-1,0,1\}} ma^m b^k \mathcal{S}_{B,B+me_1+ne_2} \\ \sum\limits_{m\in\{-1,1\},n\in\{-1,0,1\}} ma^m b^k \mathcal{T}_{B,B+me_1+ne_2} & -\sum\limits_{m\in\{-1,1\},n\in\{-1,0,1\}} ma^m b^k \mathcal{D}^*_{B,B+me_1+ne_2} \end{pmatrix}.$$

Again, due to symmetry and translational invariance the terms of the operator Q are subject to cancellation.

Finally, the operator B, which describes the effect of the auxiliary densities on the mismatch, is defined as:

$$B := \sum_{m\in[0,1],n\in\{-1,0,1\}} a^m b^n \begin{pmatrix} \mathcal{D}_{\partial D,L+me_1+ne_2} & -\mathcal{S}_{\partial D,L+me_1+ne_2} & 0 & 0 \\ \mathcal{T}_{\partial D,L+me_1+ne_2} & -\mathcal{D}^*_{\partial D,L+me_1+ne_2} & 0 & 0 \end{pmatrix}$$
$$+ \sum_{m\in\{-1,0,1\},n\in\{0,1\}} a^m b^n \begin{pmatrix} 0 & 0 & \mathcal{D}_{\partial D,B+me_1+ne_2} & -\mathcal{S}_{\partial B,L+me_1+ne_2} \\ 0 & 0 & \mathcal{T}_{\partial D,B+me_1+ne_2} & -\mathcal{D}^*_{\partial D,B+me_1+ne_2} \end{pmatrix}.$$

By avoiding the use of the quasi-periodic Green's function, the Barnett-Greengard method can be used for photonic band structure calculations that are free from the issue of empty resonance.

5.4.2.3. *Multipole Expansion Method.* When D is a circular disk of radius R, the integral equation admits an explicit representation. In this case, the solution can be represented as a sum of cylindrical waves $J_n(kr)e^{\sqrt{-1}n\theta}$ or $H^{(1)}_n(kr)e^{\sqrt{-1}n\theta}$. Here we give a multipole expansion interpretation of the integral operator $\mathcal{A}^{\alpha,k}$. It results in a numerical scheme which is much more efficient than one obtained with the usual discretization.

Recall that, for each fixed k, α, we have to find a characteristic value of $\mathcal{A}^{\alpha,k}(\omega)$ defined by (6.15) where ω in the original operator $\mathcal{A}^{\alpha,k}$ is replaced by $\sqrt{k}\omega$. The corresponding solution is associated to transverse magnetic mode and k represents the permittivity of the inclusion.

From the above expression, we see that $\mathcal{A}^{\alpha,k}$ is represented in terms of the single-layer potential only. So it is enough to derive a multipole expansion version of the single-layer potential.

Before computing $\mathcal{S}^{\alpha,\omega}[\varphi]$, let us first consider the single-layer potential $\mathcal{S}^{\omega}_D[\varphi]$ for a single disk D. We adopt the polar coordinates (r, θ). Then, since D is a circular disk, the density function $\varphi = \varphi(\theta)$ is a 2π-periodic function. So it admits the following Fourier series expansion:

$$\varphi = \sum_{n\in\mathbb{Z}} a_n e^{\sqrt{-1}n\theta},$$

for some coefficients a_n. So we only need to compute $u := \mathcal{S}_D^\omega[e^{\sqrt{-1}n\theta}]$ which satisfies

$$
(5.16) \qquad \begin{cases} \Delta u + \omega^2 u = 0 \quad \text{in } \mathbb{R}^2 \setminus \overline{D}, \\ \Delta u + \omega^2 u = 0 \quad \text{in } D, \\ u|_+ = u|_- \quad \text{on } \partial D, \\ \left.\frac{\partial u}{\partial \nu}\right|_+ - \left.\frac{\partial u}{\partial \nu}\right|_- = e^{\sqrt{-1}n\theta} \quad \text{on } \partial D, \\ u \text{ satisfies the Sommerfeld radiation condition.} \end{cases}
$$

The above equation can be easily solved by using the separation of variables technique in polar coordinates. It gives

$$
(5.17) \qquad \mathcal{S}_D^\omega[e^{\sqrt{-1}n\theta}] = \begin{cases} cJ_n(\omega R)H_n^{(1)}(\omega r)e^{\sqrt{-1}n\theta}, & |r| > R, \\ cH_n^{(1)}(\omega R)J_n(\omega r)e^{\sqrt{-1}n\theta}, & |r| \le R, \end{cases}
$$

where $c = (-\sqrt{-1}\pi R)/2$.

Now we compute the quasi-periodic single-layer potential $\mathcal{S}^{\alpha,\omega}[e^{\sqrt{-1}n\theta}]$. Since

$$
G_\sharp^{\alpha,\omega}(x,y) = -\frac{\sqrt{-1}}{4} \sum_{m\in\mathbb{Z}^2} H_0^{(1)}(\omega|x-y-m|)e^{\sqrt{-1}m\cdot\alpha},
$$

we have

$$
\begin{aligned}
\mathcal{S}^{\alpha,\omega}[e^{\sqrt{-1}n\theta}] &= \mathcal{S}_D^\omega[e^{\sqrt{-1}n\theta}] + \sum_{m\in\mathbb{Z}^2, m\neq 0} \mathcal{S}_{D+m}^\omega[e^{\sqrt{-1}n\theta}]e^{\sqrt{-1}m\cdot\alpha} \\
&= \mathcal{S}_D^\omega[e^{\sqrt{-1}n\theta}] + cJ_n(\omega R) \sum_{m\in\mathbb{Z}^2} H_n^{(1)}(\omega r_m)e^{\sqrt{-1}n\theta_m}e^{\sqrt{-1}m\cdot\alpha}.
\end{aligned}
$$

Here, $D+m$ means a translation of a disk D by m and (r_m, θ_m) are the polar coordinates with respect to the center of $D+m$. By applying the following addition theorem:

$$
H_n^{(1)}(\omega r_m)e^{\sqrt{-1}n\theta_m} = \sum_{l\in\mathbb{Z}} (-1)^{n-l} H_{n-l}^{(1)}(\omega|m|)e^{\sqrt{-1}n\arg(m)}J_l(\omega r)e^{\sqrt{-1}l\theta},
$$

we obtain

$$
(5.18) \qquad \mathcal{S}^{\alpha,\omega}[e^{\sqrt{-1}n\theta}] = \mathcal{S}_D^\omega[e^{\sqrt{-1}n\theta}] + cJ_n(\omega R)\sum_{l\in\mathbb{Z}}(-1)^{n-l}Q_{n-l}J_l(\omega r)e^{\sqrt{-1}l\theta},
$$

where Q_n is so called the lattice sum defined by

$$
Q_n := \sum_{m\in\mathbb{Z}^2, m\neq 0} H_n^{(1)}(\omega|m|)e^{\sqrt{-1}n\arg(m)}e^{\sqrt{-1}m\cdot\alpha}.
$$

So, from (5.17) and (5.18), we finally obtain the explicit representation of $\mathcal{S}^{\alpha,\omega}$.

For numerical computation, we should consider the truncated series

$$
\sum_{n=-N}^{N} a_n \mathcal{S}^{\alpha,\omega}[e^{\sqrt{-1}n\theta}],
$$

instead of $\mathcal{S}^{\alpha,\omega}[\varphi] = \sum_{n\in\mathbb{Z}} a_n \mathcal{S}^{\alpha,\omega}[e^{\sqrt{-1}n\theta}]$ for some sufficiently large $N \in \mathbb{N}$. Then, using $e^{\sqrt{-1}n\theta}$ as a basis, we have the following matrix representation of the operator

$\mathcal{S}^{\alpha,\omega}$:

$$\mathcal{S}^{\alpha,\omega}[\varphi]|_{\partial D} \approx \begin{pmatrix} S_{-N,-N} & S_{-N,-(N-1)} & \cdots & S_{-N,N} \\ S_{-(N-1),-N} & S_{-(N-1),-(N-1)} & \cdots & S_{-(N-1),N} \\ \vdots & & \ddots & \vdots \\ S_{N,-N} & \cdots & \cdots & S_{NN} \end{pmatrix} \begin{pmatrix} a_{-N} \\ a_{-(N-1)} \\ \vdots \\ a_N \end{pmatrix},$$

where $S_{m,n}$ is given by

$$S_{m,n} = cJ_n(\omega R)H_n^{(1)}(\omega R)\delta_{mn} + cJ_n(\omega R)(-1)^{n-m}Q_{n-m}J_m(\omega R).$$

Similarly, we also have the following matrix representation for $\partial\mathcal{S}^{\alpha,\omega}/\partial\nu|_{\pm}$ on ∂D:

$$\frac{\partial\mathcal{S}^{\alpha,\omega}}{\partial\nu}[\varphi]\Big|_{\pm} \approx \begin{pmatrix} S'^{\pm}_{-N,-N} & S'^{\pm}_{-N,-(N-1)} & \cdots & S'^{\pm}_{-N,N} \\ S'^{\pm}_{-(N-1),-N} & S'^{\pm}_{-(N-1),-(N-1)} & \cdots & S'^{\pm}_{-(N-1),N} \\ \vdots & & \ddots & \vdots \\ S'^{\pm}_{N,-N} & \cdots & \cdots & S'^{\pm}_{NN} \end{pmatrix} \begin{pmatrix} a_{-N} \\ a_{-(N-1)} \\ \vdots \\ a_N \end{pmatrix},$$

where $S'^{\pm}_{m,n}$ is given by

$$\begin{aligned} S'^{\pm}_{m,n} = \frac{\omega}{2}\Big[\pm 1 + c\Big(J_n \cdot (H_n^{(1)})' + J_n' \cdot H_n^{(1)}\Big)(\omega R)\Big]\delta_{mn} \\ + cJ_n(\omega R)(-1)^{n-m}Q_{n-m}\omega J_m'(\omega R). \end{aligned}$$

The matrix representation of $\mathcal{A}^{\alpha,k}(\omega)$ immediately follows.

5.4.2.4. *Computing the Lattice Sum Efficiently.* Unfortunately, the series in the definition of Q_n^α suffers from very slow convergence. Here we provide an alternative representation which converges very quickly. For $n > 0$, Q_n^α can be represented as

$$Q_n = Q_n^G + \Delta Q_n$$

where ΔQ_n is given by

$$\Delta Q_n = \sum_{m\in\mathbb{Z}} \frac{1}{\gamma_m}\left(\frac{e^{\sqrt{-1}n\theta_m}}{e^{-\sqrt{-1}\alpha(2)}e^{-\sqrt{-1}\gamma_m} - 1} + (-1)^n \frac{e^{\sqrt{-1}n\theta_m}}{e^{-\sqrt{-1}\alpha(2)}e^{-\sqrt{-1}\gamma_m} - 1}\right),$$

$$\beta_m = \alpha(1) + 2\pi m, \quad \theta_m = \sin^{-1}(\beta_m/\omega), \quad \gamma_m = \sqrt{\omega^2 - \beta_m^2},$$

and Q_n^G is given by

$$\begin{aligned} Q_0^G = -1 - \frac{2\sqrt{-1}}{\pi}(-\psi(1) + \ln\frac{\omega}{4\pi}) - \frac{2\sqrt{-1}}{\widetilde{\gamma}_0} - \frac{2\sqrt{-1}(\omega^2 + 2\beta_0^2)}{(2\pi)^3}\zeta(3) \\ - 2\sqrt{-1}\sum_{m\in\mathbb{Z}} \frac{1}{\widetilde{\gamma}_m} + \frac{1}{\widetilde{\gamma}_{-m}} - \frac{1}{m\pi} - \frac{\omega^2 + 2\beta_0^2}{(2\pi m)^3}, \end{aligned}$$

$$\begin{aligned} Q_{2l}^G = -2\sqrt{-1}\frac{e^{-2\sqrt{-1}l\theta_0}}{\widetilde{\gamma}_0} - 2\sqrt{-1}\sum_{m\in\mathbb{Z}} \frac{e^{-2\sqrt{-1}l\theta_m}}{\widetilde{\gamma}_m} + \frac{e^{2\sqrt{-1}l\theta_{-m}}}{\widetilde{\gamma}_{-m}} - \frac{(-1)^l}{m\pi}\Big(\frac{\omega}{4m\pi}\Big)^{2l} \\ - 2\sqrt{-1}\frac{(-1)^l}{\pi}(\frac{\omega}{4\pi})^{2l}\zeta(2l+1) + \frac{\sqrt{-1}}{l\pi} \\ + \frac{\sqrt{-1}}{\pi}\sum_{m=1}^{l}(-1)^m 2^{2m}\frac{(l+m-1)!}{(2m)!(l-m)!}\Big(\frac{2\pi}{\omega}\Big)^{2m} B_{2m}(\frac{\alpha(1)}{2\pi}), \end{aligned}$$

$$
\begin{aligned}
Q^G_{2l-1} = & \, 2\sqrt{-1}\sum_{m\in\mathbb{Z}} \frac{e^{-\sqrt{-1}(2l-1)\theta_m}}{\widetilde{\gamma}_m} - \frac{e^{\sqrt{-1}(2l-1)\theta_{-m}}}{\widetilde{\gamma}_{-m}} + \sqrt{-1}\frac{(-1)^l\beta_0 l}{(m\pi)^2}\Big(\frac{\omega}{4m\pi}\Big)^{2l-1} \\
& - 2\sqrt{-1}\frac{e^{-i(2l-1)\theta_0}}{\widetilde{\gamma}_0} + 2\frac{(-1)^l\beta_0 l}{\pi^2}\Big(\frac{\omega}{4\pi}\Big)^{2l-1}\zeta(2l+1) \\
& - \frac{2}{\pi}\sum_{m=0}^{l-1}(-1)^m 2^{2m}\frac{(l+m-1)!}{(2m+1)!(l-m-1)!}\Big(\frac{2\pi}{\omega}\Big)^{2m+1}B_{2m+1}\Big(\frac{\alpha(1)}{2\pi}\Big),
\end{aligned}
$$

where B_m is the Bernoulli polynomial and

$$
\widetilde{\gamma}_m = \begin{cases} \sqrt{\omega^2-\beta_m^2}, & \omega \geq \beta_m, \\ -\sqrt{-1}\sqrt{\beta_m^2-\omega^2}, & \omega < \beta_m. \end{cases}
$$

5.4.2.5. *Numerical Example.* Now we present a numerical example in which we assume D is a circular disk of radius $R = 0.42$ and $k = \infty$. The corresponding Matlab code can be downloaded at `www.ams.org/bookpages/surv-235`.

The truncation parameter for the cylindrical waves is set to be $N = 8$. The points Γ, X and M represent $\alpha = (0,0)$, $\alpha = (\pi, 0)$ and $\alpha = (\pi,\pi)$, respectively. We plot the characteristic values ω along the boundary of the triangle ΓXM. A band gap is clearly present.

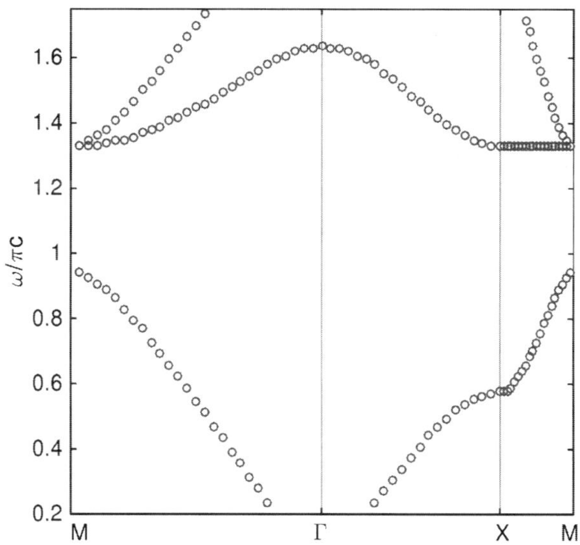

FIGURE 5.1. The band structure for a biperiodic array of circular cylinders each with radius $R = 0.42$ and $k = \infty$. The frequency is normalized to be $\omega/(\pi c)$ where c is the speed of light.

5.5. Sensitivity Analysis with Respect to the Index Ratio

Let us now turn to the sensitivity of the band gap with respect to the contrast and/or the shape of the inclusion.

Expanding the operator-valued function $\mathcal{A}^{\alpha,k}$ in terms of k and small perturbations of the shape of D, we calculate asymptotic expressions of its characteristic values with the help of the generalized Rouché theorem.

5.5.1. Preliminary Results. The following lemma, which is an immediate consequence of (2.297), gives a complete asymptotic expansion of $\mathcal{A}^{\alpha,k}$ as $k \to +\infty$.

Let $\mathcal{S}_l^{\alpha,\omega}$ and $(\mathcal{K}_l^{-\alpha,\omega})^*$ be given by (2.297). Let the operators

$$\mathcal{A}_0^\alpha(\omega) = \begin{pmatrix} \mathcal{S}^{\alpha,\omega} & -\mathcal{S}^{\alpha,0} \\ 0 & \frac{1}{2}I + (\mathcal{K}^{-\alpha,0})^* \end{pmatrix},$$

$$\mathcal{A}_1^\alpha(\omega) = \begin{pmatrix} 0 & -\mathcal{S}_1^{\alpha,\omega} \\ \left(\frac{1}{2}I - (\mathcal{K}^{-\alpha,\omega})^*\right) & (\mathcal{K}_1^{-\alpha,\omega})^* \end{pmatrix},$$

and, for $l \geq 2$,

$$\mathcal{A}_l^\alpha(\omega) = \begin{pmatrix} 0 & -\mathcal{S}_l^{\alpha,\omega} \\ 0 & (\mathcal{K}_l^{-\alpha,\omega})^* \end{pmatrix}.$$

LEMMA 5.3. *Suppose $\alpha \neq 0$. We have*

$$\mathcal{A}^{\alpha,k}(\omega) = \mathcal{A}_0^\alpha(\omega) + \sum_{l=1}^{+\infty} \frac{1}{k^l} \mathcal{A}_l^\alpha(\omega). \tag{5.19}$$

We now have the following lemma for the characteristic values of $\mathcal{A}_0^\alpha$.

LEMMA 5.4. *Suppose $\alpha \neq 0$. Then $\omega_0^\alpha \in \mathbb{R}$ is a characteristic value of $\mathcal{A}_0^\alpha$ if and only if $(\omega_0^\alpha)^2$ is either an eigenvalue of $-\Delta$ in D with the Dirichlet boundary condition or an eigenvalue of $-\Delta$ in $Y \setminus \overline{D}$ with the Dirichlet boundary condition on ∂D and the quasi-periodic condition on ∂Y.*

PROOF. Suppose that $\omega = \omega_0^\alpha \in \mathbb{R}$ is a characteristic value of $\mathcal{A}_0^\alpha$. Then there is $(\phi, \psi) \neq 0$ such that

$$\begin{cases} \mathcal{S}^{\alpha,\omega}[\phi] - \mathcal{S}^{\alpha,0}[\psi] = 0, \\ \left(\frac{1}{2}I + (\mathcal{K}^{-\alpha,0})^*\right)[\psi] = 0 \end{cases} \quad \text{on } \partial D. \tag{5.20}$$

It then follows from Lemma 2.92 that $\psi = 0$ and hence $\mathcal{S}^{\alpha,\omega}[\phi] = 0$ on ∂D. Since $\phi \neq 0$, $\mathcal{S}^{\alpha,\omega}[\phi] \neq 0$ either in D or in $Y \setminus \overline{D}$ and hence $(\omega_0^\alpha)^2$ is either an eigenvalue of $-\Delta$ in D with the Dirichlet boundary condition or an eigenvalue of $-\Delta$ in $Y \setminus \overline{D}$ with the Dirichlet boundary condition on ∂D and the quasi-periodic condition on ∂Y, and $\mathcal{S}^{\alpha,\omega}[\phi]$ is an associated eigenfunction.

Conversely if $(\omega_0^\alpha)^2$ is an eigenvalue of $-\Delta$ in D with the Dirichlet boundary condition, then by Green's representation formula, we have

$$u(x) = -\mathcal{S}^{\alpha,\omega}\left[\frac{\partial u}{\partial \nu}\bigg|_{\partial D}\right], \quad x \in D.$$

Thus (5.20) holds with $(\phi, \psi) = (\partial u/\partial \nu|_{\partial D}, 0)$. The other case can be treated similarly using (2.298). This completes the proof. □

At this moment let us invoke the results in [**295**] (see also [**246, 294, 475**]). In [**295**], it is shown, by a completely different argument which involves the convergence of quadratic forms, that the spectrum of (5.6) for $\alpha \neq 0$ accumulates near the spectrum of $-\Delta$ in D with the Dirichlet boundary condition on ∂D as $k \to +\infty$. According to this result, the eigenvalue of the exterior problem is not realized as a limit of eigenvalues of the problem (5.6). In fact, the limit of the corresponding eigenfunctions is given by

$$u(x) = \begin{cases} \mathcal{S}^{\alpha,\omega}[\phi], & x \in D, \\ \mathcal{S}^{\alpha,0}[\psi] = 0, & x \in Y \setminus \overline{D}, \end{cases}$$

where the pair (ϕ, ψ) is defined by (5.20). If $(\omega_0^\alpha)^2$ is an eigenvalue for the exterior problem and not for the interior problem, then $\mathcal{S}^{\alpha,\omega}[\phi] = 0$ in D and hence $u = 0$ in Y.

The following lemma was first proved in [**294, 295**].

LEMMA 5.5. *Let $(\omega^0)^2$ (with $\omega^0 > 0$) be a simple eigenvalue of $-\Delta$ in D with the Dirichlet boundary condition. There exists a unique eigenvalue $(\omega^{\alpha,k})^2$ (with $\omega^{\alpha,k} > 0$) of (5.6) lying in a small complex neighborhood V of ω^0. Indeed, ω^0 and $\omega^{\alpha,k}$ are simple poles of $(\mathcal{A}_0^\alpha)^{-1}$ and $(\mathcal{A}^{\alpha,k})^{-1}$, respectively.*

5.5.2. Full Asymptotic Expansion. Combining now the generalized Rouché theorem together with Lemma 5.3, we are able to derive complete asymptotic formulas for the characteristic values of $\omega \mapsto \mathcal{A}^{\alpha,k}(\omega)$. Applying Theorem 1.14 yields that

$$\omega^{\alpha,k} - \omega^0 = \frac{1}{2\sqrt{-1}\pi} \operatorname{tr} \int_{\partial V} (\omega - \omega^0)(\mathcal{A}^{\alpha,k})^{-1}(\omega) \frac{d}{d\omega} \mathcal{A}^{\alpha,k}(\omega) d\omega.$$

Suppose that the quasi-momentum $\alpha \neq 0$. We obtain the following complete asymptotic expansion for the eigenvalue perturbations $\omega^{\alpha,k} - \omega^0$.

THEOREM 5.6. *Let V be as in Lemma 5.5. Suppose $\alpha \neq 0$. Then the following asymptotic expansion holds:*

$$\omega^{\alpha,k} - \omega^0 = \frac{1}{2\sqrt{-1}\pi} \sum_{p=1}^{+\infty} \frac{1}{p} \sum_{n=p}^{+\infty} \frac{1}{k^n} \operatorname{tr} \int_{\partial V} B_{n,p}^\alpha(\omega) d\omega, \tag{5.21}$$

where

$$B_{n,p}^\alpha(\omega) = (-1)^p \sum_{\substack{n_1 + \ldots + n_p = n \\ n_i \geq 1}} (\mathcal{A}_0^\alpha)^{-1}(\omega) \mathcal{A}_{n_1}^\alpha(\omega) \ldots (\mathcal{A}_0^\alpha)(\omega)^{-1} \mathcal{A}_{n_p}^\alpha(\omega) \tag{5.22}$$

and

$$(\mathcal{A}_0^\alpha)(\omega)^{-1} = \begin{pmatrix} (\mathcal{S}^{\alpha,\omega})^{-1} & (\mathcal{S}^{\alpha,\omega})^{-1} \mathcal{S}^{\alpha,0} (\frac{1}{2} I + (\mathcal{K}^{-\alpha,0})^*)^{-1} \\ 0 & (\frac{1}{2} I + (\mathcal{K}^{-\alpha,0})^*)^{-1} \end{pmatrix}. \tag{5.23}$$

5.5.3. Leading-Order Term. Let us compute the leading-order term in the expansion of $\omega^{\alpha,k} - \omega^0$. Let u^0 be the (normalized) eigenvector associated to the simple eigenvalue $(\omega^0)^2$ and let $\varphi := \partial u^0 / \partial \nu|_-$ so that

$$u^0(x) = -\mathcal{S}^{\alpha,\omega^0}[\varphi](x) \quad \text{for } x \in D. \tag{5.24}$$

We first establish the following lemma.

LEMMA 5.7. *The following identity holds:*

$$\langle \varphi, \frac{d}{d\omega}\mathcal{S}^{\alpha,\omega}[\varphi]\big|_{\omega=\omega^0}\rangle = -2\omega^0 \int_D |u^0|^2. \tag{5.25}$$

PROOF. From (2.286), it follows that

$$\Delta \frac{d}{d\omega} G^{\alpha,\omega}(x,y) + \omega^2 \frac{d}{d\omega} G^{\alpha,\omega}(x,y) = -2\omega G^{\alpha,\omega}(x,y),$$

and therefore,

$$\frac{d}{d\omega} G^{\alpha,\omega}(x,y) = -2\omega \int_Y G^{\alpha,\omega}(x,z) G^{\alpha,\omega}(z,y) dz.$$

Consequently, for any $\psi \in L^2(\partial D)$,

$$\begin{aligned}
\frac{d\mathcal{S}^{\alpha,\omega}[\psi](x)}{d\omega} &= \frac{d}{d\omega}\int_{\partial D} G^{\alpha,\omega}(x,y)\psi(y)\, d\sigma(y) \\
&= \int_{\partial D} \frac{d}{d\omega} G^{\alpha,\omega}(x,y)\psi(y)\, d\sigma(y) \\
&= -\, 2\omega \int_Y G^{\alpha,\omega}(x,z) \int_{\partial D} G^{\alpha,\omega}(z,y)\psi(y)\, d\sigma(y) dz \\
&= -\, 2\omega \int_Y G^{\alpha,\omega}(x,z) \mathcal{S}^{\alpha,\omega}[\psi](z) dz.
\end{aligned}$$

Using the fact that

$$\mathcal{S}^{\alpha,\omega^0}[\varphi] = \begin{cases} -u^0 & \text{in } D, \\ 0 & \text{in } Y \setminus \overline{D}, \end{cases}$$

we compute

$$\begin{aligned}
\langle \varphi, \frac{d\mathcal{S}^{\alpha,\omega}[\varphi]}{d\omega}\bigg|_{\omega=\omega^0} \rangle &= \int_{\partial D} \varphi(x) \frac{d\mathcal{S}^{-\alpha,\omega}}{d\omega}[\overline{\varphi}](x)\bigg|_{\omega=\omega^0} d\sigma(x) \\
&= -2\omega^0 \int_{\partial D} \varphi(x) \int_Y G^{\alpha,\omega^0}(x,z) \mathcal{S}^{-\alpha,\omega^0}[\overline{\varphi}](z) dz\, d\sigma(x) \\
&= -2\omega^0 \int_{\partial D}\int_{\partial D}\int_Y G^{\alpha,\omega^0}(x,z) G^{-\alpha,\omega^0}(y,z)\varphi(x)\overline{\varphi}(y) dz d\sigma(x)\, d\sigma(y) \\
&= -2\omega^0 \int_Y \left| \int_{\partial D} G^{\alpha,\omega^0}(x,z)\varphi(x) d\sigma(x)\right|^2 dz \\
&= -2\omega^0 \int_Y \left| \mathcal{S}^{\alpha,\omega^0}[\varphi](z)\right|^2 dz \\
&= -2\omega^0 \int_D |u^0(z)|^2\, dz,
\end{aligned}$$

which yields the desired formula. □

We are now ready to prove the following theorem.

THEOREM 5.8. *Let v^α be the unique α-quasi-periodic solution to*

$$\begin{cases} \Delta v^\alpha = 0 & \text{in } Y \setminus \overline{D}, \\ \dfrac{\partial v^\alpha}{\partial \nu}\bigg|_+ = \dfrac{\partial u^0}{\partial \nu}\bigg|_- & \text{on } \partial D. \end{cases} \tag{5.26}$$

The following asymptotic expansion holds:

$$\omega^{\alpha,k} - \omega^0 = -\frac{1}{k}\frac{\displaystyle\int_{Y\setminus\overline{D}} |\nabla v^\alpha|^2}{\displaystyle 2\omega^0 \int_D |u^0|^2} + O(\frac{1}{k^2}) \quad \text{as } k \to +\infty. \tag{5.27}$$

PROOF. Because of (5.24), we get

$$(\frac{1}{2}I - (\mathcal{K}^{-\alpha,\omega^0})^*)[\varphi] = \varphi.$$

Moreover, since ω^0 is the only simple pole in V of the mapping $\omega \mapsto (\mathcal{S}^{\alpha,\omega})^{-1}$, we can write [**89**]

$$(\mathcal{S}^{\alpha,\omega})^{-1} = \frac{1}{\omega - \omega^0} T + \mathcal{Q}^{\alpha,\omega},$$

where the operator-valued function $\mathcal{Q}^{\alpha,\omega}$ is holomorphic in ω in V, $T : L^2(\partial D) \to \operatorname{span}\{\varphi\}$ is such that $T\mathcal{S}^{\alpha,\omega^0} = \mathcal{S}^{\alpha,\omega^0} T = 0$, and

$$T\frac{d}{d\omega}\mathcal{S}^{\alpha,\omega}\bigg|_{\omega=\omega^0} = \frac{1}{||\varphi||^2_{L^2}}\langle\varphi,\cdot\rangle\varphi$$

is the orthogonal projection from $L^2(\partial D)$ into $\operatorname{span}\{\varphi\}$. Here $(\cdot,\cdot)$ is the L^2-inner product on ∂D. It can also be shown that

$$T = \frac{1}{\langle\varphi, \frac{d}{d\omega}\mathcal{S}^{\alpha,\omega}[\varphi]\big|_{\omega=\omega^0}\rangle}\langle\varphi,\cdot\rangle\,\varphi. \tag{5.28}$$

It then follows from the residue theorem that

$$\frac{1}{2\sqrt{-1}\pi}\operatorname{tr}\int_{\partial V}(\mathcal{A}_0^\alpha)(\omega)^{-1}\mathcal{A}_1^\alpha(\omega)d\omega = \operatorname{tr}\Big[T\mathcal{S}^{\alpha,0}\big(\frac{1}{2}I + (\mathcal{K}^{-\alpha,0})^*\big)^{-1}\big(\frac{1}{2}I - (\mathcal{K}^{-\alpha,\omega^0})^*\big)\Big].$$

Let

$$v^\alpha(x) := \mathcal{S}^{\alpha,0}\big(\frac{1}{2}I + (\mathcal{K}^{-\alpha,0})^*\big)^{-1}[\varphi](x), \quad x \in Y \setminus \overline{D}.$$

Then v^α is the unique α-quasi-periodic solution to (5.26) and

$$\frac{1}{2\sqrt{-1}\pi}\operatorname{tr}\int_{\partial V}(\mathcal{A}_0^\alpha)(\omega)^{-1}\mathcal{A}_1^\alpha(\omega)d\omega = \frac{1}{||\varphi||^2_{L^2}}\langle\varphi, Tv^\alpha\rangle.$$

Therefore, we have from (5.21)

$$\omega^{\alpha,k} - \omega^0 = -\frac{1}{k||\varphi||^2_{L^2}}\langle\varphi, Tv^\alpha\rangle + O(\frac{1}{k^2}) \quad \text{as } k \to +\infty.$$

By virtue of (5.28), it follows that

$$\frac{1}{||\varphi||^2_{L^2}}\langle\varphi, Tv^\alpha\rangle = \frac{1}{\langle\varphi, \frac{d}{d\omega}\mathcal{S}^{\alpha,\omega}[\varphi]\big|_{\omega=\omega^0}\rangle}\langle\varphi, v^\alpha\rangle.$$

Integration by parts yields

$$\langle\varphi, v^\alpha\rangle = -\int_{Y\setminus\overline{D}}|\nabla v^\alpha|^2,$$

and hence we obtain (5.27) from Lemma 5.7. This completes the proof. □

Note that if u^0 is normalized, then (5.27) can be rewritten as

$$(\omega^{\alpha,k})^2 - (\omega^0)^2 = -\frac{1}{k}\int_{Y\setminus\overline{D}} |\nabla v^\alpha|^2 + O(\frac{1}{k^2}) \quad \text{as } k \to +\infty. \tag{5.29}$$

5.5.4. Periodic Case.

Turning now to the periodic case ($\alpha = 0$), we first introduce the following notation. Let χY denote the constant function 1 on Y. Let the operator $\widetilde{\Delta}$ be acting on span$\{\chi(Y), H_0^1(D)\}$, with

$$\widetilde{\Delta} u := \begin{cases} -\Delta(u|_D) & \text{in } D, \\ \dfrac{1}{|Y\setminus\overline{D}|}\displaystyle\int_{\partial D}\frac{\partial}{\partial\nu}(u|_D) & \text{in } Y\setminus\overline{D}. \end{cases} \tag{5.30}$$

See [**295**]. It is worth mentioning that the eigenvalue problem for $\widetilde{\Delta}$ can be written as

$$\begin{cases} \Delta u + \omega^2 u = 0 & \text{in } D, \\ u + \dfrac{1}{|Y\setminus\overline{D}|}\displaystyle\int_D u = 0 & \text{on } \partial D. \end{cases}$$

Define the sequence of operator-valued functions $(\widetilde{\mathcal{A}}_l^0)_{l\in\mathbb{N}}$ by

$$\widetilde{\mathcal{A}}_0^0(\omega) = \begin{pmatrix} \mathcal{S}^{0,\omega} & -\dfrac{1}{\omega^2}\displaystyle\int_{\partial D} \\ \dfrac{1}{2}I - (\mathcal{K}^{0,\omega})^* & \dfrac{1}{2}I + (\mathcal{K}^{0,0})^* \end{pmatrix}, \tag{5.31}$$

$$\widetilde{\mathcal{A}}_1^0(\omega) = \begin{pmatrix} 0 & -\mathcal{S}^{0,0} \\ 0 & (\mathcal{K}_1^{0,\omega})^* \end{pmatrix}, \quad \widetilde{\mathcal{A}}_l^0(\omega) = \begin{pmatrix} 0 & -\mathcal{S}_{l-1}^{0,\omega} \\ 0 & (\mathcal{K}_l^{0,\omega})^* \end{pmatrix} \tag{5.32}$$

for $l \geq 2$, and set

$$\widetilde{B}_{n,p}^0(\omega) = (-1)^p \sum_{\substack{n_1+\ldots+n_p=n \\ n_i\geq 1}} (\widetilde{\mathcal{A}}_0^0)(\omega)^{-1}\widetilde{\mathcal{A}}_{n_1}^0(\omega)\ldots(\widetilde{\mathcal{A}}_0^0)(\omega)^{-1}\widetilde{\mathcal{A}}_{n_p}^0(\omega). \tag{5.33}$$

Here, the operators $\mathcal{S}_l^{0,\omega}$ and $(\mathcal{K}_l^{0,\omega})^*$ are given by (2.297) with $\alpha = 0$.

The following complete asymptotic expansion of $\widetilde{\mathcal{A}}^{0,k}$ as $k \to +\infty$ holds:

$$\widetilde{\mathcal{A}}^{0,k}(\omega) = \widetilde{\mathcal{A}}_0^0(\omega) + \sum_{l=1}^{+\infty}\frac{1}{k^l}\widetilde{\mathcal{A}}_l^0(\omega).$$

On the other hand, we have the following lemma on the characteristic value of $\widetilde{\mathcal{A}}_0^0$, whose proof will be given in Section 5.9.

LEMMA 5.9. *Suppose that $(\widetilde{\omega}^0)^2$ (with $\widetilde{\omega}^0 > 0$) is not an eigenvalue of $-\Delta$ in $Y\setminus\overline{D}$ with Dirichlet boundary condition on ∂D and the periodic condition on ∂Y. Then $(\widetilde{\omega}^0)^2$ is an eigenvalue of $\widetilde{\Delta}$ if and only if $\widetilde{\omega}^0$ is a characteristic value of the operator-valued function $\widetilde{\mathcal{A}}_0^0$.*

Analogously to Theorem 5.6, the asymptotic formula for $\alpha = 0$ follows from a direct application of Theorem 1.14.

THEOREM 5.10. *Suppose $\alpha = 0$. Let $(\widetilde{\omega}^0)^2$ (with $\widetilde{\omega}^0 > 0$) be a simple eigenvalue of $\widetilde{\Delta}$. There exists a unique eigenvalue $(\omega^{0,k})^2$ (with $\omega^{0,k} > 0$) of (5.6) lying in a small complex neighborhood of $(\widetilde{\omega}^0)^2$ and the following asymptotic expansion holds:*

$$\omega^{0,k} - \widetilde{\omega}^0 = \frac{1}{2\sqrt{-1}\pi} \sum_{p=1}^{+\infty} \frac{1}{p} \sum_{n=p}^{+\infty} \frac{1}{k^n} \operatorname{tr} \int_{\partial V} \widetilde{B}^0_{n,p}(\omega) d\omega, \tag{5.34}$$

where V is a small complex neighborhood of $\widetilde{\omega}^0$ and $\widetilde{B}^0_{n,p}(\omega)$ is given by (5.33).

5.5.5. The case when $|\alpha|$ is of order $1/\sqrt{k}$. In this subsection we derive an asymptotic expansion which is valid for $|\alpha|$ of order $O(1/\sqrt{k})$, not just for fixed $\alpha \neq 0$ or $\alpha = 0$, as has been considered in the previous subsections. We give the limiting behavior of $\omega^{\alpha,k}$ in this case.

Recall that we seek for the characteristic value of the operator-valued function $\omega \mapsto \mathcal{A}^{\alpha,k}(\omega)$ where $\mathcal{A}^{\alpha,k}(\omega)$ is given in (5.13). One of the difficulties in dealing with the operator when $|\alpha|$ is of order $1/\sqrt{k}$ is that $\mathcal{A}^{\alpha,k}(\omega)$ has a singularity at $\omega^2 = |\alpha|^2/k$ as one can see from the first formula in Subsection 2.13.1. In order to avoid this difficulty, we use an argument different from those in the previous sections.

Note that finding a characteristic value of $\mathcal{A}^{\alpha,k}(\omega)$ is equivalent to finding a nonzero (φ, ψ) satisfying

$$\begin{cases} \mathcal{S}^{\alpha,\omega}[\varphi] - \mathcal{S}^{\alpha,\frac{\omega}{\sqrt{k}}}[\psi] = 0, \\ \dfrac{1}{k}\Big(\dfrac{1}{2}I - (\mathcal{K}^{-\alpha,\omega})^*\Big)[\varphi] + \Big(\dfrac{1}{2}I + (\mathcal{K}^{-\alpha,\frac{\omega}{\sqrt{k}}})^*\Big)[\psi] = 0 \end{cases} \tag{5.35}$$

on ∂D. If such a pair (φ, ψ) exists, then $\varphi \neq 0$. In fact, if $\varphi = 0$, then

$$\mathcal{S}^{\alpha,\frac{\omega}{\sqrt{k}}}[\psi] = 0 \quad \text{and} \quad \frac{\partial \mathcal{S}^{\alpha,\frac{\omega}{\sqrt{k}}}[\psi]}{\partial \nu}\bigg|_+ = \Big(\frac{1}{2}I + (\mathcal{K}^{-\alpha,\frac{\omega}{\sqrt{k}}})^*\Big)[\psi] = 0 \quad \text{on } \partial D.$$

If k is so large that ω^2/k is not a Dirichlet eigenvalue on D, then it follows that $\mathcal{S}^{\alpha,\frac{\omega}{\sqrt{k}}}[\psi] = 0$ in D and $Y \setminus \overline{D}$, and hence $\psi = 0$. Therefore finding a nonzero (φ, ψ) satisfying (5.35) amounts to finding a nonzero φ satisfying

$$\left[\Big(\frac{1}{2}I - (\mathcal{K}^{-\alpha,\omega})^*\Big) + k\Big(\frac{1}{2}I + (\mathcal{K}^{-\alpha,\frac{\omega}{\sqrt{k}}})^*\Big)(\mathcal{S}^{\alpha,\frac{\omega}{\sqrt{k}}})^{-1}\mathcal{S}^{\alpha,\omega}\right][\varphi] = 0$$

on ∂D. Thus finding a characteristic value of $\mathcal{A}^{\alpha,k}(\omega)$ is equivalent to finding a characteristic value of the operator-valued function

$$\omega \mapsto \Big(\frac{1}{2}I - (\mathcal{K}^{-\alpha,\omega})^*\Big) + kN^{\alpha,\frac{\omega}{\sqrt{k}}}\mathcal{S}^{\alpha,\omega}, \tag{5.36}$$

where we put

$$N^{\alpha,\frac{\omega}{\sqrt{k}}} := \Big(\frac{1}{2}I + (\mathcal{K}^{-\alpha,\frac{\omega}{\sqrt{k}}})^*\Big)(\mathcal{S}^{\alpha,\frac{\omega}{\sqrt{k}}})^{-1}. \tag{5.37}$$

Note that $N^{\alpha,\frac{\omega}{\sqrt{k}}}$ can be extended to the Dirichlet-to-Neumann map for

$$\Delta + \frac{\omega^2}{k} \quad \text{on } Y \setminus \overline{D}$$

with α-quasi-periodic condition on ∂Y, which is defined for

$$\frac{\omega^2}{k} < \min_{\alpha \in]-\pi,\pi]^2} \kappa(\alpha).$$

Here $\kappa(\alpha)$ is the smallest eigenvalue of $-\Delta$ with the Dirichlet boundary condition on ∂D and quasi-periodicity on ∂Y. Furthermore, $N^{\alpha,\frac{\omega}{\sqrt{k}}}$ depends smoothly both on ω and α. Therefore, we have the expansion

$$N^{\alpha,\frac{\omega}{\sqrt{k}}} = N^{\alpha,0} + \frac{\omega^2}{k}\frac{d}{d(t^2)}N^{\alpha,t}\big|_{t=0} + O\left(\frac{1}{k^2}\right).$$

A further expansion in terms of α yields

$$N^{\alpha,\frac{\omega}{\sqrt{k}}} = N^{\alpha,0} + \frac{\omega^2}{k}\dot{N} + O\left(\frac{|\alpha|}{k}\right) + \left(\frac{1}{k^2}\right), \tag{5.38}$$

where

$$\dot{N} := \frac{d}{d(t^2)}N^{0,t}\big|_{t=0}. \tag{5.39}$$

The expansion (5.38) was first obtained by Friedlander [**246**].

In order to obtain a better understanding of the operator $\dot{N}$, let us consider the following problem for t small:

$$\begin{cases} \Delta u_t + t^2 u_t = 0 & \text{in } Y \setminus \overline{D}, \\ u_t = f & \text{on } \partial D, \\ u_t \text{ and } \dfrac{\partial u_t}{\partial \nu} \text{ are periodic on } \partial Y. \end{cases}$$

Since

$$N^{0,t}[f] = \frac{\partial u_t}{\partial \nu}|_{\partial D},$$

one can see that

$$\dot{N}[f] = \frac{\partial w}{\partial \nu}\bigg|_{\partial D}, \tag{5.40}$$

where $w = \partial u_t/\partial(t^2)|_{\partial D}$, which is the solution to

$$\begin{cases} \Delta w + u_0 = 0 & \text{in } Y \setminus \overline{D}, \\ w = 0 & \text{on } \partial D, \\ w \text{ and } \dfrac{\partial w}{\partial \nu} \text{ are periodic on } \partial Y. \end{cases}$$

Using (5.40), we can derive relevant estimates for $\dot{N}$. We have

$$\|\dot{N}[f]\|_{H^{-1/2}(\partial D)} = \left\|\frac{\partial w}{\partial \nu}\right\|_{H^{-1/2}(\partial D)} \leq C\|w\|_{H^1(D)} \leq C'\|u_0\|_{H^{-1}(D)}.$$

Therefore, we have for example

$$\|\dot{N}[f]\|_{H^{-1/2}(\partial D)} \leq C\|f\|_{H^{1/2}(\partial D)}. \tag{5.41}$$

It should be noted that the estimate (5.41) is not optimal.

The following lemma will be useful.

LEMMA 5.11. *Let $u_1, u_2, \ldots$ be the eigenfunctions corresponding to $0 \leq \omega_1^{\alpha,k} \leq \omega_2^{\alpha,k} \leq \ldots$. For a given constant M there exists C such that*

$$\left\| u_j - \frac{1}{|\partial D|} \int_{\partial D} u_j \right\|_{H^{1/2}(\partial D)} \leq C \left(|\alpha| + \frac{1}{k} \right) \|u_j\|_{H^1(D)} \tag{5.42}$$

for all j satisfying $\omega_j^{\alpha,k} \leq M$. Furthermore,

$$\left| \int_D u_i \overline{u_j} + \int_{\partial D} \dot{N}[u_i|_{\partial D}] \overline{u_j} \right| \leq C \left(|\alpha| + \frac{1}{k} \right) \|u_i\|_{H^1(D)} \|u_j\|_{H^1(D)}, \tag{5.43}$$

provided that $\omega_i^{\alpha,k} \neq \omega_j^{\alpha,k}$. If $\omega_i^{\alpha,k} = \omega_j^{\alpha,k}$ for some $i \neq j$, then we can choose u_i and u_j in such a way that (5.43) holds.

PROOF. We get from (5.38) that

$$N^{0,0} \left[u_j - \frac{1}{|\partial D|} \int_{\partial D} u_j \right] = \frac{1}{k} \left(\frac{\partial u_j}{\partial \nu} - (\omega_j^{\alpha})^2 \dot{N}[u_j|_{\partial D}] \right) + (N^{0,0} - N^{\alpha,0})[u_j|_{\partial D}] + O(|\alpha|) + O\left(\frac{1}{k}\right).$$

Note that $N^{0,0}$ is the Dirichlet-to-Neuman map defined on ∂D for the Laplacian in $Y \setminus \overline{D}$ with the periodic boundary condition on ∂Y, and hence it is invertible as an operator from $H_0^{1/2}(\partial D)$ into $H_0^{-1/2}(\partial D)$, where the subscript 0 indicates the zero-mean value (in a weak sense for $H_0^{-1/2}(\partial D)$). Since $N^{\alpha,0} - N^{0,0} = O(|\alpha|)$ as an operator from $H^{1/2}(\partial D)$ into $H^{-1/2}(\partial D)$, (5.41) leads to

$$\left\| u_j - \frac{1}{|\partial D|} \int_{\partial D} u_j \right\|_{H^{1/2}(\partial D)} \leq C \left(|\alpha| + \frac{1}{k} \right) \|u_j\|_{H^{1/2}(\partial D)},$$

from which (5.42) follows.

To prove (5.43), we introduce a notation for the quadratic form: Let

$$E(u, v) := \int_D \nabla u \cdot \overline{\nabla v} \, dx. \tag{5.44}$$

Since $N^{\alpha,\omega_i^{\alpha,k}}$ is the Dirichlet-to-Neuman map for the exterior problem, it follows from the divergence theorem that

$$\begin{aligned}
&(\omega_i^{\alpha,k})^2 \left(\int_D u_i \overline{u}_j + \int_{\partial D} \dot{N}[u_i|_{\partial D}] \overline{u}_j \right) \\
&\quad = E(u_i, u_j) - \int_{\partial D} \left(k N^{\alpha,\omega_i^{\alpha,k}} - (\omega_i^{\alpha,k})^2 \dot{N} \right) [u_i|_{\partial D}] \overline{u}_j.
\end{aligned}$$

We also have

$$\begin{aligned}
&(\omega_j^{\alpha,k})^2 \left(\int_D u_i \cdot \overline{u}_j + \int_{\partial D} \dot{N}[u_i|_{\partial D}] \overline{u}_j \right) \\
&= E(u_i, u_j) - \int_{\partial D} \left(u_i \, k N^{\alpha,\omega_j^{\alpha,k}} [u_j|_{\partial D}] - (\omega_i^{\alpha,k})^2 \dot{N}[u_i|_{\partial D}] \right) \overline{u}_j \\
&= E(u_i, u_j) - \int_{\partial D} \left(k N^{\alpha,\omega_j^{\alpha,k}} - (\omega_j^{\alpha,k})^2 \dot{N} \right) [u_i|_{\partial D}] \overline{u}_j,
\end{aligned}$$

where the last equality holds thanks to the fact that the Dirichlet-to-Neuman map is self-adjoint. Consequently,

$$
\begin{aligned}
&((\omega_i^{\alpha,k})^2 - (\omega_j^{\alpha,k})^2)\left(\int_D u_i \overline{u}_j + \int_{\partial D} \dot{N}[u_i|_{\partial D}]\overline{u}_j\right) \\
&= \int_{\partial D} \left((kN^{\alpha,\omega_j^{\alpha,k}} - (\omega_j^{\alpha,k})^2 \dot{N}) - (kN^{\alpha,\omega_i^{\alpha,k}} - (\omega_i^{\alpha,k})^2 \dot{N})\right)[u_i|_{\partial D}]\overline{u}_j \\
&= \int_{\partial D} \Big((kN^{\alpha,\omega_j^{\alpha,k}} - kN^{\alpha,0} - (\omega_j^{\alpha,k})^2 \dot{N}) \\
&\qquad - (kN^{\alpha,\omega_i^{\alpha,k}} - kN^{\alpha,0} - (\omega_i^{\alpha,k})^2 \dot{N})\Big)[u_i|_{\partial D}]\overline{u}_j .
\end{aligned}
$$

Hence, (5.43) follows from (5.38), and the proof is complete. □

The estimate (5.42) shows that if $|\alpha|$ and $1/k$ are small enough, then u_j is almost constant on ∂D, which is in good agreement with the case when $\alpha = 0$.

5.6. Photonic Band Gap Opening

In this section we discuss the photonic band gap opening in the limiting case as k tends to $+\infty$. We will not include proofs in this section since very similar ones will be given in Section 6.3.

Let ω_j be the eigenvalues of $-\Delta$ in D with Dirichlet conditions and let $\widetilde{\omega}_j$ be the eigenvalues of $\widetilde{\Delta}$ defined in (5.30). Then the following min-max characterization of ω_j and $\widetilde{\omega}_j$ is proved in [**295**] (see also Lemma 6.16):

$$
\omega_j^2 = \min_{N_j} \max_{u \in N_j, \|u\|_{L^2(D)}=1} E(u,u), \tag{5.45}
$$

and

$$
\widetilde{\omega}_j^2 = \min_{N_j} \max_{u \in N_j, \|u\|_{L^2(D)}=1} \frac{E(u,u)}{1 - |\int_D u|^2}, \tag{5.46}
$$

where the minimum is taken over all j -dimensional subspaces N_j of $H_0^1(D)$ and the quadratic form E is defined by (5.44). Using the min-max characterization, one can show the following interlacing relation:

$$
\omega_j \le \widetilde{\omega}_j \le \omega_{j+1}, \quad j = 1, 2, \ldots. \tag{5.47}
$$

One can also show the following: For any $\varepsilon > 0$ and j, there exist c_1 and c_2 sufficiently small such that we have

$$
\widetilde{\omega}_j - \epsilon \le \omega_{j+1}^{\alpha,k} \le \omega_{j+1} \tag{5.48}
$$

for $|\alpha| \le c_1$ and $k > 1/c_2$. See Lemma 6.18.

Since 0 is an eigenvalue of the periodic problem with multiplicity 1, combining formulas (5.21), (5.34), and (5.48) shows that the spectral bands converge, as $k \to +\infty$, to

$$
[0, \omega_1] \cup [\widetilde{\omega}_1, \omega_2] \cup [\widetilde{\omega}_2, \omega_3] \cup \ldots, \tag{5.49}
$$

and hence we have a band gap if and only if the following holds:

$$
\omega_j < \widetilde{\omega}_j \quad \text{for some } j. \tag{5.50}
$$

It is proved in [**295**] that the spectral bands converge to (5.49) in a somewhat different way and (5.50) holds provided that $\int_D u_j \neq 0$ where u_j is an eigenfunction corresponding to ω_j^2.

As we will see in the next chapter, the situation for the phononic crystal is more subtle and complicated. Among other reasons, it is because, unlike the case of the Laplace operator, 0 is an eigenvalue of the periodic problem for the Lamé system with multiplicity 2 (in two dimensions).

5.7. Sensitivity Analysis with Respect to Small Perturbations in the Geometry of the Holes

Suppose that D is of class $\mathcal{C}^2$. Let D_ϵ be an ϵ-perturbation of D; *i.e.*, let $h \in \mathcal{C}^1(\partial D)$ and ∂D_ϵ be given by

$$\partial D_\epsilon = \Big\{ \widetilde{x} : \widetilde{x} = x + \epsilon h(x)\nu(x), \ x \in \partial D \Big\}.$$

Define the operator-valued function $\mathcal{A}_\epsilon^\alpha$ by

$$\mathcal{A}_\epsilon^\alpha : \omega \mapsto \begin{pmatrix} \mathcal{S}_{D_\epsilon}^{\alpha,\omega} & -\mathcal{S}_{D_\epsilon}^{\alpha,\frac{\omega}{\sqrt{k}}} \\ \frac{1}{k}\left(\frac{1}{2}I - (\mathcal{K}_{D_\epsilon}^{-\alpha,\omega})^*\right) & \frac{1}{2}I + (\mathcal{K}_{D_\epsilon}^{-\alpha,\frac{\omega}{\sqrt{k}}})^* \end{pmatrix}.$$

Write

$$\frac{\partial G^{\alpha,\omega}}{\partial \nu(x)}(x,y) = \frac{1}{2\pi}\frac{\langle x-y, \nu(x)\rangle}{|x-y|^2} + R^{\alpha,\omega}(x,y),$$

where $R^{\alpha,\omega}(x,y)$ is smooth for all x and y. Following Subsection 3.2.2, we have a uniformly convergent expansion for the length element $d\sigma_\epsilon(\widetilde{y})$ on ∂D_ϵ; *i.e.*,

$$d\sigma_\epsilon(\widetilde{y}) = \sum_{n=0}^{+\infty} \epsilon^n \sigma^{(n)}(y) d\sigma(y),$$

where $\sigma^{(n)}$ are bounded functions, and easily prove that the following lemma holds.

LEMMA 5.12. *Let Ψ_ϵ be the diffeomorphism from ∂D onto ∂D_ϵ given by*

$$\Psi_\epsilon(x) = x + \epsilon h(x)\nu(x).$$

Let $N \in \mathbb{N}$. There exist C depending only on N, the $\mathcal{C}^2$-norm of D, and $\|h\|_{\mathcal{C}^1(\partial D)}$ such that for any $\widetilde{\varphi} \in L^2(\partial D_\epsilon)$,

$$\left\| \mathcal{S}_{D_\epsilon}^{\alpha,\omega}[\widetilde{\varphi}] \circ \Psi_\epsilon - \mathcal{S}^{\alpha,\omega}[\varphi] - \sum_{n=1}^N \epsilon^n \mathcal{S}_{\alpha,\omega}^{(n)}[\varphi] \right\|_{L^2(\partial D)} \leq C\epsilon^{N+1} \|\varphi\|_{L^2(\partial D)}$$

and

$$\left\| ((\mathcal{K}_{D_\epsilon}^{-\alpha,\omega})^*[\widetilde{\varphi}]) \circ \Psi_\epsilon - (\mathcal{K}^{\alpha,\omega})^*\varphi - \sum_{n=1}^N \epsilon^n \mathcal{K}_{\alpha,\omega}^{(n)}[\varphi] \right\|_{L^2(\partial D)} \leq C\epsilon^{N+1} \|\varphi\|_{L^2(\partial D)},$$

where $\varphi := \widetilde{\varphi} \circ \Psi_\epsilon$. Here

$$\mathcal{S}_{\alpha,\omega}^{(n)}[\varphi](x) = \sum_{l=0}^n \sum_{|\beta|=l} \frac{1}{\beta!} \int_{\partial D} \partial^\beta G^{\alpha,\omega}(x,y)(h(x)\nu(x) - h(y)\nu(y))^l \sigma^{(n-l)}(y)\varphi(y) d\sigma(y),$$

and

$$\mathcal{K}_{\alpha,\omega}^{(n)}[\varphi](x) = \mathcal{K}_{\omega}^{(n)}[\varphi](x) + \sum_{l=0}^{n} \sum_{|\beta|=l} \frac{1}{\beta!} \int_{\partial D} \partial^{\beta} R^{\alpha,\omega}(x,y)(h(x)\nu(x) - h(y)\nu(y))^{l} \sigma^{(n-l)}(y)\varphi(y)\, d\sigma(y),$$

and the bounded operators $\mathcal{K}_{\omega}^{(n)} = \mathcal{K}_{D,\omega}^{(n)}$ *are defined in (*2.150*).*

The sensitivity analysis with respect to small perturbations in the geometry of the holes consists of expanding, based on Lemma 5.12, $\mathcal{A}_{\epsilon}^{\alpha}$ in terms of ϵ to calculate asymptotic expressions of its characteristic values. This can be done in exactly the same manner as in Theorem 5.6.

5.8. Proof of the Representation Formula

In this section we provide a proof of representation formula (5.10) which plays a central role in our analysis.

THEOREM 5.13. *Suppose that* ω^2 *is not an eigenvalue for* $-\Delta$ *in* $Y \setminus \overline{D}$ *with Dirichlet boundary condition on* ∂D *and quasi-periodic boundary condition on* ∂Y *and assume* ω^2/k *is not an eigenvalue for* $-\Delta$ *in* D *with Dirichlet boundary condition on* ∂D*. Then, for any eigenfunction* u *of (*5.6*), there exists one and only one pair* $(\phi, \psi) \in L^2(\partial D) \times L^2(\partial D)$ *such that* u *has the representation (*5.11*). Moreover,* (ϕ, ψ) *is the solution to the integral equation (*5.12*). The mapping* $u \longmapsto (\phi, \psi)$ *from solutions of (*5.6*) to solutions of the system of integral equations (*5.12*) is one-to-one.*

We first prove the following lemma.

LEMMA 5.14. *Suppose that* u *is an eigenfunction of (*5.6*). Then*

$$u|_{\partial D} \perp \mathrm{Ker}(\mathcal{S}^{-\alpha,\omega}).$$

Here $\mathcal{S}^{-\alpha,\omega}$ *is considered an operator from* $L^2(\partial D)$ *into* $H^1(\partial D)$*.*

PROOF. To prove this lemma, we observe that, since $(\Delta + \omega^2)u = 0$ in D,

$$u(x) = \mathcal{D}^{\alpha,\omega}\left[u|_{\partial D}\right](x) - \mathcal{S}^{\alpha,\omega}\left[\frac{\partial u}{\partial \nu}\Big|_{-}\right](x), \quad x \in D,$$

and consequently,

$$\frac{1}{2}u|_{\partial D} = \mathcal{K}^{\alpha,\omega}\left[u|_{\partial D}\right] - \mathcal{S}^{\alpha,\omega}\left[\frac{\partial u}{\partial \nu}\Big|_{-}\right].$$

Let $\phi \in \mathrm{Ker}(\mathcal{S}^{-\alpha,\omega})$. Because of the assumption on ω^2, we immediately deduce that $\mathcal{S}^{-\alpha,\omega}[\phi] = 0$ in $Y \setminus \overline{D}$, and hence

$$\begin{cases} \mathcal{S}^{-\alpha,\omega}[\phi] = 0, \\ \frac{1}{2}\phi + (\mathcal{K}^{\alpha,\omega})^{*}[\phi] = 0 \end{cases} \quad \text{on } \partial D. \tag{5.51}$$

Then, we have

$$\begin{aligned}\frac{1}{2}\langle u|_{\partial D},\phi\rangle &= \langle \mathcal{K}^{\alpha,\omega}\left[u|_{\partial D}\right],\phi\rangle - \left\langle \mathcal{S}^{\alpha,\omega}\left[\frac{\partial u}{\partial \nu}\Big|_{-}\right],\phi\right\rangle \\ &= \langle u|_{\partial D},(\mathcal{K}^{\alpha,\omega})^{*}[\phi]\rangle - \left\langle \frac{\partial u}{\partial \nu}\Big|_{-},\mathcal{S}^{-\alpha,\omega}[\phi]\right\rangle \\ &= -\frac{1}{2}\langle u|_{\partial D},\phi\rangle - 0,\end{aligned}$$

which proves the lemma. □

PROOF OF THEOREM 5.13. We first note that the problem of finding (ϕ,ψ) is equivalent to solving the two equations

$$(5.52)\qquad \begin{cases} \mathcal{S}^{\alpha,\omega}[\phi] = u|_{\partial D} & \text{on } \partial D,\\ \left(-\dfrac{1}{2}I + (\mathcal{K}^{-\alpha,\omega})^{*}\right)[\phi] = \dfrac{\partial u}{\partial \nu}\Big|_{-} & \text{on } \partial D \end{cases}$$

and

$$(5.53)\qquad \begin{cases} \mathcal{S}^{\alpha,\frac{\omega}{\sqrt{k}}}[\psi] = u|_{\partial D} & \text{on } \partial D,\\ \left(\dfrac{1}{2}I + (\mathcal{K}^{-\alpha,\frac{\omega}{\sqrt{k}}})^{*}\right)[\psi] = \dfrac{\partial u}{\partial \nu}\Big|_{+} & \text{on } \partial D. \end{cases}$$

Here we only consider the problem of finding ϕ, the solution to (5.52); the problem of finding ψ can be solved in the same way.

From Lemma 5.14 it follows that there exists $\phi_0 \in L^2(\partial D)$ such that

$$\mathcal{S}^{\alpha,\omega}[\phi_0+\phi] = u|_{\partial D} \quad \text{on } \partial D, \quad \forall \phi \in \operatorname{Ker}(\mathcal{S}^{\alpha,\omega}).$$

Hence, to show existence of a solution to (5.52), it suffices to prove that there exists $\phi \in \operatorname{Ker}(\mathcal{S}^{\alpha,\omega})$ such that the second equation in (5.52) is satisfied by $\phi_0+\phi$. Thanks to the second equation in (5.51), this equation becomes

$$(5.54)\qquad \phi = \frac{\partial\left(\mathcal{S}^{\alpha,\omega}[\phi_0]-u\right)}{\partial \nu}\Big|_{-},$$

and then, we only need to show that

$$\frac{\partial\left(\mathcal{S}^{\alpha,\omega}[\phi_0]-u\right)}{\partial \nu}\Big|_{-} \in \operatorname{Ker}(\mathcal{S}^{\alpha,\omega}),$$

which is an immediate consequence of the fact that $\mathcal{S}^{\alpha,\omega}[\phi_0]-u$ is a solution to $\Delta+\omega^2$ in D with the Dirichlet boundary condition. We have then proved the existence of a solution to (5.52).

Suppose now that we have two solutions ϕ_1 and ϕ_2 to (5.52). Then, because of the assumption on ω^2, we have $\mathcal{S}^{\alpha,\omega}[\phi_1-\phi_2]=0$ in $Y\setminus\overline{D}$, and hence

$$\left(\frac{1}{2}I + (\mathcal{K}^{-\alpha,\omega})^{*}\right)[\phi_1-\phi_2] = 0 \quad \text{on } \partial D.$$

By the second equation in (5.52), we have $\phi_1=\phi_2$.

So far, we have shown that there are unique ϕ and ψ satisfying (5.52) and (5.53), respectively. The jump conditions satisfied by u immediately show that the pair (ϕ,ψ) satisfies the system of integral equations (5.12).

Conversely, suppose that (ϕ,ψ) is a nontrivial solution to the system of integral equations (5.12). Then defining u by (5.11), we only need to show that u is not trivial to conclude that u is an eigenfunction of (5.6). Suppose that $u=0$ in Y.

Then $\mathcal{S}^{\alpha,\omega}[\phi] = 0$ in D, and by the assumption on ω^2, we deduce that $\mathcal{S}^{\alpha,\omega}[\phi] = 0$ in $Y \setminus \overline{D}$. Finally, from the jump of the normal derivative of $\mathcal{S}^{\alpha,\omega}[\phi]$ on ∂D, we deduce that $\phi = 0$. The assumption on ω^2/k leads to $\psi = 0$. This is in contradiction to the fact that $(\phi, \psi) \neq (0, 0)$. This completes the proof. $\square$

5.9. Characterization of the Eigenvalues of $\widetilde{\Delta}$

Let $\widetilde{\omega}_0$ be a characteristic value of the operator-valued function $\widetilde{\mathcal{A}}_0^0$. Let (ϕ, ψ) be a root function associated with $\widetilde{\omega}_0$. Set

$$u = \mathcal{S}^{0,\omega}[\phi] - \frac{1}{\widetilde{\omega}_0^2} \int_{\partial D} \psi$$

and

$$c = \frac{1}{\widetilde{\omega}_0^2 |Y \setminus \overline{D}|} \int_{\partial D} (-\frac{1}{2} I + (\mathcal{K}^{0,\omega})^*)[\phi].$$

It follows from (2.299) that

$$c = \frac{1}{\widetilde{\omega}_0^2} \int_{\partial D} \psi,$$

and therefore, $-\Delta(u + c) = \widetilde{\omega}_0^2 (u + c)$ in D and $u = 0$ on ∂D. Thus, we conclude that $\widetilde{\omega}_0^2$ is an eigenvalue of $\widetilde{\Delta}$.

Conversely, assume that $\widetilde{\omega}_0^2$ (with $\widetilde{\omega}_0 > 0$) is an eigenvalue of $\widetilde{\Delta}$ associated with $u + c$, where $u \in H_0^1(D)$, and

$$c = \frac{1}{|Y \setminus \overline{D}| \widetilde{\omega}_0^2} \int_{\partial D} \frac{\partial u}{\partial \nu}.$$

Let ϕ be a solution to

$$(\frac{1}{2} I - (\mathcal{K}^{0,\widetilde{\omega}_0})^*)[\phi] = \frac{\partial u}{\partial \nu} \quad \text{on } \partial D. \tag{5.55}$$

Such a solution exits even though $\widetilde{\omega}_0^2$ is an eigenvalue of the Laplacian in D with Neumann boundary condition since in this case $\partial u / \partial \nu$ is orthogonal in L^2 to the associated Neumann eigenvector. Set

$$\psi = -(\frac{1}{2} I + (\mathcal{K}^{0,0})^*)^{-1} \left[\frac{\partial u}{\partial \nu} \right]. \tag{5.56}$$

Then, (ϕ, ψ) satisfies

$$\begin{pmatrix} \mathcal{S}^{0,\widetilde{\omega}_0} & -\dfrac{1}{\widetilde{\omega}_0^2} \displaystyle\int_{\partial D} \\ \dfrac{1}{2} I - (\mathcal{K}^{0,\widetilde{\omega}_0})^* & \dfrac{1}{2} I + (\mathcal{K}^{0,0})^* \end{pmatrix} \begin{pmatrix} \phi \\ \psi \end{pmatrix} = 0,$$

which proves that $\widetilde{\omega}_0$ is a characteristic value of $\widetilde{\mathcal{A}}_0^0$.

5.10. Maximizing Band Gaps in Photonic Crystals

Let Y denote the periodic unit cell $[0, 1)^2$. To study the optimal design of photonic band gaps, we consider the quasi-periodic eigenvalue problems:

$$-(\nabla + \sqrt{-1}\alpha) \cdot (\nabla + \sqrt{-1}\alpha) u_\alpha = (\omega^\alpha)^2 q(x) u_\alpha \tag{5.57}$$

and

$$-(\nabla + \sqrt{-1}\alpha) \frac{1}{q(x)} \cdot (\nabla + \sqrt{-1}\alpha) v_\alpha = (\omega^\alpha)^2 v_\alpha, \tag{5.58}$$

where α is in the Brillouin zone and the density function $q(x)$ is given by

$$q(x) = \begin{cases} q_1 & \text{for } x \in \Omega \setminus \overline{D}, \\ q_2 & \text{for } x \in D. \end{cases} \tag{5.59}$$

The eigenvalue problem (5.57) is for the transverse magnetic polarization while (5.58) is for the transverse electric polarization.

The spectrum of (5.57) (resp. (5.58)) is composed of a sequence of nonnegative eigenvalues, each of finite multiplicity. Repeating them according to multiplicity, we denote them

$$0 \leq \omega_1^\alpha \leq \omega_2^\alpha \leq \omega_3^\alpha \ldots \to +\infty.$$

If we use the level set method (see Appendix B.2) to represent the interface ∂D, then

$$q(x) = \begin{cases} q_1 & \text{for } \{x, \phi(x) < 0\}, \\ q_2 & \text{for } \{x, \phi(x) > 0\}. \end{cases}$$

A typical design goal is to maximize the band gap in the transverse magnetic or the transverse electric case. In both cases, we write [**183**, **210**, **287**, **322**]

$$\mathcal{J}[D] = \inf_\alpha \omega_{j+1}^\alpha - \sup_\alpha \omega_j^\alpha \tag{5.60}$$

and maximize $\mathcal{J}[D]$ with respect to the level set function ϕ. An analysis of the problem shows that it may be nonsmooth, *i.e.*, Lipschitz continuous but not differentiable with respect to ϕ, for several reasons [**210**]. First of all, the inf and sup in the definition of $\mathcal{J}$ are nonsmooth functions. Moreover, multiple eigenvalues introduce a nondifferentiability with respect to ϕ. However, one can still use generalized gradients and bundle optimization techniques to overcome this difficulty.

The generalized gradient of a locally Lipschitz function is defined as follows [**201**, Chapter 2]. Let X be a real Banach space and let $f : X \to \mathbb{R}$ be Lipschitz near a given point $x \in X$. Define

$$\widetilde{f}(x, v) := \limsup_{y \to x, t \to 0^+} \frac{f(y + tv) - f(y)}{t}.$$

The generalized gradient of f at x, denoted $\partial f(x)$, is the (nonempty) weak*-compact subset of X^* (the dual space of continuous linear functionals on X) whose support function is $\widetilde{f}(x, \cdot)$. Thus $\zeta \in \partial f(x)$ if and only if $\widetilde{f}(x, v) \geq (\zeta, v)$ for all v in X. It is also worth noticing that if f admits a Gâteaux derivative $f'_G(x)$ at x, then $f'_G(x) \in \partial f(x)$. Moreover, if f is continuously differentiable at x, then $\partial f(x)$ reduces to a singleton: $\partial f(x) = \{f'(x)\}$.

Let co denote the convex hull, *i.e.*, the set of all convex combinations of elements in the given set. Returning now to our optimal design problem, the generalized gradient of ω_j^α with respect to ϕ can be written as follows [**210**, **322**]:

$$\partial_\phi \omega_j^\alpha \subset \mathrm{co} \left\{ -\frac{1}{2}(q_2 - q_1)\omega_j^\alpha |u_\alpha|^2 : u_\alpha \in \mathcal{V}_{TM}^{(j)}(q, \alpha) \right\},$$

in the transverse magnetic case, and

$$\partial_\phi \omega_j^\alpha \subset \mathrm{co} \left\{ \frac{1}{2\omega_j^\alpha}(\frac{1}{q_2} - \frac{1}{q_1})|(\nabla + \sqrt{-1}\alpha)v_\alpha|^2 : v_\alpha \in \mathcal{V}_{TE}^{(j)}(q, \alpha) \right\},$$

in the TE case, where $\mathcal{V}_{TM}^{(j)}$ (and $\mathcal{V}_{TE}^{(j)}$) are the span of all eigenfunctions u_α (and v_α) associated with the eigenvalue ω_j^α, respectively, and satisfying the normalization $\int_Y q|u_\alpha|^2 = 1$ and $\int_Y |v_\alpha|^2 = 1$. The shape derivatives of $\mathcal{J}[D]$ are given by

$$d_S \mathcal{J}[D] = \int_{\partial D} V_{TM}(\text{or } V_{TE})\, \theta \cdot \nu \, d\sigma,$$

where the velocities which give the ascent direction for the optimization are [**210**, **322**]

$$(5.61) \qquad \begin{aligned} V_{TM} \quad \subset \quad & \text{co}\left\{ -\frac{1}{2}(q_2 - q_1)\omega_{j+1}^\alpha |u_\alpha|^2 : u_\alpha \in \mathcal{V}_{TM}^{(j+1)}(q,\alpha) \right\} \\ & -\text{co}\left\{ -\frac{1}{2}(q_2 - q_1)\omega_j^\alpha |u_\alpha|^2 : u_\alpha \in \mathcal{V}_{TM}^{(j)}(q,\alpha) \right\} \end{aligned}$$

and

$$(5.62) \qquad \begin{aligned} V_{TE} \quad \subset \quad & \text{co}\left\{ \frac{1}{2\omega_{j+1}^\alpha}\Big(\frac{1}{q_2} - \frac{1}{q_1}\Big)|(\nabla + \sqrt{-1}\alpha)v_\alpha|^2 : v_\alpha \in \mathcal{V}_{TE}^{(j+1)}(q,\alpha) \right\} \\ & -\text{co}\left\{ \frac{1}{2\omega_j^\alpha}\Big(\frac{1}{q_2} - \frac{1}{q_1}\Big)|(\nabla + \sqrt{-1}\alpha)v_\alpha|^2 : v_\alpha \in \mathcal{V}_{TE}^{(j)}(q,\alpha) \right\}. \end{aligned}$$

5.11. Photonic Cavities

Let k be a positive constant and let $k(x)$ be a periodic function with period 1 in x_1 and x_2 such that

$$k(x) = \begin{cases} k & \text{in } Y \setminus \overline{D}, \\ 1 & \text{in } D. \end{cases}$$

Consider the solution u to the following problem:

$$(5.63) \qquad \nabla \cdot k(x) \nabla u + \omega^2 n(x) u = 0.$$

Suppose that $n(x) - 1$ is compactly supported in a bounded domain $\Omega \subset \mathbb{R}^2$, and is assumed to be known. Assume that Ω is a localized defect inserted into the photonic crystal. It can be proved that the introduction of a localized defect does not change the essential spectrum of the operator.

Assume that the operator $\nabla \cdot k(x)\nabla$ has a gap in the spectrum and seek for ω inside the bandgap such that (5.63) has a nontrivial solution. As in Section 3.3, we can use an integral formulation to compute ω. We can formally rewrite the solution to (5.63) in integral form

$$(5.64) \qquad u(x) + \omega^2 \int_\Omega (n(y) - 1) G_\omega(x,y) u(y) dy = 0\,, \quad x \in \mathbb{R}^2,$$

where G_ω is the Green's function of $\nabla \cdot k(x)\nabla + \omega^2$ in $\mathbb{R}^2$. Notice that for frequencies in the band gap, G_ω is exponentially decaying. We have [**205**]

$$(5.65) \qquad |G_\omega(x,y)| = O\big(e^{-C\text{dist}(\omega^2, \sigma(-\nabla \cdot k(x)\nabla))}\big) \quad \text{as } |x-y| \to \infty$$

with C being a positive constant and $\sigma(-\nabla \cdot k(x)\nabla)$ is the spectrum of $-\nabla \cdot k(x)\nabla$.

We call ω a defect mode if its value in (5.64) yields nontrivial solutions $u(x)$ of (5.64). Hence, in view of (5.65), a defect mode is a solution to the wave equation which is exponentially localized in the defect while its time dependence is harmonic and can be computed by the same approach as the one presented in Section 3.3. Moreover, using the generalized argument principle, we can also compute the sensitivity of the defect modes with respect to n and Ω.

5.12. Concluding Remarks

In this chapter we have first discussed the structure of spectra of periodic elliptic operators. The main tool of the theory is the Floquet transform. We have also performed a high-order sensitivity analysis of the spectral properties of high contrast band gap materials, consisting of a background medium which is perforated by a periodic array of holes, with respect to the index ratio and small perturbations in the geometry of the holes. The asymptotic expansions have been obtained by transforming the spectral problem into a system of equations involving singular integral operators, a Taylor expansion of the associated kernels, and the generalized Rouché theorem. The leading-order terms in our expansions have been explicitly computed.

The boundary integral formulation we developed can be extended to compute electromagnetic modes of photonic crystal fibers. As has been shown in [**197**], an appropriate boundary integral formulation produces very accurate results for the propagative modes. On the other hand, in [**279**, **503**], a boundary integral formulation, in which the Ewald method was combined with a fast multipole approach, has been developed to compute the transmission of electromagnetic waves through photonic crystal slabs.

Our approach in this chapter will be extended in the next chapter to the equations of linear elasticity.

CHAPTER 6

Phononic Band Gaps

6.1. Introduction

In the past decade there has been a steady growth of interest in the propagation of elastic waves through inhomogeneous materials. The ultimate objective of these investigations has been the design of the so-called phononic band gap materials or phononic crystals. The most recent research in this field has focused on theoretical and experimental demonstrations of band gaps in two-dimensional and three-dimensional structures constructed of high contrast elastic materials arranged in a periodic array. This type of structure prevents elastic waves in certain frequency ranges from propagating and could be used to generate frequency filters with control of pass or stop bands, as beam splitters, as sound or vibration protection devices, or as elastic waveguides. See, for example, [**211**, **343**, **480**, **502**].

To mathematically formulate the problem investigated in this chapter, set D to be a connected domain with Lipschitz boundary lying inside the open square $(0,1)^2$. As in Chapter 5, an important example of phononic crystals consists of a background elastic medium of constant Lamé parameters λ and μ which is perforated by an array of arbitrary-shaped inclusions $\Omega = \bigcup_{n\in\mathbb{Z}^2}(D+n)$ periodic along each of the two orthogonal coordinate axes in the plane. These inclusions have Lamé constants $\widetilde{\lambda}$, $\widetilde{\mu}$. The shear modulus μ of the background medium is assumed to be larger than that of the inclusion $\widetilde{\mu}$. Then we investigate the spectrum of the self-adjoint operator defined by

$$u \mapsto -\nabla \cdot (\mathbb{C}\nabla^s u) = -\sum_{j,k,l=1}^{2} \frac{\partial}{\partial x_j}\left(C_{ijkl}\frac{\partial u_k}{\partial x_l}\right), \tag{6.1}$$

which is densely defined on $L^2(\mathbb{R}^2)^2$. Here the elasticity tensor $\mathbb{C}$ is given by

$$C_{ijkl} := \left(\lambda\,\chi(\mathbb{R}^2\backslash\overline{\Omega})+\widetilde{\lambda}\,\chi(\Omega)\right)\delta_{ij}\delta_{kl}+\left(\mu\,\chi(\mathbb{R}^2\backslash\overline{\Omega})+\widetilde{\mu}\,\chi(\Omega)\right)(\delta_{ik}\delta_{jl}+\delta_{il}\delta_{jk}), \tag{6.2}$$

where $\chi(\Omega)$ is the indicator function of Ω.

In this chapter we adopt this specific two-dimensional model to understand the relationship between the contrast of the shear modulus and the band gap structure of the phononic crystal. We will also consider the case of two materials with different densities in order to investigate the relation between the density contrast and the band gap structure.

By Floquet theory in Section 5.2, the spectrum of the Lamé system with periodic coefficients is represented as a union of bands, called the phononic band structure. Carrying out a band structure calculation for a given phononic crystal involves a family of eigenvalue problems, as the quasi-momentum is varied over the first Brillouin zone. The problem of finding the spectrum of (6.1) is reduced to a family of eigenvalue problems with quasi-periodicity condition; *i.e.*,

$$\nabla \cdot (\mathbb{C}\nabla^s u) + \omega^2 u = 0 \text{ in } \mathbb{R}^2, \tag{6.3}$$

with the periodicity condition

$$u(x+n) = e^{\sqrt{-1}\alpha\cdot n} u(x) \quad \text{for every } n \in \mathbb{Z}^2. \tag{6.4}$$

Here the quasi-momentum α varies over the Brillouin zone $[0, 2\pi)^2$. Each of these operators has compact resolvent so that its spectrum consists of discrete eigenvalues of finite multiplicity. We show that these eigenvalues are the characteristic values of meromorphic operator-valued functions that are of Fredholm type of index zero. As in Chapter 5, this yields a natural and efficient approach to the computation of the band gap phononic structure which is based on a combination of boundary element methods and Muller's method for finding complex roots of scalar equations. See Section 1.6. Following Chapter 5, we proceed from the generalized Rouché theorem to construct complete asymptotic expressions for the characteristic values as the Lamé parameter μ of the background goes to infinity. For $\alpha \neq 0$, we prove that the discrete spectrum of (6.3) accumulates near the Dirichlet eigenvalues of the Lamé system in D as μ goes to infinity. We then obtain a full asymptotic formula for the eigenvalues. The leading-order term is of order μ^{-1} and can be calculated explicitly. For the periodic case $\alpha = 0$, we establish a formula for the asymptotic behavior of the eigenvalues. It turns out that their limiting set is generically different from that for $\alpha \neq 0$. We also consider the case when $|\alpha|$ is of order $1/\sqrt{\mu}$ and derive an asymptotic expansion for the eigenvalues in this case as well. Not surprisingly, this formula tends continuously to the previous ones as $\alpha\sqrt{\mu}$ goes to zero or to infinity. We finally provide a criterion for exhibiting gaps in the band structure. As we said before, the existence of those spectral gaps implies that the elastic waves in those frequency ranges are prohibited from travelling through the elastic body. Our criterion shows that the smaller the density of the matrix, the wider the band gap, provided that the criterion is fulfilled. This phenomenon was reported in [**227**] where it was observed that periodic elastic composites, whose matrix has lower density and higher shear modulus compared to those of inclusions, yield better open gaps.

6.2. Asymptotic Behavior of Phononic Band Gaps

The phononic crystal we consider in this chapter is a homogeneous elastic medium of Lamé constants λ, μ which contains a periodic (with respect to the lattice $\mathbb{Z}^2$) array of arbitrary-shaped inclusions $\Omega = \bigcup_{n\in\mathbb{Z}^2}(D+n)$. These inclusions have Lamé constants $\widetilde{\lambda}$, $\widetilde{\mu}$.

We use the same notation as in Chapter 5. Let $Y = (0,1)^2$ denote the fundamental periodic cell. For each quasi-momentum $\alpha \in [0, 2\pi)^2$, set $\sigma_\alpha(D)$ to be the (discrete) spectrum of the operator defined by (6.1) with the condition that

$e^{-\sqrt{-1}\alpha\cdot x}u$ is periodic. In other words, $\sigma_\alpha(D)$ is the spectrum of the problem

$$\begin{cases} \mathcal{L}^{\lambda,\mu}u + \omega^2 u = 0 & \text{in } Y \setminus \overline{D}, \\ \mathcal{L}^{\widetilde{\lambda},\widetilde{\mu}}u + \omega^2 u = 0 & \text{in } D, \\ u\big|_+ - u\big|_- = 0 & \text{on } \partial D, \\ \dfrac{\partial u}{\partial \nu}\Big|_+ - \dfrac{\partial u}{\partial \widetilde{\nu}}\Big|_- = 0 & \text{on } \partial D, \\ e^{-\sqrt{-1}\alpha\cdot x}u \text{ is periodic in the whole space.} \end{cases} \tag{6.5}$$

Recall that $\mathcal{L}^{\widetilde{\lambda},\widetilde{\mu}}$ denotes the elastostatic system corresponding to the Lamé constants $\widetilde{\lambda}$ and $\widetilde{\mu}$ and $\partial/\partial\widetilde{\nu}$ is the corresponding conormal derivative.

By the standard Floquet theory briefly described in Subsection 5.2, the spectrum of (6.5) has the band structure given by

$$\bigcup_{\alpha\in[0,2\pi)^2} \sigma_\alpha(D). \tag{6.6}$$

The main objective of this section is to investigate the behavior of $\sigma_\alpha(D)$ as $\mu \to +\infty$.

6.2.1. Integral Representation of Quasi-Periodic Solutions. In this subsection, we obtain the integral representation formula for the solution to (6.5). We denote by $\widetilde{\mathcal{S}}^\omega, \widetilde{\mathcal{D}}^\omega$, and $\widetilde{\mathcal{K}}^\omega$ the layer potentials on ∂D associated with the Lamé parameters $(\widetilde{\lambda}, \widetilde{\mu})$.

We first prove the following lemma.

LEMMA 6.1. *Suppose that ω^2 is not an eigenvalue for $-\mathcal{L}^{\lambda,\mu}$ in D with the Dirichlet boundary condition on ∂D. Let u be a solution to (6.5). Then we have*

$$u|_{\partial D} \perp \operatorname{Ker} \widetilde{\mathcal{S}}^\omega \quad \textit{and} \quad u|_{\partial D} \perp \operatorname{Ker}(\mathcal{S}^{\alpha,\omega})^*.$$

Here $\widetilde{\mathcal{S}}^\omega$ and $\mathcal{S}^{\alpha,\omega}$ are considered to be operators on $L^2(\partial D)^2$.

PROOF. We first observe that, since $(\mathcal{L}^{\widetilde{\lambda},\widetilde{\mu}} + \omega^2)u = 0$ in D, we have

$$u(x) = \widetilde{\mathcal{D}}^\omega\left[u|_{\partial D}\right](x) - \widetilde{\mathcal{S}}^\omega\left[\frac{\partial u}{\partial\widetilde{\nu}}\Big|_-\right](x), \quad x \in D, \tag{6.7}$$

and consequently by (2.161) it follows that

$$\frac{1}{2}u|_{\partial D} = \widetilde{\mathcal{K}}^\omega\left[u|_{\partial D}\right] - \widetilde{\mathcal{S}}^\omega\left[\frac{\partial u}{\partial\widetilde{\nu}}\Big|_-\right]. \tag{6.8}$$

Let $\phi \in \operatorname{Ker}(\widetilde{\mathcal{S}}^\omega)$; *i.e.*, $\widetilde{\mathcal{S}}^\omega[\phi] = 0$ on ∂D. By Lemma 2.121, we have $\widetilde{\mathcal{S}}^\omega[\phi] = 0$ in $\mathbb{R}^2 \setminus \overline{D}$ and hence $(1/2)\phi + (\widetilde{\mathcal{K}}^\omega)^*[\phi] = 0$ by (2.160). Then we have from (6.8)

$$\begin{aligned} \frac{1}{2}\langle u|_{\partial D}, \phi\rangle &= \left\langle \widetilde{\mathcal{K}}^\omega\left[u|_{\partial D}\right], \phi \right\rangle - \left\langle \widetilde{\mathcal{S}}^\omega\left[\frac{\partial u}{\partial\widetilde{\nu}}\Big|_-\right], \phi \right\rangle \\ &= \left\langle u|_{\partial D}, (\widetilde{\mathcal{K}}^\omega)^*[\phi] \right\rangle - \left\langle \frac{\partial u}{\partial\widetilde{\nu}}\Big|_-, \widetilde{\mathcal{S}}^\omega[\phi] \right\rangle \\ &= -\frac{1}{2}\langle u|_{\partial D}, \phi\rangle, \end{aligned}$$

which implies $\langle u|_{\partial D}, \phi\rangle = 0$, and hence $u|_{\partial D} \perp \operatorname{Ker} \widetilde{\mathcal{S}}^\omega$.

Observe that if u is α-quasi-periodic, then

$$\mathcal{D}_Y^{\alpha,\omega}\left[u|_{\partial Y}\right] = 0 \quad \text{and} \quad \mathcal{S}_Y^{\alpha,\omega}\left[\frac{\partial u}{\partial \nu}\Big|_+\right] = 0 \quad \text{on } \partial Y,$$

where $\mathcal{D}_Y^{\alpha,\omega}$ and $\mathcal{S}_Y^{\alpha,\omega}$ are the (α-quasi-periodic) double- and single-layer potentials on ∂Y. Thus we have

$$u(x) = -\mathcal{D}^{\alpha,\omega}\left[u|_{\partial D}\right](x) + \mathcal{S}^{\alpha,\omega}\left[\frac{\partial u}{\partial \nu}\Big|_+\right](x), \quad x \in Y \setminus \overline{D},$$

and consequently,

$$\frac{1}{2}u|_{\partial D} = -\mathcal{K}^{\alpha,\omega}\left[u|_{\partial D}\right] + \mathcal{S}^{\alpha,\omega}\left[\frac{\partial u}{\partial \nu}\Big|_+\right].$$

Let $\phi \in \mathrm{Ker}(\mathcal{S}^{\alpha,\omega})^*$. Since $(\mathcal{S}^{\alpha,\omega})^* = \mathcal{S}^{-\alpha,\omega}$, we have

$$\mathcal{S}^{-\alpha,\omega}[\phi] = 0 \quad \text{on } \partial D.$$

Since ω^2 is not a Dirichlet eigenvalue of $-\mathcal{L}^{\lambda,\mu}$ in D, we immediately get

$$\mathcal{S}^{-\alpha,\omega}[\phi] = 0 \quad \text{in } D,$$

and hence

$$-\frac{1}{2}\phi + \left(\mathcal{K}^{\alpha,\omega}\right)^*[\phi] = 0 \quad \text{on } \partial D.$$

Therefore, we can deduce that

$$\begin{aligned}\frac{1}{2}\langle u|_{\partial D}, \phi\rangle &= -\left\langle \mathcal{K}^{\alpha,\omega}\left[u|_{\partial D}\right], \phi\right\rangle + \left\langle \mathcal{S}^{\alpha,\omega}\left[\frac{\partial u}{\partial \nu}\Big|_+\right], \phi\right\rangle \\ &= -\left\langle u|_{\partial D}, (\mathcal{K}^{\alpha,\omega})^*[\phi]\right\rangle + \left\langle \frac{\partial u}{\partial \nu}\Big|_+, \mathcal{S}^{-\alpha,\omega}[\phi]\right\rangle \\ &= -\frac{1}{2}\left\langle u|_{\partial D}, \phi\right\rangle,\end{aligned}$$

which implies $\langle u|_{\partial D}, \phi\rangle = 0$. This completes the proof. □

We now establish a representation formula for solutions of (6.5).

THEOREM 6.2. *Suppose that ω^2 is not an eigenvalue for $-\mathcal{L}^{\lambda,\mu}$ in D with the Dirichlet boundary condition on ∂D. Then, for any solution u of (6.5), there exists one and only one pair $(\phi, \psi) \in L^2(\partial D)^2 \times L^2(\partial D)^2$ such that*

$$u(x) = \begin{cases} \widetilde{\mathcal{S}}^{\omega}[\phi](x), & x \in D, \\ \mathcal{S}^{\alpha,\omega}[\psi](x), & x \in Y \setminus \overline{D}. \end{cases} \tag{6.9}$$

Moreover, (ϕ, ψ) satisfies

$$\begin{cases} \widetilde{\mathcal{S}}^{\omega}[\phi] - \mathcal{S}^{\alpha,\omega}[\psi] = 0 & \text{on } \partial D, \\ \left(\frac{1}{2}I - (\widetilde{\mathcal{K}}^{\omega})^*\right)[\phi] + \left(\frac{1}{2}I + (\mathcal{K}^{-\alpha,\omega})^*\right)[\psi] = 0 & \text{on } \partial D, \end{cases} \tag{6.10}$$

and the mapping $u \longmapsto (\phi, \psi)$ from solutions of (6.5) in $H^1(Y)^2$ to solutions to the system of integral equations (6.10) in $L^2(\partial D)^2 \times L^2(\partial D)^2$ is a one-to-one correspondence.

PROOF. We first note that the problem of finding (ϕ, ψ) satisfying (6.9) and (6.10) is equivalent to solving the following two systems of equations:

$$\begin{cases} \widetilde{\mathcal{S}}^\omega[\phi] = u|_{\partial D} & \text{on } \partial D, \\ \Big(-\dfrac{1}{2}I + (\widetilde{\mathcal{K}}^\omega)^*\Big)[\phi] = \dfrac{\partial u}{\partial \widetilde{\nu}}\Big|_- & \text{on } \partial D \end{cases} \tag{6.11}$$

and

$$\begin{cases} \mathcal{S}^{\alpha,\omega}[\psi] = u|_{\partial D} & \text{on } \partial D, \\ \Big(\dfrac{1}{2}I + (\mathcal{K}^{-\alpha,\omega})^*\Big)[\psi] = \dfrac{\partial u}{\partial \nu}\Big|_+ & \text{on } \partial D. \end{cases} \tag{6.12}$$

In order to find ϕ satisfying (6.11), it suffices to find ϕ satisfying $\widetilde{\mathcal{S}}^\omega[\phi] = u$ in D. Suppose for a moment that the following holds:

$$\operatorname{Im} \widetilde{\mathcal{S}}^\omega = \Big\{ \phi : \phi \perp \operatorname{Ker} \widetilde{\mathcal{S}}^\omega \Big\}. \tag{6.13}$$

It then follows from Lemma 6.1 that there exists $\phi_0 \in L^2(\partial D)^2$ such that

$$\widetilde{\mathcal{S}}^\omega[\phi_0] = u|_{\partial D} \quad \text{on } \partial D. \tag{6.14}$$

Observe that if $\omega \neq 0$, then the solution to the Dirichlet problem for $\mathcal{L}^{\lambda,\mu} + \omega^2$ may not be unique, and hence (6.14) does not imply $\widetilde{\mathcal{S}}^\omega[\phi_0] = u$ in D. However, since $(\mathcal{L}^{\lambda,\mu} + \omega^2)(u - \widetilde{\mathcal{S}}^\omega[\phi_0]) = 0$ in D, we get by Green's formula

$$u - \widetilde{\mathcal{S}}^\omega[\phi_0] = -\widetilde{\mathcal{S}}^\omega \left[\frac{\partial}{\partial \widetilde{\nu}} \Big(u - \widetilde{\mathcal{S}}^\omega[\phi_0] \Big) \Big|_- \right] \quad \text{in } D,$$

and therefore,

$$u = \widetilde{\mathcal{S}}^\omega \left[\phi_0 - \frac{\partial}{\partial \widetilde{\nu}} \Big(u - \widetilde{\mathcal{S}}^\omega[\phi_0] \Big) \Big|_- \right] \quad \text{in } D.$$

To prove the uniqueness of ϕ satisfying (6.11), it suffices to show that the solution to

$$\begin{cases} \widetilde{\mathcal{S}}^\omega[\phi] = 0 & \text{on } \partial D, \\ \Big(-\dfrac{1}{2}I + (\widetilde{\mathcal{K}}^\omega)^*\Big)[\phi] = 0 & \text{on } \partial D \end{cases}$$

is zero. By the first equation in the above and Lemma 2.121, $\widetilde{\mathcal{S}}^\omega[\phi] = 0$ in $\mathbb{R}^2 \setminus \overline{D}$ and hence

$$\phi = \frac{\partial}{\partial \widetilde{\nu}} \widetilde{\mathcal{S}}^\omega[\phi]\Big|_+ - \frac{\partial}{\partial \widetilde{\nu}} \widetilde{\mathcal{S}}^\omega[\phi]\Big|_- = 0.$$

Similarly, we can show existence and uniqueness of ψ satisfying

$$u = \mathcal{S}^{\alpha,\omega}[\psi] \quad \text{in } Y \setminus \overline{D},$$

which yields (6.12). To complete the proof, we shall verify that (6.13) holds. Let G be a subspace of $H^1(\partial\Omega)$ such that

$$G := \Big\{ \phi : \phi \perp \operatorname{Ker} \widetilde{\mathcal{S}}^\omega \Big\}.$$

Since

$$\langle \widetilde{\mathcal{S}}^\omega \phi, \psi \rangle = \langle \phi, \widetilde{\mathcal{S}}^\omega \psi \rangle, \qquad \forall \phi, \psi \in L^2(\partial\Omega),$$

it is easy to see that $\operatorname{Im} \widetilde{\mathcal{S}}^\omega \subset G$. It remains then to show that

$$\dim H^1(\partial\Omega) / \operatorname{Im} \widetilde{\mathcal{S}}^\omega \leq \dim H^1(\partial\Omega)/G.$$

Let $\phi_1, \phi_2, \ldots, \phi_n$ be an orthonormal basis of Ker $\widetilde{\mathcal{S}}^\omega$. Since $H^1(\partial\Omega)$ is dense in $L^2(\partial\Omega)$, we can take $\widetilde{\phi}_1, \widetilde{\phi}_2, \ldots, \widetilde{\phi}_n$ in $H^1(\partial\Omega)$ such that $\|\phi_j - \widetilde{\phi}_j\|_{L^2(\partial\Omega)} \leq \epsilon$ for all j. Then $\widetilde{\phi}_1, \widetilde{\phi}_2, \ldots, \widetilde{\phi}_n$ is linearly independent in $H^1(\partial\Omega)/G$. To see this, suppose that

$$a_1\widetilde{\phi}_1 + \cdots + a_n\widetilde{\phi}_n \in G.$$

By taking the inner products with the ϕ_j's, we obtain

$$\begin{pmatrix} \phi_1 \cdot \widetilde{\phi}_1 & \ldots & \phi_1 \cdot \widetilde{\phi}_n \\ & \vdots & \\ \phi_n \cdot \widetilde{\phi}_1 & \ldots & \phi_n \cdot \widetilde{\phi}_n \end{pmatrix} \begin{pmatrix} a_1 \\ \vdots \\ a_n \end{pmatrix} = 0.$$

Since

$$\begin{pmatrix} \phi_1 \cdot \widetilde{\phi}_1 & \ldots & \phi_1 \cdot \widetilde{\phi}_n \\ & \vdots & \\ \phi_n \cdot \widetilde{\phi}_1 & \ldots & \phi_n \cdot \widetilde{\phi}_n \end{pmatrix}$$

is invertible (as it is a small perturbation of the identity), we have $a_1 = \ldots = a_n = 0$. Hence,

$$\dim H^1(\partial\Omega)/\operatorname{Im} \widetilde{\mathcal{S}}^\omega = \dim \operatorname{Ker} \widetilde{\mathcal{S}}^\omega \leq \dim H^1(\partial\Omega)/G,$$

and we can conclude that $G = \operatorname{Im} \widetilde{\mathcal{S}}^\omega$. □

Let $\mathcal{A}^{\alpha,\mu}(\omega)$ be the operator-valued function of ω defined by

$$\mathcal{A}^{\alpha,\mu}(\omega) := \begin{pmatrix} \widetilde{\mathcal{S}}^\omega & -\mathcal{S}^{\alpha,\omega} \\ \frac{1}{2}I - (\widetilde{\mathcal{K}}^\omega)^* & \frac{1}{2}I + (\mathcal{K}^{-\alpha,\omega})^* \end{pmatrix}. \tag{6.15}$$

By Theorem 6.2, ω^2 is an eigenvalue corresponding to the quasi-momentum α if and only if ω is a characteristic value of $\mathcal{A}^{\alpha,\mu}(\omega)$. Consequently, we now have a new way of computing the spectrum of (6.5) parallel to our formulation in the previous chapter of the band structure problem for photonic crystals. This way consists of examining the characteristic values of $\mathcal{A}^{\alpha,\mu}(\omega)$. Based on Muller's method for finding complex roots of scalar equations, a boundary element method similar to the one developed in Subsection 5.4.2 for photonic crystals can be designed for computing phononic band gaps.

6.2.2. Full Asymptotic Expansions. Expanding the operator-valued functions $\mathcal{A}^{\alpha,\mu}(\omega)$ in terms of μ as $\mu \to +\infty$, we can calculate asymptotic expressions of their characteristic values with the help of the generalized Rouché theorem, and this is what we do in this subsection.

We begin with the following asymptotic expansion of $G_{ij}^{\alpha,\omega}(x,y)$ in (2.530).

LEMMA 6.3. *Let* $\tau_l = 1 - (c_T/c_L)^{2l}$. *As* $\mu \to +\infty$,

$$\begin{aligned} G_{ij}^{\alpha,\omega}(x,y) &= \sum_{l=1}^{+\infty} \frac{\omega^{2(l-1)}}{\mu^l} \sum_{n\in\mathbb{Z}^2} e^{\sqrt{-1}(2\pi n+\alpha)\cdot(x-y)} \bigg(\frac{-\delta_{ij}}{|2\pi n+\alpha|^{2l}} \\ &\qquad + \tau_l \frac{(2\pi n+\alpha)_i(2\pi n+\alpha)_j}{|2\pi n+\alpha|^{2(l+1)}} \bigg), \end{aligned} \tag{6.16}$$

for fixed $\alpha \neq 0$, while for $\alpha = 0$,

(6.17)

$$G_{ij}^{0,\omega}(x,y) = \frac{\delta_{ij}}{\omega^2} + \sum_{l=1}^{+\infty} \frac{\omega^{2(l-1)}}{(2\pi)^{2l}\mu^l} \sum_{n\in\mathbb{Z}^2\setminus\{0\}} e^{2\pi\sqrt{-1}n\cdot(x-y)} \left(-\frac{\delta_{ij}}{|n|^{2l}} + \tau_l \frac{n_i n_j}{|n|^{2(l+1)}} \right).$$

The derivations of (6.16) and (6.17) are straightforward. In fact, since

$$\frac{1}{k_T^2 - |2\pi n + \alpha|^2} = \frac{1}{\dfrac{\omega^2}{\mu} - |2\pi n + \alpha|^2} = -\sum_{k=0}^{+\infty} \frac{\omega^{2k}}{\mu^k |2\pi n + \alpha|^{2(k+1)}},$$

one easily shows that (6.16) and (6.17) hold.

We can write (6.16) and (6.17) as

$$\mathbf{G}^{\alpha,\omega}(x,y) = \sum_{l=1}^{+\infty} \frac{1}{\mu^l} \mathbf{G}_l^{\alpha,\omega}(x,y) \tag{6.18}$$

and

$$\mathbf{G}^{0,\omega}(x,y) = \frac{1}{\omega^2} I + \sum_{l=1}^{+\infty} \frac{1}{\mu^l} \mathbf{G}_l^{0,\omega}(x,y), \tag{6.19}$$

where the definitions of $\mathbf{G}_l^{\alpha,\omega}(x,y)$ and $\mathbf{G}_l^{0,\omega}(x,y)$ are obvious from (6.16) and (6.17). We note that $\mathbf{G}_l^{\alpha,\omega}(x,y)$ and $\mathbf{G}_l^{0,\omega}(x,y)$ are dependent upon μ because of the factor τ_l. However, since $|\tau_l| \leq C$ for some constant C independent of μ and l, this will not affect our subsequent analysis. We also note that $\mathbf{G}_1^{\alpha,\omega}(x,y)$ is independent of ω and

$$\mathbf{G}_1^{\alpha,\omega}(x,y) = \mu \mathbf{G}^{\alpha,0}(x,y), \tag{6.20}$$

where $\mathbf{G}^{\alpha,0}(x,y)$ is the quasi-periodic fundamental function defined in (2.531).

Denote by $\mathcal{S}_l^{\alpha,\omega}$ and $(\mathcal{K}_l^{-\alpha,\omega})^*$, for $l \geq 1$ and $\alpha \in [0,2\pi)^2$, the single-layer potential and the boundary integral operator associated with the kernel $\mathbf{G}_l^{\alpha,\omega}(x,y)$ as defined in (2.538) so that

$$\mathcal{S}^{\alpha,\omega} = \sum_{l=1}^{+\infty} \frac{1}{\mu^l} \mathcal{S}_l^{\alpha,\omega} \quad \text{and} \quad (\mathcal{K}^{-\alpha,\omega})^* = \sum_{l=1}^{+\infty} \frac{1}{\mu^l} (\mathcal{K}_l^{-\alpha,\omega})^*. \tag{6.21}$$

LEMMA 6.4. *The operator $(1/2)\, I + (\mathcal{K}^{-\alpha,0})^* : L^2(\partial D)^2 \to L^2(\partial D)^2$ is invertible.*

Before proving Lemma 6.4, let us make a note of the following simple fact: If u and v are α-quasi-periodic, then

$$\int_{\partial Y} \frac{\partial u}{\partial \nu} \cdot \overline{v} d\sigma = 0. \tag{6.22}$$

To prove this, we observe that

$$\int_{\partial Y} \frac{\partial u}{\partial \nu} \cdot \overline{v} = \int_{\partial Y} \frac{\partial (e^{-\sqrt{-1}\alpha\cdot x} u)}{\partial \nu} \cdot \overline{e^{-\sqrt{-1}\alpha\cdot x} v} + \sqrt{-1} \int_{\partial Y} \Big[\lambda \alpha \cdot (e^{-\sqrt{-1}\alpha\cdot x} u) N$$
$$+ \mu \begin{pmatrix} 2\alpha_1 N_1 + \alpha_2 N_2 & \alpha_1 N_2 \\ \alpha_2 N_1 & \alpha_1 N_1 + 2\alpha_2 N_2 \end{pmatrix} (e^{-\sqrt{-1}\alpha\cdot x} u) \Big] \cdot \overline{e^{-\sqrt{-1}\alpha\cdot x} v}.$$

Here $N = (N_1, N_2)$ is the outward unit normal to the unit cell Y. Then the integrands over the opposite sides of ∂Y have the same absolute values with different signs and therefore the integration over ∂Y is zero.

PROOF OF LEMMA 6.4. For $\alpha \neq 0$, we show the injectivity of $(1/2)\, I + (\mathcal{K}^{-\alpha,0})^*$. Since $(1/2)\, I + (\mathcal{K}^0)^*$ and $(\mathcal{K}^{-\alpha,0})^* - (\mathcal{K}^0)^*$ are compact, then from the Fredholm alternative, the result follows. Suppose $\phi \in L^2(\partial D)^2$ satisfies

$$(\frac{1}{2} I + (\mathcal{K}^{-\alpha,0})^*)[\phi] = 0 \quad \text{on } \partial D.$$

Then by (2.535), $u := \mathcal{S}^{\alpha,0}[\phi]$ satisfies

$$\begin{cases} \mathcal{L}^{\lambda,\mu} u = 0 & \text{in } Y \setminus \overline{D}, \\ \dfrac{\partial u}{\partial \nu}\Big|_+ = 0 & \text{on } \partial D, \\ u \text{ is } \alpha\text{-quasi-periodic in the whole space.} \end{cases}$$

Therefore, it follows from (6.22) that

$$\int_{Y \setminus \overline{D}} \Big(\lambda |\nabla \cdot u|^2 + \frac{\mu}{2} |\nabla u + \nabla u^t|^2 \Big) = \int_{\partial Y} \frac{\partial u}{\partial \nu} \cdot \overline{u} - \int_{\partial D} \frac{\partial u}{\partial \nu}\Big|_+ \cdot \overline{u} = 0.$$

Thus, u is constant in $Y \setminus \overline{D}$ and hence in D. Thus, we get

$$\phi = \frac{\partial u}{\partial \nu}\Big|_+ - \frac{\partial u}{\partial \nu}\Big|_- = 0.$$

For the periodic case $\alpha = 0$, the proof follows the same lines. Since $(\mathcal{K}^{0,0})^* - (\mathcal{K}^0)^*$ is compact, it suffices to show the injectivity of $(1/2)\, I + \mathcal{K}^{0,0}$. Let $\phi \in L^2(\partial D)^2$ satisfying $((1/2)\, I + \mathcal{K}^{0,0})[\phi] = 0$ on ∂D. Then $u := \mathcal{D}^{0,0}[\phi]$ satisfies

$$\begin{cases} \mathcal{L}^{\lambda,\mu} u = 0 & \text{in } D, \\ u|_- = 0 & \text{on } \partial D, \end{cases}$$

and therefore $u = 0$ in D. Furthermore, if $((1/2)\, I + \mathcal{K}^{0,0})[\phi] = 0$, we can show that $\phi \in H^1(\partial D)^2$ and $\partial(\mathcal{D}^{0,0}[\phi])/\partial \nu\Big|_+ = \partial(\mathcal{D}^{0,0}[\phi])/\partial \nu\Big|_-$. See [**7**] for the details. Then we have

$$\begin{cases} \mathcal{L}^{\lambda,\mu} u = 0 & \text{in } Y \setminus \overline{D}, \\ \dfrac{\partial u}{\partial \nu}\Big|_+ = 0 & \text{on } \partial D, \\ u \text{ is periodic in the whole space.} \end{cases}$$

Therefore, it follows that

$$\int_{Y \setminus \overline{D}} \Big(\lambda |\nabla \cdot u|^2 + \frac{\mu}{2} |\nabla u + \nabla u^t|^2 \Big) = \int_{\partial Y} \frac{\partial u}{\partial \nu} \cdot \overline{u} - \int_{\partial D} \frac{\partial u}{\partial \nu}\Big|_+ \cdot \overline{u} = 0.$$

Thus, u is constant in $Y \setminus \overline{D}$, and hence $\phi = u|_- - u|_+$ is constant. By (2.539), we obtain that

$$0 = (\frac{1}{2} I + \mathcal{K}^{0,0})[\phi] = |Y \setminus \overline{D}| \phi,$$

which implies that ϕ must be zero. This completes the proof. □

We now derive complete asymptotic expansions of eigenvalues as $\mu \to +\infty$. We deal with three cases separately: $\alpha \neq 0$ (not of order $O(1/\sqrt{\mu})$), $\alpha = 0$, and $|\alpha|$ of order $O(1/\sqrt{\mu})$.

6.2.2.1. *The case* $\alpha \neq 0$. The following lemma, which is an immediate consequence of (6.21), gives a complete asymptotic expansion of $\mathcal{A}^{\alpha,\mu}(\omega)$ defined in (6.15) as $\mu \to +\infty$.

LEMMA 6.5. *Suppose* $\alpha \neq 0$. *Let*

$$\mathcal{A}_0^{\alpha}(\omega) = \begin{pmatrix} \widetilde{\mathcal{S}}^{\omega} & 0 \\ \frac{1}{2}I - (\widetilde{\mathcal{K}}^{\omega})^* & \frac{1}{2}I + (\mathcal{K}^{-\alpha,0})^* \end{pmatrix}, \tag{6.23}$$

and, for $l \geq 1$,

$$\mathcal{A}_l^{\alpha}(\omega) = \begin{pmatrix} 0 & -\mathcal{S}_l^{\alpha,\omega} \\ 0 & \frac{1}{\mu}(\mathcal{K}_{l+1}^{-\alpha,\omega})^* \end{pmatrix}. \tag{6.24}$$

Then we have

$$\mathcal{A}^{\alpha,\mu}(\omega) = \mathcal{A}_0^{\alpha}(\omega) + \sum_{l=1}^{+\infty} \frac{1}{\mu^l} \mathcal{A}_l^{\alpha}(\omega). \tag{6.25}$$

All the operators are defined on $L^2(\partial D)^2 \times L^2(\partial D)^2$.

Note that it is just for convenience that there is $1/\mu$ in the definition of $\mathcal{A}_l^{\alpha}(\omega)$. This of course does not affect any of our asymptotic results.

LEMMA 6.6. *Suppose* $\alpha \neq 0$. *Then the following assertions are equivalent:*

(i) $\omega^{\alpha,0} \in \mathbb{R}$ *is a characteristic value of* $\mathcal{A}_0^{\alpha}(\omega)$,

(ii) $\omega^{\alpha,0} \in \mathbb{R}$ *is a characteristic value of* $\widetilde{\mathcal{S}}^{\omega}$,

(iii) $(\omega^{\alpha,0})^2$ *is an eigenvalue of* $-\mathcal{L}^{\widetilde{\lambda},\widetilde{\mu}}$ *in* D *with the Dirichlet boundary condition.*

Moreover, if u *is an eigenfunction of* $-\mathcal{L}^{\widetilde{\lambda},\widetilde{\mu}}$ *in* D *with the Dirichlet boundary condition, then* $\varphi := \partial u/\partial\nu|_-$ *is a root function of* $\widetilde{\mathcal{S}}^{\omega}$. *Conversely, if* φ *is a root function of* $\widetilde{\mathcal{S}}^{\omega}$, *then* $u := -\widetilde{\mathcal{S}}^{\omega}[\varphi]$ *is an eigenfunction of* $-\mathcal{L}^{\widetilde{\lambda},\widetilde{\mu}}$ *in* D *with the Dirichlet boundary condition.*

PROOF. By Lemma 6.4, $(1/2)\,I + (\mathcal{K}^{-\alpha,0})^*$ is invertible. Thus characteristic values of $\mathcal{A}_0^{\alpha}(\omega)$ coincide with those of $\widetilde{\mathcal{S}}^{\omega}$. On the other hand, Green's identity (6.7) shows that the characteristic values of $\widetilde{\mathcal{S}}^{\omega}$ are exactly the eigenvalues of $-\mathcal{L}^{\widetilde{\lambda},\widetilde{\mu}}$ in D with the Dirichlet boundary condition. The last statements of Lemma 6.6 also follow from (6.7). □

LEMMA 6.7. *Every eigenvector of* $\widetilde{\mathcal{S}}^{\omega}$ *has rank one.*

PROOF. Let ϕ be an eigenvector of $\widetilde{\mathcal{S}}^{\omega}$ associated with the characteristic value ω^0; *i.e.*, $\widetilde{\mathcal{S}}^{\omega^0}[\phi] = 0$ on ∂D. Suppose that there exists ϕ^{ω}, holomorphic in a neighborhood of ω^0 as a function of ω, such that $\phi^{\omega^0} = \phi$ and

$$\widetilde{\mathcal{S}}^{\omega}[\phi^{\omega}] = (\omega^2 - (\omega^0)^2)\psi^{\omega}$$

for some ψ^ω. Let $u^\omega(x) := \widetilde{\mathcal{S}}^\omega[\phi^\omega](x)$, $x \in D$. Then u^ω satisfies

$$\begin{cases} (\mathcal{L}^{\widetilde{\lambda},\widetilde{\mu}} + \omega^2) u^\omega = 0 & \text{in } D, \\ u^\omega = (\omega^2 - (\omega^0)^2)\psi^\omega & \text{on } \partial D. \end{cases}$$

By Green's formula, we have

$$\begin{aligned} (\omega^2 - (\omega^0)^2) \int_D u^\omega \cdot \overline{u^{\omega^0}} &= \int_D u^\omega \cdot \overline{\mathcal{L}^{\widetilde{\lambda},\widetilde{\mu}} u^{\omega^0}} - \mathcal{L}^{\widetilde{\lambda},\widetilde{\mu}} u^\omega \cdot \overline{u^{\omega^0}} \\ &= \int_{\partial D} u^\omega \cdot \frac{\partial \overline{u^{\omega^0}}}{\partial \widetilde{\nu}} = (\omega^2 - (\omega^0)^2) \int_{\partial D} \psi^\omega \cdot \frac{\partial \overline{u^{\omega^0}}}{\partial \widetilde{\nu}}. \end{aligned}$$

Dividing by $\omega^2 - (\omega^0)^2$ and letting $\omega \to \omega^0$, we arrive at

$$\int_D |u^{\omega^0}|^2 = \int_{\partial D} \psi^{\omega^0} \cdot \frac{\partial \overline{u^{\omega^0}}}{\partial \widetilde{\nu}}.$$

Therefore, we conclude that ψ^{ω^0} is not identically zero. This completes the proof. □

By Lemma 6.4 and the fact that $\widetilde{\mathcal{S}}^\omega$ is Fredholm, we know that $\mathcal{A}_0^\alpha(\omega)$ is normal. Moreover, Lemma 6.7 says that the multiplicity of $\mathcal{A}_0^\alpha(\omega)$ at each eigenvalue $(\omega^0)^2$ of $-\mathcal{L}^{\widetilde{\lambda},\widetilde{\mu}}$ is equal to the dimension of $\operatorname{Ker} \widetilde{\mathcal{S}}^{\omega^0}$. Combining this fact with Theorem 1.15, we obtain the following lemma.

LEMMA 6.8. *For each eigenvalue $(\omega^0)^2$ of $-\mathcal{L}^{\widetilde{\lambda},\widetilde{\mu}}$ and sufficiently large μ, there exists a small neighborhood V of $\omega^0 > 0$ such that $\mathcal{A}^{\alpha,\omega}$ is normal with respect to ∂V and $\mathcal{M}(\mathcal{A}^{\alpha,\omega}, \partial V) = \dim \operatorname{Ker} \widetilde{\mathcal{S}}^{\omega^0}$.*

Let $(\omega^0)^2$ (with $\omega^0 > 0$) be a simple eigenvalue of $-\mathcal{L}^{\widetilde{\lambda},\widetilde{\mu}}$ in D with the Dirichlet boundary condition. There exists a unique eigenvalue $(\omega^{\alpha,\mu})^2$ (with $\omega^{\alpha,\mu} > 0$) of (6.5) lying in a small complex neighborhood V of ω^0. Combining the generalized Rouché theorem with Lemma 6.5 we are now able to derive complete asymptotic formulas for the characteristic values of $\omega \mapsto \mathcal{A}^{\alpha,\mu}(\omega)$. Theorem 1.14 yields that

$$\omega^{\alpha,\mu} - \omega^0 = \frac{1}{2\sqrt{-1}\pi} \operatorname{tr} \int_{\partial V} (\omega - \omega^0) \mathcal{A}^{\alpha,\mu}(\omega)^{-1} \frac{d}{d\omega} \mathcal{A}^{\alpha,\mu}(\omega) d\omega. \tag{6.26}$$

Then we obtain the following complete asymptotic expansion for the eigenvalue perturbations $\omega^{\alpha,\mu} - \omega^0$. It is proof is similar to that of Theorem 3.7.

THEOREM 6.9. *Suppose $\alpha \neq 0$. Then, for sufficiently large μ, the following asymptotic expansion holds:*

$$\omega^{\alpha,\mu} - \omega^0 = \frac{1}{2\pi\sqrt{-1}} \sum_{p=1}^{+\infty} \frac{1}{p} \sum_{n=p}^{+\infty} \frac{1}{\mu^n} \operatorname{tr} \int_{\partial V} \mathcal{B}_{n,p}^\alpha(\omega)\, d\omega, \tag{6.27}$$

where

$$\mathcal{B}_{n,p}^\alpha(\omega) = (-1)^p \sum_{\substack{n_1 + \cdots + n_p = n \\ n_i \geq 1}} \mathcal{A}_0^\alpha(\omega)^{-1} \mathcal{A}_{n_1}^\alpha(\omega) \ldots \mathcal{A}_0^\alpha(\omega)^{-1} \mathcal{A}_{n_p}^\alpha(\omega). \tag{6.28}$$

6.2.2.2. *The case* $\alpha = 0$. We now deal with the periodic case ($\alpha = 0$). By (6.17) we have

$$\mathcal{A}^{0,\mu}(\omega) = \mathcal{A}_0^0(\omega) + \sum_{l=1}^{+\infty} \frac{1}{\mu^l} \mathcal{A}_l^0(\omega), \tag{6.29}$$

where

$$\mathcal{A}_0^0(\omega) = \begin{pmatrix} \widetilde{\mathcal{S}}^\omega & -\dfrac{1}{\omega^2}\displaystyle\int_{\partial D} \cdot \, d\sigma \\ \dfrac{1}{2}I - (\widetilde{\mathcal{K}}^\omega)^* & \dfrac{1}{2}I + (\mathcal{K}^{0,0})^* \end{pmatrix} \tag{6.30}$$

and, for $l \geq 1$,

$$\mathcal{A}_l^0(\omega) = \begin{pmatrix} 0 & -\mathcal{S}_l^{0,\omega} \\ 0 & \dfrac{1}{\mu}(\mathcal{K}_{l+1}^{0,\omega})^* \end{pmatrix}. \tag{6.31}$$

Here we consider the following eigenvalue problem:

$$\begin{cases} (\mathcal{L}^{\widetilde{\lambda},\widetilde{\mu}} + \omega^2)u = 0 & \text{in } D, \\ u + \dfrac{1}{|Y \setminus \overline{D}|}\displaystyle\int_D u = 0 & \text{on } \partial D. \end{cases} \tag{6.32}$$

We note that it has a discrete spectrum and its eigenvalues are nonnegative since we have

$$\begin{aligned} \int_D \widetilde{\lambda}|\nabla \cdot u|^2 + \frac{\widetilde{\mu}}{2}|\nabla u + \nabla u^t|^2 &= \int_{\partial D} u \cdot \frac{\partial \overline{u}}{\partial \widetilde{\nu}} - \int_D u \cdot \mathcal{L}^{\widetilde{\lambda},\widetilde{\mu}}\overline{u} \\ &= -\frac{1}{|Y \setminus \overline{D}|}\int_D u \cdot \int_{\partial D} \frac{\partial \overline{u}}{\partial \widetilde{\nu}} + \overline{\omega}^2 \int_D |u|^2 \\ &= \frac{\overline{\omega}^2}{|Y \setminus \overline{D}|}\left|\int_D u\right|^2 + \overline{\omega}^2 \int_D |u|^2. \end{aligned}$$

The eigenvalue of (6.32) is related to the characteristic value of $\mathcal{A}^0(\omega)$ as follows.

LEMMA 6.10. *Equation (6.32) has a nonzero solution if and only if ω is a characteristic value of the operator-valued function $\mathcal{A}_0^0(\omega)$.*

PROOF. Suppose that there exists a nonzero pair (ϕ, ψ) such that

$$\mathcal{A}_0^0(\omega)\begin{pmatrix} \phi \\ \psi \end{pmatrix} = 0,$$

or equivalently,

$$\widetilde{\mathcal{S}}^\omega[\phi] - \frac{1}{\omega^2}\int_{\partial D} \psi \, d\sigma = 0 \quad \text{on } \partial D, \tag{6.33}$$

$$\left(\frac{1}{2}I - (\widetilde{\mathcal{K}}^\omega)^*\right)[\phi] + \left(\frac{1}{2}I + (\mathcal{K}^{0,0})^*\right)[\psi] = 0 \quad \text{on } \partial D. \tag{6.34}$$

In particular, ϕ is nonzero by the invertibility of $(1/2)\,I + (\mathcal{K}^{0,0})^*$. Let $u := \widetilde{\mathcal{S}}^\omega[\phi]$. Then we have

$$
\begin{aligned}
\frac{1}{|Y \setminus \overline{D}|} \int_D u &= -\frac{1}{\omega^2 |Y \setminus \overline{D}|} \int_{\partial D} \frac{\partial u}{\partial \widetilde{\nu}} \\
&= -\frac{1}{\omega^2 |Y \setminus \overline{D}|} \int_{\partial D} \big(-\frac{1}{2} I + (\widetilde{\mathcal{K}}^\omega)^*\big)[\phi] \\
&= -\frac{1}{\omega^2 |Y \setminus \overline{D}|} \int_{\partial D} \left(\frac{1}{2} I + (\mathcal{K}^{0,0})^*\right)[\psi] \\
&= -\frac{1}{\omega^2} \int_{\partial D} \psi,
\end{aligned}
$$

where the last equality follows from (2.540). Therefore, by (6.33), u is a nonzero solution to (6.32).

Suppose that (6.32) has a nonzero solution u. Following the same argument as in the proof of Theorem 6.2, we can see that there exists ϕ such that

$$
\begin{cases}
\widetilde{\mathcal{S}}^\omega[\phi] = u|_{\partial D} & \text{on } \partial D, \\
\big(-\frac{1}{2} I + (\widetilde{\mathcal{K}}^\omega)^*\big)[\phi] = \dfrac{\partial u}{\partial \widetilde{\nu}} & \text{on } \partial D.
\end{cases}
\tag{6.35}
$$

If we set

$$
\psi = \big(\frac{1}{2} I + (\mathcal{K}^{0,0})^*\big)^{-1} \left[\frac{\partial u}{\partial \widetilde{\nu}}\right],
$$

then (ϕ, ψ) satisfies

$$
\mathcal{A}_0^0(\omega) \begin{pmatrix} \phi \\ \psi \end{pmatrix} = 0.
$$

This completes the proof. □

We also have the following lemma.

LEMMA 6.11. *Every eigenvector of $\mathcal{A}_0^0(\omega)$ has rank one.*

PROOF. Suppose that $\begin{pmatrix} \phi \\ \psi \end{pmatrix}$ is an eigenvector of $\mathcal{A}_0^0(\omega)$ with rank m associated with characteristic value ω^0; *i.e.*, there exist ϕ^ω and ψ^ω, holomorphic as functions of ω, such that $\phi^{\omega^0} = \phi$, $\psi^{\omega^0} = \psi$, and

$$
\mathcal{A}_0^0(\omega) \begin{pmatrix} \phi^\omega \\ \psi^\omega \end{pmatrix} = (\omega - \omega^0)^m \begin{pmatrix} \widetilde{\phi}^\omega \\ \widetilde{\psi}^\omega \end{pmatrix},
$$

for some $\begin{pmatrix} \widetilde{\phi}^\omega \\ \widetilde{\psi}^\omega \end{pmatrix} \in L^2(\partial D)^2$. In other words, the following identities hold on ∂D:

$$
\begin{aligned}
\widetilde{\mathcal{S}}^\omega[\phi^\omega] - \frac{1}{\omega^2} \int_{\partial D} \psi^\omega \, d\sigma &= (\omega - \omega^0)^m \widetilde{\phi}^\omega, \\
\left(\frac{1}{2} I - (\widetilde{\mathcal{K}}^\omega)^*\right)[\phi^\omega] + \left(\frac{1}{2} I + (\mathcal{K}^{0,0})^*\right)[\psi^\omega] &= (\omega - \omega^0)^m \widetilde{\psi}^\omega.
\end{aligned}
$$

It then follows from (2.540) that

$$
\begin{aligned}
&\widetilde{\mathcal{S}}^\omega[\phi^\omega] - \frac{1}{|Y \setminus \overline{D}|\omega^2} \int_{\partial D} \big(-\frac{1}{2} I + (\widetilde{\mathcal{K}}^\omega)^*\big)[\phi^\omega]\, d\sigma \\
&\quad = \widetilde{\mathcal{S}}^\omega[\phi^\omega] - \frac{1}{|Y \setminus \overline{D}|\omega^2} \int_{\partial D} \big(\frac{1}{2} I + (\mathcal{K}^{0,0})^*\big)[\psi^\omega]\, d\sigma + \frac{(\omega-\omega^0)^m}{|Y \setminus \overline{D}|\omega^2} \int_{\partial D} \widetilde{\psi}^\omega\, d\sigma \\
&\quad = \widetilde{\mathcal{S}}^\omega[\phi^\omega] - \frac{1}{\omega^2} \int_{\partial D} \psi^\omega\, d\sigma + \frac{(\omega-\omega^0)^m}{|Y \setminus \overline{D}|\omega^2} \int_{\partial D} \widetilde{\psi}^\omega\, d\sigma \\
&\quad = (\omega-\omega^0)^m \left(\widetilde{\phi}^\omega + \frac{1}{|Y \setminus \overline{D}|\omega^2} \int_{\partial D} \widetilde{\psi}^\omega\, d\sigma \right).
\end{aligned}
$$

Let

$$
\eta^\omega := \left(\widetilde{\phi}^\omega + \frac{1}{|Y \setminus \overline{D}|\omega^2} \int_{\partial D} \widetilde{\psi}^\omega\, d\sigma \right) \quad \text{and} \quad u^\omega := \widetilde{\mathcal{S}}^\omega[\phi^\omega].
$$

Then u^ω satisfies

$$
\begin{cases}
(\mathcal{L}^{\widetilde{\lambda},\widetilde{\mu}} + \omega^2) u^\omega = 0 & \text{in } D, \\
u^\omega = \dfrac{1}{|Y \setminus \overline{D}|\omega^2} \displaystyle\int_{\partial D} \frac{\partial u^\omega}{\partial \widetilde{\nu}}\Big|_- d\sigma + (\omega-\omega^0)^m \eta^\omega & \text{on } \partial D.
\end{cases}
$$

By Green's formula, we have

$$
\begin{aligned}
&(\omega^2 - (\omega^0)^2) \int_D u^\omega \cdot \overline{u^{\omega^0}} = \int_{\partial D} u^\omega \cdot \frac{\partial \overline{u^{\omega^0}}}{\partial \widetilde{\nu}} - \overline{u^{\omega^0}} \cdot \frac{\partial u^\omega}{\partial \widetilde{\nu}} d\sigma \\
&= \left(\frac{1}{\omega^2} - \frac{1}{(\omega^0)^2} \right) \frac{1}{|Y \setminus \overline{D}|} \int_{\partial D} \frac{\partial u^\omega}{\partial \widetilde{\nu}} d\sigma \cdot \int_{\partial D} \frac{\partial \overline{u^{\omega^0}}}{\partial \widetilde{\nu}} d\sigma + (\omega-\omega^0)^m \int_{\partial D} \eta^\omega \cdot \frac{\partial \overline{u^{\omega^0}}}{\partial \widetilde{\nu}} d\sigma.
\end{aligned}
$$

Dividing by $\omega^2 - (\omega^0)^2$ and letting $\omega \to \omega^0$, we obtain

$$
\int_D |u^{\omega^0}|^2 + \frac{1}{|Y \setminus \overline{D}|(\omega^0)^4} \left| \int_{\partial D} \frac{\partial u^{\omega^0}}{\partial \widetilde{\nu}} d\sigma \right|^2 = \lim_{\omega \to \omega^0} \frac{(\omega-\omega^0)^m}{\omega^2 - (\omega^0)^2} \int_{\partial D} \eta^{\omega^0} \cdot \frac{\partial \overline{u^{\omega^0}}}{\partial \widetilde{\nu}} d\sigma.
$$

Since the term on the left is nonzero, we conclude that $m = 1$. This completes the proof. □

Analogously to Theorem 6.9, the following asymptotic formula for $\alpha = 0$ holds.

THEOREM 6.12. *Suppose $\alpha = 0$. Let $(\widetilde{\omega}^0)^2$ (with $\widetilde{\omega}^0 > 0$) be a simple eigenvalue of (6.32). Then there exists a unique characteristic value $\omega^{0,\mu} > 0$ of $\mathcal{A}^0(\omega)$ lying in a small complex neighborhood V of $\widetilde{\omega}^0$ and the following asymptotic expansion holds:*

$$
\omega^{0,\mu} - \widetilde{\omega}^0 = \frac{1}{2\pi\sqrt{-1}} \sum_{p=1}^{+\infty} \frac{1}{p} \sum_{n=p}^{+\infty} \frac{1}{\mu^n} \operatorname{tr} \int_{\partial V} \mathcal{B}_{n,p}(\omega) d\omega, \tag{6.36}
$$

where

$$
\mathcal{B}_{n,p}(\omega) = (-1)^p \sum_{\substack{n_1+\cdots+n_p=n \\ n_i \geq 1}} \mathcal{A}_0^0(\omega)^{-1} \mathcal{A}_{n_1}^0(\omega) \ldots \mathcal{A}_0^0(\omega)^{-1} \mathcal{A}_{n_p}^0(\omega). \tag{6.37}
$$

6.2.3. The case when $|\alpha|$ is of order $1/\sqrt{\mu}$. In this subsection we derive an asymptotic expansion which is valid for $|\alpha|$ of order $O(1/\sqrt{\mu})$, not just for fixed $\alpha \neq 0$ or $\alpha = 0$, as was considered in the previous subsections. We give the limiting behavior of $\omega^{\alpha,\mu}$ in this case. The argument of this subsection is similar to that of Subsection 5.5.5. The only difference is that while the operators in Subsection 5.5.5 are scaled by k, we deal with unscaled operators here, since there are two parameters μ and λ.

For exactly the same reason as in Section 5.5.5, we consider the operator

$$N^{\alpha,\omega} := \left(\frac{1}{2}I + (\mathcal{K}^{\alpha,\omega})^*\right)(\mathcal{S}^{\alpha,\omega})^{-1},$$

which can be extended to the Dirichlet-to-Neumann map on ∂D for $\mathcal{L}^{\lambda,\mu} + \omega^2$ in $Y \setminus \overline{D}$ with the Dirichlet boundary condition on ∂D and the quasi-periodicity condition on ∂Y. Note that the Dirichlet-to-Neumann map is defined for

$$\omega^2 < \min_{\alpha \in]-\pi,\pi]^2} \kappa(\alpha),$$

where $\kappa(\alpha)$ is the smallest eigenvalue of $-\mathcal{L}^{\lambda,\mu}$ with the Dirichlet boundary condition ∂D and quasi-periodicity on ∂Y. It is easy to see that $\kappa(\alpha)$ behaves like $O(\mu)$ as $\mu \to +\infty$. Furthermore, $N^{\alpha,\omega}$ depends smoothly both on ω and α. In particular, since $(1/\mu)\,\mathcal{S}^{\alpha,\omega}$ and $(1/2)\,I + (\mathcal{K}^{\alpha,\omega})^*$ depends on ω^2/μ, so does $(1/\mu)\,N^{\alpha,\omega}$, and hence we have the expansion

$$N^{\alpha,\omega} = N^{\alpha,0} + \omega^2 \dot{N} + O(|\alpha|) + O\left(\frac{1}{\mu}\right), \tag{6.38}$$

where

$$\dot{N} := \frac{d}{d(\omega^2)} N^{0,\omega}\Big|_{\omega=0}.$$

As for (5.41), we can show that $\dot{N} : H^{1/2}(\partial D) \mapsto H^{-1/2}(\partial D)$ is bounded. Note that

$$N^{\alpha,0} = O(\mu) \tag{6.39}$$

in the operator-norm from $H^{1/2}(\partial D)$ into $H^{-1/2}(\partial D)$, as $\mu \to +\infty$.

The following lemma, which is an analogous to Lemma 5.11 for the photonic band gap, will be used later.

LEMMA 6.13. *Let $u_1, u_2, \ldots$ be the eigenfunctions corresponding to $0 \leq \omega_1^{\alpha,\mu} \leq \omega_2^{\alpha,\mu} \leq \ldots$. For a given constant M there exists C such that*

$$\left\| u_j - \frac{1}{|\partial D|}\int_{\partial D} u_j \right\|_{H^{1/2}(\partial D)^2} \leq C\|u_j\|_{H^1(D)^2}\left(|\alpha| + \frac{1}{\mu}\right), \tag{6.40}$$

for all j satisfying $\omega_j^{\alpha,\mu} \leq M$. Furthermore,

$$\left| \int_D u_i \cdot \overline{u}_j + \int_{\partial D} \dot{N}[u_i|_{\partial D}] \cdot \overline{u}_j \right| \leq C\|u_i\|_{H^1(D)^2}\|u_j\|_{H^1(D)^2}\left(|\alpha| + \frac{1}{\mu}\right), \tag{6.41}$$

provided that $\omega_i^{\alpha,\mu} \neq \omega_j^{\alpha,\mu}$. If $\omega_i^{\alpha,\mu} = \omega_j^{\alpha,\mu}$ for some $i \neq j$, then we can choose u_i and u_j in such a way that (6.41) holds.

PROOF. In view of (6.38), we have

$$\frac{1}{\mu}N^{0,0}\left[u_j - \frac{1}{|\partial D|}\int_{\partial D} u_j\right] = \frac{1}{\mu}\frac{\partial u_j}{\partial \widetilde{\nu}} - \frac{1}{\mu}(\omega_j^{\alpha,\mu})^2 \dot{N}[u_j|_{\partial D}] + \frac{1}{\mu}(N^{0,0} - N^{\alpha,0})[u_j|_{\partial D}] + O(|\alpha|) + O(\frac{1}{\mu}).$$

Since $N^{\alpha,0} = N^{0,0} + O(\mu|\alpha|)$, we can derive (6.40) in exactly the same way as in Lemma 5.11.

To prove (6.41), let $\widetilde{E}$ be given by (2.403) with (λ,μ) replaced with $(\widetilde{\lambda},\widetilde{\mu})$; namely

$$\widetilde{E}(u,v) = \int_D \widetilde{\lambda}(\nabla\cdot u)(\overline{\nabla\cdot v}) + \frac{\widetilde{\mu}}{2}(\nabla u + \nabla u^t)\cdot(\overline{\nabla v} + \overline{\nabla v}^t). \tag{6.42}$$

We then obtain from the divergence theorem, as we did in the proof of Lemma 5.11,

$$\begin{aligned}(\omega_i^{\alpha,\mu})^2&\left(\int_D u_i\cdot\overline{u}_j + \int_{\partial D}\dot{N}[u_i|_{\partial D}]\cdot\overline{u}_j\right)\\ &= \widetilde{E}(u_i,u_j) - \int_{\partial D}\left(N^{\alpha,\omega_i^{\alpha,\mu}} - (\omega_i^{\alpha,\mu})^2\dot{N}\right)[u_i|_{\partial D}]\cdot\overline{u}_j\end{aligned}$$

and

$$\begin{aligned}(\omega_j^{\alpha,\mu})^2&\left(\int_D u_i\cdot\overline{u}_j + \int_{\partial D}\dot{N}[u_i|_{\partial D}]\cdot\overline{u}_j\right)\\ &= \widetilde{E}(u_i,u_j) - \int_{\partial D}(N^{\alpha,\omega_j^{\alpha,\mu}} - (\omega_j^{\alpha,\mu})^2\dot{N})[u_i|_{\partial D}]\cdot\overline{u}_j.\end{aligned}$$

It then follows that

$$\begin{aligned}&\left((\omega_i^{\alpha,\mu})^2 - (\omega_j^{\alpha,\mu})^2\right)\left(\int_D u_i\cdot\overline{u}_j + \int_{\partial D}\dot{N}[u_i|_{\partial D}]\cdot\overline{u}_j\right)\\ &= \int_{\partial D}\left((N^{\alpha,\omega_j^{\alpha,\mu}} - (\omega_j^{\alpha,\mu})^2\dot{N}) - (N^{\alpha,\omega_i^{\alpha,\mu}} - (\omega_i^{\alpha,\mu})^2\dot{N})\right)[u_i|_{\partial D}]\cdot\overline{u}_j\\ &= \int_{\partial D}\left((N^{\alpha,\omega_j^{\alpha,\mu}} - N^{\alpha,0} - (\omega_j^{\alpha,\mu})^2\dot{N}) - (N^{\alpha,\omega_i^{\alpha,\mu}} - N^{\alpha,0} - (\omega_i^{\alpha,\mu})^2\dot{N})\right)[u_i|_{\partial D}]\cdot\overline{u}_j.\end{aligned}$$

Hence, (6.41) holds and the proof is complete. □

The estimate (6.40) shows that u_j is almost constant on ∂D and there is a function $\widetilde{u}_j$ with a constant value on ∂D satisfying

$$\|\,\widetilde{u}_j - u_j\|_{H^1(D)^2} \le C\|u_j\|_{H^1(D)^2}\left(|\alpha| + \frac{1}{\mu}\right). \tag{6.43}$$

In fact, it is quite easy to find such a function. Let w_j be the solution to

$$\begin{cases}\mathcal{L}^{\widetilde{\lambda},\widetilde{\mu}} w_j = 0 & \text{in } D,\\ w_j = u_j - \dfrac{1}{|\partial D|}\displaystyle\int_{\partial D} u_j & \text{on } \partial D.\end{cases}$$

Then, thanks to (6.40), we have

$$\|w_j\|_{H^1(D)^2} \le C\|u_j\|_{H^1(D)^2}\left(|\alpha| + \frac{1}{\mu}\right),$$

and hence $\widetilde{u}_j = u_j - w_j$ does the job.

6.2.4. Derivation of the Leading-Order Terms. For $\alpha \neq 0$, let us write down explicitly the leading-order term in the expansion of $\omega^{\alpha,\mu} - \omega^0$. We first observe that

$$\mathcal{A}_0^\alpha(\omega)^{-1} = \begin{pmatrix} (\widetilde{\mathcal{S}}^\omega)^{-1} & 0 \\ \left(\frac{1}{2}I + (\mathcal{K}^{-\alpha,0})^*\right)^{-1}\left(\frac{1}{2}I - (\widetilde{\mathcal{K}}^\omega)^*\right)(\widetilde{\mathcal{S}}^\omega)^{-1} & \left(\frac{1}{2}I + (\mathcal{K}^{-\alpha,0})^*\right)^{-1} \end{pmatrix}.$$

Next, we prove the following lemma.

LEMMA 6.14. *Let u^0 be an eigenvector associated to the simple eigenvalue $(\omega^0)^2$ and let $\varphi := \partial u^0/\partial\widetilde{\nu}|_-$ on ∂D. Then we have, in a neighborhood of ω^0,*

$$(\widetilde{\mathcal{S}}^\omega)^{-1} = \frac{1}{\omega - \omega^0}T + \mathcal{Q}^\omega, \tag{6.44}$$

where $\mathcal{Q}^\omega$ are operators in $\mathcal{L}(H^2(\partial D)^2, L^2(\partial D)^2)$ holomorphic in ω and T is defined by

$$T(f) := -\frac{\langle f, \varphi\rangle\varphi}{2\omega^0 \displaystyle\int_D |u^0|^2}, \tag{6.45}$$

where $\langle\ ,\ \rangle$ is the inner product on $L^2(\partial D)^2$.

PROOF. By Lemma 6.7, there are operators T and Q^ω in $\mathcal{L}(H^2(\partial D)^2, L^2(\partial D)^2)$ such that $(\widetilde{\mathcal{S}}_D^\omega)^{-1}$ takes the form

$$(\widetilde{\mathcal{S}}^\omega)^{-1} = \frac{1}{\omega - \omega^0}T + \mathcal{Q}^\omega, \tag{6.46}$$

where $\mathcal{Q}^\omega$ is holomorphic in ω. Since

$$I = \widetilde{\mathcal{S}}^\omega(\widetilde{\mathcal{S}}^\omega)^{-1} = \frac{1}{\omega - \omega^0}\widetilde{\mathcal{S}}^\omega T + \widetilde{\mathcal{S}}^\omega\mathcal{Q}^\omega, \tag{6.47}$$

by letting $\omega \to \omega^0$, we have

$$\widetilde{\mathcal{S}}_D^{\omega^0} T = 0. \tag{6.48}$$

Similarly, we can show that

$$T\widetilde{\mathcal{S}}_D^{\omega^0} = 0. \tag{6.49}$$

It then follows from (6.48) and (6.49) that

$$\operatorname{Im} A = \operatorname{Ker}\widetilde{\mathcal{S}}^{\omega^0} = \operatorname{span}\{\varphi\} \quad \text{and} \quad \operatorname{Ker} A = \operatorname{Im}\widetilde{\mathcal{S}}^{\omega^0} = \operatorname{span}\{\varphi\}^\perp.$$

Here span$\{\varphi\}$ denotes the vector space spanned by φ. Therefore

$$T = C\langle\cdot, \varphi\rangle\,\varphi, \tag{6.50}$$

for some constant C.

By Green's formula, we have for $x \in D$,

$$\begin{aligned} \widetilde{\mathcal{S}}^\omega[\varphi](x) &= \widetilde{\mathcal{S}}^\omega\left[\frac{\partial u^0}{\partial \nu}\Big|_-\right](x) - \widetilde{\mathcal{D}}^\omega\left[u^0\right](x) \\ &= (\omega^2 - (\omega^0)^2)\int_D \widetilde{\Gamma}^\omega(x-y)u^0(y)dy - u^0(x). \end{aligned} \tag{6.51}$$

In particular, we get

$$\widetilde{\mathcal{S}}^\omega[\varphi](x) = (\omega^2 - (\omega^0)^2) \int_D \widetilde{\Gamma}^\omega(x-y)u^0(y)dy, \quad x \in \partial D. \tag{6.52}$$

Expanding $\widetilde{\Gamma}^\omega(x-y)$ in ω gives

$$\widetilde{\mathcal{S}}^\omega[\varphi](x) = 2\omega^0(\omega - \omega^0) \int_D \widetilde{\Gamma}^{\omega^0}(x-y)u^0(y)dy + (\omega-\omega^0)^2 A^\omega, \tag{6.53}$$

for some function A^ω holomorphic in ω. Therefore, it follows that

$$(\widetilde{\mathcal{S}}^\omega)^{-1}\left[2\omega^0 \int_D \widetilde{\Gamma}^{\omega^0}(x-y)u^0(y)dy\right] = \frac{1}{\omega-\omega^0}\varphi + B^\omega, \tag{6.54}$$

where B^ω is holomorphic in ω, which together with (6.46) implies that

$$T\left[2\omega^0 \int_D \widetilde{\Gamma}^{\omega^0}(\cdot - y)u^0(y)dy\right] = \varphi. \tag{6.55}$$

Note that if we take $\omega = \omega^0$ in (6.51), then

$$u^0(x) = -\widetilde{\mathcal{S}}^{\omega^0}[\varphi](x), \quad x \in D. \tag{6.56}$$

It then follows from (6.50) and (6.55) that

$$\begin{aligned} 1 &= C\left\langle 2\omega^0 \int_D \widetilde{\Gamma}^{\omega^0}(x-y)u^0(y)dy, \varphi\right\rangle \\ &= 2C\omega^0\left\langle u^0, \widetilde{\mathcal{S}}^\omega[\varphi]\right\rangle = -2C\omega^0\int_D |u^0|^2. \end{aligned}$$

This completes the proof. □

Because of (6.56), we have

$$(\frac{1}{2}I - (\widetilde{\mathcal{K}}^{\omega^0})^*)[\varphi] = \varphi \quad \text{on } \partial D. \tag{6.57}$$

Observe from (6.23) and (6.24) that the diagonal elements of $\mathcal{A}_0^\alpha(\omega)^{-1}\mathcal{A}_1^\alpha(\omega)$ are 0 and

$$-\big(\frac{1}{2}I + (\mathcal{K}^{-\alpha,0})^*\big)^{-1} = \big(\frac{1}{2}I - (\widetilde{\mathcal{K}}^{\omega^0})^*\big)(\widetilde{\mathcal{S}}^\omega)^{-1}\mathcal{S}_1^{\alpha,\omega} + \text{an operator holomorphic in } \omega.$$

Identity (6.20) implies that $\mathcal{S}_1^{\alpha,\omega} = \mu\mathcal{S}^{\alpha,0}$, and hence (6.44) yields

$$\frac{1}{2\pi\sqrt{-1}} \operatorname{tr} \int_{\partial V} \mathcal{A}_0^\alpha(\omega)^{-1}\mathcal{A}_1^\alpha(\omega)d\omega = -\mu \operatorname{tr}\left[T\mathcal{S}^{\alpha,0}\big(\frac{1}{2}I + (\mathcal{K}^{-\alpha,0})^*\big)^{-1}\big(\frac{1}{2}I - (\widetilde{\mathcal{K}}^{\omega^0})^*\big)\right].$$

Since $\operatorname{Im} T = \operatorname{span}\{\varphi\}$, it follows from (6.57) that

$$\begin{aligned} &\operatorname{tr}\left[T\mathcal{S}^{\alpha,0}\big(\frac{1}{2}I + (\mathcal{K}^{-\alpha,0})^*\big)^{-1}\big(\frac{1}{2}I - (\widetilde{\mathcal{K}}^{\omega^0})^*\big)\right] \\ &\qquad = \frac{\left\langle \left[T\mathcal{S}^{\alpha,0}\big(\frac{1}{2}I + (\mathcal{K}^{-\alpha,0})^*\big)^{-1}\big(\frac{1}{2}I - (\widetilde{\mathcal{K}}^{\omega^0})^*\big)\right][\varphi], \varphi\right\rangle}{||\varphi||^2_{L^2(\partial D)}} \\ &\qquad = \frac{\left\langle \left[T\mathcal{S}^{\alpha,0}\big(\frac{1}{2}I + (\mathcal{K}^{-\alpha,0})^*\big)^{-1}\right][\varphi], \varphi\right\rangle}{||\varphi||^2_{L^2(\partial D)}}. \end{aligned}$$

We set

$$v^\alpha(x) := \mu \mathcal{S}^{\alpha,0} \left(\frac{1}{2} I + (\mathcal{K}^{-\alpha,0})^* \right)^{-1} [\varphi](x), \quad x \in Y \setminus \overline{D}.$$

Then v^α is the unique α-quasi-periodic solution to

$$\begin{cases} \mathcal{L}^{\lambda,\mu} v^\alpha = 0 & \text{in } Y \setminus \overline{D}, \\ \dfrac{\partial v^\alpha}{\partial \nu}\Big|_+ = \mu \dfrac{\partial u^0}{\partial \widetilde{\nu}}\Big|_- & \text{on } \partial D, \end{cases}$$

and

$$\frac{1}{2\pi\sqrt{-1}} \operatorname{tr} \int_{\partial V} \mathcal{A}_0^\alpha(\omega)^{-1} \mathcal{A}_1^\alpha(\omega) d\omega = \frac{1}{||\varphi||_{L^2}^2} \langle \varphi, T v^\alpha \rangle$$
$$= \frac{1}{\mu} \frac{\displaystyle\int_{Y\setminus\overline{D}} \lambda |\nabla \cdot v^\alpha|^2 + \frac{\mu}{2} |\nabla v^\alpha + (\nabla v^\alpha)^t|^2}{\displaystyle 2\omega^0 \int_D |u^0|^2}.$$

Thus the following corollary holds.

COROLLARY 6.15. *Suppose $\alpha \neq 0$. Then the following asymptotic formula holds:*

$$\omega^{\alpha,\mu} - \omega^0 = -\frac{1}{\mu} \frac{\displaystyle\int_{Y\setminus\overline{D}} \frac{\lambda}{\mu} |\nabla \cdot v^\alpha|^2 + \frac{1}{2} |\nabla v^\alpha + (\nabla v^\alpha)^t|^2}{\displaystyle 2\omega^0 \int_D |u^0|^2} + O\left(\frac{1}{\mu^2}\right) \tag{6.58}$$

as $\mu \to +\infty$.

The formula (6.58) may be rephrased, like (5.29), as

$$(\omega^{\alpha,\mu})^2 - (\omega^0)^2 = -\frac{1}{\mu} \int_{Y\setminus\overline{D}} \frac{\lambda}{\mu} |\nabla \cdot v^\alpha|^2 + \frac{1}{2} |\nabla v^\alpha + (\nabla v^\alpha)^t|^2 + O\left(\frac{1}{\mu^2}\right),$$

assuming that u^0 is normalized.

When $\alpha = 0$, it does not seem to be likely that we can explicitly compute the leading-order term in a closed form as in the case $\alpha \neq 0$. However, we can compute the leading-order term in the asymptotic expansion of $\omega^{0,\mu} - \widetilde{\omega}^0$.

Let u^0 be the (normalized) eigenvector of (6.32) associated with the simple eigenvalue $\widetilde{\omega}_0$. Let $(\widetilde{\phi}_0, \widetilde{\psi}_0)$ satisfy (6.35) with u replaced by u^0 and $\omega = \widetilde{\omega}^0$. Since $\widetilde{\omega}^0$ is the only simple pole in V of the mapping $\omega \mapsto \mathcal{A}_0^0(\omega)^{-1}$, one can prove that

$$\mathcal{A}_0^0(\omega)^{-1} = \frac{1}{\omega - \widetilde{\omega}_0} \left(\frac{d}{d\omega} \mathcal{A}_0^0(\omega)\Big|_{\omega=\widetilde{\omega}^0} \begin{pmatrix} \widetilde{\phi}_0 \\ \widetilde{\psi}_0 \end{pmatrix} \cdot \begin{pmatrix} \widetilde{\phi}_0 \\ \widetilde{\psi}_0 \end{pmatrix} \right)^{-1} \begin{pmatrix} (\cdot, \widetilde{\phi}_0)\, \widetilde{\phi}_0 & 0 \\ 0 & (\cdot, \widetilde{\psi}_0)\, \widetilde{\psi}_0 \end{pmatrix}$$
$$+ \text{ an operator-valued function holomorphic in } \omega,$$

which allows us to make explicit the leading-order term in the expansion of $\omega^{0,\mu} - \widetilde{\omega}^0$. Similar calculations and expressions in the transition region ($|\alpha| = O(1/\sqrt{\mu})$) can be derived as well.

6.3. Criterion for Gap Opening

Following Chapter 5, we provide in this subsection a criterion for gap opening in the spectrum of the operator given by (6.1) as $\mu \to +\infty$.

Let ω_j be the eigenvalues of $-\mathcal{L}^{\widetilde{\lambda},\widetilde{\mu}}$ in D with the Dirichlet boundary condition. Let $\widetilde{\omega}_j$ denote the eigenvalues of (6.32). We first prove the following min-max characterization of ω_j and $\widetilde{\omega}_j$.

LEMMA 6.16. *The following min-max characterizations of ω_j^2 and $\widetilde{\omega}_j^2$ hold:*

$$\omega_j^2 = \min_{N_j} \max_{u \in N_j, ||u||=1} \widetilde{E}(u,u) \tag{6.59}$$

and

$$\widetilde{\omega}_j^2 = \min_{N_j} \max_{u \in N_j, ||u||=1} \frac{\widetilde{E}(u,u)}{1 - \left| \int_D u \right|^2}, \tag{6.60}$$

where the minimum is taken over all j-dimensional subspaces N_j of $(H_0^1(D))^2$. Recall that $H_0^1(D)$ is the set of all functions in $H^1(D)$ with zero-trace on ∂D and $\widetilde{E}$ is given by (6.42).

PROOF. The identity (6.59) is well known [**326**]. Note that if v satisfies the Dirichlet condition on ∂D, then $\widetilde{v} := v - \int_D v$ satisfies the boundary condition

$$\widetilde{v} + \frac{1}{|Y \setminus \overline{D}|} \int_D \widetilde{v} = 0 \quad \text{on } \partial D. \tag{6.61}$$

Conversely, if $\widetilde{v}$ satisfies (6.61), then

$$v := \widetilde{v} + \frac{1}{|Y \setminus \overline{D}|} \int_D \widetilde{v}$$

obviously satisfies the Dirichlet boundary condition.

Observe that the operator with the boundary condition in (6.32) is not self-adjoint, and hence Poincaré's min-max principle cannot be applied. So we now introduce an eigenvalue problem whose eigenvalues are exactly those of (6.32). Let $\mathcal{H} = \operatorname{span}\{H_0^2(D), \chi(Y)\}$ in $L^2(Y)$ where $H_0^2(D)$ is regarded as a subspace of $L^2(Y)$ by extending the functions to be 0 in $Y \setminus \overline{D}$. Let $\mathcal{G}$ be the closure of $\mathcal{H}$ in $L^2(Y)$. Define the operator $T : \mathcal{H} \times \mathcal{H} \to \mathcal{G} \times \mathcal{G}$ by

$$Tu = \begin{cases} -\mathcal{L}^{\widetilde{\lambda},\widetilde{\mu}} u & \text{on } D, \\ \dfrac{1}{|Y \setminus \overline{D}|} \displaystyle\int_D \mathcal{L}^{\widetilde{\lambda},\widetilde{\mu}} u & \text{on } Y \setminus \overline{D}. \end{cases}$$

The constant value of Tu in $Y \setminus \overline{D}$ was chosen so that $\int_Y Tu = 0$. Then one can easily see that T is a densely defined self-adjoint operator on $\mathcal{H} \times \mathcal{H}$ and

$$\langle Tu, v \rangle_Y = \widetilde{E}(u,v) \quad \text{for } u, v \in \mathcal{H} \times \mathcal{H}. \tag{6.62}$$

Here $\langle \, , \, \rangle_Y$ denotes the inner-product on $L^2(Y)^2$. One can also show that nonzero eigenvalues of T are eigenvalues of (6.32) and vice versa.

Let M_j be a j-dimensional subspace of $\mathcal{H} \times \mathcal{H}$ perpendicular to constant vectors which are eigenvectors corresponding to the eigenvalue zero. Then by Poincaré's min-max principle, we have

$$\begin{aligned}
\widetilde{\omega}_j^2 &= \min_{M_j} \max_{u \in M_j} \frac{\langle Tu, u \rangle_Y}{\langle u, u \rangle_Y} \\
&= \min_{M_j} \max_{u \in M_j} \frac{\widetilde{E}(u, u)}{\langle u, u \rangle_Y} \\
&= \min_{N_j} \max_{v \in N_j} \frac{\widetilde{E}(v - \int_D v, v - \int_D v)}{\langle v - \int_D v, v - \int_D v \rangle_Y} \\
&= \min_{N_j} \max_{v \in N_j} \frac{\widetilde{E}(v, v)}{\langle v, v \rangle_D - |\int_D v|^2},
\end{aligned}$$

where $\langle \, , \, \rangle_D$ denotes the inner-product on $L^2(D)^2$, which completes the proof of the lemma. □

LEMMA 6.17. *The eigenvalues* ω_j *and* $\widetilde{\omega}_j$ *interlace in the following way:*

$$\omega_j \leq \widetilde{\omega}_j \leq \omega_{j+2}, \quad j = 1, 2, \ldots . \tag{6.63}$$

PROOF. Lemma 6.16 ensures that the first inequality in (6.63) is trivial. Then we only have to prove the second one. Let u_j denote the normalized eigenvector associated with ω_j. Let N_{j+2} denote the span of the eigenvectors $u_1, \ldots, u_{j+2}$ and let $\widetilde{N}$ be the subspace of N_{j+2} composed of all the elements in N_{j+2} which have zero integral over D. Since the set of constant vectors has dimension 2, $\widetilde{N}$ is of dimension greater than j. Therefore, we have $\widetilde{\omega}_j \leq \omega_{j+2}$, as desired. □

We will also need the following lemma.

LEMMA 6.18. *For any* $\epsilon > 0$ *and* j*, there exist* c_1 *and* c_2 *sufficiently small such that we have*

$$\widetilde{\omega}_j - \epsilon \leq \omega_{j+2}^{\alpha,\mu} \leq \omega_{j+2} \tag{6.64}$$

for $|\alpha| \leq c_1$ *and* $\mu > 1/c_2$.

PROOF. The second inequality easily follows from the min-max principle for eigenvalues with Dirichlet boundary condition on ∂D and with quasi-periodicity on ∂Y.

To prove the first inequality, let $u_1, \ldots, u_j$ be eigenfunctions corresponding to $(\omega_1^{\alpha,\mu})^2, \ldots, (\omega_j^{\alpha,\mu})^2$, respectively, satisfying

$$\int_D |u_i|^2 + \int_{\partial D} \dot{N}[u_i|_{\partial D}] \cdot \overline{u}_i = 1,$$

together with the orthogonality condition (6.41). For $u = \sum_{i=1}^{j} c_i u_i$, we have, with the aid of (6.41) and the divergence theorem, that

$$\frac{\widetilde{E}(u,u) - \int_{\partial D} N^{\alpha,0}[u|_{\partial D}] \cdot \overline{u}}{\int_D |u|^2 + \int_{\partial D} \dot{N}[u|_{\partial D}] \cdot \overline{u}}$$

$$= \frac{\sum_{i=1}^{j} c_i^2 (\omega_j^{\alpha,\mu})^2 \left(\int_D |u_i|^2 + \int_{\partial D} \dot{N}[u_i|_{\partial D}] \cdot \overline{u}_i \right)}{\sum_{i=1}^{j} c_i^2 \left(\int_D |u_i|^2 + \int_{\partial D} \dot{N}[u_i|_{\partial D}] \cdot \overline{u}_i \right)} + O(|\alpha|) + O(\frac{1}{\mu}).$$

Hence, we have

$$(6.65) \qquad (\omega_j^{\alpha,\mu})^2 = \max_{u \in \langle u_1, \ldots, u_j \rangle} \frac{\widetilde{E}(u,u) - \int_{\partial D} N^{\alpha,0}[u|_{\partial D}] \cdot \overline{u}}{\int_D |u|^2 + \int_{\partial D} \dot{N}[u|_{\partial D}] \cdot \overline{u}} + O(|\alpha|) + O(\frac{1}{\mu}),$$

where $\langle u_1, \ldots, u_j \rangle$ denotes the span of the eigenvectors $u_1, \ldots, u_j$ (*i.e.*, N_j).

Since

$$- \int_{\partial D} N^{\alpha,0}[u|_{\partial D}] \cdot \overline{u} \geq 0,$$

we get

$$(6.66) \qquad (\omega_j^{\alpha,\mu})^2 \geq \max_{u \in \langle u_1, \ldots, u_j \rangle} \frac{\widetilde{E}(u,u)}{\int_D |u|^2 + \int_{\partial D} \dot{N}[u|_{\partial D}] \cdot \overline{u}} + O(|\alpha|) + O(\frac{1}{\mu}).$$

Let $\widetilde{u}_i$, $i = 1, \ldots, j$, be an approximation of u_j with constant values on ∂D satisfying (6.43). Then, one can see that

$$(6.67) \qquad (\omega_j^{\alpha,\mu})^2 \geq \max_{u \in \langle \widetilde{u}_1, \ldots, \widetilde{u}_j \rangle} \frac{\widetilde{E}(u,u)}{\int_D |u|^2 + \int_{\partial D} \dot{N}[u|_{\partial D}] \cdot \overline{u}} + O(|\alpha|) + O\left(\frac{1}{\mu}\right).$$

By the definition of $\dot{N}$, we can easily check that

$$\int_{\partial D} \dot{N}[U] \cdot \overline{U} = |Y \setminus D| |U|^2$$

for any constant vector U, and hence we obtain

$$(\omega_j^{\alpha,\mu})^2 \geq \max_{u \in \langle \widetilde{v}_1, \ldots, \widetilde{v}_j \rangle} \frac{\widetilde{E}(u,u)}{\int_D |u|^2 + |Y \setminus D| |u|_{\partial D}|^2} + O(|\alpha|) + O\left(\frac{1}{\mu}\right).$$

Thus we get

$$(6.68) \qquad \omega_j(\alpha)^2 \geq \min_{\mathcal{N}_j} \max_{u \in \mathcal{N}_j} \frac{\widetilde{E}(u,u)}{\int_D |u|^2 + |Y \setminus D| |u|_{\partial D}|^2} + O(|\alpha|) + O\left(\frac{1}{\mu}\right),$$

where $\mathcal{N}_j$ is a j-dimensional subspace of $H^1(D)^2$ of elements with constant values a.e. on ∂D. Recalling that

$$\widetilde{\omega}_{j-2}^2 = \min_{\mathcal{N}_j} \max_{u \in \mathcal{N}_j} \frac{\widetilde{E}(u,u)}{\displaystyle\int_D |u|^2 + |Y \setminus D||u|_{\partial D}|^2},$$

we finally arrive at

$$(\omega_j^\alpha)^2 \geq \widetilde{\omega}_{j-2}^2 + O(|\alpha|) + O\left(\frac{1}{\mu}\right),$$

which gives the desired result. □

Since 0 is an eigenvalue of the periodic problem with multiplicity 2, combining formulas (6.27), (6.36), and Lemma 6.18 shows that the spectral bands converge, as $\mu \to +\infty$, to

$$[0, \omega_1] \cup [0, \omega_2] \bigcup_{j \geq 1} [\widetilde{\omega}_j, \omega_{j+2}], \tag{6.69}$$

and hence we have a band gap if and only if the following holds:

$$\omega_{j+1} < \widetilde{\omega}_j \quad \text{for some } j \quad (\text{ criterion for gap opening}). \tag{6.70}$$

Observe that by (6.59) and (6.60) the gap opening criterion is equivalent to

$$\min_{\mathcal{N}_{j+1}} \max_{u \in \mathcal{N}_{j+1}, ||u||=1} \widetilde{E}(u,u) < \min_{\mathcal{N}_j} \max_{u \in \mathcal{N}_j, ||u||=1} \frac{\widetilde{E}(u,u)}{1 - \left|\displaystyle\int_D u\right|^2}, \tag{6.71}$$

where $\mathcal{N}_j$ is a j-dimensional subspace of $H_0^1(\partial D)^2$.

To find conditions on the inclusion D so that the gap opening criterion is satisfied by rigorous analysis is unlikely. However, finding such conditions by means of numerical computations will be of great importance. It should be emphasized that the criterion (6.70) is for the case when the matrix and the inclusion have the same density, assumed to be equal to 1.

6.4. Gap Opening Criterion When Densities Are Different

We now consider periodic elastic composites such that the matrix and the inclusion have different densities.

Suppose that the density of the matrix is ρ while that of the inclusion is 1 (after normalization). The Lamé parameters are the same as before. In this case, the first equation of the eigenvalue problem (6.5) has to be replaced by

$$\mathcal{L}^{\lambda,\mu} u + \rho\omega^2 u = 0 \quad \text{in } Y \setminus \overline{D}. \tag{6.72}$$

Hence we can show by exactly the same analysis that the asymptotic expansions (6.27) and (6.36) hold if we replace the operators (6.30) and (6.24) (and (6.31)) with the new operators (depending on the density ρ) given by

$$\mathcal{A}_0^0(\omega) = \begin{pmatrix} \widetilde{\mathcal{S}}^\omega & -\dfrac{1}{\rho\omega^2}\displaystyle\int_{\partial D} \cdot \, d\sigma \\ \dfrac{1}{2}I - (\widetilde{\mathcal{K}}^\omega)^* & \dfrac{1}{2}I + (\mathcal{K}^{0,0})^* \end{pmatrix} \tag{6.73}$$

and

$$\mathcal{A}_l^\alpha(\omega) = \rho^{l-1}\begin{pmatrix} 0 & -\mathcal{S}_l^{\alpha,\omega} \\ 0 & \dfrac{\rho}{\mu}(\mathcal{K}_{l+1}^{-\alpha,\omega})^* \end{pmatrix}, \quad l \geq 1, \tag{6.74}$$

respectively, and the eigenvalue problem (6.32) with the eigenvalue problem

$$\begin{cases} (\mathcal{L}^{\widetilde{\lambda},\widetilde{\mu}} + \omega^2)u = 0 & \text{in } D, \\ u + \dfrac{1}{\rho|Y \setminus \overline{D}|}\displaystyle\int_D u = 0 & \text{on } \partial D. \end{cases} \tag{6.75}$$

Let $\{\widetilde{\omega}_j\}$ be the set of eigenvalues of (6.75). In order to express $\widetilde{\omega}_j$ using the min-max principle, we define $\langle\ ,\ \rangle_Y$ by

$$\langle u, v\rangle_Y = \int_D u\cdot v + \rho\int_{Y\setminus\overline{D}} u\cdot v. \tag{6.76}$$

We also define T, as before, by

$$Tu = \begin{cases} -\mathcal{L}^{\widetilde{\lambda},\widetilde{\mu}}u & \text{on } D, \\ \dfrac{1}{\rho|Y\setminus\overline{D}|}\displaystyle\int_D \mathcal{L}^{\widetilde{\lambda},\widetilde{\mu}}u & \text{on } Y\setminus\overline{D}. \end{cases} \tag{6.77}$$

Then T is self-adjoint with respect to $\langle\ ,\ \rangle_Y$. By Poincaré's min-max principle again, we have

$$\begin{aligned} \widetilde{\omega}_j^2 &= \min_{M_j}\max_{u\in M_j}\frac{\langle Tu,u\rangle_Y}{\langle u,u\rangle_Y} \\ &= \min_{M_j}\max_{u\in M_j}\frac{\widetilde{E}(u,u)}{\langle u,u\rangle_Y} \\ &= \min_{N_j}\max_{v\in N_j}\frac{\widetilde{E}(v,v)}{\left\langle v - \dfrac{1}{|D|+\rho|Y\setminus\overline{D}|}\displaystyle\int_D v, v - \dfrac{1}{|D|+\rho|Y\setminus\overline{D}|}\displaystyle\int_D v\right\rangle_Y} \\ &= \min_{N_j}\max_{v\in N_j}\frac{\widetilde{E}(v,v)}{\langle v,v\rangle_D - \dfrac{1}{|D|+\rho|Y\setminus\overline{D}|}\left|\displaystyle\int_D v\right|^2}, \end{aligned}$$

where M_j and N_j are the same as in the proof of Lemma 6.16. Therefore, we have the following min-max characterization of the eigenvalues of problem (6.75):

$$\widetilde{\omega}_j^2 = \min_{N_j}\max_{u\in N_j}\frac{\widetilde{E}(u,u)}{\langle u,u\rangle_D - \dfrac{1}{|D|+\rho|Y\setminus\overline{D}|}\left|\displaystyle\int_D u\right|^2}. \tag{6.78}$$

We then get a band gap criterion for the different density case which is equivalent to (6.70):

$$\min_{N_{j+1}}\max_{u\in N_{j+1},||u||=1}\widetilde{E}(u,u) < \min_{N_j}\max_{u\in N_j,||u||=1}\frac{\widetilde{E}(u,u)}{1 - \dfrac{1}{|D|+\rho|Y\setminus\overline{D}|}\left|\displaystyle\int_D u\right|^2}. \tag{6.79}$$

It is quite interesting to compare (6.79) with (6.71). If $\rho < 1$, then

$$(6.80)\quad \min_{N_j} \max_{u \in N_j, ||u||=1} \frac{\widetilde{E}(u,u)}{1 - |\int_D u|^2} < \min_{N_j} \max_{u \in N_j, ||u||=1} \frac{\widetilde{E}(u,u)}{1 - \frac{1}{|D| + \rho|Y \setminus \overline{D}|} |\int_D u|^2},$$

which shows that the smaller the density ρ, the wider the band gap, provided that (6.70) is fulfilled. This phenomenon was reported by Economou and Sigalas [**227**] who observed that periodic elastic composites whose matrix has lower density and higher shear modulus compared to those of inclusions yield better open gaps. The analysis of this chapter agrees with these experimental findings.

6.5. Concluding Remarks

In this chapter we have reduced band structure calculations for phononic crystals to the problem of finding the characteristic values of a family of meromorphic integral operators. We have also provided complete asymptotic expansions of these characteristic values as the shear modulus goes to infinity, established a connection between the band gap structure and the Dirichlet eigenvalue problem for the Lamé operator, and given a criterion for gap opening as the shear modulus becomes large. The leading-order terms in the expansions of the characteristic values were explicitly computed. An asymptotic analysis for the band gap structure in three-dimensions can be provided with only minor modifications of the techniques presented here. Our results in this chapter open the road to numerous numerical and analytical investigations on phononic crystals and could, in particular, be used for systematic optimal design of phononic structures as well as for efficient computations of the band structure problem.

Part 3

Subwavelength Resonant Structures and Super-resolution

CHAPTER 7

Plasmonic Resonances for Nanoparticles

7.1. Introduction

Driven by the search for new materials with interesting and unique properties, the field of nanoparticle research has grown immensely in recent decades [**372**]. Plasmon resonant nanoparticles have unique capabilities such as enhancing the brightness and directivity of light, confining strong electromagnetic fields, and the outcoupling of light into advantageous directions [**469**]. Recent advances in nanofabrication techniques have made it possible to construct complex nanostructures such as arrays using plasmonic nanoparticles as components. A reason for the thriving interest in optical studies of plasmon resonant nanoparticles is due to their recently proposed use as labels in molecular biology [**307**]. New types of cancer diagnostic nanoparticles are constantly being developed. Nanoparticles are also being used in thermotherapy as nanometric heat-generators that can be activated remotely by external electromagnetic fields [**114**]. Furthermore, photothermal effects could be used to boost the efficiency of solar cells [**111**].

The optical response of plasmon resonant nanoparticles is dominated by the appearance of plasmon resonances over a wide range of wavelengths [**372**]. For individual particles or very low concentrations of non-interacting nanoparticles in a solvent, separated from one another by distances larger than the wavelength, these resonances depend on the electromagnetic parameters of the nanoparticle, those of the surrounding material, and the particle shape and size. High scattering and absorption cross sections and strong near-fields are unique effects of plasmonic resonant nanoparticles. In order to profit from them, a rigorous understanding of the interactive effects between the particle size and shape and the contrasts in the electromagnetic parameters is required. One of the most important parameters in the context of applications is the position of the resonances in terms of wavelength or frequency. A longstanding problem is to tune this position by changing the particle size or the concentration of the nanoparticles in a solvent [**263, 372**]. It was experimentally observed, for instance, in [**263, 470**] that the scaling behavior of nanoparticles is critical. The question of how the resonant properties of plasmonic nanoparticles develops with increasing size or/and concentration is therefore fundamental.

In this chapter we use the full Maxwell equations for light propagation in order to analyze plasmonic resonances for nanoparticles. We mathematically define the notion of plasmonic resonance. At the quasi-static limit, we show that plasmon resonances in nanoparticles can be treated as an eigenvalue problem for the Neumann–Poincaré integral operator and unfortunately, they are size-independent. Then we analyze the plasmon resonance shift and broadening with respect to changes in size and shape, using the layer potential techniques associated with the full Maxwell equations. We give a rigorous detailed description of the scaling behavior of plasmonic resonances to improve our understanding of light scattering by plasmonic nanoparticles beyond the quasi-static regime. On the other hand, we present an effective medium theory for resonant plasmonic systems. We treat a composite material in which plasmonic nanoparticles are embedded and isolated from each other. The particle dimension and interparticle distances are considered to be infinitely small compared with the wavelength of the interacting light. We extend the validity of the Maxwell-Garnett effective medium theory in order to describe the behavior of a system of plasmonic resonant nanoparticles. We show that by homogenizing plasmonic nanoparticles one can obtain high-contrast or negative parameter materials, depending on how the frequencies used correspond to the plasmonic resonant frequency.

7.2. Quasi-Static Plasmonic Resonances

7.2.1. Uniform Validity of Small-Volume Expansions. We consider the scattering problem of a time-harmonic electromagnetic wave incident on a particle D. The homogeneous medium is characterized by electric permittivity ε_m and magnetic permeability μ_m, while D is characterized by electric permittivity ε_c and magnetic permeability μ_c, both of which depend on the frequency. Define

$$k_m = \omega\sqrt{\varepsilon_m \mu_m}, \quad k_c = \omega\sqrt{\varepsilon_c \mu_c},$$

and

$$\varepsilon_D = \varepsilon_m \chi(\mathbb{R}^3 \setminus \overline{D}) + \varepsilon_c \chi(D), \quad \mu_D = \varepsilon_m \chi(\mathbb{R}^3 \setminus \overline{D}) + \varepsilon_c \chi(D).$$

For a given incident plane wave (E^i, H^i), solution to the Maxwell equations in free space (2.330), the scattering problem can be modeled by the system of equations (2.331) subject to the Silver-Müller radiation condition (2.332).

Let $D = z + \delta B$ where B contains the origin and $|B| = O(1)$. For any $x \in \partial D$, let $\widetilde{x} = \frac{x-z}{\delta} \in \partial B$ and define for each function f defined on ∂D, a corresponding function defined on B as follows

$$\eta(f)(\widetilde{x}) = f(z + \delta\widetilde{x}). \tag{7.1}$$

The following result follows from [**37**]. It is a refinement of Theorem 2.120. Its proof is sketched at the end of this chapter.

THEOREM 7.1. *Let*

$$d_\sigma = \min\left\{\text{dist}\left(\lambda_\mu, \sigma((\mathcal{K}_D^0)^*) \cup -\sigma((\mathcal{K}_D^0)^*)\right), \text{dist}\left(\lambda_\varepsilon, \sigma((\mathcal{K}_D^0)^*) \cup -\sigma((\mathcal{K}_D^0)^*)\right)\right\}.$$

Then, for $D = z + \delta B \Subset \mathbb{R}^3$ of class $\mathcal{C}^{1,\alpha}$ for $\alpha > 0$, the following uniform far-field expansion holds

$$\begin{aligned} E^s(x) &= -\frac{\sqrt{-1}\omega\mu_m}{\varepsilon_m}\nabla \times \mathbf{G}_{k_m}(x-z)M(\lambda_\mu, D)H^i(z) \\ &\quad -\omega^2\mu_m \mathbf{G}_{k_m}(x-z)M(\lambda_\varepsilon, D)E^i(z) + O(\frac{\delta^4}{d_\sigma}), \end{aligned} \tag{7.2}$$

where $\mathbf{G}_{k_m}(x-z)$ is the Dyadic Green (matrix valued) function for the full Maxwell equations defined by (2.380) and $M(\lambda_\mu, D)$ and $M(\lambda_\varepsilon, D)$ are the polarization tensors associated with D and the contrasts λ_μ and λ_ε given by (2.72) with $k = \mu_c/\mu_c$ and $k = \varepsilon_c/\varepsilon_m$, respectively.

Suppose that ε_c and μ_c are changing with respect to the angular frequency ω while ε_m and μ_m are independent of ω. Because of causality, the real and imaginary parts of ε_c and μ_c obey Kramers-Kronig relations

$$
\begin{aligned}
\Re F(\omega) &= -\frac{1}{\pi}\text{p.v.}\int_{-\infty}^{+\infty}\frac{\Im F(\omega')}{\omega-\omega'}\,d\omega', \\
\Im F(\omega) &= \frac{1}{\pi}\text{p.v.}\int_{-\infty}^{+\infty}\frac{\Re F(\omega')}{\omega-\omega'}\,d\omega',
\end{aligned}
\tag{7.3}
$$

$F(\omega) = \varepsilon_c(\omega)$ or $\mu_c(\omega)$. The permittivity and permeability of plasmonic nanoparticles in the infrared spectral regime can be described by the Drude model given by

$$
\varepsilon_c(\omega) = \varepsilon_m\big(1-\frac{\omega_p^2}{\omega(\omega+\sqrt{-1}\tau^{-1})}\big), \quad \mu_c(\omega) = \mu_m\big(1-F\frac{\omega^2}{\omega^2-\omega_0^2+\sqrt{-1}\tau^{-1}}\big),
\tag{7.4}
$$

where ω_p is the plasma frequency of the bulk material, τ^{-1} is the damping coefficient, F is a filling factor, and ω_0 is a localized plasmon frequency.

DEFINITION 7.2. *Let* $d'_\sigma(\omega) = \min\left\{\operatorname{dist}\big(\lambda_\mu(\omega),\sigma((\mathcal{K}_D^0)^*)\big),\operatorname{dist}\big(\lambda_\varepsilon(\omega),\sigma((\mathcal{K}_D^0)^*)\big)\right\}$. *Then we call* ω *a quasi-static plasmonic resonance if* $d'_\sigma(\omega) \ll 1$.

Notice that, in view of (2.72), if ω is a quasi-static plasmonic resonance, then at least one of the polarization tensors $M(\lambda_\varepsilon, D)$ and $M(\lambda_\mu, D)$ blows up.

Assume that the incident fields are plane waves given by

$$
\begin{aligned}
E^i(x) &= p e^{\sqrt{-1}k_m d\cdot x}, \\
H^i(x) &= d\times p e^{\sqrt{-1}k_m d\cdot x},
\end{aligned}
$$

where $p\in\mathbb{R}^3$ and $d\in\mathbb{R}^3$ with $|d|=1$ are such that $p\cdot d = 0$.

From Taylor expansions on the formula of Theorem 7.1, it follows that the following far-field asymptotic expansion holds:

$$
\begin{aligned}
E^s(x) &= -\frac{e^{\sqrt{-1}k_m|x|}}{4\pi|x|}\Big(\omega\mu_m k_m e^{\sqrt{-1}k_m(d-\hat{x})\cdot z}\big(\hat{x}\times I\big)M(\lambda_\mu,D)(d\times p) \\
&\quad -k_m^2 e^{\sqrt{-1}k_m(d-\hat{x})\cdot z}\big(I-\hat{x}\hat{x}^t\big)M(\lambda_\varepsilon,D)p\Big) + O(\frac{1}{|x|^2}) + O(\frac{\delta^4}{d_\sigma})
\end{aligned}
$$

as $|x|\to+\infty$, where $\hat{x} = x/|x|$ and t denotes the transpose. Therefore, up to an error term of order $O(\frac{\delta^4}{d_\sigma})$, the scattering amplitude A_∞ defined by (2.356) is given by

$$
\begin{aligned}
A_\infty(\hat{x}) &= \omega\mu_m k_m e^{\sqrt{-1}k_m(d-\hat{x})\cdot z}\big(\hat{x}\times I\big)M(\lambda_\mu,D)(d\times p) \\
&\quad -k_m^2 e^{\sqrt{-1}k_m(d-\hat{x})\cdot z}\big(I-\hat{x}\hat{x}^t\big)M(\lambda_\varepsilon,D)p.
\end{aligned}
\tag{7.5}
$$

Formula (7.5) allows us to compute the extinction cross-section Q^{ext} in terms of the polarization tensors associated with the particle D and the material parameter contrasts. Moreover, an estimate for the blow up of the extinction cross-section Q^{ext} at the plasmonic resonances follows immediately from (2.72).

THEOREM 7.3. *We have*

$$
Q^{\text{ext}} = \frac{4\pi}{k_m|p|^2}\Im\left[p\cdot\left[\omega\mu_m k_m\big(d\times I\big)M(\lambda_\mu,D)(d\times p) - k_m^2\big(I-dd^t\big)M(\lambda_\varepsilon,D)p\right]\right].
$$

7.2.1.1. *Shape Derivative of Quasi-Static Plasmonic Resonances.* In order to compute the shape derivative of quasi-static plasmonic resonances, it suffices to compute the shape derivative of eigenvalues of the Neumann–Poincaré operator $(\mathcal{K}_D^0)^*$.

Let D_ϵ be given by

$$\partial D_\epsilon = \left\{ \tilde{x} : \tilde{x} = x + \epsilon h(x)\nu(x), \; x \in \partial D \right\},$$

where $h \in \mathcal{C}^1(\partial D)$ and $0 < \epsilon \ll 1$. From Lemma 2.42, we have

$$((\mathcal{K}_{D_\epsilon}^0)^*[\cdot]) \circ \Psi_\epsilon = (\mathcal{K}_D^0)^*[\cdot] + \epsilon \mathcal{K}_D^{(1)}[\cdot] + O(\epsilon^2),$$

where $\mathcal{K}_D^{(1)} : \mathcal{H}^*(\partial D) \to \mathcal{H}^*(\partial D)$ is defined by (2.148).

Assume that λ_j is a simple eigenvalue of $(\mathcal{K}_D^0)^*$ associated to the eigenfunction φ_j. Then, there exists an eigenvalue λ_j^ϵ of $(\mathcal{K}_{D_\epsilon}^0)^*$ in a small neighborhood of λ_j and the following asymptotic formula for λ_j^ϵ as $\epsilon \to 0$ holds:

$$\lambda_j^\epsilon = \lambda_j + \epsilon \langle \mathcal{K}_D^{(1)}[\varphi_j], \varphi_j \rangle_{L^2(\partial D)} + O(\epsilon^2).$$

The shape derivative of λ_j is therefore given by $\langle \mathcal{K}_D^{(1)}[\varphi_j], \varphi_j \rangle_{L^2(\partial D)}$. It can be used dor the design of nanoparticles that resonate at specific frequencies [**29**, **274**, **386**].

7.3. Effective Medium Theory for Suspensions of Plasmonic Nanoparticles

In this section we derive effective properties of a system of plasmonic nanoparticles. To begin with, we consider a bounded and simply connected domain $\Omega \Subset \mathbb{R}^3$ of class $\mathcal{C}^{1,\alpha}$ for $\alpha > 0$, filled with a composite material that consists of a matrix of constant electric permittivity ε_m and a set of periodically distributed plasmonic nanoparticles with (small) period η and electric permittivity ε_c; see Figure 7.1.

FIGURE 7.1. Periodic composite material.

Let $Y = (-1/2, 1/2)^3$ be the unit cell and denote $\delta = \eta^\beta$ for $\beta > 0$. We set the (rescaled) periodic function

$$\gamma = \varepsilon_m \chi(Y \backslash \bar{D}) + \varepsilon_c \chi(D),$$

where $D = \delta B$ with $B \Subset \mathbb{R}^3$ being of class $\mathcal{C}^{1,\alpha}$ and the volume of B, $|B|$, is assumed to be equal to 1. Thus, the electric permittivity of the composite is given by the periodic function

$$\gamma_\eta(x) = \gamma(x/\eta),$$

which has period η. Now, consider the problem

$$\nabla \cdot \gamma_\eta \nabla u_\eta = 0 \quad \text{in } \Omega \tag{7.6}$$

with an appropriate boundary condition on $\partial\Omega$. Then, there exists a homogeneous, generally anisotropic, permittivity γ^*, such that the replacement, as $\eta \to 0$, of the original equation (7.6) by

$$\nabla \cdot \gamma^* \nabla u_0 = 0 \quad \text{in } \Omega$$

is a valid approximation in a certain sense. The coefficient γ^* is called an effective permittivity. It represents the overall macroscopic material property of the periodic composite made of plasmonic nanoparticles embedded in an isotropic matrix.

The (effective) matrix $\gamma^* = (\gamma^*_{pq})_{p,q=1,2,3}$ is defined by [**52**]

$$\gamma^*_{pq} = \int_Y \gamma(x) \nabla u_p(x) \cdot \nabla u_q(x) dx,$$

where u_p, for $p = 1, 2, 3$, is the unique solution to the cell problem

$$\begin{cases} \nabla \cdot \gamma \nabla u_p = 0 \quad \text{in } Y, \\ u_p - x_p \quad \text{periodic (in each direction) with period 1}, \\ \displaystyle\int_Y u_p(x) dx = 0. \end{cases} \tag{7.7}$$

Using Green's formula, we can rewrite γ^* in the following form:

$$\gamma^*_{pq} = \varepsilon_m \int_{\partial Y} u_q(x) \frac{\partial u_p}{\partial \nu}(x) d\sigma(x). \tag{7.8}$$

The matrix γ^* depends on η as a parameter and cannot be written explicitly.

Let $\mathcal{S}^0_{D,\sharp}$ and $(\mathcal{K}^0_{D,\sharp})^*$ be the single-layer potential and the Neumann–Poincaré operator, respectively, associated with the periodic Green's function $G_\sharp$ defined in (2.119) for $d = 3$.

From Theorem 2.38, we get

$$\gamma^*_{pq} = \varepsilon_m \int_{\partial Y} \big(y_q + C_q + \mathcal{S}^0_{D,\sharp}[\phi_q](y)\big) \frac{\partial\big(y_p + \mathcal{S}^0_{D,\sharp}[\phi_p](y)\big)}{\partial \nu} d\sigma(y),$$

where

$$\phi_p(y) = (\lambda_\varepsilon I - (\mathcal{K}^0_{D,\sharp})^*)^{-1}[\nu_p](y) \quad \text{for } y \text{ in } \partial D, \tag{7.9}$$

and $p = 1, 2, 3$.

Because of the periodicity of $\mathcal{S}^0_{D,\sharp}[\phi_p]$, we get

$$\gamma^*_{pq} = \varepsilon_m \Big(\delta_{pq} + \int_{\partial Y} y_q \frac{\partial \mathcal{S}^0_{D,\sharp}[\phi_p]}{\partial \nu}(y) d\sigma(y)\Big). \tag{7.10}$$

In view of the periodicity of $\mathcal{S}^0_{D,\sharp}[\phi_p]$, the divergence theorem applied on $Y \setminus \bar{D}$ and Lemma 2.37 yields

$$\int_{\partial Y} y_q \frac{\partial \mathcal{S}^0_{D,\sharp}[\phi_p]}{\partial \nu}(y) = \int_{\partial D} y_q \phi_p(y) d\sigma(y).$$

Let

$$\psi_p(y) = \phi_p(\delta y) \quad \text{for } y \in \partial B.$$

Then, by (7.10), we obtain

$$\gamma^* = \varepsilon_m(I + fP), \tag{7.11}$$

where $f = |D| = \delta^3 (= \eta^{3\beta})$ is the volume fraction of D and $P = (P_{pq})_{p,q=1,2,3}$ is given by

$$P_{pq} = \int_{\partial B} y_q \psi_p(y) d\sigma(y). \tag{7.12}$$

Now we proceed with the computation of P and prove the main result of this section, which shows the validity of the Maxwell-Garnett theory uniformly with respect to the frequency under the assumptions that

$$f \ll \operatorname{dist}(\lambda_\varepsilon(\omega), \sigma((\mathcal{K}_B^0)^*))^{3/5} \quad \text{and } (I - \delta^3 R^{-1}_{\lambda_\varepsilon(\omega)} T_0)^{-1} = O(1), \tag{7.13}$$

where $R^{-1}_{\lambda_\varepsilon(\omega)}$ and T_0 are to be defined and $\operatorname{dist}(\lambda_\varepsilon(\omega), \sigma((\mathcal{K}_D^0)^*))$ is the distance between $\lambda_\varepsilon(\omega)$ and the spectrum of $(\mathcal{K}_B^0)^*$.

THEOREM 7.4. *Assume that (7.13) holds. Then we have*

$$\gamma^* = \varepsilon_m\big(I + fM(I - \frac{f}{3}M)^{-1}\big) + O\Big(\frac{f^{8/3}}{\operatorname{dist}(\lambda_\varepsilon(\omega), \sigma((\mathcal{K}_B^0)^*))^2}\Big), \tag{7.14}$$

uniformly in ω. Here, $M = M(\lambda_\varepsilon(\omega), B)$ is the polarization tensor (2.71) associated with B and $\lambda_\varepsilon(\omega)$.

PROOF. In view of Lemma 2.36 and (7.9), we can write, for $x \in \partial D$,

$$(\lambda_\varepsilon(\omega)I - (\mathcal{K}_D^0)^*)[\phi_p](x) - \int_{\partial D} \frac{\partial R(x-y)}{\partial \nu(x)} \phi_p(y) d\sigma(y) = \nu_p(x),$$

which yields, for $x \in \partial B$,

$$(\lambda_\varepsilon(\omega)I - (\mathcal{K}_B^0)^*)[\psi_p](x) - \delta^2 \int_{\partial B} \frac{\partial R(\delta(x-y))}{\partial \nu(x)} \psi_p(y) d\sigma(y) = \nu_p(x).$$

By virtue of Lemma 2.36, we get

$$\nabla R(\delta(x-y)) = -\frac{\delta}{3}(x-y) + O(\delta^3)$$

uniformly in $x, y \in \partial B$. Since

$$\int_{\partial B} \psi_p(y) d\sigma(y) = 0,$$

we now have

$$(R_{\lambda_\varepsilon(\omega)} - \delta^3 T_0 + \delta^5 T_1)[\psi_p](x) = \nu_p(x),$$

and so

$$(I - \delta^3 R^{-1}_{\lambda_\varepsilon(\omega)} T_0 + \delta^5 R^{-1}_{\lambda_\varepsilon(\omega)} T_1)[\psi_p](x) = R^{-1}_{\lambda_\varepsilon(\omega)}[\nu_p](x), \tag{7.15}$$

where

$$
(7.16)\qquad \begin{aligned}
R_{\lambda_\varepsilon(\omega)}[\psi_p](x) &= (\lambda_\varepsilon(\omega)I - (\mathcal{K}_B^0)^*)[\psi_p](x),\\
T_0[\psi_p](x) &= \frac{\nu(x)}{3}\cdot \int_{\partial B} y\psi_p(y)d\sigma(y),\\
\|T_1\|_{\mathcal{L}(\mathcal{H}^*(\partial B),\mathcal{H}^*(\partial B))} &= O(1).
\end{aligned}
$$

Here, $\mathcal{H}^*(\partial B)$ is defined by (2.18) with Ω replaced with B. Since

$$
(\mathcal{K}_B^0)^* : \mathcal{H}^*(\partial B) \to \mathcal{H}^*(\partial B)
$$

is a self-adjoint, compact operator (see Theorem 2.8), it follows that

$$
(7.17)\qquad \|(\lambda_\varepsilon(\omega)I - (\mathcal{K}_B^0)^*)^{-1}\|_{\mathcal{L}(\mathcal{H}^*(\partial B),\mathcal{H}^*(\partial B))} \le \frac{c}{\operatorname{dist}(\lambda_\varepsilon(\omega),\sigma((\mathcal{K}_B^0)^*))}
$$

for a constant c.

In view of (7.16), T_0 is a compact operator. From the fact that the imaginary part of $R_{\lambda_\varepsilon(\omega)}$ is nonzero, it follows that $I-\delta^3 R^{-1}_{\lambda_\varepsilon(\omega)}T_0$ is invertible.

Under the assumption that

$$
\begin{aligned}
&(I-\delta^3 R^{-1}_{\lambda_\varepsilon(\omega)}T_0)^{-1} = O(1),\\
&\delta^5 \ll \operatorname{dist}(\lambda_\varepsilon(\omega),\sigma((\mathcal{K}_B^0)^*)),
\end{aligned}
$$

we get from (7.15) and (7.17)

$$
\begin{aligned}
\psi_p(x) &= (I-\delta^3 R^{-1}_{\lambda_\varepsilon(\omega)}T_0 + \delta^5 R^{-1}_{\lambda_\varepsilon(\omega)}T_1)^{-1} R^{-1}_{\lambda_\varepsilon(\omega)}[\nu_p](x),\\
&= (I-\delta^3 R^{-1}_{\lambda_\varepsilon(\omega)}T_0)^{-1} R^{-1}_{\lambda_\varepsilon(\omega)}[\nu_p](x) + O\Big(\frac{\delta^5}{\operatorname{dist}(\lambda_\varepsilon(\omega),\sigma((\mathcal{K}_B^0)^*))}\Big).
\end{aligned}
$$

Therefore, we obtain the estimate for ψ_p

$$
\psi_p = O\Big(\frac{1}{\operatorname{dist}(\lambda_\varepsilon(\omega),\sigma((\mathcal{K}_B^0)^*))}\Big).
$$

Now, we multiply (7.15) by y_q and integrate over ∂B. We can derive from the estimate of ψ_p that

$$
P(I-\frac{f}{3}M) = M + O\Big(\frac{\delta^5}{\operatorname{dist}(\lambda_\varepsilon(\omega),\sigma((\mathcal{K}_B^0)^*))^2}\Big),
$$

and therefore,

$$
P = M(I+\frac{f}{3}M)^{-1} + O\Big(\frac{\delta^5}{\operatorname{dist}(\lambda_\varepsilon(\omega),\sigma((\mathcal{K}_B^0)^*)^2}\Big)
$$

with P being defined by (7.12). Since $f=\delta^3$ and

$$
M = O\Big(\frac{\delta^3}{\operatorname{dist}(\lambda_\varepsilon(\omega),\sigma((\mathcal{K}_B^0)^*))}\Big),
$$

it follows from (7.11) that the Maxwell-Garnett formula (7.14) holds (uniformly in the frequency ω) under the assumption (7.13) on the volume fraction f. □

REMARK 7.5. *As a corollary of Theorem 7.4, we see that in the case when* $fM = O(1)$, *which is equivalent to the scale* $f = O\Big(\operatorname{dist}(\lambda_\varepsilon(\omega),\sigma((\mathcal{K}_B^0)^*))\Big)$, *the* *matrix* $fM(I-\frac{f}{3}M)^{-1}$ *may have a negative-definite symmetric real part. On the other hand, if*

$$\mathrm{dist}(\lambda_\varepsilon(\omega),\sigma((\mathcal{K}_B^0)^*))=O(f^{1+\beta})\quad \textit{for } 0<\beta<2/3,$$

then the effective matrix γ^* *may be very large. This provides evidence of the possibility of constructing both negative and high-contrast materials using plasmonic nanoparticles in appropriate regimes. In view of (2.74), one can prove that for an appropriate volume fraction and* ω *near and above a plasmonic resonance,* γ^* *may be large, while for* ω *below the plasmonic resonance,* γ^* *may be negative.*

7.4. Shift in Plasmonic Resonances Due to the Particle Size

As shown in Section 7.2, according to the quasi-static approximation for small particles, the surface plasmon resonance peak occurs when the particle's polarization tensor is maximized. Plasmon resonances in nanoparticles can be treated at the quasi-static limit as an eigenvalue problem for the Neumann-Poincaré integral operator. At this limit, they are size-independent. However, as the particle size increases, they are determined from scattering and absorption blow up and become size-dependent. This was experimentally observed, for instance, in [**263**, **387**, **437**, **470**].

In this section we analyze the shift of the plasmon resonance with changes in size of the nanoparticle.

Let $\widetilde{\mathcal{M}}_B$ be defined by (2.323) with D replaced with B and let $\sigma_j, j=1,2,3$, be given by (2.328). For simplicity we assume that Conditions 2.103 and 2.105 hold.

We consider the original system of integral equations (2.335) for a given incident plane wave (E^i,H^i). With the same notation as in Section 2.14, the following result holds by using Lemmas 2.110 and 2.111.

LEMMA 7.6. *Let* η *be defined by (7.1). The system of equations* (2.335) *can be rewritten as follows:*

$$\mathcal{W}_B(\delta)\begin{pmatrix}\eta(\psi)\\ \omega\eta(\phi)\end{pmatrix}=\begin{pmatrix}\dfrac{\eta(\nu\times E^i)}{\mu_m-\mu_c}\\ \dfrac{\eta(\sqrt{-1}\nu\times H^i)}{\varepsilon_m-\varepsilon_c}\end{pmatrix}\Bigg|_{\partial B},\tag{7.18}$$

with

$$\mathcal{W}_B(\delta)=\begin{pmatrix}\mathcal{W}_B(\delta)_{11} & \mathcal{W}_B(\delta)_{12}\\ \mathcal{W}_B(\delta)_{21} & \mathcal{W}_B(\delta)_{22}\end{pmatrix},\tag{7.19}$$

where

$$\begin{aligned}
\mathcal{W}_B(\delta)_{11}&=\lambda_\mu I-\mathcal{M}_B+\delta^2\frac{\mu_m\mathcal{M}_{B,2}^{k_m}-\mu_c\mathcal{M}_{B,2}^{k_c}}{\mu_m-\mu_c}+O(\delta^3),\\
\mathcal{W}_B(\delta)_{12}&=\frac{1}{\mu_m-\mu_c}(\delta\mathcal{L}_{B,1}+\delta^2\mathcal{L}_{B,2})+O(\delta^3),\\
\mathcal{W}_B(\delta)_{21}&=\frac{1}{\varepsilon_m-\varepsilon_c}(\delta\mathcal{L}_{B,1}+\delta^2\mathcal{L}_{B,2})+O(\delta^3),\\
\mathcal{W}_B(\delta)_{22}&=\lambda_\varepsilon I-\mathcal{M}_B+\delta^2\frac{\varepsilon_m\mathcal{M}_{B,2}^{k_m}-\varepsilon_c\mathcal{M}_{B,2}^{k_c}}{\varepsilon_m-\varepsilon_c}+O(\delta^3),
\end{aligned}$$

and the material parameter contrasts λ_μ and λ_ε are given by

$$\lambda_\mu = \frac{\mu_c + \mu_m}{2(\mu_m - \mu_c)}, \quad \lambda_\varepsilon = \frac{\varepsilon_c + \varepsilon_m}{2(\varepsilon_m - \varepsilon_c)}. \tag{7.20}$$

It is clear that

$$\mathcal{W}_B(0) = \mathcal{W}_{B,0} = \begin{pmatrix} \lambda_\mu I - \mathcal{M}_B & 0 \\ 0 & \lambda_\varepsilon I - \mathcal{M}_B \end{pmatrix}.$$

Moreover,

$$\mathcal{W}_B(\delta) = \mathcal{W}_{B,0} + \delta \mathcal{W}_{B,1} + \delta^2 \mathcal{W}_{B,2} + O(\delta^3),$$

in the sense that

$$\|\mathcal{W}_B(\delta) - \mathcal{W}_{B,0} - \delta \mathcal{W}_{B,1} - \delta^2 \mathcal{W}_{B,2}\| \leq C\delta^3$$

for a constant C independent of δ. Here $\|A\| = \sup_{i,j} \|A_{i,j}\|_{H_T^{-\frac{1}{2}}(\mathrm{div},\partial B)}$ for any operator-valued matrix A with entries $A_{i,j}$.

We are now interested in finding $\mathcal{W}_B^{-1}(\delta)$. The following result holds.

LEMMA 7.7. *The system of equations* (2.335) *is equivalent to*

$$W_B(\delta) \begin{pmatrix} \eta(\psi)^{(1)} \\ \eta(\psi)^{(2)} \\ \omega\eta(\phi)^{(1)} \\ \omega\eta(\phi)^{(2)} \end{pmatrix} = \begin{pmatrix} \dfrac{\eta(\nu \times E^i)^{(1)}}{\mu_m - \mu_c} \\ \dfrac{\eta(\nu \times E^i)^{(2)}}{\mu_m - \mu_c} \\ \dfrac{\eta(\sqrt{-1}\nu \times H^i)^{(1)}}{\varepsilon_m - \varepsilon_c} \\ \dfrac{\eta(\sqrt{-1}\nu \times H^i)^{(2)}}{\varepsilon_m - \varepsilon_c} \end{pmatrix}\Bigg|_{\partial B}, \tag{7.21}$$

where

$$W_B(\delta) = W_{B,0} + \delta W_{B,1} + \delta^2 W_{B,2} + O(\delta^3)$$

with

$$W_{B,0} = \begin{pmatrix} \lambda_\mu I - \widetilde{\mathcal{M}}_B & O \\ O & \lambda_\varepsilon I - \widetilde{\mathcal{M}}_B \end{pmatrix},$$

$$W_{B,1} = \begin{pmatrix} O & \dfrac{1}{\mu_m - \mu_c}\widetilde{\mathcal{L}}_{B,1} \\ \dfrac{1}{\varepsilon_m - \varepsilon_c}\widetilde{\mathcal{L}}_{B,1} & O \end{pmatrix},$$

$$W_{B,2} = \begin{pmatrix} \dfrac{1}{\mu_m - \mu_c}\widetilde{\mathcal{M}}^\mu_{B,2} & \dfrac{1}{\mu_m - \mu_c}\widetilde{\mathcal{L}}_{B,2} \\ \dfrac{1}{\varepsilon_m - \varepsilon_c}\widetilde{\mathcal{L}}_{B,2} & \dfrac{1}{\varepsilon_m - \varepsilon_c}\widetilde{\mathcal{M}}^\varepsilon_{B,2} \end{pmatrix},$$

$$\widetilde{\mathcal{M}}^\mu_{B,2} = \begin{pmatrix} (\widetilde{\mathcal{M}}^\mu_{B,2})_{11} & (\widetilde{\mathcal{M}}^\mu_{B,2})_{12} \\ (\widetilde{\mathcal{M}}^\mu_{B,2})_{21} & (\widetilde{\mathcal{M}}^\mu_{B,2})_{22} \end{pmatrix},$$

$$\widetilde{\mathcal{M}}^\varepsilon_{B,2} = \begin{pmatrix} (\widetilde{\mathcal{M}}^\varepsilon_{B,2})_{11} & (\widetilde{\mathcal{M}}^\varepsilon_{B,2})_{12} \\ (\widetilde{\mathcal{M}}^\varepsilon_{B,2})_{21} & (\widetilde{\mathcal{M}}^\varepsilon_{B,2})_{22} \end{pmatrix},$$

and

$$\widetilde{\mathcal{M}}_B = \begin{pmatrix} -\Delta_{\partial B}^{-1}(\mathcal{K}_B^0)^*\Delta_{\partial B} & 0 \\ \mathcal{R}_B & \mathcal{K}_B^0 \end{pmatrix},$$

$$\begin{aligned}
(\widetilde{\mathcal{M}}^{\mu}_{B,2})_{11} &= \Delta_{\partial B}^{-1}\nabla_{\partial B}\cdot(\mu_m\mathcal{M}^{k_m}_{B,2} - \mu_c\mathcal{M}^{k_c}_{B,2})\nabla_{\partial B},\\
(\widetilde{\mathcal{M}}^{\mu}_{B,2})_{12} &= \Delta_{\partial B}^{-1}\nabla_{\partial B}\cdot(\mu_m\mathcal{M}^{k_m}_{B,2} - \mu_c\mathcal{M}^{k_c}_{B,2})\vec{\mathrm{curl}}_{\partial B},\\
(\widetilde{\mathcal{M}}^{\mu}_{B,2})_{21} &= -\Delta_{\partial B}^{-1}\mathrm{curl}_{\partial B}(\mu_m\mathcal{M}^{k_m}_{B,2} - \mu_c\mathcal{M}^{k_c}_{B,2})\nabla_{\partial B},\\
(\widetilde{\mathcal{M}}^{\mu}_{B,2})_{22} &= -\Delta_{\partial B}^{-1}\mathrm{curl}_{\partial B}(\mu_m\mathcal{M}^{k_m}_{B,2} - \mu_c\mathcal{M}^{k_c}_{B,2})\vec{\mathrm{curl}}_{\partial B},
\end{aligned}$$

$$\begin{aligned}
(\widetilde{\mathcal{M}}^{\varepsilon}_{B,2})_{11} &= \Delta_{\partial B}^{-1}\nabla_{\partial B}\cdot(\varepsilon_m\mathcal{M}^{k_m}_{B,2} - \varepsilon_c\mathcal{M}^{k_c}_{B,2})\nabla_{\partial B},\\
(\widetilde{\mathcal{M}}^{\varepsilon}_{B,2})_{12} &= \Delta_{\partial B}^{-1}\nabla_{\partial B}\cdot(\varepsilon_m\mathcal{M}^{k_m}_{B,2} - \varepsilon_c\mathcal{M}^{k_c}_{B,2})\vec{\mathrm{curl}}_{\partial B},\\
(\widetilde{\mathcal{M}}^{\varepsilon}_{B,2})_{21} &= -\Delta_{\partial B}^{-1}\mathrm{curl}_{\partial B}(\varepsilon_m\mathcal{M}^{k_m}_{B,2} - \varepsilon_c\mathcal{M}^{k_c}_{B,2})\nabla_{\partial B},\\
(\widetilde{\mathcal{M}}^{\varepsilon}_{B,2})_{22} &= -\Delta_{\partial B}^{-1}\mathrm{curl}_{\partial B}(\varepsilon_m\mathcal{M}^{k_m}_{B,2} - \varepsilon_c\mathcal{M}^{k_c}_{B,2})\vec{\mathrm{curl}}_{\partial B},
\end{aligned}$$

$$\widetilde{\mathcal{L}}_{B,s} = \begin{pmatrix} \Delta_{\partial B}^{-1}\nabla_{\partial B}\cdot\mathcal{L}_{B,s}\nabla_{\partial B} & \Delta_{\partial B}^{-1}\nabla_{\partial B}\cdot\mathcal{L}_{B,s}\vec{\mathrm{curl}}_{\partial B} \\ -\Delta_{\partial B}^{-1}\mathrm{curl}_{\partial B}\mathcal{L}_{B,s}\nabla_{\partial B} & -\Delta_{\partial B}^{-1}\mathrm{curl}_{\partial B}\mathcal{L}_{B,s}\vec{\mathrm{curl}}_{\partial B} \end{pmatrix}$$

for $s = 1, 2$.
Moreover, the eigenfunctions of $W_{B,0}$ *in* $H(\partial B)^2$ *are given by*

$$\begin{aligned}
\Psi_{1,j,i} &= \begin{pmatrix} \psi_{j,i} \\ O \end{pmatrix}, \quad j = 0, 1, 2, \ldots; i = 1, 2, 3,\\
\Psi_{2,j,i} &= \begin{pmatrix} O \\ \psi_{j,i} \end{pmatrix}, \quad j = 0, 1, 2, \ldots; i = 1, 2, 3,
\end{aligned}$$

associated to the eigenvalues $\lambda_\mu - \lambda_{j,i}$ *and* $\lambda_\varepsilon - \lambda_{j,i}$, *respectively, and generalized eigenfunctions of order one*

$$\begin{aligned}
\Psi_{1,j,3,g} &= \begin{pmatrix} \psi_{j,3,g} \\ O \end{pmatrix},\\
\Psi_{2,j,3,g} &= \begin{pmatrix} O \\ \psi_{j,3,g} \end{pmatrix},
\end{aligned}$$

associated to eigenvalues $\lambda_\mu - \lambda_{j,3}$ *and* $\lambda_\varepsilon - \lambda_{j,3}$, *respectively, all of which form a non-orthogonal basis of* $H(\partial B)^2$. *Here,* $H(\partial B)$ *is defined by (*2.327*) with* D *replaced with* B.

PROOF. The proof follows directly from Lemmas 2.102 and 2.104. □

We regard the operator $W_B(\delta)$ as a perturbation of the operator $W_{B,0}$ for small δ. Using perturbation theory, we can derive the perturbed eigenvalues and their associated eigenfunctions in $H(\partial B)^2$.
We denote by $\Gamma = \{(k, j, i) : k = 1, 2; j = 1, 2, \ldots; i = 1, 2, 3\}$ the set of indices for the eigenfunctions of $W_{B,0}$ and by $\Gamma_g = \{(k, j, 3, g) : k = 1, 2; j = 1, 2, \ldots\}$ the set of indices for the generalized eigenfunctions. We denote by γ_g the generalized

eigenfunction index corresponding to eigenfunction index γ and vice-versa. We also denote by

$$\tau_\gamma = \begin{cases} \lambda_\mu - \lambda_{j,i}, & k = 1, \\ \lambda_\varepsilon - \lambda_{j,i}, & k = 2. \end{cases} \tag{7.22}$$

We suppose that the following technical assumption holds.

CONDITION 7.8. $\lambda_\mu \neq \lambda_\varepsilon$.

In the following we will only consider $\gamma \in \Gamma$ for which there is no generalized eigenfunction index associated. In other words, we only consider $\gamma = (k, i, j) \in \Gamma$ such that $\lambda_{j,i} \in \sigma_1 \cup \sigma_2$ (see (2.328) for the definitons). We call this subset Γ_{sim}. Note that Conditions 2.103 and 7.8 imply that the eigenvalues of $W_{B,0}$ indexed by $\gamma \in \Gamma_{\text{sim}}$ are simple.

THEOREM 7.9. *As $\delta \to 0$, the perturbed eigenvalues and eigenfunctions indexed by $\gamma \in \Gamma_{\text{sim}}$ have the following asymptotic expansions:*

$$\begin{aligned} \tau_\gamma(\delta) &= \tau_\gamma + \delta\tau_{\gamma,1} + \delta^2\tau_{\gamma,2} + O(\delta^3), \\ \Psi_\gamma(\delta) &= \Psi_\gamma + \delta\Psi_{\gamma,1} + O(\delta^2), \end{aligned} \tag{7.23}$$

where

$$\begin{aligned} \tau_{\gamma,1} &= \frac{\langle W_{B,1}\Psi_\gamma, \widetilde{\Psi}_\gamma\rangle_{H(\partial B)^2}}{\langle \Psi_\gamma, \widetilde{\Psi}_\gamma\rangle_{H(\partial B)^2}} = 0, \\ \tau_{\gamma,2} &= \frac{\langle W_{B,2}\Psi_\gamma, \widetilde{\Psi}_\gamma\rangle_{H(\partial B)^2} - \langle W_{B,1}\Psi_{\gamma,1}, \widetilde{\Psi}_\gamma\rangle_{H(\partial B)^2}}{\langle \Psi_\gamma, \widetilde{\Psi}_\gamma\rangle_{H(\partial B)^2}}, \\ (\tau_\gamma - W_{B,0})\Psi_{\gamma,1} &= -W_{B,1}\Psi_\gamma. \end{aligned} \tag{7.24}$$

Here, $\widetilde{\Psi}_{\gamma'} \in \operatorname{Ker}(\bar{\tau}_{\gamma'} - W_{B,0}^*)$ and $W_{B,0}^*$ is the $H(\partial B)^2$ adjoint of $W_{B,0}$.

Using Lemma 2.107 and Remark 2.109 we can solve $\Psi_{\gamma,1}$. Indeed,

$$\begin{aligned} \Psi_{\gamma,1} &= \sum_{\substack{\gamma' \in \Gamma \\ \gamma' \neq \gamma}} \frac{\alpha(-W_{B,1}\Psi_\gamma, \Psi_{\gamma'})\Psi_{\gamma'}}{\tau_\gamma - \tau_{\gamma'}} \\ &\quad + \sum_{\substack{\gamma'_g \in \Gamma_g \\ \gamma' \neq \gamma}} \alpha(-W_{B,1}\Psi_\gamma, \Psi_{\gamma'_g}) \left(\frac{\Psi_{\gamma'_g}}{\tau_\gamma - \tau_{\gamma'}} + \frac{\Psi_{\gamma'}}{(\tau_\gamma - \tau_{\gamma'})^2} \right) \\ &\quad + \alpha(-W_{B,1}\Psi_\gamma, \Psi_\gamma)\Psi_\gamma. \end{aligned}$$

By abuse of notation,

$$\alpha(x, \Psi_\gamma) = \begin{cases} \alpha(x_1, \psi_\kappa) & \gamma = (1, j, i),\ \kappa = (j, i), \\ \alpha(x_2, \psi_\kappa) & \gamma = (2, j, i),\ \kappa = (j, i), \end{cases} \tag{7.25}$$

for

$$x = \begin{pmatrix} x_1 \\ x_2 \end{pmatrix} \in H(\partial B)^2$$

with α being introduced in Definition 2.106.

Consider now the degenerate case $\gamma \in \Gamma \backslash \Gamma_{\text{sim}} =: \Gamma_{\text{deg}} = \{\gamma = (k, i, j) \in \Gamma$ s.t $\lambda_{j,i} \in \sigma_3\}$. It is clear that, for $\gamma \in \Gamma_{\text{deg}}$, the algebraic multiplicity of the eigenvalue τ_γ is 2 while the geometric multiplicity is 1. In this case every eigenvalue

τ_γ and associated eigenfunction Ψ_γ will split into two branches, as δ goes to zero, represented by a convergent Puiseux series as

$$(7.26)\qquad \begin{aligned} \tau_{\gamma,h}(\delta) &= \tau_\gamma + (-1)^h \delta^{1/2}\tau_{\gamma,1} + (-1)^{2h}\delta^{2/2}\tau_{\gamma,2} + O(\delta^{3/2}),\\ \Psi_{\gamma,h}(\delta) &= \Psi_\gamma + (-1)^h \delta^{1/2}\Psi_{\gamma,1} + (-1)^{2h}\delta^{2/2}\Psi_{\gamma,2} + O(\delta^{3/2}), \end{aligned}$$

for $h = 0, 1$, where $\tau_{\gamma,j}$ and $\Psi_{\gamma,j}$ can be recovered by recurrence formulas; see Section 2.9.6.

Recall that the electric and magnetic parameters, ε_c and μ_c, depend on the frequency of the incident field, ω, following a Drude model. Therefore, the eigenvalues of the operator $W_{B,0}$ and the perturbation in the eigenvalues depend on the frequency as well, that is,

$$\begin{aligned} \tau_\gamma(\delta,\omega) &= \tau_\gamma(\omega) + \delta^2\tau_{\gamma,2}(\omega) + O(\delta^3), \quad \gamma \in \Gamma_{\text{sim}},\\ \tau_{\gamma,h}(\delta,\omega) &= \tau_\gamma + \delta^{1/2}(-1)^h\tau_{\gamma,1}(\omega) + \delta^{2/2}(-1)^{2h}\tau_{\gamma,2}(\omega) + O(\delta^{3/2}), \quad \gamma \in \Gamma_{\text{deg}} \end{aligned}$$

for $h = 0, 1$.

In the sequel, we will omit frequency dependence to simplify the notation. However, we will keep in mind that all these quantities are frequency dependent.

We first state the following result.

PROPOSITION 7.10. *If ω is a quasi-static plasmonic resonance (as stated in Definition 7.2), then $|\tau_\gamma| \ll 1$ and is locally minimized for some $\gamma \in \Gamma$ with τ_γ being defined by (7.22).*

Then we recall two different notions of plasmonic resonance [**78**].

DEFINITION 7.11.

(i) *We say that ω is a plasmonic resonance if $|\tau_\gamma(\delta)| \ll 1$ and is locally minimized for some $\gamma \in \Gamma_{\text{sim}}$ or $|\tau_{\gamma,h}(\delta)| \ll 1$ and is locally minimized for some $\gamma \in \Gamma_{\text{deg}}$, $h = 0, 1$.*

(ii) *We say that ω is a first-order corrected quasi-static plasmonic resonance if $|\tau_\gamma + \delta^2\tau_{\gamma,2}| \ll 1$ and is locally minimized for some $\gamma \in \Gamma_{\text{sim}}$ or $|\tau_\gamma + \delta^{1/2}(-1)^h\tau_{\gamma,1}| \ll 1$ and is locally minimized for some $\gamma \in \Gamma_{\text{deg}}$, $h = 0, 1$. Here, the correction terms $\tau_{\gamma,2}$ and $\tau_{\gamma,1}$ are defined by (7.24) and (7.26).*

Note that quasi-static resonance is size independent and is therefore a zero-order approximation of the plasmonic resonance in terms of the particle size while the first-order corrected quasi-static plasmonic resonance depends on the size of the nanoparticle.

We are interested in solving equation (7.21)

$$W_B(\delta)[\Psi] = f,$$

where

$$\Psi = \begin{pmatrix} \eta(\psi)^{(1)} \\ \eta(\psi)^{(2)} \\ \omega\eta(\phi)^{(1)} \\ \omega\eta(\phi)^{(2)} \end{pmatrix}, f = \left.\begin{pmatrix} \dfrac{\eta(\nu\times E^i)^{(1)}}{\mu_m-\mu_c} \\ \dfrac{\eta(\nu\times E^i)^{(2)}}{\mu_m-\mu_c} \\ \dfrac{\eta(\sqrt{-1}\nu\times H^i)^{(1)}}{\varepsilon_m-\varepsilon_c} \\ \dfrac{\eta(\sqrt{-1}\nu\times H^i)^{(2)}}{\varepsilon_m-\varepsilon_c} \end{pmatrix}\right|_{\partial B}$$

for ω close to the resonance frequencies, *i.e.*, when $\tau_\gamma(\delta)$ is very small for some γ's $\in \Gamma_{\text{sim}}$ or $\tau_{\gamma,h}(\delta)$ is very small for some γ's $\in \Gamma_{\text{deg}}$, $h = 0, 1$. In this case, the major part of the solution would be the contributions of the excited resonance modes $\Psi_\gamma(\delta)$ and $\Psi_{\gamma,h}(\delta)$.

We introduce the following definition.

DEFINITION 7.12. *We call $J \subset \Gamma$ an index set of resonances if the τ_γ's are close to zero when $\gamma \in \Gamma$ and are bounded from below when $\gamma \in \Gamma^c$. More precisely, we choose a threshold number $\eta_0 > 0$ independent of ω such that*

$$|\tau_\gamma| \geq \eta_0 > 0 \quad \textit{for } \gamma \in J^c.$$

From now on, we shall use J as our index set of resonances. For simplicity, we assume throughout this paper that the following condition holds.

CONDITION 7.13. *We assume that $\lambda_\mu \neq 0$, $\lambda_\varepsilon \neq 0$ or equivalently, $\mu_c \neq -\mu_m$, $\varepsilon_c \neq -\varepsilon_m$.*

It follows that the set J is finite.
Consider the space $\mathcal{E}_J = \text{span}\{\Psi_\gamma(\delta), \Psi_{\gamma,h}(\delta);\ \gamma \in J,\ h = 0, 1\}$. Note that, under Condition 7.13, $\mathcal{E}_J$ is finite dimensional. Similarly, we define $\mathcal{E}_{J^c}$ as the spanned by $\Psi_\gamma(\delta), \Psi_{\gamma,h}(\delta);\ \gamma \in J^c,\ h = 0, 1$ and eventually other vectors to complete the base. We have $H(\partial B)^2 = \mathcal{E}_J \oplus \mathcal{E}_{J^c}$.

We define $P_J(\delta)$ and $P_{J^c}(\delta)$ as the (non-orthogonal) projection into the finite-dimensional space $\mathcal{E}_J$ and infinite-dimensional space $\mathcal{E}_{J^c}$, respectively. It is clear that, for any $f \in H(\partial B)^2$

$$f = P_J(\delta)[f] + P_{J^c}(\delta)[f].$$

Moreover, we have an explicit representation for $P_J(\delta)$

$$P_J(\delta)[f] = \sum_{\gamma \in J \cap \Gamma_{\text{sim}}} \alpha_\delta(f, \Psi_\gamma(\delta))\Psi_\gamma(\delta) + \sum_{\substack{\gamma \in J \cap \Gamma_{\text{deg}} \\ h=0,1}} \alpha_\delta(f, \Psi_{\gamma,h}(\delta))\Psi_{\gamma,h}(\delta). \tag{7.27}$$

Here, as in Lemma 2.107,

$$\begin{aligned}
\alpha_\delta(f, \Psi_\gamma(\delta)) &= \frac{\langle f, \tilde{\Psi}_\gamma(\delta)\rangle_{H(\partial B)^2}}{\langle \Psi_\gamma(\delta), \tilde{\Psi}_\gamma(\delta)\rangle_{H(\partial B)^2}}, \quad \gamma \in J \cap \Gamma_{\text{sim}}, \\
\alpha_\delta(f, \Psi_{\gamma,h}(\delta)) &= \frac{\langle f, \tilde{\Psi}_{\gamma,h}(\delta)\rangle_{H(\partial B)^2}}{\langle \Psi_{\gamma,h}(\delta), \tilde{\Psi}_{\gamma,h}(\delta)\rangle_{H(\partial B)^2}}, \quad \gamma \in J \cap \Gamma_{\text{deg}},\ h = 0, 1,
\end{aligned}$$

where $\widetilde{\Psi}_\gamma \in \text{Ker}(\bar{\tau}_{\gamma,h}(\delta) - W_B^*(\delta))$, $\widetilde{\Psi}_{\gamma,h} \in \text{Ker}(\bar{\tau}_{\gamma,h}(\delta) - W_B^*(\delta))$ and $W_B^*(\delta)$ is the $H(\partial B)^2$-adjoint of $W_B(\delta)$.

We are now ready to solve the equation $W_B(\delta)\Psi = f$. In view of Remark 2.109,

$$\begin{aligned}
\Psi = W_B^{-1}(\delta)[f] &= \sum_{\gamma \in J \cap \Gamma_{\text{sim}}} \frac{\alpha_\delta(f, \Psi_\gamma(\delta))\Psi_\gamma(\delta)}{\tau_\gamma(\delta)} + \sum_{\substack{\gamma \in J \cap \Gamma_{\text{deg}} \\ h=0,1}} \frac{\alpha_\delta(f, \Psi_{\gamma,h}(\delta))\Psi_{\gamma,h}(\delta)}{\tau_{\gamma,h}(\delta)} \\
&\quad + W_B^{-1}(\delta)P_{J^c}(\delta)[f].
\end{aligned} \tag{7.28}$$

The following lemma holds.

LEMMA 7.14. *The norm $\|W_B^{-1}(\delta)P_{J^c}(\delta)\|_{\mathcal{L}(H(\partial B)^2, H(\partial B)^2)}$ is uniformly bounded in ω and δ.*

PROOF. Consider the operator

$$W_B(\delta)|_{J^c} : P_{J^c}(\delta)H(\partial B)^2 \to P_{J^c}(\delta)H(\partial B)^2.$$

We can show that for every ω and δ, $\mathrm{dist}(\sigma(W_B(\delta)|_{J^c}), 0) \geq \frac{\eta_0}{2}$, where $\sigma(W_B(\delta)|_{J^c})$ is the discrete spectrum of $W_B(\delta)|_{J^c}$. Here and throughout the paper, dist denotes the distance. Then, it follows that

$$\begin{aligned} \|W_B^{-1}(\delta)P_{J^c}(\delta)[f]\| = \|W_B^{-1}(\delta)|_{J^c}P_{J^c}(\delta)[f]\| &\lesssim \frac{1}{\eta_0}\exp(\frac{C_1}{\eta_0^2})\|P_{J^c}(\delta)[f]\| \\ &\lesssim \frac{1}{\eta_0}\exp(\frac{C_1}{\eta_0^2})\|f\|, \end{aligned}$$

where the notation $A \lesssim B$ means that $A \leq CB$ for some constant C independent of A and B. □

Finally, we are ready to state our main result in this section.

THEOREM 7.15. *Let η be defined by (7.1). Under Conditions 2.103, 2.105, 7.8 and 7.13, the scattered field $E^s = E - E^i$ due to a single plasmonic particle has the following representation:*

$$E^s = \mu_m \nabla \times \vec{\mathcal{S}}_D^{k_m}[\psi](x) + \nabla \times \nabla \times \vec{\mathcal{S}}_D^{k_m}[\phi](x) \qquad x \in \mathbb{R}^3 \backslash \bar{D},$$

where

$$\begin{aligned} \psi &= \eta^{-1}\big(\nabla_{\partial B}\widetilde{\psi}^{(1)} + \vec{\mathrm{curl}}_{\partial B}\widetilde{\psi}^{(2)}\big), \\ \phi &= \frac{1}{\omega}\eta^{-1}\big(\nabla_{\partial B}\widetilde{\phi}^{(1)} + \vec{\mathrm{curl}}_{\partial B}\widetilde{\phi}^{(2)}\big), \end{aligned}$$

$$\begin{aligned} \Psi = \begin{pmatrix} \widetilde{\psi}^{(1)} \\ \widetilde{\psi}^{(2)} \\ \widetilde{\phi}^{(1)} \\ \widetilde{\phi}^{(2)} \end{pmatrix} &= \sum_{\gamma \in J \cap \Gamma_{\mathrm{sim}}} \frac{\alpha(f, \Psi_\gamma)\Psi_\gamma + O(\delta)}{\tau_\gamma(\delta)} \\ &\quad + \sum_{\gamma \in J \cap \Gamma_{\mathrm{deg}}} \frac{\zeta_1(f)\Psi_\gamma + \zeta_2(f)\Psi_{\gamma,1} + O(\delta^{1/2})}{\tau_{\gamma,0}(\delta)\tau_{\gamma,1}(\delta)} + O(1), \end{aligned}$$

and

$$\begin{aligned} \zeta_1(f) &= \frac{\langle f, \tilde{\Psi}_{\gamma,1}\rangle_{H(\partial B)^2}\tau_\gamma - \langle f, \tilde{\Psi}_\gamma\rangle_{H(\partial B)^2}(\tau_{\gamma,1} + \tau_\gamma\frac{a_2}{a_1})}{a_1}, \\ \zeta_2(f) &= \frac{\langle f, \tilde{\Psi}_\gamma\rangle_{H(\partial B)^2}}{a_1}, \\ a_1 &= \langle \Psi_\gamma, \tilde{\Psi}_{\gamma,1}\rangle_{H(\partial B)^2} + \langle \Psi_{\gamma,1}, \tilde{\Psi}_\gamma\rangle_{H(\partial B)^2}, \\ a_2 &= \langle \Psi_\gamma, \tilde{\Psi}_{\gamma,2}\rangle_{H(\partial B)^2} + \langle \Psi_{\gamma,2}, \tilde{\Psi}_\gamma\rangle_{H(\partial B)^2} + \langle \Psi_{\gamma,1}, \tilde{\Psi}_{\gamma,1}\rangle_{H(\partial B)^2}. \end{aligned}$$

PROOF. Recall that

$$\begin{aligned} \Psi &= \sum_{\gamma \in J \cap \Gamma_{\mathrm{sim}}} \frac{\alpha_\delta(f, \Psi_\gamma(\delta))\Psi_\gamma(\delta)}{\tau_\gamma(\delta)} + \sum_{\substack{\gamma \in J \cap \Gamma_{\mathrm{deg}} \\ h=0,1}} \frac{\alpha_\delta(f, \Psi_{\gamma,h}(\delta))\Psi_{\gamma,h}(\delta)}{\tau_{\gamma,h}(\delta)} \\ &\quad + W_B^{-1}(\delta)P_{J^c}(\delta)[f]. \end{aligned}$$

By Lemma 7.14, we have $W_B^{-1}(\delta)P_{J^c}(\delta)[f] = O(1)$.
If $\gamma \in J \cap \Gamma_{\text{sim}}$, an asymptotic expansion on δ yields

$$\alpha_\delta(f, \Psi_\gamma(\delta))\Psi_\gamma(\delta) = \alpha(f, \Psi_\gamma)\Psi_\gamma + O(\delta).$$

If $\gamma \in J \cap \Gamma_{\text{deg}}$ then $\langle \Psi_\gamma, \tilde{\Psi}_\gamma \rangle_{H(\partial B)^2} = 0$. Therefore, an asymptotic expansion on δ yields

$$\begin{aligned}
\alpha_\delta(f, \Psi_{\gamma,h}(\delta))\Psi_{\gamma,h}(\delta) &= \frac{(-1)^h \langle f, \tilde{\Psi}_\gamma \rangle_{H(\partial B)^2} \Psi_\gamma}{\delta^{-1/2} a_1} \\
&\quad + \frac{1}{a_1}\Bigg((\langle f, \tilde{\Psi}_{\gamma,1} \rangle_{H(\partial B)^2} - \langle f, \tilde{\Psi}_\gamma \rangle_{H(\partial B)^2} \frac{a_2}{a_1})\Psi_\gamma \\
&\qquad\qquad + \langle f, \tilde{\Psi}_\gamma \rangle_{H(\partial B)^2} \Psi_{\gamma,1} \Bigg) \\
&\quad + O(\delta^{1/2})
\end{aligned}$$

with

$$\begin{aligned}
a_1 &= \langle \Psi_\gamma, \tilde{\Psi}_{\gamma,1} \rangle_{H(\partial B)^2} + \langle \Psi_{\gamma,1}, \tilde{\Psi}_\gamma \rangle_{H(\partial B)^2}, \\
a_2 &= \langle \Psi_\gamma, \tilde{\Psi}_{\gamma,2} \rangle_{H(\partial B)^2} + \langle \Psi_{\gamma,2}, \tilde{\Psi}_\gamma \rangle_{H(\partial B)^2} + \langle \Psi_{\gamma,1}, \tilde{\Psi}_{\gamma,1} \rangle_{H(\partial B)^2}.
\end{aligned}$$

Since $\tau_{\gamma,h}(\delta) = \tau_\gamma + \delta^{1/2}(-1)^h \tau_{\gamma,1} + O(\delta)$, the result follows by adding the terms

$$\frac{\alpha_\delta(f, \Psi_{\gamma,0}(\delta))\Psi_{\gamma,0}(\delta)}{\tau_{\gamma,0}(\delta)} \quad \text{and} \quad \frac{\alpha_\delta(f, \Psi_{\gamma,1}(\delta))\Psi_{\gamma,1}(\delta)}{\tau_{\gamma,1}(\delta)}.$$

The proof is then complete. □

COROLLARY 7.16. *Assume the same conditions as in Theorem 7.15. Under the additional condition that*

$$\min_{\gamma \in J \cap \Gamma_{\text{sim}}} |\tau_\gamma(\delta)| \gg \delta^3, \quad \min_{\gamma \in J \cap \Gamma_{\text{deg}}} |\tau_\gamma(\delta)| \gg \delta, \tag{7.29}$$

we have

$$\Psi = \sum_{\gamma \in J \cap \Gamma_{\text{sim}}} \frac{\alpha(f, \Psi_\gamma)\Psi_\gamma + O(\delta)}{\tau_\gamma + \delta^2 \tau_{\gamma,2}} + \sum_{\gamma \in J \cap \Gamma_{\text{deg}}} \frac{\zeta_1(f)\Psi_\gamma + \zeta_2(f)\Psi_{\gamma,1} + O(\delta^{1/2})}{\tau_\gamma^2 - \delta \tau_{\gamma,1}^2} + O(1).$$

COROLLARY 7.17. *Assume the same conditions as in Theorem 7.15. Under the additional condition that*

$$\min_{\gamma \in J \cap \Gamma_{\text{sim}}} |\tau_\gamma(\delta)| \gg \delta^2, \quad \min_{\gamma \in J \cap \Gamma_{\text{deg}}} |\tau_\gamma(\delta)| \gg \delta^{1/2}, \tag{7.30}$$

we have

$$\begin{aligned}
\Psi &= \sum_{\gamma \in J \cap \Gamma_{\text{sim}}} \frac{\alpha(f, \Psi_\gamma)\Psi_\gamma + O(\delta)}{\tau_\gamma} + \sum_{\gamma \in J \cap \Gamma_{\text{deg}}} \frac{\alpha(f, \Psi_\gamma)\Psi_\gamma}{\tau_\gamma} \\
&\quad + \alpha(f, \Psi_{\gamma,g}) \left(\frac{\Psi_{\gamma,g}}{\tau_\gamma} + \frac{\Psi_\gamma}{\tau_\gamma^2} \right) + O(1).
\end{aligned}$$

PROOF. We have

$$\begin{aligned}\lim_{\delta\to 0} W_B^{-1}(\delta) P_{\mathrm{span}\{\Psi_{\gamma,0}(\delta),\Psi_{\gamma,1}(\delta)\}}[f] &= \lim_{\delta\to 0} \frac{\alpha_\delta(f,\Psi_{\gamma,0}(\delta))\Psi_{\gamma,0}(\delta)}{\tau_{\gamma,0}(\delta)} \\ &\quad + \frac{\alpha_\delta(f,\Psi_{\gamma,1}(\delta))\Psi_{\gamma,1}(\delta)}{\tau_{\gamma,1}(\delta)} \\ &= W_{B,0}^{-1}(\delta) P_{\mathrm{span}\{\Psi_\gamma,\Psi_{\gamma_g}\}}[f] \\ &= \frac{\alpha(f,\Psi_\gamma)\Psi_\gamma}{\tau_\gamma} + \alpha(f,\Psi_{\gamma,g})\left(\frac{\Psi_{\gamma,g}}{\tau_\gamma} + \frac{\Psi_\gamma}{\tau_\gamma^2}\right),\end{aligned}$$

where $\gamma \in J \cap \Gamma_{\mathrm{deg}}$, $f \in H(\partial B)^2$ and $P_{\mathrm{span}E}$ is the projection into the linear space generated by the elements in the set E. □

REMARK 7.18. *Note that for* $\gamma \in J$,

$$\tau_\gamma \approx \min\left\{\mathrm{dist}\big(\lambda_\mu, \sigma((\mathcal{K}_B^0)^*) \cup -\sigma((\mathcal{K}_B^0)^*)\big), \mathrm{dist}\big(\lambda_\varepsilon, \sigma((\mathcal{K}_B^0)^*) \cup -\sigma((\mathcal{K}_B^0)^*)\big)\right\}.$$

It is clear, from Remark 7.18, that resonances can occur when exciting the spectrum of $(\mathcal{K}_B^0)^*$ or/and that of $-(\mathcal{K}_B^0)^*$. We substantiate in the following that only the spectrum of $(\mathcal{K}_B^0)^*$ can be excited to create the plasmonic resonances in the quasi-static regime.
Recall that

$$f = \left.\begin{pmatrix} \dfrac{\eta(\nu\times E^i)^{(1)}}{\mu_m-\mu_c} \\ \dfrac{\eta(\nu\times E^i)^{(2)}}{\mu_m-\mu_c} \\ \dfrac{\eta(\sqrt{-1}\nu\times H^i)^{(1)}}{\varepsilon_m-\varepsilon_c} \\ \dfrac{\eta(\sqrt{-1}\nu\times H^i)^{(2)}}{\varepsilon_m-\varepsilon_c} \end{pmatrix}\right|_{\partial B},$$

and therefore,

$$f_1 := \frac{\eta(\nu\times E^i)^{(1)}}{\mu_m-\mu_c} = \frac{\Delta_{\partial B}^{-1}\nabla_{\partial B}\cdot\eta(\nu\times E^i)}{\mu_m-\mu_c}.$$

Now, suppose $\gamma = (1,j,1) \in J$ (recall that J is the index set of resonances). Then $\tau_\gamma = \lambda_\mu - \lambda_{1,j}$, where $\lambda_{1,j} \in \sigma_1 = \sigma(-(\mathcal{K}_B^0)^*)\backslash\sigma((\mathcal{K}_B^0)^*)$. From Remark 2.108,

$$\alpha(f,\Psi_\gamma) = \langle \Delta_{\partial B} f_1, \varphi_{j,1}\rangle_{\mathcal{H}^*} = \alpha(f,\Psi_\gamma) = \frac{1}{\mu_m-\mu_c}\langle \nabla_{\partial B}\cdot\eta(\nu\times E^i), \varphi_{j,1}\rangle_{\mathcal{H}^*},$$

where $\varphi_{j,1} \in \mathcal{H}_0^*(\partial B)$ is a normalized eigenfunction of $(\mathcal{K}_B^0)^*(\partial B)$.

A Taylor expansion of E^i gives, for $x \in \partial D$,

$$E^i(x) = \sum_{\beta\in\mathbb{N}^3}^{\infty} \frac{(x-z)^\beta \partial^\beta E^i(z)}{|\beta|!}.$$

Thus,

$$\eta(\nu\times E^i)(\tilde{x}) = \eta(\nu)(\tilde{x}) \times E^i(z) + O(\delta),$$

and

$$\begin{aligned}\nabla_{\partial B}\cdot\eta(\nu\times E^i)(\tilde{x}) &= -\eta(\nu)(\tilde{x})\cdot\nabla\times E^i(z) + O(\delta) \\ &= O(\delta).\end{aligned}$$

Therefore, the zeroth-order term of the expansion of $\nabla_{\partial B} \cdot \eta(\nu \times E^i)$ in δ is zero. Hence,

$$\alpha(f, \Psi_\gamma) = 0.$$

In the same way, we have

$$\begin{aligned} \alpha(f, \Psi_\gamma) &= 0, \\ \alpha(f, \Psi_{\gamma_g}) &= 0 \end{aligned}$$

for $\gamma = (2, j, 1) \in J$ and γ_g such that $\gamma \in J$.

As a result we see that the spectrum of $-(\mathcal{K}_B^0)^*$ is not excited in the zeroth-order term. However, we note that $\sigma(-(\mathcal{K}_B^0)^*)$ can be excited in higher-order terms.

Finally, we sketch a proof of Theorem 7.1. From (2.333), we have

$$E^s(x) = \mu_m \nabla \times \vec{\mathcal{S}}_D^{k_m}[\psi](x) + \nabla \times \nabla \times \vec{\mathcal{S}}_D^{k_m}[\phi](x), \quad x \in \mathbb{R}^3 \setminus \overline{D},$$

where ψ and ϕ are determined by (7.18). Since $\mathcal{W}_B(\delta) = \mathcal{W}_{B,0} + O(\delta)$, formula (7.2) follows by using the identities stated in Lemma 2.100.

7.4.1. Numerical Examples. Here we present numerical examples to demonstrate the shift of the plasmonic resonance. The corresponding Matlab code can be downloaded at `www.ams.org/bookpages/surv-235`.

The first example involves a spherical nanoparticle of radius R with permittivity ϵ_c. For the permittivity ϵ_c, we use Drude's model as follows: $\epsilon_c(\omega) = 1 - \omega_p^2/(\omega(\omega + \sqrt{-1}\gamma))$ where $\omega_p = 5.8(eV)$ and $\gamma = 0.2$. We compute the extinction cross-section Q^{ext} as a function of the operating wavelength $\lambda = 2\pi c/\omega$. Due to spherical symmetry, it can be shown that Q^{ext} has the following simple representation

$$Q^{\text{ext}} = \frac{2}{(k_m R)^2} \sum_{n=1}^{\infty} (2n+1) \Re \{ \frac{\sqrt{-1} k_m}{n(n+1)} (W_n^{TE} + W_n^{TM}) \},$$

where W_n^{TE} and W_n^{TM} are the scattering coefficients of a spherical structure. We have already seen in section 2.14 how to compute W_n^{TE} and W_n^{TM}. We now repeatedly plot Q^{ext} while changing the radius R from 5 nm to 30 nm in Figure 7.2. The shift of the plasmonic resonance is clearly shown.

We also present a numerical example of a spherical shell with outer radius R and inner radius $R/2$. We also assume the outer sphere has the permittivity ϵ_c and the inner sphere has the same permittivity as background. In Figure 7.2, we plot Q^{ext} for the shell for various values of radius R. Again, the shift of plasmon resonance is clearly shown.

7.5. Plasmonic Resonance for a System of Spheres

Confining light at the nanoscale is challenging due to the diffraction limit. Strongly localized surface plasmon modes in singular metallic structures, such as sharp tips and two nearly touching surfaces, offer a promising route to overcome this difficulty. Recently, transformation optics has been applied to analyze various structural singularities and then provides novel physical insights for a broadband nanofocusing of light.

Among three-dimensional singular structures, the system of nearly touching spheres is of fundamental importance. In the narrow gap regions, a large field enhancement occurs. The significant spectral shift of resonance mode also occurs due to the plasmon hybridization. A cluster of plasmonic spheres such as a heptamer

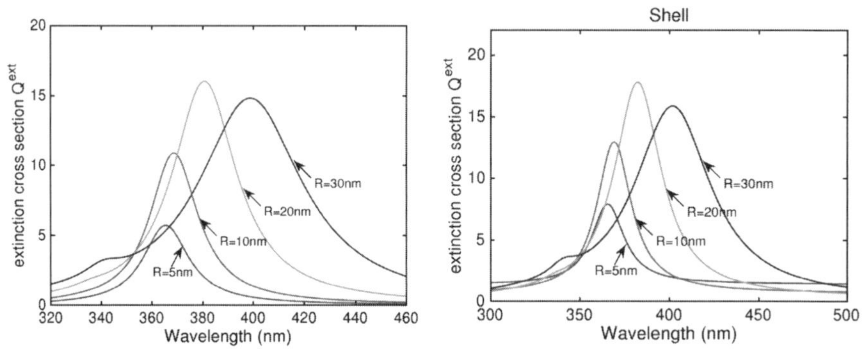

FIGURE 7.2. Extinction cross-section Q^{ext} for a spherical nanoparticle and a shell of radius R. We change the radius R from 5 (nm) to 30 (nm). The inner radius of the shell is set to be $R/2$. The shift of plasmon resonance is clearly shown.

and a octamer can support collective resonance modes such as Fano resonances. For theoretical investigations of these phenomena, the numerical computation plays a important role. Unfortunately, in the nearly touching case, it is quite challenging to compute the field distribution in the gap accurately. In fact, the required computational cost dramatically increases as the spheres get closer. The multipole expansion method requires a large number of spherical harmonics and the finite element method (or boundary element method) requires very fine mesh in the gap. Moreover, the linear systems to be solved are ill-conditioned. So conventional numerical methods are time consuming or inaccurate for this extreme case.

Here we present a hybrid numerical scheme that overcomes difficulty. The key idea of our hybrid scheme is to clarify the connection between Transformation Optics and the image charge method. The results of this section are from [**513**].

7.5.1. Two Metallic Spheres. We consider the two metallic spheres which are shown in Figure 7.3. The permittivity ϵ of each individual sphere is modeled as $\epsilon = 1 - \omega_p^2/(\omega(\omega + \sqrt{-1}\gamma))$ where ω is the operating frequency, ω_p is the plasma frequency and γ is the damping parameter. We fit Palik's data [**436**] for silver by adding a few Lorentz terms. We shall assume that the plasmonic spheres are small compared to optical wavelengths so that the quasi-static approximation can be adopted.

7.5.2. Transformation Optics. Let us briefly review the transformation optics approach by Pendry et al. [**445**]. To transform two spheres into a concentric shell, Pendry et al. introduced the inversion transformation Φ defined as

$$x' = \Phi(x) = R_T^2(x - X_0)/|x - X_0|^2 + X_0' \tag{7.31}$$

where X_0, X_0' and R_T are given parameters. This inversion mapping induces the inhomogeneous permittivity $\epsilon'(x') = R_T^2|x' - X_0'|\epsilon$ in the transformed space. Then, by taking advantage of the symmetry of the shell, the electric potential can be represented in terms of the following basis functions:

$$\mathcal{M}_{n,\pm}^m(x) = |x' - X_0'|(r')^{\pm(n+\frac{1}{2})-\frac{1}{2}} Y_n^m(\theta', \phi'), \tag{7.32}$$

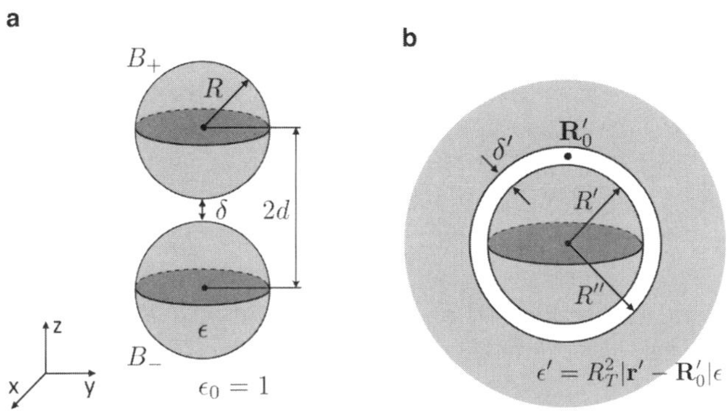

FIGURE 7.3. Two spheres and the transformation optics inversion mapping. (a) Two identical spheres, each of radius R and permittivity ϵ, are separated by a distance δ. The distance between their centers is $2d$. The background permittivity is $\epsilon_0 = 1$. (b) The transformation optics inversion mapping transforms the lower sphere B_- (or the upper sphere B_+) into a sphere of radius R' (or a hollow sphere of radius R'') centered at the origin, respectively.

where $\{Y_n^m\}$ are the spherical harmonics and (r', θ', ϕ') are the spherical coordinates. We will call $\mathcal{M}_{n,\pm}^m$ a transformation optics basis.

Let us assume that two plasmonic spheres $B_+ \cup B_-$ are placed in a uniform incident field

$$(0, 0, E_0 \Re\{e^{\sqrt{-1}\omega t}\}).$$

Then the (quasi-static) electrical potential V outside the two spheres can be represented in the following form:

$$V(x) = -E_0 x_3 + \sum_{n=0}^{\infty} A_n \big(\mathcal{M}_{n,+}^0(x) - \mathcal{M}_{n,-}^0(x)\big) \tag{7.33}$$

Here, the coefficients A_n can be determined by solving some tridiagonal system and x_1, x_2, x_3 are the Cartesian coordinates. Unfortunately, it cannot be solved analytically.

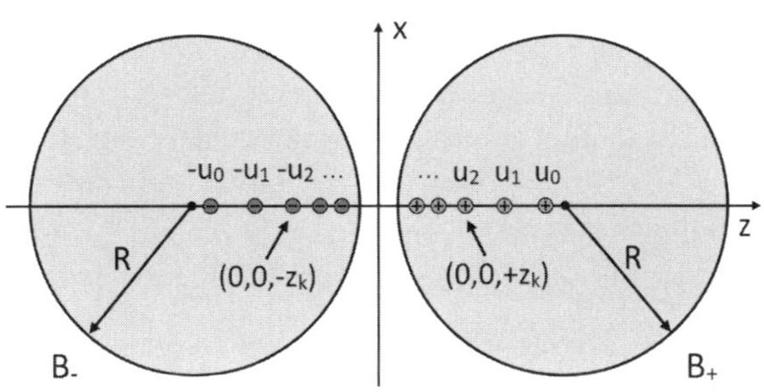

FIGURE 7.4. Image charges for two spheres. Red and green circles represent image charges placed along the x_3-axis.

7.5.3. Method of Image Charges. Now we discuss the method of images. Since the imaging rule for a pair of cylinders is simple, an exact image series solution and its properties can be easily derived. However, for two dielectric spheres, an exact solution cannot be obtained due to the appearance of a continuous line image source. Poladian observed in [**451**] that the continuous source can be well approximated by a point charge and then derived an approximate but analytic image series solution. Let us briefly review Poladian's solution for two dielectric spheres. Let $\tau = (\epsilon - 1)/(\epsilon + 1)$, $s = \cosh^{-1}(d/R)$ and $\alpha = R \sinh s$. Suppose that two point charges of strength ± 1 are located at $(0, 0, \pm z_0) \in B_\pm$, respectively. By Poladian's imaging rule, they produce an infinite series of image charges of strength $\pm u_k$ at $(0, 0, \pm z_k)$ for $k = 0, 1, 2, \ldots$, where z_k and u_k are given by

$$z_k = \alpha \coth(ks + s + t_0), \quad u_k = \tau^k \frac{\sinh(s + t_0)}{\sinh(ks + s + t_0)}. \tag{7.34}$$

Here, the parameter t_0 is such that $z_0 = \alpha \coth(s + t_0)$. See Figure 7.4. The potential $U(x)$ generated by all the above image charges is given by

$$U(x) = \sum_{k=0}^{\infty} u_k (\Gamma_0(x - x_k) - \Gamma_0(x + x_k)) \tag{7.35}$$

where $x_k = (0, 0, z_k)$ and $\Gamma_0(x) = 1/(4\pi|x|)$.

Let us consider the potential V outside the two spheres when a uniform incident field $(0, 0, E_0 \Re\{e^{\sqrt{-1}\omega t}\})$ is applied. Let p_0 be the induced polarizability when a single sphere is subjected to the uniform incident field, that is, $p_0 = E_0 R^3 2\tau/(3 - \tau)$. Using the potential $U(x)$, we can derive an approximate solution for $V(x)$. For $|\tau| \approx 1$, we have

$$V(x) \approx -E_0 x_3 + 4\pi p_0 \frac{\partial(U(x))}{\partial z_0}\bigg|_{z_0 = d} + QU(x)|_{z_0 = d}, \tag{7.36}$$

where Q is a constant chosen so that the right-hand side in equation (7.36) has no net flux on the surface of each sphere. The accuracy of the approximate formula, equation (7.36), improves as $|\epsilon|$ increases and it becomes exact when $|\epsilon| = \infty$. Moreover, its accuracy is pretty good even if the value of $|\epsilon|$ is moderate.

We now explain the difficulty involved in applying the the image series solution, equation (7.36), to the case of plasmonic spheres. In view of the expressions for u_k, equation (7.34), we can see that equation (7.36) is not convergent when $|\tau| > e^s$. For plasmonic materials such as gold and silver, the real part of the permittivity ϵ is negative over the optical frequencies and this means that the corresponding parameter $|\tau|$ can attain any value in the interval (e^s, ∞). Moreover, it turns out that all the plasmonic resonant values for τ are contained in the set $\{\tau \in \mathbb{C} : |\tau| > e^s\}$. So, equation (7.36) cannot describe the plasmonic interaction between the spheres due to the non-convergence.

7.5.4. Analytical Solution for Two Plasmonic Spheres. Here we present an analytic approximate solution for two plasmonic spheres in a uniform incident field $(0, 0, E_0 \Re\{e^{\sqrt{-1}\omega t}\})$. More importantly, we shall see that our analytical approximation completely captures the singular behavior of the exact solution. This feature is essential in developing our hybrid numerical scheme.

The solution which is valid for two plasmonic spheres can be derived by establishing the explicit connection between transformation optics and the method of image charges. We can convert the image series into a transformation optics-type solution by using the explicit connection formula. The result is shown in the following theorem.

THEOREM 7.19. *If* $|\tau| \approx 1$, *the following approximation for the electric potential* $V(x)$ *holds: for* $x \in \mathbb{R}^3 \setminus (B_+ \cup B_-)$,

$$V(x) \approx -E_0 x_3 + \sum_{n=0}^{\infty} \widetilde{A}_n \Big(\mathcal{M}^0_{n,+}(x) - \mathcal{M}^0_{n,-}(x) \Big), \tag{7.37}$$

where the coefficient $\widetilde{A}_n$ *is given by*

$$\begin{aligned} \widetilde{A}_n &= E_0 \frac{2\tau\alpha}{3-\tau} \cdot \frac{2n+1-\gamma_0}{e^{(2n+1)s} - \tau}, \\ \gamma_0 &= \sum_{n=0}^{\infty} \frac{2n+1}{e^{(2n+1)s}-\tau} \Big/ \sum_{n=0}^{\infty} \frac{1}{e^{(2n+1)s}-\tau}. \end{aligned} \tag{7.38}$$

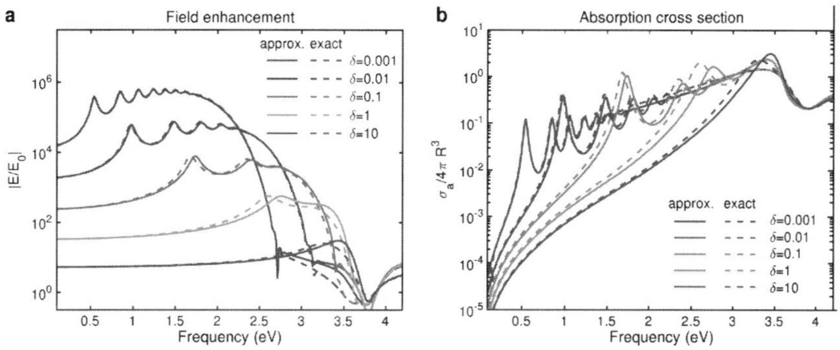

FIGURE 7.5. Exact solution versus analytic approximation. (a) Field enhancement plot as a function of frequency ω for various separation distances δ. The solid lines represent the approximate analytical solution and the dashed lines represent the exact solution. Two identical silver spheres of radius 30 nm are considered. (b) Same as (a) but for the absorption cross section.

As expected, the above approximate expression is valid even if $|\tau| > e^s$. Therefore, it can furnish useful information about the plasmonic interaction between the two spheres. As a first demonstration, let us investigate the (approximate) resonance condition, that is, the condition for τ at which the coefficients $\widetilde{A}_n$ diverge. One might conclude that the resonance condition is given by $\tau = e^{(2n+1)s}$. However, one can see that $\widetilde{A}_n$ has a removable singularity at each $\tau = e^{(2n+1)s}$. In fact, the (approximate) resonance condition turns out to be

$$\sum_{n=0}^{\infty} \frac{1}{e^{(2n+1)s} - \tau} = 0. \tag{7.39}$$

In other words, the plasmonic resonance happens when τ is one of the zeros of equation (7.39). It turns out that the zeros $\{\tau_n\}_{n=0}^{\infty}$ lie on the positive real axis and satisfy, for $n = 0, 1, 2, \ldots,$

$$e^{(2n+1)s} < \tau_n < e^{(2n+3)s}. \tag{7.40}$$

The above estimate helps us understand the asymptotic behavior of the resonance when two spheres get closer. As the separation distance δ goes to zero, the parameter s also goes to zero (in fact, $s = O(\delta^{1/2})$). Then, in view of equation (7.40), τ_n will converge to 1 and the corresponding permittivity ϵ_n goes to infinity. This means that a red-shift of the (bright) resonance modes occurs. Since the approximate analytical formula for V becomes more accurate as $|\epsilon|$ increases, we can expect that accuracy improves as the separation distance goes to zero. It also indicates that our formula captures the singular nature of the field distribution completely. Furthermore, the difference between τ_n and τ_{n+1} decreases, which means that the spectrum becomes almost continuous.

We now derive approximate formulas for the field at the gap and for the absorption cross section. From Theorem 7.19, we obtain the following:

$$\begin{aligned} E(0,0,0) \approx E_0 - E_0 \frac{8\tau}{3-\tau} \Bigg[& \sum_{n=0}^{\infty} \frac{(2n+1)^2}{e^{(2n+1)s} - \tau} (-1)^n \\ & - \gamma_0 \sum_{n=0}^{\infty} \frac{2n+1}{e^{(2n+1)s} - \tau} (-1)^n \Bigg]. \end{aligned} \tag{7.41}$$

In the quasi-static approximation, the absorption cross section σ_a is defined by $\sigma_a = \omega \Im\{p\}$, where p is the polarizability of the system of two spheres. From Theorem 7.19, σ_a is approximated as follows:

$$\begin{aligned} \sigma_a \approx \omega E_0 \frac{8\tau\alpha^3}{3-\tau} \Bigg[& \sum_{n=0}^{\infty} \frac{(2n+1)^2}{e^{(2n+1)s} - \tau} \\ & - \Bigg(\sum_{n=0}^{\infty} \frac{2n+1}{e^{(2n+1)s} - \tau} \Bigg)^2 \Bigg/ \sum_{n=0}^{\infty} \frac{1}{e^{(2n+1)s} - \tau} \Bigg]. \end{aligned} \tag{7.42}$$

We compare the above approximate formulas with the exact ones. Figure 7.5 represents respectively the field enhancement and the absorption cross section σ_a as functions of the frequency ω for various distances ranging from 0.001 nm to 10 nm. The strong accuracy of our approximate formulas over broad ranges of frequencies and gap distances is clearly shown. As mentioned previously, the accuracy improves as the spheres get closer. It is also worth highlighting the red-shift of the plasmonic resonance modes as the separation distance δ goes to zero.

7.5.5. Hybrid Numerical Scheme for a Many-Spheres System. Now we consider a system involving an arbitrary number of plasmonic spheres. If all the spheres are well separated, then the multipole expansion method is efficient and accurate for computing the field distribution. However, when the spheres are close to each other, the problem becomes very challenging since the charge densities on each sphere are nearly singular. To overcome this difficulty, Cheng and Greengard [**196, 198**] developed a hybrid numerical scheme combining the multipole expansion method and the method of images.

Let us briefly explain the main idea of Cheng and Greengard's method. In the standard multipole expansion method, the potential is represented as a sum of general multipole sources $\mathcal{Y}_{lm}(x) = Y_l^m(\theta, \phi)/r^{l+1}$ located at the center of each of the spheres. Suppose that a pair of spheres is close to touching. For convenience,

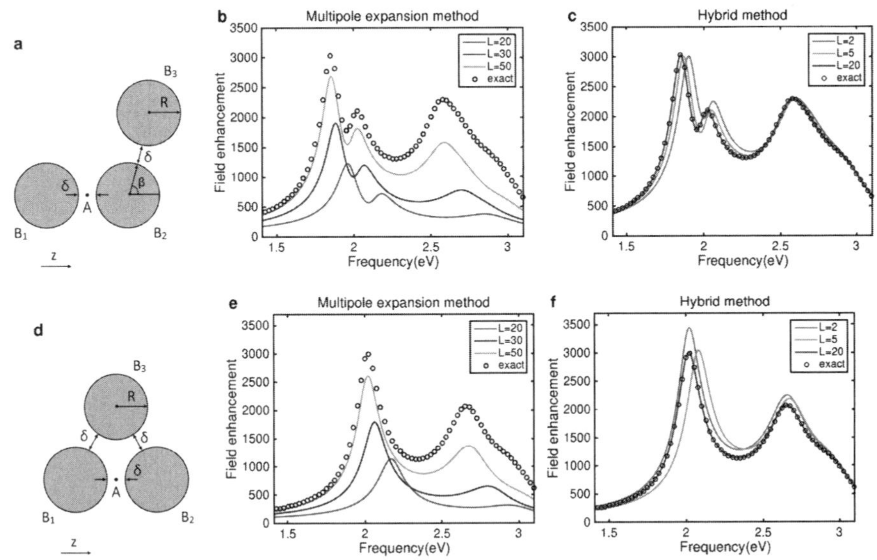

FIGURE 7.6. Multipole expansion method versus Hybrid scheme. (a) and (d) Two examples of three spheres configuration. (b) and (c) The field enhancement at point A as a function of frequency for the configuration (a) using the multipole expansion method and the hybrid method, respectively. The parameters are given as $R = 30$ nm, $\delta = 0.25$ nm and $\beta = 80^\circ$. The uniform incident field $(0, 0, \Re\{e^{\sqrt{-1}\omega t}\})$ is applied. (e) (f) Same as (b) and (c) but for the configuration (d).

let us identify the pair as $B_+ \cup B_-$. A multipole source $\mathcal{Y}_{lm}$ located at the center of B_+ generates an infinite sequence of image multipole sources by Poladian's imaging rule. Let us denote the resulting image multipole potential by U_{lm}^+. We also define U_{lm}^- in a similar way. Roughly speaking, Cheng and Greengard [**198**] modified the multipole expansion method by replacing a multipole source $\mathcal{Y}_{lm}$ with its corresponding image multipole series $U_{lm}^\pm$.

Since the image series $U_{lm}^\pm$ captures the close-to-touching interactions analytically, their scheme is very efficient and highly accurate even if the distance between the spheres is extremely small. However, the image multipole series $U_{lm}^\pm$ are not convergent for $|\tau| > e^s$. Hence it cannot be applied to cluster of plasmonic spheres. Therefore, in order to extend Cheng and Greengard's method to the plasmonic case, it is essential to establish an explicit connection between the image multipole series $U_{lm}^\pm$ and transformation optics. We develop a hybrid numerical scheme valid for plasmonics spheres by replacing the image multipole series with its transformation optics version.

Next, we present numerical examples to illustrate the hybrid method. We consider two examples of the three-spheres configuration. We provide a comparison between multipole expansion method and the hybrid method by plotting the field enhancement at the gap center A. For the numerical implementation, only a finite number of the multipoles $\mathcal{Y}_{lm}$ or hybrid multipoles $U_{lm}^\pm$ should be used. Let L be the truncation number for the order l. In Figures 7.6 (b) and 7.6 (e), the field enhancement is computed using the standard multipole expansion method. The computations give inaccurate results even if we include a large number of multipole sources with $L = 50$. On the contrary, the hybrid method gives pretty accurate results even for small values of L such as $L = 2$ and 5 (Figures 7.6 (c) and 7.6 (f)). Furthermore, 99% accuracy can be achieved using only $L = 20$. For each hybrid multipole $U_{lm}^{\pm}$, the transformation optics harmonics are included up to order $n = 300$ to ensure convergence and we note that the multipole can be evaluated very efficiently.

To achieve 99.9% accuracy at the first resonant peak, it is necessary to set $L = 150$ in the multipole expansion method which means a $68,400 \times 68,400$ linear system needs to be solved. However, the same accuracy can be achieved with only $L = 23$ in the hybrid method. The corresponding linear system has size $1,725 \times 1,725$ and it can be solved 2,000 times faster than the multipole expansion method.

7.6. Quasi-Static Plasmonic Resonances for Domains with Corners

In this section, we consider quasi-static plasmonic resonances for nonsmooth particles.

When the boundary of the particle D is smooth, the Neumann–Poincaré operator $(\mathcal{K}_D^0)^*$ is compact in $H^{-1/2}(\partial D)$. However, for domains with corners (that is, Lipschitz domains), the Neumann–Poincaré operator is not compact. So it may have a continuous spectrum. The purpose of this section is to characterize complete spectrum of Lipschitz domains. We also investigate the plasmonic resonance at the continuous spectrum. In Subsection 7.6.1, we consider the complete spectrum on the intersecting disks and its associated plasmonic resonance. As we will see, the blow-up rate of the norm of the solution at the continuous spectrum is significantly different from the discrete one. In Subsection 7.6.2, a numerical algorithm for classifying the spectrum on general Lipschitz domains using resonance is considered.

7.6.1. Neumann–Poincaré Operator on Intersecting Disks and Plasmonic Resonances.

7.6.1.1. *Intersecting Disks and Continuous Spectrum.* Let $B_1 = B_a(\mathbf{c}_1)$ and $B_2 = B_a(c_2)$ be two disks of the same radius a centered at c_1 and c_2, respectively, which are intersecting, namely, $|c_1 - c_2| < 2a$. The domain D is defined to be

$$D := B_1 \cup B_2. \tag{7.43}$$

(We may consider $B_1 \cap B_2$ instead.) Note that ∂D is Lipschitz continuous with corners at the intersecting points. We assume that c_1 is on the x_2-axis and $c_1 = -c_2$ so that the intersecting points lie on the x_1-axis and they are symmetric with respect to the origin. Let $2\theta_0$ ($\theta_0 > 0$) be the angle between the two circles at the intersecting point (see Figure 7.7). One can easily see that the intersecting points are given by $\pm a \sin\theta_0$.

For a given positive number α the bilinear transformation

$$\lambda(z) = \frac{z + \alpha}{z - \alpha}, \tag{7.44}$$

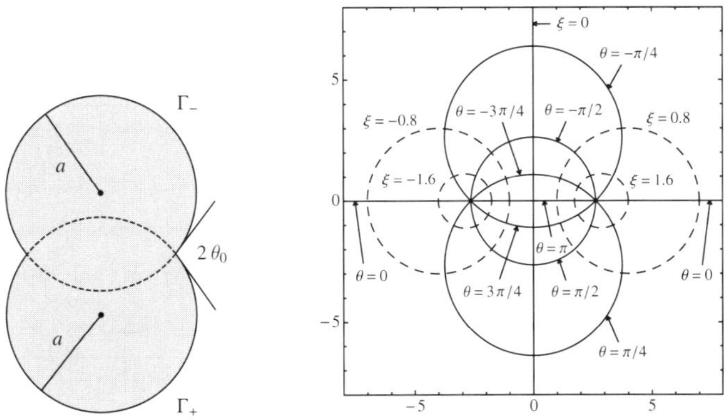

FIGURE 7.7. The left figure is the intersecting disks D. Level coordinate curves of ζ (when $\alpha = \sqrt{7}$) are illustrated in z-plane in the right figure.

is a conformal map from the Riemann sphere onto itself, and maps the interval $[-\alpha, \alpha]$ onto the negative real axis. So,

$$\zeta = \xi + \sqrt{-1}\theta := \ln\, \lambda(z), \tag{7.45}$$

is well defined for $z \in \mathbb{C} \setminus [-\alpha, \alpha]$, where ln is the logarithm with the principal branch. The coordinate ζ or (ξ, θ) is the bipolar coordinates we will use. The level coordinate curves for ζ are illustrated in Figure 7.7. Here and thereafter, we take

$$\alpha = a \sin \theta_0, \tag{7.46}$$

so that $(\alpha, 0)$ and $(-\alpha, 0)$ are two intersecting points of the circles. Note that $\ln \lambda$ maps $\mathbb{C} \setminus [-\alpha, \alpha]$ onto $\{\lambda = \xi + \sqrt{-1}\theta,\ -\infty < \xi < \infty, -\pi < \theta < \pi\}$. If we let Γ_+ and Γ_- be the lower and upper halves of ∂D, respectively, one can see that $\ln \lambda$ maps Γ_+ onto $\{\theta = \theta_0\}$, and Γ_- onto $\{\theta = -\theta_0\}$, and the exterior of D onto

$$\{\zeta = \xi + \sqrt{-1}\theta,\ -\infty < \xi < \infty, -\theta_0 < \theta < \theta_0\}$$

(see Figure 7.8). Note that we denote the lower half by Γ_+ because it is mapped into $\{\theta = +\theta_0\}$. For later use, we define

$$\ln \lambda(z) =: \Psi_1(z) + \sqrt{-1}\Psi_2(z).$$

Using the bipolar coordinates, it is possible to compute the single-layer potential operator explicitly. Then we can derive the complete spectral resolution of the Neumann–Poincaré operator. Consequently, we completely characterize the spectrum on the intersecting disks with the following theorem (see [**318**] for the details)

THEOREM 7.20. *Let $\sigma_{ac}((\mathcal{K}_D^0)^*)$, $\sigma_{sc}((\mathcal{K}_D^0)^*)$ and $\sigma_{sc}((\mathcal{K}_D^0)^*)$ be the absolutely continuous, singularly continuous, and pure point spectrum of $(\mathcal{K}_D^0)^*$ on $\mathcal{H}_0^*$, respectively. Then we have*

$$\sigma_{ac}((\mathcal{K}_D^0)^*) = [-b, b], \quad \sigma_{sc}((\mathcal{K}_D^0)^*) = \emptyset, \quad \sigma_{pp}((\mathcal{K}_D^0)^*) = \emptyset.$$

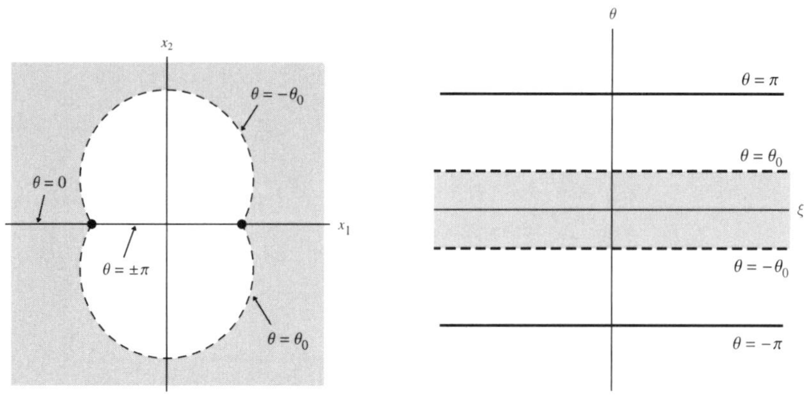

FIGURE 7.8. The exterior of the intersecting disks (the gray region in the left figure) is the image of the infinite strip (the gray region in the right figure) via the conformal map $z = x_1 + \sqrt{-1}x_2 = \Phi(\xi + \sqrt{-1}\theta)$.

The analysis of this section goes through with slight modification for the case when $\theta_0 > \pi/2$. In this case the spectral bound is given by

$$b = \frac{1}{2} - \frac{\pi - \theta_0}{\pi} = \frac{\theta_0}{\pi} - \frac{1}{2}. \tag{7.47}$$

In [**448**], a bound on the essential spectrum of the Neumann–Poincaré operator on curvilinear polygonal domains has been obtained. Recently, it was proved in [**449**] that the interval determined by the bound on the essential spectrum obtained in [**448**] is actually the essential spectrum. In [**153**], the same characterization of the essential spectrum is obtained by using Weyl sequences.

7.6.1.2. *Quasi-Static Plasmonic Resonance at the Continuous Spectrum.* Here we consider plasmonic resonance on the intersecting disks. Suppose that the intersecting disks has dielectric constant $\epsilon_c + \sqrt{-1}\delta$ (δ is the dissipation) and the matrix $\mathbb{R}^2 \setminus \overline{D}$ have the dielectric constant ϵ_m. So, the distribution of the dielectric constant is given by

$$\epsilon = (\epsilon_c + \sqrt{-1}\delta)\chi(D) + \epsilon_m \chi(\mathbb{R}^2 \setminus \overline{D}),$$

where χ denotes the indicator function. After normalization we assume that $\epsilon_m = 1$. To investigate the resonance on the plasmonic structure, we consider the following problem:

$$\begin{cases} \nabla \cdot \epsilon \nabla u_\delta = f & \text{in } \mathbb{R}^2, \\ u_\delta(x) = O(|x|^{-1}) & \text{as } |x| \to \infty, \end{cases} \tag{7.48}$$

where f is a source function compactly supported in $\mathbb{R}^2 \setminus \overline{D}$ satisfying

$$\int_{\mathbb{R}^2} f = 0. \tag{7.49}$$

A typical such source functions are polarized dipoles, namely, $f(x) = a \cdot \nabla \delta_z(x)$ for some $z \in \mathbb{R}^2 \setminus \overline{D}$, where a is a constant vector and δ_z is the Dirac mass. It is worth mentioning that (7.49) is a necessary condition for the existence of a solution to (7.48). For a given ϵ_c, the resonances are characterized by the fact that

$$\|\nabla u_\delta\|_{L^2(D)} \to \infty \quad \text{as } \delta \to 0. \tag{7.50}$$

Let

$$\lambda := \frac{\epsilon_c + 1 + \sqrt{-1}\delta}{2(\epsilon_c - 1) + 2\sqrt{-1}\delta}. \tag{7.51}$$

Then, the solution u_δ to (7.48) can be represented as

$$u_\delta(x) = q(x) + \mathcal{S}_D^0[\varphi_\delta](x), \quad x \in \mathbb{R}^2, \tag{7.52}$$

where q is the Newtonian potential of f, *i.e.*,

$$q(x) = \int_{\mathbb{R}^2} \Gamma(x-y) f(y) dy,$$

and the potential $\varphi_\delta \in \mathcal{H}_0^*(\partial D)$ is the solution to the integral equation

$$(\lambda I - (\mathcal{K}_D^0)^*)[\varphi_\delta] = \partial_\nu q \quad \text{on } \partial D. \tag{7.53}$$

From (7.53), we then obtain

$$\varphi_\delta = \int_{-b}^{b} \frac{1}{\lambda - t}\, d\mathcal{E}_t[\partial_\nu q], \tag{7.54}$$

where $\mathcal{E}_t$ is the resolution of identity.

Since we can prove that $\|\nabla u\|_{L^2(D)} \approx \|\varphi_\delta\|_{\mathcal{H}^*}$, (7.50) is equivalent to

$$\|\varphi_\delta\|_{\mathcal{H}^*} \to \infty \quad \text{as } \delta \to 0. \tag{7.55}$$

It follows from (7.54) that

$$\|\varphi_\delta\|_{\mathcal{H}^*}^2 = \int_{-b}^{b} \frac{1}{|\lambda - t|^2}\, d\langle \partial_\nu q, \mathcal{E}_t[\partial_\nu q]\rangle_{\mathcal{H}^*}. \tag{7.56}$$

As $\delta \to 0$, λ tends to λ_0, which is defined by

$$\lambda_0 := \frac{\epsilon_c + 1}{2(\epsilon_c - 1)}.$$

Formula (7.56) shows that the resonance (7.55) occurs only when $\lambda_0 \in [-b, b] \subset]-1/2, 1/2[$. This condition can be fulfilled only if $\epsilon_c < 0$. So, a resonance occurs only when the dielectric constant of the inclusion is negative, which is the case for plasmonic materials.

Suppose that the source f is given by

$$f(x) = a \cdot \nabla \delta_z(x), \tag{7.57}$$

where a is a constant vector and z is a point outside D. Choose $z \in \mathbb{R}^2 \setminus \overline{D}$ such that

$$\Psi_2(z) \neq 0. \tag{7.58}$$

We also choose a such that

$$a \cdot \nabla \Psi_1(z) \neq 0 \quad \text{and} \quad a \cdot \nabla \Psi_2(z) \neq 0. \tag{7.59}$$

By using an explicit formula for the spectral resolution of the Neumann–Poincaré operator, the blow-up rate of $\|\varphi_\delta\|_{\mathcal{H}^*}$ can be derived and the following result holds (for a more complete statement, see again [**318**]).

THEOREM 7.21. *Suppose f is the dipole source given in (7.57) with a and z satisfying (7.58) and (7.59). Then, the blow-up rate of the norm $\|\varphi_\delta\|_{\mathcal{H}^*}$ depends on λ_0 as follows:*

(i) *If $\lambda_0 \in]-b, 0[\cup]0, b[$, then $\|\varphi_\delta\|_{\mathcal{H}^*} \approx \delta^{-1/2}$.*
(ii) *If $\lambda_0 = \pm b$, then $\|\varphi_\delta\|_{\mathcal{H}^*} \approx \delta^{-3/4}$.*
(iii) *If $\lambda_0 = 0$, then*

$$\|\varphi_\delta\|_{\mathcal{H}^*} \approx |\log \delta|^{1/2} \delta^{-1/2-|\Psi_2(z)|/(2\theta_0)}.$$

It is known that $\|\varphi_\delta\|_{\mathcal{H}^*} \approx \delta^{-1}$ when λ_0 belongs to the pure point spectrum (that is, λ_0 is an eigenvalue). So we infer that the blow-up rate of the norm $\|\varphi_\delta\|_{\mathcal{H}^*}$ becomes significantly different when λ_0 is in the continuous spectrum.

7.6.2. Classification of the Spectrum by Resonances for General Lipschitz Domains. In view of Theorem 7.21, one may expect that the spectrum can be classified using the blow-up rate of the norm $\|\varphi_\delta\|_{\mathcal{H}^*}$. Indeed, it turns out to be true. Here we present a method based on resonance to distinguish between the continuous spectrum and the pure point spectrum.

Let $f \in \mathcal{H}_0^*$. For $t \in]-1/2, 1/2[$ and $\delta > 0$, let $\varphi_{t,\delta}$ be the solution of the integral equation

$$((t + \sqrt{-1}\delta)I - (\mathcal{K}_D^0)^*)[\varphi_{t,\delta}] = f \quad \text{on } \partial D. \tag{7.60}$$

By the spectral resolution theorem [**512**], there is a family of projection operators $\mathcal{E}_s$ (called the resolution identity) such that

$$(\mathcal{K}_D^0)^* = \int_{\sigma((\mathcal{K}_D^0)^*)} s \, d\mathcal{E}_s.$$

We then obtain from (7.60) that

$$\varphi_{t,\delta} = \int_{\sigma((\mathcal{K}_D^0)^*)} \frac{1}{t + \sqrt{-1}\delta - s} \, d\mathcal{E}_s,$$

and hence

$$\|\varphi_{t,\delta}\|_{\mathcal{H}^*} = \int_{\sigma((\mathcal{K}_D^0)^*)} \frac{1}{t + \sqrt{-1}\delta - s} \, d(f, \mathcal{E}_s)_{\mathcal{H}^*}. \tag{7.61}$$

If $t \notin \sigma((\mathcal{K}_D^0)^*)$, we can immediately see from (7.61) that

$$\lim_{\delta \to 0} \delta^\alpha \|\varphi_{t,\delta}\|_{\mathcal{H}^*} = 0 \tag{7.62}$$

for any $\alpha > 0$.

Suppose that $t \in \sigma((\mathcal{K}_D^0)^*)$. An eigenvalue t of $(\mathcal{K}_D^0)^*$ is characterized by the discontinuity $\mathcal{E}_{t+} - \mathcal{E}_t \neq 0$ (and the fact that t is isolated). So, if f satisfies

$$(f, \mathcal{E}_{t+}[f])_{\mathcal{H}^*} - (f, \mathcal{E}_t[f])_{\mathcal{H}^*} > 0,$$

then

$$\|\varphi_{t,\delta}\|_{\mathcal{H}^*}^2 \geq \frac{(f, \mathcal{E}_{t+}[f])_{\mathcal{H}^*} - (f, \mathcal{E}_t[f])_{\mathcal{H}^*}}{\delta^2},$$

and hence

$$\|\varphi_{t,\delta}\|_{\mathcal{H}^*}^2 \approx \delta^{-2}. \tag{7.63}$$

Suppose that the spectral measure $d(f, \mathcal{E}_s)_{\mathcal{H}^*}$ is absolutely continuous near t, namely, there exists an $\epsilon > 0$ and a function $\mu_f(s)$ which is integrable on $[t-\epsilon, t+\epsilon]$ such that

$$d(f, \mathcal{E}_s)_{\mathcal{H}^*} = \mu_f(s)ds, \quad s \in [t-\epsilon, t+\epsilon].$$

Then we obtain from (7.61)

$$\|\varphi_{t,\delta}\|^2_{\mathcal{H}^*} = \int_{\sigma((\mathcal{K}_D^0)^*)\setminus[t-\epsilon,t+\epsilon]} \frac{1}{(s-t)^2+\delta^2}\, d(f, \mathcal{E}_s)_{\mathcal{H}^*} + \int_{[t-\epsilon,t+\epsilon]} \frac{\mu_f(s)ds}{(s-t)^2+\delta^2}.$$

From the boundary behavior of the Poisson integral, we have

$$\lim_{\delta\to 0} \delta \int_{[t-\epsilon,t+\epsilon]} \frac{\mu_f(s)ds}{(s-t)^2+\delta^2} = \frac{\pi}{2}(\mu_f(t+) + \mu_f(t-)).$$

On the other hand, it is proved in [**318**] that

$$\lim_{\delta\to 0} \delta^2 \int_{[t-\epsilon,t+\epsilon]} \frac{\mu_f(s)ds}{(s-t)^2+\delta^2} = 0.$$

So, we have

$$\lim_{\delta\to 0} \delta \|\varphi_{t,\delta}\|^2_{\mathcal{H}^*} = \frac{\pi}{2}(\mu_f(t+) + \mu_f(t-)),$$

and hence,

$$\lim_{\delta\to 0} \delta^2 \|\varphi_{t,\delta}\|^2_{\mathcal{H}^*} = 0. \tag{7.64}$$

Define the indicator function $\alpha_f(t)$ by

$$\alpha_f(t) := \sup\Big\{\alpha \,|\, \limsup_{\to 0} \delta^\alpha \|\varphi_{t,\delta}\|_{\mathcal{H}^*} = \infty\Big\}, \quad t \in]-1/2, 1/2[.$$

We see that $0 \le \alpha_f(t) \le 1$ for all t. The classification of spectra of the Neumann–Poincaré operator is based on the following theorem.

THEOREM 7.22. *Let $f \in \mathcal{H}_0^*$.*
(i) *If $\alpha_f(t) > 0$, then $t \in \sigma((\mathcal{K}_D^0)^*)$.*
(ii) *If $\alpha_f(t) = 1$ and t is isolated, then $t \in \sigma_{pp}((\mathcal{K}_D^0)^*)$.*
(iii) *If $1/2 \le \alpha_f(t) < 1$, then $t \in \sigma_c((\mathcal{K}_D^0)^*)$.*

PROOF. The assertion (i) is an immediate consequence of (7.62), while (ii) follows from (7.63) and (7.64). (iii) is a consequence of (i) and (ii). □

Because of the complicated nature of the singularly continuous spectrum, it is hard to classify it from the continuous spectrum. However, if $\alpha_f(t) = 1$ and t is not isolated (embedded in the continuous spectrum), then we may infer that it is in the singularly continuous spectrum. The indicator function $\alpha_f(t)$ can be computed using the following lemma.

LEMMA 7.23. *For $t \in]-1/2, 1/2[$ and $\delta > 0$ define*

$$\alpha_f(t, \delta) = -\frac{\log \|\varphi_{t,\delta}\|_{\mathcal{H}^*}}{\log \delta}.$$

It holds that

$$\alpha_f(t) = \lim_{\delta\to 0} \alpha_f(t, \delta)$$

provided that the limit exists.

PROOF. Let $l = \lim_{\delta \to 0} \alpha_f(t, \delta)$ under the assumption that the limit exists. If $\alpha < l$, then

$$\limsup_{\delta \to 0} \delta^\alpha \|\varphi_{t,\delta}\|_{\mathcal{H}^*} = \infty,$$

and hence $\alpha \leq \alpha_f(t)$. If $\alpha > l$, then

$$\lim_{\delta \to 0} \delta^\alpha \|\varphi_{t,\delta}\|_{\mathcal{H}^*} = 0,$$

and hence, $\alpha \geq \alpha_f(t)$. So, $\alpha_f(t) = l$. □

Lemma 7.23 allows us to approximate $\alpha_f(t)$ by $\alpha_f(t, \delta)$ for small δ.

In order to classify the spectra using α_f, it is quite important to choose the source function properly. For example, to have $\alpha_f(t) = 1$ for classification of eigenvalues, f needs to have a non-zero eigenmode (the component of the corresponding eigenfunctions). In [**291**], it was proposed to use the following source function:

$$f_z(x) = \nu(x) \cdot \nabla q_z(x), \quad q_z(x) = d \cdot \nabla_x \left(\frac{1}{2\pi} \ln |x - z| \right),$$

where d is a constant unit vector and $z \in \mathbb{R}^2 \setminus \overline{D}$. In fact, q_z is the Newtonian potential of the polarizable dipole $d \cdot \nabla \delta_z(x)$ located at z.

Using f_z as a source function has several advantages. First of all, q_z is a harmonic function in D, and hence

$$\|f_z\|_{H^{-1/2}(\partial D)} \approx \|q_z\|_{H^1(D)},$$

follows from the standard regularity estimates for the Laplace equation with Neumann boundary conditions. Moreover, we can see easily that there are constants C_1 and C_2 such that

$$C_1 \leq \|q_z\|_{H^1(D)} \leq C_2,$$

as long as the location z of the dipole is at some distance from D and ∞, namely, there are constants C_3 and C_4 such that

$$C_3 \leq \operatorname{dist}(z, D) \leq C_4. \tag{7.65}$$

This means that, for all z satisfying (7.65), we have

$$\|f_z\|_{H^{-1/2}(\partial D)} \approx 1.$$

In particular, we do not have to normalize f_z.

Another advantage of using f_z as a source function is that for any $\varphi \in \mathcal{H}_0^*$ f_z contains a component of φ, namely, $(f_z, \varphi)_{\mathcal{H}^*} \neq 0$ for most z's. To see this, we first observe that $(f_z, \varphi)_{\mathcal{H}^*}$ is harmonic as a function of the z variable, and it is non-vanishing. In fact, if $(f_z, \varphi)_{\mathcal{H}^*} = 0$ for all $z \in \mathbb{R}^2 \setminus D$, then $\mathcal{S}_D^0[\partial_\nu \mathcal{S}_D^0[\varphi]] = 0$, and hence, $\partial_\nu \mathcal{S}_D^0[\varphi] = 0$ on ∂D. Thus $\varphi = 0$. As a non-vanishing harmonic function, $(f_z, \varphi)_{\mathcal{H}^*}$ cannot be zero for z in an open set. So, we infer that $(f_z, \varphi)_{\mathcal{H}^*} \neq 0$ for almost all z.

Still $(f_z, \varphi)_{\mathcal{H}^*}$ can be small for z in a large set. To avoid such a case, we choose several z's, say $z_1, \ldots, z_N$, satisfying (7.65), and consider the new indicator function

$$\alpha_\sharp(t) := \max_{1 \leq j \leq N} \{\alpha_{f_{z_j}}(t)\}.$$

We emphasize that Theorem 7.22 is still valid with $\alpha_\sharp(t)$ replaced with $\alpha_f(t)$. The indicator function $\alpha_\sharp(t)$ is approximated by $\alpha_\sharp(t, \delta)$ for small δ, which is defined by

$$\alpha_\sharp(t, \delta) := \max_{1 \leq j \leq N} \{\alpha_{f_{z_j}}(t, \delta)\}.$$

In [**291**], the authors perform numerical experiments for the spectrum of the Neumann–Poincaré operator on various domains as revealed by the indicator function $\alpha_\sharp(t)$.

For classifying spectra using plasmonic resonances, quite high precision computations are required. However, when the domain has corners, it is a considerable challenge to compute the Neumann–Poincaré operator with high accuracy. Here, we use the Recursively Compressed Inverse Preconditioning (RCIP) method [**293**], explained in detail in [**289, 290**], which is a high precision method for solving integral equations on piecewise smooth boundaries. It is worth mentioning that this method has been successfully adapted for computing the polarizabilities of domains with corners [**292**].

7.7. Concluding Remarks

In this chapter we have analyzed plasmonic resonances for nanoparticles. We have estimated the plasmon resonance shift due to changes in the size and shape of the nanoparticles. We have derived effective electromagnetic parameters of a composite material in which plasmonic nanoparticles are embedded. We have shown that by homogenizing plasmonic nanoparticles one can obtain high-contrast or negative material parameters, depending on how the frequencies used correspond to the plasmonic resonant frequency. These results will play a key role in later chapters in the analysis of super-resolution imaging mechanisms and sub-wavelength bandgap crystals. Finally, we have also presented a hybrid numerical scheme to compute plasmonic resonances of systems of spheres and discussed the quasi-static plasmonic resonances for particles with nonsmooth boundaries.

The results of this chapter can be used to provide a mathematical foundation for plasmonic sensing. Consider the inverse problem of reconstructing a small object from far-field measurements. As it will be shown in Chapter 8, this inverse problem is severely ill-posed because of the diffraction limit and low signal to noise ratio. In [**83, 84**], a novel methodology to solve this type of inverse problem based on an idea from plasmonic sensing is proposed. By using the field interaction with a known plasmonic particle, the fine detail information of the small object can be encoded into the shift of the resonant frequencies of the two particle system in the far field. Two interaction regimes are considered. In the intermediate interaction regime, it is shown that this information is exactly the generalized polarization tensors associated with the small object, from which one can perform the reconstruction. In the strong interaction regime, a well chosen conformal mapping, which transforms the two particle system into a coated structure, is used. The shifts in the resonance frequencies for a fixed relative position of the particles are expressed in terms of the generalized polarization tensors of the transformed domain.

Purely optical effects are only one aspect discussed in this chapter. As mentioned in the introduction, plasmonic nanoparticles can also be used to create thermal effects, because under light illumination the plasmonic nanoparticles heat-up locally owing to absorption. Indeed, heating effects are already under investigations as a treatment for cancer, where tumor cells are destroyed by plasmonic nanoparticles excited by an infrared laser beam [**114**]. In [**81**], layer potentials and asymptotic analysis techniques are used to analyze the heat generation due to nanoparticles when illuminated at their plasmonic resonances. Arbitrary-shaped particles and both the case of a single and multiple particles are considered. An asymptotic expansion formula for the temperature elevation due to plasmonic nanoparticles is derived. For close-to-touching nanoparticles, it is shown that the temperature field deviates significantly from the one generated by two single nanoparticles.

CHAPTER 8

Imaging of Small Particles

8.1. Introduction

In this chapter we consider, in the presence of noise, the detection and localization of small particles from multi-static measurements. Multistatic imaging usually involves two steps. The first consists in recording the waves generated by the particles on an array of receivers. The second step consists in processing the recorded data in order to estimate some relevant features of the particles. In its most general form, this inverse problem is severely ill-posed and nonlinear. The recorded data is entangled in the global structure of the geometric and material parameters of the target in a highly nonlinear way. This is the main fundamental obstacle to finding efficient reconstruction algorithms. If, however, in advance we have additional information about the target, then we may be able to determine specific features with a good accuracy.

In this chapter, we discuss direct (non-iterative) reconstruction algorithms that take advantage of the smallness of the particles. We apply the asymptotic formulas derived in Theorems 2.86, 2.120, and 2.143 for the purpose of identifying the locations and certain properties of small particles. In particular, we introduce the MUltiple Signal Classification algorithm (MUSIC), reverse-time migration, and Kirchhoff migration. We analyze their resolution and stability with respect to noise in the measurements. Resolution analysis is to estimate the size of the finest detail that can be reconstructed from the data while stability analysis is to quantify the localization error in the presence of noise. We refer the reader to [**43**] for more details on these direct imaging algorithms.

Taking into account the sparsity of the problem of imaging small particles, we show that it can be recast to a joint sparse recovery problem and outline the algorithm proposed in [**354**]. In [**193**], other l_1 minimization-based imaging methods are designed for locating small particles. In particular, a hybrid approach combining the use of singular value decomposition with l_1 norm minimization is introduced.

8.2. Scalar Wave Imaging of Small Particles

8.2.1. MUSIC-type Method. Let $B_R := \{|x| < R\}$. Let D be a small particle with location at $z \in B_R$ and electromagnetic parameters ε_c and μ_c. Let $x_i, i = 1, \ldots, N$ be equi-distributed points along the boundary ∂B_R for $N \gg 1$. The array of N elements $\{x_1, \ldots, x_N\}$ is used to detect the particle. The array of elements $\{x_1, \ldots, x_N\}$ is operating both in transmission and in reception. Let u_j^s be the scattered wave by D corresponding to the incident wave $\Gamma_{k_m}(x - x_j)$. From Theorem 2.86 it follows that

$$u_j^s(x) = \delta^d \Big(\nabla_z \Gamma_{k_m}(z - x_j) M \nabla_z \Gamma_{k_m}(x - z) + k_m^2 (\frac{\varepsilon_c}{\varepsilon_m} - 1)|B| \Gamma_{k_m}(z - x_j)\Gamma_{k_m}(x - z) \Big) \\ + O(\delta^{d+1}),$$

where M is the polarization tensor defined in (2.71) with λ given by (2.272).

Suppose for the sake of simplicity that D is a disk. Define the $N \times N$ data matrix by

$$A^\omega := \left(u_j^s(x_i)\right)_{i,j}, \tag{8.1}$$

and introduce the N-dimensional vector fields $g^{(j)}(z^S)$, for $z^S \in B_R$ and $j = 1, \ldots, d+1$, by

$$g^{(j)}(z^S) = \frac{1}{\sqrt{\sum_{i=1}^N |e_j \cdot \nabla_z \Gamma_{k_m}(z^S - x_i)|^2}} \left(e_j \cdot \nabla_z \Gamma_{k_m}(z^S - x_1), \ldots, e_j \cdot \nabla_z \Gamma_{k_m}(z - x_N)\right)^t, \tag{8.2}$$

for $j = 1, \ldots, d$, and

$$g^{(d+1)}(z^S) = \frac{1}{\sqrt{\sum_{i=1}^N |\Gamma_{k_m}(z^S - x_i)|^2}} \left(\Gamma_{k_m}(z^S - x_1), \ldots, \Gamma_{k_m}(z^S - x_N)\right)^t, \tag{8.3}$$

where $\{e_1, \ldots, e_d\}$ is an orthonormal basis of $\mathbb{R}^d$.

Let $g(z^S)$ be the $N \times d$ matrix whose columns are $g^{(1)}(z^S), \ldots, g^{(d)}(z^S)$. Then, from (2.86), (8.1) can be written as

$$A^\omega \approx \tau_\mu g(z) \overline{g(z)}^t + \tau_\varepsilon g^{(d+1)}(z) \overline{g^{(d+1)}(z)}^t,$$

where

$$\tau_\mu := 2|D| \frac{\mu_m - \mu_c}{\mu_m + \mu_c} \Big(\sum_{i=1}^N |\nabla_z \Gamma_{k_m}(z - x_i)|^2\Big),$$

$$\tau_\varepsilon := |D| k_m^2 \Big(\frac{\varepsilon_c}{\varepsilon_m} - 1\Big) \Big(\sum_{i=1}^N |\Gamma_{k_m}(z - x_i)|^2\Big).$$

Let P be the orthogonal projection onto the range $\operatorname{Im} A^\omega$ of A^ω. The MUSIC-type imaging functional is defined by

$$\mathcal{I}_{\mathrm{MU}}(z^S, \omega) := \Big(\sum_{j=1}^{d+1} \|(I - P)[g^{(j)}](z^S)\|^2\Big)^{-1/2}. \tag{8.4}$$

This functional has large peaks only at the locations of the particles [**43**].

8.2.2. Reverse-Time Migration and Kirchhoff Imaging. A backpropagation-type imaging functional at a single frequency ω for the particle D is given for $z^S \in B_R$ by

$$\mathcal{I}_{\mathrm{BP}}(z^S, \omega) := \sum_{j=1}^{d+1} u_j^s(x_j) \overline{g^{(j)}(z^S)} \cdot g^{(j)}(z^S), \tag{8.5}$$

where $g^{(j)}$ are defined by (8.2) and (8.3).

For sufficiently large N, we have

$$\frac{1}{N} \sum_{j=1}^N \overline{\Gamma_{k_m}}(x_j - z^S) \Gamma_{k_m}(x_j - z^S) \sim \Im \Gamma_{k_m}(z - z^S),$$

and

$$\frac{1}{N}\sum_{j=1}^{N}\nabla_z\overline{\Gamma_{k_m}}(x_j-z^S)\cdot\nabla_z\Gamma_{k_m}(x_j-z^S)^t\sim\Im\Gamma_{k_m}(z-z^S)\,\frac{z-z^S}{|z-z^S|}\Big(\frac{z-z^S}{|z-z^S|}\Big)^t,$$

where $A\sim B$ means $A\approx CB$ for some constant C.

Therefore,

$$\mathcal{I}_{\mathrm{BP}}(z^S,\omega)\sim\begin{cases}\mathrm{sinc}(k_m|z-z^S|) & \text{for } d=3,\\ J_0(k_m|z-z^S|) & \text{for } d=2,\end{cases}$$

where $\mathrm{sinc}(s)=\sin(s)/s$ is the sinc function and J_0 is the Bessel function of the first kind and of order zero.

These formulas show that the resolution of the imaging functional is the standard diffraction limit. It is of the order of half the wavelength $2\pi/k_m$.

Note that $\mathcal{I}_{\mathrm{BP}}$ uses only the diagonal terms of the response matrix A^ω, defined by (8.1). Using the whole matrix, we arrive at the Kirchhoff migration functional:

$$\mathcal{I}_{\mathrm{KM}}(z^S,\omega)=\sum_{j=1}^{d+1}\overline{g^{(j)}(z^S)}\cdot A^\omega g^{(j)}(z^S)\,. \tag{8.6}$$

REMARK 8.1. *Suppose for the sake of simplicity that $\mu_c=\mu_m$. In this case the response matrix is*

$$A^\omega=\tau_\varepsilon g^{(d+1)}(z)\overline{g^{(d+1)}(z)}^t$$

and we can prove that $\mathcal{I}_{\mathrm{MU}}$ is a nonlinear function of $\mathcal{I}_{\mathrm{KM}}$ [**46**]. *In fact, we have*

$$\mathcal{I}_{\mathrm{KM}}(z^S,\omega)=\tau_\varepsilon\Big(1-\mathcal{I}_{\mathrm{MU}}^{-2}(z^S,\omega)\Big)\,.$$

It is worth pointing out that this transformation improves neither the stability nor the resolution.

Moreover, in the presence of additive measurement noise with variance $k_m^2\sigma_{\mathrm{noise}}^2$, the response matrix can be written as

$$A^\omega=\tau_\varepsilon g^{(d+1)}(z)\overline{g^{(d+1)}(z)}^t+\sigma_{\mathrm{noise}}k_m W\,,$$

where W is a complex symmetric Gaussian matrix with mean zero and variance 1.

Let $\mathbb{E}$ *and* Var *denote the mean and the variance, respectively. According to* [**46**], *the Signal-to-Noise Ratio (SNR) of the imaging functional $\mathcal{I}_{\mathrm{KM}}$, defined by*

$$\mathrm{SNR}(\mathcal{I}_{\mathrm{KM}})=\frac{\mathbb{E}[\mathcal{I}_{\mathrm{KM}}(z,\omega)]}{\mathrm{Var}(\mathcal{I}_{\mathrm{KM}}(z,\omega))^{1/2}}\,,$$

is then equal to

$$\mathrm{SNR}(\mathcal{I}_{\mathrm{KM}})=\frac{\tau_\varepsilon}{k_m\sigma_{\mathrm{noise}}}\,. \tag{8.7}$$

For the MUSIC algorithm, the peak of $\mathcal{I}_{\mathrm{MU}}$ is affected by measurement noise. We have

$$\mathcal{I}_{\mathrm{MU}}(z,\omega)=\begin{cases}\frac{|\tau_\varepsilon|}{\sigma_{\mathrm{noise}}} & \textit{if } \tau_\varepsilon\gg\sigma_{\mathrm{noise}},\\ 1 & \textit{if } |\tau_\varepsilon|\ll k_m\sigma_{\mathrm{noise}}\,.\end{cases}$$

REMARK 8.2. *Consider m closely spaced particles $\cup_{s=1}^m(\delta B_s + z)$ and let the magnetic permeability and electric permittivity of the particle $\delta B_s + z$, for $s = 1, \ldots, m$, be given by $\mu_c^{(s)}$ and $\varepsilon_c^{(s)}$. In view of Theorem 2.88, only the position z, the overall polarization tensor M defined by (2.284), and $\sum_{s=1}^m \left((\frac{\varepsilon_c^{(s)}}{\varepsilon_m} - 1)|B_s|\right)$ can be reconstructed from measured far-field data.*

8.2.3. Joint Sparse Recovery. In this subsection we show that the problem of imaging small particles can be recast to a joint sparse recovery problem. The following algorithm was proposed in [**354**].

Let us first recall the Lippmann-Schwinger representation of u_j^s. We have

$$u_j^s(x) = \int_D \left((\frac{1}{\mu_c} - \frac{1}{\mu_m})\nabla_y \Gamma_{k_m}(x-y) \cdot \nabla_z u_j^s(y) + k_m^2(\frac{\varepsilon_c}{\varepsilon_m} - 1)\Gamma_{k_m}(x-y)u_j^s(y) \right) dy, \tag{8.8}$$

for $x \in \mathbb{R}^d$.

Then we approximate ∇u_j^s and u_j^s in the search domain Ω^S by either piecewise constant functions or splines as

$$\nabla u_j^s(y) = \begin{bmatrix} \sum_{l=1}^L \alpha_{l,j}^{(1)} \phi^{(1)}(y, y_l) \\ \vdots \\ \sum_{l=1}^L \alpha_{l,j}^{(d)} \phi^{(d)}(y, y_l) \end{bmatrix},$$

and

$$u_j^s(y) = \sum_{l=1}^L \alpha_{l,j}^{(d+1)} \phi^{(d+1)}(y, y_l),$$

where $\{y_l\}_{l=1}^L$, for some $L \in \mathbb{N}$, are finite sampling points of Ω^S and $\phi^{(n)}(y, y_l)$ is the basis function of the nth coordinate with $n \in \{1, \ldots, d+1\}$.

With these definitions at hand, we obtain from (8.8) the following matrix equation:

$$A^\omega = [S^{(1)}, \ldots, S^{(d+1)}] \begin{bmatrix} (\alpha_{l,j}^{(1)})_{l,j} \\ \vdots \\ (\alpha_{l,j}^{(d+1)})_{l,j} \end{bmatrix}, \tag{8.9}$$

where A^ω is the data matrix and $S = [S^{(1)}, \ldots, S^{(d+1)}]$ is the sensing matrix with

$$(S^{(n)})_{i,l} = (\frac{1}{\mu_c} - \frac{1}{\mu_m}) \int_{\Omega^S} (\nabla_y \Gamma_{k_m}(x_i - y) \cdot e_n) \phi^{(n)}(y, y_l) dy$$

for $n = 1, \ldots, d$, and

$$(S^{(d+1)})_{i,l} = k_m^2(\frac{\varepsilon_c}{\varepsilon_m} - 1) \int_{\Omega^S} \Gamma_{k_m}(x_i - y)\phi^{(d+1)}(y, y_l) dy.$$

Here, $(e_1, \ldots, e_d)$ is an orthonormal basis of $\mathbb{R}^d$. Let $X = \begin{bmatrix} (\alpha_{l,j}^{(1)})_{l,j} \\ \vdots \\ (\alpha_{l,j}^{(d+1)})_{l,j} \end{bmatrix}$. The solution X to (8.9) has a pairwise joint sparsity meaning that $(\alpha_{l,j}^{(1)}), \ldots, (\alpha_{l,j}^{(d+1)})$ are nonzero at the rows corresponding to the particle's location. Based on (8.9), we can formulate in the presence of measurement noise the following joint sparse recovery problem:

$$\min_X \|X\|_0 \quad \text{subject to } \|A^\omega - SX\|_F^2 \leq \eta,$$

where $\|X\|_0$ denotes the number of rows that have nonzero elements in the matrix X, η is a small (regularization) parameter, and $\| \ \|_F$ denotes the Frobenius norm. We refer to [**354**] for an implementation of this algorithm.

8.3. Electromagnetic Imaging

Let $B_R := \{|x| < R\}$. With the notation of Subsection 2.14.9, we denote by D a small elastic particle (with location at $z \in B_R$ and electromagnetic parameters ε_c and ε_c). Let $x_i, i = 1, \ldots, N$ be equi-distributed points along the boundary ∂B_R for $N \gg 1$. The array of N elements $\{x_1, \ldots, x_N\}$ is used to detect the particle. Let $\theta_1, \ldots, \theta_N$ be the corresponding unit directions of the incident fields/observation directions. For the sake of simplicity, we suppose that D is a ball and $\mu_c = \mu_m$. We choose the incident electric field to be such that

$$E^i(x) = \mathbf{G}_{k_m}(x - x_j)\theta_j, \quad x \in \mathbb{R}^3, \tag{8.10}$$

where $\mathbf{G}_{k_m}(x-z)$ is the Dyadic Green (matrix valued) function for the full Maxwell equations defined by (2.380). Let E_j^s denote the solution to (2.331) corresponding to the incident field E^i given by (8.10).

The asymptotic expansion (2.120) yields

$$\begin{aligned} E_j^s(x) &= 3k_m^2 \frac{\varepsilon_m - \varepsilon_c}{2\varepsilon_m + \varepsilon_c} |D| \mathbf{G}_{k_m}(x - z)\mathbf{G}_{k_m}(z - x_j)\theta_j \\ &\quad + O(\delta^4). \end{aligned} \tag{8.11}$$

Here we have used the explicit formula (2.88) of the polarization tensor for a ball in three dimensions.

The measured data is the $N \times N$ matrix given by

$$A^\omega := \Big(E_j^s(x_i) \cdot \theta_i \Big)_{i,j}. \tag{8.12}$$

Introduce the N-dimensional vector fields $g^{(j)}(z^S)$, for $z^S \in B_R$ and $j = 1, 2, 3$, by

$$\begin{aligned} &g^{(j)}(z^S) \\ &= \frac{1}{\sqrt{\sum_{i=1}^N |e_j \cdot \mathbf{G}_{k_m}(z^S - x_i)\theta_i|^2}} \Big(e_j \cdot \mathbf{G}_{k_m}(z^S - x_1)\theta_1, \ldots, e_j \cdot \mathbf{G}_{k_m}(z^S - x_N)\theta_N \Big)^t. \end{aligned} \tag{8.13}$$

With this at hand, the MUSIC, reverse-time migration, Kirchhoff, and joint sparse recovery algorithms described in the previous subsection can be easily extended to the electromagnetic case [**49**]. The performance of reverse-time migration and Kirchhoff algorithms in the presence of measurement noise is investigated in [**154**].

8.4. Elasticity Imaging

8.4.1. MUSIC-type Method. Let $d = 2$ and let $B_R := \{|x| < R\}$. Let D be a small elastic particle (with location at $z \in B_R$ and Lamé parameters $\widetilde{\lambda}$ and $\widetilde{\mu}$). Let $x_i, i = 1, \ldots, N$ be equi-distributed points along the boundary ∂B_R for $N \gg 1$.

The array of N elements $\{x_1, \dots, x_N\}$ is used to detect the particle. Let $\theta_1, \dots, \theta_N$ be the corresponding unit directions of incident fields/observation directions. The array of elements $\{x_1, \dots, x_N\}$ is operating both in transmission and in reception. For the sake of simplicity, we take $d = 2$. We choose the incident displacement field to be such that

$$u_0^{(j)}(x) = \mathbf{\Gamma}^\omega(x, x_j)\theta_j, \quad x \in \mathbb{R}^2, \tag{8.14}$$

and denote by $u_\delta^{(j)}$ the solution to (2.521) corresponding to the incident field $u_0^{(j)}$.

From the asymptotic expansion (2.528), we have

$$\begin{aligned}(u_\delta^{(j)} - u_0^{(j)})(x) = -\delta^2 \Big(& \nabla_z \mathbf{\Gamma}^\omega(x, z) : \mathbb{M} \nabla_z (\mathbf{\Gamma}^\omega(z, x_j)\theta_j) \\ & + \omega^2 (\rho - \widetilde{\rho}) |B| \mathbf{\Gamma}^\omega(x, z) \mathbf{\Gamma}^\omega(z, x_j)\theta_j \Big) + O(\delta^3).\end{aligned} \tag{8.15}$$

The measured data is the $N \times N$ matrix given by

$$A^\omega := \left((u_\delta^{(j)} - u_0^{(j)})(x_i) \cdot \theta_i \right)_{i,j}. \tag{8.16}$$

For any point $x \in \mathbb{R}^2$, let us introduce the $N \times 2$ matrix of the incident field emitted by the array of N transmitters $G(x, \omega)$, which will be called the Green matrix, and the $N \times 3$ matrix of the corresponding independent components of the stress tensors $S(x, \omega)$, which will be called the stress matrix:

$$G(x, \omega) = \left(\mathbf{\Gamma}^\omega(x, x_1)\theta_1, \dots, \mathbf{\Gamma}^\omega(x, x_N)\theta_N \right)^t, \tag{8.17}$$

$$S(x, \omega) = \left(s_1(x), \dots, s_N(x) \right)^t, \tag{8.18}$$

where

$$s_j(x) = [\sigma_{11}^{(j)}(x), \sigma_{22}^{(j)}(x), \sigma_{12}^{(j)}(x)]^t, \quad \sigma^{(j)}(x) = \mathbb{C} \nabla^s (\mathbf{\Gamma}^\omega(x, x_j)\theta_j),$$

where $\mathbb{C}$ is the elasticity tensor defined by (2.388).

One can see from (8.15) and (8.16) that the data matrix A^ω is factorized as follows:

$$A^\omega = -\delta^2 H(z, \omega)\, D(\omega)\, H^t(z, \omega), \tag{8.19}$$

where

$$H(x, \omega) = [S(x, \omega),\, G(x, \omega)] \tag{8.20}$$

and $D(\omega)$ is a symmetric 5×5 matrix given by

$$D(\omega) = \begin{pmatrix} \mathcal{L}[\mathbb{M}] & 0 \\ 0 & \omega^2 (\rho - \widetilde{\rho}) |B| I \end{pmatrix} \tag{8.21}$$

for some linear operator $\mathcal{L}$.

Consequently, the data matrix A^ω is the product of three matrices $H^t(z, \omega)$, $D(\omega)$ and $H(z, \omega)$. The physical meaning of the above factorization is the following: the matrix $H^t(z, \omega)$ is the propagation matrix from the transmitter points toward the particle located at the point z, the matrix $D(\omega)$ is the scattering matrix and $H(z, \omega)$ is the propagation matrix from the particle toward the receiver points.

Recall that MUSIC is essentially based on characterizing the range of the data matrix A^ω, which is the so-called signal space, forming projections onto its null (noise) spaces, and computing its singular value decomposition.

From the factorization (8.19) of A^ω and the fact that the scattering matrix D is nonsingular (so, it has rank 5), the standard argument from linear algebra yields that, if $N \geq 5$ and if the propagation matrix $H(z, \omega)$ has maximal rank 5 then the ranges $\operatorname{Im} H(z,\omega)$ and $\operatorname{Im} A$ coincide.

The following is a MUSIC characterization of the location of the elastic particle and is valid if N is sufficiently large: Suppose that $N \geq 5$. Let $a \in \mathbb{C}^5 \setminus \{0\}$, then

$$H(z^S)a \in \operatorname{Im} A^\omega \qquad \text{if and only if} \qquad z^S = z.$$

In other words, any linear combination of the column vectors of the propagation matrix $H(z^S, \omega)$ defined by (8.20) belongs to the range of A^ω (signal space) if and only if the points z^S and z coincide.

If the dimension of the signal space, $s\,(\leq 5)$, is known or is estimated from the singular value decomposition of A^ω, defined by $A^\omega = V\Sigma\overline{U}^t$, then the MUSIC algorithm applies. Furthermore, if v_i denote the column vectors of the matrix V then for any vector $a \in \mathbb{C}^5 \setminus \{0\}$ and for any space point z^S within the search domain, a map of the estimator $\mathcal{I}_{\mathrm{MU}}(z^S, \omega)$ defined as the inverse of the Euclidean distance from the vector $H(z^S,\omega)a$ to the signal space by

$$\mathcal{I}_{\mathrm{MU}}(z^S,\omega) = \left(\sqrt{\sum_{i=s+1}^{N} |v_i \cdot H(z^S,\omega)a|^2}\right)^{-1/2} \tag{8.22}$$

peaks (to infinity, in theory) at the center z of the particle. The visual aspect of the peak of $\mathcal{I}_{\mathrm{MU}}$ at z depends upon the choice of the vector a. A common choice which means that we are working with all the significant singular vectors is $a = (1, 1, \ldots, 1)^t$. However, we emphasize the fact that a choice of the vector a in (8.22) with dimension (number of nonzero components) much lower than 5 still permits one to image the elastic particle with our MUSIC-type algorithm. See the numerical results below. It is worth mentioning that the estimator $\mathcal{I}_{\mathrm{MU}}(z^S, \omega)$ is obtained via the projection of the linear combination of the column vectors of the Green matrix $G(z^S)$ onto the noise subspace of the A^ω for a signal space of dimension l if the dimension of a is l.

Let us also point out here that the function $\mathcal{I}_{\mathrm{MU}}(z^S, \omega)$ does not contain any information about the shape and the orientation of the particle. Yet, if the position of the particle is found (approximately at least) via observation of the map of $\mathcal{I}_{\mathrm{MU}}(z^S, \omega)$, then one could attempt, using the decomposition (8.19), to retrieve the elastic moment tensor of the particle (which is of order δ^2).

Finally, it is worth emphasizing that in dimension 3, the matrix D is 9×9 and is of rank 9. For locating the particle, the number N then has to be larger than 9. We also mention that the MUSIC algorithm developed here applies to the crack location problem in the time-harmonic regime.

8.4.2. Reverse-Time Migration and Kirchhoff Imaging. Suppose for simplicity that a small elastic particle (with location at $z \in B_R$) has only a density contrast and set $\theta_j = \theta$ for all j. Formula (8.15) simplifies to

$$(u_\delta^{(j)} - u_0^{(j)})(x) = -\delta^2\omega^2(\rho - \widetilde{\rho})|B|\mathbf{\Gamma}^\omega(x,z)\mathbf{\Gamma}^\omega(z,x_j)\theta + O(\delta^3).$$

Thus, for a search point $z^S \in B_R$, it follows by using (2.483) that

$$\frac{1}{N}\sum_{i=1}^{N}\overline{\mathbf{\Gamma}_\alpha^\omega}(z^S, x_i)(u_\delta^{(j)} - u_0^{(j)})(x_i) \simeq \frac{\delta^2}{c_\alpha}\omega(\rho - \widetilde{\rho})|B|(\Im\,\mathbf{\Gamma}_\alpha^\omega(z^S, z))\mathbf{\Gamma}^\omega(z, x_j)\theta.$$

We introduce the reverse-time migration imaging functional $\mathcal{I}_{\mathrm{RM},\alpha}(z^S,\omega)$ for $\alpha = p$ or s given by

$$\frac{1}{N^2}\sum_{i,j=1}^{N}\overline{\mathbf{\Gamma}_\alpha^\omega}(z^S, x_j)\theta \cdot \sum_{i=1}^{N}\overline{\mathbf{\Gamma}_\alpha^\omega}(z^S, x_i)(u_\delta^{(j)} - u_0^{(j)})(x_i). \tag{8.23}$$

$\mathcal{I}_{\mathrm{RM},\alpha}(z^S,\omega)$ consists in backpropagating with the α-Green function the data set $\{(u_\delta^{(j)} - u_0^{(j)})(x_i)\}$ both from the source point x_j and the receiver point x_i.

Using (2.483) and the reciprocity property (2.482) we obtain that

$$\mathcal{I}_{\mathrm{RM},\alpha}(z^S,\omega) \simeq -\frac{\delta^2}{c_\alpha^2}(\rho - \widetilde{\rho})|B||\Im\,\mathbf{\Gamma}_\alpha^\omega(z^S, z)\theta|^2.$$

The imaging functional $\mathcal{I}_{\mathrm{RM},\alpha}(z^S,\omega)$ attains then its maximum (if $\rho < \widetilde{\rho}$) or minimum (if $\rho > \widetilde{\rho}$) at $z^S = z$.

The imaging functional $\mathcal{I}_{\mathrm{RM},\alpha}(z^S,\omega)$ can be simplified as follows to yield the so-called Kirchhoff migration imaging functional $\mathcal{I}_{\mathrm{KM},\alpha}(z^S,\omega)$ given by

$$\frac{1}{N^2}\sum_{i,j=1}^{N} e^{-\sqrt{-1}\kappa_\alpha(|x_j - z^S| + |z^S - x_i|)}\theta \cdot (u_\delta^{(j)} - u_0^{(j)})(x_i). \tag{8.24}$$

The function $\mathcal{I}_{\mathrm{KM},\alpha}$ also attains its maximum at $z^S = z$. In this simplified version, backpropagation is approximated by travel time migration.

REMARK 8.3. *The joint sparse recovery framework in Subsection* 8.2.3 *was extended to allow for the accurate reconstruction of elastic particles in* [**511**].

8.5. Numerical Illustrations

In this section we present numerical examples for direct imaging of small particles with MUSIC. The corresponding Matlab code can be downloaded at `www.ams.org/bookpages/surv-235`.

We consider two particles: one magnetic and the second dielectric, as shown in Figures 8.1(a) and 8.1(b). We give in Figure 8.1(c) a map of $\mathcal{I}_{\mathrm{MU}}(z^S,\omega)$ obtained via the projection of the linear combination of the matrix $g(z^S)$ onto the noise space of A^ω.

8.6. Concluding Remarks

In this chapter we have introduced direct algorithms for small particle detection. We have seen that the MUSIC algorithm is based on characterizing the range of the data while reverse-time migration and Kirchhoff imaging are based on extracting phase information from the data in order to locate the particles. Based on Lippmann-Schwinger integral formulations, we have also provided a joint sparse recovery framework for small particles.

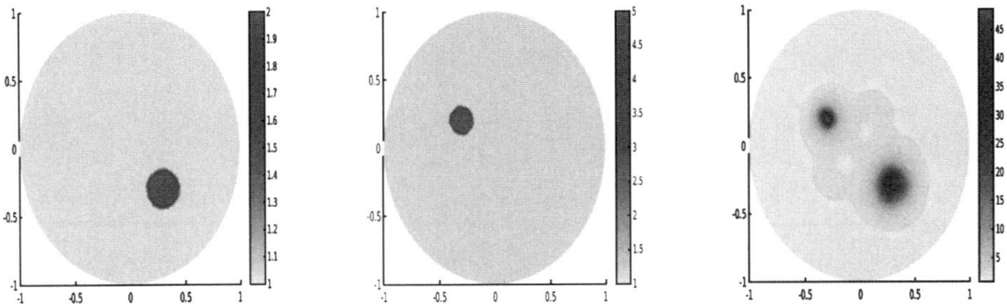

(a) μ (b) ε (c) MUSIC reconstruction

FIGURE 8.1. From left to right: The magnetic particle with coefficient μ, the electrical particle with coefficient ε, and the reconstructed fields using MUSIC algorithm.

As a direct consequence of Helmholtz-Kirchhoff identities, we have also shown that the resolution limit, defined as the minimum distance required between two small particles to distinguish between them, is of the order of half the operating wavelength. The purpose of the next chapter is to investigate super-resolution imaging mechanisms.

CHAPTER 9

Super-Resolution Imaging

9.1. Introduction

Super-resolution has many applications in nanophotonics. It is being intensively investigated as a technique that can potentially focus electromagnetic radiation in a region of the order of a few nanometers beyond the diffraction limit of light and thereby cause an extraordinary enhancement of the electromagnetic fields [**307**].

As shown in the previous chapter, the resolution in the homogeneous space for far-field imaging systems is limited by half the operating wavelength, which is a direct consequence of Helmholtz-Kirchhoff identities. In order to differentiate point sources or small particles which are located less than half the wavelength apart, super-resolution techniques have to be used.

While many techniques exist in practice, here we are only interested in the one using resonant media. The resolution enhancement in resonant media has been demonstrated in various recent experiments [**105, 356–359**]. The basic idea is the following: suppose that we have sources or particles that are placed inside a domain of typical size of order of the wavelength of the wave the sources can emit, and we want to differentiate them by making measurements in the far-field. While this is impossible in the homogeneous space, it is possible if the medium around these sources or particles is changed so that the point spread function, which is the imaginary part of the Green function in the new medium, displays a much sharper peak than the homogeneous one and thus can resolve sub-wavelength details. The key issue in such an approach is to design the surrounding medium so that the corresponding Green function has the tailored property.

9.2. Super-Resolution Imaging in High-Contrast Media

In this section, we present the mathematical theory for realizing this approach by using high-contrast media. We show that in high-contrast media, the super-resolution is due to the propagating sub-wavelength resonant modes excited in the media and is limited by the finest structure in these modes. For the sake of simplicity, we consider inverse source problems. The problem of imaging small particles can be handled by a similar approach.

9.2.1. Inverse Source Problems. We consider the following inverse source problem in a general medium characterized by refractive index $n(x)$:

$$\Delta u + k^2 n(x) u = f,$$
$$u \text{ satisfies the Sommerfeld radiation condition.}$$

We assume that $n - 1$ is compactly supported in a bounded domain $D \Subset \mathbb{R}^d$ for $d = 2, 3$, and is assumed to be known. We are interested in imaging f, which can be either a function in $L^2(D)$ or consists of a finite number of point sources supported in D, from the scattered field u in the far-field. Denote by $\Phi_k(x,y)$ the corresponding Green function for the media, that is, the solution to

$$\Delta_x \Phi_k(x,y) + k^2 n(x)\Phi_k(x,y) = \delta_y(x),$$
$$\Phi_k \text{ satisfies the Sommerfeld radiation condition}$$

with δ_y being the Dirac mass at y, we have

$$u(x) = \mathtt{K}_D[f](x) := \int_D \Phi_k(x,y) f(y)\, dy.$$

The inverse source problem of reconstructing f from u for fixed frequency is well-known to be ill-posed for general sources; see, for instance, [**20**, **52**, **133**]. While there are many methods of reconstructing f from u, we concentrate on the following three most common ones in the literature:

(i) Time reversal based method;
(ii) Minimum L^2-norm solution; and
(iii) Minimum L^1-norm solution.

9.2.2. Time Reversal Based Method. We first present some basics about the time-reversal-based method. The imaging functional is given as follows:

$$I(x) = \int_\Gamma \overline{\Phi_k(x,z)} u(z)\, ds(z) = \mathtt{K}_D^* \mathtt{K}_D[f](x), \tag{9.1}$$

where Γ is a closed surface in the far-field where the measurements are taken, and $\mathtt{K}_D^*$ is the adjoint of $\mathtt{K}_D$ viewed as a linear operator from the space $L^2(D)$ to $L^2(\Gamma)$. Physically, the operator $\mathtt{K}_D^*$ corresponds to time-reversing the observed field. This imaging method is the simplest and perhaps the mostly used one in practice.

The resolution of this imaging method can be derived from the Helmholtz-Kirchhoff identity. As a corollary of Theorem 2.74, the following result holds.

Corollary 9.1. *We have*

$$I(x) = K_D^* K_D[f](x) \approx -\frac{1}{k}\int_D \Im \Phi_k(x,y) f(y)\, dy.$$

If we take f to be a point source, we obtain the point spread function of the imaging functional, which shows that the time-reversal based method has resolution limited by $\Im\Phi_k(x,y)$.

9.2.3. Minimum L^2-Norm Solution. We now consider the second method which is based on L^2-minimization. We assume that the source $f \in L^2(D)$. The method is given as follows:

$$\min \|g\|_{L^2(D)} \text{ subject to } \quad K_D[g] = u, \tag{9.2}$$

which can be relaxed in the presence of noise as follows:

$$\min \|g\|_{L^2(D)} \text{ subject to } \quad \|\mathtt{K}_D[g] - u\|^2_{L^2(\Gamma)} < \delta \tag{9.3}$$

with $\delta > 0$ being a given small parameter.

In order to obtain an explicit formula for this method, we consider the singular value decomposition for the operator

$$\mathtt{K}_D : L^2(D) \to L^2(\Gamma).$$

We have

$$\mathtt{K}_D = \sum_{l \geq 0} \sigma_l P_l,$$

where σ_l is the lth singular value and P_l is the associated projection. The ill-posedness of the inverse source problem is due to the fast decay of the singular values to zero; see, for instance, [**52, 485**].

By a direct calculation, one can show that the minimum L^2-norm solution to (9.2) is given by

$$I(x) = \sum_{l \geq 0} \frac{P_l^* P_l}{\sigma_l^2} \mathtt{K}_D^* \mathtt{K}_D[f](x), \tag{9.4}$$

while the regularized one, which is the solution to (9.3) is given by

$$I_\alpha(x) = \sum_{l \geq 0} \frac{P_l^* P_l}{\sigma_l^2 + \alpha} \mathtt{K}_D^* \mathtt{K}_D[f](x), \tag{9.5}$$

with α as a function of δ introduced in (9.3) being chosen by Morozov's discrepancy principle; see, for instance, [**270**].

9.2.4. Minimum L^1-Norm Solution. The method of minimum L^1-norm solution is proposed in [**186, 187**]. Assume that f is equal to a superposition of separate point sources. The method of minimum L^1-norm solution is to solve the minimization problem

$$\min \|g\|_{L^1(D)} \quad \text{subject to} \quad \mathtt{K}_D^* \mathtt{K}_D[g] = \mathtt{K}_D^*[u],$$

or its relaxed version, which reads as

$$\min \|g\|_{L^1(D)} \quad \text{subject to} \quad \|\mathtt{K}_D^* \mathtt{K}_D[g] - \mathtt{K}_D^*[u]\|_{L^2(\Gamma)}^2 < \delta.$$

In [**186, 187**], it is shown that under a minimum separation condition for the point sources, the inverse source problem is well posed. A main feature of their approach is that the L^1-minimization can pull out small spikes even though they may be completely buried in the side lobes of large ones.

It is worth emphasizing that without any *a priori* information, the resolution of the raw image, which is obtained by time-reversal method, is determined by the imaginary part of the Green function in the associated media.

9.2.5. The Special Case of Homogeneous Medium. In a homogeneous medium, we have $n \equiv 1$. For simplicity, we consider the case $d = 3$ and recall that

$$\Phi_k(x, y) = \Gamma_k(x - y) = -\frac{e^{\sqrt{-1}k|x-y|}}{4\pi|x - y|}.$$

In the far-field, where $k|y| = O(1)$ and $k|x| \gg 1$, we have $|x - y| \approx |x| - \hat{x} \cdot y$, where $\hat{x} = \frac{x}{|x|}$. Thus,

$$u(x) = -\int_D \frac{e^{\sqrt{-1}k|x-y|}}{4\pi|x - y|} f(y)\, dy \approx -\frac{e^{\sqrt{-1}k|x|}}{4\pi|x|} \hat{f}(k\hat{x}),$$

where $\hat{f}$ is the Fourier transform of f.

If we make measurements on the surface ∂B_R, the sphere of radius R and center the origin, then we have

$$u(x) = -\frac{e^{\sqrt{-1}kR}}{4\pi R} \hat{f}(k\hat{x}).$$

Using the time-reversal method, we have for R large enough

$$I(z) \approx \frac{1}{16\pi^2 R^2} \int_{\partial B_R} \int_D e^{\sqrt{-1}k\hat{x}\cdot(y-z)} f(y)\, dy\, ds(x) = \frac{1}{4\pi} \int_D f(y) \frac{\sin k|z-y|}{k|z-y|}\, dy,$$

where the imaging functional I is defined by (9.1) with $\Gamma = \partial B_R$.

9.2.6. Green Function in High-Contrast Media. Throughout this section, we set the wavenumber k to be the unit and suppress its presence in what follows. We assume that the wave speed in the free space is one. The free-space wavelength is given by 2π. We consider the following Helmholtz equation with a delta source term:

$$\Delta_x \Phi(x, x_0) + \Phi(x, x_0) + \tau n(x)\chi(D)(x)\Phi(x, x_0) = \delta(x - x_0) \quad \text{in } \mathbb{R}^d, \tag{9.6}$$

where $\chi(D)$ is the characteristic function of D, which has size of order of the free-space wavelength, $n(x)$ is a positive function of order one in the space of $\mathcal{C}^1(\overline{D})$ and $\tau \gg 1$ is the contrast. We denote by $\Phi_0(x, x_0)$ the free-space Green's function $\Gamma_1(x - x_0)$.

Write $\Phi = v + \Phi_0$, we can show that

$$\Delta v + v = -\tau n(x)\chi(D)(v + \Phi_0). \tag{9.7}$$

Thus,

$$v(x, x_0) = -\tau \int_D n(y)\Phi_0(x, y)\Big(v(y, x_0) + \Phi_0(y, x_0)\Big)\, dy.$$

Define

$$\mathrm{K}_D[f](x) = -\int_D n(y)\Phi_0(x, y) f(y)\, dy. \tag{9.8}$$

Then, $v = v(x) = v(x, x_0)$ satisfies the following integral equation:

$$(I - \tau \mathrm{K}_D)[v] = \tau \mathrm{K}_D[\Phi(\cdot, x_0)], \tag{9.9}$$

and hence,

$$v(x) = (\frac{1}{\tau} - \mathrm{K}_D)^{-1}\mathrm{K}_D[\Phi(\cdot, x_0)].$$

In what follows, we present properties of the integral operator K_D.

LEMMA 9.2. *The operator K_D is compact from $L^2(D)$ to $L^2(D)$. In fact, K_D is bounded from $L^2(D)$ to $H^2(D)$. Moreover, K_D is a Hilbert-Schmidt operator.*

LEMMA 9.3. *Let $\sigma(K_D)$ be the spectrum of K_D defined by (9.8). We have*

(i) *$\sigma(K_D) = \{0, \lambda_1, \lambda_2, \dots, \lambda_n, \dots\}$, where $|\lambda_1| \geq |\lambda_2| \geq |\lambda_3| \geq \dots$ and $\lambda_n \to 0$;*

(ii) *$\{0\} = \sigma(K_D)\backslash\sigma_p(K_D)$ with $\sigma_p(K_D)$ being the point spectrum of K_D.*

PROOF. We need only to prove the second assertion. Assume that

$$\mathrm{K}_D[u] = \int_D \Phi_0(x, y) n(y) u(y)\, dy = 0.$$

We have $0 = (\Delta + 1)\mathrm{K}_D[u] = nu$, which shows that $u = 0$. The assertion is then proved. □

LEMMA 9.4. *Let K_D be defined by (9.8). Then, $\lambda \in \sigma(K_D)$ if and only if there is a non-trivial solution in $H^2_{\text{loc}}(\mathbb{R}^d)$ to the following problem:*

$$(\Delta + 1)u(x) = \tfrac{1}{\lambda} n(x)u(x) \quad \textit{in } D, \tag{9.10}$$

$$(\Delta + 1)u = 0 \quad \textit{in } \mathbb{R}^d \backslash D, \tag{9.11}$$

$$u \textit{ satisfies the Sommerfeld radiation condition.} \tag{9.12}$$

PROOF. Assume that $\mathtt{K}_D[u] = \lambda u$. We define $\tilde{u}(x) = \int_D \Phi_0(x,y)n(y)u(y)\,dy$, where $x \in \mathbb{R}^d$. Then $\tilde{u}$ satisfies the required equations. □

Notice that the resonant modes have sub-wavelength structures in D for $|\lambda| < 1$ and can propagate into the far-field. It is these sub-wavelength propagating modes that cause super-resolution.

LEMMA 9.5. *Let $\mathcal{H}_j$ denote the generalized eigenspace of the operator K_D for the eigenvalue λ_j. The following decomposition holds:*

$$L^2(D) = \overline{\bigcup_{j=1}^{\infty} \mathcal{H}_j}.$$

PROOF. By the same argument as the one in the proof of Lemma 9.3, we can show that Ker $\mathtt{K}_D^* = \{0\}$. As a result, we have

$$\overline{\mathtt{K}_D\big(L^2(D)\big)} = \big(\text{Ker } \mathtt{K}_D^*\big)^{\perp} = L^2(D),$$

and the lemma is proved. □

LEMMA 9.6. *There exists a basis $\{u_{j,l,k}\}$, $1 \le l \le m_j, 1 \le k \le n_{j,l}$ for $\mathcal{H}_j$ such that*

$$K_D(u_{j,1,1}, \ldots, u_{j,m_j,n_{j,m_j}}) = (u_{j,1,1}, \ldots, u_{j,m_j,n_{j,m_j}}) \begin{pmatrix} J_{j,1} & & \\ & \ddots & \\ & & J_{j,m_j} \end{pmatrix},$$

where $J_{j,l}$ is the canonical Jordan matrix of size $n_{j,l}$ in the form

$$J_{j,l} = \begin{pmatrix} \lambda_j & 1 & & \\ & \ddots & \ddots & \\ & & \lambda_j & 1 \\ & & & \lambda_j \end{pmatrix}.$$

PROOF. This follows from the Jordan theory applied to the linear operator $\mathtt{K}_D|_{\mathcal{H}_j} : \mathcal{H}_j \to \mathcal{H}_j$ on the finite dimensional space $\mathcal{H}_j$. □

We denote $\Gamma = \{(j,l,k) \in \mathbb{N} \times \mathbb{N} \times \mathbb{N}; 1 \le l \le m_j, 1 \le k \le n_{j,l}\}$ the set of indices for the basis functions. We introduce a partial order on $\mathbb{N} \times \mathbb{N} \times \mathbb{N}$. Let $\gamma = (j,k,l) \in \Gamma, \gamma' = (j',l',k') \in \Gamma$, we say that $\gamma' \preceq \gamma$ if one of the following conditions are satisfied:

(i) $j > j'$;
(ii) $j = j'$, $l > l'$;
(iii) $j = j'$, $l = l'$, $k \ge k'$.

By the Gram-Schmidt orthonormalization process, the following result is obvious.

LEMMA 9.7. *There exists an orthonormal basis $\{e_\gamma : \gamma \in \Gamma\}$ for $L^2(D)$ such that*

$$e_\gamma = \sum_{\gamma' \preceq \gamma} a_{\gamma,\gamma'} u_{\gamma'},$$

where $a_{\gamma,\gamma'}$ are constants and $a_{\gamma,\gamma} \neq 0$.

We can regard $A = \{a_{\gamma,\gamma'}\}_{\gamma,\gamma' \in \Gamma}$ as a matrix. It is clear that A is upper-triangular and has non-zero diagonal elements. Its inverse is denoted by $B = \{b_{\gamma,\gamma'}\}_{\gamma,\gamma' \in \Gamma}$ which is also upper-triangular and has non-zero diagonal elements. We have

$$u_\gamma = \sum_{\gamma' \preceq \gamma} b_{\gamma,\gamma'} e_{\gamma'}.$$

LEMMA 9.8. *The functions $\{e_\gamma(x)\overline{e_{\gamma'}(y)}\}$ form a normal basis for the Hilbert space $L^2(D \times D)$. Moreover, the following completeness relation holds:*

$$\delta(x-y) = \sum_\gamma e_\gamma(x)\overline{e_\gamma(y)}.$$

By standard elliptic theory, we have $\Phi(x, x_0) \in L^2(D \times D)$ for fixed τ. Thus we have

$$\Phi(x, x_0) = \sum_{\gamma,\gamma'} \alpha_{\gamma,\gamma'} e_\gamma(x)\overline{e_{\gamma'}}(x_0), \tag{9.13}$$

for some constants $\alpha_{\gamma,\gamma'}$ satisfying

$$\sum_{\gamma,\gamma'} |\alpha_{\gamma,\gamma'}|^2 = \|\Phi(x, x_0)\|^2_{L^2(D\times D)} < \infty.$$

To analyze the Green function Φ, we need to find the constants $\alpha_{\gamma,\gamma'}$. To do so, we first note that

$$\Phi_0(x, x_0) = \frac{1}{n(x_0)} \mathrm{K}_D[\delta(\cdot - x_0)].$$

Thus,

$$\begin{aligned} \Phi(x, x_0) &= \Phi_0(x, x_0) + (\frac{1}{\tau} - \mathrm{K}_D)^{-1}\mathrm{K}_D^2[\delta(\cdot - x_0)] \\ &= \Phi_0(x, x_0) + \frac{1}{n(x_0)} \sum_\gamma \overline{e_\gamma}(x_0)(\frac{1}{\tau} - \mathrm{K}_D)^{-1}\mathrm{K}_D^2[e_\gamma]. \end{aligned}$$

We next compute $(\frac{1}{\tau} - \mathrm{K}_D)^{-1}\mathrm{K}_D^2[e_\gamma]$. For ease of notation, we define $u_{j,l,k} = 0$ for $k \leq 0$. We have

$$\mathrm{K}_D[u_{j,l,k}] = \lambda_j u_{j,l,k} + u_{j,l,k-1} \quad \text{for all } j, l, k,$$

and

$$\mathrm{K}_D^2[u_{j,l,k}] = \lambda_j^2 u_{j,l,k} + 2\lambda_j u_{j,l,k-1} + u_{j,l,k-2} \quad \text{for all } j, l, k.$$

On the other hand, for $z \notin \sigma(\mathrm{K}_D)$, we have

$$(z - \mathrm{K}_D)^{-1}[u_{j,l,k}] = \frac{1}{z-\lambda_j} u_{j,l,k} + \frac{1}{(z-\lambda_j)^2} u_{j,l,k-1} + \ldots + \frac{1}{(z-\lambda_j)^k} u_{j,l,1},$$

and therefore, it follows that

$$
\begin{aligned}
(z-\mathrm{K}_D)^{-1}\mathrm{K}_D^2[u_{j,l,k}] &= \frac{\lambda_j^2}{z-\lambda_j}u_{j,l,k} + \frac{\lambda_j^2}{(z-\lambda_j)^2}u_{j,l,k-1}\cdots + \frac{\lambda_j^2}{(z-\lambda_j)^k}u_{j,l,1} \\
&\quad + \frac{2\lambda_j}{z-\lambda_j}u_{j,l,k-1} + \frac{2\lambda_j}{(z-\lambda_j)^2}u_{j,l,k-2}\cdots + \frac{2\lambda_j}{(z-\lambda_j)^{k-1}}u_{j,l,1} \\
&\quad + \frac{1}{z-\lambda_j}u_{j,l,k-2} + \frac{1}{(z-\lambda_j)^2}u_{j,l,k-3}\cdots + \frac{1}{(z-\lambda_j)^{k-2}}u_{j,l,1} \\
&= \frac{\lambda_j^2}{z-\lambda_j}u_{j,l,k} + \Big(\frac{\lambda_j^2}{(z-\lambda_j)^2} + \frac{2\lambda_j}{z-\lambda_j}\Big)u_{j,l,k-1} \\
&\quad + \Big(\frac{\lambda_j^2}{(z-\lambda_j)^3} + \frac{2\lambda_j}{z-\lambda_j} + \frac{1}{z-\lambda_j}\Big)u_{j,l,k-2} \\
&\quad + \ldots + \Big(\frac{\lambda_j^2}{(z-\lambda_j)^k} + \frac{2\lambda_j}{(z-\lambda_j)^{k-1}} + \frac{1}{(z-\lambda_j)^{k-2}}\Big)u_{j,l,1} \\
&= \sum_{\gamma'} d_{\gamma,\gamma'}u_{\gamma'},
\end{aligned}
$$

where we have introduced the matrix $D=\{d_{\gamma,\gamma'}\}_{\gamma,\gamma'\in\Gamma}$, which is upper-triangular and has block-structure.

With these calculations, by taking $z=1/\tau$, we arrive at the following result.

THEOREM 9.9. *The following expansion holds for the Green function*

$$
\Phi(x,x_0) = \Phi_0(x,x_0) + \sum_{\gamma\in\Gamma}\sum_{\gamma'''\in\Gamma}\alpha_{\gamma,\gamma'''}\overline{e_\gamma}(x_0)e_{\gamma'''}(x), \tag{9.14}
$$

where

$$
\alpha_{\gamma,\gamma'''} = \frac{1}{n(x_0)}\sum_{\gamma'\preceq\gamma}\sum_{\gamma''\preceq\gamma'} a_{\gamma,\gamma'}d_{\gamma',\gamma''}b_{\gamma'',\gamma'''}.
$$

Moreover, for τ^{-1} belonging to a compact subset of $\mathbb{R}\setminus\big(\mathbb{R}\cap\sigma(K_D)\big)$, we have the following uniform bound:

$$
\sum_{\gamma,\gamma'}|\alpha_{\gamma,\gamma'}|^2 < \infty.
$$

Alternatively, if we start from the identity,

$$
\begin{aligned}
\delta(x-x_0) &= \sum_{\gamma''} e_{\gamma''}(x)\overline{e_{\gamma''}(x_0)} \\
&= \sum_{\gamma''}\sum_{\gamma'\preceq\gamma''}\sum_{\gamma'''\preceq\gamma''} a_{\gamma'',\gamma'}\overline{a}_{\gamma'',\gamma'''}u_{\gamma'}(x)\overline{u_{\gamma'''}}(x_0),
\end{aligned}
$$

then we can obtain an equivalent expansion for the Green function in terms of the basis of resonant modes.

THEOREM 9.10. *The following expansion holds for the Green function:*

$$
\Phi(x,x_0) = \Phi_0(x,x_0) + \sum_{\gamma''\in\Gamma}\sum_{\gamma'''\preceq\gamma''}\sum_{\gamma\preceq\gamma''}\beta_{\gamma'',\gamma,\gamma'''}u_\gamma(x)\overline{u_{\gamma'''}}(x_0), \tag{9.15}
$$

where

$$\beta_{\gamma'',\gamma,\gamma'''} = \frac{1}{n(x_0)} \sum_{\gamma' \preceq \gamma''} \overline{a}_{\gamma'',\gamma'''} a_{\gamma'',\gamma'} d_{\gamma',\gamma}. \tag{9.16}$$

Here, the infinite summation can be interpreted as follows:

$$\lim_{\gamma_0 \to \infty} \sum_{\gamma'' \leq \gamma_0} \sum_{\gamma' \preceq \gamma''} \sum_{\gamma''' \preceq \gamma''} \beta_{\gamma'',\gamma,\gamma'''} u_\gamma(x) \overline{u_{\gamma'''}}(x_0) = \Phi(x, x_0) - \Phi_0(x, x_0) \quad in \ L^2(D \times D). \tag{9.17}$$

In order to have some idea of the expansions of the Green function $\Phi(x, y)$, we compare them to the expansion of the Green function in the homogeneous space, *i.e.* $\Phi_0(x, y)$. For this purpose, we introduce the matrix $H = \{h_{\gamma,\gamma'}\}_{\gamma,\gamma' \in \Gamma}$, which is defined by

$$\mathrm{K}_D[u_\gamma] = \sum_{\gamma'} h_{\gamma,\gamma'} u_{\gamma'}.$$

In fact, we have

$$h_{j,l,k,j',l',k'} = \lambda_j \delta_{j,j'} \delta_{l,l'} \delta_{k,k'} + \delta_{j,j'} \delta_{l,l'} \delta_{k-1,k'},$$

where δ denotes the Kronecker symbol.

LEMMA 9.11. (i) *In the normal basis $\{e_\gamma\}_{\gamma \in \Gamma}$, the following expansion holds for the Green function $\Phi_0(x, x_0)$:*

$$\Phi_0(x, x_0) = \sum_{\gamma \in \Gamma} \sum_{\gamma''' \in \Gamma} \tilde{\alpha}_{\gamma,\gamma'''} \overline{e_\gamma}(x_0) e_{\gamma'''}(x), \tag{9.18}$$

where

$$\tilde{\alpha}_{\gamma,\gamma'''} = \frac{1}{n(x_0)} \sum_{\gamma' \preceq \gamma} \sum_{\gamma'' \preceq \gamma'} a_{\gamma,\gamma'} h_{\gamma',\gamma''} b_{\gamma'',\gamma'''}.$$

Moreover, we have the following uniform bound:

$$\sum_{\gamma,\gamma'} |\tilde{\alpha}_{\gamma,\gamma'}|^2 < C < \infty.$$

(ii) *In the basis of resonant modes $\{u_\gamma\}_{\gamma \in \Gamma}$, the following expansion holds for the Green function $\Phi_0(x, x_0)$:*

$$\Phi_0(x, x_0) = \sum_{\gamma'' \in \Gamma} \sum_{\gamma''' \preceq \gamma''} \sum_{\gamma \preceq \gamma''} \tilde{\beta}_{\gamma'',\gamma,\gamma'''} u_\gamma(x) \overline{u_{\gamma'''}}(x_0), \tag{9.19}$$

where

$$\tilde{\beta}_{\gamma'',\gamma,\gamma'''} = \frac{1}{n(x_0)} \sum_{\gamma' \preceq \gamma''} \overline{a}_{\gamma'',\gamma'''} a_{\gamma'',\gamma'} h_{\gamma',\gamma}. \tag{9.20}$$

Here, the infinite summation can be interpreted as follows:

$$\lim_{\gamma_0 \to \infty} \sum_{\gamma'' \leq \gamma_0} \sum_{\gamma''' \preceq \gamma''} \sum_{\gamma \preceq \gamma''} \tilde{\beta}_{\gamma'',\gamma,\gamma'''} u_\gamma(x) \overline{u_{\gamma'''}}(x_0) = \Phi_0(x, x_0) \quad in \ L^2(D \times D).$$

Based on the resonance expansions of the Green functions in high-contrast media and in the free space, we can now propose an explanation for the super-resolution phenomenon. Observe that the difference between the coefficients $\beta_{\gamma'',\gamma,\gamma'''}$ and

$\tilde{\beta}_{\gamma'',\gamma,\gamma'''}$ in (9.16) and (9.20) are the quantities $d_{\gamma,\gamma'}$ and $h_{\gamma,\gamma'}$ ($a_{\gamma'',\gamma'}$ are constants). If, for example, we consider the special case where the spaces $\mathcal{H}_j$ are of dimension one, then we have

$$d_{\gamma,\gamma'} = \delta_{\gamma,\gamma'} \frac{\lambda_j^2}{z - \lambda_j}, \quad h_{\gamma,\gamma'} = \delta_{\gamma,\gamma'} \lambda_j,$$

and therefore,

$$d_{\gamma,\gamma'} = \frac{1}{\frac{z}{\lambda_j} - 1} h_{\gamma,\gamma'},$$

which shows that the contribution to the Green function Φ of the sub-wavelength resonant mode u_γ is amplified when z is close to λ_j.

Therefore, we can see that the imaginary part of Φ may have sharper peak than that of Φ_0 due to the excited sub-wavelength resonant modes. When the high contrast is properly chosen (the frequency is fixed), one or several of these sub-wavelength resonance modes can be excited, and they dominate over the other ones in the expansion of the Green function Φ. It is those sub-wavelength modes that essentially determine the behavior of Φ and hence the associated resolution in the media. Therefore, we can expect super-resolution to occur in this case.

REMARK 9.12. *Using the Maxwell-Garnett effective medium theory for Maxwell's equations derived in Section* 7.3 *for dilute periodic distributions of plasmonic nanoparticles, one can see that near plasmonic resonances the effective (or overall) permittivity ε^* is high. By investigating the spectral properties of the operator*

$$E|_\Omega \mapsto \int_\Omega (\varepsilon^*(y) - \varepsilon_m) \mathbf{G}_{k_m}(x, y) dy, \quad x \in \Omega,$$

where $\mathbf{G}_{k_m}$ is defined by (2.380) *with $k_m = \omega\sqrt{\varepsilon_m \mu_m}$, one can extend our results in this section to the case of the full Maxwell equations and give evidence of the super-resolution phenomenon for electromagnetic waves in composite materials made of plasmonic nanoparticles.*

9.2.7. Numerical Illustrations. Here we consider a more general situation than in the previous theoretical analysis and explain how to compute the Green's function numerically. We also present a numerical example in which a high-contrast medium is represented as a disk.

9.2.7.1. *Solving an Integral Equation for the Green's Function.* The Green's function Φ is the solution to the following problem:

$$(9.21) \quad \begin{cases} \nabla \cdot \dfrac{1}{\mu} \nabla \Phi(\cdot, x_0) + \omega^2 \varepsilon \Phi(\cdot, x_0) = \dfrac{1}{\mu_c} \delta_{x_0} & \text{in } \mathbb{R}^d, \\ \Phi(\cdot, x_0) \text{ satisfies the Sommerfeld radiation condition.} \end{cases}$$

It can be shown that the above problem is equivalent to the following system of equations:

$$(9.22) \quad \begin{cases} (\Delta + k_c^2) \Phi(\cdot, x_0) = \delta_{x_0} & \text{in } D, \\ (\Delta + k_m^2) \Phi(\cdot, x_0) = 0 & \text{in } \mathbb{R}^d \setminus \overline{D}, \\ \Phi(\cdot, x_0)|_+ = \Phi(\cdot, x_0)|_- & \text{on } \partial D, \\ \dfrac{1}{\mu_m} \dfrac{\partial \Phi(\cdot, x_0)}{\partial \nu}\Big|_+ = \dfrac{1}{\mu_c} \dfrac{\partial \Phi(\cdot, x_0)}{\partial \nu}\Big|_- & \text{on } \partial D, \\ \Phi(\cdot, x_0) \text{ satisfies the Sommerfeld radiation condition.} \end{cases}$$

Note that the wave number k_c plays the role of the high contrast parameter analogously to τ in the previous theoretical analysis.

Let $\Phi_0^{k_c}$ be the free space Green's function with wave number k_c. Since $\Phi_0^{k_c}$ satisfies

$$(\Delta + k_c^2)\Phi_0^{k_c}(\cdot, x_0) = \delta_{x_0} \quad \text{in } \mathbb{R}^d, \tag{9.23}$$

we see that $v := \Phi - \Phi_0^{k_c}$ satisfies $\Delta v + k_c^2 v = 0$ in D. Therefore we can represent the Green's function Φ using the single-layer potential as follows:

$$\Phi(x, x_0) = \begin{cases} \Phi_0^{k_c}(x, x_0) + \mathcal{S}_D^{k_c}[\varphi](x), & x \in D, \\ \mathcal{S}_D^{k_m}[\psi](x), & x \in \mathbb{R}^d \setminus D. \end{cases} \tag{9.24}$$

Next we determine the densities φ and ψ. From the transmission conditions on ∂D and the jump relations for the single-layer potentials, we get

$$\begin{cases} \mathcal{S}_D^{k_c}[\varphi] - \mathcal{S}_D^{k_m}[\psi] = -\Phi_0^{k_c}(\cdot, x_0) \\ \displaystyle\frac{1}{\mu_c}(-\frac{1}{2}I + (\mathcal{K}_D^{k_c})^*)[\varphi]\Big|_- - \frac{1}{\mu_m}(\frac{1}{2}I + (\mathcal{K}_D^{k_m})^*)[\psi]\Big|_+ = -\frac{1}{\mu_c}\frac{\partial \Phi_0^{k_c}(\cdot, x_0)}{\partial \nu} \end{cases} \quad \text{on } \partial D. \tag{9.25}$$

The above system of integral equations has the same form as that of (2.178). We have already discussed how to solve that system of equations numerically in Chapter 2.

9.2.7.2. *Explicit Expression of Green's Function for a Disk.* Let D be a disk of radius R located at the origin. Then it can be shown that the explicit solution is given by

$$\Phi(r, \theta) = \begin{cases} -\dfrac{\sqrt{-1}}{4}H_0^{(1)}(k_c r) + aJ_0(k_c r), & |r| \le R, \\ bH_0^{(1)}(k_m r), & |r| > R, \end{cases} \tag{9.26}$$

where (r, θ) are the polar coordinates and the constants a and b are given by

$$a = -\frac{\sqrt{-1}}{4}\frac{\frac{k_m}{\mu_m}H_0^{(1)}(k_cR)(H_0^{(1)})'(k_mR) - \frac{k_c}{\mu_c}H_0^{(1)}(k_mR)(H_0^{(1)})'(k_cR)}{\frac{k_c}{\mu_c}H_0^{(1)}(k_mR)J_0'(k_cR) - \frac{k_m}{\mu_m}J_0(k_cR)(H_0^{(1)})'(k_mR)},$$

$$b = -\frac{\sqrt{-1}}{4}\frac{\frac{k_c}{\mu_c}H_0^{(1)}(k_cR)J_0'(k_cR) - \frac{k_c}{\mu_c}(H_0^{(1)})'(k_cR)J_0(k_cR)}{\frac{k_c}{\mu_c}H_0^{(1)}(k_mR)J_0'(k_cR) - \frac{k_m}{\mu_m}J_0(k_cR)(H_0^{(1)})'(k_mR)}.$$

9.2.7.3. *Resonant Wavenumber k_c for a Disk.* It is also worth emphasizing that we can derive resonant values for k_c. From the expressions for a and b, we can immediately see that the nth resonant value $k_{c,n}$ is nth zero of

$$\frac{k_c}{\mu_c}H_0^{(1)}(k_mR)J_0'(k_cR) - \frac{k_m}{\mu_m}J_0(k_cR)(H_0^{(1)})'(k_mR) = 0. \tag{9.27}$$

So the resonant values for k_c can be computed using Muller's method. When we solve the above equation, we need to be careful because μ_c depends on k_c via $\mu_c = k_c^2/(\omega^2\epsilon_c)$.

9.2.7.4. *Numerical Example.* Let D be a circular disk of radius $R = 2$ centered at the origin O in $\mathbb{R}^2$. We fix $\omega = 1$, $\epsilon_c = \epsilon_m = 1$ and $\mu_m = 1$. Then μ_c is determined by $\mu_c = k_c^2$.

First, let us compute the distribution of the resonant values for k_c. To do this, we plot the term on the left-hand side of (9.27) as a function of k_c. The plot is shown in Figure 9.1 and it shows that there are many local maximum points which converge to zero as their corresponding wave number k_c increases. This reflects the fact that the resonant values $k_{c,n}$ (or the corresponding eigenvalues of the operator $\mathtt{K}_D$) are complex numbers and $1/k_{c,n}$ converges to zero as $n \to \infty$. This is in accordance with our previous theoretical analysis of the super-resolution phenomenon because a large wavenumber k_c plays the role of the high contrast parameter τ.

n	$k_c(A_n)$	$k_c(B_n)$
1	1.86	2.74
2	3.48	4.32
3	5.08	5.9

TABLE 9.1. Corresponding value of k_c to the points A_n and B_n.

Next we determine how the shape of $\Im\Phi$ changes as a function of k_c. We choose three local maximum (or minimum) points A_1, A_2 and A_3 (or B_1, B_2 and B_3) as shown in Figure 9.1. At the point A_1, A_2 or A_3, we expect that the corresponding $\Im\Phi$ does not have a sharp peak because the term on the left-hand side of (9.27) is not small, which means k_c is not close to a resonant value. On the other hand, we expect that $\Im\Phi$ has a sharper peak than that of $\Im\Phi_0^{k_m}$ at the points B_1, B_2 and B_3. The (approximate) numerical values of k_c corresponding to the points A_n and B_n are shown in Table 9.1.

First we consider the non-resonant case. In Figure 9.2, we plot $\Im\Phi$ when $k_c = k_c(A_n), n = 1, 2, 3$ over the line segment from $(-R, 0)$ to $(R, 0)$. The dotted line represents the imaginary part of $\Phi_0^{k_m}$. The circles and the lines represent the exact values and the numerically computed values, respectively. We note that in this case the peak is not sharper than that of the free space Green's function, as shown in Figure 9.2.

Next, we consider the resonant case. In Figure 9.3, we plot $\Im\Phi$ when $k_c = k_c(B_n), n = 1, 2, 3$ over the line segment from $(-R, 0)$ to $(R, 0)$. In contrast to the previous case, in the case of a resonant k_c the peak is sharper than that of the free space Green's function. Also the sub-wavelength structure of the resonant mode is clearly shown in Figure 9.3.

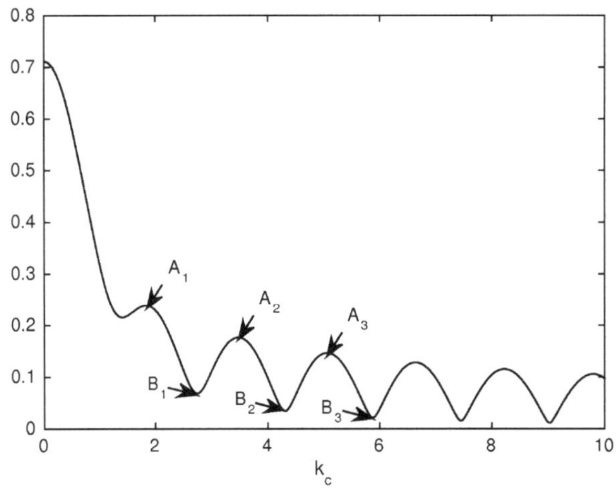

FIGURE 9.1. The plot for the term on the left-hand side of (9.27) as a function of k_c. D is a disk with radius $R = 2$. The parameters are $\omega = 1, \varepsilon_m = 1, \varepsilon_c = 1, \mu_m = 1$ and μ_c is determined by $\mu_c = k_c^2/(\varepsilon_c \omega^2)$. Three local maximums (or minimums) are marked as A_n (or B_n), respectively.

The Matlab code used to generate the numerical results shown in Table 9.1 and Figures 9.1, 9.2, and 9.3 can be downloaded at `www.ams.org/bookpages/surv-235`.

9.3. Super-Resolution in Resonant Structures

By modifying the homogeneous spaces with sub-wavelength resonators, we can introduce propagating sub-wavelength resonance modes to the space which encode sub-wavelength information in a neighborhood of the space embedded by the sub-wavelength resonators, thus yielding a Green's function whose imaginary part exhibits sub-wavelength peaks, thereby breaking the resolution limit (or diffraction limit) in the homogeneous space. In this section, using the fact that plasmonic particles are ideal sub-wavelength resonators, we consider the possibility of super-resolution by using a system of identical plasmonic particles.

9.3.1. Multiple Plasmonic Nanoparticles. We consider the scattering of an incident time harmonic wave u^i by multiple weakly coupled plasmonic nanoparticles in three dimensions. For ease of exposition, we consider the case of L particles with an identical shape and use the Helmholtz equation for light propagation.

We write $D_l = z_l + \delta \widetilde{D}$, $l = 1, 2, \ldots, L$, where $\widetilde{D}$ is centered at the origin. Moreover, we denote $D_0 = \delta \widetilde{D}$ as our reference nanoparticle and let

$$D = \bigcup_{l=1}^{L} D_l, \quad \varepsilon_D = \varepsilon_m \chi(\mathbb{R}^3 \backslash \bar{D}) + \varepsilon_c \chi(\bar{D}), \quad \mu_D = \mu_m \chi(\mathbb{R}^3 \backslash \bar{D}) + \mu_c \chi(D).$$

We assume that the following conditions hold.

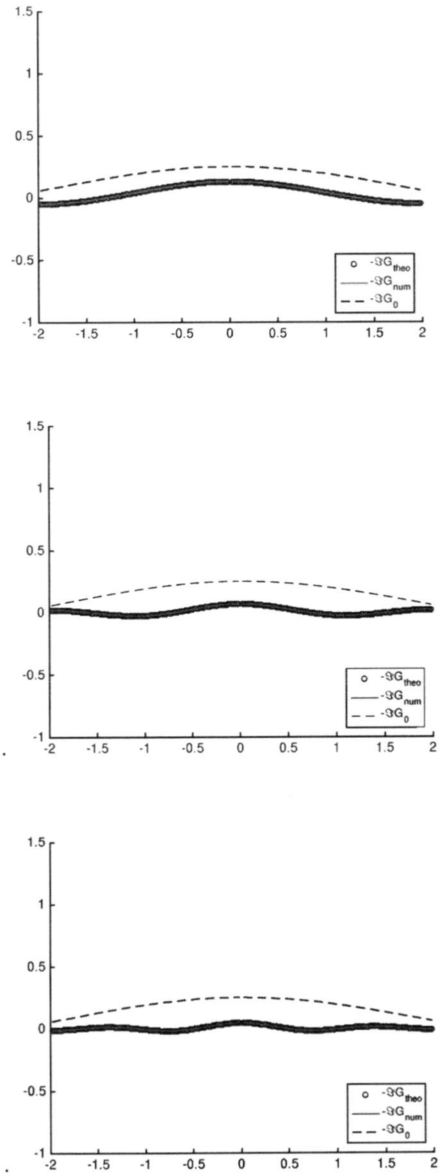

FIGURE 9.2. The plot for $\Im\Phi$ when $k_c = k_c(A_n), n = 1, 2, 3$ over the line segment from $(-R, 0)$ to $(R, 0)$. The dotted line represents the imaginary part of $\Phi_0^{k_m}$. The circles and the lines represent the exact values and the numerically computed values, respectively. In this case the peak is not sharper than that of the free space Green's function.

CONDITION 9.13. *We assume that the numbers $\varepsilon_m, \mu_m, \varepsilon_c, \mu_c$ are dimensionless and are of order one. We also assume that the particle $\widetilde{D}$ has size of order one and ω is dimensionless and is of order $o(1)$.*

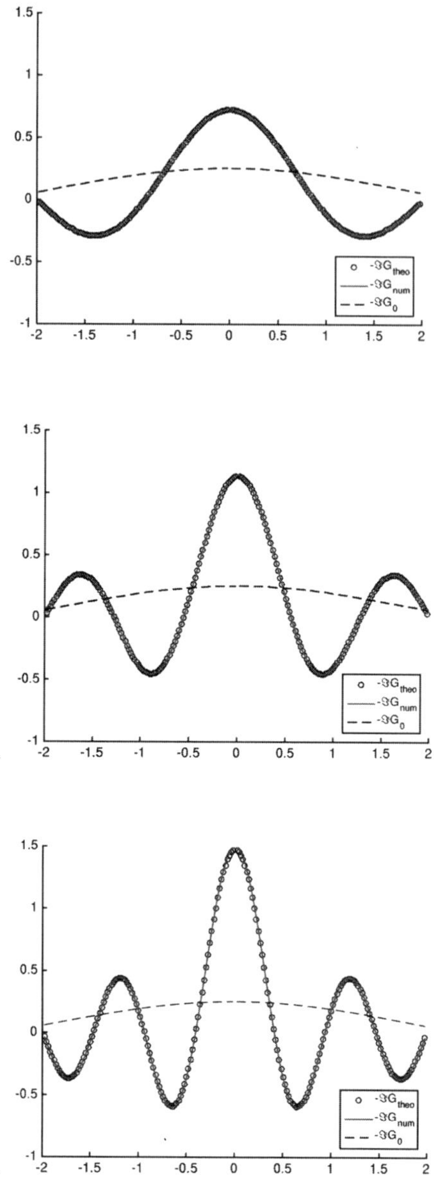

FIGURE 9.3. The plot for $\Im\Phi$ when $k_c = k_c(B_n), n = 1, 2, 3$ over the line segment from $(-R, 0)$ to $(R, 0)$. The dotted line represents the imaginary part of $\Phi_0^{k_m}$. The circles and the lines represent the exact values and the numerically computed values, respectively. In this case the peak is sharper than that of the free space Green's function. Also the sub-wavelength structure of the resonant mode is clearly shown.

CONDITION 9.14. *Let*

$$\lambda = \frac{\mu_m + \mu_c}{2(\mu_m - \mu_c)}. \tag{9.28}$$

We assume that $\lambda \neq 0$ *or equivalently,* $\mu_c \neq -\mu_m$.

CONDITION 9.15. *The size* δ *of the particles is a small parameter and the distances between neighboring particles are of order one.*

The scattering problem can be modeled by the following Helmholtz equation:

$$\begin{cases} \nabla \cdot \dfrac{1}{\mu_D} \nabla u + \omega^2 \varepsilon_D u = 0 \quad \text{in } \mathbb{R}^3 \backslash \partial D, \\ u_+ - u_- = 0 \quad \text{on } \partial D, \\ \dfrac{1}{\mu_m} \dfrac{\partial u}{\partial \nu}\Big|_+ - \dfrac{1}{\mu_c} \dfrac{\partial u}{\partial \nu}\Big|_- = 0 \quad \text{on } \partial D, \\ u^s := u - u^i \ \text{ satisfies the Sommerfeld radiation condition.} \end{cases} \tag{9.29}$$

Let

$$\begin{aligned} u^i(x) &= e^{\sqrt{-1} k_m d \cdot x}, \\ F_{l,1}(x) &= -u^i(x)\big|_{\partial D_l} = -e^{\sqrt{-1} k_m d \cdot x}\big|_{\partial D_l}, \\ F_{l,2}(x) &= -\frac{\partial u^i}{\partial \nu}(x)\Big|_{\partial D_l} = -\sqrt{-1} k_m e^{\sqrt{-1} k_m d \cdot x} d \cdot \nu(x)\big|_{\partial D_l}, \end{aligned}$$

and define the operator $\mathcal{K}^k_{D_p, D_l}$ by

$$\mathcal{K}^k_{D_p, D_l}[\psi](x) = \int_{\partial D_p} \frac{\partial \Gamma_k(x-y)}{\partial \nu(x)} \psi(y) d\sigma(y), \quad x \in \partial D_l.$$

Analogously, we define

$$\mathcal{S}^k_{D_p, D_l}[\psi](x) = \int_{\partial D_p} \Gamma_k(x-y) \psi(y) d\sigma(y), \quad x \in \partial D_l.$$

The solution u of (9.29) can be represented as follows:

$$u(x) = \begin{cases} u^i(x) + \displaystyle\sum_{l=1}^{L} \mathcal{S}^{k_m}_{D_l}[\psi_l](x), & x \in \mathbb{R}^3 \backslash \bar{D}, \\ \displaystyle\sum_{l=1}^{L} \mathcal{S}^{k_c}_{D_l}[\phi_l](x), & x \in D, \end{cases}$$

where $\phi_l, \psi_l \in H^{-\frac{1}{2}}(\partial D_l)$ satisfy the following system of integral equations

$$\begin{cases} \mathcal{S}^{k_m}_{D_l}[\psi_l] - \mathcal{S}^{k_c}_{D_l}[\phi_l] + \displaystyle\sum_{p \neq l} \mathcal{S}^{k_m}_{D_p, D_l}[\psi_p] = F_{l,1}, \\ \dfrac{1}{\mu_m}\big(\dfrac{1}{2} I + (\mathcal{K}^{k_m}_{D_l})^*\big)[\psi_l] + \dfrac{1}{\mu_c}\big(\dfrac{1}{2} I - (\mathcal{K}^{k_c}_{D_l})^*\big)[\phi_l] \\ \qquad + \dfrac{1}{\mu_m} \displaystyle\sum_{p \neq l} \mathcal{K}^{k_m}_{D_p, D_l}[\psi_p] = F_{l,2}, \end{cases}$$

and

$$\begin{cases} F_{l,1} = -u^i & \text{on } \partial D_l, \\ F_{l,2} = -\dfrac{1}{\mu_m}\dfrac{\partial u^i}{\partial \nu} & \text{on } \partial D_l. \end{cases}$$

9.3.2. First-Order Correction to Plasmonic Resonances and Field Behavior at Plasmonic Resonances in the Multi-Particle Case. We consider the scattering in the quasi-static regime, *i.e.*, when the incident wavelength is much greater than one. With proper dimensionless analysis, we can assume that $\omega \ll 1$. As a consequence, $\mathcal{S}_D^{k_c}$ is invertible. Note that

$$\phi_l = (\mathcal{S}_{D_l}^{k_c})^{-1}\big(\mathcal{S}_{D_l}^{k_m}[\psi_l] + \sum_{p\neq l} \mathcal{S}_{D_p,D_l}^{k_m}[\psi_p] - F_{l,1}\big).$$

We obtain the following equation for ψ_l's,

$$\mathcal{A}_D(w)[\psi] = f,$$

where

$$\begin{aligned} \mathcal{A}_D(w) &= \begin{pmatrix} \mathcal{A}_{D_1}(\omega) & & & \\ & \mathcal{A}_{D_2}(\omega) & & \\ & & \ddots & \\ & & & \mathcal{A}_{D_L}(\omega) \end{pmatrix} \\ &\quad + \begin{pmatrix} 0 & \mathcal{A}_{1,2}(\omega) & \cdots & \mathcal{A}_{1,L}(\omega) \\ \mathcal{A}_{2,1}(\omega) & 0 & \cdots & \mathcal{A}_{2,L}(\omega) \\ \vdots & \cdots & 0 & \vdots \\ \mathcal{A}_{L,1}(\omega) & \cdots & \mathcal{A}_{L,L-1}(\omega) & 0 \end{pmatrix}, \end{aligned}$$

$$\psi = \begin{pmatrix} \psi_1 \\ \psi_2 \\ \vdots \\ \psi_L \end{pmatrix}, \quad f = \begin{pmatrix} f_1 \\ f_2 \\ \vdots \\ f_L \end{pmatrix},$$

with

$$\mathcal{A}_{D_l}(\omega) = \frac{1}{\mu_m}\big(\frac{1}{2}I + (\mathcal{K}_{D_l}^{k_m})^*\big) + \frac{1}{\mu_c}\big(\frac{1}{2}I - (\mathcal{K}_{D_l}^{k_c})^*\big)(\mathcal{S}_{D_l}^{k_c})^{-1}\mathcal{S}_{D_l}^{k_m}, \tag{9.30}$$

and

$$\begin{aligned} \mathcal{A}_{l,p}(\omega) &= \frac{1}{\mu_c}\big(\frac{1}{2}I - (\mathcal{K}_{D_l}^{k_c})^*\big)(\mathcal{S}_{D_l}^{k_c})^{-1}\mathcal{S}_{D_p,D_l}^{k_m} + \frac{1}{\mu_m}\mathcal{K}_{D_p,D_l}^{k_m}, \\ f_l &= F_{l,2} + \frac{1}{\mu_c}\big(\frac{1}{2}I - (\mathcal{K}_{D_l}^{k_c})^*\big)(\mathcal{S}_{D_l}^{k_c})^{-1}[F_{l,1}]. \end{aligned}$$

For $j = 1, \ldots, L$, let $\mathcal{H}^*(\partial D_j)$ and $\mathcal{H}(\partial D_j)$ be respectively defined by (2.18) and (2.20) with Ω replaced with D_j. We first consider the operators $\mathcal{S}_{D_j}^k$ and $(\mathcal{K}_{D_j}^k)^*$. The following asymptotic expansions hold. Its proof is immediate.

LEMMA 9.16. (i) *Regarded as operators from $\mathcal{H}^*(\partial D_j)$ into $\mathcal{H}(\partial D_j)$, we have*

$$\mathcal{S}_{D_j}^k = \mathcal{S}_{D_j}^0 + k\mathcal{S}_{D_j,1} + k^2\mathcal{S}_{D_j,2} + O(k^3\delta^3),$$

where $\mathcal{S}_{D_j}^0 = O(1)$ and $\mathcal{S}_{D_j,m} = O(\delta^m)$;

(ii) *Regarded as operators from* $\mathcal{H}(\partial D_j)$ *into* $\mathcal{H}^*(\partial D_j)$, *we have*

$$(\mathcal{S}_{D_j}^k)^{-1} = (\mathcal{S}_{D_j}^0)^{-1} + k\mathcal{B}_{D_j,1} + k^2\mathcal{B}_{D_j,2} + O(k^3\delta^3),$$

where $\mathcal{S}_{D_j}^{-1} = O(1)$ *and* $\mathcal{B}_{D_j,m} = O(\delta^m)$;

(iii) *Regarded as operators from* $\mathcal{H}^*(\partial D_j)$ *into* $\mathcal{H}^*(\partial D_j)$, *we have*

$$(\mathcal{K}_{D_j}^k)^* = (\mathcal{K}_{D_j}^0)^* + k^2 O(\delta^2),$$

where $(\mathcal{K}_{D_j}^0)^* = O(1)$.

We also need the following result.

LEMMA 9.17.

(i) *Regarded as an operator from* $\mathcal{H}^*(\partial D_j)$ *into* $\mathcal{H}(\partial D_l)$ *we have,*

$$\mathcal{S}_{D_j,D_l}^k = \mathcal{S}_{j,l,0,0} + \mathcal{S}_{j,l,0,1} + \mathcal{S}_{j,l,0,2} + k\mathcal{S}_{j,l,1} + k^2\mathcal{S}_{j,l,2,0} + O(\delta^4) + O(k^2\delta^2).$$

Moreover,

$$\mathcal{S}_{j,l,m,n} = O(\delta^{n+1}).$$

(ii) *Regarded as an operator from* $\mathcal{H}^*(\partial D_j)$ *into* $\mathcal{H}^*(\partial D_l)$, *we have*

$$\mathcal{K}_{D_j,D_l}^k = \mathcal{K}_{j,l,0,0} + O(k^2\delta^2).$$

Moreover,

$$\mathcal{K}_{j,l,0,0} = O(\delta^2).$$

Finally, the following asymptotic expansions hold.

LEMMA 9.18.

(i) *Regarded as operators from* $\mathcal{H}^*(\partial D_j)$ *into* $\mathcal{H}^*(\partial D_j)$, *we have*

$$\mathcal{A}_{D_j}(\omega) = \mathcal{A}_{D_j,0} + O(\delta^2\omega^2),$$

(ii) *Regarded as operators from* $\mathcal{H}^*(\partial D_l)$ *into* $\mathcal{H}^*(\partial D_p)$, *we have*

$$\mathcal{A}_{l,p}(\omega) = \frac{1}{\mu_c}\big(\frac{1}{2}I-(\mathcal{K}_{D_l}^0)^*\big)(\mathcal{S}_{D_l}^0)^{-1}\big(\mathcal{S}_{p,l,0,1}+\mathcal{S}_{p,l,0,2}\big)+\frac{1}{\mu_m}\mathcal{K}_{p,l,0,0}+O(\delta^2\omega^2)+O(\delta^4).$$

Moreover,

$$\begin{aligned}
\big(\frac{1}{2}I - (\mathcal{K}_{D_l}^0)^*\big) \circ (\mathcal{S}_{D_l}^0)^{-1} \circ \mathcal{S}_{p,l,0,1} &= O(\delta^2),\\
\big(\frac{1}{2}I - (\mathcal{K}_{D_l}^0)^*\big) \circ (\mathcal{S}_{D_l}^0)^{-1} \circ \mathcal{S}_{p,l,0,2} &= O(\delta^3),\\
\mathcal{K}_{p,l,0,0} &= O(\delta^2).
\end{aligned}$$

PROOF. The proof of (i) follows from Lemma 9.16. We now prove (ii). Recall that

$$\begin{aligned}
\frac{1}{2}I - (\mathcal{K}_{D_l}^{k_c})^* &= \frac{1}{2}I - (\mathcal{K}_{D_l}^0)^* + O(\delta^2\omega^2),\\
(\mathcal{S}_{D_l}^{k_c})^{-1} &= (\mathcal{S}_{D_l}^0)^{-1} - k_c(\mathcal{S}_{D_l}^0)^{-1}\mathcal{S}_{D_l,1}(\mathcal{S}_{D_l}^0)^{-1} + O(\delta^2\omega^2),\\
\mathcal{S}_{D_p,D_l}^{k_m} &= \mathcal{S}_{p,l,0,0} + \mathcal{S}_{p,l,0,1} + \mathcal{S}_{p,l,0,2} + k_m\mathcal{S}_{p,l,1} + k_m^2\mathcal{S}_{p,l,2,0} + O(\delta^4) + O(\omega^2\delta^2)\\
\mathcal{K}_{D_p,D_l}^{k_m} &= \mathcal{K}_{p,l,0,0} + O(\omega^2\delta^2).
\end{aligned}$$

Using the identity

$$\big(\frac{1}{2}I - (\mathcal{K}_{D_l}^0)^*\big)(\mathcal{S}_{D_l}^0)^{-1}[\chi(D_l)] = 0,$$

we can derive that

$$\begin{aligned}\mathcal{A}_{l,p}(\omega) &= \frac{1}{\mu_c}(\frac{1}{2}I - (\mathcal{K}_{D_l}^0)^*)(\mathcal{S}_{D_l}^{k_c})^{-1}\mathcal{S}_{D_p,D_l}^{k_m} + \frac{1}{\mu_m}\mathcal{K}_{p,l,0,0} + O(\delta^2\omega^2)\\ &= \frac{1}{\mu_c}(\frac{1}{2}I - (\mathcal{K}_{D_l}^0)^*)(\mathcal{S}_{D_l}^{0})^{-1}\mathcal{S}_{D_p,D_l}^{k_m} + \frac{1}{\mu_m}\mathcal{K}_{p,l,0,0} + O(\delta^2\omega^2)\\ &= \frac{1}{\mu_c}(\frac{1}{2}I - (\mathcal{K}_{D_l}^0)^*)(\mathcal{S}_{D_l}^{0})^{-1}\big(\mathcal{S}_{p,l,0,0} + \mathcal{S}_{p,l,0,1} + \mathcal{S}_{p,l,0,2} + k_m\mathcal{S}_{p,l,1}\\ &\quad +k_m^2\mathcal{S}_{p,l,2,0} + O(\delta^4)\big) + \frac{1}{\mu_m}\mathcal{K}_{p,l,0,0} + O(\delta^2\omega^2)\\ &= \frac{1}{\mu_c}(\frac{1}{2}I - (\mathcal{K}_{D_l}^0)^*)(\mathcal{S}_{D_l}^{0})^{-1}\big(\mathcal{S}_{p,l,0,1} + \mathcal{S}_{p,l,0,2}\big) + \frac{1}{\mu_m}\mathcal{K}_{p,l,0,0} + O(\delta^2\omega^2)\\ &\quad +O(\delta^4).\end{aligned}$$

The rest of the lemma follows from Lemmas 9.16 and 9.17. □

Denote by $\mathcal{H}^*(\partial D) = \mathcal{H}^*(\partial D_1) \times \ldots \times \mathcal{H}^*(\partial D_L)$, which is equipped with the inner product

$$\langle \psi, \phi \rangle_{\mathcal{H}^*} = \sum_{l=1}^{L} \langle \psi_l, \phi_l \rangle_{\mathcal{H}^*(\partial D_l)}.$$

With the help of Lemma 9.18, the following result is obvious.

LEMMA 9.19. *Regarded as an operator from $\mathcal{H}^*(\partial D)$ into $\mathcal{H}^*(\partial D)$, we have*

$$\mathcal{A}(\omega) = \mathcal{A}_{D,0} + \mathcal{A}_{D,1} + O(\omega^2\delta^2) + O(\delta^4),$$

where

$$\mathcal{A}_{D,0} = \begin{pmatrix} \mathcal{A}_{D_1,0} & & & \\ & \mathcal{A}_{D_2,0} & & \\ & & \cdots & \\ & & & \mathcal{A}_{D_L,0} \end{pmatrix},$$

and

$$\mathcal{A}_{D,1} = \begin{pmatrix} 0 & \mathcal{A}_{D,1,12} & \mathcal{A}_{D,1,13} & \cdots \\ \mathcal{A}_{D,1,21} & 0 & \mathcal{A}_{D,1,23} & \cdots \\ & & \cdots & \\ \mathcal{A}_{D,1,L1} & \cdots & \mathcal{A}_{D,1,LL-1} & 0 \end{pmatrix}$$

with

$$\begin{aligned}\mathcal{A}_{D_l,0} &= (\frac{1}{2\mu_m} + \frac{1}{2\mu_c})I - (\frac{1}{\mu_c} - \frac{1}{\mu_m})(\mathcal{K}_{D_l}^0)^*,\\ \mathcal{A}_{D,1,pq} &= \frac{1}{\mu_c}(\frac{1}{2}I - (\mathcal{K}_{D_p}^0)^*)(\mathcal{S}_{D_p}^0)^{-1}\big(\mathcal{S}_{q,p,0,1} + \mathcal{S}_{q,p,0,2}\big) + \frac{1}{\mu_m}\mathcal{K}_{q,p,0,0}.\end{aligned}$$

It is evident that

$$\mathcal{A}_{D,0}[\psi] = \sum_{j=0}^{\infty}\sum_{l=1}^{L} \tau_j \langle \psi, \varphi_{j,l}\rangle_{\mathcal{H}^*}\varphi_{j,l}, \tag{9.31}$$

where

$$\tau_j = \frac{1}{2\mu_m} + \frac{1}{2\mu_c} - (\frac{1}{\mu_c} - \frac{1}{\mu_m})\lambda_j, \tag{9.32}$$

$$\varphi_{j,l} = \varphi_j e_l \tag{9.33}$$

with e_l being the standard basis of $\mathbb{R}^L$ and (λ_j, φ_j) being the eigenvalues and associated eigenfunctions of the operator $(\mathcal{K}_{D_1}^0)^*$.

We consider $\mathcal{A}(\omega)$ as a perturbation to the operator $\mathcal{A}_{D,0}$ for small ω and small δ. Using a standard perturbation argument, we can derive the perturbed eigenvalues and eigenfunctions.

In what follows, we only use the first order perturbation theory and derive the leading order term, *i.e.*, the perturbation due to the term $\mathcal{A}_{D,1}$. For each l, we define an $L \times L$ matrix R_l by letting

$$\begin{aligned} R_{l,pq} &= \langle \mathcal{A}_{D,1}[\varphi_{l,p}], \varphi_{l,q} \rangle_{\mathcal{H}^*}, \\ &= \left\langle \mathcal{A}_{D,1}[\varphi_l e_p], \varphi_l e_q \right\rangle_{\mathcal{H}^*}, \\ &= \langle \mathcal{A}_{D,1,pq}[\varphi_l], \varphi_l \rangle_{\mathcal{H}^*}. \end{aligned}$$

LEMMA 9.20. *The matrix $R_l = (R_{l,pq})_{p,q=1,\dots,L}$ has the following explicit expression:*

$$\begin{aligned} R_{l,pp} &= 0, \\ R_{l,pq} &= \frac{3}{4\pi\mu_c}(\lambda_j - \frac{1}{2}) \sum_{|\alpha|=|\beta|=1} \int_{\partial D_0} \int_{\partial D_0} \frac{(z_p - z_q)^{\alpha+\beta}}{|z_p - z_q|^5} x^\alpha y^\beta \varphi_l(x)\varphi_l(y) d\sigma(x) d\sigma(y) \\ &\quad + (\frac{1}{4\pi\mu_c} - \frac{1}{4\pi\mu_m})(\lambda_j - \frac{1}{2}) \int_{\partial D_0} \int_{\partial D_0} \frac{x \cdot y}{|z_p - z_q|^3} \varphi_l(x)\varphi_l(y) d\sigma(x) d\sigma(y) \\ &= O(\delta^3), \quad p \neq q. \end{aligned}$$

PROOF. It is clear that $R_{l,pp} = 0$. For $p \neq q$, we have

$$R_{l,pq} = R_{l,pq}^I + R_{l,pq}^{II} + R_{l,pq}^{III},$$

where

$$\begin{aligned} R_{l,pq}^I &= \frac{1}{\mu_c} \left\langle (\frac{1}{2}I - (\mathcal{K}_{D_p}^0)^*)(\mathcal{S}_{D_p}^0)^{-1} \mathcal{S}_{q,p,0,1}[\varphi_l], \varphi_l \right\rangle_{\mathcal{H}^*(\partial D_l)}, \\ R_{l,pq}^{II} &= \frac{1}{\mu_c} \left\langle (\frac{1}{2}I - (\mathcal{K}_{D_p}^0)^*)(\mathcal{S}_{D_p}^0)^{-1} \mathcal{S}_{q,p,0,2}[\varphi_l], \varphi_l \right\rangle_{\mathcal{H}^*(\partial D_l)}, \\ R_{l,pq}^{III} &= \frac{1}{\mu_m} \langle \mathcal{K}_{q,p,0,0}[\varphi_l], \varphi_l \rangle_{\mathcal{H}^*(\partial D_l)}. \end{aligned}$$

We first consider $R_{l,pq}^I$. By the following identity

$$(\frac{1}{2}I - (\mathcal{K}_{D_p}^0)^*)\mathcal{S}_{D_l}[\varphi_l] = \mathcal{S}_{D_l}^0(\frac{1}{2}I - \mathcal{K}_{D_p}^0)[\varphi_l] = (\lambda_j - \frac{1}{2})\varphi_l,$$

we obtain

$$\begin{aligned} R_{l,pq}^I &= -\frac{1}{\mu_c} \left\langle (\frac{1}{2}I - (\mathcal{K}_{D_p}^0)^*)(\mathcal{S}_{D_p}^0)^{-1} \mathcal{S}_{q,p,0,1}[\varphi_l], \mathcal{S}_{D_l}^0[\varphi_l] \right\rangle_{L^2(\partial D_l)}, \\ &= \frac{1}{\mu_c}(\lambda_j - \frac{1}{2}) \langle \mathcal{S}_{q,p,0,1}[\varphi_l], \mathcal{S}_{D_l}^0[\varphi_l] \rangle_{L^2(\partial D_l)}. \end{aligned}$$

Using the explicit representation of $\mathcal{S}_{q,p,0,1}$, we further conclude that

$$R_{l,pq}^I = 0.$$

Similarly, we have

$$\begin{aligned}
\frac{\mu_c}{\lambda_j-\frac{1}{2}}R^{II}_{l,pq} &= \big\langle \mathcal{S}_{q,p,0,2}[\varphi_l], \mathcal{S}^0_{D_l}[\varphi_l]\big\rangle_{L^2(\partial D_l)},\\
&= \sum_{|\alpha|=|\beta|=1}\int_{\partial D_0}\int_{\partial D_0}\Big(\frac{3(z_p-z_q)^{\alpha+\beta}}{4\pi|z_p-z_q|^5}x^\alpha y^\beta+\frac{\delta_{\alpha\beta}x^\alpha y^\beta}{4\pi|z_p-z_q|^3}\Big)\varphi_l(x)\varphi_l(y)d\sigma(x)d\sigma(y)\\
&= \frac{3}{4\pi}\sum_{|\alpha|=|\beta|=1}\int_{\partial D_0}\int_{\partial D_0}\frac{(z_p-z_q)^{\alpha+\beta}}{|z_p-z_q|^5}x^\alpha y^\beta\varphi_l(x)\varphi_l(y)d\sigma(x)d\sigma(y)\\
&\quad+\frac{1}{4\pi}\sum_{|\alpha|=1}\int_{\partial D_0}\int_{\partial D_0}\frac{1}{|z_p-z_q|^3}x^\alpha y^\alpha\varphi_l(x)\varphi_l(y)d\sigma(x)d\sigma(y).
\end{aligned}$$

Finally, note that

$$\mathcal{K}_{q,p,0,0}[\varphi_l]=\frac{1}{4\pi|z_p-z_q|^3}a\cdot\nu(x)=\frac{1}{4\pi|z_p-z_q|^3}\sum_{m=1}^{3}a_m\nu_m(x),$$

where $a_m=\big\langle (y-z_q)_m,\varphi_l\big\rangle_{L^2(\partial D_q)}$, and $a=(a_1,a_2,a_3)^t$.

Therefore, we have

$$\begin{aligned}
R^{III}_{l,pq} &= -\frac{1}{\mu_m}\langle \mathcal{K}_{q,p,0,0}[\varphi_l],\varphi_l\rangle_{\mathcal{H}^*(\partial D_l)}\\
&= -\frac{1}{4\pi|z_p-z_q|^3\mu_m}\langle a\cdot\nu(x),\varphi_l\rangle_{\mathcal{H}^*(\partial D_l)}\\
&= -\frac{1}{4\pi|z_p-z_q|^3\mu_m}\Big\langle\big(\frac{1}{2}I-(\mathcal{K}^0_{D_p})^*\big)(\mathcal{S}^0_{D_p})^{-1}(a\cdot(x-z_p)),\varphi_l\Big\rangle_{\mathcal{H}^*(\partial D_l)}\\
&= -\frac{1}{4\pi|z_p-z_q|^3\mu_m}(\lambda_j-\frac{1}{2})\langle a\cdot(x-z_p),\varphi_l\rangle_{L^2(\partial D_p)}\\
&= -\frac{1}{4\pi|z_p-z_q|^3\mu_m}(\lambda_j-\frac{1}{2})\int_{\partial D_0}\int_{\partial D_0}x\cdot y\varphi_l(x)\varphi_l(y)d\sigma(x)d\sigma(y).
\end{aligned}$$

This completes the proof of the lemma. □

We now have an explicit formula for the matrix R_l. It is clear that R_l is symmetric, but not self-adjoint. For ease of presentation, we assume the following condition.

CONDITION 9.21. *R_l has L-distinct eigenvalues.*

We remark that Condition 9.21 is not essential for our analysis. Without this condition, the perturbation argument is still applicable, but the results may be quite complicated. We refer to [**326**] for a complete description of the perturbation theory.

Let $\tau_{j,l}$ and $X_{j,l}=(X_{j,l,1},\dots,X_{j,l,L})^t$, $l=1,2,\dots,L$, be the eigenvalues and normalized eigenvectors of the matrix R_j. We remark that each $X_{j,l}$ may be complex valued and may not be orthogonal to other eigenvectors.

Under perturbation, each τ_j splits into the following L eigenvalues of $\mathcal{A}(\omega)$,

$$\tau_{j,l}(\omega)=\tau_j+\tau_{j,l}+O(\delta^4)+O(\omega^2\delta^2). \tag{9.34}$$

The associated perturbed eigenfunctions have the following form

$$\varphi_{j,l}(\omega) = \sum_{p=1}^{L} X_{j,l,p} e_p \varphi_j + O(\delta^4) + O(\omega^2\delta^2). \tag{9.35}$$

We are interested in solving the equation $\mathcal{A}_D(\omega)[\psi] = f$ when ω is close to the resonance frequencies, *i.e.*, when $\tau_j(\omega)$ are very small for some j's. In this case, the major part of the solution would be based on the excited resonance modes $\varphi_{j,l}(\omega)$.

For this purpose, we introduce the index set of resonances J just as we did in (7.12) for the single particle case.

DEFINITION 9.22. *We call $J \subset \mathbb{N}$ index set of resonances if the $\tau_{j,l}(\omega)$'s are close to zero when $j \in J$ and are bounded from below when $j \in J^c$. More precisely, we choose a threshold number $\eta_0 > 0$ independent of ω such that*

$$|\tau_{j,l}(\omega)| \geq \eta_0 > 0 \quad \textit{for } j \in J^c.$$

For simplicity, we assume that the following conditions hold.

CONDITION 9.23. *Each eigenvalue λ_j, $j \in J$, of the operator $(\mathcal{K}_{D_1}^0)^*$ is simple. Moreover, we have $\omega^2 \ll \delta$.*

We define

$$P_J(\omega)\varphi_{j,m}(\omega) = \begin{cases} \varphi_{j,m}(\omega), & j \in J, \\ 0, & j \in J^c. \end{cases}$$

In fact,

$$P_J(\omega) = \sum_{j\in J} P_j(\omega) = \sum_{j\in J} \frac{1}{2\pi\sqrt{-1}} \int_{\gamma_j} (\xi - \mathcal{A}_D(\omega))^{-1} d\xi, \tag{9.36}$$

where γ_j is a Jordan curve in the complex plane enclosing only the eigenvalues $\tau_{j,l}(\omega)$ for $l = 1, 2, \ldots, L$ among all the eigenvalues.

To obtain an explicit representation of $P_J(\omega)$, we consider the adjoint operator $\mathcal{A}_D(\omega)^*$. By a similar perturbation argument, we can obtain its perturbed eigenvalue and eigenfunctions. Note that the adjoint matrix $\bar{R}_j^t = \bar{R}_j$ has eigenvalues $\overline{\tau_{j,l}}$ and corresponding eigenfunctions $\overline{X_{j,l}}$. Then the eigenvalues and eigenfunctions of $\mathcal{A}_D(\omega)^*$ have the following form

$$\begin{aligned} \widetilde{\tau}_{j,l}(\omega) &= \tau_j + \overline{\tau_{j,l}} + O(\delta^4) + O(\omega^2\delta^2), \\ \widetilde{\varphi}_{j,l}(\omega) &= \widetilde{\varphi}_{j,l} + O(\delta^4) + O(\omega^2\delta^2), \end{aligned}$$

where

$$\widetilde{\varphi}_{j,l} = \sum_{p=1}^{L} \widetilde{X}_{j,l,p} e_p \varphi_j$$

with $\widetilde{X}_{j,l,p}$ being a multiple of $\overline{X_{j,l,p}}$.

We normalize $\widetilde{\varphi}_{j,l}$ in a way such that the following holds:

$$\langle \varphi_{j,p}, \widetilde{\varphi}_{j,q} \rangle_{\mathcal{H}^*(\partial D)} = \delta_{pq},$$

which is also equivalent to the following condition

$$\overline{X_{j,p}}^t \widetilde{X}_{j,q} = \delta_{pq}.$$

Then, we can show that the following result holds.

LEMMA 9.24. *In the space $\mathcal{H}^*(\partial D)$, as ω goes to zero, we have*

$$f = \omega f_0 + O(\omega^2 \delta^{\frac{3}{2}}),$$

where $f_0 = (f_{0,1}, \dots, f_{0,L})^t$ with

$$\begin{aligned} f_{0,l} &= -\sqrt{-1}\sqrt{\varepsilon_m \mu_m} e^{\sqrt{-1}k_m d \cdot z_l} \Big(\frac{1}{\mu_m} d \cdot \nu(x) + \frac{1}{\mu_c} \big(\frac{1}{2} I - (\mathcal{K}_{D_l}^0)^*\big)(\mathcal{S}_{D_l}^0)^{-1}[d \cdot (x - z)] \Big) \\ &= O(\delta^{\frac{3}{2}}). \end{aligned}$$

PROOF. We first show that

$$\|u\|_{\mathcal{H}^*(\partial D_0)} = \delta^{\frac{3}{2}+m} \|u\|_{\mathcal{H}^*(\partial \tilde{D})}, \quad \|u\|_{\mathcal{H}(\partial D_0)} = \delta^{\frac{1}{2}+m} \|u\|_{\mathcal{H}(\partial \tilde{D})}$$

for any homogeneous function u such that $u(\delta x) = \delta^m u(x)$. Indeed, we have $\eta(u)(x) = \delta^m u(x)$. Since $\|\eta(u)\|_{\mathcal{H}^*(\partial \tilde{D})} = \delta^{-\frac{3}{2}} \|u\|_{\mathcal{H}^*(\partial D_0)}$, we obtain

$$\|u\|_{\mathcal{H}^*(\partial D_0)} = \delta^{\frac{3}{2}} \|\eta(u)\|_{\mathcal{H}^*(\partial \tilde{D})} = \delta^{\frac{3}{2}+m} \|u\|_{\mathcal{H}^*(\partial \tilde{D})},$$

which proves our first claim. The second claim follows in a similar way. Using this result, we arrive at the desired asymptotic result. □

Denote by $Z = (Z_1, \dots, Z_L)$, where $Z_j = \sqrt{-1} k_m e^{\sqrt{-1} k_m d \cdot z_j}$. We are ready to present our main result in this section.

THEOREM 9.25. *Under Conditions 9.13, 9.14, 9.15, and 9.21, the scattered field due to L plasmonic particles in the quasi-static regime has the following representation*

$$u^s = \sum_l^L \mathcal{S}_{D_l}^{k_m}[\psi_l],$$

where $\psi = (\psi_1, \dots, \psi_L)^t$ has the following asymptotic expansion:

$$\begin{aligned} \psi &= \sum_{j \in J} \sum_{l=1}^L \frac{\langle f, \widetilde{\varphi}_{j,l}(\omega) \rangle_{\mathcal{H}^*} \varphi_{j,l}(\omega)}{\tau_{j,l}(\omega)} + \mathcal{A}_D(\omega)^{-1}(P_{J^c}(\omega) f) \\ &= \sum_{j \in J} \sum_{l=1}^L \frac{\langle d \cdot \nu(x), \varphi_j \rangle_{\mathcal{H}^*(\partial D_0)} Z \overline{\widetilde{X}_{j,l}} \varphi_{j,l} + O(\omega^2 \delta^{\frac{3}{2}})}{\lambda - \lambda_j + \big(\frac{1}{\mu_c} - \frac{1}{\mu_m}\big)^{-1} \tau_{j,l} + O(\delta^4) + O(\delta^2 \omega^2)} + O(\omega \delta^{\frac{3}{2}}). \end{aligned}$$

PROOF. The proof is similar to that of Theorem 7.15. □

As a consequence, the following result holds.

COROLLARY 9.26. *With the same notation as in Theorem 9.25 and under the additional condition that*

$$\min_{j \in J} |\tau_{j,l}(\omega)| \gg \omega^q \delta^p,$$

for some integer p and q, and

$$\tau_{j,l}(\omega) = \tau_{j,l,p,q} + o(\omega^q \delta^p),$$

we have

$$\psi = \sum_{j \in J} \sum_{l=1}^L \frac{\langle d \cdot \nu(x), \varphi_j \rangle_{\mathcal{H}^*(\partial D_0)} Z \overline{\widetilde{X}_{j,l}} \varphi_{j,l} + O(\omega^2 \delta^{\frac{3}{2}})}{\tau_{j,l,p,q}} + O(\omega \delta^{\frac{3}{2}}).$$

9.3.3. Super-Resolution by Using Plasmonic Nanoparticles.

9.3.3.1. *Asymptotic Expansion of the Scattered Field.* In order to illustrate the super-focusing phenomenon, we set

$$u^i(x) = \Gamma_{k_m}(x - x_0) = -\frac{e^{\sqrt{-1}k_m|x-x_0|}}{4\pi|x - x_0|}.$$

LEMMA 9.27. *In the space $\mathcal{H}^*(\partial D)$, as ω goes to zero, we have*

$$f = f_0 + O(\omega\delta^{\frac{3}{2}}) + O(\delta^{\frac{5}{2}}),$$

where $f_0 = (f_{0,1}, \dots, f_{0,L})^t$ with

$$\begin{aligned} f_{0,l} &= -\frac{1}{4\pi|z_l - x_0|^3}\Big(\frac{1}{\mu_m}(z_l - x_0)\cdot\nu(x) + \frac{1}{\mu_c}(\frac{1}{2}I - (\mathcal{K}^0_{D_l})^*)(\mathcal{S}^0_{D_l})^{-1}[(z_l - x_0)\cdot(x - z_l)]\Big) \\ &= O(\delta^{\frac{3}{2}}). \end{aligned}$$

PROOF. Recall that

$$f_l = F_{l,2} + \frac{1}{\mu_c}(\frac{1}{2}I - (\mathcal{K}^{k_c}_{D_l})^*)(\mathcal{S}^{k_c}_{D_l})^{-1}[F_{l,1}].$$

We can show that

$$F_{l,2} = -\frac{1}{\mu_m}\frac{\partial u^i}{\partial \nu} = -\frac{1}{4\pi\mu_m|z_l - x_0|^3}(z_l - x_0)\cdot\nu(x) + O(\delta^{\frac{5}{2}}) + O(\omega\delta^{\frac{3}{2}}) \quad \text{in } \mathcal{H}^*(\partial D_l).$$

Besides,

$$u^i(x)|_{\partial D_l} = -\frac{e^{\sqrt{-1}k_m|z_l-x_0|}}{4\pi|z_l - x_0|}\chi(\partial D_l) + \frac{1}{4\pi|z_l - x_0|^3}(z_l - x_0)\cdot(x - z_l) + O(\delta^{\frac{5}{2}}) + O(\omega\delta^{\frac{3}{2}})$$

in $\mathcal{H}(\partial D_l)$.

Using the identity $(\frac{1}{2}I - (\mathcal{K}^0_{D_l})^*)(\mathcal{S}^0_{D_l})^{-1}[\chi(\partial D_l)] = 0$, we obtain that

$$\frac{1}{\mu_c}(\frac{1}{2}I - (\mathcal{K}^{k_c}_{D_l})^*)(\mathcal{S}^{k_c}_{D_l})^{-1}[F_{l,1}] = -\frac{1}{4\pi|z_l - x_0|^3\mu_c}(\frac{1}{2}I - (\mathcal{K}^0_{D_l})^*)(\mathcal{S}^0_{D_l})^{-1}[(z_l - x_0)\cdot(x - z_l)].$$

This completes the proof of the lemma. □

We now derive an asymptotic expansion of the scattered field in an intermediate regime which is neither too close to the plasmonic particles nor too far away. More precisely, we consider the following domain

$$\mathrm{D}_{\delta,k} = \{x \in \mathbb{R}^3;\ \min_{1\le l\le L}|x - z_l| \gg \delta,\ \max_{1\le l\le L}|x - z_l| \ll \frac{1}{k}\}.$$

LEMMA 9.28. *Let $\psi_l \in \mathcal{H}^*(\partial D_l)$ and let $v(x) = \mathcal{S}^k_{D_l}[\psi_l](x)$. Then we have for $x \in \mathrm{D}_{\delta,k}$,*

$$\begin{aligned} v(x) &= \Gamma_k(x - z_l)\Big(\frac{1}{|x - z_l|} - \sqrt{-1}k\Big)\frac{x - z_l}{|x - z_l|}\cdot\int_{\partial D_0} y\psi_l(y)d\sigma(y) + O(\delta^{\frac{5}{2}})\|\psi_l\|_{\mathcal{H}^*(\partial D_l)} \\ &\quad + \Gamma_k(x - z_l)\int_{\partial D_0}\psi_l(y)d\sigma(y). \end{aligned}$$

Moreover, the following estimates hold

$$\begin{aligned} v(x) &= O(\delta^{\frac{3}{2}}) \quad \text{if } \int_{\partial D_0}\psi_l(y)d\sigma(y) = 0, \\ v(x) &= O(\delta^{\frac{1}{2}}) \quad \text{if } \int_{\partial D_0}\psi_l(y)d\sigma(y) \neq 0. \end{aligned}$$

PROOF. We only consider the case when $l = 0$. The other case follows similarly or by coordinate translation. We have

$$v(x) = \mathcal{S}_D^k[\psi](x) = \int_{\partial D_0} \Gamma_k(x-y)\psi(y)d\sigma(y) = -\int_{\partial D_0} \frac{e^{\sqrt{-1}k|x-y|}}{4\pi|x-y|}\psi(y)d\sigma(y).$$

Since

$$\Gamma_k(x-y) = \Gamma_k(x) + \sum_{|\alpha=1|} \frac{\partial \Gamma_k(x)}{\partial y^\alpha} y^\alpha + \sum_{m\geq 2}\sum_{|\alpha=m|} \frac{\partial^m \Gamma_k(x)}{\partial y^\alpha} y^\alpha,$$

and

$$\frac{\partial \Gamma_k(x)}{\partial y^\alpha} = -\frac{e^{\sqrt{-1}k|x|}}{4\pi|x|}\Big(\frac{1}{|x|} - \sqrt{-1}k\Big)\frac{x}{|x|} = \Gamma_k(x)\Big(\frac{1}{|x|} - \sqrt{-1}k\Big)\frac{x^\alpha}{|x|},$$

we obtain the required identity for the case $l = 0$. The estimate follows from the fact that

$$\|y^\alpha\|_{\mathcal{H}(\partial D_0)} = O(\delta^{\frac{2|\alpha|+1}{2}}).$$

This completes the proof of the lemma. □

Denote by

$$\begin{aligned}
S_{j,l}(x,k) &= \Gamma_k(x-z_l)\frac{x-z_l}{|x-z_l|^2} \cdot \int_{\partial D_0} y\varphi_j(y)d\sigma(y),\\
S_l(x,k) &= \Gamma_k(x-z_l)\int_{\partial D_0} \varphi_0(y)d\sigma(y),\\
H_{j,l}(x_0) &= -\frac{1}{4\pi|z_l-x_0|^3}\langle (z_l-x_0)\cdot\nu(x), \varphi_j\rangle_{\mathcal{H}^*(\partial D_0)}.
\end{aligned}$$

It is clear that the following size estimates hold

$$S_{j,l}(x,k) = O(\delta^{\frac{3}{2}}), \quad S_l(x,k) = O(\delta^{\frac{1}{2}}), \quad H_{j,l}(x_0) = O(\delta^{\frac{3}{2}}) \quad \text{for } j\neq 0, \quad H_{0,l}(x_0) = 0.$$

THEOREM 9.29. *Under Conditions 9.13, 9.14, 9.15, and 9.21, the Green function $\Phi_{k_m}(x,x_0)$ in the presence of L plasmonic particles has the following representation in the quasi-static regime: for $x \in \mathrm{D}_{\delta,k_m}$,*

$$\begin{aligned}
\Phi_{k_m}(x,x_0) &= \Gamma_{k_m}(x-x_0)\\
&\quad + \sum_{j\in J}\sum_{l=1}^{L} \frac{H_{j,p}(x_0)\widetilde{X}_{j,l,p}X_{j,l,q}S_{j,q}(x,k_m) + O(\delta^4) + O(\omega\delta^3)}{\lambda - \lambda_j + \big(\frac{1}{\mu_c} - \frac{1}{\mu_m}\big)^{-1}\tau_{j,l} + O(\delta^4) + O(\delta^2\omega^2)} + O(\delta^3).
\end{aligned}$$

PROOF. With $u^i(x) = \Gamma_{k_m}(x-x_0)$, we have

$$\psi = \sum_{j\in J}\sum_{1\leq l\leq L} a_{j,l}\varphi_{j,l} + \sum_{1\leq l\leq L} a_{0,l}\varphi_{0,l} + O(\delta^{\frac{3}{2}}),$$

where

$$\begin{aligned}
a_{j,l} &= \langle f, \widetilde{\varphi}_{j,l}\rangle_{\mathcal{H}^*(\partial D)} = \langle f_0, \widetilde{\varphi}_{j,l}\rangle_{\mathcal{H}^*(\partial D)} + O(\omega\delta^{\frac{3}{2}}) + O(\delta^{\frac{5}{2}}),\\
&= \Big(\frac{1}{\mu_c} - \frac{1}{\mu_m}\Big)\widetilde{X}_{j,l,p}H_{j,p}(x_0) + O(\omega\delta^{\frac{3}{2}}) + O(\delta^{\frac{5}{2}}),\\
a_{0,l} &= \langle f, \widetilde{\varphi}_{0,l}\rangle_{\mathcal{H}^*(\partial D)} = O(\delta^{\frac{5}{2}}).
\end{aligned}$$

By Lemma 9.28,

$$\begin{aligned}\mathcal{S}_D^{k_m}[\varphi_{j,l}](x) &= \sum_{1\le p\le L} \mathcal{S}_D^{k_m}[X_{j,l,p}\varphi_j e_p](x) = \sum_{1\le p\le L} X_{j,l,p}\mathcal{S}_{D_p}^{k_m}[\varphi_j](x)\\ &= \sum_{1\le p\le L} X_{j,l,p}S_{j,p}(x,k_m) + O(\delta^{\frac{5}{2}}) + O(\omega\delta^{\frac{3}{2}}).\end{aligned}$$

On the other hand, for $j=0$, we have

$$\begin{aligned}\mathcal{S}_D^{k_m}[\varphi_{0,l}](x) &= O(\delta^{\frac{1}{2}}),\\ \tau_{0,l}(\omega) &= \tau_0 + O(\delta^4) + O(\delta^2\omega^2) = O(1).\end{aligned}$$

Therefore, we can deduce that

$$\begin{aligned}u^s &= \mathcal{S}_D^{k_m}[\psi](x) = \sum_{j\in J}\sum_{1\le l\le L} a_{j,l}\mathcal{S}_D^{k_m}[\varphi_{j,l}] + \sum_{1\le l\le L} a_{0,l}\mathcal{S}_D^{k_m}[\varphi_{0,l}] + O(\delta^3),\\ &= \sum_{j\in J}\sum_{l=1}^{L} \frac{1}{\tau_{j,l}(\omega)}\Big((\frac{1}{\mu_c}-\frac{1}{\mu_m})H_{j,p}(x_0)\widetilde{X}_{j,l,p}X_{j,l,q}S_{j,q}(x,k_m) + O(\omega\delta^3)\\ &\quad +O(\delta^4)\Big) + O(\delta^3),\\ &= \sum_{j\in J}\sum_{l=1}^{L} \frac{H_{j,p}(x_0)\widetilde{X}_{j,l,p}X_{j,l,q}S_{j,q}(x,k_m) + O(\omega\delta^3) + O(\delta^4)}{\lambda-\lambda_j + \big(\frac{1}{\mu_c}-\frac{1}{\mu_m}\big)^{-1}\tau_{j,l} + O(\delta^4) + O(\delta^2\omega^2)} + O(\delta^3).\end{aligned}$$

□

9.3.4. Asymptotic Expansion of the Imaginary Part of the Green's Function. As a consequence of Theorem 9.29, we obtain the following result on the imaginary part of the Green function.

THEOREM 9.30. *Assume the same conditions as in Theorem 9.29. Under the additional assumption that*

$$\begin{aligned}\lambda-\lambda_j + (\frac{1}{\mu_c}-\frac{1}{\mu_m})^{-1}\tau_{j,l} &\gg O(\delta^4) + O(\delta^2\omega^2),\\ \Re\left(\lambda-\lambda_j + (\frac{1}{\mu_c}-\frac{1}{\mu_m})^{-1}\tau_{j,l}\right) &\lesssim \Im\left(\lambda-\lambda_j + (\frac{1}{\mu_c}-\frac{1}{\mu_m})^{-1}\tau_{j,l}\right)\end{aligned}$$

for each l and $j\in J$, we have

$$\begin{aligned}\Im\Phi_{k_m}(x,x_0) &= \Im\Gamma_{k_m}(x-x_0) + O(\delta^3) +\\ &\quad \sum_{j\in J}\sum_{l=1}^{L} \Re\left(H_{j,p}(x_0)\widetilde{X}_{j,l,p}X_{j,l,q}S_{j,q}(x,0) + O(\omega\delta^3) + O(\delta^4)\right)\\ &\quad \times\Im\left(\frac{1}{\lambda-\lambda_j + \big(\frac{1}{\mu_c}-\frac{1}{\mu_m}\big)^{-1}\tau_{j,l}}\right),\end{aligned}$$

where $x, x_0 \in \mathrm{D}_{\delta,k_m}$.

Note that $\Re\left(H_{j,p}(x_0)\widetilde{X}_{j,l,p}X_{j,l,q}S_{j,q}(x,0)\right) = O(\delta^3)$. Under the conditions in Theorem 9.30, if we have additionally that

$$\Im\left(\frac{1}{\lambda-\lambda_j+\left(\frac{1}{\mu_c}-\frac{1}{\mu_m}\right)^{-1}\tau_{j,l}}\right)=O(\frac{1}{\delta^3})$$

for some plasmonic frequency ω, then the term in the expansion of $\Im\Phi_{k_m}(x,x_0)$ which is due to resonance has size one and exhibits a sub-wavelength peak with width of order one. This breaks the diffraction limit $1/k_m$ in the free space. We also note that the term $\Im\Gamma_{k_m}(x-x_0)$ has size $O(\omega)$. Thus, we can conclude that super-focusing (or super-resolution) can indeed be achieved by using a system of plasmonic particles.

REMARK 9.31. *The results of this section on the super-resolution effect obtained by using a finite system of plasmonic nanoparticles can be extended to the full Maxwell equations.*

9.4. Super-Resolution Based on Scattering Tensors

9.4.1. Multipolar Expansion. In this subsection we use the same notation as in Section 8.2. We consider for the sake of simplicity the two-dimensional case. We let $B_R := \{|x| < R\}$ and D be a small particle located at $z \in B_R$ with electromagnetic parameters ε_c and μ_c. We set $x_i, i = 1,\dots,N$ to be equi-distributed points along the boundary ∂B_R for $N \gg 1$. The array of N elements $\{x_1,\dots,x_N\}$ is used to detect the particle. The array of elements $\{x_1,\dots,x_N\}$ is operating both in transmission and in reception. Let u_j^s be the wave scattered by $D = z+\delta B$ corresponding to the incident wave $\Gamma_{k_m}(x-x_j)$.

From the multipolar expansion (2.266), it follows that

$$u_j^s(x) = \sum_{|l'|=0}^{n+1}\sum_{|l|=0}^{n-|l'|+1}\frac{\delta^{|l|+|l'|}}{l!l'!}\partial^l\Gamma_{k_m}(x-z)\partial_z^{l'}\Gamma_{k_m}(x_j-z)\widetilde{W}_{ll'} + O(\delta^{n+2}), \tag{9.37}$$

where $\widetilde{W}_{ll'}$ is the scattering tensor defined by (2.265).

Therefore, the entries a_{ij}^{ω} of the data matrix A^{ω} introduced in (8.1) can be approximated as follows:

$$a_{ij}^{\omega} = g(x_i,z)\mathcal{W}g(x_j,z)^t + O(\delta^{n+2}), \tag{9.38}$$

where $g(x_i,z)$ is a row vector of size $(n+1)\times(n+2)/2$, which is given by

$$g(x_i,z) = \left(\frac{1}{l!}\partial^l\Gamma_{k_m}(x_i-z)\right)_{|l|\le n}, \tag{9.39}$$

and $\mathcal{W}$ is defined by

$$\mathcal{W} = \left(\delta^{|l|+|l'|}\widetilde{W}_{ll'}\right)_{|l|,|l'|\le n}. \tag{9.40}$$

If δ is small, then high-order terms in (9.37) can be neglected. In this case, the analysis of the data matrix reduces to the one in Section 8.2, which is based on the dipolar approximation (2.282). As δ increases, more and more multipolar terms could be included in formula (9.38) in order to approximate the data matrix. For fixed δ, the number of multipolar terms (or the maximal resolving order) which can be robustly reconstructed from the measured data depends only on the signal-to-noise ratio and can be estimated as a function of the signal-to-noise ratio [**87**].

In view of (9.39), the signal space of the data matrix A^ω becomes richer. The set of the singular vectors consists of the Green function and its derivatives. In order to locate the particle, exactly the same imaging functionals constructed in Section 8.2 can be used. They peak at the location of the particle. However, the significant singular values are perturbed, even those associated with the dipolar approximation (2.282). Indeed, when δ is increasing, new significant singular values can merge. Those are related to higher-order multipolar terms. They can be expressed in terms of the scattering tensors $\widetilde{W}_{ll'}$. These new singular values, which are intermediate between the three larger ones (in the case of a single particle) and zero, contain some information on the particle and give a better approximation of its shape and electromagnetic parameters.

9.4.2. Reconstruction Procedure. In this subsection we first present formulas for the reconstruction of the size and material parameters of the particle from the data matrix A^ω. Then we introduce optimal control and dictionary-based matching approaches to identify the shape of the particle.

Once the location z of the particle is estimated for instance by using MUSIC algorithm, the matrix $\mathcal{W}$ can be recovered from the data matrix A^ω. The size $|D|$ and the electric permittivity ε can be estimated as follows:

$$|D| = \frac{1}{k_m^2} |\widetilde{W}_{(2,0),(0,0)}|$$

and

$$\varepsilon_c = \frac{|\widetilde{W}_{(2,0),(0,0)}|}{\omega^2 \mu_m |D|}.$$

By (2.274) we know that $(\widetilde{W}_{ll'})_{|l|=|l'|=1}$ is approximately the polarization tensor associated with D and λ given by (2.272). Therefore, by (2.107) and (2.108) an equivalent ellipse can be computed and an estimate of μ_c can be found.

Once the equivalent ellipse is reconstructed we can use it as an initial guess and minimize using an optimal control scheme the discrepancy between the computed and measured matrix $\mathcal{W}$; see [**56**]. The level set method can be implemented in order to reconstruct separately closely spaced particles [**45**].

In [**87**], a dictionary matching approach is proposed. It is an alternative to the optimal control approach. It relies on learning the geometric features contained in the matrix data. In the dictionary matching approach, we identify and classify a particle, knowing in advance that the latter belongs to a certain collection of particles. The method relies on computing the invariants under rigid motions from the extracted scattering tensors and allows us to handle the scaling within certain ranges. A particle is classified by comparing its invariants with those of a set of learned shapes at multiple-frequencies. The larger the frequency band used, the better the classification performance of the dictionary matching algorithm.

The reconstruction results obtained by the optimal control method and the dictionary matching approach are far beyond the resolution limit.

9.5. Concluding Remarks

In this chapter, we have provided a mathematical theory to explain the super-resolution mechanism in high-contrast media. We have investigated the behavior of the Green's functions of high-contrast media. Our resonance expansions of the Green's functions, which were first derived in [**95**], are the key to mathematically explaining the super-resolution mechanism in high-contrast media. From (9.15) and (9.17), we have proved that the super-resolution is due to propagating subwavelength resonant modes. It is worth mentioning that in (9.15) and (9.17), we have observed that a phenomenon of mixing of modes occurs. This is essentially due to the non-hermitian nature of the operator $\mathtt{K}_D$. Finally, it is worth mentioning that the results in this chapter can be used to analyse the Mie resonances of dielectric particles [**517**].

Part 4

Metamaterials

CHAPTER 10

Near-Cloaking

10.1. Introduction

To cloak a target is to make it invisible with respect to probing by electromagnetic or elastic waves. Extensive work has been produced on cloaking in the context of electromagnetic and elastic waves. Many schemes for cloaking are currently under active investigation. These include interior cloaking, where the cloaking region is inside the cloaking device, and exterior cloaking in which the cloaking region is outside the cloaking device.

In this chapter, we focus on interior cloaking and describe effective near-cloaking structures for electromagnetic and elastic scattering problems. The focus of the next chapter will be placed on exterior cloaking.

In interior cloaking, the difficulty is to construct material parameter distributions of a cloaking structure such that any target placed inside the structure is undetectable to waves. One approach is to use transformation optics (also called the scheme of changing variables) [**179, 273, 284, 339, 459, 500**]. The principle behind transformation optics is to use a coordinate transformation to derive the spatial dependent material parameters to guide the wave. Transformation optics takes advantage of the fact that the equations governing electromagnetic and acoustic wave propagation have transformation laws under change of variables; see Subsection 2.14.4. They are form invariant under coordinate transformations. This allows one to design structures that bend waves around a hidden region, returning them to their original path on the far side. The change of variables based cloaking method uses a singular transformation to boost the material properties so that it makes a cloaking region look like a point to outside measurements. However, this transformation induces the singularity of material constants in the transversal direction (also in the tangential direction in two dimensions), which causes difficulty both in the theory and applications. To overcome this weakness, so called 'near-cloaking' is naturally considered, which is a regularization or an approximation of singular cloaking. Instead of the singular transformation, one can use a regular one to push forward the material parameters, in which a small ball is blown up to the cloaking region [**334, 335**]. Enhanced cloaking can be achieved by using a cancellation technique [**63, 64**]. The first stage of this approach involves designing a multi-coated structure around a small perfect insulator to significantly reduce its effect on boundary or scattering cross-section measurements. The multi-coating cancels the generalized polarization tensors or the scattering coefficients of the cloaking device. One then obtains a near-cloaking structure by pushing forward the multi-coated structure around a small object via the standard blow-up transformation technique.

The purpose of this chapter is to review the cancellation technique. We first design a structure coated around a particle to have vanishing generalized polarization tensors of lower orders and show that the order of perturbation due to a small particle can be reduced significantly. We then obtain a near-cloaking structure by pushing forward the multi-coated structure around a small object via the usual blow-up transformation.

When considering near-cloaking for the Helmholtz equation, we construct structures such that their first scattering coefficients vanish. Analogously to the quasi-static limit, we prove that, after applying transformation optics, structures with vanishing scattering coefficients enhance near-cloaking. We emphasize that such a structure achieves near-cloaking for a band of frequencies. We also show that near-cloaking for the Helmholtz equation becomes increasingly difficult as the cloaked object becomes bigger or the operating frequency becomes higher. The difficulty scales inversely proportionally to the object diameter or the frequency.

Finally, the cancellation technique is extended to the full Maxwell equations and the Lamé system.

The results of this chapter are from [**1**, **63**–**65**].

10.2. Near-Cloaking in the Quasi-Static Limit

To explain the principle of construction of cloaking structures, we review the results on the quasi-static model obtained in [**63**].

Let Ω be a domain in $\mathbb{R}^2$ containing 0 possibly with multiple components with smooth boundary. For a contrast λ, recall that the generalized polarization tensors $M_{\alpha\beta}(\lambda, \Omega)$ associated with Ω and λ are defined in (2.69).

Let $P_m(x)$ be the complex-valued polynomial

$$P_m(x) = (x_1 + \sqrt{-1}x_2)^m := \sum_{|\alpha|=m} a_\alpha^m x^\alpha + \sqrt{-1} \sum_{|\beta|=m} b_\beta^m x^\beta \,. \tag{10.1}$$

Using polar coordinates $x = re^{\sqrt{-1}\theta}$, the above coefficients a_α^m and b_β^m can also be characterized by

$$\sum_{|\alpha|=m} a_\alpha^m x^\alpha = r^m \cos m\theta, \text{ and } \sum_{|\beta|=m} b_\beta^m x^\beta = r^m \sin m\theta \,. \tag{10.2}$$

We introduce the following combination of generalized polarization tensors using the coefficients in (10.1):

$$M_{mn}^{cc} = \sum_{|\alpha|=m} \sum_{|\beta|=n} a_\alpha^m a_\beta^n M_{\alpha\beta} \,, \tag{10.3}$$

$$M_{mn}^{cs} = \sum_{|\alpha|=m} \sum_{|\beta|=n} a_\alpha^m b_\beta^n M_{\alpha\beta} \,, \tag{10.4}$$

$$M_{mn}^{sc} = \sum_{|\alpha|=m} \sum_{|\beta|=n} b_\alpha^m a_\beta^n M_{\alpha\beta} \,, \tag{10.5}$$

$$M_{mn}^{ss} = \sum_{|\alpha|=m} \sum_{|\beta|=n} b_\alpha^m b_\beta^n M_{\alpha\beta} \,. \tag{10.6}$$

For a given harmonic function H in $\mathbb{R}^2$, consider

$$\begin{cases} \nabla\cdot\Big(\sigma_0\chi(\mathbb{R}^2\setminus\overline{\Omega})+\sigma\chi(\Omega)\Big)\nabla u=0 & \text{in } \mathbb{R}^2\,, \\ u(x)-H(x)=O(|x|^{-1}) & \text{as } |x|\to\infty\,, \end{cases} \tag{10.7}$$

where σ_0 and σ are conductivities (positive constants) of $\mathbb{R}^2\setminus\overline{\Omega}$ and Ω, respectively.

If the harmonic function H admits the expansion

$$H(x)=H(0)+\sum_{n=1}^{\infty} r^n\big(a_n^c(H)\cos n\theta+a_n^s(H)\sin n\theta\big)$$

with $x=(r\cos\theta, r\sin\theta)$, then, we have the following formula

$$\begin{aligned}(u-H)(x)=&-\sum_{m=1}^{\infty}\frac{\cos m\theta}{2\pi m r^m}\sum_{n=1}^{\infty}\big(M_{mn}^{cc}a_n^c(H)+M_{mn}^{cs}a_n^s(H)\big)\\ &-\sum_{m=1}^{\infty}\frac{\sin m\theta}{2\pi m r^m}\sum_{n=1}^{\infty}\big(M_{mn}^{sc}a_n^c(H)+M_{mn}^{ss}a_n^s(H)\big)\quad \text{as } r=|x|\to\infty\,,\end{aligned} \tag{10.8}$$

where $M_{mn}^{cc}, M_{mn}^{cs}, M_{mn}^{sc}$, and M_{mn}^{sc} are defined by (10.3)–(10.6).

In order to make u look like H for large $|x|$, we construct structures with vanishing generalized polarization tensors for all $|n|, |m| \le N$. We call such structures GPT-vanishing structures of order N. To do so, we use a disc with multiple coatings. Let Ω be a disc of radius r_1. For a positive integer N, let $0<r_{N+1}<r_N<\cdots<r_1$ and define

$$A_j := \{r_{j+1}<r=|x|\le r_j\},\quad j=1,2,\ldots,N\,. \tag{10.9}$$

Let $A_0=\mathbb{R}^2\setminus\overline{\Omega}$ and $A_{N+1}=\{r\le r_{N+1}\}$. Set σ_j to be the conductivity of A_j for $j=1,2,\ldots,N+1$, and $\sigma_0=1$. Let

$$\sigma=\sum_{j=0}^{N+1}\sigma_j\chi(A_j)\,. \tag{10.10}$$

Because of the symmetry of the disc, one can easily see that

$$M_{mn}^{cs}[\sigma]=M_{mn}^{sc}[\sigma]=0 \quad \text{for all } m,n\,, \tag{10.11}$$

$$M_{mn}^{cc}[\sigma]=M_{mn}^{ss}[\sigma]=0 \quad \text{if } m\ne n\,, \tag{10.12}$$

and

$$M_{nn}^{cc}[\sigma]=M_{nn}^{ss}[\sigma] \quad \text{for all } n\,. \tag{10.13}$$

Let $M_n=M_{nn}^{cc}$, $n=1,2,\ldots$, for the simplicity of notation. Let

$$\zeta_j := \frac{\sigma_j-\sigma_{j-1}}{\sigma_j+\sigma_{j-1}},\quad j=1,\ldots,N+1\,. \tag{10.14}$$

One can prove that [**63**]

$$|M_n|\le 2\pi n r_1^{2n} \quad \text{for all } n\in\mathbb{N}. \tag{10.15}$$

The following is a characterization of GPT-vanishing structures. Again, see [**63**].

PROPOSITION 10.1. *If there are nonzero constants* $\zeta_1, \dots, \zeta_{N+1}$ ($|\zeta_j| < 1$) *and* $r_1 > \cdots > r_{N+1} > 0$ *such that*

$$\prod_{j=1}^{N+1} \begin{bmatrix} 1 & \zeta_j r_j^{-2l} \\ \zeta_j r_j^{2l} & 1 \end{bmatrix} \text{ is an upper triangular matrix for } l = 1, 2, \dots, N, \tag{10.16}$$

then (Ω, σ), *given by* (10.9), (10.10), *and* (10.14), *is a GPT-vanishing structure of order* N, *i.e.,* $M_l = 0$ *for* $l \leq N$. *More generally, if there are nonzero constants* $\zeta_1, \zeta_2, \zeta_3, \dots$ ($|\zeta_j| < 1$) *and* $r_1 > r_2 > r_3 > \dots$ *such that* r_n *converges to a positive number, say* $r_\infty > 0$, *and*

$$\prod_{j=1}^{\infty} \begin{bmatrix} 1 & \zeta_j r_j^{-2l} \\ \zeta_j r_j^{2l} & 1 \end{bmatrix} \text{ is an upper triangular matrix for every } l, \tag{10.17}$$

then (Ω, σ), *given by* (10.9), (10.10), *and* (10.14), *is a GPT-vanishing structure with* $M_l = 0$ *for all* l.

Let (Ω, σ) be a GPT-vanishing structure of order N of the form (10.10). We take $r_1 = 2$ so that Ω is the disk of radius 2, and $r_{N+1} = 1$. We assume that $\sigma_{N+1} = 0$ which amounts to the structure being insulated along ∂B_1. For small $\delta > 0$, let

$$\Psi_{\frac{1}{\delta}}(x) = \frac{1}{\delta} x, \quad x \in \mathbb{R}^2 . \tag{10.18}$$

Then, $(B_{2\delta}, \sigma \circ \Psi_{\frac{1}{\delta}})$ is a GPT-vanishing structure of order N and it is insulated on ∂B_δ.

For a given domain Ω and a subdomain $B \Subset \Omega$, we introduce the Dirichlet-to-Neumann map $\Lambda_{\Omega,B}[\sigma]$ as

$$\Lambda_{\Omega,B}[\sigma](f) = \sigma \frac{\partial u}{\partial \nu}\bigg|_{\partial\Omega}, \tag{10.19}$$

where u is the solution to

$$\begin{cases} \nabla \cdot \sigma \nabla u = 0 & \text{in } \Omega \setminus \overline{B}, \\ \dfrac{\partial u}{\partial \nu} = 0 & \text{on } \partial B, \\ u = f & \text{on } \partial \Omega, \end{cases} \tag{10.20}$$

where ν is the outward normal to ∂B. Note that with $\Omega = B_2$, $\Lambda_{\Omega,B_\delta}[\sigma \circ \Psi_{\frac{1}{\delta}}]$ may be regarded as small perturbation of $\Lambda_{\Omega,\emptyset}[1]$ if $M_l = 0$ for all $l \leq N$. A complete asymptotic expansion of $\Lambda_{\Omega,B_\delta}[\sigma \circ \Psi_{\frac{1}{\delta}}]$ as $\delta \to 0$ can be obtained and it can be proved that

$$\left\| \Lambda_{B_2,B_\delta}\left[\sigma \circ \Psi_{\frac{1}{\delta}}\right] - \Lambda_{B_2,\emptyset}[1] \right\| \leq C\delta^{2N+2} \tag{10.21}$$

for some constant C independent of δ, where the norm is the operator norm from $H^{1/2}(\partial\Omega)$ into $H^{-1/2}(\partial\Omega)$. In fact, if f admits the Fourier expansion $f(\theta) = \sum_{n\in\mathbb{Z}} f_n e^{\sqrt{-1}n\theta}$, then we have

$$\left(\Lambda_{B_2,B_\delta}\left[\sigma \circ \Psi_{\frac{1}{\delta}}\right] - \Lambda_{B_2,\emptyset}[1]\right)(f) = \sum_{n\in\mathbb{Z}} \frac{|n|(\delta/2)^{2|n|} M_{|n|}}{2\pi|n| - (\delta/2)^{2|n|} M_{|n|}} f_n e^{\sqrt{-1}n\theta}.$$

From (10.15) , it follows that as $\delta \to 0$,

$$|M_n|\delta^{2n} \leq C\delta^{2N+2} \quad \text{for all } n \in \mathbb{N}$$

for some constant C independent of n and hence, (10.21) holds; see [**63**]. We then push forward $\sigma \circ \Psi_{\frac{1}{\delta}}$ by the change of variables F_δ,

$$F_\delta(x) := \begin{cases} \Big(\dfrac{3-4\delta}{2(1-\delta)} + \dfrac{1}{4(1-\delta)}|x|\Big)\dfrac{x}{|x|} & \text{for } 2\delta \leq |x| \leq 2\,, \\ \Big(\dfrac{1}{2} + \dfrac{1}{2\delta}|x|\Big)\dfrac{x}{|x|} & \text{for } \delta \leq |x| \leq 2\delta\,, \\ \dfrac{x}{\delta} & \text{for } |x| \leq \delta\,, \end{cases} \tag{10.22}$$

in other words,

$$(F_\delta)_\star(\sigma \circ \Psi_{\frac{1}{\delta}}) = \frac{(DF_\delta)(\sigma \circ \Psi_{\frac{1}{\delta}})(DF_\delta)^t}{|\det(DF_\delta)|} \circ F_\delta^{-1}\,. \tag{10.23}$$

Note that F_δ maps $|x| = \delta$ onto $|x| = 1$, and is the identity on $|x| = 2$. So by invariance of the Dirichlet-to-Neumann map, we have

$$\Lambda_{B_2,B_1}\Big[(F_\delta)_\star(\sigma \circ \Psi_{\frac{1}{\delta}})\Big] = \Lambda_{B_2,B_\delta}\Big[\sigma \circ \Psi_{\frac{1}{\delta}}\Big]\,. \tag{10.24}$$

Identity (10.24) can be proved using the divergence theorem [**179**]. Thus we obtain the following theorem from [**63**], which shows that, using GPT-vanishing structures we achieve enhanced near-cloaking.

THEOREM 10.2. *Let the conductivity profile σ be a GPT-vanishing structure of order N such that $\sigma_{N+1} = 0$. There exists a constant C independent of δ such that*

$$\Big\|\Lambda_{B_2,B_1}\Big[(F_\delta)_\star(\sigma \circ \Psi_{\frac{1}{\delta}})\Big] - \Lambda_{B_2,\emptyset}[1]\Big\| \leq C\delta^{2N+2}\,. \tag{10.25}$$

REMARK 10.3. *It is worth emphasizing that the conductivities of the constructed near-cloaking devices are anisotropic. Nevertheless, they can be approximated by concentric isotropic homogeneous coatings* [**44, 459**]. *The approximation is in the sense that it minimizes the discrepancy between the associated Dirichlet-to-Neumann maps for only the first few eigenvectors.*

REMARK 10.4. *If we consider the spectra of the Laplacian with Dirichlet or Neumann boundary conditions inside on one hand the near-cloaking device of order N and on the other hand the homogeneous disk of conductivity one, then one can show that the first eigenvalues are approximately the same (up to an error of order of δ^{2N+2}).*

10.3. Near-Cloaking for the Helmholtz Equation

Analogously to the quasi-static case, in order to achieve enhanced near-cloaking for the Helmholtz equation, we construct multi-coated structures such that their first scattering coefficients vanish. Then by pushing forward the multi-coated structures via the transformation optics, we obtain enhanced cloaking with respect to scattering cross-section measurements.

10.3.1. Scattering Coefficients. Let D be a bounded domain in $\mathbb{R}^2$ with smooth boundary ∂D, and let (ε_0, μ_0) be the pair of electromagnetic parameters (permittivity and permeability) of $\mathbb{R}^2 \setminus \bar{D}$ and (ε_c, μ_c) be that of D. Then the permittivity and permeability distributions are given by

$$\varepsilon = \varepsilon_0 \chi(\mathbb{R}^2 \setminus \bar{D}) + \varepsilon_c \chi(D) \quad \text{and} \quad \mu = \mu_0 \chi(\mathbb{R}^2 \setminus \bar{D}) + \mu_c \chi(D) . \tag{10.26}$$

Given a frequency ω, set $k_c = \omega\sqrt{\varepsilon_c \mu_c}$ and $k_0 = \omega\sqrt{\varepsilon_0 \mu_0}$. For a function u^i satisfying $(\Delta + k_0^2)u^i = 0$ in $\mathbb{R}^2$, we consider the scattered wave u, *i.e.*, the solution to (2.172).

Suppose that u^i is given by a plane wave $e^{\sqrt{-1}k_0 \xi \cdot x}$ with ξ being on the unit circle, then (2.228) yields (2.233), where W_{nm}, given by (2.219), are the scattering coefficients, $\xi = (\cos\theta_\xi, \sin\theta_\xi)$, and $x = (|x|, \theta_x)$.

Let far-field pattern $A_\infty[\varepsilon, \mu, \omega]$, when the incident field is given by $e^{\sqrt{-1}k_0 \xi \cdot x}$, be defined by (2.234). As shown in Theorem 2.79, the scattering coefficients are the Fourier coefficients of $A_\infty[\varepsilon, \mu, \omega]$.

The scattering cross-section $Q^s[\varepsilon, \mu, \omega]$ is defined by

$$Q^s[\varepsilon, \mu, \omega](\theta') := \int_0^{2\pi} \left| A_\infty[\varepsilon, \mu, \omega](\theta, \theta') \right|^2 d\theta . \tag{10.27}$$

It is worth recalling that the optical theorem (Theorem 2.81) leads to a natural constraint on W_{nm}. In fact, (2.247) or equivalently (2.248) holds.

In the next subsection, we compute the scattering coefficients of multiply coated inclusions and provide structures whose scattering coefficients vanish. Such structures will be used to enhance near-cloaking. Any target placed inside such structures will have nearly vanishing scattering cross-section Q^s, uniformly in the direction θ'.

10.3.2. S-Vanishing Structures. The purpose of this subsection is to construct multiply layered structures whose scattering coefficients vanish. We call such structures *S-vanishing structures.* We design a multi-coating around an insulated inclusion D, for which the scattering coefficients vanish. The computations of the scattering coefficients of multi-layered structures (with multiple phase electromagnetic materials) follow in exactly the same way as in Subsection 2.10.4. The system of two equations (2.178) should be replaced by a system of $2\times$ the number of phase interfaces (-1 if the core is perfectly insulating).

For positive numbers $r_1, \ldots, r_{L+1}$ with $2 = r_1 > r_2 > \cdots > r_{L+1} = 1$, let

$$A_j := \{x : r_{j+1} \le |x| < r_j\}, \quad j = 1, \ldots, L, \quad A_0 := \mathbb{R}^2 \setminus \overline{A_1} ,$$

and

$$A_{L+1}(= D) := \{x : |x| < 1\} .$$

Let (μ_j, ε_j) be the pair of permeability and permittivity of A_j for $j = 0, 1, \ldots, L+1$. Set $\mu_0 = 1$ and $\varepsilon_0 = 1$. Let

$$\mu = \sum_{j=0}^{L+1} \mu_j \chi(A_j) \quad \text{and} \quad \varepsilon = \sum_{j=0}^{L+1} \varepsilon_j \chi(A_j) . \tag{10.28}$$

In this case the scattering coefficient $W_{nm} = W_{nm}[\mu, \varepsilon, \omega]$ can be defined using (2.227). In fact, if u is the solution to

$$\nabla \cdot \frac{1}{\mu} \nabla u + \omega^2 \varepsilon u = 0 \qquad \text{in } \mathbb{R}^2 \tag{10.29}$$

with the outgoing radiation condition on $u - U$ where U is given by (2.226), then $u - U$ admits the asymptotic expansion (2.227) with $k_0 = \omega\sqrt{\varepsilon_0\mu_0}$.

Exactly like the conductivity case, one can show using symmetry that

$$W_{nm} = 0 \quad \text{if } m \neq n. \tag{10.30}$$

Let us define W_n by

$$W_n := W_{nn}\,. \tag{10.31}$$

Our purpose is to design, given N and ω, material parameters μ and ε so that $W_n[\mu, \varepsilon, \omega] = 0$ for $|n| \leq N$. We call such a structure (μ, ε) an *S-vanishing structure of order N at frequency ω*. Since $H^{(1)}_{-n} = (-1)^n H^{(1)}_n$ and $J_{-n} = (-1)^n J_n$, we have

$$W_{-n} = W_n\,, \tag{10.32}$$

and hence it suffices to consider W_n only for $n \geq 0$.

Moreover, from Lemma 2.76 it follows that there exists δ_0 such that, for all $\delta \leq \delta_0$,

$$|W_n[\varepsilon, \mu, \delta\omega]| \leq \frac{C^{2n}}{n^{2n}}\delta^{2n} \quad \text{for all } n \in \mathbb{N} \setminus \{0\}\,, \tag{10.33}$$

where the constant C depends on $(\varepsilon, \mu, \omega)$ but is independent of δ.

Furthermore, note that (2.248) leads to

$$\Im \sum_{n\in\mathbb{Z}} W_n[\varepsilon, \mu, \omega] = -\sqrt{\frac{\pi\omega}{2}} \sum_{n\in\mathbb{Z}} \left| W_n[\varepsilon, \mu, \omega] \right|^2. \tag{10.34}$$

Let $k_j := \omega\sqrt{\mu_j\varepsilon_j}$ for $j = 0, 1, \ldots, L$. We assume that $\mu_{L+1} = +\infty$, which amounts to the solution satisfying the zero Neumann condition on $|x| = r_{L+1} (= 1)$. To compute W_n for $n \geq 0$, we look for solutions u_n to (10.29) of the form

(10.35)
$$u_n(x) = a_j^{(n)} J_n(k_j r) e^{\sqrt{-1}n\theta} + b_j^{(n)} H_n^{(1)}(k_j r) e^{\sqrt{-1}n\theta}, \quad x \in A_j, \quad j = 0, \ldots, L\,,$$

with $a_0^{(n)} = 1$. Note that

$$W_n = 4\sqrt{-1}\, b_0^{(n)}\,. \tag{10.36}$$

The solution u_n satisfies the transmission conditions

$$u_n|_+ = u_n|_- \quad \text{and} \quad \frac{1}{\mu_{j-1}} \frac{\partial u_n}{\partial \nu}\Big|_+ = \frac{1}{\mu_j} \frac{\partial u_n}{\partial \nu}\Big|_- \quad \text{on } |x| = r_j$$

for $j = 1, \ldots, L$, which reads

$$\begin{bmatrix} J_n(k_j r_j) & H_n^{(1)}(k_j r_j) \\ \sqrt{\dfrac{\varepsilon_j}{\mu_j}} J_n'(k_j r_j) & \sqrt{\dfrac{\varepsilon_j}{\mu_j}} \left(H_n^{(1)}\right)'(k_j r_j) \end{bmatrix} \begin{bmatrix} a_j^{(n)} \\ b_j^{(n)} \end{bmatrix}$$
$$= \begin{bmatrix} J_n(k_{j-1} r_j) & H_n^{(1)}(k_{j-1} r_j) \\ \sqrt{\dfrac{\varepsilon_{j-1}}{\mu_{j-1}}} J_n'(k_{j-1} r_j) & \sqrt{\dfrac{\varepsilon_{j-1}}{\mu_{j-1}}} \left(H_n^{(1)}\right)'(k_{j-1} r_j) \end{bmatrix} \begin{bmatrix} a_{j-1}^{(n)} \\ b_{j-1}^{(n)} \end{bmatrix}. \tag{10.37}$$

The Neumann condition $\frac{\partial u_n}{\partial \nu}\big|_+ = 0$ on $|x| = r_{L+1}$ amounts to

$$\begin{bmatrix} 0 & 0 \\ J_n'(k_L) & \left(H_n^{(1)}\right)'(k_L) \end{bmatrix} \begin{bmatrix} a_L^{(n)} \\ b_L^{(n)} \end{bmatrix} = \begin{bmatrix} 0 \\ 0 \end{bmatrix}. \tag{10.38}$$

Combining (10.37) and (10.38), we obtain

$$\begin{bmatrix} 0 \\ 0 \end{bmatrix} = P^{(n)}[\varepsilon, \mu, \omega] \begin{bmatrix} a_0^{(n)} \\ b_0^{(n)} \end{bmatrix}, \tag{10.39}$$

where

$$P^{(n)}[\varepsilon, \mu, \omega] := \begin{bmatrix} 0 & 0 \\ p_{21}^{(n)} & p_{22}^{(n)} \end{bmatrix} = (-\frac{\pi}{2}\sqrt{-1}\omega)^L \left(\prod_{j=1}^L \mu_j r_j\right) \begin{bmatrix} 0 & 0 \\ J_n'(k_L) & \left(H_n^{(1)}\right)'(k_L) \end{bmatrix}$$
$$\times \prod_{j=1}^L \begin{bmatrix} \sqrt{\frac{\varepsilon_j}{\mu_j}}\left(H_n^{(1)}\right)'(k_j r_j) & -H_n^{(1)}(k_j r_j) \\ -\sqrt{\frac{\varepsilon_j}{\mu_j}} J_n'(k_j r_j) & J_n(k_j r_j) \end{bmatrix}$$
$$\times \begin{bmatrix} J_n(k_{j-1} r_j) & H_n^{(1)}(k_{j-1} r_j) \\ \sqrt{\frac{\varepsilon_{j-1}}{\mu_{j-1}}} J_n'(k_{j-1} r_j) & \sqrt{\frac{\varepsilon_{j-1}}{\mu_{j-1}}}\left(H_n^{(1)}\right)'(k_{j-1} r_j) \end{bmatrix}.$$

In order to have a structure whose scattering coefficients W_n vanishes up to the Nth order, we need to have $b_0^{(n)} = 0$ (when $a_0^{(n)} = 1$) for $n = 0, \ldots, N$, which amounts to

$$p_{21}^{(n)} = 0 \quad \text{for } n = 0, \ldots, N, \tag{10.40}$$

because of (10.39). We emphasize that $p_{22}^{(n)} \neq 0$. In fact, if $p_{22}^{(n)} = 0$, then (10.39) can be fulfilled with $a_0^{(n)} = 0$ and $b_0^{(n)} = 1$. It means that there exists (μ, ε) on $\mathbb{R}^2 \setminus D$ such that the following problem has a solution:

$$\begin{cases} \nabla \cdot \dfrac{1}{\mu} \nabla u + \omega^2 \varepsilon u = 0 & \text{in } \mathbb{R}^2 \setminus \overline{D}, \\ \dfrac{\partial u}{\partial \nu}\Big|_+ = 0 \quad \text{on } \partial D, \\ u(x) = H_n^{(1)}(k_0 r) e^{\sqrt{-1} n \theta} & \text{for } |x| = r > 2, \end{cases} \tag{10.41}$$

which is not possible.

We note that (10.40) is a set of conditions on (μ_j, ε_j) and r_j for $j = 1, \ldots, L$. In fact, $p_{21}^{(n)}$ is a nonlinear algebraic function of μ_j, ε_j and r_j, $j = 1, \ldots, L$. We are not able to show existence of (μ_j, ε_j) and r_j, $j = 1, \ldots, L$, satisfying (10.40) even if it is quite important to do so. But the solutions (at fixed frequency) can be computed numerically in the same way as in the conductivity case.

We now consider the S-vanishing structure for all (low) frequencies. Let ω be fixed and we look for a structure (μ, ε) such that

$$W_n[\mu, \varepsilon, \delta\omega] = 0 \quad \text{for all } |n| \leq N \text{ and } \delta \leq \delta_0 \tag{10.42}$$

for some δ_0. Such a structure may not exist. Even numerically, it does not seem to exist. So instead we look for a structure such that

$$W_n[\mu, \varepsilon, \delta\omega] = o(\delta^{2N}) \quad \text{for all } |n| \leq N \text{ and } \delta \to 0. \tag{10.43}$$

We call such a structure an *S-vanishing structure of order N at low frequencies.*

To investigate the behavior of $W_n[\mu, \varepsilon, \delta\omega]$ as $\delta \to 0$, we need the asymptotic expansions of Bessel functions for small arguments. We have

$$H_\nu^{(1)}(x) \approx -\frac{\sqrt{-1}2^\nu\Gamma(\nu)}{\pi} x^{-\nu} \quad \text{for fixed } \nu \text{ and } x \to 0\,. \tag{10.44}$$

For $n \in \mathbb{N}$, we also recall that, as $x \to 0$,

$$J_n(x) = \frac{x^n}{2^n}\left(\frac{1}{\Gamma(n+1)} - \frac{\frac{1}{4}x^2}{\Gamma(n+2)} + \frac{(\frac{1}{4}x^2)^2}{2!\Gamma(n+3)} - \frac{(\frac{1}{4}x^2)^3}{3!\Gamma(n+4)} + \dots\right), \tag{10.45}$$

$$\begin{aligned} Y_n(x) = &-\frac{(\frac{1}{2}x)^{-n}}{\pi}\sum_{l=0}^{n-1}\frac{(n-l-1)!}{l!}(\frac{1}{4}x^2)^l + \frac{2}{\pi}\ln(\frac{1}{2}x)J_n(x) \\ &- \frac{(\frac{1}{2}x)^n}{\pi}\sum_{l=0}^{\infty}(\psi(l+1)+\psi(n+l+1))\frac{(-\frac{1}{4}x^2)^l}{l!(n+l)!}, \end{aligned} \tag{10.46}$$

where $\psi(1) = -\gamma$ and

$$\psi(n) = -\gamma + \sum_{l=1}^{n-1}\frac{1}{l} \quad \text{for } n \geq 2$$

with γ being the Euler constant.

Plugging formulas (10.45) and (10.46) into (10.39), we obtain

$$\begin{aligned} P^{(0)}[\varepsilon, \mu, \delta\omega] &= (-\frac{\pi}{2}\sqrt{-1}\delta\omega)^L\left(\prod_{j=1}^L \mu_j r_j\right)\begin{bmatrix} 0 & 0 \\ -\frac{k_L}{2}\delta + O(\delta^3) & \frac{2\sqrt{-1}}{\pi k_L}\delta^{-1} + O(\delta\ln\delta)\end{bmatrix} \\ &\times \prod_{j=1}^L \begin{bmatrix} \frac{2\sqrt{-1}}{\pi\omega\mu_j r_j}\delta^{-1} + O(\delta\ln\delta) & \frac{4}{\pi^2}\left(\frac{1}{\omega\mu_{j-1}r_j} - \frac{1}{\omega\mu_j r_j}\right)\frac{\ln\delta}{\delta} + O(\delta^{-1}) \\ \frac{r_j}{2}\omega\varepsilon_j\left(1 - \frac{\varepsilon_{j-1}}{\varepsilon_j}\right)\delta + O(\delta^3) & \frac{2i}{\pi\omega\mu_{j-1}r_j}\delta^{-1} + O(\delta\ln\delta)\end{bmatrix} \\ &= \delta^{-1}\begin{bmatrix} 0 & 0 \\ O(\delta^2) & \frac{2\sqrt{-1}}{\pi k_L}\prod_{j=1}^L\frac{\mu_j}{\mu_{j-1}} + O(\delta)\end{bmatrix}, \end{aligned} \tag{10.47}$$

and, for $n \geq 1$,

$$\begin{aligned} P^{(n)}[\varepsilon, \mu, \delta\omega] &= (-\sqrt{-1}\tfrac{\pi}{2}\delta\omega)^L \\ &\times \left(\prod_{j=1}^L \mu_j r_j\right)\begin{bmatrix} 0 & 0 \\ \frac{nk_L^{n-1}}{2^n\Gamma(n+1)}\delta^{n-1} + O(\delta^n) & \frac{\sqrt{-1}2^n\Gamma(n+1)}{\pi k_L^{n+1}}\delta^{-n-1} + O(\delta^{-n})\end{bmatrix} \\ &\times \prod_{j=1}^L \begin{bmatrix} \sqrt{\frac{\varepsilon_j}{\mu_j}}\frac{\sqrt{-1}2^n\Gamma(n+1)}{\pi(k_jr_j)^{n+1}}\delta^{-n-1} + O(\delta^{-n}) & \frac{\sqrt{-1}2^n\Gamma(n)}{\pi(k_jr_j)^n}\delta^{-n} + O(\delta^{-n+1}) \\ -\sqrt{\frac{\varepsilon_j}{\mu_j}}\frac{n(k_jr_j)^{n-1}}{2^n\Gamma(n+1)}\delta^{n-1} + O(\delta^n) & \frac{(k_jr_j)^n}{2^n\Gamma(n+1)}\delta^n + O(\delta^{n+1})\end{bmatrix} \\ &\times \begin{bmatrix} \frac{(k_{j-1}r_j)^n}{2^n\Gamma(n+1)}\delta^n + O(\delta^{n+1}) & -\frac{\sqrt{-1}2^n\Gamma(n)}{\pi(k_{j-1}r_j)^n}\delta^{-n} + O(\delta^{-n+1}) \\ \sqrt{\frac{\varepsilon_{j-1}}{\mu_{j-1}}}\frac{n(k_{j-1}r_j)^{n-1}}{2^n\Gamma(n+1)}\delta^{n-1} + O(\delta^n) & \sqrt{\frac{\varepsilon_{j-1}}{\mu_{j-1}}}\frac{\sqrt{-1}2^n\Gamma(n+1)}{\pi(k_{j-1}r_j)^{n+1}}\delta^{-n-1} + O(\delta^{-n})\end{bmatrix}, \end{aligned}$$

and hence

$$
\begin{aligned}
P^{(n)}[\varepsilon,\mu,\delta\omega] &= \frac{1}{2^L}\begin{bmatrix} 0 & 0 \\ \dfrac{nk_L^{n-1}}{2^n\Gamma(n+1)}\delta^{n-1}+O(\delta^n) & \dfrac{\sqrt{-1}2^n\Gamma(n+1)}{\pi k_L^{n+1}}\delta^{-n-1}+O(\delta^{-n})\end{bmatrix} \\
&\quad\times\prod_{j=1}^{L}\begin{bmatrix} a_j(b_j+1)+o(1) & c_j(b_j-1)\delta^{-2n}+o(\delta^{-2n}) \\ \dfrac{b_j-1}{c_j}\delta^{2n}+o(\delta^{2n}) & \dfrac{b_j+1}{a_j}+o(1)\end{bmatrix},
\end{aligned}
\tag{10.48}
$$

where

$$
a_j := \left(\frac{k_{j-1}}{k_j}\right)^n,\quad b_j := \frac{\mu_j}{\mu_{j-1}},\quad c_j := \frac{\sqrt{-1}2^{2n}\Gamma(n)\Gamma(n+1)}{\pi(k_{j-1}k_jr_j^2)^n},
$$

with $k_j = \omega\sqrt{\varepsilon_j\mu_j}$.

From the above calculations of the leading order terms of $P^{(n)}[\varepsilon,\mu,\delta\omega]$ and the expansion formula of $J_n(t)$ and $Y_n(t)$, we see that $p_{21}^{(n)}$ and $p_{22}^{(n)}$ admit the following expansions:

$$
p_{21}^{(n)}(\mu,\varepsilon,t) = t^{n-1}\left(f_0^{(n)}(\mu,\varepsilon)+\sum_{l=1}^{(N-n)}\sum_{j=0}^{L+1}f_{l,j}^{(n)}(\mu,\varepsilon)t^{2l}(\ln t)^j+o(t^{2N-2n})\right)
\tag{10.49}
$$

and

$$
p_{22}^{(n)}(\mu,\varepsilon,t) = t^{-n-1}\left(g_0^{(n)}(\mu,\varepsilon)+\sum_{l=1}^{(N-n)}\sum_{j=0}^{L+1}g_{l,j}^{(n)}(\mu,\varepsilon)t^{2l}(\ln t)^j+o(t^{2N-2n})\right)
\tag{10.50}
$$

for $t=\delta\omega$ and some functions $f_0^{(n)}, g_0^{(n)}, f_{l,j}^{(n)}$, and $g_{l,j}^{(n)}$ independent of t.

LEMMA 10.5. *For any pair of (μ,ε), we have*

$$
g_0^{(n)}(\mu,\varepsilon)\neq 0\,.
\tag{10.51}
$$

PROOF. For $n=0$, it follows from (10.47) that

$$
g_0^{(0)}(\mu,\varepsilon) = \frac{2\sqrt{-1}}{\pi\sqrt{\varepsilon_L\mu_L}}\prod_{j=1}^{L}\frac{\mu_j}{\mu_{j-1}}\neq 0\,.
$$

Suppose $n>0$. Assume that there exists a pair of (μ,ε) such that $g_0^{(n)}(\mu,\varepsilon)=0$. Then the solution given by (10.35) with $a_0^{(n)}=0$ and $b_0^{(n)}=1$ satisfies

$$
\begin{cases}
\nabla\cdot\dfrac{1}{\mu}\nabla u+\delta^2\omega^2\varepsilon u = 0 & \text{in } \mathbb{R}^2\setminus\overline{D}\,,\\
\dfrac{\partial u}{\partial\nu}\Big|_+ = o(\delta^{-n}) & \text{on } \partial D\,,\\
u(x) = H_n^{(1)}(\delta k_0 r)e^{\sqrt{-1}n\theta} & \text{for } |x|=r>2\,.
\end{cases}
\tag{10.52}
$$

Let $v(x) := \lim_{\delta \to 0} \delta^n u(x)$. Then using (10.44) it follows that v satisfies

$$\begin{cases} \nabla \cdot \dfrac{1}{\mu} \nabla v = 0 \qquad \text{in } \mathbb{R}^2 \setminus \overline{D}\,, \\ \dfrac{\partial v}{\partial \nu}\Big|_+ = 0 \quad \text{on } \partial D\,, \\ v(x) = -\dfrac{\sqrt{-1} 2^n \Gamma(n)}{\pi k_0^n} r^{-n} e^{\sqrt{-1} n\theta} \quad \text{for } |x| = r > 2\,, \end{cases} \tag{10.53}$$

which is impossible. Thus $g_0^{(n)}(\mu, \varepsilon) \neq 0$, as desired and the proof is complete. □

Equations (10.49) and (10.50) together with the above lemma give us the following proposition.

PROPOSITION 10.6. *For $n \geq 1$, let W_n be defined by (10.31). We have*

$$W_n[\mu, \varepsilon, t] = t^{2n} \left(W_n^0[\mu, \varepsilon] + \sum_{l=1}^{(N-n)} \sum_{j=0}^{M_l} W_n^{l,j}[\mu, \varepsilon] t^{2l} (\ln t)^j \right) + o(t^{2N})\,, \tag{10.54}$$

where $t = \delta\omega$, $M_l := (L+1)l$ (L being the number of layers), and the coefficients $W_n^0[\mu, \varepsilon]$ and $W_n^{l,j}[\mu, \varepsilon]$ are independent of t.

To construct an S-vanishing structure of order N at low frequencies, we need to have a pair (μ, ε) of the form (10.28) satisfying

$$W_n^0[\mu, \varepsilon] = 0, \text{ and } W_n^{l,j}[\mu, \varepsilon] = 0 \quad \text{for } 0 \leq n \leq N,\ 1 \leq l \leq (N-n),\ 1 \leq j \leq M_l\,. \tag{10.55}$$

As in the conductivity case, it should be emphasized that one does not know if a solution exists for any order N. Nevertheless, numerical constructions of such structures for small N are given in the last subsection.

10.3.3. Enhancement of Near-Cloaking. In this subsection we show that the S-vanishing structures (after applying transformation optics) enhance the near-cloaking.

Let (μ, ε) be an S-vanishing structure of order N at low frequencies, *i.e.*, (10.55) holds, and it is of the form (10.28). It follows from (2.221), Theorem 2.79, and Proposition 10.6 that

$$A_\infty[\mu, \varepsilon, \delta\omega](\theta, \theta') = o(\delta^{2N}) \tag{10.56}$$

uniformly in (θ, θ') if $\delta \leq \delta_0$ for some δ_0.

Let

$$\Psi_\delta(x) = \frac{1}{\delta} x, \quad x \in \mathbb{R}^2\,. \tag{10.57}$$

Then we have

$$A_\infty\Big[\mu \circ \Psi_\delta, \varepsilon \circ \Psi_\delta, \omega\Big] = A_\infty[\mu, \varepsilon, \delta\omega]\,. \tag{10.58}$$

To see this, let u be the solution to

$$\begin{cases} \nabla \cdot \dfrac{1}{(\mu \circ \Psi_\delta)(x)} \nabla u(x) + \omega^2 (\varepsilon \circ \Psi_\delta)(x) u(x) = 0 \quad \text{in } \mathbb{R}^2 \setminus \overline{B_\delta}\,, \\ \dfrac{\partial u}{\partial \nu} = 0 \quad \text{on } \partial B_\delta\,, \\ (u - U) \text{ satisfies the outgoing radiation condition,} \end{cases} \tag{10.59}$$

where $U(x) = e^{\sqrt{-1}k_0(\cos\theta,\sin\theta)\cdot x}$. Here B_δ is the disk of radius δ centered at 0. Define for $y = x/\delta$

$$\widetilde{u}(y) := \Big(u \circ \Psi_\delta^{-1}\Big)(y) = \Big(u \circ \Psi_{\frac{1}{\delta}}\Big)(y) \quad \text{and} \quad \widetilde{U}(y) = \Big(U \circ \Psi_{\frac{1}{\delta}}\Big)(y).$$

Then, we have

$$\begin{cases} \nabla \cdot \dfrac{1}{\mu(y)} \nabla_y \widetilde{u}(y) + \delta^2 \omega^2 \varepsilon(y) \widetilde{u}(y) = 0 \quad \text{in } \mathbb{R}^2, \\ \dfrac{\partial \widetilde{u}}{\partial \nu} = 0 \quad \text{on } \partial B_1, \\ (\widetilde{u} - \widetilde{U}) \text{ satisfies the outgoing radiation condition.} \end{cases} \tag{10.60}$$

From the definition (2.237) of the far-field pattern A_∞, we get

$$(u - U)(x) \sim -\sqrt{-1} e^{-\frac{\pi\sqrt{-1}}{4}} \frac{e^{\sqrt{-1}k_0|x|}}{\sqrt{|x|}} A_\infty\Big[\mu \circ \Psi_\delta, \varepsilon \circ \Psi_\delta, \omega\Big](\theta, \theta') \quad \text{as } |x| \to \infty,$$

and

$$(\widetilde{u} - \widetilde{U})(y) \sim -\sqrt{-1} e^{-\frac{\pi\sqrt{-1}}{4}} \frac{e^{\sqrt{-1}\delta k_0|y|}}{\sqrt{|y|}} A_\infty[\mu, \varepsilon, \delta\omega](\theta, \theta') \quad \text{as } |y| \to \infty,$$

where $x = |x|(\cos\theta', \sin\theta')$. So, we have (10.58). It then follows from (10.56) that

$$A_\infty\Big[\mu \circ \Psi_\delta, \varepsilon \circ \Psi_\delta, \omega\Big](\theta, \theta') = o(\delta^{2N}). \tag{10.61}$$

We also obtain from (2.246)

$$Q^s\Big[\mu \circ \Psi_\delta, \varepsilon \circ \Psi_\delta, \omega\Big](\theta') = o(\delta^{4N}). \tag{10.62}$$

It is worth emphasizing that $(\mu \circ \Psi_\delta, \varepsilon \circ \Psi_\delta)$ is a multi-coated structure of radius 2δ.

We now apply a transformation to the structure $(\mu \circ \Psi_\delta, \varepsilon \circ \Psi_\delta)$ to blow up the small disk of radius δ.

For a small number δ, let F_δ be the diffeomorphism defined by

$$F_\delta(x) := \begin{cases} x & \text{for } |x| \geq 2, \\ \Big(\dfrac{3 - 4\delta}{2(1-\delta)} + \dfrac{1}{4(1-\delta)}|x|\Big) \dfrac{x}{|x|} & \text{for } 2\delta \leq |x| \leq 2, \\ \Big(\dfrac{1}{2} + \dfrac{1}{2\delta}|x|\Big) \dfrac{x}{|x|} & \text{for } \delta \leq |x| \leq 2\delta, \\ \dfrac{x}{\delta} & \text{for } |x| \leq \delta. \end{cases} \tag{10.63}$$

We then get from (10.61), (10.62), and Lemma 2.112 the main result of this section.

THEOREM 10.7. *If (μ, ε) is an S-vanishing structure of order N at low frequencies, then there exists δ_0 such that*

$$A_\infty\Big[(F_\delta)_\star(\mu \circ \Psi_\delta), (F_\delta)_\star(\varepsilon \circ \Psi_\delta), \omega\Big](\theta, \theta') = o(\delta^{2N}) \tag{10.64}$$

and

$$Q^s\Big[(F_\delta)_\star(\mu \circ \Psi_\delta), (F_\delta)_\star(\varepsilon \circ \Psi_\delta), \omega\Big](\theta') = o(\delta^{4N}) \tag{10.65}$$

*for all $\delta \leq \delta_0$, uniformly in θ and θ'. Moreover, the cloaking enhancement, given by (*10.64*) and (*10.65*), is achieved for all frequencies smaller than ω.*

Since (10.43) holds if we replace ω by $\omega' \leq \omega$, the cloaking enhancement is achieved for all the frequencies smaller than ω. Then it is worth comparing (10.64) with (10.56). In (10.56), (μ, ε) is a multiply layered structure between radius 1 and 2 in which each layer is filled with an isotropic material, and enhanced near-cloaking is achieved for low frequencies $\delta\omega$ with $\delta \leq \delta_0$. On the other hand, in (10.64) the frequency ω does not have to be small. In fact, (10.64) says that for any frequency ω there is a radius δ which yields the enhanced near-cloaking up to $o(\delta^{2N})$.

10.4. Near-Cloaking for the Full Maxwell Equations

In this section, the scattering coefficients vanishing approach is used to consider near-cloaking for the full Maxwell equations. As in the previous section, these S-vanishing structures are, prior to using the transformation optics in Subsection 2.14.4, layered-structures designed so that their first scattering coefficients W_n^{TE} and W_n^{TM} defined in Subsection 2.14.6.2 vanish. We therefore construct multilayered structures whose scattering coefficients vanish, which are called *S-vanishing structures* for the full Maxwell equations.

10.4.1. Scattering Coefficients of Multilayered Structures. The scattering coefficients $\left(W^{TE,TE}_{(n,m)(p,q)}, W^{TE,TM}_{(n,m)(p,q)}, W^{TM,TE}_{(n,m)(p,q)}, W^{TM,TM}_{(n,m)(p,q)}\right)$ are defined in Subsection 2.14.6.2, namely, if E^i given as in (2.353), the scattered field $E - E^i$ can be expanded as (2.354) and (2.355). The transmission condition on each interface Σ_j is given by (2.361).

Assume that the core A_{L+1} is perfectly conducting, namely,

$$E \times \nu = 0 \quad \text{on } \Sigma_{L+1} = \partial A_{L+1}. \tag{10.66}$$

In Subsection 2.14.6.2, using the symmetry of the layered radial structure, the scattering coefficients are reduced to W_n^{TE} and W_n^{TM}, given by (2.371) and (2.377).

The multi-layered structure is defined as follows: For positive numbers

$$r_1, \ldots, r_{L+1} \quad \text{with} \quad 2 = r_1 > r_2 > \cdots r_{L+1} = 1,$$

let

$$A_j := \{x : r_{j+1} \leq |x| < r_j\}, \quad j = 1, \ldots, L,$$
$$A_0 := \mathbb{R}^3 \setminus \overline{B_2},$$

and

$$A_{L+1}(= D) := \{x : |x| < 1\},$$

where B_2 denotes the central ball of radius 2 and

$$\Sigma_j = \{|x| = r_j\}, \quad j = 1, \ldots, L+1.$$

For $j = 1, \ldots, L+1$, let (μ_j, ϵ_j) be the pair of permeability and permittivity parameters of A_j. Set $\mu_0 = 1$ and $\epsilon_0 = 1$. Then define

$$\mu = \sum_{j=0}^{L+1} \mu_j \chi(A_j) \quad \text{and} \quad \epsilon = \sum_{j=0}^{L+1} \epsilon_j \chi(A_j), \tag{10.67}$$

which are permeability and permittivity distributions of the layered structure.

To construct the S-vanishing structure at a fixed frequency ω, one looks for (μ, ϵ) such that

$$W_n^{TE}[\epsilon, \mu, \omega] = 0, \; W_n^{TM}[\epsilon, \mu, \omega] = 0, \quad n = 1, \ldots, N,$$

for some N. More ambitiously one may look for a structure (μ, ϵ) for a fixed ω such that

$$W_n^{TE}[\mu, \epsilon, \delta\omega] = 0, \quad W_n^{TM}[\mu, \epsilon, \delta\omega] = 0$$

for all $1 \leq n \leq N$ and $\delta \leq \delta_0$ for some δ_0. Such a structure may not exist. So instead one looks for a structure such that

$$W_n^{TE}[\mu, \epsilon, \delta\omega] = o(\delta^{2N+1}), \quad W_n^{TM}[\mu, \epsilon, \delta\omega] = o(\delta^{2N+1}), \tag{10.68}$$

for all $1 \leq n \leq N$ and $\delta \leq \delta_0$ for some δ_0. Such a structure is called an *S-vanishing structure of order N at low frequencies.* In the following subsection, the scattering coefficients are expanded at low frequencies and conditions for the magnetic permeability and the electric permittivity to be an S-vanishing structure are derived.

Recall that from Lemma 2.117 it follows that there exists $\delta_0 > 0$ such that, for all $\delta \leq \delta_0$,

$$\left|W_n^{TE}[\varepsilon, \mu, \delta\omega]\right| \leq \frac{C^{2n}}{n^{2n}} \delta^{2n+1}, \tag{10.69}$$

for all $n \in \mathbb{N} \setminus \{0\}$, where the constant C depends on $(\varepsilon, \mu, \omega)$ but is independent of δ. The same estimate holds for W_n^{TM}.

Suppose now that (μ, ϵ) is an S-vanishing structure of order N at low frequencies. Let the incident wave E^i be given by a plane wave $e^{\sqrt{-1}\delta k \cdot x} c$ with $|k| = k_0 (= \omega\sqrt{\varepsilon_0 \mu_0})$ and $k \cdot c = 0$. From (2.358), the corresponding scattering amplitude, $A_\infty[\mu, \epsilon, \delta\omega](c, \hat{k} := k/|k|; \hat{x} := x/|x|)$, is given by (2.357) with the following $\alpha_{n,m}$ and $\beta_{n,m}$:

$$\begin{cases} \alpha_{n,m} = \dfrac{4\pi(\sqrt{-1})^n}{\sqrt{n(n+1)}} (V_{n,m}(\hat{k}) \cdot c) W_n^{TE}[\mu, \epsilon, \delta\omega], \\ \beta_{n,m} = -\dfrac{4\pi(\sqrt{-1})^n}{\sqrt{n(n+1)}} \dfrac{1}{\sqrt{-1}\omega\mu_0} (U_{n,m}(\hat{k}) \cdot c) W_n^{TM}[\mu, \epsilon, \delta\omega]. \end{cases}$$

Applying (2.359) and (10.68),

$$A_\infty[\mu, \epsilon, \delta\omega](c, \hat{k}; \hat{x}) = o(\delta^{2N+1}) \tag{10.70}$$

uniformly in $(\hat{k}, \hat{x})$ if $\delta \leq \delta_0$. Thus using such a structure, the visibility of the scattering amplitude is greatly reduced.

10.4.2. Asymptotic Expansion of the Scattering Coefficients. The spherical Bessel functions of the first and second kinds have the series expansions

$$j_n(t) = \sum_{l=0}^{\infty} \frac{(-1)^l t^{n+2l}}{2^l l! 1 \times 3 \times \cdots \times (2n+2l+1)}$$

and

$$y_n(t) = -\frac{(2n)!}{2^n n!} \sum_{l=0}^{\infty} \frac{(-1)^l t^{2l-n-1}}{2^l l! (-2n+1)(-2n+3) \ldots (-2n+2l-1)}.$$

So, using the notation of double factorials, which is defined by

$$n!! := \begin{cases} n \times (n-2) \times \cdots \times 3 \times 1 & \text{if } n > 0 \text{ is odd,} \\ n \times (n-2) \times \cdots \times 4 \times 2 & \text{if } n > 0 \text{ is even,} \\ 1 & \text{if } n = -1, 0, \end{cases}$$

one has

$$j_n(t) = \frac{t^n}{(2n+1)!!}\big(1+o(t)\big) \quad \text{for } t \ll 1 \tag{10.71}$$

and

$$y_n(t) = -\big((2n-1)!!\big)t^{-n+1}\big(1+o(t)\big) \quad \text{for } t \ll 1. \tag{10.72}$$

Let $P_n^{TE}[\varepsilon,\mu,\omega], p_{n,1}^{TE}$, and $p_{n,2}^{TE}$ be defined by ((2.370)). One now computes $P_n^{TE}[\epsilon,\mu,t]$ for small t. For $n \geq 1$,

$$P_n^{TE}[\epsilon,\mu,t] = (-\sqrt{-1}t)^L\left(\prod_{j=1}^L \mu_j^{\frac{3}{2}}\epsilon_j^{\frac{1}{2}}r_j\right)\begin{bmatrix}\frac{z_L^n}{(2n+1)!!}t^n+o(t^n) & \frac{-\sqrt{-1}Q(n)}{z_L^{n+1}}t^{-n-1}\\ 0 & 0\end{bmatrix}$$

$$\times\prod_{j=1}^L\left(\begin{bmatrix}\frac{\sqrt{-1}Q(n)n}{\mu_j(z_jr_j)^{n+1}}t^{-n-1}+o(t^{-n-1}) & \frac{\sqrt{-1}Q(n)}{(z_jr_j)^{n+1}}t^{-n-1}+o(t^{-n-1})\\ \frac{-(n+1)(z_jr_j)^n}{\mu_j(2n+1)!!}t^n+o(t^n) & \frac{(z_jr_j)^n}{(2n+1)!!}t^n+o(t^n)\end{bmatrix}\right.$$

$$\left.\begin{bmatrix}\frac{(z_{j-1}r_j)^n}{(2n+1)!!}t^n+o(t^n) & \frac{-\sqrt{-1}Q(n)}{(z_{j-1}r_j)^{n+1}}t^{-n-1}+o(t^{-n-1})\\ \frac{(n+1)(z_{j-1}r_j)^n}{\mu_{j-1}(2n+1)!!}t^n+o(t^n) & \frac{\sqrt{-1}Q(n)n}{\mu_{j-1}(z_{j-1}r_j)^{n+1}}t^{-n-1}+o(t^{-n-1})\end{bmatrix}\right),$$

where $z_j = \sqrt{\epsilon_j\mu_j}$ and $Q(n) = (2n-1)!!$. One then has

$$P_n^{TE}[\epsilon,\mu,t] = \begin{bmatrix}\frac{z_L^n}{(2n+1)!!}t^n+o(t^n) & \frac{-\sqrt{-1}Q(n)}{z_L^{n+1}}t^{-n-1}+o(t^{-n-1})\\ 0 & 0\end{bmatrix}\times$$

$$\times\prod_{j=1}^L\begin{bmatrix}h_{11}^{TE,n,j} & h_{12}^{TE,n,j}\\ h_{21}^{TE,n,j} & h_{22}^{TE,n,j}\end{bmatrix}$$

with

$$\begin{aligned}
h_{11}^{TE,n} &= \frac{Q(n)z_{j-1}^n}{(2n+1)!!z_j^n}\big(n+\frac{(n+1)\mu_j}{\mu_{j-1}}\big)\big(1+o(1)\big),\\
h_{12}^{TE,n} &= (-\sqrt{-1})\frac{(Q(n))^2n}{z_j^nz_{j-1}^{n+1}r_j^{2n+1}}\big(1-\frac{\mu_j}{\mu_{j-1}}\big)t^{-2n-1}\big(1+o(1)\big),\\
h_{21}^{TE,n} &= \sqrt{-1}\frac{z_{j-1}^nz_j^{n+1}r_j^{2n+1}(n+1)}{((2n+1)!!)^2}\big(1-\frac{\mu_j}{\mu_{j-1}}\big)t^{2n+1}\big(1+o(1)\big),\\
h_{22}^{TE,n} &= \frac{Q(n)z_j^{n+1}}{(2n+1)!!z_{j-1}^{n+1}}\big(n+1+\frac{n\mu_j}{\mu_{j-1}}\big)\big(1+o(1)\big).
\end{aligned}$$

Similarly, for the transverse magnetic case, let $P_n^{TM}[\varepsilon,\mu,\omega], p_{n,1}^{TM}$, and $p_{n,2}^{TM}$ be defined by ((2.376)). One has

$$P_n^{TM}[\epsilon,\mu,t] = \begin{bmatrix}\frac{(n+1)z_L^n}{(2n+1)!!}t^n+o(t^n) & \frac{-\sqrt{-1}nQ(n)}{z_L^{n+1}}t^{-n-1}+o(t^{-n-1})\\ 0 & 0\end{bmatrix}$$

$$\times\prod_{j=1}^L\begin{bmatrix}h_{11}^{TM,n,j} & h_{12}^{TM,n,j}\\ h_{21}^{TM,n,j} & h_{22}^{TM,n,j}\end{bmatrix}$$

with

$$\begin{aligned} h_{11}^{TM,n,j} &= \frac{Q(n)z_{j-1}^n}{(2n+1)!!z_j^n}\left((n+\frac{\epsilon_j}{\epsilon_{j-1}}(n+1)\right)(1+o(1)), \\ h_{12}^{TM,n,j} &= (-\sqrt{-1})\frac{(Q(n))^2 n}{z_j^n z_{j-1}^{n+1} r_j^{2n+1}}\left(1-\frac{\epsilon_j}{\epsilon_{j-1}}\right)t^{-2n-1}(1+o(1)), \\ h_{21}^{TM,n,j} &= \sqrt{-1}\frac{z_{j-1}^n z_j^{n+1} r_j^{2n+1}(n+1)}{((2n+1)!!)^2}\left(1-\frac{\epsilon_j}{\epsilon_{j-1}}\right)t^{2n+1}(1+o(1)), \\ h_{22}^{TM,n,j} &= \frac{Q(n)z_j^{n+1}}{(2n+1)!!z_{j-1}^{n+1}}\left(n+1+\frac{\epsilon_j}{\epsilon_{j-1}}n\right)(1+o(1)). \end{aligned}$$

Using the behavior of spherical Bessel functions for small arguments, one can see that $p_{n,1}^{TE}$ and $p_{n,2}^{TE}$ admit the following expansions:

$$p_{n,1}^{TE}[\mu,\epsilon,t] = t^n\left(\sum_{l=0}^{N-n} f_{n,l}^{TE}(\mu,\epsilon)t^{2l} + o(t^{2N-2n})\right) \tag{10.73}$$

and

$$p_{n,2}^{TE}[\mu,\epsilon,t] = t^{-n-1}\left(\sum_{l=0}^{N-n} g_{n,l}^{TE}(\mu,\epsilon)t^{2l} + o(t^{2N-2n})\right). \tag{10.74}$$

Similarly, $p_{n,1}^{TM}$ and $p_{n,2}^{TM}$ have the following expansions:

$$p_{n,1}^{TM}[\mu,\epsilon,t] = t^n\left(\sum_{l=0}^{N-n} f_{n,l}^{TM}(\mu,\epsilon)t^{2l} + o(t^{2N-2n})\right) \tag{10.75}$$

and

$$p_{n,2}^{TM}[\mu,\epsilon,t] = t^{-n-1}\left(\sum_{l=0}^{N-n} g_{n,l}^{TM}(\mu,\epsilon)t^{2l} + o(t^{2N-2n})\right) \tag{10.76}$$

for $t=\delta\omega$ and some functions $f_{n,l}^{TE}, g_{n,l}^{TE}, f_{n,l}^{TM}$, and $g_{n,l}^{TM}$ independent of t.

LEMMA 10.8. *For any pair of* (μ,ϵ), *one has*

$$g_{n,0}^{TE}(\mu,\epsilon) \neq 0 \tag{10.77}$$

and

$$g_{n,0}^{TM}(\mu,\epsilon) \neq 0. \tag{10.78}$$

PROOF. Assume that there exists a pair of (μ,ϵ) such that $g_{n,0}^{TE}(\mu,\epsilon)=0$. Since $p_{n,2}^{TE}[\mu,\epsilon,\delta\omega] = o(\delta^{-n-1})$, the solution given by (2.363) with $a_0=1$ and $\tilde{a}_0=0$ satisfies

$$\begin{cases} \nabla\times\left(\dfrac{1}{\mu}\nabla\times E\right) - \delta^2\omega^2\epsilon E = 0 & \text{in } \mathbb{R}^3\setminus\overline{D}, \\ \nabla\cdot E = 0 & \text{in } \mathbb{R}^3\setminus\overline{D}, \\ (\nu\times E)\big|_+ = o(\delta^{-(n+1)}) & \text{on } \partial D, \\ E(x) = h_n^{(1)}(\delta k_0|x|)V_{n,0}(\hat{x}) & \text{for } |x|>2. \end{cases}$$

Let $V(x) = \lim_{\delta \to 0} \delta^{n+1} E(x)$. Using (10.72) one knows that the limit V satisfies

$$\begin{cases} \nabla \times \left(\dfrac{1}{\mu} \nabla \times V \right) = 0 & \text{in } \mathbb{R}^3 \setminus \overline{D}, \\ \nabla \cdot V = 0 & \text{in } \mathbb{R}^3 \setminus \overline{D}, \\ (\nu \times V)\big|_{+} = 0 & \text{on } \partial D, \\ V(x) = -\big((2n-1)!!\big) V_{n,0}(\hat{x}) & \text{for } |x| > 2. \end{cases}$$

Since $V_{n,0}(\hat{x}) = O(|x|^{-1})$, one gets $V(x) = 0$ by Green's formula, which is a contradiction. Thus $g_{n,0}^{TE}(\mu, \epsilon) \neq 0$. In a similar way, (10.78) can be proved. □

From Lemma 10.8, one obtains the following result.

PROPOSITION 10.9. *One has*

$$W_n^{TE}[\mu, \epsilon, t] = t^{2n+1} \sum_{l=0}^{N-n} W_{n,l}^{TE}[\mu, \epsilon] t^{2l} + o(t^{2N+1})$$

and

$$W_n^{TM}[\mu, \epsilon, t] = t^{2n+1} \sum_{l=0}^{N-n} W_{n,l}^{TM}[\mu, \epsilon] t^{2l} + o(t^{2N+1}),$$

where $t = \delta\omega$ and the coefficients $W_{n,l}^{TE}[\mu, \epsilon]$ and $W_{n,l}^{TM}[\mu, \epsilon]$ are independent of t.

Hence, if one has (μ, ϵ) such that

$$W_{n,l}^{TE}[\mu, \epsilon] = W_{n,l}^{TM}[\mu, \epsilon] = 0, \quad \text{for all } 1 \leq n \leq N,\ 0 \leq l \leq (N-n), \tag{10.79}$$

(μ, ϵ) satisfies (10.68); in other words, it is an *S-vanishing structure of order N at low frequencies.* It is quite challenging to construct (μ, ϵ) analytically satisfying (10.79). The next subsection presents some numerical examples of such structures.

10.4.3. Enhancement of Near Cloaking. In this section one constructs a cloaking structure based on the following lemma.

LEMMA 10.10. *Let F be an orientation-preserving diffeomorphism of $\mathbb{R}^3$ onto $\mathbb{R}^3$ such that $F(x)$ is identity for $|x|$ large enough. If (E, H) is a solution to*

$$\begin{cases} \nabla \times E = \sqrt{-1}\omega\mu H & \textit{in } \mathbb{R}^3, \\ \nabla \times H = -\sqrt{-1}\omega\epsilon E & \textit{in } \mathbb{R}^3, \\ (E - E^i, H - H^i) \textit{ is radiating,} \end{cases} \tag{10.80}$$

then $(\widetilde{E}, \widetilde{H})$ defined by $(\widetilde{E}(y), \widetilde{H}(y)) = \big((DF)^{-T} E(F^{-1}(y)), (DF)^{-T} H(F^{-1}(y))\big)$ satisfies

$$\begin{cases} \nabla \times \widetilde{E} = \sqrt{-1}\omega(F_*\mu)\widetilde{H} & \textit{in } \mathbb{R}^3, \\ \nabla \times \widetilde{H} = -\sqrt{-1}\omega(F_*\epsilon)\widetilde{E} & \textit{in } \mathbb{R}^3, \\ (\widetilde{E} - \widetilde{E}^i, \widetilde{H} - \widetilde{H}^i) \textit{ is radiating,} \end{cases}$$

where $(\widetilde{E}^i(y), \widetilde{H}^i(y)) = \big((DF)^{-T} E^i(F^{-1}(y)), (DF)^{-T} H^i(F^{-1}(y))\big)$,

$$(F_*\mu)(y) = \frac{DF(x)\mu(x)DF^T(x)}{\det(DF(x))}, \quad \textit{and} \quad (F_*\epsilon)(y) = \frac{DF(x)\epsilon(x)DF^T(x)}{\det(DF(x))},$$

with $x = F^{-1}(y)$ and DF is the Jacobian matrix of F.

Hence,

$$A[\mu, \epsilon, \omega] = A[F_*\mu, F_*\epsilon, \omega].$$

To compute the scattering amplitude which corresponds to the material parameters before the transformation, one considers the following scaling function, for small parameter δ,

$$\Psi_{\frac{1}{\delta}}(x) = \frac{1}{\delta}x, \quad x \in \mathbb{R}^3.$$

Then one has the following relation between the scattering amplitudes which correspond to two sets of differently scaled material parameters and frequency:

$$A_\infty\left[\mu \circ \Psi_{\frac{1}{\delta}}, \epsilon \circ \Psi_{\frac{1}{\delta}}, \omega\right] = A_\infty[\mu, \epsilon, \delta\omega]. \tag{10.81}$$

To see this, consider (E, H) which satisfies

$$\begin{cases} (\nabla \times E)(x) = \sqrt{-1}\omega\big(\mu \circ \Psi_{\frac{1}{\delta}}\big)(x)H(x) & \text{for } x \in \mathbb{R}^3 \setminus \overline{B_\delta}, \\ (\nabla \times H)(x) = -\sqrt{-1}\omega\big(\epsilon \circ \Psi_{\frac{1}{\delta}}\big)(x)E(x) & \text{for } x \in \mathbb{R}^3 \setminus \overline{B_\delta}, \\ \hat{x} \times E(x) = 0 & \text{on } \partial B_\delta, \\ (E - E^i, H - H^i) \text{ is radiating}, \end{cases}$$

with the incident wave $E^i(x) = e^{\sqrt{-1}k\cdot x}\hat{c}$ and $H^i = \frac{1}{\sqrt{-1}\omega\mu_0}\nabla \times E^i$ with $k \cdot \hat{c} = 0$ and $|k| = k_0$. Here B_δ is the ball of radius δ centered at the origin. Set $y = \frac{1}{\delta}x$ and define

$$\big(\widetilde{E}(y), \widetilde{H}(y)\big) := \Big(\big(E \circ \Psi_{\frac{1}{\delta}}^{-1}\big)(y), \big(H \circ \Psi_{\frac{1}{\delta}}^{-1}\big)(y)\Big) = \Big(\big(E \circ \Psi_\delta\big)(y), \big(H \circ \Psi_\delta\big)(y)\Big)$$

and

$$\big(\widetilde{E^i}(y), \widetilde{H^i}(y)\big) := \Big(\big(E^i \circ \Psi_\delta\big)(y), \big(H^i \circ \Psi_\delta\big)(y)\Big).$$

Then, one has

$$\begin{cases} \Big(\nabla_y \times \widetilde{E}\Big)(y) = \sqrt{-1}\delta\omega\mu(y)\widetilde{H}(y) & \text{for } y \in \mathbb{R}^3 \setminus \overline{B_1} \\ \Big(\nabla_y \times \widetilde{H}\Big)(y) = -\sqrt{-1}\delta\omega\epsilon(y)\widetilde{E}(y) & \text{for } y \in \mathbb{R}^3 \setminus \overline{B_1}, \\ \hat{y} \times \widetilde{E}(y) = 0 & \text{on } \partial B_1, \\ (\widetilde{E} - \widetilde{E^i}, \widetilde{H} - \widetilde{H^i}) \text{ is radiating} \end{cases}$$

Remind that the scattered wave can be represented using the scattering amplitude as follows:

$$(E - E^i)(x) \sim \frac{e^{\sqrt{-1}k_0|x|}}{k_0|x|}A_\infty\left[\mu \circ \Psi_{\frac{1}{\delta}}, \epsilon \circ \Psi_{\frac{1}{\delta}}, \omega\right](c, \hat{k}; \hat{x}) \quad \text{as } |x| \to \infty,$$

and

$$(\widetilde{E} - \widetilde{E^i})(y) \sim \frac{e^{\sqrt{-1}k_0\delta|y|}}{k_0\delta|y|}A_\infty[\mu, \epsilon, \omega](c, \hat{k}; \hat{x}) \quad \text{as } |y| \to \infty.$$

Since the left-hand sides of the previous equations are coincident, one has (10.81).

Suppose that (μ, ϵ) is an S-vanishing structure of order N at low frequencies as in Section 10.4. From (10.70) and (10.81), one has

$$A_\infty\left[\mu \circ \Psi_{\frac{1}{\delta}}, \epsilon \circ \Psi_{\frac{1}{\delta}}, \omega\right](c, \hat{k}; \hat{x}) = o(\delta^{2N+1}) \tag{10.82}$$

Then, one defines the diffeomorphism F_δ as

$$F_\delta(x) := \begin{cases} x & \text{for } |x| \geq 2, \\ \Big(\dfrac{3-4\delta}{2(1-\delta)} + \dfrac{1}{4(1-\delta)}|x|\Big)\dfrac{x}{|x|} & \text{for } 2\delta \leq |x| \leq 2, \\ \Big(\dfrac{1}{2} + \dfrac{1}{2\delta}|x|\Big)\dfrac{x}{|x|} & \text{for } \delta \leq |x| \leq 2\delta, \\ \dfrac{x}{\delta} & \text{for } |x| \leq \delta. \end{cases}$$

One then gets from (10.82) and Lemma 10.10 the main result of this chapter.

THEOREM 10.11. *If (μ, ϵ) is an S-vanishing structure of order N at low frequencies, then there exists δ_0 such that*

$$A_\infty\left[(F_\delta)_*(\mu \circ \Psi_{\frac{1}{\delta}}), (F_\delta)_*(\epsilon \circ \Psi_{\frac{1}{\delta}}), \omega\right](c, \hat{k}; \hat{x}) = o(\delta^{2N+1}),$$

for all $\delta \leq \delta_0$, uniformly in $(\hat{k}, \hat{x})$.

Remark that the cloaking structure $\big((F_\delta)_*(\mu \circ \Psi_{\frac{1}{\delta}}), (F_\delta)_*(\epsilon \circ \Psi_{\frac{1}{\delta}})\big)$ in Theorem 10.11 satisfies the perfect electric conductor boundary condition on $|x| = 1$.

10.4.4. Numerical Implementation. This section provides numerical examples of S-vanishing structures of order N at low frequencies based on (10.79). As in previous sections, we use a gradient descent method for a suitable energy functional. We symbolically compute the scattering coefficients. In the place of spherical Bessel functions and spherical Hankel functions, we use their low-frequency expansions and symbolically compute W_n^{TE} and W_n^{TM} to obtain $W_{n,l}^{TE}$ and $W_{n,l}^{TM}$. The corresponding Matlab code can be downloaded at `www.ams.org/bookpages/surv-235`.

The following example is an S-vanishing structure of order $N = 2$ made of 6 multilayers. The radii of the concentric disks are $r_j = 2 - \frac{j-1}{6}$ for $j = 1, \dots, 7$. From Proposition 10.9, the nonzero leading terms of $W_n^{TE}[\mu, \epsilon, t]$ and $W_n^{TM}[\mu, \epsilon, t]$ up to t^5 are

- $[t^3, t^5]$ terms in $W_1^{TE}[\mu, \epsilon, t]$, *i.e.*, $W_{1,0}^{TE}, W_{1,1}^{TE}$,
- $[t^3, t^5]$ terms in $W_1^{TM}[\mu, \epsilon, t]$, *i.e.*, $W_{1,0}^{TM}, W_{1,1}^{TM}$,
- $[t^5]$ term in $W_2^{TE}[\mu, \epsilon, t]$, *i.e.*, $W_{2,0}^{TE}$,
- $[t^5]$ term in $W_2^{TM}[\mu, \epsilon, t]$, *i.e.*, $W_{2,0}^{TM}$.

Consider the mapping

$$(\mu, \epsilon) \longrightarrow (W_{1,0}^{TE}, W_{1,1}^{TE}, W_{1,0}^{TM}, W_{1,1}^{TM}, W_{2,0}^{TE}, W_{2,0}^{TM}), \tag{10.83}$$

where, $\mu = (\mu_1, \dots, \mu_6)$ and $\epsilon = (\epsilon_1, \dots, \epsilon_6)$. One looks for (μ, ϵ) which has the right-hand side of (10.83) as small as possible. Since (10.83) is a nonlinear equation, one solves it iteratively. Initially, one wets $\mu = \mu^{(0)}$ and $\epsilon = \epsilon^{(0)}$. One iteratively modifies $(\mu^{(i)}, \epsilon^{(i)})$

$$[\mu^{(i+1)}\ \epsilon^{(i+1)}]^T = [\mu^{(i)}\ \epsilon^{(i)}]^T - A_i^\dagger b^{(i)}, \tag{10.84}$$

where $A_i^\dagger$ is the pseudo-inverse of

$$A_i := \frac{\partial(W_{1,0}^{TE}, W_{1,1}^{TE}, \dots, W_{2,0}^{TM})}{\partial(\mu, \epsilon)}\Big|_{(\mu,\epsilon)=(\mu^{(i)},\epsilon^{(i)})},$$

and

$$b^{(i)} = \begin{bmatrix} W_{1,0}^{TE} \\ W_{1,1}^{TE} \\ \vdots \\ W_{2,0}^{TM} \end{bmatrix} \Bigg|_{(\mu,\epsilon)=(\mu^{(i)},\epsilon^{(i)})}.$$

Figure 10.1 and Figure 10.2 show computational results of 6-layers S-vanishing structure of order $N = 2$. One sets $r = (2, \frac{11}{6}, \ldots, \frac{7}{6})$, $\mu^{(0)} = (3, 6, 3, 6, 3, 6)$ and $\epsilon^{(0)} = (3, 6, 3, 6, 3, 6)$ and modify them following (10.84) with the constraints that μ and ϵ belongs to the interval between 0.1 and 10. The obtained material parameters are $\mu = (0.1000, 1.1113, 0.2977, 2.0436, 0.1000, 1.8260)$ and $\epsilon = (0.4356, 1.1461, 0.2899, 1.8199, 0.1000, 3.1233)$, respectively. Differently from the no-layer structure with the perfect electric conductor condition at $|x| = 1$, the obtained multilayer structure has the nearly zero coefficients of $W_n^{TE}[\mu, \epsilon, t]$ and $W_n^{TM}[\mu, \epsilon, t]$ up to t^5.

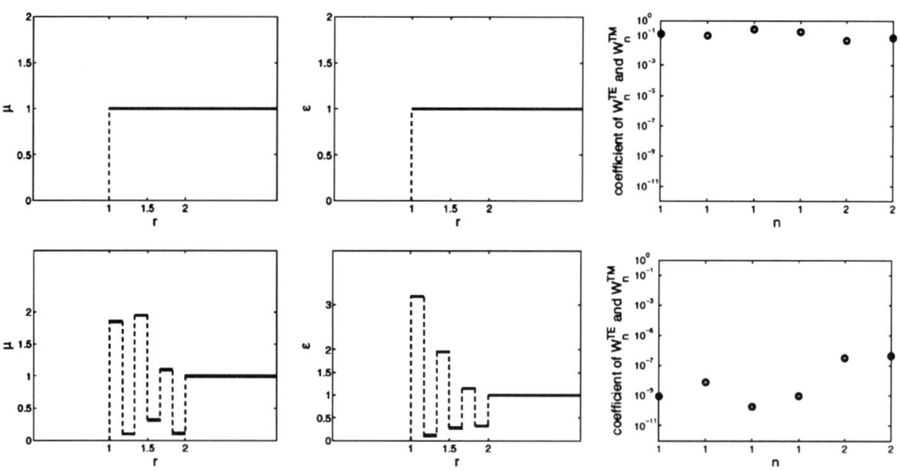

FIGURE 10.1. This figure shows the graph of the material parameters and the corresponding coefficients in $W_n^{TE}[\mu, \epsilon, t]$ and $W_n^{TM}[\mu, \epsilon, t]$ up to t^5. The first row is of the no-layer case, and the second row is of 6-layers S-vanishing structure of order $N = 2$ which is explained in Example 1. In the third column, the y-axis shows $(W_{1,0}^{TE}, W_{1,1}^{TE}, W_{1,0}^{TM}, W_{1,1}^{TM}, W_{2,0}^{TE}, W_{2,0}^{TM})$ from the left to the right.

10.5. Near-Cloaking for the Elasticity System

As an application of the elastic scattering coefficients introduced in Subsection 2.15.15.2 we consider the elastic cloaking problem. The aim here is to construct an effective near cloaking structure at a fixed frequency to make the objects inside the unit disk invisible. We extend the approach of the previous sections to the elasticity system. Towards this end, we first design S-vanishing structures in the next subsection by canceling the first elastic scattering coefficients.

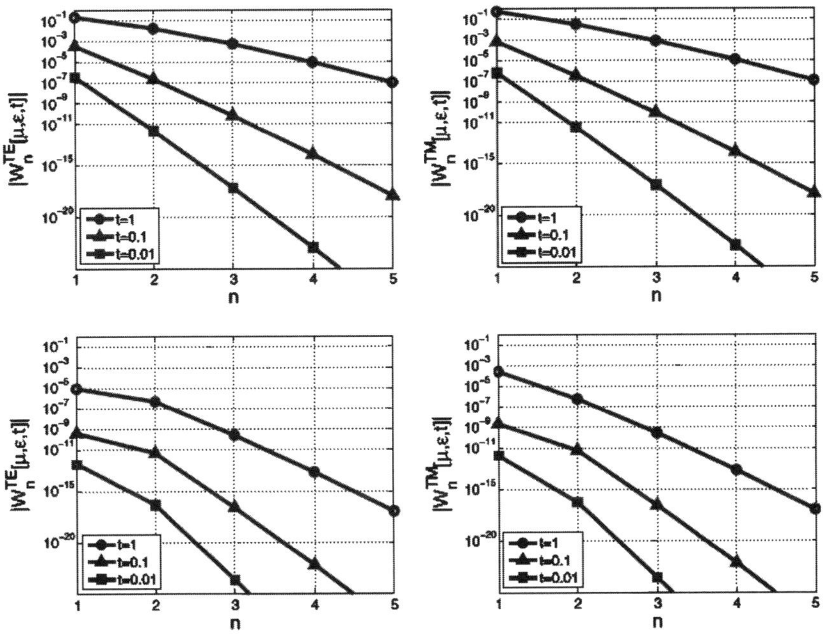

FIGURE 10.2. This figure shows the graph of $W_n^{TE}[\mu, \epsilon, t]$ and $W_n^{TM}[\mu, \epsilon, t]$ for various values of t. The first row is of the no-layer case, and the second row if of 6-layers S-vanishing structure of order $N = 2$ which is explained in Example 1. The values of W_n^{TE} and W_n^{TM} are much smaller in the S-vanishing structure than in the no-layer structure.

10.5.1. S-vanishing Structures. For positive numbers r_j $(j = 1, 2, \dots, L + 1)$ with $2 = r_1 > r_2 > \cdots > r_{L+1} = 1$ we construct a multi-layered structure by defining

$$\begin{aligned} A_0 :=& \{x \in \mathbb{R}^2 : \quad |x| > 2\} \\ A_j :=& \{x \in \mathbb{R}^2 : \quad r_{j+1} \leq |x| < r_j\}, \quad j = 1, \dots, L \\ A_{L+1} :=& \{x \in \mathbb{R}^2 : \quad |x| < 1\}. \end{aligned}$$

Let $(\lambda_j, \mu_j, \rho_j)$ be the Lamé parameters and densities of A_j for $j = 0, \dots, L + 1$. In particular, λ_0, μ_0 and ρ_0 are the parameters of the background medium. In the sequel, the piecewise constant parameters λ, μ and ρ are defined as

$$\lambda = \sum_{j=0}^{L+1} \lambda_j \, \chi(A_j), \mu = \sum_{j=0}^{L+1} \mu_j \, \chi(A_j), \text{ and } \rho = \sum_{j=0}^{L+1} \rho_j \, \chi(A_j), \tag{10.85}$$

in accordance with the aforementioned multi-layered structure. The scattering coefficients $W_{m,n}^{\alpha,\beta} = W_{m,n}^{\alpha,\beta}(\lambda, \mu, \rho, \omega)$ can be defined analogously to (2.506) and the total field $u = (u_1, u_2)^t$ solves the equation

$$\mathcal{L}_{\lambda,\mu} u + \rho \omega^2 u = 0 \quad \text{in} \quad \mathbb{R}^2. \tag{10.86}$$

Since the multi-layered structure is circularly symmetric it is easy to check that

$$W_{m,n}^{\alpha,\beta} = 0 \qquad \text{for all} \quad \alpha, \beta \in \{p, s\} \quad \text{and} \quad n \neq m.$$

Therefore, we have the following definition of the S-vanishing structures.

DEFINITION 10.12 (S-vanishing Structure). *The medium (λ, μ, ρ) defined by (10.85) is called an S-vanishing structure of order N at frequency ω if $W_{n,n}^{\alpha,\beta} = 0$ for all $|n| \leq N$ and $\alpha, \beta \in \{p, s\}$.*

In the rest of this subsection, we aim to construct an S-vanishing structure for general elastic waves. To facilitate the later analysis, the notation $T_{\lambda,\mu}$ is adopted for the surface traction operator $\partial/\partial\nu$ associated with elastic moduli λ and μ. In order to design envisioned structure, it suffices to construct (λ, μ, ρ) such that $W_n^{\alpha,\beta} := W_{n,n}^{\alpha,\beta} = 0$ for all $0 \leq n \leq N$ and $\alpha, \beta \in \{p, s\}$. We assume that the cloaked region $\{|x| < 1\}$ is a cavity, so that the scattered field u satisfies the traction-free boundary condition $T_{\lambda_{L+1},\mu_{L+1}} u := \partial u/\partial\nu = 0$ on $|x| = 1$. Note that the two-dimensional surface traction admits the expression

$$T_{\lambda,\mu} w = 2\mu(\nu \cdot \nabla w_1, \nu \cdot \nabla w_2) + \lambda N \, \nabla \cdot w + \mu T \, (\partial_2 w_1 - \partial_1 w_2), \quad w = (w_1, w_2),$$

in terms of the surface normal and tangent vectors $N = (n_1, n_2)$ and $T = (-n_1, n_2)$ on the surface respectively. The solutions u_n to (10.86) of the form

$$u_n(x) = \widehat{a}_j^{n,p} J_n^p(x) + \widehat{a}_j^{n,s} J_n^s(x) + a_j^{n,p} H_n^p(x) + a_j^{n,s} H_n^s(x), \quad x \in A_j, \; j = 0, \ldots, L,$$

are sought with unknown coefficients $\widehat{a}_j^{n,\alpha}, a_j^{n,\alpha} \in \mathbb{C}$, to be determined later. Intuitively, one should look for solutions u_n whose coefficients fulfill the relations

$$\widehat{a}_0^{n,p} \, \widehat{a}_0^{n,s} \neq 0 \quad \text{and} \quad a_0^{n,p} = a_0^{n,s} = 0 \quad \text{for all} \quad n = 0, \ldots, N. \tag{10.87}$$

Note that by (2.504) and (2.505), the scattering coefficients in this case turn out to be

$$\begin{cases} W_n^{\alpha,p} = \sqrt{-1} 4\rho_0 \omega^2 a_0^{n,\alpha} = 0 & \text{when } \widehat{a}_0^{n,p} = 1 \text{ and } \widehat{a}_0^{n,s} = 0, \\ W_n^{\alpha,s} = \sqrt{-1} 4\rho_0 \omega^2 a_0^{n,\alpha} = 0 & \text{when } \widehat{a}_0^{n,p} = 0 \text{ and } \widehat{a}_0^{n,s} = 1. \end{cases} \tag{10.88}$$

The solution u_n satisfies the transmission conditions

$$u_n|_+ = u_n|_- \quad \text{and} \quad T_{\lambda_{j-1},\mu_{j-1}} u_n|_+ = T_{\lambda_j,\mu_j} u_n|_- \quad \text{on} \quad |x| = r_j, \, \forall j = 1, \ldots, L. \tag{10.89}$$

Fairly easy calculations indicate that on $|x| = r$,

$$\begin{aligned} \widehat{e}_r \cdot [T_{\lambda,\mu} H_n^p(x)] &= 2\mu \frac{\partial^2 v_n(x, \kappa_p)}{\partial r^2} + \lambda \Delta v_n(x, \kappa_p) \\ &= 2\mu\kappa_p^2 (H_n^{(1)})''(r\kappa_p) e^{\sqrt{-1} n \varphi_x} - \lambda \kappa_p^2 H_n^{(1)}(r\kappa_p) e^{\sqrt{-1} n \varphi_x} \\ &= \frac{e^{\sqrt{-1} n \varphi_x}}{r^2} \left(-2\mu r \kappa_p (H_n^{(1)})'(r\kappa_p) + (2\mu n^2 - (\lambda + 2\mu) r^2 \kappa_p^2) H_n^{(1)}(r\kappa_p) \right), \\ &=: \frac{e^{\sqrt{-1} n \varphi_x}}{r^2} B_n^p(r\kappa_p, \lambda, \mu), \end{aligned}$$

and

$$\begin{aligned}\widehat{e}_\theta \cdot [T_{\lambda,\mu} H_n^p(x)] &= 2\mu \left(-\frac{1}{r^2}\frac{\partial v_n(x,\kappa_p)}{\partial \varphi_x} + \frac{1}{r}\frac{\partial^2 v_n(x,\kappa_p)}{\partial r \partial \varphi_x} \right) \\ &= \frac{e^{\sqrt{-1}n\varphi_x}}{r^2}(2\sqrt{-1}\mu n)\left(-H_n^{(1)}(r\kappa_p) + r\kappa_p (H_n^{(1)})'(r\kappa_p) \right) \\ &=: \frac{e^{\sqrt{-1}n\varphi_x}}{r^2} C_n^p(r\kappa_p,\lambda,\mu),\end{aligned}$$

where

$$\begin{aligned}B_n^p(t,\lambda,\mu) &:= -2\mu t (H_n^{(1)})'(t) + (2\mu n^2 - (\lambda+2\mu)t^2) H_n^{(1)}(t), \\ C_n^p(t,\lambda,\mu) &:= (2\sqrt{-1}\mu n)\left(-H_n^{(1)}(t) + t(H_n^{(1)})'(t) \right).\end{aligned}$$

In the sequel, the shorthand notation

$$B_{n,j}^p = B_n^p(r_j\kappa_p,\lambda_j,\mu_j) \quad \text{and} \quad C_{n,j}^p = C_n^p(r_j\kappa_p,\lambda_j,\mu_j)$$

is used for simplicity. It holds that

$$T_{\lambda_j,\mu_j} H_n^p(x) = \frac{1}{r_j^2}\left(B_{n,j}^p \, P_n(\hat{x}) + C_{n,j}^p \, S_n(\hat{x}) \right) \quad \text{on } |x| = r_j.$$

By arguing as for H_n^p, we obtain that

$$T_{\lambda_j,\mu_j} H_n^s(x) = \frac{1}{r_j^2}\left(B_{n,j}^s \, P_n(\hat{x}) + C_{n,j}^s \, S_n(\hat{x}) \right) \quad \text{on } |x| = r_j,$$

where

$$\begin{aligned}B_{n,j}^s &= B_n^s(t)\big|_{t=r_j\kappa_s} := (2\sqrt{-1}\mu n)\left(H_n^{(1)}(r_j\kappa_s) + r_j\kappa_s \left(H_n^{(1)}\right)'(r_j\kappa_s) \right), \\ C_{n,j}^s &= C_n^s(t)\big|_{t=r_j\kappa_s} := 2\mu(r_j\mathrm{K}_s)\left(H_n^{(1)}\right)'(r_j\kappa_s) + \left(-2\mu n^2 + 3\mu(r_j\kappa_s)^2 H_n^{(1)}(r_j\kappa_s) \right).\end{aligned}$$

Note that $B_{n,j}^s = C_{n,j}^p$. Analogously, we obtain

$$T_{\lambda_j,\mu_j} J_n^\alpha(x) = \frac{1}{r_j^2}\left(\widehat{B}_{n,j}^\alpha(r_j) \, P_n(\hat{x}) + \widehat{C}_{n,j}^\alpha(r_j) \, S_n(\hat{x}) \right), \quad \alpha = p, s,$$

where $\widehat{B}_{n,j}^\alpha$ and $\widehat{C}_{n,j}^\alpha$ are defined in the same way as $B_{n,j}^\alpha$ and $C_{n,j}^\alpha$ with $H_n^{(1)}$ replaced by J_n. Hence, the transmission conditions in (10.89) can be written as

$$M_{n,j-1}(r_j)(\widehat{a}_{j-1}^{n,p}, \widehat{a}_{j-1}^{n,s}, a_{j-1}^{n,p}, a_{j-1}^{n,s})^t = M_{n,j}(r_j)(\widehat{a}_j^{n,p}, \widehat{a}_j^{n,s}, a_j^{n,p}, a_j^{n,s})^t, \tag{10.90}$$

for $j = 1, \ldots, L$. Here $M_{n,j}$, $j = 0, \ldots, L$, $n = 0, \ldots, N$, is the 4×4 matrix defined by

$$M_{n,j}(r) := \begin{pmatrix} t_p J_n'(t_p) & \sqrt{-1}nJ_n(t_s) & t_p(H_n^{(1)})'(t_p) & \sqrt{-1}nH_n^{(1)}(t_s) \\ \sqrt{-1}nJ_n(t_p) & -t_sJ_n'(t_s) & \sqrt{-1}nH_n^{(1)}(t_p) & -t_sH_n^{(1)}(t_s) \\ \widehat{B}_{n,j}^p(t_p) & \widehat{B}_{n,j}^s(t_s) & B_{n,j}^p(t_p) & B_{n,j}^s(t_s) \\ \widehat{C}_{n,j}^p(t_p) & \widehat{C}_{n,j}^s(t_p) & C_{n,j}^s(t_s) & C_{n,j}^s(t_s) \end{pmatrix},$$

where $t_\alpha := r\kappa_\alpha$.

The traction-free boundary condition on $|x| = r_{L+1} = 1$ amounts to

$$M_{n,L+1}(\widehat{a}_L^{n,p}, \widehat{a}_L^{n,s}, a_L^{n,p}, a_L^{n,s})^t = (0,0,0,0)^t, \tag{10.91}$$

for $n = 0, \dots, N$ with

$$M_{n,L+1} := \begin{pmatrix} 0 & 0 & 0 & 0 \\ 0 & 0 & 0 & 0 \\ \widehat{B}^p_{n,L}(r_{L+1}\kappa_p) & \widehat{B}^s_{n,L}(r_{L+1}\kappa_s) & B^p_{n,L}(r_{L+1}\kappa_p) & B^s_{n,L}(r_{L+1}\kappa_s) \\ \widehat{C}^p_{n,L}(r_{L+1}\kappa_p) & \widehat{C}^s_{n,L}(r_{L+1}\kappa_s) & C^p_{n,L}(r_{L+1}\kappa_p) & C^s_{n,L}(r_{L+1}\kappa_s) \end{pmatrix}.$$

Combining (10.90) and (10.91) we obtain

$$(10.92) \quad \begin{cases} Q^{(n)}(\widehat{a}_0^{n,p}, \widehat{a}_0^{n,s}, a_0^{n,p}, a_0^{n,s})^t = (0,0,0,0)^t, \\ Q^{(n)} = Q^{(n)}(\lambda, \mu, \rho\omega^2) := M_{n,L+1} \\ \qquad \times \prod_{j=1}^L M_{n,j}^{-1}(r_j) M_{n,j-1}(r_j) = \begin{pmatrix} 0 & 0 \\ Q_{21}^{(n)} & Q_{22}^{(n)} \end{pmatrix}, \end{cases}$$

where $Q_{21}^{(n)}, Q_{22}^{(n)}$ are 2×2 matrix functions of λ, μ and $\rho\omega^2$.

Exactly as in the acoustic case in Section 10.3, one can show that the determinant of $Q_{22}^{(n)}$ is non-vanishing. In fact, if $\det(Q_{22}^{(n)}) = 0$, then one can derive a contradiction to the uniqueness of our forward scattering problems. Therefore, it suffices to look for the parameters λ_j, μ_j, ρ_j ($j = 1, 2, \dots, L$) from the nonlinear algebraic equations

$$(Q_{21}^{(n)})_{i,k}(\lambda, \mu, \rho\omega^2) = 0, \qquad i,k = 1,2, \quad n = 1, 2 \dots.$$

We are interested in a nearly S-vanishing structure of order N at low frequencies, that is, a structure (λ, μ, ρ) such that

$$W_n^{\alpha,\beta}(\lambda, \mu, \rho, \omega) = o(\omega^{2N}) \quad \text{for all} \quad \alpha, \beta \in \{p, s\}, \ |n| \le N, \quad \text{as} \quad \omega \to 0.$$

Towards this end, one needs to study the asymptotic behavior of $W_n^{\alpha,\beta}(\lambda, \mu, \rho, \omega)$ as ω tends to zero. In view of (10.88) and (10.92), it is found that

$$(10.93) \quad \begin{aligned} (W_n^{\alpha,p}, W_n^{\alpha,s})^t &= \sqrt{-1}4\rho_0\omega^2 \, (a_0^{n,p}, a_0^{n,s})^t \\ &= -\sqrt{-1}4\rho_0\omega^2 \, (Q_{22}^{(n)})^{-1} \, Q_{21}^{(n)} \, (\widehat{a}_0^{n,p}, \widehat{a}_0^{n,s})^t \end{aligned}$$

where $\widehat{a}_0^{n,p}$ and $\widehat{a}_0^{n,s}$ are selected depending on (10.88).

Let W_n denote the 2×2 matrix

$$W_n = \begin{pmatrix} W_n^{p,p} & W_n^{s,p} \\ W_n^{p,s} & W_n^{s,s} \end{pmatrix}.$$

Then, the following result based on relation (10.93) elucidates the low-frequency asymptotic behavior of W_n [**1**].

THEOREM 10.13. *For all $n \in \mathbb{N}$, we have*

(10.94)

$$W_n(\lambda, \mu, \rho, \omega) = \omega^{2n+2} \left(V_{n,0}(\lambda, \mu, \rho) + \sum_{l=0}^{N-n} \sum_{j=0}^{(L+1)l} \omega^{2l} \, (\ln \omega)^j \, V_{n,l,j}(\lambda, \mu, \rho) \right) + \Upsilon_n$$

as $\omega \to 0$, where the matrices $V_{n,0}$ and $V_{n,l,j}$ are defined by

$$V_{n,0} = \begin{pmatrix} V_{n,0}^{p,p} & V_{n,0}^{s,p} \\ V_{n,0}^{p,s} & V_{n,0}^{s,s} \end{pmatrix} \quad \text{and} \quad V_{n,l,j} = \begin{pmatrix} V_{n,l,j}^{p,p} & V_{n,l,j}^{s,p} \\ V_{n,l,j}^{p,s} & V_{n,l,j}^{s,s} \end{pmatrix},$$

in terms of some $V_{n,0}^{\alpha,\beta}$ *and* $V_{n,l,j}^{\alpha,\beta}$ *dependent on* λ, μ, ρ *but independent of* ω. *The residual matrix* $\Upsilon_n = (\Upsilon_{ik}^n)_{i,k=1,2}$ *is such that* $|\Upsilon_{ik}^n| \leq C\omega^{2N+2}$, *for all* $i, k = 1, 2$, *where the constant* $C \in \mathbb{R}_+$ *is independent of* ω.

The analytic expressions of the quantities $V_{n,0}^{\alpha,\beta}$ and $V_{n,l,j}^{\alpha,\beta}$ in terms of λ_j, μ_j and ρ_j are very complicated, but can be extracted, for example, by using the symbolic toolbox of MATLAB. Theorem 10.13 follows from (10.93) and the low-frequency asymptotics of $Q_{22}^{(n)}(\lambda, \mu, \rho\omega^2)$ and $Q_{21}^{(n)}(\lambda, \mu, \rho\omega^2)$ as $\omega \to 0$. The latter can be derived based on the definition given in (10.92) in combination with the expansion formulas for Bessel and Neumann functions and their derivatives for small arguments. We refer the reader to [**1**, Appendix D] for the proof of Theorem 10.13.

In order to construct a nearly S-vanishing structure of order N at low frequencies, thanks to Theorem 10.13 we need to determine the parameters λ_j, μ_j and ρ_j from the equations

$$V_{n,0}^{\alpha,\beta}(\lambda, \mu, \rho) = V_{n,l,j}^{\alpha,\beta}(\lambda, \mu, \rho) = 0,$$

for all $0 \leq n \leq N$, $1 \leq l \leq (N-n)$, $1 \leq j \leq (L+1)l$ and $\alpha, \beta \in \{p, s\}$. Numerically, this can be achieved by applying, for example, the gradient descent method to the minimization problem

$$\min_{\lambda_j, \mu_j, \rho_j} \sum_{\alpha,\beta \in \{p,s\}} \left\{ \left| V_{n,0}^{\alpha,\beta} \right|^2 + \sum_{l=0}^{N-n} \sum_{j=0}^{(L+1)l} \left| V_{n,l,j}^{\alpha,\beta} \right|^2 \right\}.$$

10.5.2. Enhancement of Near Cloaking. The aim of this section is to show that the nearly S-vanishing structures constructed in Subsection 10.5.1 can be used to enhance the cloaking effect in elasticity. The enhancement of near cloaking is based on the idea of transformation optics used in the previous sections. Let (λ, μ, ρ) be a nearly S-vanishing structure of order N at low frequencies, taking the form of (10.85). This implies that for some fixed $\omega > 0$ there exists $\delta_0 > 0$ such that

$$\left| W_{m,n}^{\alpha,\beta}[\lambda, \mu, \rho, \delta\omega] \right| = o(\delta^{2N+2}), \quad |n| \leq N, \quad \delta \leq \delta_0.$$

On the other hand, recall from the proof of Lemma 2.139 that

$$\left| W_n^{\alpha,\beta}[\lambda, \mu, \rho, \delta\omega] \right| \leq C\delta^{2N} \quad \text{for all} \quad |n| \geq N, \ \delta \leq \delta_0. \tag{10.95}$$

Hence, by Theorem 2.140, the far-field elastic scattering amplitudes can be estimated by

$$u_\alpha^\infty[\lambda, \mu, \rho, \delta\omega](\hat{x}, \hat{x}') = o(\delta^{2N-1}), \quad \alpha = p, s, \quad \text{as} \quad \delta \to 0 \tag{10.96}$$

uniformly in all observation directions $\hat{x}$ and incident directions $\hat{x}'$. Introduce the transformation on $\mathbb{R}^2$:

$$\Psi_\delta(x) := \frac{1}{\delta} x, \quad x \in \mathbb{R}^2.$$

Then, by arguing as in the acoustic and electromagnetic case, we have

$$u_\alpha^\infty[\lambda \circ \Psi_\delta, \mu \circ \Psi_\delta, \rho \circ \Psi_\delta, \omega] = u_\alpha^\infty[\lambda, \mu, \rho, \delta\omega] = o(\delta^{2N-1}) \quad \text{for all} \quad \delta \leq \delta_0.$$

Note that the medium $(\lambda \circ \Psi_\delta, \mu \circ \Psi_\delta, \rho \circ \Psi_\delta)$ is a homogeneous multi-coated structure of radius 2δ.

We now apply the transformation invariance of the Lamé system to the medium $(\lambda \circ \Psi_\delta, \mu \circ \Psi_\delta, \rho \circ \Psi_\delta)$. Recall that the elastic wave propagation in such a homogeneous isotropic medium can be restated as

$$\nabla \cdot (\mathfrak{C} : \nabla u) + \omega^2 (\rho \circ \Psi_\delta) u = 0 \quad \text{in} \quad \mathbb{R}^2,$$

where $\mathfrak{C} = (C_{ijkl})_{i,j,k,l=1}^N$ is the rank-four stiffness tensor defined by

$$C_{ijkl}(x) = (\lambda \circ \Psi_\delta)\, \delta_{i,j}\delta_{k,l} + (\mu \circ \Psi_\delta)\,(\delta_{i,k}\delta_{j,l} + \delta_{i,l}\delta_{j,k}), \tag{10.97}$$

and the action of $\mathfrak{C}$ on a matrix $A = (a_{ij})_{i,j=1,2}$ is defined as

$$\mathfrak{C} : A = (\mathfrak{C} : A)_{i,j=1}^2 = \left(\sum_{k,l=1,2} C_{ijkl}\, a_{kl} \right)_{i,j=1,2}. \tag{10.98}$$

In the case of a generic anisotropic elastic material, the stiffness tensor satisfies the following symmetries
(10.99)

$$\text{major symmetry:} \quad C_{ijkl} = C_{klij}, \qquad \text{minor symmetry:} \quad C_{ijkl} = C_{jikl} = C_{ijlk},$$

for all $i,j,k,l = 1,2$. Let $\tilde{x} = (\tilde{x}_1, \tilde{x}_2) = F_\delta(x) : \mathbb{R}^2 \to \mathbb{R}^2$ be a bi-Lipschitz and orientation-preserving transformation such that $F_\delta(\{|x| < \delta\}) = \{|\tilde{x}| < 1\}$ and the region $|x| \geq 2$ remains invariant under the transformation. This implies that we have blown up a small traction-free disk of radius $\delta < 1$ to the unit disk centered at the origin. The push-forwards of $\mathfrak{C}$ and ρ are defined respectively by

$$\begin{aligned}
(F_\delta)_* \mathfrak{C} := \widehat{\mathfrak{C}} &= \left(\widehat{C}_{iqkp}(\tilde{x}) \right)_{i,q,k,p=1}^2 \\
&= \left(\frac{1}{\det(M)} \left\{ \sum_{l,j=1,2} C_{ijkl} \frac{\partial \tilde{x}_p}{\partial x_l} \frac{\partial \tilde{x}_q}{\partial x_j} \right\} \Bigg|_{x = F_\delta^{-1}(\tilde{x})} \right)_{i,q,k,p=1,2}, \\
(F_\delta)_* \rho := \widehat{\rho} &= \left(\frac{\rho}{\det(M)} \right) \Bigg|_{x = F_\delta^{-1}(\tilde{x})}, \qquad M = \left(\frac{\partial \tilde{x}_i}{\partial x_j} \right)_{i,j=1,2}.
\end{aligned}$$

We need the following lemma (see, for instance, [**301, 398**]).

LEMMA 10.14. *The function u is a solution to $\nabla \cdot (\mathfrak{C} : \nabla u) + \omega^2 \rho u = 0$ in $\mathbb{R}^2$ if and only if $\widehat{u} = u \circ (F_\delta)^{-1}$ satisfies $\widehat{\nabla} \cdot (\widehat{\mathfrak{C}} : \widehat{\nabla} \widehat{u}) + \omega^2 \widehat{\rho} \widehat{u} = 0$ in $\mathbb{R}^2$, where $\widehat{\nabla}$ denotes the gradient operator with respect to the transformed variable $\tilde{x}$.*

Applying the above lemma to the Lamé system (10.97) we obtain the following result [**1**].

THEOREM 10.15. *If (λ, μ, ρ) is a nearly S-vanishing structure of order N at low frequencies, there exists $\delta_0 > 0$ such that*

$$u_\alpha^\infty[(F_\delta)_* \mathfrak{C}, (F_\delta)_*(\rho \circ \Psi_\delta), \omega](x, x') = o(\delta^{2N-1}), \quad \alpha = p, s,$$

for all $\delta < \delta_0$, uniformly in all x and x'. Here the stiffness tensor $\mathfrak{C}$ is defined by (10.97). Moreover, an elastic medium $((F_\delta)_* \mathfrak{C}, (F_\delta)_*(\rho \circ \Psi_\delta))$ in $1 < |x| < 2$ is a nearly cloaking device for the hidden *region $|x| < 1$.*

Theorem 10.15 implies that for any frequency ω and any integer number N there exist $\delta_0 = \delta_0(\omega, N) > 0$ and the elastic medium $((F_\delta)_* \mathfrak{C}, (F_\delta)_*(\rho \circ \Psi_\delta))$ with $\delta < \delta_0$ such that the nearly cloaking enhancement can be achieved at the order $o(\delta^{2N-2})$. We finish this section with the following remarks.

REMARK 10.16. *Unlike the acoustic and electromagnetic case, the transformed elastic tensor* $(F_\delta)_*\mathfrak{C}$ *is not anisotropic, since it possesses the major symmetry only. Note that the transformed mass density* $(F_\delta)_*(\rho \circ \Psi_\delta)$ *is still isotropic. In fact, it has been pointed out by Milton, Briane and Willis* [**398**] *that the invariance of the Lamé system can be achieved only if one relaxes the assumption on the minor symmetry of the transformed elastic tensor. This has led Norris and Shuvalov* [**418**] *and Parnell* [**441**] *to explore the elastic cloaking by using Cosserat material or by employing non-linear pre-stress in a neo-Hookean elastomeric material.*

REMARK 10.17. *We have designed an enhanced nearly cloaking device for general incoming elastic plane waves. A device for cloaking only compressional or shear waves can be analogously constructed by using the corresponding elastic scattering coefficients.*

10.6. Concluding Remarks

In this chapter, near-cloaking examples for electromagnetic and elastic waves have been shown. Based on the scattering coefficients vanishing approach, cloaking devices that achieve an enhanced interior cloaking effect have been designed. Such cloaking devices have been obtained via blow-up using the transformation optics of multicoated domains. The cloaking devices have anisotropic material parameters. Nevertheless, they can be approximated by concentric isotropic homogeneous coatings [**459**]. For wave propagation problems, when considering near-cloaking for the Helmholtz or Maxwell equations, it was proved in [**44**] that cloaking is increasingly difficult as the cloaked object becomes bigger or the operating frequency becomes higher. The difficulty scales proportionally to the object diameter of the frequency. Another important observation made in [**44**] is that the reduction factor of the scattering cross-section is higher in the backscattering region than in the forwarded one. This is due to the creeping waves propagating in the shadow region. As a consequence, the cloaking problem becomes easier if only scattered waves at certain angles are visible. The constructions proposed in this chapter can be extended to the enhanced reshaping problem. In [**44**], it was also shown how to make any target look like a disc with homogeneous physical parameters.

CHAPTER 11

Anomalous Resonance Cloaking and Shielding

11.1. Introduction

In this chapter, we consider the dielectric problem with a source term, which models the quasi-static (zero-frequency) transverse magnetic regime. The cloaking of the source is achieved in a region external to a plasmonic structure. The plasmonic structure consists of a shell having relative permittivity $-1 + \sqrt{-1}\delta$ with δ modeling losses.

The cloaking issue is directly linked to the existence of anomalous localized resonance (ALR), which is tied to the fact that an elliptic system of equations can exhibit localization effects near the boundary of ellipticity. The plasmonic structure exhibits ALR if, as the loss parameter δ goes to zero, the magnitude of the quasi-static in-plane electric field diverges throughout a specific region (with sharp boundary not defined by any discontinuities in the relative permittivity), called the anomalous resonance region, but converges to a smooth field outside that region. The anomalous feature of the resonance is that it is not associated to a finite dimensional eigenvalue and a forcing term at or near the resonant frequency. Instead, the resonance here is associated to an infinite dimensional kernel of the limiting (non-elliptic) operator. The localized feature of the resonance refers to the fact that the resonance is spatially localized.

To state the problem, let Ω_e be a bounded domain in $\mathbb{R}^2$ and let Ω_i be a domain whose closure is contained in Ω_e. Throughout this chapter, we assume that Ω_e and Ω_i are smooth. For a given loss parameter $\delta > 0$, the permittivity distribution in $\mathbb{R}^2$ is given by

$$\varepsilon_\delta = \begin{cases} 1 & \text{in } \mathbb{R}^2 \setminus \overline{\Omega_e}\,, \\ -1 + \sqrt{-1}\delta & \text{in } \Omega_e \setminus \overline{\Omega_i}\,, \\ 1 & \text{in } \Omega_i\,. \end{cases} \tag{11.1}$$

We may consider the configuration as a core with permittivity 1 coated by the shell $\Omega_e \setminus \overline{\Omega_i}$ with permittivity $-1 + \sqrt{-1}\delta$. This structure is called a superlens. It is inserted into a medium with permittivity 1. It turns out that quite interesting behavior happens in the limit as $\delta \to 0$. The superlens acts as an exterior cloaking device for certain sources since the resonance cancels the effect of those sources.

For a given function f compactly supported in $\mathbb{R}^2$ satisfying

$$\int_{\mathbb{R}^2} f\,dx = 0 \tag{11.2}$$

(which physically is required by conservation of charge), we consider the following dielectric problem:

$$\nabla \cdot \varepsilon_\delta \nabla V_\delta = \alpha f \quad \text{in } \mathbb{R}^2\,, \tag{11.3}$$

with the decay condition $V_\delta(x) \to 0$ as $|x| \to \infty$.

A fundamental problem is to identify those sources f such that when $\alpha = 1$ then first

$$E_\delta := \int_{\Omega_e \setminus \overline{\Omega_i}} \delta |\nabla V_\delta|^2 dx \to \infty \quad \text{as } \delta \to 0 . \tag{11.4}$$

and second V_δ remains bounded outside some radius a:

$$|V_\delta(x)| < C, \text{ when } |x| > a \tag{11.5}$$

for some constants C and a independent of δ (which requires that the ball $\Omega_a := \{|x| < a\}$ contains the entire region of anomalous localized resonance). The quantity E_δ is proportional to the electromagnetic power dissipated into heat by the time harmonic electrical field averaged over time. Hence (11.4) implies an infinite amount of energy dissipated per unit time in the limit $\delta \to 0$ which is unphysical. If instead we choose $\alpha = 1/\sqrt{E_\delta}$ then the source αf will produce the same power independent of δ and the new associated solution V_δ (which is the previous solution V_δ multiplied by α) will approach zero outside the radius a: cloaking due to anomalous localized resonance (CALR) occurs. The conditions (11.4) and (11.5) are sufficient to ensure CALR: a necessary and sufficient condition is that (with $\alpha = 1$) $V_\delta/\sqrt{E_\delta}$ goes to zero outside some radius as $\delta \to 0$. We also consider a weaker blow-up of the energy dissipation, namely,

$$\limsup_{\delta \to 0} E_\delta = \infty . \tag{11.6}$$

We say that weak CALR takes place if (11.6) holds (in addition to (11.5)). Then the (renormalized) source $f/\sqrt{E_\delta}$ will be essentially invisible for an infinite sequence of small values of δ tending to zero (but would be visible for values of δ interspersed between this sequence if CALR does not additionaly hold).

The aim of this chapter is to review a general method based on the potential theory to study cloaking due to anomalous resonance. Using layer potential techniques, we reduce the problem to a singularly perturbed system of integral equations. The system is non-self-adjoint. A symmetrization technique can be applied in the general case. In the case of an annulus (Ω_i is the disk of radius ρ_i and Ω_e is the concentric disk of radius ρ_e), it is known [**399**] that there exists a critical radius (the cloaking radius)

$$\rho_\star = \sqrt{\rho_e^3 {\rho_i}^{-1}} . \tag{11.7}$$

such that any finite collection of dipole sources located at fixed positions within the annulus $\Omega_{\rho_\star} \setminus \overline{\Omega_e}$ is cloaked. We show that if f is an integrable function supported in $E \subset \Omega_{\rho_\star} \setminus \overline{\Omega_e}$ satisfying (11.2) and the Newtonian potential of f does not extend as a harmonic function in $\Omega_{\rho_\star}$, then weak CALR takes place. Moreover, we show that if the Fourier coefficients of the Newtonian potential of f satisfy a mild gap condition, then CALR takes place. Conversely we show that if the source function f is supported outside $\Omega_{\rho_\star}$ then (11.4) does not happen and no cloaking occurs.

In this chapter, we also show that a cylindrical superlens can also act as a new kind of shielding device if the core is eccentric to the shell. Shielding is the phenomenon that is observed when a Faraday cage operates to block the effects of an electric field. Such a cage can block the effects of an external field on its internal contents, or the effects of an internal field on the outside environment. While such a conventional device shields a region enclosed by the device, a superlens with an eccentric core can shield a non-coated region which is located outside the device. We call this phenomenon *shielding at a distance.* The key element to study in the eccentric case is the Möbius transformation via which a concentric annulus is transformed into an eccentric one. We also provide various numerical examples to show the cloaking and shielding effects due to anomalous resonance.

This chapter is organized as follows. In Section 11.2 we transform the problem into a system of integral equations using layer potentials. In Section 11.3, we treat the special case of an annulus. In Section 11.4 we investigate the conditions required for shielding at a distance and geometric features such as the location and size of the shielded region. The results on cloaking are from [**34**] and those on shielding at a distance are from [**514**]. As shown in this chapter, plasmonic resonance effects have many applications in cloaking and shielding. This is one of the reasons why the development of negative index metamaterials is another very much-studied research area [**162**, **163**, **365**, **486**].

11.2. Layer Potential Formulation

As in Chapter 2, for $\partial\Omega_i$ or $\partial\Omega_e$, we denote, respectively, the single- and double-layer potentials of a function $\phi \in L^2$ as $\mathcal{S}^0_{\Omega_i}[\phi]$ and $\mathcal{D}^0_{\Omega_e}[\phi]$. We also introduce the associated Neumann–Poincaré operators $\mathcal{K}^0_{\Omega_i}$ and $\mathcal{K}^0_{\Omega_e}$.

Let F be the Newtonian potential of f, *i.e.*,

$$F(x) = \int_{\mathbb{R}^2} \Gamma(x, y) f(y) dy, \quad x \in \mathbb{R}^2 . \tag{11.8}$$

Then F satisfies $\Delta F = f$ in $\mathbb{R}^2$, and the solution V_δ to (11.3) may be represented as

$$V_\delta(x) = F(x) + \mathcal{S}^0_{\Omega_i}[\phi_i](x) + \mathcal{S}^0_{\Omega_e}[\phi_e](x) \tag{11.9}$$

for some functions $\phi_i \in L^2_0(\partial\Omega_i)$ and $\phi_e \in L^2_0(\partial\Omega_e)$ (L^2_0 is the collection of all square integrable functions with the integral zero). The transmission conditions along the interfaces $\partial\Omega_e$ and $\partial\Omega_i$ satisfied by V_δ read

$$\begin{aligned} (-1+\sqrt{-1}\delta)\frac{\partial V_\delta}{\partial \nu}\Big|_+ &= \frac{\partial V_\delta}{\partial \nu}\Big|_- \quad \text{on } \partial\Omega_i , \\ \frac{\partial V_\delta}{\partial \nu}\Big|_+ &= (-1+\sqrt{-1}\delta)\frac{\partial V_\delta}{\partial \nu}\Big|_- \quad \text{on } \partial\Omega_e . \end{aligned}$$

Hence the pair of potentials (ϕ_i, ϕ_e) is the solution to the following system of integral equations:

$$\begin{cases} (-1+\sqrt{-1}\delta)\dfrac{\partial \mathcal{S}^0_{\Omega_i}[\phi_i]}{\partial \nu_i}\Big|_+ - \dfrac{\partial \mathcal{S}^0_{\Omega_i}[\phi_i]}{\partial \nu_i}\Big|_- + (-2+\sqrt{-1}\delta)\dfrac{\partial \mathcal{S}^0_{\Omega_e}[\phi_e]}{\partial \nu_i} \\ \quad = (2-\sqrt{-1}\delta)\dfrac{\partial F}{\partial \nu_i} \quad \text{on } \partial\Omega_i , \\ (2-\sqrt{-1}\delta)\dfrac{\partial \mathcal{S}^0_{\Omega_i}[\phi_i]}{\partial \nu_e} + \dfrac{\partial \mathcal{S}^0_{\Omega_e}[\phi_e]}{\partial \nu_e}\Big|_+ - (-1+\sqrt{-1}\delta)\dfrac{\partial \mathcal{S}^0_{\Omega_e}[\phi_e]}{\partial \nu_e}\Big|_- \\ \quad = (-2+\sqrt{-1}\delta)\dfrac{\partial F}{\partial \nu_e} \quad \text{on } \partial\Omega_e . \end{cases}$$

Note that we have used the notation ν_i and ν_e to indicate the outward normal on $\partial\Omega_i$ and $\partial\Omega_e$, respectively. Using the jump formula (2.8) for the normal derivative of the single-layer potentials, the above equations can be rewritten as

$$\begin{bmatrix} -z_\delta I + (\mathcal{K}^0_{\Omega_i})^* & \dfrac{\partial}{\partial \nu_i} \mathcal{S}^0_{\Omega_e} \\ \dfrac{\partial}{\partial \nu_e} \mathcal{S}^0_{\Omega_i} & z_\delta I + (\mathcal{K}^0_{\Omega_e})^* \end{bmatrix} \begin{bmatrix} \phi_i \\ \phi_e \end{bmatrix} = - \begin{bmatrix} \dfrac{\partial F}{\partial \nu_i} \\ \dfrac{\partial F}{\partial \nu_e} \end{bmatrix} \tag{11.10}$$

on $L^2_0(\partial \Omega_i) \times L^2_0(\partial \Omega_e)$, where we set

$$z_\delta = \frac{\sqrt{-1}\delta}{2(2 - \sqrt{-1}\delta)}. \tag{11.11}$$

Note that the operator in (11.10) can be viewed as a compact perturbation of the operator

$$R_\delta := \begin{bmatrix} -z_\delta I + (\mathcal{K}^0_{\Omega_i})^* & 0 \\ 0 & z_\delta I + (\mathcal{K}^0_{\Omega_e})^* \end{bmatrix}. \tag{11.12}$$

From Lemma 2.2, it follows that the eigenvalues of $(\mathcal{K}^0_{\Omega_i})^*$ and $(\mathcal{K}^0_{\Omega_e})^*$ lie in the interval $(-\frac{1}{2}, \frac{1}{2}]$. Observe that $z_\delta \to 0$ as $\delta \to 0$ and that there are sequences of eigenvalues of $(\mathcal{K}^0_{\Omega_i})^*$ and $(\mathcal{K}^0_{\Omega_e})^*$ approaching 0 since $(\mathcal{K}^0_{\Omega_i})^*$ and $(\mathcal{K}^0_{\Omega_e})^*$ are compact. So 0 is the essential singularity of the operator valued meromorphic function

$$\lambda \in \mathbb{C} \mapsto (\lambda I + (\mathcal{K}^0_{\Omega})^*)^{-1}.$$

This causes a serious difficulty in dealing with (11.10). We emphasize that $(\mathcal{K}^0_{\Omega_e})^*$ is not self-adjoint in general. In fact, $(\mathcal{K}^0_{\Omega_e})^*$ is self-adjoint only when $\partial \Omega_e$ is a circle (or a sphere in three dimensions).

Let $\mathcal{H} = L^2(\partial \Omega_i) \times L^2(\partial \Omega_e)$. We write (11.10) in a slightly different form. We first apply the operator

$$\begin{bmatrix} -I & 0 \\ 0 & I \end{bmatrix} : \mathcal{H} \to \mathcal{H}$$

to (11.10). Then the equation becomes

$$\begin{bmatrix} z_\delta I - (\mathcal{K}^0_{\Omega_i})^* & -\dfrac{\partial}{\partial \nu_i} \mathcal{S}^0_{\Omega_e} \\ \dfrac{\partial}{\partial \nu_e} \mathcal{S}^0_{\Omega_i} & z_\delta I + (\mathcal{K}^0_{\Omega_e})^* \end{bmatrix} \begin{bmatrix} \phi_i \\ \phi_e \end{bmatrix} = \begin{bmatrix} \dfrac{\partial F}{\partial \nu_i} \\ -\dfrac{\partial F}{\partial \nu_e} \end{bmatrix}. \tag{11.13}$$

Let the Neumann–Poincaré-type operator $\mathbb{K}^* : \mathcal{H} \to \mathcal{H}$ be defined by

$$\mathbb{K}^* := \begin{bmatrix} -(\mathcal{K}^0_{\Omega_i})^* & -\dfrac{\partial}{\partial \nu_i} \mathcal{S}^0_{\Omega_e} \\ \dfrac{\partial}{\partial \nu_e} \mathcal{S}^0_{\Omega_e} & (\mathcal{K}^0_{\Omega_e})^* \end{bmatrix}, \tag{11.14}$$

and let

$$\Phi := \begin{bmatrix} \phi_i \\ \phi_e \end{bmatrix}, \quad g := \begin{bmatrix} \dfrac{\partial F}{\partial \nu_i} \\ -\dfrac{\partial F}{\partial \nu_e} \end{bmatrix}. \tag{11.15}$$

Then, (11.13) can be rewritten in the form

$$(z_\delta \mathbb{I} + \mathbb{K}^*)[\Phi] = g, \tag{11.16}$$

where $\mathbb{I}$ is given by

$$\mathbb{I} = \begin{bmatrix} I & 0 \\ 0 & I \end{bmatrix}.$$

11.3. Anomalous Resonance in an Annulus

In this section we consider the anomalous resonance when the domains Ω_e and Ω_i are concentric disks. We calculate the explicit form of the limiting solution. Throughout this section, we set $\Omega_e = \{|x| < \rho_e\}$ and $\Omega_i = \{|x| < \rho_i\}$, where $\rho_e > \rho_i$.

According to (2.45) and (2.46), if Φ is given by

$$\Phi = \sum_{n \neq 0} \begin{bmatrix} \phi_i^n \\ \phi_e^n \end{bmatrix} e^{\sqrt{-1}n\theta},$$

then

$$\mathbb{K}^*[\Phi] = \sum_{n \neq 0} \begin{bmatrix} \dfrac{(\rho_i/\rho_e)^{|n|-1}}{2} \phi_e^n \\ \dfrac{(\rho_i/\rho_e)^{|n|+1}}{2} \phi_i^n \end{bmatrix} e^{\sqrt{-1}n\theta}.$$

Thus, if g is given by

$$g = \sum_{n \neq 0} \begin{bmatrix} g_i^n \\ g_e^n \end{bmatrix} e^{\sqrt{-1}n\theta},$$

the integral equations (11.16) are equivalent to

$$\begin{cases} z_\delta \phi_i^n + \dfrac{(\rho_i/\rho_e)^{|n|-1}}{2} \phi_e^n = g_i^n, \\ z_\delta \phi_e^n + \dfrac{(\rho_i/\rho_e)^{|n|+1}}{2} \phi_i^n = g_e^n, \end{cases} \tag{11.17}$$

for every $|n| \geq 1$. It is readily seen that the solution $\Phi = (\phi_i, \phi_e)$ to (11.17) is given by

$$\phi_i = 2 \sum_{n \neq 0} \frac{2z_\delta g_i^n - (\rho_i/\rho_e)^{|n|-1} g_e^n}{4z_\delta^2 - (\rho_i/\rho_e)^{2|n|}} e^{\sqrt{-1}n\theta},$$

$$\phi_e = 2 \sum_{n \neq 0} \frac{2z_\delta g_e^n - (\rho_i/\rho_e)^{|n|+1} g_i^n}{4z_\delta^2 - (\rho_i/\rho_e)^{2|n|}} e^{\sqrt{-1}n\theta}.$$

If the source is located outside the structure, *i.e.*, f is supported in $\mathbb{R}^2 \setminus \overline{\Omega_e}$, then the Newtonian potential of f, F, is harmonic in Ω_e and

$$F(x) = c - \sum_{n \neq 0} \frac{g_e^n}{|n| \rho_e^{|n|-1}} r^{|n|} e^{\sqrt{-1}n\theta}, \tag{11.18}$$

for $|x| \leq \rho_e$, where g is defined by (11.15). Thus we have

$$g_i^n = -g_e^n (\rho_i/\rho_e)^{|n|-1}. \tag{11.19}$$

Here, g_e^n is the Fourier coefficient of $-\frac{\partial F}{\partial \nu_e}$ on Γ_e, or in other words,

$$-\frac{\partial F}{\partial \nu_e} = \sum_{n \neq 0} g_e^n e^{\sqrt{-1}n\theta}. \tag{11.20}$$

We then get

$$
\begin{cases}
\phi_i = -2\displaystyle\sum_{n\neq 0} \frac{(2z_\delta+1)(\rho_i/\rho_e)^{|n|-1} g_e^n}{4z_\delta^2 - (\rho_i/\rho_e)^{2|n|}} e^{\sqrt{-1}n\theta}, \\
\phi_e = 2\displaystyle\sum_{n\neq 0} \frac{(2z_\delta + (\rho_i/\rho_e)^{2|n|}) g_e^n}{4z_\delta^2 - (\rho_i/\rho_e)^{2|n|}} e^{\sqrt{-1}n\theta}.
\end{cases}
\tag{11.21}
$$

Therefore, from (2.38) we find that

$$
\mathcal{S}_D^0[\phi_i](x) + \mathcal{S}_\Omega^0[\phi_e](x) = \sum_{n\neq 0} \frac{2(\rho_i^{2|n|} - \rho_e^{2|n|}) z_\delta}{|n|\rho_e^{|n|-1}(4z_\delta^2 - (\rho_i/\rho_e)^{2|n|})} \frac{g_e^n}{r^{|n|}} e^{\sqrt{-1}n\theta}, \quad \rho_e < r = |x|,
\tag{11.22}
$$

and

$$
\mathcal{S}_D^0[\phi_i](x) = -\sum_{n\neq 0} \frac{\rho_i^{2|n|}(2z_\delta+1)}{|n|\rho_e^{|n|-1}((\rho_i/\rho_e)^{2|n|} - 4z_\delta^2)} \frac{g_e^n}{r^{|n|}} e^{\sqrt{-1}n\theta}, \quad \rho_i < r = |x| < \rho_e,
\tag{11.23}
$$

$$
\mathcal{S}_\Omega^0[\phi_e](x) = \sum_{n\neq 0} \frac{(2z_\delta + (\rho_i/\rho_e)^{2|n|})}{|n|\rho_e^{|n|-1}((\rho_i/\rho_e)^{2|n|} - 4z_\delta^2)} g_e^n r^{|n|} e^{\sqrt{-1}n\theta}, \quad \rho_i < r = |x| < \rho_e.
\tag{11.24}
$$

We next obtain the following lemma which provides essential estimates for the investigation of this section.

LEMMA 11.1. *There exists δ_0 such that*

$$
E_\delta := \int_{\Omega_e \setminus \overline{\Omega_i}} \delta |\nabla V_\delta|^2 \approx \sum_{n\neq 0} \frac{\delta |g_e^n|^2}{|n|(\frac{\delta^2}{4} + (\rho_i/\rho_e)^{2|n|})}
\tag{11.25}
$$

uniformly in $\delta \le \delta_0$.

PROOF. Using (11.18), (11.23), and (11.24), one can see that

$$
V_\delta(x) = c + \rho_e \sum_{n\neq 0} \left[\frac{\rho_i^{2|n|}}{r^{|n|}}(2z_\delta+1) - (4z_\delta^2 + 2z_\delta) r^{|n|} \right] \frac{g_e^n e^{\sqrt{-1}n\theta}}{|n|\rho_e^{|n|}(4z_\delta^2 - (\rho_i/\rho_e)^{2|n|})}.
$$

Then straightforward computations yield that

$$
E_\delta \approx \rho_e^2 \sum_{n\neq 0} \delta(1 - (\rho_i/\rho_e)^{2|n|}) \left| \frac{2z_\delta + 1}{4z_\delta^2 + (\rho_i/\rho_e)^{2|n|}} \right|^2 (4|z_\delta|^2 - (\rho_i/\rho_e)^{2|n|}) \frac{|g_e^n|^2}{|n|}.
$$

If δ is sufficiently small, then one can also easily show that

$$
|4z_\delta^2 - (\rho_i/\rho_e)^{2|n|}| \approx \frac{\delta^2}{4} + (\rho_i/\rho_e)^{2|n|}.
$$

Therefore we get (11.25) and the proof is complete. □

We next investigate the behavior of the series in the right-hand side of (11.25). Let

$$
N_\delta = \frac{\ln(\delta/2)}{\ln(\rho_i/\rho_e)}.
\tag{11.26}
$$

If $|n| \le N_\delta$, then $(\delta/2) \le (\rho_i/\rho_e)^{|n|}$, and hence

$$
(11.27)\quad \begin{aligned}
\sum_{n\neq 0} \frac{\delta |g_e^n|^2}{|n|(\frac{\delta^2}{4} + (\rho_i/\rho_e)^{2|n|})} &\geq \sum_{0\neq |n|\leq N_\delta} \frac{\delta |g_e^n|^2}{|n|(\frac{\delta^2}{4} + (\rho_i/\rho_e)^{2|n|})} \\
&\geq \frac{1}{2} \sum_{0\neq |n|\leq N_\delta} \frac{\delta |g_e^n|^2}{|n|(\rho_i/\rho_e)^{2|n|}} .
\end{aligned}
$$

Suppose that

$$
(11.28)\quad \limsup_{|n|\to\infty} \frac{|g_e^n|^2}{|n|(\rho_i/\rho_e)^{|n|}} = \infty.
$$

Then there is a subsequence $\{n_k\}$ with $|n_1| < |n_2| < \dots$ such that

$$
(11.29)\quad \lim_{k\to\infty} \frac{|g_e^{n_k}|^2}{|n_k|(\rho_i/\rho_e)^{|n_k|}} = \infty.
$$

If we take $\delta = 2(\rho_i/\rho_e)^{|n_k|}$, then $N_\delta = |n_k|$ and
(11.30)

$$
\sum_{0\neq |n|\leq N_\delta} \frac{\delta |g_e^n|^2}{|n|(\rho_i/\rho_e)^{2|n|}} = (\rho_i/\rho_e)^{|n_k|} \sum_{0\neq |n|\leq |n_k|} \frac{|g_e^n|^2}{|n|(\rho_i/\rho_e)^{2|n|}} \geq \frac{|g_e^{|n_k|}|^2}{|n_k|(\rho_i/\rho_e)^{|n_k|}} .
$$

Thus we obtain from (11.25) that

$$
(11.31)\quad \lim_{k\to\infty} E_{(\rho_i/\rho_e)^{|n_k|}} = \infty .
$$

We emphasize that (11.28) is not enough to guarantee (11.4). We now impose an additional condition for CALR to occur. We assume that $\{g_e^n\}$ satisfies the following gap property:

GP : There exists a sequence $\{n_k\}$ with $|n_1| < |n_2| < \dots$ such that

$$
\lim_{k\to\infty} (\rho_i/\rho_e)^{|n_{k+1}|-|n_k|} \frac{|g_e^{n_k}|^2}{|n_k|(\rho_i/\rho_e)^{|n_k|}} = \infty .
$$

If GP holds, then we immediately see that (11.28) holds, but the converse is not true. If (11.28) holds, *i.e.*, there is a subsequence $\{n_k\}$ with $|n_1| < |n_2| < \dots$ satisfying (11.29) and the gap $|n_{k+1}| - |n_k|$ is bounded, then GP holds. In particular, if

$$
(11.32)\quad \lim_{n\to\infty} \frac{|g_e^n|^2}{|n|(\rho_i/\rho_e)^{|n|}} = \infty ,
$$

then GP holds.

Assume that $\{g_e^n\}$ satisfies GP and $\{n_k\}$ is such a sequence. Let $\delta = 2(\rho_i/\rho_e)^\alpha$ for some α and let $k(\alpha)$ be the number such that

$$
|n_{k(\alpha)}| \leq \alpha < |n_{k(\alpha)+1}|.
$$

Then, we have

$$
(11.33)\quad \begin{aligned}
\sum_{0\neq |n|\leq N_\delta} \frac{\delta |g_e^n|^2}{|n|(\rho_i/\rho_e)^{2|n|}} &= (\rho_i/\rho_e)^\alpha \sum_{0\neq |n|\leq \alpha} \frac{|g_e^n|^2}{|n|(\rho_i/\rho_e)^{2|n|}} \\
&\geq (\rho_i/\rho_e)^{|n_{k(\alpha)+1}|-|n_{k(\alpha)}|} \frac{|g_e^{n_{k(\alpha)}}|^2}{|n_{k(\alpha)}|(\rho_i/\rho_e)^{|n_{k(\alpha)}|}} \to \infty ,
\end{aligned}
$$

as $\alpha \to \infty$.

We obtain the following lemma.

LEMMA 11.2. *If (11.28) holds, then*

$$\limsup_{\delta \to 0} E_\delta = \infty \,. \tag{11.34}$$

If $\{g_e^n\}$ satisfies the condition GP, then

$$\lim_{\delta \to 0} E_\delta = \infty \,. \tag{11.35}$$

Suppose that the source function is supported inside the radius $\rho_\star = \sqrt{\rho_e^3 \rho_i^{-1}}$. Then its Newtonian potential cannot be extended harmonically in $|x| < \rho_\star$ in general. So, if F is given by

$$F = c - \sum_{n \neq 0} a_n r^{|n|} e^{\sqrt{-1} n \theta}, \quad r < \rho_e \,, \tag{11.36}$$

then the radius of convergence is less than $\rho_\star$. Thus we have

$$\limsup_{|n| \to \infty} |n| |a_n|^2 \rho_\star^{2|n|} = \infty \,, \tag{11.37}$$

i.e., (11.28) holds. The GP condition is equivalent to the existence of $\{n_k\}$ with

$$|n_1| < |n_2| < \dots$$

such that

$$\lim_{k \to \infty} (\rho_i / \rho_e)^{|n_{k+1}| - |n_k|} |n_k| |a_{n_k}|^2 \rho_\star^{2|n_k|} = +\infty \,. \tag{11.38}$$

The following is the main theorem of this section.

THEOREM 11.3. *Let f be a source function supported in $\mathbb{R}^2 \setminus \overline{\Omega_e}$ and F be the Newtonian potential of f.*

(i) *If F does not extend as a harmonic function in $\Omega_{\rho_\star} := \{|x| < \rho_\star\}$, then weak CALR occurs, i.e.,*

$$\limsup_{\delta \to 0} E_\delta = \infty \tag{11.39}$$

and (11.5) holds with $a = \rho_e^2 / \rho_i$.

(ii) *If the Fourier coefficients of F satisfy (11.38), then CALR occurs, i.e.,*

$$\lim_{\delta \to 0} E_\delta = \infty \tag{11.40}$$

and (11.5) holds with $a = \rho_e^2 / \rho_i$.

(iii) *If F extends as a harmonic function in a neighborhood of $\overline{\Omega_{\rho_\star}}$, then CALR does not occur, i.e.,*

$$E_\delta < C \tag{11.41}$$

for some C independent of δ.

PROOF. If F does not extend as a harmonic function in $\Omega_{\rho_\star}$, then (11.28) holds. Thus we have (11.39). If (11.38) holds, then (11.40) holds by Lemma 11.2.

Moreover, by (11.22), we see that

$$\begin{aligned}|V_\delta| &\le |F| + \sum_{n\neq 0} \left| \frac{2(\rho_i^{2|n|} - \rho_e^{2|n|}) z_\delta}{|n|\rho_e^{|n|-1}(4z_\delta^2 - (\rho_i/\rho_e)^{2|n|})} \frac{g_e^n}{r^{|n|}} \right| \le |F| \\ &\quad + C \sum_{n\neq 0} \frac{\delta \rho_e^{|n|}}{(\frac{\delta^2}{4} + (\rho_i/\rho_e)^{2|n|})|n| r^{|n|}} \\ &\le |F| + C \sum_{n\neq 0} \frac{\rho_e^{2|n|}}{|n|\rho_i^{|n|} r^{|n|}} < C, \quad \text{if} \quad r = |x| > \frac{\rho_e^2}{\rho_i}\end{aligned}$$

for some constants C which may differ at each occurrence.

If F extends as a harmonic function in a neighborhood of $\overline{\Omega_{\rho_\star}}$, then the power series of F, which is given by (11.18), converges for $r < \rho_\star + 2\epsilon$ for some $\epsilon > 0$. Therefore there exists a constant C such that

$$\frac{|g_e^n|}{|n|\rho_e^{|n|-1}} \le C \frac{1}{(\rho_\star + \epsilon)^{|n|}}$$

for all n. It then follows that

$$|g_e^n| \le C(\rho_e^2(\rho_i/\rho_e)^{-1} + \rho_e \epsilon)^{-|n|/2} \rho_e^{|n|} \le ((\rho_i/\rho_e)^{-1} + \epsilon)^{-|n|/2} \tag{11.42}$$

for all n. This tells us that

$$\sum_{n\neq 0} \frac{\delta |g_e^n|^2}{|n|(\delta^2 + (\rho_i/\rho_e)^{2|n|})} \le \sum_{n\neq 0} \frac{|g_e^n|^2}{2|n|(\rho_i/\rho_e)^{|n|}} \le \sum_{n\neq 0} \frac{1}{2|n|(1 + \epsilon(\rho_i/\rho_e))^{|n|}}.$$

This completes the proof. □

If f is a dipole in $\Omega_{\rho_\star} \setminus \overline{\Omega_e}$, *i.e.*, $f(x) = a \cdot \nabla \delta_y(x)$ for a vector a and $y \in \Omega_{\rho_\star} \setminus \overline{\Omega_e}$ where δ_y is the Dirac delta function at y, then $F(x) = a \cdot \nabla \Gamma(x, y)$. From the following expansion of the fundamental solution of the Laplacian:

$$\frac{(-1)^{|\alpha|}}{\alpha!} \partial^\alpha \Gamma(x, 0) = \frac{-1}{2\pi|\alpha|} \left[a_\alpha^{|\alpha|} \frac{\cos|\alpha|\theta}{r^{|\alpha|}} + b_\alpha^{|\alpha|} \frac{\sin|\alpha|\theta}{r^{|\alpha|}} \right], \tag{11.43}$$

we have

$$\Gamma(x, y) = \sum_{n=1}^{\infty} \frac{-1}{2\pi n} \left[\frac{\cos n\theta_y}{r_y^n} r^n \cos n\theta + \frac{\sin n\theta_y}{r_y^n} r^n \sin n\theta \right] + C\,. \tag{11.44}$$

Then we see that the Fourier coefficients of F have the growth rate r_y^{-n} and satisfy (11.38), and hence CALR takes place. Similarly CALR takes place for a sum of dipole sources at different fixed positions in $\Omega_{\rho_\star} \setminus \overline{\Omega_e}$. We mention that this fact was found in [**399**].

If f is a quadrapole, *i.e.*,

$$f(x) = A : \nabla\nabla \delta_y(x) = \sum_{i,j=1}^{2} a_{ij} \frac{\partial^2}{\partial x_i \partial x_j} \delta_y(x)$$

for a 2×2 matrix $A = (a_{ij})$ and $y \in \Omega_{\rho_\star} \setminus \overline{\Omega_e}$. Then

$$F(x) = \sum_{i,j=1}^{2} a_{ij} \frac{\partial^2 \Gamma(x, y)}{\partial x_i \partial x_j}\,.$$

Thus CALR takes place. This is in agreement with the numerical result in [**416**].

If f is supported in $\mathbb{R}^2 \setminus \overline{\Omega_{\rho_\star}}$, then F is harmonic in a neighborhood of $\overline{\Omega_{\rho_\star}}$, and hence CALR does not occur by Theorem 11.3. In fact, we can say more about the behavior of the solution V_δ as $\delta \to 0$ which is related to the observation in [**400, 415**] that in the limit $\delta \to 0$ the annulus itself becomes invisible to sources that are sufficiently far away.

THEOREM 11.4. *If f is supported in $\mathbb{R}^2 \setminus \overline{\Omega_{\rho_\star}}$, then (11.41) holds (with $\alpha = 1$ in (11.3)). Moreover, we have*

$$\sup_{|x| \geq \rho_\star} |V_\delta(x) - F(x)| \to 0 \quad as \quad \delta \to 0\,. \tag{11.45}$$

PROOF. Since $\operatorname{supp} f \subset \mathbb{R}^2 \setminus \overline{\Omega_{\rho_\star}}$, the power series of F, which is given by (11.18), converges for $r < \rho_\star + 2\epsilon$ for some $\epsilon > 0$.

According to (11.22), if $\rho_e < r = |x|$, then we have

$$V_\delta(x) - F(x) = \sum_{n \neq 0} \frac{2(\rho_e^{2|n|} - \rho_i^{2|n|}) z_\delta}{|n| \rho_e^{|n|-1} ((\rho_i/\rho_e)^{2|n|} - 4 z_\delta^2)} \frac{g_e^n}{r^{|n|}} e^{\sqrt{-1} n \theta}\,.$$

If $|x| = \rho_\star$, then the identity

$$\frac{(\rho_e^{2|n|} - \rho_i^{2|n|}) z_\delta}{|n| \rho_e^{|n|-1} ((\rho_i/\rho_e)^{2|n|} - 4 z_\delta^2)} \frac{g_e^n}{\rho_\star^{|n|}} = \frac{(1 - (\rho_i/\rho_e)^{2|n|}) z_\delta}{((\rho_i/\rho_e)^{|n|} - 4 z_\delta^2 (\rho_i/\rho_e)^{-|n|})} \frac{g_e^n \rho_\star^{|n|}}{|n| \rho_e^{|n|-1}}$$

holds and

$$\begin{aligned}
&\left| \frac{(1 - (\rho_i/\rho_e)^{2|n|}) z_\delta}{((\rho_i/\rho_e)^{|n|} - 4 z_\delta^2 (\rho_i/\rho_e)^{-|n|})} \right| \leq \left| \frac{1}{(z_\delta^{-1} (\rho_i/\rho_e)^{|n|} - z_\delta (\rho_i/\rho_e)^{-|n|})} \right| \\
&\leq \left| \frac{1}{\Im (z_\delta^{-1} (\rho_i/\rho_e)^{|n|} - z_\delta (\rho_i/\rho_e)^{-|n|})} \right| = \left(\frac{\delta}{4 + \delta^2} (\rho_i/\rho_e)^{-|n|} + \frac{1}{\delta} (\rho_i/\rho_e)^{|n|} \right)^{-1}.
\end{aligned}$$

It then follows from (11.42) that

$$\begin{aligned}
|V_\delta(x) - F(x)| \leq 2 \sum_{n \neq 0} & \left(\frac{\delta}{4 + \delta^2} (\rho_i/\rho_e)^{-|n|} + \frac{1}{\delta} (\rho_i/\rho_e)^{|n|} \right)^{-1} \\
& \times \frac{\rho_e}{|n|} \left(\frac{(\rho_i/\rho_e)^{-1}}{(\rho_i/\rho_e)^{-1} + \epsilon} \right)^{|n|/2},
\end{aligned}$$

and hence

$$|V_\delta(x) - F(x)| \to 0 \quad \text{as } \delta \to 0\,.$$

Since $V_\delta - F$ is harmonic in $|x| > \rho_e$ and tends to 0 as $|x| \to \infty$, we obtain (11.45) by the maximum principle. This completes the proof. □

Theorem 11.4 shows that any source supported outside $\Omega_{\rho_\star}$ cannot make the blow-up of the power dissipation happen and hence is not cloaked. In fact, it is known that we can recover the source f from its Newtonian potential F outside $\Omega_{\rho_\star}$ since f is supported outside $\overline{\Omega_{\rho_\star}}$ (see [**305**]). Therefore we infer from (11.45) that f may be recovered approximately by observing V_δ outside $\Omega_{\rho_\star}$.

11.4. Shielding at a Distance

The aim of this section is to investigate the conditions required for shielding at a distance and geometric features such as the location and size of the shielded region. The key element to study in the eccentric case is the Möbius transformation via which a concentric annulus is transformed into an eccentric one. The quasi-static properties of the eccentric superlens can be derived in a straightforward way from those of the concentric case since the Möbius transformation is a conformal mapping.

We let Ω_i and Ω_e denote circular disks centered at the origin with the radii ρ_i and ρ_e, respectively. We assume that $0 < \rho_i < \rho_e < 1$.

As in the previous section, the core (Ω_i) and the background ($\mathbb{R}^2 \setminus \overline{\Omega_e}$) are assumed to be occupied by the isotropic material of permittivity 1 and the shell ($\Omega_e \setminus \overline{\Omega_i}$ by the plasmonic material of permittivity $-1 + \sqrt{-1}\delta$. The obtained concentric superlens geometry is described in Figure 11.1(a).

Identifying $\mathbb{R}^2$ as $\mathbb{C}$, Ω_i and Ω_e can be represented as

$$\Omega_i = \big\{z \in \mathbb{C} : |z| < \rho_i\big\} \quad \text{and} \quad \Omega_e = \big\{z \in \mathbb{C} : |z| < \rho_e\big\}.$$

We also assume the annulus structure to be small compared to the operating wavelength so that it can adopt the quasi-static approximation. Then (the quasi-static) electric potential V_δ satisfies

$$\nabla \cdot \epsilon_\delta \nabla V_\delta = f \quad \text{in } \mathbb{C}, \tag{11.46}$$

where f represents an electrical source. We assume that f is a point multipole source of order n located at a location $z_0 \in \mathbb{R}^2 \setminus \overline{\Omega_e}$. Then the potential F generated by the source f can be represented as

$$F(z) = \sum_{k=1}^{n} \Re\{c_k (z - z_0)^{-k}\}, \quad z \in \mathbb{C},$$

with complex coefficients c_k's. When $n = 1$, the source f (or the potential F) means a point dipole source.

Then, from the previous section, the anomalous localized resonance can be summarized as follows.

(i) the dissipation energy W_δ diverges as the loss parameter δ goes to zero if and only if a point source f is located inside the region $\Omega_* := \{|z| < \rho_*\}$, where $\rho_* := \sqrt{\rho_e^3/\rho_i}$ and W_δ is given by

$$W_\delta := \Im \int_{\mathbb{R}^2} \epsilon_\delta |\nabla V_\delta|^2 \, dx = \delta \int_{\Omega_e \setminus \overline{\Omega_i}} |\nabla V_\delta|^2. \tag{11.47}$$

Let us call Ω_* (or ρ_*) *the critical region* (or *the critical radius*), respectively.

(ii) the electric field $-\nabla V_\delta$ stays bounded outside some circular region regardless of δ. More precisely, we have

$$|\nabla V_\delta(z)| \leq C, \quad z \in \Omega_b := \{|z| > \rho_e^2/\rho_i\}, \tag{11.48}$$

for some constant C independent of δ. Here, the subscript 'b' in Ω_b indicates the boundedness of the electric field. Let us call Ω_b *the calm region.*

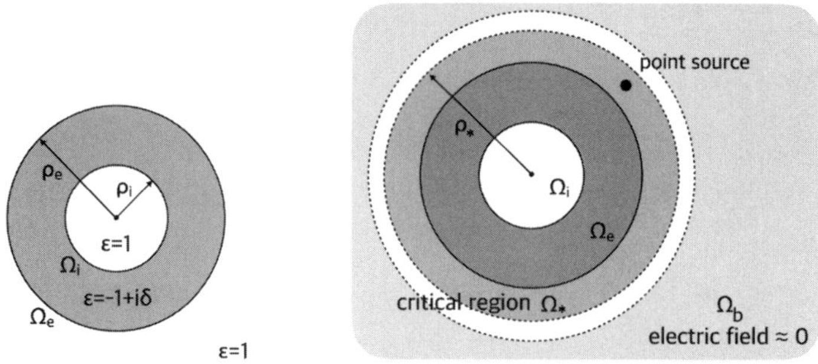

FIGURE 11.1. Cloaking due to the anomalous localized resonance: (a) shows the structure of the superlens with concentric core; (b) illustrates the cloaking effect.

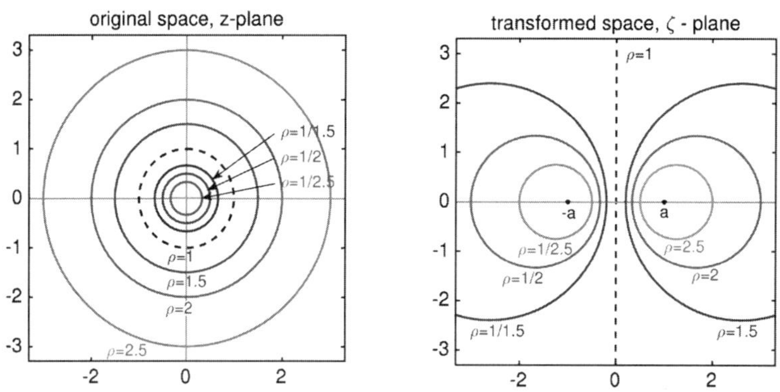

FIGURE 11.2. The Möbius transformation Φ defined in (11.49) maps 0, ∞, 1 to $-a$, $+a$, ∞, respectively. The left figure shows radial coordinate curves $\{|z| = \rho\}$, $\rho > 0$, and the right figure their images transformed by Φ with $a = 1$. Concentric circles satisfying $\rho \neq 1$ are transformed into eccentric ones.

11.4.1. Möbius Transformation. In this subsection, we show that the concentric annulus can be transformed into an eccentric one by applying the Möbius transformation Φ defined as

$$\zeta = \Phi(z) := a\frac{z+1}{z-1} \tag{11.49}$$

with a given positive number a. We shall also discuss how the critical region is transformed depending on the critical parameter ρ_*.

The function Φ is a conformal mapping from $\mathbb{C} \setminus \{1\}$ to $\mathbb{C} \setminus \{a\}$. It maps the point $z = 1$ to infinity, infinity to $\zeta = a$, and $z = 0$ to $\zeta = -a$. It maps a circle centered at the origin, say $S_\rho := \{z \in \mathbb{C} : |z| = \rho\}$, to the circle given by

$$\Phi(S_\rho) = \{z \in \mathbb{C} : |z - c| = r\}, \quad \text{where } c = a\frac{\rho^2+1}{\rho^2-1} \text{ and } r = \frac{2a}{|\rho - \rho^{-1}|}. \tag{11.50}$$

So the concentric circles S_ρ's with $\rho \neq 1$ are transformed to eccentric ones in ζ-plane; see Figure 11.2.

Let us discuss how the concentric superlens described in Section 11.3 is geometrically transformed by the mapping Φ. Note that for $0 < \rho < 1$, the transformed circle $\Phi(S_\rho)$ always lies in the left half-plane of $\mathbb{C}$. Since we assume that $0 < \rho_i < \rho_e < 1$, the concentric annulus in z-plane is changed to an eccentric one contained in the left half ζ-plane. We let $\widetilde{\Omega}_i$ (or $\widetilde{\Omega}_e$) denote the transformed disk of Ω_i (or Ω_e), respectively.

Now we consider the critical region $\Omega_* = \{|z| < \rho_*\}$ and the calm region Ω_b. Let us denote the transformed critical region (or calm region) by $\widetilde{\Omega}_*$ (or $\widetilde{\Omega}_b$), respectively. The shape of $\widetilde{\Omega}_*$ can be very different depending on the value of ρ_*. Suppose $0 < \rho_* < 1$ for a moment. Then the region $\widetilde{\Omega}_*$ is a circular disk contained in the left half ζ-plane. Next, assume that $\rho_* > 1$. In this case, $\widetilde{\Omega}_*$ becomes the region outside a disk which is disjoint from the eccentric annulus. Contrary to the case when $\rho_* < 1$, the region $\widetilde{\Omega}_*$ is now unbounded. Similarly, the shape of $\widetilde{\Omega}_b$ depends on the parameter $\rho_b := \rho_e^2/\rho_i$. If $0 < \rho_b < 1$, $\widetilde{\Omega}_b$ is a region outside a circle. But, if $\rho_b > 1$, $\widetilde{\Omega}_b$ becomes a bounded circular region which does not intersect with the eccentric superlens. This unbounded (or bounded) feature of the shape of $\widetilde{\Omega}_*$ (or $\widetilde{\Omega}_b$) will be essentially used to design a new shielding device.

11.4.2. Potential in the Transformed Space. Here, we will transform the potential V_δ via the Möbius map Φ and then show that the resulting potential describes the physics of the eccentric superlens. Let us define the transformed potential $\widetilde{V}_\delta$ by $\widetilde{V}_\delta(\zeta) := V_\delta \circ \Phi^{-1}(\zeta)$. Since the Möbius transformation Φ is a conformal mapping, it preserves the harmonicity of the potential and interface conditions. It can be easily shown that the transformed potential V_δ satisfies

$$\nabla \cdot \widetilde{\epsilon}_\delta \nabla \widetilde{V}_\delta = \widetilde{f} \quad \text{in } \mathbb{C}, \tag{11.51}$$

where $\widetilde{f}(\zeta) = \frac{1}{|\Phi'|^2}(f \circ \Phi^{-1})(\zeta)$ and the permittivity $\widetilde{\epsilon}_\delta$ is given by

$$\widetilde{\epsilon}_\delta(\zeta) = \begin{cases} 1 & \text{in } \widetilde{\Omega}_i, \\ -1 + \sqrt{-1}\delta & \text{in } \widetilde{\Omega}_e \setminus \overline{\widetilde{\Omega}_i}, \\ 1 & \text{in the background.} \end{cases} \tag{11.52}$$

Therefore, the transformed potential $\widetilde{V}_\delta$ represents the quasi-static electrical potential of the eccentric superlens (11.52) induced by the source $\tilde{f}(\zeta)$.

Now we consider some physical properties in the transformed space. The dissipation energy $\widetilde{W}_\delta$ in the transformed space turns out to be the same as the original one W_δ as follows:

$$\widetilde{W}_\delta = \delta \int_{\partial(\widetilde{\Omega}_e \setminus \widetilde{\Omega}_i)} \widetilde{V}_\delta \frac{\partial \widetilde{V}_\delta}{\partial \widetilde{n}} \, d\widetilde{l} = \delta \int_{\partial(\Omega_e \setminus \Omega_i)} V_\delta \frac{1}{|\Phi'|} \frac{\partial V_\delta}{\partial n} |\Phi'| \, dl = W_\delta. \tag{11.53}$$

In the derivation we have used the Green's identity and the harmonicity of the potentials V_δ and $\widetilde{V}_\delta$.

The point source f is transformed into another point source at a different location. To see this, we recall that the source f is located at $z = z_0$ in the original space. It generates the potential $F(z) = \sum_{k=1}^n \Re\{c_k(z - z_0)^{-k}\}$. By the map Φ, the potential F becomes $\widetilde{F} := F \circ \Phi^{-1}$ which is of the following form:

$$\widetilde{F}(\zeta) = \sum_{k=1}^{n} \Re\left\{d_n(\zeta - \zeta_0)^{-k}\right\}, \tag{11.54}$$

where d_k's are complex constants and $\zeta_0 := \Phi(z_0)$. So the transformed source $\widetilde{f}$ is a point multipole source of order n located at $\zeta = \zeta_0$. It is also worth remarking that, if the point source f is located at $z_0 = 1$ in the original space, then $\widetilde{f}$ becomes a multipole source at infinity in the transformed space. In fact, its corresponding potential $\widetilde{F}$ is of the following form:

$$\widetilde{F}(\zeta) = \sum_{k=1}^{n} \Re\left\{e_k \zeta^k\right\}$$

for some complex constants e_k. For example, if $n = 1$, then the source $\widetilde{f}$ (or potential $\widetilde{F}$) represents a uniform incident field.

11.4.3. Shielding at a Distance Due to Anomalous Resonance. In this subsection, we analyze the anomalous resonance in the eccentric annulus and explain how a new kind of shielding effect can arise. In view of the previous subsection, the mathematical description of anomalous resonance in the eccentric case can be directly obtained from that in the concentric case as follows:

(i) the dissipation energy $\widetilde{W_\delta}$ diverges as the loss parameter δ goes to zero if and only if a point source $\widetilde{f}$ is located inside the region $\widetilde{\Omega}_*$.

(ii) the electric field $-\nabla \widetilde{V}_\delta$ stays bounded in the calm region $\widetilde{\Omega}_b$ regardless of δ, *i.e.*,

$$|\nabla \widetilde{V}_\delta(\zeta)| \leq C, \quad \zeta \in \widetilde{\Omega}_b, \tag{11.55}$$

for some constant C independent of δ.

Now we discuss a new shielding effect. Suppose the parameters ρ_i and ρ_e satisfy $\rho_* = \sqrt{\rho_e^3/\rho_i} > 1$. Then, as explained in Subsection 11.4.1, the calm region $\widetilde{\Omega}_b$ becomes a bounded circular region which does not intersect with the eccentric structure. If a point source is located within the critical region $\widetilde{\Omega}_*$, then the anomalous resonance occurs and the normalized electric field $-\nabla \widetilde{V}_\delta/\sqrt{E}_\delta$ is nearly zero inside the calm region $\widetilde{\Omega}_b$. So the bounded circular region $\widetilde{\Omega}_b$ is not affected by any surrounding point source located in $\widetilde{\Omega}_*$. In other words, the shielding effect does occur in $\widetilde{\Omega}_b$, but there is a significant difference in this shielding effect compared to the standard one. There is no additional material enclosing the region $\widetilde{\Omega}_b$; the eccentric structure is located disjointly. So we call this effect 'shielding at a distance' and $\widetilde{\Omega}_b$ 'the shielding region'. The condition for its occurrence can be summarized as follows: shielding at a distance happens in $\widetilde{\Omega}_b$ if and only if the critical parameter ρ_* and the source location ζ_0 satisfy

$$\rho_* > 1 \text{ and } \zeta_0 \in \widetilde{\Omega}_*. \tag{11.56}$$

The shielding effect occurs for not only a point source but also an external field like a uniform incident field $\widetilde{F}(\zeta) = -\Re\{E_0\zeta\}$ for a complex constant E_0. As mentioned previously, an external field of the form $\Re\{\sum_{k=1}^n e_k\zeta^k\}$ can be considered as a point source at $\zeta = \infty$. Since the critical region $\widetilde{\Omega}_*$ contains the point at infinity when $\rho_* > 1$, the anomalous resonance will happen and then the circular bounded

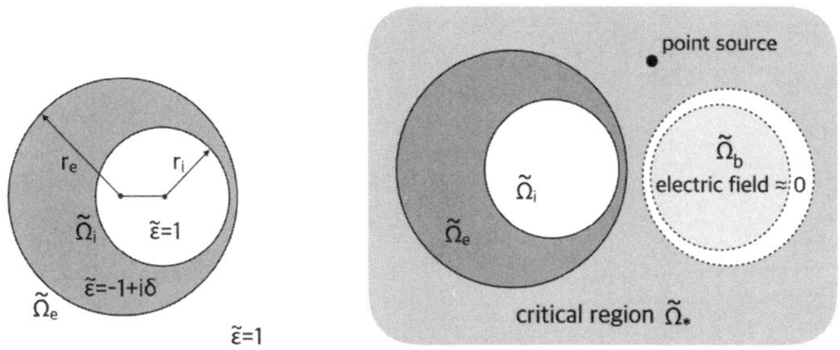

FIGURE 11.3. Shielding at a distance due to the anomalous localized resonance: (left) shows the structure of the superlens with the eccentric core; (right) illustrates shielding at a distance.

region $\widetilde{\Omega}_b$ will be shielded. It is worth remarking that, unlike in the eccentric case, the anomalous resonance cannot result from any external field with source at infinity for the concentric case.

We now discuss the range of validity of the new shielding effect. Recall that we assume the eccentric superlens has a negative permittivity. For this, we may use plasmonic materials (such as gold and silver) whose permittivities have negative real part in the infrared to visible regime. In these regimes, the operating wavelength lies in the range of several hundreds of nanometers. Then the eccentric structure and the shielded region should have less than a few tens of nanometers in size because our analysis is based on the quasi-static approximation. Also, they should be relatively close to each other. We, however, emphasize that the sources don't have to be located near the eccentric structure. In fact, the shielding can happen for external electromagnetic waves surrounding the structure and the shielded region. Suppose, for instance, that a plane wave is incident. Since the eccentric structure is small compared to the wavelength, the incident field is nearly uniform across the structure. As already explained, the anomalous resonance can happen for the uniform field, which is a point dipole at infinity. So the shielding at a distance will occur for the plane wave. Similarly, it can happen for general surrounding waves.

11.4.4. Numerical Illustration. In this subsection we illustrate shielding at a distance by showing several examples of the field distribution generated by an eccentric annulus and a point source. To compute the field distribution, we use an analytic solution derived by applying a separation of variables method in the polar coordinates to the concentric case and then using the Möbius transformation Φ. The corresponding Matlab code can be downloaded at `www.ams.org/bookpages/surv-235`.

For all the examples below, we fix $\rho_e = 0.7$ for the concentric shell and $a = 1$ for the Möbius transformation. We also fix the loss parameter as $\delta = 10^{-12}$.

Example 1 (Cloaking of a dipole source) We first present an eccentric annulus which acts as a cloaking device (Figure 11.4). Since we want to make a 'cloaking' device, we need ρ_b to satisfy the condition $\rho_b < 1$. Setting $\rho_i = 0.55$ for this example, we have $\rho_b = \rho_e^2/\rho_i = 0.89 < 1$ ($\rho_* = (\rho_e^3/\rho_i)^{1/2} = 0.79$). Then by applying the Möbius transformation Φ, the concentric annulus is transformed to the following eccentric structure from (11.50): the outer region $\widetilde{\Omega}_e = \Phi(\Omega_e)$ is the circular disk of radius 2.75 centered at $(-2.92, 0)$ and the core $\widetilde{\Omega}_i = \Phi(\Omega_i)$ is of

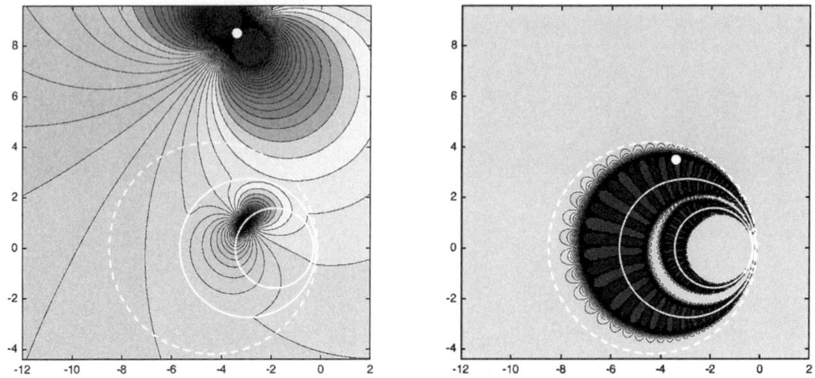

FIGURE 11.4. Cloaking for a dipole source. We set $\rho_i = 0.55$, $\rho_e = 0.7$ and $a = 1$. The dipole source is located at $\zeta_0 = (-3.4, 8.5)$ in the left figure and at $\zeta_0 = (-3.4, 3.5)$ in the right figure. (left) A dipole source (small solid disk in white) is located outside the critical region $\widetilde{\Omega}_*$ (white dashed circle). The field outside $\widetilde{\Omega}_*$ is significantly perturbed by the source. (right) A dipole source is located inside the critical region $\widetilde{\Omega}_*$. The anomalous resonance happens near the superlens but the field outside $\widetilde{\Omega}_*$ becomes nearly zero. The source becomes almost cloaked.

radius 1.58 centered at $(-1.87, 0)$. The boundaries of the physical regions $\partial\widetilde{\Omega}_i$ and $\partial\widetilde{\Omega}_e$ are plotted as solid white curves in Figure 11.4. On the other hand, the critical region's boundary $\partial\widetilde{\Omega}_*$, which is not a material interface, and is the circle of radius 4.08 centered at $(-4.55, 0)$, is plotted as a dashed white circle. We refrain from plotting the calm region's boundary $\partial\widetilde{\Omega}_b$ in the figure for the sake of the simplicity; it is relatively close to $\partial\widetilde{\Omega}_*$. Note that the calm region $\widetilde{\Omega}_b$ is an unbounded region whose boundary is slightly outside of $\partial\widetilde{\Omega}_*$.

In Figure 11.4(a), we assume that a dipole source $\widetilde{F}(\zeta) = \Re\{\overline{b}(\zeta - \zeta_0)^{-1}\}$ is located at $\zeta_0 = (-3.4, 8.5)$ with the dipole moment $b = (3, -3)$. The point source is plotted as a small solid disk (in white). It is clearly seen that the field distribution is smooth over the entire region except at the dipole source. That is, the anomalous resonance does not occur. We can detect the dipole source by measuring the perturbation of the electric field.

In Figure 11.4(b), we change the location of the source to $\zeta_0 = (-3.4, 3.5)$ so that the source's location belongs to the critical region $\widetilde{\Omega}_*$. Then the anomalous resonance does occur, as shown in the figure. As a result, the potential outside the white dashed circle becomes nearly constant. In other words, the dipole source is almost cloaked.

Example 2 (Shielding at a distance for a dipole source) Next we show that changing the size of the core allows for shielding at a distance to happen for a dipole source (Figure 11.5).

In Figure 11.5(a), we let $\rho_i = 0.55$ as in Example 1. We also assume that a dipole source $\widetilde{F}(\zeta) = \Re\{\overline{b}(\zeta - \zeta_0)^{-1}\}$ is located at $\zeta_0 = (5, 5)$ with the dipole moment

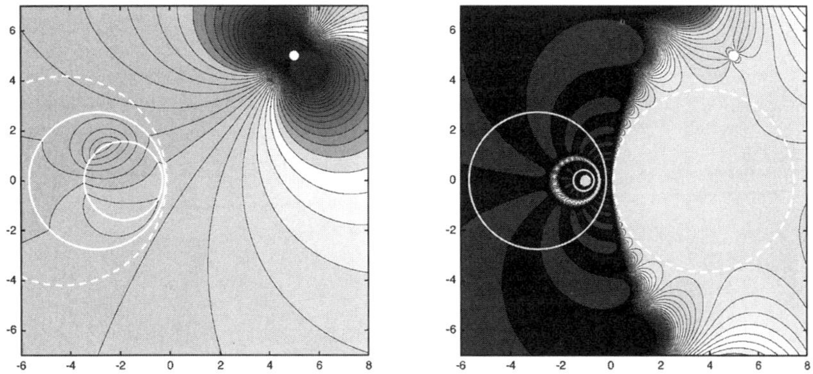

FIGURE 11.5. Shielding at a distance for a dipole source. We set $\rho_e = 0.7$, $a = 1$ and $\zeta_0 = (5,5)$. We also set $\rho_i = 0.55$ in the left figure and $\rho_i = 0.2$ in the right figure. (left) The critical region $\widetilde{\Omega}_*$ (white dashed line) contains the eccentric superlens (white solid lines). The field outside the white dashed circle is significantly perturbed by the source. (right) The critical region is now the region outside the white dashed circle which does not contain the superlens any longer. The field inside the white dashed circle is nearly zero and so the shielding occurs.

$b = (3,3)$. Since the source is located outside the critical region, the anomalous resonance does not happen.

Now let us change the size of the core. To make the shielding at distance occur, the critical radius ρ_* satisfies the condition $\rho_* > 1$. We set $\rho_i = 0.2$ so that $\rho_* = \sqrt{\rho_e^3/\rho_i} = 1.31 > 1$. Then, the core $\widetilde{\Omega}_i = \Phi(\Omega_i)$ becomes the circular disk of radius 0.42 centered at $(-1.08, 0)$. The critical region $\widetilde{\Omega}_*$ becomes the region outside the circle of radius 3.53 centered at $(4.06, 0)$. The resulting eccentric annulus and the critical region are illustrated in Figure 11.5(b). Note that the source is contained in the new critical region $\widetilde{\Omega}_*$ and $\rho_* > 1$. In other words, the condition (11.56) for shielding at a distance is satisfied. Indeed, inside the white dashed circle, the potential becomes nearly constant while there is an anomalous resonance outside. Thus, the shielding at a distance occurs.

Example 3 (Shielding at a distance for a uniform field) Finally, we consider shielding at a distance for a uniform field (Figure 11.6). We keep the parameters a, ρ_i and ρ_e as in the previous example but change the dipole source to a uniform field $\widetilde{F}(\zeta) = -\Re\{E_0\zeta\}$ with $E_0 = 1$. As mentioned previously, an external field can be considered as a point source located at infinity.

In the left figure, the critical region does not contain infinity. So the anomalous resonance does not happen. The uniform field can be easily detected. In the right figure, we change the core as in the previous example. Now the critical region (the region outside the white dashed circle) does contain infinity. So the anomalous resonance does happen. Again, the potential becomes nearly constant in the region inside the dashed circle. This means there is shielding at a distance for a uniform field.

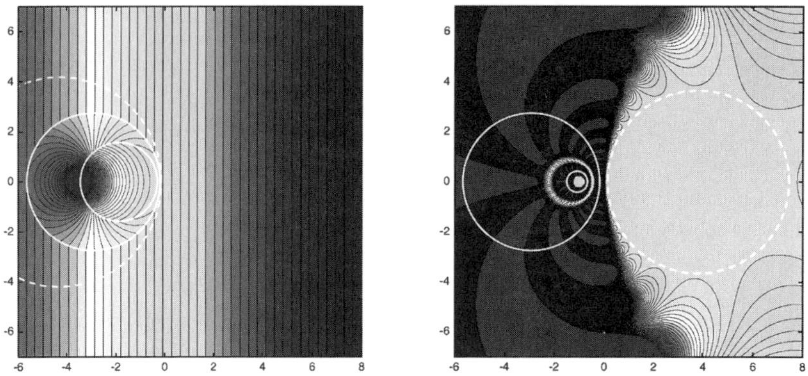

FIGURE 11.6. Shielding at a distance for a uniform field. We set $\rho_e = 0.7$ and $a = 1$. We also set $\rho_i = 0.55$ in the left figure and $\rho_i = 0.2$ in the right figure. (left) The critical region $\widetilde{\Omega}_*$ (white dashed line) contains the eccentric superlens (white solid lines). The uniform incident field is nearly unperturbed outside the white dashed circle. (right) The critical region $\widetilde{\Omega}_*$ is the region outside the white dashed circle which does not contain the superlens any longer. The field inside the white dashed circle is nearly zero and so the shielding occurs.

11.5. Concluding Remarks

The convergence to a smooth field outside the region was shown in [**415**], where the first numerical evidence for ALR was also presented. A proof of ALR for a dipolar source outside a plasmonic annulus was given in [**400**]. The condition for CALR in the annulus case was also derived in [**34**]. A necessary and sufficient condition on the source term to be cloaked in the general case was derived in [**34**]. It is based on a symmetrization principle for the associated boundary integral formulation. It is worth mentioning that if the real part of the permittivity of the shell is different from -1, then CALR does not occur [**35**]. On the other hand, if the cylindrical structure has an eccentric core, then a new kind of shielding effect can happen [**514**]. Using the Möbius transformation, the shielding at distance has been investigated both analytically and numerically. In contrast with conventional shielding, the anomalous resonance shielding effect does not require any material which encloses the region to be shielded.

The results of this chapter on anomalous resonance cloaking were extended in [**333**] to the case when the core D is not radial by a different method based on a variational approach. In [**393**], the cloaking due to anomalous localized resonance in the quasi-static regime in the case when a general charge density distribution is brought near a slab superlens is analyzed.

On the other hand, it was shown in [**35**] that in three dimensions CALR does not occur. The occurrence of CALR is in fact determined by the eigenvalue distribution of the Neumann–Poincaré-type operator associated with the structure [**35**]. However, using a shell with a specially designed anisotropic dielectric constant, it is possible to make CALR occur in three dimensions [**36**].

In [**169**], various examples including an elliptical core in an elliptical shell are considered via numerical simulation. The results are similar to those of the radial case considered here; in particular, the structure seemed to cloak a polarizable dipole placed sufficiently near the shell. When the core and the shell are confocal ellipses, the critical elliptic radius such that, for any source inside it, CALR takes place and, for any source outside it, CALR does not take place was computed in [**200**] by again using the spectral properties of the Neumann–Poincaré-type operator associated with the two elliptic interfaces.

Using the spectral properties of the Neumann–Poincaré operator for the Lamé system of elasto-statics (see Subsection 2.15.8), it was shown in [**97**] that CALR takes place at accumulation points of eigenvalues of the Neumann–Poincaré operator for the elasto-static system. In [**99**], the anomalous localized resonance on the circular coated structure and cloaking related to it in the context of elasto-static systems are investigated. The structure consists of the circular core with constant Lamé parameters and the circular shell of negative Lamé parameters proportional to those of the core. As in this chapter, it is shown that cloaking by anomalous localized resonance takes place if and only if the dipole type source lies inside critical radii determined by the radii of the core and the shell. This result has been obtained in [**366**] using a variational approach similar to that in [**333**].

In [**330**], it is shown that anomalous localized resonance may appear only for bodies so small such that the quasi-static approximation is realistic. This gives limits for size of the objects for which CALR may be used. Such limits may also apply for shielding at a distance for the full Maxwell equations.

CHAPTER 12

Plasmonic Metasurfaces

12.1. Introduction

A metasurface is a composite material layer, designed and optimized in order to control and transform electromagnetic fields. The layer thickness is negligible with respect to the wavelength in the surrounding space. The composite structure forming the metasurface is assumed to behave as a material in the electromagnetic sense, meaning that it can be homogenized on the wavelength scale, and the metasurface can be adequately characterized by its effective, surface-averaged properties [**498**].

In this chapter, we consider the scattering by a thin layer of periodic plasmonic nanoparticles mounted on a perfectly conducting sheet. We design the thin layer to have anomalous reflection properties and therefore to be viewed as a metasurface. As the thickness of the layer, which is of the same order as the diameter of the individual nanoparticles, is negligible compared to the wavelength, it can be approximated by an impedance boundary condition. Our main result is to show that at some resonant frequencies the impedance blows up, allowing for a significant reduction of the scattering from the plate. Using the spectral properties of the periodic Neumann–Poincaré operator defined in (12.6), we investigate the dependency of the impedance with respect to changes in the nanoparticle geometry and configuration. We fully characterize the resonant frequencies in terms of the periodicity, the shape and the material parameters of the nanoparticles. As the period of the array is much smaller than the wavelength, the resonant frequencies of the array of nanoparticles differ significantly from those of single nanoparticles. As shown in this chapter, they are associated with eigenvalues of a periodic Neumann–Poincaré type operator. In contrast with quasi-static plasmonic resonances of single nanoparticles, they depend on the particle size. For simplicity, only one-dimensional arrays embedded in $\mathbb{R}^2$ are considered in this chapter. The extension to the two-dimensional case is straightforward and the dependence of the plasmonic resonances on the parameters of the lattice is easy to derive.

We present numerical results to illustrate our main results in this chapter, which open a door for a mathematical and numerical framework for realizing full control of waves using metasurfaces [**18**, **405**, **498**]. Our approach applies to any example of periodic distributions of resonators having (sub-wavelength) resonances in the quasi-static regime. It provides a framework for explaining the observed extraordinary or meta-properties of such structures and for optimizing these properties.

The chapter is organized as follows. We first formulate the problem of approximating the effect of a thin layer with impedance boundary conditions. Then using the results of Subsection 2.6.1 on the one-dimensional periodic Green's function and layer potentials, we derive an explicit formula for the equivalent boundary condition in terms of the eigenvalues and eigenvectors of the one-dimensional periodic Neumann–Poincaré operator defined by (12.6), and give the shape derivative of the impedance parameter. Finally, we illustrate with a few numerical experiments the anomalous change in the equivalent impedance boundary condition due to the plasmonic resonances of the periodic array of nanoparticles. For simplicity, we only consider the scalar wave equation and use a two-dimensional setup. The results of this chapter can be readily generalized to higher dimensions as well as to the full Maxwell equations. Our results in this chapter are from [**82**].

12.2. Setting of the Problem

We use the Helmholtz equation to model the propagation of light. As said before, this approximation can be viewed as a special case of Maxwell's equations, when the incident wave u^i is transverse magnetic or transverse electric polarized.

Consider a particle occupying a bounded domain $D \Subset \mathbb{R}^2$ of class $\mathcal{C}^{1,\eta}$ for some $\eta > 0$ and with size of order $\delta \ll 1$. The particle is characterized by electric permittivity ε_c and magnetic permeability μ_c, both of which may depend on the frequency of the incident wave. Assume that $\Im \varepsilon_c > 0, \Re \mu_c < 0, \Im \mu_c > 0$ and define

$$k_m = \omega\sqrt{\varepsilon_m \mu_m}, \quad k_c = \omega\sqrt{\varepsilon_c \mu_c},$$

where ε_m and μ_m are the permittivity and permeability of free space, respectively, and ω is the frequency. Throughout this chapter, we assume that ε_m and μ_m are real and positive and k_m is of order 1.

We consider the configuration shown in Figure 12.1, where a particle D is repeated periodically in the x_1-axis with period δ, and is of a distance of order δ from the boundary $x_2 = 0$ of the half-space $\mathbb{R}^2_+ := \{(x_1, x_2) \in \mathbb{R}^2,\ x_2 > 0\}$. We denote by $\mathcal{D}$ this collection of periodically arranged particles.

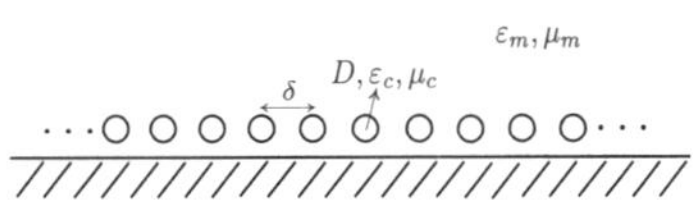

FIGURE 12.1. Thin layer of nanoparticles in the half-space.

Let $u^i(x) = e^{\sqrt{-1}k_m d\cdot x}$ be the incident wave. Here, d is the unit incidence direction. The scattering problem is modeled as follows:

$$(12.1) \qquad \begin{cases} \nabla \cdot \dfrac{1}{\mu_{\mathcal{D}}} \nabla u + \omega^2 \varepsilon_{\mathcal{D}} u = 0 \quad \text{in } \mathbb{R}^2_+ \setminus \partial\mathcal{D}, \\ u_+ - u_- = 0 \quad \text{on } \partial\mathcal{D}, \\ \dfrac{1}{\mu_m} \dfrac{\partial u}{\partial \nu}\bigg|_+ - \dfrac{1}{\mu_c} \dfrac{\partial u}{\partial \nu}\bigg|_- = 0 \quad \text{on } \partial\mathcal{D}, \\ u - u^i \text{ satisfies an outgoing radiation condition at infinity}, \\ u = 0 \quad \text{on } \partial\mathbb{R}^2_+ = \{(x_1, 0),\ x_1 \in \mathbb{R}\}, \end{cases}$$

where

$$\varepsilon_{\mathcal{D}} = \varepsilon_m \chi(\mathbb{R}^2_+ \setminus \overline{\mathcal{D}}) + \varepsilon_c \chi(\mathcal{D}), \quad \mu_{\mathcal{D}} = \varepsilon_m \chi(\mathbb{R}^2_+ \setminus \overline{\mathcal{D}}) + \varepsilon_c \chi(\mathcal{D}),$$

and $\partial/\partial\nu$ denotes the outward normal derivative on $\partial\mathcal{D}$.

Following [**2**], under the assumption that the wavelength of the incident wave is much larger than the size of the nanoparticle, a certain homogenization occurs, and we can construct $z \in \mathbb{C}$ such that the solution to

$$
\begin{cases}
\Delta u_{\text{app}} + k_m^2 u_{\text{app}} = 0 \quad \text{in } \mathbb{R}^2_+, \\
u_{\text{app}} + \delta z \frac{\partial u_{\text{app}}}{\partial x_2} = 0 \quad \text{on } \partial\mathbb{R}^2_+, \\
u_{\text{app}} - u^i \text{ satisfies outgoing radiation condition at infinity,}
\end{cases}
\tag{12.2}
$$

gives the leading order approximation for u. We refer to $u_{\text{app}} + \delta z \partial u_{\text{app}}/\partial x_2 = 0$ as the equivalent impedance boundary condition for problem (12.1). Note that if $\Im z < 0$, then (12.2) has a unique solution.

12.3. Boundary-Layer Corrector and Effective Impedance

In order to compute z, we introduce the following asymptotic expansion [**2, 6**]:

$$
u = u^{(0)} + u^{(0)}_{BL} + \delta(u^{(1)} + u^{(1)}_{BL}) + \dots,
\tag{12.3}
$$

where the leading-order term $u^{(0)}$ is solution to

$$
\begin{cases}
\Delta u^{(0)} + k_m^2 u^{(0)} = 0 \quad \text{in } \mathbb{R}^2_+, \\
u^{(0)} = 0 \quad \text{on } \partial\mathbb{R}^2_+, \\
u^{(0)} - u^i \text{ satisfies an outgoing radiation condition at infinity.}
\end{cases}
$$

The boundary-layer correctors $u^{(0)}_{BL}$ and $u^{(1)}_{BL}$ have to be exponentially decaying in the x_2-direction. Note that, according to [**2, 6**], $u^{(0)}_{BL}$ is introduced in order to correct (up to the first order in δ) the transmission condition on the boundary of the nanoparticles, which is not satisfied by the leading-order term $u^{(0)}$ in the asymptotic expansion of u, while $u^{(1)}_{BL}$ is a higher-order correction term and does not contribute to the first-order equivalent boundary condition in (12.2).

We next construct the corrector $u^{(0)}_{BL}$. We first introduce a function α and a complex constant α_∞ such that they satisfy the rescaled problem

$$
\begin{cases}
\Delta \alpha = 0 \quad \text{in } \left(\mathbb{R}^2_+ \backslash \overline{\mathcal{B}}\right) \cup \mathcal{B}, \\
\alpha|_+ - \alpha|_- = 0 \quad \text{on } \partial\mathcal{B}, \\
\left.\dfrac{1}{\mu_m}\dfrac{\partial \alpha}{\partial \nu}\right|_+ - \left.\dfrac{1}{\mu_c}\dfrac{\partial \alpha}{\partial \nu}\right|_- = \left(\dfrac{1}{\mu_c} - \dfrac{1}{\mu_m}\right)\nu_2 \quad \text{on } \partial\mathcal{B}, \\
\alpha = 0 \quad \text{on } \partial\mathbb{R}^2_+, \\
\alpha - \alpha_\infty \text{ is exponentially decaying as } x_2 \to +\infty.
\end{cases}
\tag{12.4}
$$

Here, $\nu = (\nu_1, \nu_2)$ and $B = D/\delta$ is repeated periodically in the x_1-axis with period 1 and $\mathcal{B}$ is the collection of these periodically arranged particles.

Then $u^{(0)}_{BL}$ is defined by

$$
u^{(0)}_{BL}(x) := \delta \frac{\partial u^{(0)}}{\partial x_2}(x_1, 0)\left(\alpha(\frac{x}{\delta}) - \alpha_\infty\right).
$$

The corrector $u^{(1)}$ can be found to be the solution to

$$\begin{cases} \Delta u^{(1)} + k_m^2 u^{(1)} = 0 \quad \text{in } \mathbb{R}^2_+, \\ u^{(1)} = \alpha_\infty \frac{\partial u^{(0)}}{\partial x_2} \quad \text{on } \partial\mathbb{R}^2_+, \\ u^{(1)} \text{ satisfies an outgoing radiation condition at infinity.} \end{cases}$$

By writing

$$u_{\text{app}} = u^{(0)} + \delta u^{(1)}, \tag{12.5}$$

we arrive at (12.2) with $z = -\alpha_\infty$, up to a second-order term in δ. We summarize the above results in the following theorem.

THEOREM 12.1. *The solution u_{app} to (12.2) with $z = -\alpha_\infty$ approximates pointwisely (for $x_2 > 0$) the exact solution u to (12.1) as $\delta \to 0$, up to a second-order term in δ.*

In order to compute α_∞, we derive an integral representation for the solution α to (12.4). We make use of the periodic Green function $G_\sharp$ defined by (2.110). Let

$$G_\sharp^+(x, y) = G_\sharp\big((x_1 - y_1, x_2 - y_2)\big) - G_\sharp\big((x_1 - y_1, -x_2 - y_2)\big),$$

which is the periodic Green's function in the upper half-space with Dirichlet boundary conditions, and define

$$\begin{aligned} \mathcal{S}_{B\sharp}^+ : H^{-\frac{1}{2}}(\partial B) \quad &\longrightarrow \quad H^1_{\text{loc}}(\mathbb{R}^2), H^{\frac{1}{2}}(\partial B) \\ \varphi \quad &\longmapsto \quad \mathcal{S}_{B,\sharp}^+[\varphi](x) = \int_{\partial B} G_\sharp^+(x, y)\varphi(y)d\sigma(y) \end{aligned}$$

for $x \in \mathbb{R}^2_+, x \in \partial B$ and

$$\begin{aligned} &(\mathcal{K}_{B\sharp}^*)^+ : H^{-\frac{1}{2}}(\partial B) \longrightarrow H^{-\frac{1}{2}}(\partial B) \\ &\varphi \longmapsto (\mathcal{K}_{B,\sharp}^*)^+[\varphi](x) = \int_{\partial B} \frac{\partial G_\sharp^+(x, y)}{\partial \nu(x)}\varphi(y)d\sigma(y) \end{aligned} \tag{12.6}$$

for $x \in \partial B$.

It is clear that the results of Lemma 2.35 hold true for $\mathcal{S}_{B\sharp}^+$ and $(\mathcal{K}_{B\sharp}^*)^+$. Moreover, for any $\varphi \in H^{-\frac{1}{2}}(\partial B)$, we have

$$\mathcal{S}_{B,\sharp}^+[\varphi](x) = 0 \quad \text{for } x \in \partial\mathbb{R}^2_+.$$

Now, we can readily see that α can be represented as $\alpha = \mathcal{S}_{B,\sharp}^+[\varphi]$, where $\varphi \in H^{-\frac{1}{2}}(\partial B)$ satisfies

$$\frac{1}{\mu_m} \frac{\partial \mathcal{S}_{B,\sharp}^+[\varphi]}{\partial \nu}\bigg|_+ - \frac{1}{\mu_c} \frac{\partial \mathcal{S}_{B,\sharp}^+[\varphi]}{\partial \nu}\bigg|_- = \Big(\frac{1}{\mu_c} - \frac{1}{\mu_m}\Big)\nu_2 \quad \text{on } \partial B.$$

Using the jump formula from Lemma 2.35, we arrive at

$$\big(\lambda_\mu I - (\mathcal{K}_{B\sharp}^*)^+\big)[\varphi] = \nu_2,$$

where

$$\lambda_\mu = \frac{\mu_c + \mu_m}{2(\mu_c - \mu_m)}.$$

Therefore, using item (v) in Lemma 2.35 on the characterization of the spectrum of $\mathcal{K}_{B\sharp}^*$ and the fact that the spectra of $(\mathcal{K}_{B\sharp}^*)^+$ and $\mathcal{K}_{B\sharp}^*$ are the same, we obtain that

$$\alpha = \mathcal{S}_{B,\sharp}^+ \big(\lambda_\mu I - (\mathcal{K}_{B\sharp}^*)^+\big)^{-1}[\nu_2].$$

LEMMA 12.2. *Let $x = (x_1, x_2)$. Then, for $x_2 \to +\infty$, the following asymptotic expansion holds:*

$$\alpha = \alpha_\infty + O(e^{-x_2}),$$

with

$$\alpha_\infty = -\int_{\partial B} y_2 \big(\lambda_\mu I - (\mathcal{K}_{B\sharp}^*)^+\big)^{-1}[\nu_2](y)d\sigma(y).$$

PROOF. The result follows from an asymptotic analysis of $G_\sharp^+(x,y)$. Indeed, suppose that $x_2 \to +\infty$, we have

$$\begin{aligned}
&G_\sharp^+(x,y) = \tfrac{1}{4\pi} \ln\big(\sinh^2(\pi(x_2 - y_2)) + \sin^2(\pi(x_1 - y_1))\big)\\
&-\frac{1}{4\pi} \ln\big(\sinh^2(\pi(x_2 + y_2)) + \sin^2(\pi(x_1 - y_1))\big)\\
&= \frac{1}{4\pi} \ln\big(\sinh^2(\pi(x_2 - y_2))\big)\\
&-\frac{1}{4\pi} \ln\big(\sinh^2(\pi(x_2 + y_2))\big)\\
&+O\Big(\ln\Big(1 + \frac{1}{\sinh^2(x_2)}\Big)\Big)\\
&= \frac{1}{2\pi}\Big(\ln\Big(\frac{e^{\pi(x_2-y_2)} - e^{-\pi(x_2+y_2)}}{2}\Big)\\
&-\ln\Big(\frac{e^{\pi(x_2+y_2)} - e^{-\pi(x_2-y_2)}}{2}\Big)\Big) + O\Big(\ln\Big(1+e^{-x_2^2}\Big)\Big)\\
&= -y_2 + O(e^{-x_2}),
\end{aligned}$$

which yields the desired result. □

Finally, it is important to note that α_∞ depends on the geometry and size of the particle B.

As $(\mathcal{K}_{B\sharp}^*)^+ : \mathcal{H}_0^* \to \mathcal{H}_0^*$ is a compact self-adjoint operator, where $\mathcal{H}_0^*$ is defined as in Lemma 2.35, we can write

$$\begin{aligned}
\alpha_\infty &= -\int_{\partial B} y_2 \big(\lambda_\mu I - (\mathcal{K}_{B\sharp}^*)^+\big)^{-1}[\nu_2](y)d\sigma(y),\\
&= -\int_{\partial B} y_2 \sum_{j=1}^\infty \frac{\langle \varphi_j, \nu_2\rangle_{\mathcal{H}_0^*} \varphi_j(y)}{\lambda_\mu - \lambda_j} d\sigma(y),\\
&= \sum_{j=1}^\infty \frac{\langle \varphi_j, \nu_2\rangle_{\mathcal{H}_0^*} \langle \varphi_j, y_2\rangle_{-\frac12,\frac12}}{\lambda_\mu - \lambda_j},
\end{aligned}$$

where $\lambda_1, \lambda_2, \ldots$ are the eigenvalues of $(\mathcal{K}_{B\sharp}^*)^+$ and $\varphi_1, \varphi_2, \ldots$ is a corresponding orthornormal basis of eigenvectors.

On the other hand, by integrating by parts we get

$$\langle \varphi_j, y_2\rangle_{-\frac{1}{2},\frac{1}{2}} = \frac{1}{\frac{1}{2}-\lambda_j}\langle \varphi_j, \nu_2\rangle_{\mathcal{H}_0^*}.$$

This, together with the fact that $\Im \lambda_\mu < 0$ (by the Drude model (7.4)), yields the following lemma.

LEMMA 12.3. *We have* $\Im \alpha_\infty > 0$.

Note that as a consequence of Lemma 12.3 it follows that (12.2) with $z = -\alpha_\infty$ is well-posed.

Finally, we give a formula for the shape derivative of α_∞. This formula can be used to optimize $|\alpha_\infty|$, for a given frequency ω, in terms of the shape B of the nanoparticle. Let B_η be an η-perturbation of B; *i.e.*, let $h \in \mathcal{C}^1(\partial B)$ and ∂B_η be given by

$$\partial B_\eta = \Big\{x + \eta h(x)\nu(x), x \in \partial B\Big\}.$$

Following Subsection 7.2.1.1, we can prove that

$$\begin{aligned}\alpha_\infty(B_\eta) &= \alpha_\infty(B) + \eta(\frac{\mu_m}{\mu_c} - 1)\\ &\quad \times \int_{\partial B} h\left[\frac{\partial v}{\partial \nu}\Big|_- \frac{\partial w}{\partial \nu}\Big|_- + \frac{\mu_c}{\mu_m}\frac{\partial v}{\partial T}\Big|_- \frac{\partial w}{\partial T}\Big|_-\right] d\sigma,\end{aligned}$$

where $\partial/\partial T$ is the tangential derivative on ∂B, v and w periodic with respect to x_1 of period 1 and satisfy

$$\begin{cases} \Delta v = 0 \quad \text{in } \left(\mathbb{R}^2_+\backslash\overline{\mathcal{B}}\right)\cup\mathcal{B}, \\ v|_+ - v|_- = 0 \quad \text{on } \partial\mathcal{B}, \\ \dfrac{\partial v}{\partial \nu}\Big|_+ - \dfrac{\mu_m}{\mu_c}\dfrac{\partial v}{\partial \nu}\Big|_- = 0 \quad \text{on } \partial\mathcal{B}, \\ v - x_2 \to 0 \quad \text{as } x_2 \to +\infty, \end{cases}$$

and

$$\begin{cases} \Delta w = 0 \quad \text{in } \left(\mathbb{R}^2_+\backslash\overline{\mathcal{B}}\right)\cup\mathcal{B}, \\ \dfrac{\mu_m}{\mu_c} w|_+ - w|_- = 0 \quad \text{on } \partial\mathcal{B}, \\ \dfrac{\partial w}{\partial \nu}\Big|_+ - \dfrac{\partial w}{\partial \nu}\Big|_- = 0 \quad \text{on } \partial\mathcal{B}, \\ w - x_2 \to 0 \quad \text{as } x_2 \to +\infty, \end{cases}$$

respectively. Therefore, the following proposition holds.

PROPOSITION 12.4. *The shape derivative* $d_\mathcal{S}\alpha_\infty(B)$ *of* α_∞ *is given by*

$$d_\mathcal{S}\alpha_\infty(B) = (\frac{\mu_m}{\mu_c} - 1)\left[\frac{\partial v}{\partial \nu}\Big|_- \frac{\partial w}{\partial \nu}\Big|_- + \frac{\mu_c}{\mu_m}\frac{\partial v}{\partial T}\Big|_- \frac{\partial w}{\partial T}\Big|_-\right].$$

If we aim to maximize the functional $J := \frac{1}{2}|\alpha_\infty|^2$ over B, then it can be easily seen that J is Fréchet differentiable and its Fréchet derivative is given by

$$\Re\, d_\mathcal{S}\alpha_\infty(B)\overline{\alpha_\infty(B)}.$$

In order to include cases where topology changes and multiple components are allowed, a level-set version of the optimization procedure described below can be developed (see Appendix B).

12.4. Numerical Illustrations

We now demonstrate the dependence of the equivalent boundary condition parameter α_∞ on the incident wavelength for various nanoparticle configurations. We use the Drude model for the permeability of background material, which is water, and the nanoparticles which are gold. The Drude model for the permeability μ is given by

$$\mu(\omega) = 1 - \frac{\omega_p^2}{\omega^2 + \sqrt{-1}\tau\omega}.$$

In particular, to model gold nanoparticles we choose the plasma frequency ω_p to be

$$\omega_p = 9.03 \times 2\pi \times \frac{1.6 \times 10^{-19}}{6.6 \times 10^{-34}},$$

and the damping coefficient τ to be

$$\tau = 0.053 \times 2\pi \times \frac{1.6 \times 10^{-19}}{6.6 \times 10^{-34}}.$$

The discretization of the boundary of the nanoparticle, along with the computation of the Neumann–Poincaré operator $(\mathcal{K}_{B\sharp}^*)^+$, where B is a disk, is performed in the same fashion as in Section 2.4.5 We then calculate

$$\alpha_\infty = -\int_{\partial B} y_2 \big(\lambda_\mu I - (\mathcal{K}_{B\sharp}^*)^+\big)^{-1}[\nu_2](y)d\sigma(y),$$

and plot its modulus $|\alpha_\infty|$ for a range of wavelengths in the interval $[150\times 10^{-9}, 350\times 10^{-9}]$. The corresponding Matlab code can be downloaded at:
`www.ams.org/bookpages/surv-235`.

In Figure 12.2 we place the row of nanoparticles a distance of 0.5 from the surface $\partial\mathbb{R}^2_+$ and vary the radii from 0.1 to 0.4. In Figure 12.3 we set the nanoparticle radius to be 0.2 and observe the change in $|\alpha_\infty|$ when we first position the nanoparticles a distance of 0.25 from the surface, and then a distance of 0.75.

In Figures 12.4 and 12.5 we demonstrate that in the case of a single row of nanoparticles we have a distinct resonance peak, whereas in in the case of three well-separated nanoparticles (all in the unit cell) we have delocalized resonances.

12.5. Concluding Remarks

In this chapter, we have considered the scattering by an array of plasmonic nanoparticles mounted on a perfectly conducting plate and showed both analytically and numerically the significant change in the boundary condition induced by the nanoparticles at their periodic plasmonic frequencies. We have also proposed an optimization approach to maximize this change in terms of the shape of the nanoparticles. Our results in this chapter can be generalized in many directions. Different boundary conditions on the plate as well as curved plates can be considered. Our approach can be easily extended to two-dimensional arrays embedded in $\mathbb{R}^3$ and the lattice effect can be included by using the Green's function for a general lattice (as in (2.132)). Full Maxwell's equations to model the light propagation can be used. The observed extraordinary or meta-properties of periodic distributions of sub-wavelength resonators can be explained by the approach proposed in this chapter.

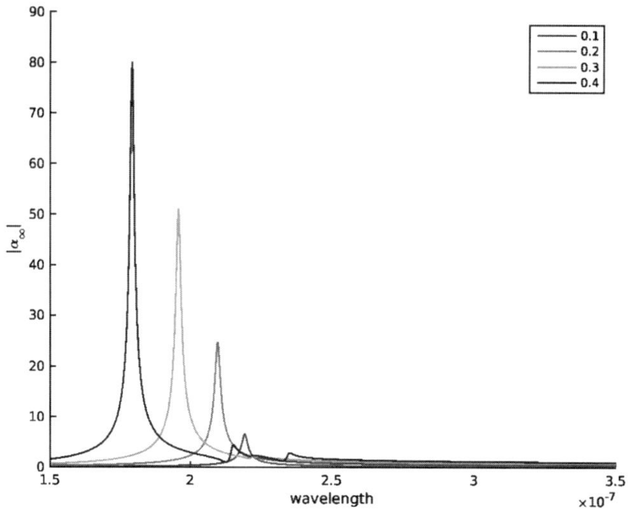

FIGURE 12.2. $|\alpha_\infty|$ as a function of wavelength for a set of radii varying from 0.1 to 0.4.

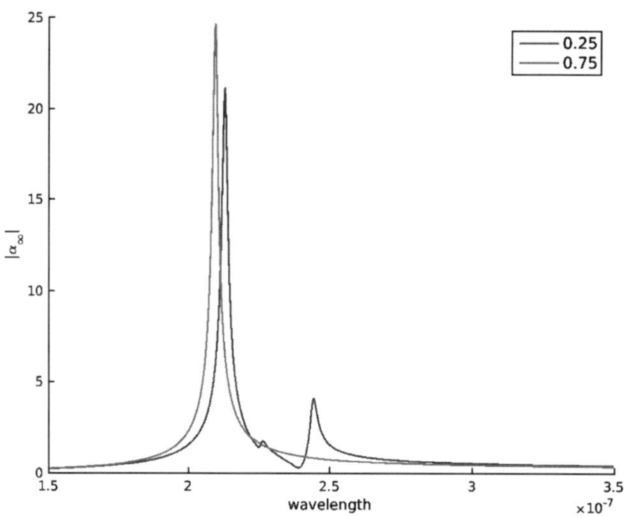

FIGURE 12.3. $|\alpha_\infty|$ as a function of wavelength for a set of radii for a disk of radius 0.2 as for distances of 0.25 and 0.75 from the boundary at $x_2 = 0$.

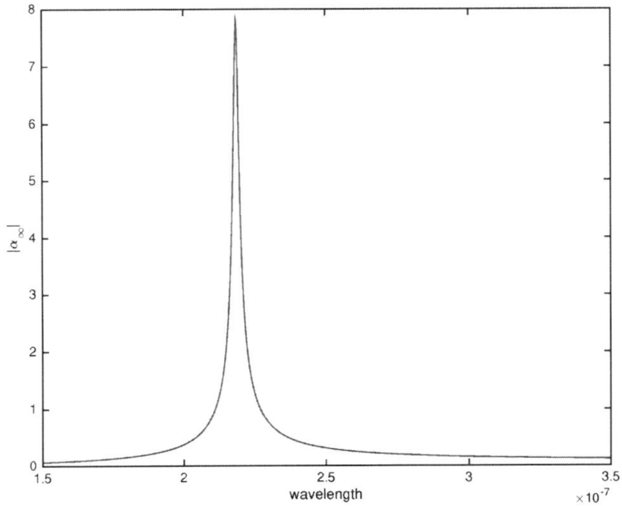

FIGURE 12.4. We observe a strong localized resonant peak in the case of a single row of nanoparticles.

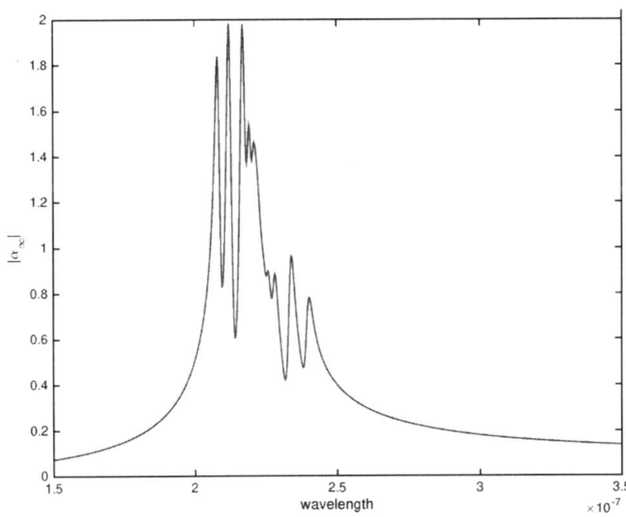

FIGURE 12.5. When we have three nanoparticles in each cell of the array we observe delocalized resonance.

Part 5

Sub-wavelength Phononics

CHAPTER 13

Helmholtz Resonator

13.1. Introduction

The Helmholtz resonator is an acoustic device which has many important applications in phononics. It consists of a closed cavity connected to the exterior domain by an opening hole.

In this section we study perturbations of scattering frequencies of Helmholtz resonators with small openings. We provide on the one hand results on the existence and localization of the scattering frequencies and on the other hand the leading-order terms in their asymptotic expansions in terms of the characteristic width of the openings.

We show that the spectrum of the Helmholtz resonator essentially coincides with the spectrum of the Laplacian with Neumann boundary condition in the closed cavity, but there is an additional resonant frequency which is a sub-wavelength resonance. Its associated eigenfunction is essentially constant in the cavity and it essentially vanishes in the exterior domain. It is the key to super-resolution and super-focusing for acoustic waves in systems of Helmholtz resonators.

As in the previous chapters, we transform the problem of finding the scattering frequencies into that of the determination of the characteristic values of certain integral operator-valued functions in the complex plane. The generalization of the Steinberg theorem given in Theorem 1.16 yields the discreteness of the set of resonant frequencies. The generalized Rouché theorem shows the existence of resonant frequencies close to the eigenfrequencies of the unperturbed resonator. In principle, the general form of the argument principle in Theorem 1.14 can be applied to derive full asymptotic expansions for the scattering frequencies. As will be shown in this chapter, the leading-order terms can be obtained by a simpler method based on pole-pencil decomposition of integral operator-valued functions and the use of the Hilbert transform.

13.2. Hilbert Transform

In order to study resonance frequencies of Helmholtz resonators, the Hilbert transform will be needed.

Let the set $\mathcal{X}^\epsilon$, for small $\epsilon > 0$, be defined by

$$\mathcal{X}^\epsilon = \left\{ \varphi : \int_{-\epsilon}^{\epsilon} \sqrt{\epsilon^2 - x^2}\, |\varphi(x)|^2 \, dx < +\infty \right\}, \tag{13.1}$$

Equipped with the norm

$$\|\varphi\|_{\mathcal{X}^\epsilon} = \left(\int_{-\epsilon}^{\epsilon} \sqrt{\epsilon^2 - x^2}\, |\varphi(x)|^2 \, dx \right)^{1/2},$$

$\mathcal{X}^\epsilon$ is a Hilbert space. Introduce

$$\mathcal{Y}^\epsilon = \Big\{\psi \in \mathcal{C}^0\left([-\epsilon,\epsilon]\right) : \psi' \in \mathcal{X}^\epsilon\Big\}, \tag{13.2}$$

where ψ' is the distribution derivative of ψ. The set $\mathcal{Y}^\epsilon$ is a Hilbert space with the norm

$$\|\psi\|_{\mathcal{Y}^\epsilon} = \left(\|\psi\|^2_{\mathcal{X}^\epsilon} + \|\psi'\|^2_{\mathcal{X}^\epsilon}\right)^{1/2}.$$

Let $\mathcal{L}_\epsilon : \mathcal{X}^\epsilon \to \mathcal{Y}^\epsilon$ be defined by

$$\mathcal{L}_\epsilon[\varphi](x) = \int_{-\epsilon}^{\epsilon} \ln|x-y|\,\varphi(y)\,dy. \tag{13.3}$$

We establish two results concerning the integral operator $\mathcal{L}_\epsilon$. These results are proved in [**468**].

LEMMA 13.1. *For all $0 < \epsilon < 2$, the integral operator $\mathcal{L}_\epsilon : \mathcal{X}^\epsilon \mapsto \mathcal{Y}^\epsilon$ is invertible.*

Let φ be a function of $\mathcal{X}^\epsilon$. The function

$$\psi(x) = \int_{-\epsilon}^{\epsilon} \ln|x-y|\,\varphi(y)\,dy$$

is differentiable and its derivative on $(-\epsilon,\epsilon)$ is given by (see for instance [**410**, p. 30])

$$\psi'(x) = \mathcal{H}_\epsilon[\varphi](x), \tag{13.4}$$

where $\mathcal{H}_\epsilon$ denotes the finite Hilbert transform (or Tricomi's operator)

$$\mathcal{H}_\epsilon[\varphi](x) = \int_{-\epsilon}^{\epsilon} \frac{\varphi(y)}{x-y}\,dy \quad \text{for } x \in (-\epsilon,\epsilon)\,. \tag{13.5}$$

The following explicit expressions hold. For any $x \in (-\epsilon,\epsilon)$, we have

$$\mathcal{H}_\epsilon[\frac{1}{\sqrt{\epsilon^2-y^2}}](x) = 0, \tag{13.6}$$

$$\mathcal{H}_\epsilon[\sqrt{\epsilon^2-y^2}](x) = \pi x, \tag{13.7}$$

and

$$\mathcal{H}_\epsilon[\frac{y}{\sqrt{\epsilon^2-y^2}}](x) = -\pi. \tag{13.8}$$

The main difficulty in studying the finite Hilbert transform $\mathcal{H}_\epsilon$ on Hölder continuous functions (with requirements at the endpoints) is that it has no smoothness preserving property, as shown by the following formulas:

$$\mathcal{H}_\epsilon[1](x) = \ln\frac{\epsilon+x}{\epsilon-x}$$

and

$$\mathcal{H}'_\epsilon[\varphi](x) - \frac{\varphi(-\epsilon)}{\epsilon+x} + \frac{\varphi(\epsilon)}{\epsilon-x} + \mathcal{H}_\epsilon[\varphi'](x).$$

The development of such a theory is a rather long and complicated process. See for instance [**422**].

Here the weighted space $\mathcal{X}^\epsilon$ is introduced to make the theory relatively simple and yet general enough for applications. The following mapping properties of the finite Hilbert transform hold. See [**468**] for a proof.

LEMMA 13.2 (Finite Hilbert transform). *The operator* $\mathcal{H}_\epsilon : \mathcal{X}^\epsilon \to \mathcal{X}^\epsilon$ *satisfies* $\dim \operatorname{Ker}(\mathcal{H}_\epsilon) = 1$ *and* $\operatorname{Im} \mathcal{H}_\epsilon = \mathcal{X}^\epsilon$.

As shown by formula (13.6), $\operatorname{Ker}(\mathcal{H}_\epsilon)$ is spanned by $1/\sqrt{\epsilon^2 - y^2}$. We refer the reader to [**107**, **228**] for the mapping properties of the finite Hilbert transform in more general weighted spaces. We will need the following Hölder estimate [**410**]

$$\left\| \int_{-1}^{1} \frac{\varphi(y)}{x-y}\, dy \right\|_{L^\infty([-1,1])} \leq C \, \|\varphi\|_{\mathcal{C}^{0,\eta}([-1,1])} \tag{13.9}$$

for $\varphi \in \mathcal{C}^{0,\eta}([-1,1])$ with $\eta > 0$.

We now solve explicitly the integral equation

$$\mathcal{L}_\epsilon[\varphi](x) = \psi(x), \quad \forall\, x \in (-\epsilon, \epsilon), \tag{13.10}$$

where ψ is a given function in $\mathcal{Y}^\epsilon$ and φ is the unknown function. Differentiating (13.10) with respect to the x-variable, we obtain the singular integral equation (13.4). The general solution of equation (13.4) is given by the Hilbert inversion formula (see for instance [**410**]):

$$\varphi_\lambda(x) = -\frac{1}{\pi^2\sqrt{\epsilon^2 - x^2}} \int_{-\epsilon}^{\epsilon} \frac{\sqrt{\epsilon^2 - y^2}\,\psi'(y)}{x-y}\, dy + \frac{\lambda}{\sqrt{\epsilon^2 - x^2}}, \tag{13.11}$$

where λ is a complex constant. Therefore, the solution φ in $\mathcal{X}^\epsilon$ of (13.10) is necessarily one of the φ_λ given by (13.11), where λ is chosen appropriately. Denote by $\lambda(\psi)$ the appropriate choice of λ and consider

$$a(\psi) = \psi(x) - \mathcal{L}_\epsilon[\varphi_{\lambda=0}](x). \tag{13.12}$$

We first observe that the quantity $a(\psi)$ is a constant, since its derivative with respect to x is identically equal to zero on $(-\epsilon, \epsilon)$. Now, substitute $\varphi_{\lambda(\psi)}$ into (13.10) to get

$$\lambda(\psi)\, \mathcal{L}_\epsilon \left[y \mapsto \frac{1}{\sqrt{\epsilon^2 - y^2}} \right] = a(\psi).$$

But, a straightforward calculation shows that

$$\mathcal{L}_\epsilon \left[y \mapsto \frac{1}{\sqrt{\epsilon^2 - y^2}} \right](x) = \pi \ln \frac{\epsilon}{2} \quad \text{for all } x \in (-\epsilon, \epsilon),$$

and therefore,

$$\lambda(\psi) = \frac{a(\psi)}{\pi \ln(\epsilon/2)}.$$

Thus,

$$\mathcal{L}_\epsilon^{-1}[\psi](x) = -\frac{1}{\pi^2\sqrt{\epsilon^2 - x^2}} \int_{-\epsilon}^{\epsilon} \frac{\sqrt{\epsilon^2 - y^2}\,\psi'(y)}{x-y}\, dy + \frac{a(\psi)}{(\pi \ln(\epsilon/2))\sqrt{\epsilon^2 - x^2}}, \tag{13.13}$$

where $a(\psi)$ is given by (13.12). This calculation has been done by Carleman in [**191**].

Note that for $\epsilon = 2$, $\mathcal{L}_2$ has a nontrivial kernel. However, for $0 < \epsilon < 2$, the solution to (13.10) is clearly unique. In fact, by (13.11) and (13.12), it follows that if $\psi \equiv 0$, then $\mathcal{L}_\epsilon^{-1}[\psi] \equiv 0$.

We will also need the following lemma.

LEMMA 13.3. *Let $\mathcal{R}_\epsilon$ be the integral operator defined from $\mathcal{X}^\epsilon$ into $\mathcal{Y}^\epsilon$ by*

$$\mathcal{R}_\epsilon[\varphi](x) = \int_{-\epsilon}^{\epsilon} R(x,y)\,\varphi(y)\,dy,$$

with $R(x,y)$ of class $\mathcal{C}^{1,\eta}$ in x and y, for $\eta > 0$. There exists a positive constant C, independent of ϵ, such that

$$\|\mathcal{L}_\epsilon^{-1}\,\mathcal{R}_\epsilon\|_{\mathcal{L}(\mathcal{X}^\epsilon,\mathcal{X}^\epsilon)} \leq \frac{C}{|\ln\epsilon|}, \tag{13.14}$$

where

$$\|\mathcal{L}_\epsilon^{-1}\,\mathcal{R}_\epsilon\|_{\mathcal{L}(\mathcal{X}^\epsilon,\mathcal{X}^\epsilon)} = \sup_{\varphi\in\mathcal{X}^\epsilon,\|\varphi\|_{\mathcal{X}^\epsilon}=1} \|\mathcal{L}_\epsilon^{-1}\mathcal{R}_\epsilon[\varphi]\|_{\mathcal{X}^\epsilon}.$$

PROOF. Let $\varphi \in \mathcal{X}^\epsilon$. By the Hilbert inversion formula (13.11), we have

$$\begin{aligned}\mathcal{L}_\epsilon^{-1}\mathcal{R}_\epsilon[\varphi](x) &= -\frac{1}{\pi^2\sqrt{\epsilon^2-x^2}}\int_{-\epsilon}^{\epsilon}\frac{\sqrt{\epsilon^2-y^2}(\mathcal{R}_\epsilon[\varphi])'(y)}{x-y}\,dy + \frac{\lambda(\mathcal{R}_\epsilon[\varphi])}{\sqrt{\epsilon^2-x^2}}\\ &:= I_1^\epsilon[\varphi](x) + I_2^\epsilon[\varphi](x),\end{aligned} \tag{13.15}$$

where

$$\lambda(\mathcal{R}_\epsilon[\varphi]) = \frac{a(\mathcal{R}_\epsilon[\varphi])}{\pi\ln(\epsilon/2)}.$$

We estimate $\|I_1^\epsilon[\varphi]\|_{\mathcal{X}^\epsilon}$ and $\|I_2^\epsilon[\varphi]\|_{\mathcal{X}^\epsilon}$ separately.

For $\|I_1^\epsilon[\varphi]\|_{\mathcal{X}^\epsilon}$, we have

$$\begin{aligned}\|I_1^\epsilon[\varphi]\|_{\mathcal{X}^\epsilon} &\left(\int_{-\epsilon}^{\epsilon}\frac{1}{\sqrt{\epsilon^2-x^2}}\left(\int_{-\epsilon}^{\epsilon}\frac{\sqrt{\epsilon^2-y^2}\,(\mathcal{R}_\epsilon[\varphi])'(y)}{x-y}\,dy\right)^2 dx\right)^{\frac{1}{2}}\\ &\leq C\left\|\int_{-\epsilon}^{\epsilon}\frac{\sqrt{\epsilon^2-y^2}\,(\mathcal{R}_\epsilon[\varphi])'(y)}{x-y}\,dy\right\|_{L^\infty([-\epsilon,\epsilon])}\\ &= C\epsilon\left\|\int_{-1}^{1}\frac{\sqrt{1-y^2}\,(\mathcal{R}_\epsilon[\varphi])'(\epsilon y)}{x-y}\,dy\right\|_{L^\infty([-1,1])}.\end{aligned}$$

We then have from the Hölder estimate (13.9) for the Hilbert transform

$$\begin{aligned}\left\|\int_{-1}^{1}\frac{\sqrt{1-y^2}\,(\mathcal{R}_\epsilon[\varphi])'(\epsilon y)}{x-y}\,dy\right\|_{L^\infty([-1,1])} &\leq C\left\|\sqrt{1-y^2}\,(\mathcal{R}_\epsilon[\varphi])'(\epsilon y)\right\|_{\mathcal{C}^{0,\eta}([-1,1])}\\ &\leq C\,\|(\mathcal{R}_\epsilon[\varphi])'(\epsilon y)\|_{\mathcal{C}^{0,\eta}([-1,1])}\\ &\leq C\,\|(\mathcal{R}_\epsilon[\varphi])'(y)\|_{\mathcal{C}^{0,\eta}([-\epsilon,\epsilon])}\\ &\leq C\int_{-\epsilon}^{\epsilon}|\varphi(y)|\,dy \leq C\|\varphi\|_{\mathcal{X}^\epsilon},\end{aligned}$$

since the kernel R is of class $\mathcal{C}^{1,\eta}$. Thus, we obtain

$$\|I_1^\epsilon[\varphi]\|_{\mathcal{X}^\epsilon} \leq C\epsilon\|\varphi\|_{\mathcal{X}^\epsilon}. \tag{13.16}$$

To estimate $\|I_2^\epsilon[\varphi]\|_{\mathcal{X}^\epsilon}$, we first observe that

$$\begin{aligned}\|I_2^\epsilon[\varphi]\|_{\mathcal{X}^\epsilon} &= \left(\int_{-\epsilon}^{\epsilon}\sqrt{\epsilon^2-x^2}|I_2^\epsilon[\varphi](x)|^2dx\right)^{1/2}\\ &= \sqrt{\pi}|\lambda(\mathcal{R}_\epsilon[\varphi])| = \left|\frac{a(\mathcal{R}_\epsilon[\varphi])}{\sqrt{\pi}\ln(\epsilon/2)}\right|.\end{aligned}$$

At this point, let us recall that

$$a(\mathcal{R}_\epsilon[\varphi]) = \mathcal{R}_\epsilon[\varphi] - \mathcal{L}_\epsilon\left[t\mapsto \frac{1}{\sqrt{\epsilon^2-t^2}}\int_{-\epsilon}^{\epsilon}\frac{\sqrt{\epsilon^2-y^2}}{t-y}(\mathcal{R}_\epsilon[\varphi])'(y)\,dy\right].$$

Once again, from the smoothness of the kernel R in x and y, we have

$$\sup_{-\epsilon\le x\le\epsilon}|\mathcal{R}_\epsilon[\varphi](x)| \le C\,\|\varphi\|_{\mathcal{X}^\epsilon}. \tag{13.17}$$

On the other hand, we get from (13.7) that

$$\begin{aligned}&\int_{-\epsilon}^{\epsilon}\frac{\sqrt{\epsilon^2-y^2}}{t-y}(\mathcal{R}_\epsilon[\varphi])'(y)\,dy\\ &= (\mathcal{R}_\epsilon[\varphi])'(t)\int_{-\epsilon}^{\epsilon}\frac{\sqrt{\epsilon^2-y^2}}{t-y}\,dy + \int_{-\epsilon}^{\epsilon}\frac{\sqrt{\epsilon^2-y^2}}{t-y}[(\mathcal{R}_\epsilon[\varphi])'(y)-(\mathcal{R}_\epsilon[\varphi])'(t)]\,dy\\ &= \pi t(\mathcal{R}_\epsilon[\varphi])'(t) + \int_{-\epsilon}^{\epsilon}\frac{\sqrt{\epsilon^2-y^2}}{t-y}[(\mathcal{R}_\epsilon[\varphi])'(y)-(\mathcal{R}_\epsilon[\varphi])'(t)]\,dy.\end{aligned}$$

Put

$$I_\epsilon(t) := \int_{-\epsilon}^{\epsilon}\frac{\sqrt{\epsilon^2-y^2}}{t-y}[(\mathcal{R}_\epsilon[\varphi])'(y)-(\mathcal{R}_\epsilon[\varphi])'(t)]\,dy.$$

Then, we have

$$\begin{aligned}|I_\epsilon(t)| &\le C\epsilon^{1+\eta}\sup_{t,y}\left|\frac{(\mathcal{R}_\epsilon[\varphi])'(y)-(\mathcal{R}_\epsilon[\varphi])'(t)}{|y-t|^\eta}\right|\\ &\le C\epsilon^{1+\eta}\int_{-\epsilon}^{\epsilon}|\varphi(y)|dy \le C\epsilon^{1+\eta}\|\varphi\|_{\mathcal{X}^\epsilon}.\end{aligned}$$

Therefore, it follows that

$$\begin{aligned}&\left|\mathcal{L}_\epsilon\left[t\mapsto\frac{1}{\sqrt{\epsilon^2-t^2}}\int_{-\epsilon}^{\epsilon}\frac{\sqrt{\epsilon^2-y^2}}{t-y}(\mathcal{R}_\epsilon[\varphi])'(y)\,dy\right]\right|\\ &\quad\le \left|\mathcal{L}_\epsilon\left[\frac{\pi t(\mathcal{R}_\epsilon[\varphi])'(t)}{\sqrt{\epsilon^2-t^2}}\right]\right| + \left|\mathcal{L}_\epsilon\left[\frac{I_\epsilon(t)}{\sqrt{\epsilon^2-t^2}}\right]\right|\\ &\quad\le\left(\sup_t|\pi t(\mathcal{R}_\epsilon[\varphi])'(t)| + \sup_t|I_\epsilon(t)|\right)\left|\mathcal{L}_\epsilon\left[\frac{1}{\sqrt{\epsilon^2-t^2}}\right]\right| \le C\epsilon|\ln\epsilon|\|\varphi\|_{\mathcal{X}^\epsilon},\end{aligned}$$

which, combined with (13.17), gives

$$\|I_2^\epsilon[\varphi]\|_{\mathcal{X}^\epsilon} \le \frac{C}{|\ln\epsilon|}\|\varphi\|_{\mathcal{X}^\epsilon}. \tag{13.18}$$

Combining (13.16) and (13.18) yields the desired estimate (13.14). □

13.3. Perturbations of Scattering Frequencies of a Helmholtz Resonator

Let $\Omega \subset \mathbb{R}^2$ be a bounded simply connected domain with boundary $\partial\Omega$ of class $\mathcal{C}^2$. Let μ_0 be a simple eigenvalues of $-\Delta$ in Ω with Neumann conditions, namely, a simple eigenvalue of (3.1) and let V be a neighborhood of μ_0 in the complex plane such that μ_0 is the only eigenvalue of (3.1) in V.

The acoustic Helmholtz resonator we consider is a surface $\partial\Omega_\epsilon = \partial\Omega \setminus \Sigma_\epsilon$, where $\partial\Omega_\epsilon$ is obtained from $\partial\Omega$ by making a small opening Σ_ϵ in the boundary with diameter tending to zero as $\epsilon \to 0$. This opening connects the interior and the exterior parts of the resonator. If μ_0 is an eigenvalue of (3.1), the corresponding scattering problem is to find μ^ϵ (with $\Im m\, \mu^\epsilon \geq 0$) close to μ_0 such that there exists a nontrivial solution to

$$
\begin{cases}
(\Delta + \mu^\epsilon)u^\epsilon = 0 \quad \text{in } \Omega \cup (\mathbb{R}^2 \setminus \overline{\Omega}), \\
\dfrac{\partial u^\epsilon}{\partial \nu} = 0 \quad \text{on } \partial\Omega_\epsilon, \\
\left| \dfrac{\partial u^\epsilon}{\partial r} - \sqrt{-1\mu^\epsilon} u^\epsilon \right| = O(r^{-1}) \quad \text{as } r = |x| \to +\infty.
\end{cases}
\tag{13.19}
$$

As in the previous chapters, we reduce the scattering problem (13.19) to the study of characteristic values of a certain operator-valued function, and by means of the generalized Rouché theorem we prove the existence of a scattering frequency μ^ϵ with small imaginary part which converges to μ_0 as $\epsilon \to 0$. We then construct the leading-order term in its asymptotic expansion.

To simplify the exposition, we shall assume that 0 is the center to which the opening can be contracted and the opening Σ_ϵ is flat: $\Sigma_\epsilon = (-\epsilon, \epsilon)$. It can be shown that the curvature of the opening does not influence the leading-order term in the asymptotic expansion of the scattering frequencies [**252**]. Following the arguments presented in the previous section, we only outline the derivation of an asymptotic expansion of μ^ϵ, leaving the details to the reader.

13.3.1. Problem Formulation.

We say that $\mu \in \mathbb{C}$ (with $\Im m\, \mu \geq 0$) is a scattering pole if there exists a nontrivial solution to the exterior problem

$$
\begin{cases}
(\Delta + \mu)v = 0 \quad \text{in } \mathbb{R}^2 \setminus \overline{\Omega}, \\
\dfrac{\partial v}{\partial \nu} = 0 \quad \text{on } \partial\Omega, \\
\displaystyle\int_{\mathbb{R}^2 \setminus \overline{\Omega}} |v|^2 < +\infty.
\end{cases}
\tag{13.20}
$$

Introduce the exterior Neumann function $N^{\sqrt{\mu}}_{\mathbb{R}^2\setminus\overline{\Omega}}$, that is, the unique solution to

$$
\begin{cases}
(\Delta_x + \mu) N^{\sqrt{\mu}}_{\mathbb{R}^2\setminus\overline{\Omega}}(x, z) = -\delta_z & \text{in } \mathbb{R}^2 \setminus \overline{\Omega}, \\
\dfrac{\partial N^{\sqrt{\mu}}_{\mathbb{R}^2\setminus\overline{\Omega}}}{\partial \nu}\bigg|_{\partial\Omega} = 0 & \text{on } \partial\Omega, \\
\left| \dfrac{\partial N^{\sqrt{\mu}}_{\mathbb{R}^2\setminus\overline{\Omega}}}{\partial r} - \sqrt{-1\mu} N^{\sqrt{\mu}}_{\mathbb{R}^2\setminus\overline{\Omega}} \right| = O(r^{-1}) \quad \text{as } r = |x| \to +\infty.
\end{cases}
\tag{13.21}
$$

Set φ^ϵ to be the normal derivative of u^ϵ on the opening Σ_ϵ:

$$\varphi^\epsilon = \frac{\partial u^\epsilon}{\partial \nu} \quad \text{on } \Sigma_\epsilon.$$

By Green's formula, φ^ϵ satisfies the integral equation

$$\int_{\Sigma_\epsilon} \left(N^{\sqrt{\mu^\epsilon}}_{\mathbb{R}^2 \setminus \overline{\Omega}} + N^{\sqrt{\mu^\epsilon}}_{\Omega} \right)(x, y)\, \varphi^\epsilon(y)\, dy = 0 \quad \text{on } \Sigma_\epsilon, \tag{13.22}$$

where the interior Neumann function $N^{\sqrt{\mu^\epsilon}}_{\Omega}$ is defined by (2.186) and the exterior Neumann function $N^{\sqrt{\mu^\epsilon}}_{\mathbb{R}^2 \setminus \overline{\Omega}}$ is defined by (13.21).

Define the operator-valued function $\mu \mapsto \mathcal{A}_\epsilon(\mu)$ by

$$\mathcal{A}_\epsilon(\mu)[\varphi](x) := \int_{-\epsilon}^{\epsilon} \left(N^{\sqrt{\mu^\epsilon}}_{\mathbb{R}^2 \setminus \overline{\Omega}} + N^{\sqrt{\mu^\epsilon}}_{\Omega} \right)(x, y)\, \varphi(y)\, dy.$$

By virtue of (13.22), the problem of finding the scattering frequencies can be reduced to that of finding the characteristic values of $\mathcal{A}_\epsilon(\mu)$.

13.3.2. Asymptotic Formula for Perturbations in Scattering Frequencies. Let μ_0 be a simple eigenvalue of (3.1) associated with the normalized eigenfunction u_{j_0} and let V be a complex neighborhood of μ_0 such that (i) μ_0 is the only eigenvalue of (3.1) in V and (ii) there is no scattering pole of (13.20) in V.

Writing

$$N^{\sqrt{\mu}}_{\mathbb{R}^2 \setminus \overline{\Omega}}(x, z) = -\frac{1}{2\pi} \ln|x - z| + r(x, z, \mu),$$

where $r(x, z, \mu)$ is holomorphic with respect to μ in V and smooth in x and z, we obtain the following pole-pencil decomposition of $\mathcal{A}_\epsilon$ in V.

LEMMA 13.4. *The following pole-pencil decomposition of $\mathcal{A}_\epsilon(\mu) : \mathcal{X}^\epsilon \to \mathcal{Y}^\epsilon$ holds for any $\mu \in V \setminus \{\mu_0\}$:*

$$\mathcal{A}_\epsilon(\mu) = -\frac{1}{\pi} \mathcal{L}_\epsilon + \frac{\mathcal{K}_\epsilon}{\mu_0 - \mu} + \mathcal{R}_\epsilon(\mu), \tag{13.23}$$

where

$$\mathcal{L}_\epsilon[\varphi](x) = \int_{-\epsilon}^{\epsilon} \ln|x - y|\, \varphi(y)\, dy,$$

$\mathcal{K}_\epsilon$ is the one-dimensional operator given by

$$\mathcal{K}_\epsilon[\varphi](x) = \langle \varphi, u_{j_0} \rangle_{L^2(\Sigma_\epsilon)}\, u_{j_0},$$

and

$$\mathcal{R}_\epsilon(\mu)[\varphi](x) = \int_{-\epsilon}^{\epsilon} R(\mu, x, y)\, \varphi(y)\, dy,$$

with $(\mu, x, y) \mapsto R(\mu, x, y)$ holomorphic in μ and smooth in x and y.

We now prove that the set of characteristic values of $\mathcal{A}_\epsilon$ is discrete.

LEMMA 13.5. *The set of characteristic values of $\mu \mapsto \mathcal{A}_\epsilon(\mu)$ is discrete.*

PROOF. We only give a proof for the discreteness of the set of characteristic values of $\mathcal{A}_\epsilon$ in V. The same arguments apply in neighborhoods V_j of $\mu_j, j \geq 1$, and in $\mathbb{C} \setminus \bigcup_j V_j$. Recall that $\mathcal{L}_\epsilon : \mathcal{X}^\epsilon \to \mathcal{Y}^\epsilon$ is invertible and $\|\mathcal{L}_\epsilon^{-1} \mathcal{R}_\epsilon\|_{\mathcal{L}(\mathcal{X}^\epsilon, \mathcal{X}^\epsilon)} \to 0$ as $\epsilon \to 0$. Therefore, $-(1/(\pi))\, \mathcal{L}_\epsilon + \mathcal{R}_\epsilon : \mathcal{X}^\epsilon \to \mathcal{Y}^\epsilon$ is invertible for ϵ small enough. It then follows from the pole-pencil decomposition (13.23) that $\mathcal{A}_\epsilon$ is finitely meromorphic and of Fredholm type in V. Moreover, since $\mathcal{K}_\epsilon$ is of finite-dimension, there exists $\mu^* \in V$ such that $\mathcal{A}_\epsilon(\mu^*)$ is invertible. Therefore, the generalization of the Steinberg theorem (Theorem 1.16) gives the discreteness of the set of characteristic values of $\mathcal{A}_\epsilon$ in V. □

Next, we prove that there exists exactly one characteristic value of $\mathcal{A}_\epsilon$ located in the neighborhood V of μ_0 and compute its asymptotic expansion as ϵ goes to zero. The method is based on the pole-pencil decomposition (13.23) of the operator-valued function $\mathcal{A}_\epsilon$, followed by an application of the generalized Rouché theorem.

LEMMA 13.6. *The operator-valued function $\mathcal{A}_\epsilon(\mu)$ has exactly one characteristic value in V.*

PROOF. We first study the principal part of $\mathcal{A}_\epsilon$, that is, the integral operator-valued function defined by

$$\mathcal{N}_\epsilon : \mu \mapsto \mathcal{N}_\epsilon(\mu) = -\frac{1}{\pi}\mathcal{L}_\epsilon + \frac{\mathcal{K}_\epsilon}{\mu_0 - \mu},$$

and show that its multiplicity in V is equal to zero. Let us find the characteristic values of $\mathcal{N}_\epsilon$ in V, that is, the complex numbers $\hat{\mu}$, such that there exists $\hat{\varphi} \not\equiv 0$ satisfying $\mathcal{N}_\epsilon(\hat{\mu})[\hat{\varphi}] \equiv 0$ on $(-\epsilon, \epsilon)$. Equivalently, we have

$$\frac{1}{\pi}\mathcal{L}_\epsilon[\hat{\varphi}] + \frac{\langle \hat{\varphi}, u_{j_0}\rangle}{\hat{\mu} - \mu_0}\, u_{j_0} = 0.$$

Since the operator $\mathcal{L}_\epsilon$ is invertible, it follows that

$$\frac{1}{\pi}\,\hat{\varphi} + \frac{\langle \hat{\varphi}, u_{j_0}\rangle}{\hat{\mu} - \mu_0}\,\mathcal{L}_\epsilon^{-1}[u_{j_0}] = 0, \tag{13.24}$$

and, by multiplying (13.24) by u_{j_0}, we find

$$\langle \hat{\varphi}, u_{j_0}\rangle \left(\frac{1}{\pi} + \frac{\langle \mathcal{L}_\epsilon^{-1}[u_{j_0}], u_{j_0}\rangle}{\hat{\mu} - \mu_0} \right) = 0.$$

Hence,

$$\hat{\mu} = \mu_0 - \pi \langle \mathcal{L}_\epsilon^{-1}[u_{j_0}], u_{j_0}\rangle, \tag{13.25}$$

since by (13.24), $\langle \hat{\varphi}, u_{j_0}\rangle = 0$ would imply that $\hat{\varphi} \equiv 0$.

Moreover, from

$$|\langle \mathcal{L}_\epsilon^{-1}[u_{j_0}], u_{j_0}\rangle| \longrightarrow 0 \quad \text{as } \epsilon \to 0,$$

it follows that

$$|\hat{\mu} - \mu_0| \longrightarrow 0 \quad \text{as } \epsilon \to 0.$$

If the normalization condition $\langle \hat{\varphi}, u_{j_0}\rangle = 1$ is chosen, then (13.24) and (13.25) show that the root function associated to this characteristic value $\hat{\mu}$ is given by

$$\hat{\varphi} = \frac{\mathcal{L}_\epsilon^{-1}[u_{j_0}]}{\langle \mathcal{L}_\epsilon^{-1}[u_{j_0}], u_{j_0}\rangle}.$$

The last point to investigate is the multiplicity of $\hat{\mu}$ as a characteristic value of $\mathcal{N}_\epsilon$, that is, the order of $\hat{\mu}$ as a pole of $\mathcal{N}_\epsilon^{-1}$. A straightforward calculation shows that $\mathcal{N}_\epsilon(\mu)[\varphi] = f$ is equivalent to

$$-\frac{1}{\pi}\varphi + \frac{\langle \varphi, u_{j_0}\rangle \mathcal{L}_\epsilon^{-1}[u_{j_0}]}{\mu_0 - \mu} = \mathcal{L}_\epsilon^{-1}[f].$$

But

$$\langle \varphi, u_{j_0} \rangle \left[-\frac{1}{\pi} + \frac{\langle \mathcal{L}_\epsilon^{-1}[u_{j_0}], u_{j_0} \rangle}{\mu_0 - \mu} \right] = \langle \mathcal{L}_\epsilon^{-1}[f], u_{j_0} \rangle,$$

which yields

$$\langle \varphi, u_{j_0} \rangle = \frac{\pi(\mu_0 - \mu)}{\mu - \hat{\mu}} \langle \mathcal{L}_\epsilon^{-1}[f], u_{j_0} \rangle,$$

and therefore,
(13.26)

$$\mathcal{N}_\epsilon(\mu)[\varphi] = f \quad \text{if and only if} \quad \varphi = -\pi\, \mathcal{L}_\epsilon^{-1}[f] + \frac{\pi^2 \langle \mathcal{L}_\epsilon^{-1}[f], u_{j_0} \rangle}{\mu - \hat{\mu}} \mathcal{L}_\epsilon^{-1}[u_{j_0}],$$

which justifies that $\hat{\mu}$ is a characteristic value of order one of $\mathcal{N}_\epsilon$. Therefore, $\mathcal{N}_\epsilon$ has exactly one pole μ_0 and one characteristic value $\hat{\mu}$ in V, each of order one, and its full multiplicity is equal to zero.

Now we estimate the multiplicity of $\mathcal{A}_\epsilon$ in V. The function $\mu \mapsto \mathcal{N}_\epsilon(\mu)$ is clearly finitely meromorphic and of Fredholm type at $\mu = \mu_0$. For all $\mu \in V \setminus \{\mu_0, \hat{\mu}\}$, $\mathcal{N}_\epsilon$ is invertible. Thus, $\mathcal{N}_\epsilon$ is normal in V. Moreover, $\mathcal{A}_\epsilon(\mu) - \mathcal{N}_\epsilon(\mu) = \mathcal{R}_\epsilon(\mu)$ is holomorphic in V and, by (13.26), it satisfies

$$\lim_{\epsilon \to 0} \|\mathcal{N}_\epsilon^{-1}(\mu)\, \mathcal{R}_\epsilon(\mu)\|_{\mathcal{L}(\mathcal{X}^\epsilon, \mathcal{X}^\epsilon)} = 0, \quad \forall\, \mu \in \partial V.$$

Consequently, the integral operator-valued function $\mu \mapsto \mathcal{A}_\epsilon(\mu)$ has, by the generalized Rouché theorem, the same full multiplicity as $\mathcal{N}_\epsilon$ in V. Since it already admits μ_0 as a pole, it admits in this neighborhood exactly one characteristic value μ^ϵ. □

Now in view of Lemmas 13.4 and 13.6, we obtain the following theorem.

THEOREM 13.7. *The operator-valued function $\mathcal{A}_\epsilon(\mu)$ has exactly one characteristic value μ^ϵ in V. Moreover, the following asymptotic expansion of μ^ϵ holds:*

$$\mu^\epsilon \approx \mu_0 - \pi \langle \mathcal{L}_\epsilon^{-1}[u_{j_0}], u_{j_0} \rangle + \pi^2 \langle \mathcal{L}_\epsilon^{-1} \mathcal{R}_\epsilon(\mu_0) \mathcal{L}_\epsilon^{-1}[u_{j_0}], u_{j_0} \rangle, \tag{13.27}$$

which yields

$$\mu^\epsilon \approx \mu_0 - \frac{\pi}{\ln \epsilon} |u_{j_0}(0)|^2.$$

PROOF. Recall that if μ^ϵ is the eigenvalue of (13.19) in V, then it is the characteristic value of $\mathcal{A}_\epsilon$ in V. Let φ^ϵ be an associated root function to μ^ϵ. Since $I + \mathcal{L}_\epsilon^{-1} \mathcal{R}_\epsilon$ is invertible for ϵ small enough, one can see as in the proof of Lemma 13.6 that $\langle \varphi^\epsilon, u_{j_0} \rangle \neq 0$, and hence we can choose φ^ϵ such that $\langle \varphi^\epsilon, u_{j_0} \rangle = 1$. With this choice, we have

$$\frac{1}{2\pi} + \frac{\langle \mathcal{L}_\epsilon^{-1}[u_{j_0}], u_{j_0} \rangle}{\mu^\epsilon - \mu_0} - \langle \mathcal{L}_\epsilon^{-1} \mathcal{R}_\epsilon(\mu^\epsilon)[\varphi^\epsilon], u_{j_0} \rangle = 0, \tag{13.28}$$

from which it follows by using (13.14) that

$$\mu^\epsilon = \mu_0 - 2\pi \langle \mathcal{L}_\epsilon^{-1}[u_{j_0}], u_{j_0} \rangle + O(|\ln \epsilon|^{-2}).$$

But,

$$(-\frac{1}{2\pi} I + \mathcal{L}_\epsilon^{-1} \mathcal{R}_\epsilon(\mu^\epsilon))[\varphi^\epsilon] + \frac{\mathcal{L}_\epsilon^{-1}[u_{j_0}]}{\mu_0 - \mu^\epsilon} = 0,$$

and thus,

$$\varphi^\epsilon \approx \frac{\mathcal{L}_\epsilon^{-1}[u_{j_0}]}{\langle \mathcal{L}_\epsilon^{-1}[u_{j_0}], u_{j_0} \rangle}.$$

Inserting the above approximation of φ^ϵ into (13.28) yields (13.27), as desired. $\square$

Theorem 13.7 shows in particular the existence of a sub-wavelength resonance for the Helmholtz resonator Ω_ϵ and lets us determine its asymptotic expansion in terms of the size ϵ of the opening. Choose $\mu_0 = 0$ (with the associated eigenfunction $= 1/\sqrt{|\Omega|}$ in Ω). It follows that there exists a unique characteristic value μ^ϵ of $\mathcal{A}_\epsilon$ in a small neighborhood of 0. Moreover,

$$\mu^\epsilon \approx -\frac{\pi}{|\Omega| \ln \epsilon}. \tag{13.29}$$

In the three-dimensional case, we can prove that [**94**]

$$\mu^\epsilon \approx \frac{1}{|\Omega|} \epsilon \mathrm{cap}(\Sigma), \tag{13.30}$$

where

$$\mathrm{cap}(\Sigma) := -\langle \mathcal{L}_1^{-1}[1], 1 \rangle_{L^2(\Sigma)} \tag{13.31}$$

is the capacity of $\Sigma := \Sigma_\epsilon$ in the rescaled opening (of arbitrary smooth shape) and $\mathcal{L}_1$ is the three-dimensional analog to $\mathcal{L}_\epsilon$ defined by (13.3) with $\epsilon = 1$:

$$\mathcal{L}_1 : \varphi \mapsto \frac{1}{\pi} \int_\Sigma \frac{\varphi(y)}{|x-y|} dy, \quad x \in \mathbb{R}^3 \backslash \overline{\Sigma}.$$

We refer to formulas (13.29) and (13.30) as the frequency formula for respectively the two and three-dimensional Helmholtz resonator. Formulas (13.29) and (13.30) indicate that it is possible to construct acoustic resonant structures that are much smaller than the wavelength of the corresponding acoustic wave. Furthermore, the frequency formulas can be generalized to a system of L disjoint Helmholtz resonators separated by a distance of the order of their characteristic size. It was shown in [**94**] that L (counting multiplicity) sub-wavelength resonances do exist and their asymptotic expansions as the characteristic size of the openings goes to zero were derived. Note that the term "sub-wavelength resonator" is associated with scattering in the quasi-stationary regime. In fact, in the case of the Helmholtz resonator, it is in that regime that the free space wavelength is significantly greater than the size of the resonator. We also remark that the resonance in the quasi-stationary regime results from the perturbations of the zero-eigenvalue of the Neumann problem in the closed resonator that are due to small openings. In the next subsection, we briefly outline the frequency formula and its consequence on super-resolution.

13.4. Resonances of a System of Helmholtz Resonators and Super-Resolution

Let the single three-dimensional Helmholtz resonator Ω be such that $\Omega = S(0,1) \times [-h, 0]$, where $S(0,1) = \{(x_1, x_2) : x_1^2 + x_2^2 \leq 1\}$ and h is the height of Ω, which is of order one. Let $\Sigma \subset S(0,1) \subset \mathbb{R}^2$ be a simply connected smooth domain which is of size one and let $\epsilon > 0$ be a small number. We assume that $0 \in \Sigma$ without loss of generality.

Consider a system of such resonators in three dimensions which consists of L disjoint Ω_j's ($1 \leq j \leq L$), where $\Omega_j = \Omega + z^{(j)}$ and $z^{(j)} = (z_1^{(j)}, z_2^{(j)}, 0)$ is the center of the opening for jth resonator. We denote by $\Omega^{in} = \bigcup_{j=1}^L \Omega_j$, $\Omega^{ex} = \{(x_1, x_2, x_3) \in \mathbb{R}^3 : x_3 > 0\}$ and $\Omega_\epsilon = \Omega^{in} \bigcup \Omega^{ex} \bigcup \Sigma_\epsilon$ with $\Sigma_\epsilon = \bigcup_{j=1}^L \Sigma_{\epsilon,j}$.

We first introduce two auxiliary Green's functions. Let N^{ex} be the Green function for the following exterior scattering problem:

$$\begin{cases} (\Delta+\omega^2)N_\omega^{ex}(x,y)=\delta_y(x), \quad x\in\Omega^{ex}, \\ \dfrac{\partial N_\omega^{ex}}{\partial\nu}(x,y)=0, \quad x\in\partial\Omega^{ex}, \\ N_\omega^{ex} \text{ satisfies the Sommerfeld radiation condition,} \end{cases}$$

and N_ω^{in} be the Green function for the following interior problem:

$$\begin{cases} (\Delta+\omega^2)N_\omega^{in}(x,y)=\delta_y(x), \quad x\in\Omega, \\ \dfrac{\partial N_\omega^{in}}{\partial\nu}(x,y)=0, \quad x\in\partial\Omega. \end{cases}$$

Throughout this section, we denote by

$$W=\{\omega\in\mathbb{C}:|\omega|\le\frac{1}{2}\sqrt{\mu_2}\},$$

where μ_2 is the first nonzero eigenvalue of the Neumann problem in Ω ($\mu_1=0$).

We have the following result.

LEMMA 13.8. *Let $y\in\{x_3=0\}$ and $\omega\in W$. Then,*

$$N_\omega^{ex}(x,y) = \frac{1}{2\pi|x-y|}+R^{ex}(x,y,\omega), \quad x\in\Omega^{ex}, \tag{13.32}$$

$$N_\omega^{in}(x,y) = \frac{1}{2\pi|x-y|}-\frac{\psi(x)\psi(y)}{\omega^2}+R^{in}(x,y,\omega), \quad x\in\Omega, \tag{13.33}$$

where $\psi=\chi(\Omega)$ and

$$\begin{aligned} R^{ex}(x,y,\omega) &= \frac{\sqrt{-1}\omega}{2\pi}\int_0^1 e^{\sqrt{-1}\omega|x-y|t}dt, \\ R^{in}(x,y,\omega) &= \omega\int_0^1 \sin\sqrt{-1}\omega|x-y|tdt+r(x,y,\omega) \end{aligned}$$

for some function r which is holomorphic in W with respect to ω and is smooth in a neighborhood of Σ in the plane $\{x_3=0\}$ with respect to both the variables x and y.

We denote by

$$R(x,y,\omega)=R^{ex}(x,y,\omega)+R^{in}(x,y,\omega),$$

and

$$\alpha_0=R(0,0,0), \quad \alpha_1=\frac{\partial R}{\partial\omega}(0,0,0). \tag{13.34}$$

It is clear that

$$\alpha_0\in\mathbb{R}, \quad \Im\alpha_1=\Im\frac{\partial R^{ex}}{\partial\omega}(0,0,0)=\frac{1}{2\pi}. \tag{13.35}$$

We now introduce the matrices $T=(T_{ij})_{L\times L}$ and $S=(S_{ij})_{L\times L}$ with

$$\begin{cases} T_{ij} = \frac{1}{2\pi|z^{(i)}-z^{(j)}|} \quad \text{for } i\ne j, \quad \text{and } T_{ii}=0, \\ S_{ij} = \frac{\sqrt{-1}}{2\pi}+\delta_{ij}\Re\alpha_1. \end{cases} \tag{13.36}$$

Observe that T is symmetric, thus T has L real eigenvalues, which are denoted by $\beta_1,\beta_2,\ldots,\beta_L$. For the ease of exposition, we assume that $\beta_1,\ldots,\beta_L$ are mutually distinct. This is the generic case among all the possible arrangements of the resonators. The corresponding normalized eigenvectors are denoted by $Y_1, Y_2, \ldots, Y_L$, respectively. Then $Y_1, Y_2, \ldots, Y_L$ form a normal basis for $\mathbb{R}^L$. We also denote by Y the matrix

$$Y = (Y_1, Y_2, \ldots, Y_L).$$

For convenience, we write

$$\mathcal{N}(x,\omega) = \left(N_\omega^{ex}(x, z^{(1)}), N_\omega^{ex}(x, z^{(2)}), \ldots, N_\omega^{ex}(x, z^{(L)})\right)^t \tag{13.37}$$

with the subscript t denoting the transpose. For each $1 \le j \le L$, we denote by

$$\zeta_j(x, x_0, \omega) = \mathcal{N}(x,\omega)^t Y_j Y_j^t \mathcal{N}(x_0,\omega). \tag{13.38}$$

It is clear that $\zeta_j = \zeta_j(x, x_0, \omega)$ is holomorphic in ω for fixed x and x_0.

The following result on the resonances of the above scattering problem was proved in [**94**].

PROPOSITION 13.9. *There exist exactly* $2L$ *resonances of order one in the domain* W *for the system of resonators, given by*

$$\omega_{0,\epsilon,j,1} = \tau_1 \epsilon^{\frac{1}{2}} + \tau_{3,j}\epsilon^{\frac{3}{2}} + \tau_{4,j}\epsilon^2 + O(\epsilon^{\frac{5}{2}}), \tag{13.39}$$

$$\omega_{0,\epsilon,j,2} = -\tau_1 \epsilon^{\frac{1}{2}} - \tau_{3,j}\epsilon^{\frac{3}{2}} + \tau_{4,j}\eta^2 + O(\epsilon^{\frac{5}{2}}), \tag{13.40}$$

where

$$\tau_1 = \sqrt{\frac{\mathrm{cap}(\Sigma)}{|\Omega|}}, \tag{13.41}$$

$$\tau_{3,j} = -\frac{1}{2}(\alpha_0 + \beta_j)\left(\frac{\mathrm{cap}(\Sigma)}{|\Omega|}\right)^{\frac{1}{2}} \mathrm{cap}(\Sigma), \tag{13.42}$$

and

$$\tau_{4,j} = -\frac{1}{2}\frac{\mathrm{cap}(\Sigma)^2}{|\Omega|} Y_j^t S Y_j \tag{13.43}$$

with $\mathrm{cap}(\Sigma)$ *being the capacity of the set* Σ *defined by (*13.31*).*

From Chapter 9, the super-resolution relies in analysis of the following Green function in the frequency domain

$$\begin{cases} (\Delta + \omega^2) N_\epsilon(x, x_0, \omega) = \delta_{x_0}(x), & x \in \Omega_\epsilon, \\ \frac{\partial N_\epsilon}{\partial \nu}(x, x_0, \omega) = 0, & x \in \partial\Omega_\epsilon, \\ N_\epsilon \text{satisfies the Sommerfeld radiation condition.} \end{cases}$$

The following result on N_ϵ in Ω_ϵ holds. We refer to [**94**] for its proof.

THEOREM 13.10. *Assume that* $\omega \in \mathbb{R} \bigcap W$. *Then for* ϵ *sufficiently small. Then* N_ϵ^{ex} *has the following asymptotic expansion*

$$\begin{aligned} N_\epsilon^{ex}(x, x_0, \omega) &= N_\omega^{ex}(x, x_0) - \epsilon \mathrm{cap}(\Sigma) \sum_{1 \le j \le L} N_\omega^{ex}(z^{(j)}, x_0, k) N_\omega^{ex}(x, z^{(j)}) \\ &\quad - \sum_{j=1}^{L}\left(\frac{1}{\omega - \omega_{0,\epsilon,j,2}} - \frac{1}{\omega - \omega_{0,\epsilon,j,1}}\right)\frac{(\mathrm{cap}(\Sigma)\epsilon)^{\frac{3}{2}}}{\sqrt{|D|}} \mathcal{N}(x,\omega)^t Y_j Y_j^t \mathcal{N}(x_0,\omega) \\ &\quad + \sum_{1 \le j \le L}\left(\frac{O(\epsilon^2)}{\omega - \omega_{0,\epsilon,j,2}} + \frac{O(\epsilon^2)}{\omega - \omega_{0,\epsilon,j,1}}\right) + O(\epsilon^2). \end{aligned}$$

As a consequence of Theorem 13.10, we can establish the following result on super-resolution (or super-focusing), which shows that super-resolution can be achieved with a single specific frequency.

THEOREM 13.11. *Let τ_1 be given by (13.41), where* cap(Σ) *is the capacity of the set Σ defined by (13.31). For $\omega = \tau_1\sqrt{\epsilon}$, the resolution function $\Im N_\epsilon^{ex}$ has the following estimate:*

$$\Im N_\epsilon^{ex}(x, x_0, \omega) = \frac{\sin \tau_1 \sqrt{\epsilon}|x - x_0|}{2\pi|x - x_0|} + \frac{\mathrm{cap}(\Sigma)^{\frac{3}{2}}}{|\Omega|^{\frac{1}{2}}}\epsilon^{\frac{1}{2}} \sum_{j=1}^{L} \frac{\Im \tau_{4,j}}{\tau_{3,j}^2}\zeta_j(x, x_0, 0) + O(\epsilon),$$

where $\zeta_j(x, x_0, 0)$ is given by (13.38) and $\tau_{3,j}$ and $\tau_{4,j}$ are defined by (13.42) and (13.43), respectively.

13.5. Concluding Remarks

In this chapter, we have derived asymptotic expansions of perturbations of scattering frequencies of sub-wavelength acoustic resonators. As in the previous chapters, we have transformed the problem of finding the scattering frequencies into the determination of the characteristic values of certain integral operator-valued functions in the complex plane. Our method in this chapter is based on pole-pencil type decomposition, followed by an application of the generalized Rouché theorem. The techniques developed in this chapter can be extended to electromagnetic and elastic analogues of the Helmholtz resonator [**238, 254, 257, 419**].

CHAPTER 14

Minnaert Resonances for Bubbles

14.1. Introduction

In this chapter, we consider acoustic wave propagation in bubbly media. At particular low frequencies known as Minnaert resonances [**401**], bubbles behave as strong sound scatterers. Using layer potential techniques and Gohberg-Sigal theory, we first derive a formula for the Minnaert resonances of bubbles of arbitrary shapes. The Minnaert resonance is a low frequency resonance in which the wavelength is much larger than the size of the bubble. Our formula for the Minnaert resonance is expressed in terms of the bulk modulus of the air, the density of the fluid, and the capacity and the volume of the bubbles.

Then, we derive an effective medium theory for acoustic wave propagation in bubbly media near the Minnaert resonant frequency. We consider the dilute regime. We start with a multiple scattering formulation of the scattering problem in which an incident wave impinges on a large number of small, identical bubbles in a homogeneous fluid. Under certain conditions on the configuration of the bubbles distribution, we justify the point interaction approximation and establish an effective medium theory for the bubbly fluid as the number of bubbles tends to infinity. The convergence rate is also derived. As a consequence, we show that just below the Minnaert resonant frequency, the obtained effective media may be highly refractive, making the super-focusing of acoustic waves achievable.

On the other hand, we show that a sub-wavelength bandgap opening may occur in bubble phononic crystals. To demonstrate the opening of a sub-wavelength phononic bandgap, we consider a periodic arrangement of bubbles and exploit their Minnaert resonance. As shown in Chapter 5, for bandgaps to occur, the period of the structure must be of the order of the wavelength and the contrast in the material parameters must be large. This limits the use of phononic crystals in applications targeting low frequencies, because phononic crystals would require impractically large geometries [**385**, **496**]. In this chapter, we provide a mathematical and numerical framework for analyzing bandgap opening in bubble phononic crystals at low-frequencies. We derive a formula for the quasi-periodic Minnaert resonance frequencies of an arbitrarily shaped bubble, along with proving the existence of a sub-wavelength bandgap. In the dilute regime, above the Minnaert resonant frequency, the real part of the effective modulus is negative and consequently, the bubbly fluid behaves as a diffusive media for the acoustic waves.

We remark that such behavior near the resonant frequency is rather analogous to the coupling of electromagnetic waves with plasmonic nanoparticles, which results in effective negative or high contrast dielectric constants for frequencies near the plasmonic resonance frequencies as shown in Chapter 7.

Finally, we provide a mathematical theory for understanding the mechanism behind the double-negative refractive index phenomenon in bubbly fluids. The design of double-negative metamaterials generally requires the use of two different kinds of sub-wavelength resonators, which may limit the applicability of double-negative metamaterials. Herein we rely on media that consists of only a single type of resonant element, and show how to turn the acoustic metamaterial with a single negative effective property into a negative refractive index metamaterial, which refracts waves negatively, hence acting as a superlens. Using bubble dimers made of two identical bubbles, we show that both the effective mass density and the bulk modulus of the bubbly fluid can be negative near the anti-resonance of the two hybridized Minnaert resonances for a single constituent bubble dimer.

The results of this chapter, which are from [**38**–**42**, **77**], formally explain the experimental observations reported in [**350**, **360**, **362**].

14.2. Derivation of Minnaert Resonance Formula

We consider the scattering of acoustic waves in a homogeneous three-dimensional acoustic medium by a bubble embedded inside. Assume that the bubble occupies a bounded and simply connected domain D with $\partial D \in \mathcal{C}^{1,\eta}$ for some $\eta > 0$. We denote by ρ_b and κ_b the density and the bulk modulus of the air inside the bubble, respectively. We let ρ and κ be the corresponding parameters for the background media $\mathbb{R}^3 \backslash D$.

Let u^i be an incident plane wave. The scattering problem can be modeled by the following equations:

$$
\left\{
\begin{array}{l}
\nabla \cdot \dfrac{1}{\rho} \nabla u + \dfrac{\omega^2}{\kappa} u = 0 \quad \text{in } \mathbb{R}^3 \backslash D, \\[2mm]
\nabla \cdot \dfrac{1}{\rho_b} \nabla u + \dfrac{\omega^2}{\kappa_b} u = 0 \quad \text{in } D, \\[2mm]
u_+ - u_- = 0 \quad \text{on } \partial D, \\[2mm]
\dfrac{1}{\rho} \dfrac{\partial u}{\partial \nu} \bigg|_+ - \dfrac{1}{\rho_b} \dfrac{\partial u}{\partial \nu} \bigg|_- = 0 \quad \text{on } \partial D, \\[2mm]
u - u^i \ \text{satisfies the Sommerfeld radiation condition.}
\end{array}
\right. \tag{14.1}
$$

We introduce four auxiliary parameters to facilitate our analysis:

$$
v = \sqrt{\frac{\kappa}{\rho}}, \quad v_b = \sqrt{\frac{\kappa_b}{\rho_b}}, \quad k = \frac{\omega}{v}, \quad k_b = \frac{\omega}{v_b}. \tag{14.2}
$$

We also introduce two dimensionless contrast parameters:

$$
\delta = \frac{\rho_b}{\rho}, \quad \tau = \frac{k_b}{k} = \frac{v}{v_b} = \sqrt{\frac{\rho_b \kappa}{\rho \kappa_b}}. \tag{14.3}
$$

By choosing appropriate physical units, we may assume that the bubble size is of order one and that the wave speeds outside and inside the bubble are both of order one. Thus the contrast between the wave speeds is not significant. We assume, however, that there is a large contrast in the bulk modulii. In summary, we assume that

$$
\delta \ll 1 \quad \text{and} \quad \tau = O(1). \tag{14.4}
$$

Then the solution u can be written as

$$u(x) = \begin{cases} u^i + \mathcal{S}_D^k[\psi], & x \in \mathbb{R}^3 \backslash \overline{D}, \\ \mathcal{S}_D^{k_b}[\psi_b], & x \in D, \end{cases} \tag{14.5}$$

for some surface potentials $\psi, \psi_b \in L^2(\partial D)$. Using the jump relations for the single-layer potentials, it is easy to derive that ψ and ψ_b satisfy the following system of boundary integral equations:

$$\mathcal{A}(\omega, \delta)[\Psi] = F, \tag{14.6}$$

where

(14.7)

$$\mathcal{A}(\omega, \delta) = \begin{pmatrix} \mathcal{S}_D^{k_b} & -\mathcal{S}_D^k \\ -\frac{1}{2}I + (\mathcal{K}_D^{k_b})^* & -\delta(\frac{1}{2}I + (\mathcal{K}_D^k)^*) \end{pmatrix}, \quad \Psi = \begin{pmatrix} \psi_b \\ \psi \end{pmatrix}, \quad F = \begin{pmatrix} u^i \\ \delta \frac{\partial u^i}{\partial \nu} \end{pmatrix}.$$

One can show that the scattering problem (14.1) is equivalent to the system of boundary integral equations (14.6).

It is clear that $\mathcal{A}(\omega, \delta)$ is a bounded linear operator from $\mathcal{H} := L^2(\partial D) \times L^2(\partial D)$ to $\mathcal{H}_1 := H^1(\partial D) \times L^2(\partial D)$, *i.e.*, $\mathcal{A}(\omega, \delta) \in \mathcal{L}(\mathcal{H}, \mathcal{H}_1)$.

The resonance of the bubble in the scattering problem (14.1) can be defined as all the complex numbers ω with positive imaginary part such that there exists a nontrivial solution to the following equation:

$$\mathcal{A}(\omega, \delta)[\Psi] = 0. \tag{14.8}$$

These can be viewed as the characteristic values of the holomorphic operator-valued function (with respect to ω) $\mathcal{A}(\omega, \delta)$. We are interested in the quasi-static resonance of the bubble, that is, the resonance frequency at which the size of the bubble is much smaller than the wavelength of the incident wave outside the bubble. In some physics literature, this resonance is called the Minnaert resonance. Due to our assumptions on the bubble being of size order one, and the wave speed outside of the bubble also being of order one, this resonance should lie in a small neighborhood of the origin in the complex plane. In what follows, we apply the Gohberg-Sigal theory to find this resonance.

We first look at the limiting case when $\delta = \omega = 0$. It is clear that

$$\mathcal{A}_0 := \mathcal{A}(0, 0) = \begin{pmatrix} \mathcal{S}_D^0 & -\mathcal{S}_D^0 \\ -\frac{1}{2}I + (\mathcal{K}_D^0)^* & 0 \end{pmatrix}, \tag{14.9}$$

where $\mathcal{S}_D^0$ and $(\mathcal{K}_D^0)^*$ are respectively the single-layer potential and the Neumann–Poincaré operator on ∂D associated with the Laplacian.

Let $\mathcal{A}_0^*$ be the adjoint of $\mathcal{A}$.

LEMMA 14.1. *We have*

(i) $\mathrm{Ker}(\mathcal{A}_0) = \mathrm{Im}\,\{\Psi_0\}$ *where* $\mathrm{Im}\{\Psi_0\}$ *denotes the space spanned by* Ψ_0 *and*

$$\Psi_0 = \alpha_0 \begin{pmatrix} \psi_0 \\ \psi_0 \end{pmatrix}$$

with $\psi_0 = (\mathcal{S}_D^0)^{-1}[1]$ *and the constant* α_0 *being chosen such that* $\|\Psi_0\| = 1$;

(ii) $\mathrm{Ker}(\mathcal{A}_0^*) = \mathrm{Im}\{\Phi_0\}$ *where*

$$\Phi_0 = \beta_0 \begin{pmatrix} 0 \\ \phi_0 \end{pmatrix}$$

with $\phi_0 = 1$ *on* ∂D *and the constant* β_0 *being chosen such that* $\|\Phi_0\| = 1$.

The above lemma shows that $\omega = 0$ is a characteristic value for the holomorphic operator-valued function $\mathcal{A}(\omega, \delta)$. By the Gohberg-Sigal theory, we can conclude the following result about the existence of the quasi-static resonance.

LEMMA 14.2. *For any* δ, *sufficiently small, there exists a characteristic value* $\omega_0 = \omega_0(\delta)$ *to the holomorphic operator-valued function* $\mathcal{A}(\omega, \delta)$ *such that* $\omega_0(0) = 0$ *and* ω_0 *depends on* δ *continuously. This characteristic value is also the quasi-static resonance (or Minnaert resonance).*

We next perform asymptotic analysis on the operator $\mathcal{A}(\omega, \delta)$. With the same notation as in Subsection 2.8.3.1, the following result holds.

LEMMA 14.3. *In the space* $\mathcal{L}(\mathcal{H}, \mathcal{H}_1)$, *we have*

$$\begin{aligned} \mathcal{A}(\omega, \delta) &:= \mathcal{A}_0 + \mathcal{B}(\omega, \delta) \\ &= \mathcal{A}_0 + \omega \mathcal{A}_{1,0} + \omega^2 \mathcal{A}_{2,0} + \omega^3 \mathcal{A}_{3,0} + \delta \mathcal{A}_{0,1} + \delta \omega^2 \mathcal{A}_{2,1} \\ &\quad + O(\omega^4) + O(\delta \omega^3), \end{aligned}$$

where

$$\mathcal{A}_{1,0} = \begin{pmatrix} v_b^{-1} \mathcal{S}_{D,1} & -v^{-1} \mathcal{S}_{D,1} \\ 0 & 0 \end{pmatrix}, \quad \mathcal{A}_{2,0} = \begin{pmatrix} v_b^{-2} \mathcal{S}_{D,2} & -v^{-2} \mathcal{S}_{D,2} \\ v_b^{-2} \mathcal{K}_{D,2} & 0 \end{pmatrix},$$

$$\mathcal{A}_{3,0} = \begin{pmatrix} v_b^{-3} \mathcal{S}_{D,3} & -v^{-3} \mathcal{S}_{D,3} \\ v_b^{-3} \mathcal{K}_{D,3} & 0 \end{pmatrix}, \quad \mathcal{A}_{0,1} = \begin{pmatrix} 0 & 0 \\ 0 & -(\frac{1}{2} I + (\mathcal{K}_D^0)^*) \end{pmatrix},$$

and

$$\mathcal{A}_{2,1} = \begin{pmatrix} 0 & 0 \\ 0 & -v^{-2} \mathcal{K}_{D,2} \end{pmatrix}.$$

We define a projection $\mathcal{P}_0 : \mathcal{H} \to \mathcal{H}_1$ by

$$\mathcal{P}_0[\Psi] := \langle \Psi, \Psi_0 \rangle_{\mathcal{H}} \Phi_0,$$

and denote by

$$\widetilde{\mathcal{A}_0} = \mathcal{A}_0 + \mathcal{P}_0.$$

The following results hold.

LEMMA 14.4. *We have*

(i) *The operator* $\widetilde{\mathcal{A}_0}$ *is a bijective operator in* $\mathcal{L}(\mathcal{H}_1, \mathcal{H})$. *Moreover,*

$$\widetilde{\mathcal{A}_0}[\Psi_0] = \Phi_0;$$

(ii) *The adjoint of* $\widetilde{\mathcal{A}_0}$, $\widetilde{\mathcal{A}_0}^*$, *is a bijective operator in* $\mathcal{L}(\mathcal{H}, \mathcal{H}_1)$. *Moreover,* $\widetilde{\mathcal{A}_0}^*[\Phi_0] = \Psi_0$.

PROOF. By construction, and the fact that $\mathcal{S}_D^0$ is bijective from $L^2(\partial D)$ to $H^1(\partial D)$ in three dimensions, we can show that $\widetilde{\mathcal{A}_0}$ is a bijective. So too is $\widetilde{\mathcal{A}_0}^*$. We only need to show that $\widetilde{\mathcal{A}_0}^*[\Phi_0] = \Psi_0$. Indeed, we can check that $\mathcal{P}_0^*[\theta] = \langle \theta, \Phi_0 \rangle \Psi_0$. Thus, it follows that

$$\widetilde{\mathcal{A}_0}^*[\Phi_0] = \mathcal{P}_0^*[\Phi_0] = \langle \Phi_0, \Phi_0 \rangle \Psi_0 = \Psi_0,$$

which completes the proof. □

The following theorem characterizes the Minnaert frequencies in terms of the shape of the bubbles.

THEOREM 14.5. *In the quasi-static regime, there exists two resonances for a single bubble:*

$$\begin{aligned}\omega_{0,0}(\delta) &= \sqrt{\frac{\mathrm{cap}(\partial D)}{|D|}}v_b\delta^{\frac{1}{2}} - \sqrt{-1}\frac{\mathrm{cap}(\partial D)^2 v_b^2}{8\pi v|D|}\delta + O(\delta^{\frac{3}{2}}),\\ \omega_{0,1}(\delta) &= -\sqrt{\frac{\mathrm{cap}(\partial D)}{|D|}}v_b\delta^{\frac{1}{2}} - \sqrt{-1}\frac{\mathrm{cap}(\partial D)^2 v_b^2}{8\pi v|D|}\delta + O(\delta^{\frac{3}{2}}),\end{aligned}$$

where $|D|$ is the volume of D and

$$\mathrm{cap}(\partial D) := -\langle \psi_0, 1\rangle_{L^2(\partial D)} = -\langle (\mathcal{S}_D^0)^{-1}[1], 1\rangle_{L^2(\partial D)} \tag{14.10}$$

is the capacity of D. The first resonance $\omega_{0,0}$ is called the Minnaert resonance.

PROOF. Step 1. We find the resonance by solving the following equation:

$$\mathcal{A}(\omega,\delta)[\Psi_\delta] = 0. \tag{14.11}$$

Since $\mathcal{A}(0,0)[\Psi_0] = 0$, we may view Ψ_δ as a perturbation of Ψ_0 and write it as $\Psi_\delta = \Psi_0 + \Psi_1$. In order to uniquely determine Ψ_1, we assume that

$$\langle \Psi_1, \Psi_0\rangle = 0, \tag{14.12}$$

where here and throughout this chapter, $\langle\ ,\ \rangle$ denotes the L^2-scalar product.

Note that we let the coefficient of Ψ_0 be one for the purpose of normalization. Since Ψ_δ is defined up to multiplicative constant, (14.12) holds without loss of generality by changing $\Psi_0 + \Psi_1$ to $\Psi_0 + \big(\Psi_1 - \langle \Psi_0, \Psi_1\rangle\Psi_0\big)/(1 + \langle \Psi_0, \Psi_1\rangle)$.

Step 2. Since $\widetilde{\mathcal{A}_0} = \mathcal{A}_0 + \mathcal{P}_0$, (14.11) is equivalent to the following:

$$(\widetilde{\mathcal{A}_0} - \mathcal{P}_0 + \mathcal{B})[\Psi_0 + \Psi_1] = 0.$$

Observe that as the operator $\widetilde{\mathcal{A}_0} + \mathcal{B}$ is invertible for sufficiently small δ and ω, we can apply $(\widetilde{\mathcal{A}_0} + \mathcal{B})^{-1}$ to both sides of the above equation to deduce that

$$\Psi_1 = (\widetilde{\mathcal{A}_0} + \mathcal{B})^{-1}\mathcal{P}_0[\Psi_0] - \Psi_0 = (\widetilde{\mathcal{A}_0} + \mathcal{B})^{-1}[\Phi_0] - \Psi_0. \tag{14.13}$$

Step 3. Using the orthogonality condition (14.12), we arrive at the following equation:

$$A(\omega,\delta) := \Big\langle (\widetilde{\mathcal{A}_0} + \mathcal{B})^{-1}[\Phi_0], \Psi_0\Big\rangle - 1 = 0 \tag{14.14}$$

Step 4. We calculate $A(\omega,\delta)$. Using the identity

$$(\widetilde{\mathcal{A}_0} + \mathcal{B})^{-1} = \Big(I + \widetilde{\mathcal{A}_0}^{-1}\mathcal{B}\Big)^{-1}\widetilde{\mathcal{A}_0}^{-1} = \Big(I - \widetilde{\mathcal{A}_0}^{-1}\mathcal{B} + \widetilde{\mathcal{A}_0}^{-1}\mathcal{B}\widetilde{\mathcal{A}_0}^{-1}\mathcal{B} + \dots\Big)\widetilde{\mathcal{A}_0}^{-1},$$

and the fact that

$$\widetilde{\mathcal{A}_0}^{-1}[\Phi_0] = \Psi_0,$$

we obtain

$$\begin{aligned}
A(\omega,\delta) &= -\omega\left\langle \mathcal{A}_{1,0}[\Psi_0],\Phi_0\right\rangle - \omega^2\left\langle \mathcal{A}_{2,0}[\Psi_0],\Phi_0\right\rangle - \omega^3\left\langle \mathcal{A}_{3,0}[\Psi_0],\Phi_0\right\rangle\\
&\quad -\delta\left\langle \mathcal{A}_{0,1}[\Psi_0],\Phi_0\right\rangle + \omega^2\left\langle \mathcal{A}_{1,0}\widetilde{\mathcal{A}_0}^{-1}\mathcal{A}_{1,0}[\Psi_0],\Phi_0\right\rangle\\
&\quad +\omega^3\left\langle \mathcal{A}_{1,0}\widetilde{\mathcal{A}_0}^{-1}\mathcal{A}_{2,0}[\Psi_0],\Phi_0\right\rangle + \omega^3\left\langle \mathcal{A}_{2,0}\widetilde{\mathcal{A}_0}^{-1}\mathcal{A}_{1,0}[\Psi_0],\Phi_0\right\rangle\\
&\quad +\omega\delta\left\langle \mathcal{A}_{1,0}\widetilde{\mathcal{A}_0}^{-1}\mathcal{A}_{0,1}[\Psi_0],\Phi_0\right\rangle + \omega\delta\left\langle \mathcal{A}_{0,1}\widetilde{\mathcal{A}_0}^{-1}\mathcal{A}_{1,0}[\Psi_0],\Phi_0\right\rangle\\
&\quad +\omega^3\left\langle \mathcal{A}_{1,0}\widetilde{\mathcal{A}_0}^{-1}\mathcal{A}_{1,0}\widetilde{\mathcal{A}_0}^{-1}\mathcal{A}_{1,0}[\Psi_0],\Phi_0\right\rangle + O(\omega^4) + O(\delta^2).
\end{aligned}$$

It is clear that $\mathcal{A}_{1,0}^*[\Phi_0]=0$. Consequently, we get

$$\begin{aligned}
A(\omega,\delta) &= -\omega^2\left\langle \mathcal{A}_{2,0}[\Psi_0],\Phi_0\right\rangle - \omega^3\left\langle \mathcal{A}_{3,0}[\Psi_0],\Phi_0\right\rangle - \delta\left\langle \mathcal{A}_{0,1}[\Psi_0],\Phi_0\right\rangle\\
&\quad +\omega^3\left\langle \mathcal{A}_{2,0}\widetilde{\mathcal{A}_0}^{-1}\mathcal{A}_{1,0}[\Psi_0],\Phi_0\right\rangle + \omega\delta\left\langle \mathcal{A}_{0,1}\widetilde{\mathcal{A}_0}^{-1}\mathcal{A}_{1,0}[\Psi_0],\Phi_0\right\rangle\\
&\quad +O(\omega^4) + O(\delta^2).
\end{aligned}$$

In the next four steps, we calculate the terms $\langle \mathcal{A}_{2,0}[\Psi_0],\Phi_0\rangle$, $\langle \mathcal{A}_{3,0}[\Psi_0],\Phi_0\rangle$, $\langle \mathcal{A}_{0,1}[\Psi_0],\Phi_0\rangle$, $\left\langle \mathcal{A}_{2,0}\widetilde{\mathcal{A}_0}^{-1}\mathcal{A}_{1,0}[\Psi_0],\Phi_0\right\rangle$ and $\left\langle \mathcal{A}_{0,1}\widetilde{\mathcal{A}_0}^{-1}\mathcal{A}_{1,0}[\Psi_0],\Phi_0\right\rangle$.

Step 5. We have

$$\begin{aligned}
\langle \mathcal{A}_{2,0}[\Psi_0],\Phi_0\rangle &= \alpha_0\beta_0 v_b^{-2}\left\langle \mathcal{K}_{D,2}[\psi_0],\phi_0\right\rangle = \alpha_0\beta_0 v_b^{-2}\left\langle \psi_0,\mathcal{K}_{D,2}^*[\phi_0]\right\rangle\\
&= -\alpha_0\beta_0 v_b^{-2}\int_{\partial D} d\sigma(x)(\mathcal{S}_D^0)^{-1}[1](x)\int_D \Gamma_0(x-y)\,dy\\
&= -\alpha_0\beta_0 v_b^{-2}\int_D dy\int_{\partial D}\Gamma_0(x-y)(\mathcal{S}_D^0)^{-1}[1](x)\,d\sigma(x)\\
&= -\alpha_0\beta_0 v_b^{-2}\int_D dy\\
&= -\alpha_0\beta_0 v^{-2}|D|.
\end{aligned}$$

Here, Γ_0 is the fundamental solution of the Laplacian in $\mathbb{R}^3$, defined by (2.2).

Step 6. On the other hand, we have

$$\begin{aligned}
\langle \mathcal{A}_{3,0}[\Psi_0],\Phi_0\rangle &= \alpha_0\beta_0 v_b^{-3}\langle \psi_0,\mathcal{K}_{D,3}^*[\phi_0]\rangle = \alpha_0\beta_0 v_b^{-3}\langle \psi_0,\frac{\sqrt{-1}}{4\pi}|D|\rangle\\
&= \alpha_0\beta_0 v_b^{-3}|D|\frac{\sqrt{-1}}{4\pi}\left\langle (\mathcal{S}_D^0)^{-1}[1],1\right\rangle = -\alpha_0\beta_0 v_b^{-3}|D|\frac{\sqrt{-1}}{4\pi}\mathrm{cap}(\partial D).
\end{aligned}$$

Step 7. It is easy to see that

$$\langle \mathcal{A}_{0,1}[\Psi_0],\Phi_0\rangle = -\langle \psi_0,\phi_0\rangle = -\alpha_0\beta_0\left\langle (\mathcal{S}_D^0)^{-1}[1],1\right\rangle = \alpha_0\beta_0\mathrm{cap}(\partial D).$$

Step 8. We now calculate the term $\left\langle \mathcal{A}_{0,1}\widetilde{\mathcal{A}_0}^{-1}\mathcal{A}_{1,0}[\Psi_0],\Phi_0\right\rangle$. We have

$$\begin{aligned}
\mathcal{A}_{1,0}[\Psi_0] &= \begin{pmatrix}(\tau^{-1}-1)v^{-1}\mathcal{S}_{D,1}[\psi_0]\\ 0\end{pmatrix} = \begin{pmatrix}(\tau^{-1}-1)v^{-1}\dfrac{\sqrt{-1}}{4\pi}\mathrm{cap}(\partial D)\\ 0\end{pmatrix},\\
\mathcal{A}_{0,1}^*[\Phi_0] &= \begin{pmatrix}0\\ -\left(\dfrac{1}{2}I+\mathcal{K}_D^0\right)[\phi_0]\end{pmatrix} = \begin{pmatrix}0\\ -\phi_0\end{pmatrix} = -\begin{pmatrix}0\\ 1\end{pmatrix}.
\end{aligned}$$

We need to calculate

$$\widetilde{\mathcal{A}}_0^{-1}\begin{pmatrix}1\\0\end{pmatrix}.$$

Assume that

$$(\mathcal{A}_0+\mathcal{P}_0)\begin{pmatrix}y_b\\y\end{pmatrix}=\begin{pmatrix}\mathcal{S}_D^0[y_b-y]\\(-\frac{1}{2}I+(\mathcal{K}_D^0)^*)[y_b]\end{pmatrix}+(\langle y_b,\psi_0\rangle+\langle y,\psi_0\rangle)\begin{pmatrix}0\\\phi_0\end{pmatrix}=\begin{pmatrix}1\\0\end{pmatrix}$$

By solving the above equations directly, we obtain that

$$y_b=\frac{1}{2}\psi_0,\, y=-\frac{1}{2}\psi_0.$$

Therefore,

$$\widetilde{\mathcal{A}}_0^{-1}\begin{pmatrix}1\\0\end{pmatrix}=\begin{pmatrix}\frac{1}{2}\psi_0\\-\frac{1}{2}\psi_0\end{pmatrix}.$$

It follows that

$$\begin{aligned}\left\langle\mathcal{A}_{0,1}\widetilde{\mathcal{A}_0}^{-1}\mathcal{A}_{1,0}[\Psi_0],\Phi_0\right\rangle &= (\tau^{-1}-1)v^{-1}\frac{\sqrt{-1}}{8\pi}\mathrm{cap}(\partial D)\langle\psi_0,\phi_0\rangle\\ &= (1-\tau^{-1})v^{-1}\frac{\sqrt{-1}}{8\pi}\mathrm{cap}(\partial D)^2\alpha_0\beta_0.\end{aligned}$$

Step 9. We calculate the term $\left\langle\mathcal{A}_{2,0}\widetilde{\mathcal{A}_0}^{-1}\mathcal{A}_{1,0}[\Psi_0],\Phi_0\right\rangle$. Using the results in Step 8, we obtain

$$\begin{aligned}\left\langle\mathcal{A}_{2,0}\widetilde{\mathcal{A}_0}^{-1}\mathcal{A}_{1,0}[\Psi_0],\Phi_0\right\rangle &= \left\langle\widetilde{\mathcal{A}_0}^{-1}\mathcal{A}_{1,0}[\Psi_0],\mathcal{A}_{2,0}^*[\Phi_0]\right\rangle\\ &= \frac{\sqrt{-1}(\tau^{-1}-1)\tau v_b^{-3}}{8\pi}\mathrm{cap}(\partial D)\alpha_0\beta_0\left\langle\psi_0,\mathcal{K}_{D,2}^*[\phi_0]\right\rangle\\ &= \frac{\sqrt{-1}(1-\tau^{-1})\tau v_b^{-3}}{8\pi}\mathrm{cap}(\partial D)|D|\alpha_0\beta_0.\end{aligned}$$

Step 10. Considering the above results, we can derive

$$\begin{aligned}A(\omega,\delta) &= \alpha_0\beta_0\Bigg(v_b^{-2}|D|\omega^2+\frac{\sqrt{-1}\tau(\tau^{-1}+1)v_b^{-3}|D|\mathrm{cap}(\partial D)}{8\pi}\omega^3-\mathrm{cap}(\partial D)\delta\\ &\quad -\frac{\sqrt{-1}(\tau^{-1}-1)v^{-1}\mathrm{cap}(\partial D)^2}{8\pi}\omega\delta\Bigg)\\ &\quad +O(\omega^4)+O(\delta^2).\end{aligned}$$

We now solve $A(\omega,\delta)=0$. It is clear that $\delta=O(\omega^2)$, and thus $\omega_0(\delta)=O(\sqrt{\delta})$. Write

$$\omega_0(\delta)=a_1\delta^{\frac{1}{2}}+a_2\delta+O(\delta^{\frac{3}{2}}).$$

We get

$$\begin{aligned}&v_b^{-2}|D|\left(a_1\delta^{\frac{1}{2}}+a_2\delta+O(\delta^{\frac{3}{2}})\right)^2\\ &+\frac{\sqrt{-1}\tau(\tau^{-1}+1)v_b^{-3}|D|\mathrm{cap}(\partial D)}{8\pi}\left(a_1\delta^{\frac{1}{2}}+a_2\delta+O(\delta^{\frac{3}{2}})\right)^3\\ &-\mathrm{cap}(\partial D)\delta-\frac{\sqrt{-1}(\tau^{-1}-1)v^{-1}\mathrm{cap}(\partial D)^2}{8\pi}\left(a_1\delta^{\frac{1}{2}}+a_2\delta+O(\delta^{\frac{3}{2}})\right)\delta\\ &+O(\delta^2)=0.\end{aligned}$$

From the coefficients of the δ and $\delta^{\frac{3}{2}}$ terms, we obtain

$$\begin{aligned}
&v_b^{-2}|D|a_1^2 - \mathrm{cap}(\partial D) = 0,\\
&2v_b^{-2}|D|a_1a_2 + \frac{\sqrt{-1}\tau(\tau^{-1}+1)v_b^{-3}|D|\mathrm{cap}(\partial D)}{8\pi}a_1^3\\
&-\frac{\sqrt{-1}(\tau^{-1}-1)v^{-1}\mathrm{cap}(\partial D)^2}{8\pi}a_1 = 0,
\end{aligned}$$

which yields

$$\begin{aligned}
a_1 &= \pm\sqrt{\frac{\mathrm{cap}(\partial D)}{|D|}}v_b,\\
a_2 &= -\frac{\sqrt{-1}(\tau^{-1}+1)v^{-1}\mathrm{cap}(\partial D)}{16\pi}a_1^2 + \frac{\sqrt{-1}(\tau^{-1}-1)\mathrm{cap}(\partial D)^2 v_b^2}{16\pi v|D|}\\
&= -\frac{\sqrt{-1}(\tau^{-1}+1)\mathrm{cap}(\partial D)^2 v_b^2}{16\pi v|D|} + \frac{\sqrt{-1}(\tau^{-1}-1)\mathrm{cap}(\partial D)^2 v_b^2}{16\pi v|D|}\\
&= \frac{-\sqrt{-1}\mathrm{cap}(\partial D)^2 v_b^2}{8\pi v|D|}.
\end{aligned}$$

This completes the proof of the theorem. □

The following monopole approximation holds [**38**].

THEOREM 14.6. *In the far field, i.e., for x such that $|x-y_0| \gg 2\pi/k$, the solution to the scattering problem at x with a single bubble D located at y_0 has the following pointwise behavior near the Minnaert resonant frequency $\omega_M := \omega_{0,0}$:*

(14.15)

$$u^s(x) := u(x) - u^i(x) = g(\omega,\delta,D)\left(1+O(\omega)+O(\delta)+o(1)\right)u^i(y_0)\Gamma_k(x-y_0),$$

where the scattering coefficient g is given by

$$g(\omega,\delta,D) = \frac{\mathrm{cap}(\partial D)}{1-(\frac{\omega_M}{\omega})^2+\sqrt{-1}\gamma}, \tag{14.16}$$

with

$$\gamma = \frac{(\tau+1)v\mathrm{cap}(\partial D)\omega}{8\pi} - \frac{(\tau-1)\mathrm{cap}(\partial D)^2\delta}{8\pi\tau^2 v|D|\omega}$$

being the damping constant.

A few remarks are in order.

REMARK 14.7. *In [**38**], the following properties for the operator $\mathcal{A}_0$ in the two-dimensional case were derived. Let ψ_0 be (the real-valued function) defined by*

$$(\mathcal{K}_D^0)^*[\psi_0] = \frac{1}{2}\psi_0$$

and $\|\psi_0\|_{L^2(\partial D)} = 1$. Let the constant

$$\gamma_0 := \mathcal{S}_D^0[\psi_0]\Big|_{\partial D}.$$

With the same notation as in Subsection 2.8.3.2, we have

(i) $\mathrm{Ker}(\mathcal{A}_0) = \mathrm{Im}\{\Psi_0\}$ *where*

$$\Psi_0 = \alpha_0 \begin{pmatrix} \psi_0 \\ a\psi_0 \end{pmatrix}$$

with

$$a = \begin{cases} \dfrac{\eta_{k_b}}{\eta_k} & \text{if } \gamma_0 = 0, \\ \dfrac{\gamma_0 + \langle \psi_0, \phi_0 \rangle \eta_{k_b}}{\gamma_0 + \langle \psi_0, \phi_0 \rangle \eta_k} & \text{if } \gamma_0 \neq 0, \end{cases}$$

and the constant α_0 *being chosen such that* $\|\Psi_0\| = 1$. *Here,* η_k *is the constant defined by* (2.153) *with* ω *replaced with* k*;*

(ii) $\mathrm{Ker}(\mathcal{A}_0^*) = \mathrm{Im}\{\Phi_0\}$ *where*

$$\Phi_0 = \beta_0 \begin{pmatrix} 0 \\ \phi_0 \end{pmatrix}$$

with $\phi_0 = 1$ *on* ∂D *and the constant* β_0 *being chosen such that* $\|\Phi_0\| = 1$.

Using the method developed above, we can derive the Minnaert resonance for a single bubble in two dimensions. We can prove that there exist two Minnaert resonances for a single bubble and their leading order terms are given by the roots of the following equation in ω*:*

$$\omega^2 \ln \omega + \left[(\ln v_b^{-1} + 1 + \frac{c_1}{b_1}) - \frac{\gamma_0}{\langle \psi_0, 1 \rangle} \right] \omega^2 - \frac{1}{4|D|} \frac{a\delta}{b_1} = 0, \tag{14.17}$$

where the constants b_1, c_1 *are defined in Subsection* 2.8.3.2.

REMARK 14.8. *Using the method developed above, we can also obtain the full asymptotic expansion for the resonance with respect to the small parameter* δ.

REMARK 14.9. *In the case of a collection of* L *identical bubbles, with separation distance much larger than their characteristic sizes, the Minnaert resonance for a single bubble will be split into* L *resonances. The splitting will be related to the eigenvalues of an* L*-by-*L *matrix which encodes information on the configuration of the* L *bubbles. This can be proved by the same argument as the one for systems of Helmholtz resonators in Section* 13.4.

REMARK 14.10. *Taking into consideration Theorems* 14.5 *and* 14.6, *we can deduce that if the bubble is represented by* $D = y_0 + sB$ *for some small positive number* s *and a normalized domain* B *with size of order one, then the Minnaert resonance for* D *is approximately given by the following formula*

$$\omega_M = \frac{1}{s} \left[\sqrt{\frac{\mathrm{cap}(\partial B)}{|B|}} v_b \delta^{\frac{1}{2}} - \sqrt{-1} \frac{\mathrm{cap}(\partial B)^2 v_b^2}{8\pi v |B|} \delta + O(\delta^{\frac{3}{2}}) \right]. \tag{14.18}$$

Moreover, the monopole approximation (14.15) *holds with the scattering coefficient* g *given by*

$$g(\omega, \delta, D) = \frac{s\mathrm{cap}(\partial B)}{1 - (\frac{\omega_M}{\omega})^2 + \sqrt{-1}\gamma},$$

where

$$\gamma = \frac{(\tau^{-1} + 1) v^{-1} \mathrm{cap}(\partial B) s\omega}{8\pi} - \frac{(\tau^{-1} - 1) \mathrm{cap}(\partial B)^2 \delta v_b^2}{8\pi v |B| s\omega}.$$

REMARK 14.11. *In the special case when D is the unit sphere, we have*

$$\mathrm{cap}(\partial D) = 4\pi, \quad |D| = 4\pi/3.$$

Consequently,

$$\begin{aligned} \sqrt{\frac{\mathrm{cap}(\partial D)}{|D|}} v_b &= \sqrt{3} v_b, \\ \frac{\mathrm{cap}(\partial D)^2 v_b^2}{8\pi v |D|} &= \frac{3 v_b^2}{2v}. \end{aligned}$$

Therefore, the Minnaert resonance is given by

$$\begin{aligned} \omega_{0,0}(\delta) &= \sqrt{3} v_b \delta^{\frac{1}{2}} - \sqrt{-1} \frac{3 v_b^2}{2v} \delta + O(\delta^{\frac{3}{2}}), \\ &= \sqrt{\frac{3\kappa_b}{\rho}} - \sqrt{-1} \frac{3}{2} \kappa_b \sqrt{\frac{1}{\rho\kappa}} + O((\frac{\rho_b}{\rho})^{\frac{3}{2}}). \end{aligned}$$

REMARK 14.12. *As in Chapter 12, using the sub-wavelength resonance of the bubble, one can design bubble metascreens. An acoustic meta-screen is a thin sheet with patterned sub-wavelength structures, which nevertheless has a macroscopic effect on the acoustic wave propagation. In* [**39**]*, periodic sub-wavelength bubbles mounted on a reflective surface (with Dirichlet boundary condition) were considered. It was shown that the structure behaves as an equivalent surface with Neumann boundary condition at the Minnaert resonant frequency which corresponds to a wavelength much greater than the size of the bubbles. An analytical formula for this resonance was derived and numerically confirmed. The super-absorption behavior of the met-screen observed in* [**362**] *was formally explained.*

14.3. Effective Medium Theory for a System of Bubbles and Super-Resolution

In this section, we derive an effective medium theory for acoustic wave propagation in bubbly fluid near the Minnaert resonant frequency. We start with a multiple scattering formulation of the scattering problem of an incident wave by a large number of identical small bubbles in a homogeneous fluid. Under certain conditions on the configuration of the bubbles, we establish an effective medium theory for the bubbly fluid as the number of bubbles tends to infinity. As a consequence, we show that just below the Minnaert resonant frequency, the obtained effective media can have a high refractive index, which is the reason for the super-focusing experiment observed in [**350**].

14.3.1. Problem Formulation. Consider the scattering of acoustic waves by N identical bubbles distributed in a homogeneous fluid in $\mathbb{R}^3$. The bubbles are represented by

$$D^N := \cup_{1\le j\le N} D_j^N,$$

where $D_j^N = y_j^N + sB$ for $1 \le j \le N$ with y_j^N being the location, s being the characteristic size and B being the normalized bubble which is a smooth and simply connected domain with size of order one. We denote by ρ_b and κ_b the density and the bulk modulus of the air inside the bubble respectively, which are different from the corresponding ρ and κ in the background medium $\mathbb{R}^3 \backslash D^N$.

We assume that $0 < s \ll 1$, $N \gg 1$ and that $\{y_j^N\} \subset \Omega$. Let u^i be the incident wave which we assume to be a plane wave for simplicity. The scattering can be modeled by the following system of equations:

$$(14.19)\qquad \begin{cases} \nabla \cdot \dfrac{1}{\rho}\nabla u^N + \dfrac{\omega^2}{\kappa}u^N = 0 \quad \text{in } \mathbb{R}^3 \backslash D^N, \\ \nabla \cdot \dfrac{1}{\rho_b}\nabla u^N + \dfrac{\omega^2}{\kappa_b}u^N = 0 \quad \text{in } D^N, \\ u_+^N - u_-^N = 0 \quad \text{on } \partial D^N, \\ \dfrac{1}{\rho}\dfrac{\partial u^N}{\partial \nu}\bigg|_+ - \dfrac{1}{\rho_b}\dfrac{\partial u^N}{\partial \nu}\bigg|_- = 0 \quad \text{on } \partial D^N, \\ u^N - u^i \text{ satisfies the Sommerfeld radiation condition,} \end{cases}$$

where u^N is the total field and ω is the frequency.

As in the previous section, we introduce the four auxiliary parameters v, v_b, k, and k_b, and the two dimensionless contrast parameters δ and τ to facilitate our analysis. By choosing proper physical units, we may again assume that both the frequency ω and the wave speed outside the bubbles are of order one. As a result, the wavenumber k outside the bubbles is also of order one. We assume that there is a large contrast between both the densities and bulk moduli inside and outside the bubbles. However, the contrast between the wave speeds are small. Thus, both the wave speed and wavenumber k_b inside the bubbles are of order one. To sum up, we assume that (14.4) holds. We also assume that the domain of interest Ω has size of order one.

Using layer potentials, the solution u^N can be written as

$$(14.20)\qquad u^N(x) = \begin{cases} u^i + \mathcal{S}_{D^N}^k[\psi^N], & x \in \mathbb{R}^3 \backslash \overline{D^N}, \\ \mathcal{S}_D^{k_b}[\psi_b^N], & x \in D^N, \end{cases}$$

for some surface potentials $\psi, \psi_b \in L^2(\partial D^N)$. Here, we have used the notation

$$\begin{aligned} L^2(\partial D^N) &= L^2(\partial D_1^N) \times L^2(\partial D_2^N) \times \cdots \times L^2(\partial D_N^N), \\ \mathcal{S}_{D^N}^k[\psi^N] &= \sum_{1 \le j \le N} \mathcal{S}_{D_j^N}^k[\psi_j^N], \\ \mathcal{S}_D^{k_b}[\psi_b^N] &= \sum_{1 \le j \le N} \mathcal{S}_{D_j^N}^k[\psi_{bj}^N]. \end{aligned}$$

Using the jump relations for the single-layer potentials, it is easy to derive that ψ and ψ_b satisfy the following system of boundary integral equations:

$$(14.21)\qquad \mathcal{A}^N(\omega, \delta)[\Psi^N] = F^N,$$

where

$$\mathcal{A}^N(\omega, \delta) = \begin{pmatrix} \mathcal{S}_{D^N}^{k_b} & -\mathcal{S}_{D^N}^k \\ -\frac{1}{2}I + (\mathcal{K}_{D^N}^{k_b})^* & -\delta(\frac{1}{2}I + (\mathcal{K}_{D^N}^k)^*) \end{pmatrix},$$

$$\Psi^N = \begin{pmatrix} \psi_b^N \\ \psi^N \end{pmatrix}, \quad F^N = \begin{pmatrix} u^i \\ \delta \frac{\partial u^i}{\partial \nu} \end{pmatrix}\bigg|_{\partial D^N}.$$

One can easily show that the scattering problem (14.19) is equivalent to the system of boundary integral equations (14.21) and there exists a unique solution to the scattering problem (14.19), or equivalently to the system (14.21).

Let $\mathcal{H} = L^2(\partial D^N) \times L^2(\partial D^N)$ and $\mathcal{H}_1 = H^1(\partial D^N) \times L^2(\partial D^N)$. It is clear that $\mathcal{A}^N(\omega,\delta)$ is a bounded linear operator from $\mathcal{H}$ to $\mathcal{H}_1$, *i.e.*, $\mathcal{A}^N(\omega,\delta) \in \mathcal{L}(\mathcal{H},\mathcal{H}_1)$. Here, we also use the following convention: let a_N and b_N be two real numbers which may depend on N, then

$$a_N \lesssim b_N$$

means that $a_N \leq C \cdot b_N$ for some constant C which is independent of a_N, b_N and N.

We are interested in the case when there is a large number N of small identical bubbles distributed in a bounded domain and the incident wave has a frequency near the Minnaert resonant frequency for an individual bubble. We recall from (14.18) that for the bubble given by $D_j^N = y_j^N + sB$, its corresponding Minnaert resonant frequency ω_M is approximately given by

$$\omega_M = \frac{1}{s}\sqrt{\frac{\text{cap}(\partial B)\delta}{|B|}}v_b. \tag{14.22}$$

We shall use the idea of point interaction approximation to study the scattering by a large number of bubbles. For this, certain assumptions on the frequency, the bubble volume fraction, and the configuration of the bubble distribution are needed.

ASSUMPTION 14.13. *The frequency* $\omega = O(1)$ *and is independent of* N. *Moreover,*

$$1 - (\frac{\omega_M}{\omega})^2 = \beta_0 s^{\epsilon_1} \tag{14.23}$$

for some fixed $0 < \epsilon_1 < 1$ *and constant* β_0.

It is clear that equation (14.23) controls the deviation of frequency from the Minnaert resonant frequency. There are two cases depending on whether $\omega > \omega_M$ or $\omega < \omega_M$. In the former case, we have $\beta_0 > 0$, while in the latter case we have $\beta_0 < 0$. We shall see later on that acoustic wave propagation is quite different in these two cases. In fact, the wave field may be dissipative in the former case while highly oscillatory and propagating in the latter case.

We next make the following assumption on the bubble volume fraction.

ASSUMPTION 14.14. *The following identity holds*

$$s^{1-\epsilon_1} \cdot N = \Lambda, \tag{14.24}$$

where Λ *is a constant independent of* N. *Moreover, we will assume that* Λ *is large.*

From (14.22) and (14.23) we see that

$$\delta = \omega^2 s^2 (1 - s^{\epsilon_1}) \cdot \frac{|B|}{\text{cap}(\partial B) v_b^2}. \tag{14.25}$$

This indicates that both the bubble size s and contrast parameter δ depend on N. In the limit when $N \to \infty$, we have $s \to 0$ and $\delta \to 0$.

We now impose conditions on the bubble configuration. We first assume that the size of each bubble is much smaller than the typical distance between neighboring bubbles. More precisely, we make the following assumption.

ASSUMPTION 14.15. *The following conditions hold:*

$$\begin{cases} \min_{l\neq j} |y_l^N - y_j^N| \geq r_N, \\ s \ll r_N, \end{cases}$$

where $r_N = \zeta N^{-\frac{1}{3}}$ for some constant ζ independent of N. Here, r_N can be viewed as the minimum separation distance between neighboring bubbles.

Following [**430**, **438**], we assume that there exists $\widetilde{V} \in L^\infty(\Omega)$ such that

$$\Theta^N(A) \to \int_A \widetilde{V}(x)dx, \quad \text{as } N \to \infty, \tag{14.26}$$

for any measurable subset $A \subset \mathbb{R}^3$, where $\Theta^N(A)$ is defined by

$$\Theta^N(A) = \frac{1}{N} \times \{\text{number of points } y_j^N \text{ in } A \subset \mathbb{R}^3\}.$$

In addition, we assume that the following condition on the regularity of the "sampling" points $\{y_j^N\}$ holds.

ASSUMPTION 14.16. *There exists $0 < \epsilon_0 < 1$ such that for all $h \geq 2r_N$:*

$$\frac{1}{N} \sum_{|x-y_j^N| \geq h} \frac{1}{|x-y_j^N|^2} \lesssim |h|^{-\epsilon_0}, \quad \text{uniformly for all } x \in \Omega, \tag{14.27}$$

$$\frac{1}{N} \sum_{2r_N \leq |x-y_j^N| \leq 3h} \frac{1}{|x-y_j^N|} \lesssim |h|, \quad \text{uniformly for all } x \in \Omega. \tag{14.28}$$

ASSUMPTION 14.17. *Let Γ_k the outgoing fundamental solution to the Helmholtz operator $\Delta + k^2$. For any $f \in \mathcal{C}^{0,\eta}(\Omega)$ with $0 < \eta < 1$,*

$$\max_{1 \leq j \leq N} \Big|\frac{1}{N} \sum_{l \neq j} \Gamma_k(y_l^N - y_j^N) f(y_l^N) - \int_\Omega \Gamma_k(y_j^N - y)\widetilde{V}(y) f(y) dy\Big| \lesssim \frac{1}{N^{\frac{\eta}{3}}} \|f\|_{\mathcal{C}^{0,\eta}(\Omega)}. \tag{14.29}$$

ASSUMPTION 14.18. $\epsilon_0 < \frac{3\epsilon_1}{1-\epsilon_1}$.

We now make several remarks about our assumptions.

REMARK 14.19. *One can show that equations (14.27) and (14.28) are respectively equivalent to the following ones*

$$\max_l \Big\{\frac{1}{N} \sum_{|y_l^N - y_j^N| \geq h} \frac{1}{|y_l^N - y_j^N|^2}\Big\} \lesssim h^{-\epsilon_0}; \tag{14.30}$$

$$\max_l \Big\{\frac{1}{N} \sum_{2r_N \leq |y_l^N - y_j^N| \leq 3h} \frac{1}{|y_l^N - y_j^N|}\Big\} \lesssim h. \tag{14.31}$$

Indeed, these estimates follow from the fact that for each $x \in \Omega$ there exists a finite number of points $y_{j_1}^N, y_{j_2}^N, \ldots, y_{j_L}^N$ in the neighborhood of x with L independent of N such that

$$\frac{1}{|x-y_j^N|^2} \leq \sum_{1 \leq l \leq L} \frac{1}{|y_{j_l}^N - y_j^N|^2}, \quad \frac{1}{|x-y_j^N|} \leq \sum_{1 \leq l \leq L} \frac{1}{|y_{j_l}^N - y_j^N|},$$

for all y_j^N such that $|x - y_j^N| \geq h$.

REMARK 14.20. *By decomposing $\Gamma_k(x-y)$ into the singular part, $\Gamma_0(x-y)$, and a smooth part, one can show that Assumption 14.17 is equivalent to*

$$\max_{1 \leq j \leq N} \Big|\frac{1}{N} \sum_{l \neq j} \frac{1}{|y_l^N - y_j^N|} f(y_l^N) - \int_\Omega \frac{1}{|y - y_j^N|} \widetilde{V}(y) f(y) dy\Big| \lesssim \frac{1}{N^{\frac{\eta}{3}}} |f\|_{\mathcal{C}^{0,\eta}}. \tag{14.32}$$

REMARK 14.21. *Assumptions* 14.13 *and* 14.18 *are important in our justification of the point interaction approximation; see Theorem* 14.24. *The assumption that* $\epsilon_1 > 0$ *is critical here. For the case* $\epsilon_1 = 0$, *the frequency is away from the Minnaert resonant frequency. The scattering coefficient* g *has magnitude of order* s. *The fluctuation in the scattered field from all the other bubbles may generate multipole modes which are comparable with the monopole mode and hence invalidate the monopole point interaction approximation. We leave this case as an open question for future investigation.*

REMARK 14.22. *Assumptions* 14.13 *and* 14.14 *are important in our effective medium theory. The parameter* ϵ_1 *in Assumption* 14.13 *controls the deviation of the frequency from the Minnaert resonant frequency, which further controls the amplitude of the scattering strength of each bubble. This parameter, together with* Λ, *also controls the volume fraction of the bubbles through Assumption* 14.14. *In an informal way, if the bubble volume fraction is below the level as set by Assumption* 14.14, *say* $s^{1-\epsilon_3} \cdot N = O(1)$ *for some* $\epsilon_3 < \epsilon_1$, *then the effect of the bubbles is negligible and the effective medium would be the same as if there are no bubbles in the limit as* $N \to \infty$. *On the other hand, if* $s^{1-\epsilon_3} \cdot N = O(1)$ *for some* $\epsilon_3 > \epsilon_1$, *then the bubbles interact strongly with each other and eventually behave as a medium with infinite effective refractive index. Only at the appropriate volume fraction as in Assumption* 14.14, *do we have an effective medium theory with finite refractive index. The larger* Λ *is, the higher the effective refractive index is. These statements can be justified by the method developed in the chapter.*

REMARK 14.23. *One can easily check that Assumptions* 14.15, 14.16 *and* 14.17 *hold for periodically distributed* y_j^N*'s.*

14.3.2. Point Interaction Approximation. In this subsection, we justify the point interaction approximation under the assumptions we made in the previous subsection. For $1 \leq j \leq N$, denote by

$$\begin{aligned} u_j^{i,N} &= u^i + \sum_{l \neq j} \mathcal{S}_{D_l^N}^k [\psi_l^N], \\ u_j^{s,N} &= \mathcal{S}_{D_j^N}^k [\psi_j^N]. \end{aligned}$$

It is clear that $u_j^{i,N}$ is the total incident field which impinges on the bubble D_j^N and $u_j^{s,N}$ is the corresponding scattered field. Our main result is the following.

THEOREM 14.24. *Under Assumptions* 14.13, 14.15, 14.16 *and* 14.18, *the following relation between* $u_j^{s,N}$ *and* $u_j^{i,N}$ *holds for all* x *such that* $|x - y_j^N| \gg s$:

$$u_j^{s,N}(x) = \Gamma_k(x - y_j^N) \cdot g \cdot \left(u_j^{i,N}(y_j^N) + O(N^{\frac{\epsilon_0}{3} - \frac{\epsilon_1}{1-\epsilon_1}} + \frac{s}{|x - y_j^N|}) \cdot \max_{1 \leq l \leq N} |u_l^{i,N}(y_l^N)| \right).$$

Moreover, for $x = y_j^N$,

$$u_j^{i,N}(y_j^N) = u^i(y_j^N) + \sum_{l \neq j} u_l^{s,N}(y_j^N) = u^i(y_j^N) + \sum_{l \neq j} g \cdot \left(u_l^{i,N}(y_l^N) + p_l^N \right) \Gamma_k(y_j^N - y_l^N),$$

where

$$\begin{aligned} g &= g(\omega,\delta,D_j^N) = -\frac{s\mathrm{cap}(\partial B)}{1-(\frac{\omega_M}{\omega})^2+\sqrt{-1}\gamma}(1+O(s)+O(\delta)),\\ \gamma &= \frac{(\tau^{-1}+1)v^{-1}\mathrm{cap}(\partial B)s\omega}{8\pi}-\frac{(\tau^{-1}-1)\mathrm{cap}(\partial B)^2v_b^2\delta}{8\pi v|B|\omega s}, \end{aligned}$$

are the scattering and damping coefficients near the Minnaert resonant frequency respectively, and p_l^N *satisfies*

$$|p_l^N| = \max_{1\le j\le N}|u_j^{i,N}(y_j^N)|\cdot O(N^{\frac{\epsilon_0}{3}-\frac{\epsilon_1}{1-\epsilon_1}}).$$

PROOF. First, by Taylor series expansion of $\Gamma_k(x-y)$ with respect to y around y_j^N, we have

(14.33)
$$\begin{aligned} u_j^{s,N}(x) &= \int_{\partial D_j^N}\Gamma_k(x-y)\psi_j^N(y)d\sigma(y)\\ &= \Gamma_k(x-y_j^N)\Big(\langle\chi(\partial D_j^N),\psi_j^N\rangle_{L^2}+O(\frac{s}{|x-y_j^N|})\cdot s\cdot\|\psi_j^N\|_{L^2}\Big). \end{aligned}$$

On the other hand, one can easily obtain

$$\psi_j^N = u_j^{i,N}(y_j^N)\mathcal{S}_{D_j^N}^{-1}[1]\cdot\frac{g}{\mathrm{cap}(\partial D_j^N)}+\frac{1}{s}\cdot O(\|F_{j,2}\|_{H^1(\partial D_j^N)})\quad\text{in } L^2(\partial D_j^N),$$

where

$$F_{j,2}(y) = u_j^{i,N}(y)-u_j^{i,N}(y_j^N) = \sum_{l\ne j}\Big(\mathcal{S}_{D_l^N}^k[\psi_l^N](y)-\mathcal{S}_{D_l^N}^k[\psi_l^N](y_j^N)\Big).$$

By Lemma 14.25, we get

$$\|\mathcal{S}_{D_l^N}^k[\psi_l^N](y)-\mathcal{S}_{D_l^N}^k[\psi_l^N](y_j^N)\|_{H^1(\partial D_j^N)}\lesssim\frac{1}{|y_l^N-y_j^N|^2}\cdot s^2\cdot\|\psi_l^N\|_{L^2(\partial D_l^N)}.$$

Thus,

$$\|F_{j,2}\|_{H^1(\partial D_j^N)}\lesssim\sum_{l\ne j}\frac{1}{|y_l^N-y_j^N|^2}\cdot s^2\cdot\max_{1\le l\le N}\|\psi_l^N\|_{L^2(\partial D_l^N)}.$$

Therefore, it follows that

$$\begin{aligned} \|\psi_j^N\|_{L^2(\partial D_j^N)} &\lesssim |u_j^{i,N}(y_j^N)|\cdot\|\mathcal{S}_{D_j^N}^{-1}[1]\|_{L^2(\partial D_j^N)}\cdot|\frac{g}{\mathrm{cap}(\partial D_j^N)}|\\ &\quad+\sum_{l\ne j}\frac{1}{|y_l^N-y_j^N|^2}\cdot s\cdot\max_{1\le j\le N}\|\psi_j^N\|_{L^2(\partial D_j^N)}. \end{aligned}$$

Note that $\|\mathcal{S}_{D_j^N}^{-1}[1]\|_{L^2(\partial D_j^N)} = O(1)$ and

$$\sum_{l\ne j}\frac{1}{|y_l^N-y_j^N|^2}\cdot s\lesssim r_N^{-\epsilon_0}s\cdot N\lesssim N^{\frac{\epsilon_0}{3}-\frac{\epsilon_1}{1-\epsilon_1}},$$

where we have used Assumption 14.13 in the last inequality.

We can therefore conclude that

$$\max_{1\le j\le N}\|\psi_j^N\|_{L^2(\partial D_j^N)}\lesssim\max_{1\le j\le N}|u_j^{i,N}(y_j^N)|\cdot|\frac{g}{\mathrm{cap}(\partial D_j^N)}|. \tag{14.34}$$

Consequently, by (14.33),

$$\begin{aligned} u_j^{s,N}(x) &= \Gamma_k(x-y_j^N)\left(\langle 1,\psi_j^N\rangle_{L^2} + O(\frac{s}{|x-y_j^N|})\cdot s\cdot \|\psi_j^N\|_{L^2}\right) \\ &= \Gamma_k(x-y_j^N)\left(\langle 1,\psi_j^N\rangle_{L^2} + O(\frac{s}{|x-y_j^N|})\max_{1\le l\le N}|u_l^{i,N}(y_l^N)|\cdot|g|\right). \end{aligned}$$

Since

$$\begin{aligned} \langle 1,\psi_j^N\rangle_{L^2} &= \left\langle 1, u_j^{i,N}(y_j^N)\mathcal{S}_{D_j^N}^{-1}[1]\cdot\frac{g}{\mathrm{cap}(\partial D_j^N)}\right\rangle_{L^2} + O(\|F_{j,2}\|_{H^1(\partial D_j^N)}) \\ &= u_j^{i,N}(y_j^N)g + O(s\cdot N^{\frac{\epsilon_0}{3}-\frac{\epsilon_1}{1-\epsilon_1}})\cdot\max_{1\le l\le N}\|\psi_l^N\|_{L^2(\partial D_l^N)} \\ &= u_j^{i,N}(y_j^N)g + O(s\cdot N^{\frac{\epsilon_0}{3}-\frac{\epsilon_1}{1-\epsilon_1}})\cdot\max_{1\le l\le N}|u_l^{i,N}(y_l^N)|\cdot\left|\frac{g}{\mathrm{cap}(\partial D_l^N)}\right| \\ &= g\left(u_j^{i,N}(y_j^N) + O(N^{\frac{\epsilon_0}{3}-\frac{\epsilon_1}{1-\epsilon_1}})\cdot\max_{1\le l\le N}|u_l^{i,N}(y_l^N)|\right), \end{aligned}$$

we arrive at

$$u_j^{s,N}(x) = \Gamma_k(x-y_j^N)g\left(u_j^{i,N}(y_j^N) + O\big(N^{\frac{\epsilon_0}{3}-\frac{\epsilon_1}{1-\epsilon_1}} + \frac{s}{|x-y_j^N|}\big)\cdot\max_{1\le l\le N}|u_l^{i,N}(y_l^N)|\right).$$

Finally, recall that

$$u_j^{i,N}(x) = u^i(x) + \sum_{l\neq j} u_l^{s,N}(x).$$

By taking $x=y_l^N$ and using the assumption that

$$|y_l^N - y_j^N| \ge r_N,$$

we obtain

$$\frac{s}{|x-y_j^N|} \le \frac{s}{r_N} \lesssim \frac{1}{N}\cdot s^{\epsilon_1}\cdot N^{\frac{1}{3}} \lesssim N^{\frac{\epsilon_0}{3}-\frac{\epsilon_1}{1-\epsilon_1}}.$$

The second part of the proposition follows immediately. □

LEMMA 14.25. *The following estimate holds:*

$$\|\mathcal{S}_{D_l^N}^k[\psi_l^N](y) - \mathcal{S}_{D_l^N}^k[\psi_l^N](y_j^N)\|_{H^1(\partial D_j^N)} \lesssim \frac{1}{|y_l^N-y_j^N|^2}\cdot s^2\cdot\|\psi_l^N\|_{L^2(\partial D_l^N)}. \tag{14.35}$$

PROOF. By Taylor series expansion of $\Gamma_k(y-z)$ with respect to y around y_j^N and z around y_l^N, we have

$$\begin{aligned} \mathcal{S}_{D_l^N}^k[\psi_l^N](y) - \mathcal{S}_{D_l^N}^k[\psi_l^N](y_j^N) &= \int_{\partial D_l^N}\left(\Gamma_k(y-z)-\Gamma_k(y_j^N-z)\right)\psi_l^N(z)d\sigma(z) \\ = \sum_{|\alpha|\ge 1}(y-y_j^N)^\alpha &\sum_{|\beta|\ge 0}\int_{\partial D_l^N}\frac{\partial^{|\alpha|+|\beta|}\Gamma_k}{\partial y^\alpha\partial z^\beta}(y_j^N-y_l^N)(z-y_l^N)^\beta\psi_l^N(z)d\sigma(z). \end{aligned}$$

Using the estimate

$$\left|\frac{\partial^{|\alpha|+|\beta|}\Gamma_k}{\partial y^\alpha\partial z^\beta}(y_j^N-y_l^N)\right| \lesssim \max\left\{\frac{1}{|y_l^N-y_j^N|},\frac{1}{|y_l^N-y_j^N|^{|\alpha|+|\beta|+1}}\right\},$$

we obtain

$$\begin{aligned} |\mathcal{S}_{D_l^N}^k[\psi_l^N](y) - \mathcal{S}_{D_l^N}^k[\psi_l^N](y_j^N)| &\lesssim \frac{1}{|y_l^N - y_j^N|^2} \cdot s^2 \cdot \|\psi_l^N\|_{L^2}, \\ |\nabla \mathcal{S}_{D_l^N}^k[\psi_l^N](y)| &\lesssim \frac{1}{|y_l^N - y_j^N|^2} \cdot s \cdot \|\psi_l^N\|_{L^2}, \end{aligned}$$

whence estimate (14.35) follows. This completes the proof. □

Let us denote $x_j^N = u_j^{i,N}(y_j^N)$, $b_j^N = u^i(y_j^N)$, $T^N = (T_{lj}^N)_{1\le l,j\le N}$ with

$$T_{lj}^N = g\Gamma_k(y_l^N - y_j^N), \quad l \neq j,$$

$T_{ll}^N = 0$, and

$$q_j^N = \sum_{l\neq j} g\Gamma_k(y_j^N - y_l^N)p_l^N.$$

We obtain the following system of equations for $x^N = (x_j^N)_{1\le j\le N}$:

$$x^N - T^N x^N = b^N + q^N. \tag{14.36}$$

14.3.3. Derivation of the Effective Medium Theory of Bubbly Media. In this section we derive an effective medium theory for the acoustic wave propagation in the bubbly medium considered in Subsection 14.3.1. We first investigate the system of equations (14.36), which resulted from the point interaction approximation. We establish its well-posedness, including existence, uniqueness and stability of a solution, and derive its limiting behavior as the number of bubbles $N \to +\infty$. We then construct the wave field from the solution to (14.36) and show the convergence of the constructed micro-field to a macro-effective field.

14.3.3.1. *Well-posedness and limiting behaviour of the point interaction system.* We start from the summation $\sum_{l\neq j} g\Gamma_k(y_j^N - y_l^N)f(y_l^N)$. It is clear that

$$\sum_{l\neq j} g\Gamma_k(y_j^N - y_l^N)f(y_l^N) = \frac{1}{N}\sum_{l\neq j} \frac{\mathrm{cap}(\partial B)}{\beta_0 s^{\epsilon_1} + \sqrt{-1}\cdot O(\omega\cdot s)}(s\cdot N)\cdot\Gamma_k(y_j^N - y_l^N)f(y_l^N).$$

Denote by

$$\beta_N = \frac{\mathrm{cap}(\partial B)}{\beta_0 + \sqrt{-1}\cdot O(\omega\cdot s^{1-\epsilon_1})}(1+O(s)), \quad \beta = \frac{\mathrm{cap}(\partial B)}{\beta_0}.$$

Note that β and B are independent of N. By Assumption 14.14, we have the following identity:

$$\sum_{l\neq j} g\Gamma_k(y_j^N - y_l^N)f(y_l^N) = \frac{1}{N}\sum_{l\neq j} \beta_N\cdot\Lambda\cdot\Gamma_k(y_j^N - y_l^N)f(y_l^N).$$

Let

$$V(x) = \beta\cdot\Lambda\cdot\widetilde{V}(x). \tag{14.37}$$

We note that there are two cases depending on whether $\omega > \omega_M$ or $\omega < \omega_M$. In the former case, $\beta_0 > 0$, thus $\beta > 0$ which leads to $V(x) \ge 0$, while in the latter case we have $\beta_0 < 0$ and thus $\beta < 0$ which leads to $V(x) \le 0$.

We now present a result on the approximation of the summation

$$\sum_{l\neq j} g\Gamma_k(y_j^N - y_l^N)f(y_l^N)$$

by using volume integrals.

LEMMA 14.26. *For any $f \in \mathcal{C}^{0,\eta}(\Omega)$ with $0<\eta<1$,*

$$\max_{1\le j\le N} \Big|\frac{1}{N}\sum_{l\neq j} \beta_N \cdot \Lambda \cdot \Gamma_k(y_j^N - y_l^N) f(y_l^N) - \int_\Omega \Gamma_k(y_j^N - y) V(y) f(y) dy\Big| \lesssim \frac{1}{N^{\frac{\eta}{3}}} \|f\|_{\mathcal{C}^{0,\eta}(\Omega)}.$$

PROOF. By Assumption 14.17, we have

$$\max_{1\le j\le N} \Big|\frac{1}{N}\sum_{l\neq j} \beta \cdot \Lambda \cdot \Gamma_k(y_j^N - y_l^N) f(y_l^N) - \int_\Omega \Gamma_k(y_j^N - y) V(y) f(y) dy\Big| \lesssim \frac{1}{N^{\frac{\eta}{3}}} \|f\|_{\mathcal{C}^{0,\eta}(\Omega)}.$$

On the other hand, note that

$$|\beta_N - \beta| \lesssim s^{1-\epsilon_1} \lesssim \frac{1}{N}.$$

Thus,

$$\begin{aligned}
&\max_{1\le j\le N} \Big|\frac{1}{N}\sum_{l\neq j} (\beta - \beta_N) \cdot \Lambda \cdot \Gamma_k(y_j^N - y_l^N) f(y_l^N)\Big| \\
&\lesssim \frac{1}{N}\cdot\frac{1}{N}\sum_{l\neq j} \frac{1}{|y_j^N - y_l^N|} \|f\|_{\mathcal{C}^{0,\eta}(\Omega)} \\
&\lesssim \frac{1}{N}\|f\|_{\mathcal{C}^{0,\eta}(\Omega)} \le \frac{1}{N^{\frac{\eta}{3}}} \|f\|_{\mathcal{C}^{0,\eta}(\Omega)}.
\end{aligned}$$

The lemma then follows immediately.

□

Let $X = \mathcal{C}^{0,\eta}(\Omega)$ for some $0<\eta<1$ (later on we will take $\eta = \frac{1-\epsilon_0}{2}$). Define $\mathcal{T}$ by

$$\mathcal{T}[f](x) = \int_\Omega \Gamma_k(x-y) V(y) f(y) dy.$$

The operator $\mathcal{T}$ can be viewed as the continuum limit of T^N in some sense. One can show that $\mathcal{T}: X \to X$ is a compact linear operator. Moreover, the following properties hold.

LEMMA 14.27. *We have*

(i) *The operator $\mathcal{T}$ is bounded from $\mathcal{C}^0(\overline{\Omega})$ to $\mathcal{C}^{0,\eta}(\Omega)$ for any $0<\eta<1$.*
(ii) *The operator $\mathcal{T}$ is bounded from $\mathcal{C}^{0,\eta}(\Omega)$ to $\mathcal{C}^{1,\eta}(\Omega)$ for any $0<\eta<1$.*
(iii) *In the case when $\omega<\omega_M$, the operator $I-\mathcal{T}$ has a bounded inverse on the Banach space X. More precisely, for each $b\in X$, there exists a unique $f\in X$ such that $f-\mathcal{T}[f]=b$ and $\|f\|_X \le C\|b\|_X$, where C is a positive constant independent of b.*
(iv) *In the case when $\omega>\omega_M$, the same conclusion as in Assertion (iii) holds, provided that $V(x)>k^2$ almost everywhere in Ω.*

PROOF. Assertions (i) and (ii) follow from the general theory on integral operators. We now show Assertion (iii). Let $b\in X$ and consider the following integral equation

$$x - \mathcal{T}[x] = b.$$

Applying the operator $\triangle + k^2$ to both sides of the above equation, we obtain

$$(\triangle + k^2)x - Vx = (\triangle + k^2)b \text{ in } \Omega.$$

In the case when $\omega < \omega_M$, we have $V(x) \leq 0$. Thus the above equation yields a Lippmann-Schwinger equation with potential $k^2 - V$, for which the solution is known to be unique. This proves that the operator $I - \mathcal{T}$ has a trivial kernel. The rest of statements of Assertion (iii) follow from standard Fredholm theory for compact operators. Similarly, for Assertion (iv), we note that the operator $\triangle + k^2 - V$ is elliptic, then the statement follows from the standard theory of elliptic equations. □

REMARK 14.28. *In the case when $\omega > \omega_M$, one has $V(x) \geq 0$. The integral equation $x - \mathcal{T}[x] = b$ leads to the following partial differential operator $\triangle + (k^2 - V)$ where $k^2 - V$ may change sign in the domain Ω depending on the values of $V(x)$. In fact, in some physical situations, $\widetilde{V}$ may be zero or negligible near $\partial\Omega$ while of order one inside Ω. When $\beta \cdot \Lambda \gg 1$, we see that $k^2 - V < 0$ in the inner region of Ω. As a consequence, the wave field is attenuating therein, which implies that the effective medium is dissipative. On the other hand, the wave field is still propagating near $\partial\Omega$ where $k^2 - V(x)$ is positive. One may also observe a transition layer between the propagative region and the dissipative region near the place when $k^2 - V(x)$ is close to 0. It is not clear whether the operator $\triangle + (k^2 - V)$ with $k^2 - V$ changing sign is uniquely solvable or not.*

In view of Remark 14.28, we shall restrict our investigation to the case when $\omega < \omega_M$ from now on. However, we remark that if we assume that the kernel of the operator $I - \mathcal{T}$ is trivial in the case when $\omega > \omega_M$, then all the arguments and results which hold for the case $\omega < \omega_M$ also hold for $\omega > \omega_M$.

Note that $u^i \in X$. Let ψ be the unique solution satisfying

$$\psi - \mathcal{T}[\psi] = u^i. \tag{14.38}$$

It is clear that

$$(\triangle + k^2)\psi - V\psi = 0 \text{ in } \mathbb{R}^3.$$

We shall show that ψ is the limit of the solution x^N to (14.36) in a sense which will be made clear later on. We first present the following result concerning the well-posedness of the discrete system (14.36).

PROPOSITION 14.29. *Let $X = \mathcal{C}^{0,\eta}(\Omega)$ for $\eta = \frac{1-\epsilon_0}{2}$ and assume that $\omega < \omega_M$. Then under Assumptions* 14.15, 14.16 *and* 14.17, *there exists $N_0 > 0$ such that for all $N \geq N_0$ and $b \in X$, there is a unique solution to the equation*

$$z^N - T^N z^N = b^N$$

with $b_j^N = b(y_j^N)$. Moreover,

$$\max_{1 \leq j \leq N} |z_j^N| \leq C_1 \|b\|_X,$$

for some positive constant C_1 independent of N and b.

The proof of this proposition is technical and is given in [**96**]. As a corollary of the proposition, we can prove our main result on the limiting behavior of the solution to the system (14.36).

THEOREM 14.30. *Let $X = \mathcal{C}^{0,\eta}(\Omega)$ for $\eta = \frac{1-\epsilon_0}{2}$ and assume that $\omega < \omega_M$. Then under Assumptions* 14.13, 14.14, 14.15, 14.16, 14.17 *and* 14.18, *there exists $N_0 > 0$ such that for all $N \geq N_0$,*

$$\max_{1 \leq j \leq N} |x_j^N - \psi(y_j^N)| \lesssim N^{-\min\{\frac{1-\epsilon_0}{6}, \frac{\epsilon_0}{3}, \frac{\epsilon_1}{1-\epsilon_1}\}},$$

where x^N and ψ are the solutions to (14.36) *and* (14.38), *respectively.*

PROOF. Step 1. We have

$$\begin{aligned} x_j^N - \frac{1}{N}\sum_{l \neq j} \beta_N \cdot \Lambda \cdot \Gamma_k(y_j^N - y_l^N) x_l^N &= b_j^N + q_j^N, \\ \psi(y_j^N) - \int_\Omega \Gamma_k(y_j^N - y) V(y) f(y) dy &= b_j^N. \end{aligned}$$

Let $r_j^N = x_j^N - \psi(y_j^N)$. Then

$$r^N - T^N r^N = e^N + q^N,$$

where

$$e_j^N = \frac{1}{N}\sum_{l \neq j} \beta_N \cdot \Lambda \cdot \Gamma_k(y_j^N - y_l^N)\psi(y_j^N) - \int_\Omega \Gamma_k(y_j^N - y) V(y) \psi(y) dy.$$

Step 2. Let $G_N(x, y)$ be given by

$$G_N(x, y) = \begin{cases} -\frac{1}{4\pi|x-y|} & \text{if } |x-y| \geq r_N, \\ -\frac{1}{4\pi r_N}\frac{|x-y|}{r_N} & \text{if } 0 \leq |x-y| < r_N, \end{cases}$$

and define

$$G_N(x, y, k) = G_N(x, y) + (\Gamma_k(x-y) - \Gamma_0(x-y)) := G_{N,1}(x, y, k) + G_{N,2}(x, y, k).$$

Denote by

$$\widetilde{q}^N(y) = \sum_{l \neq j} g G_N(y, y_l^N, k) p_l^N = \widetilde{q}_1^N(y) + \widetilde{q}_2^N(y),$$

where

$$\widetilde{q}_1^N(y) = \sum_{l \neq j} g G_{N,1}(y, y_l^N, k) p_l^N, \quad \widetilde{q}_2^N(y) = \sum_{l \neq j} g G_{N,2}(y, y_l^N, k) p_l^N.$$

One can prove that $\widetilde{q}_1^N \in X$; see [**96**]. Moreover,

$$\|\widetilde{q}_1^N\|_X \lesssim \max_{1 \leq l \leq N} |p_l^N| \lesssim O(N^{\frac{\epsilon_0}{3} - \frac{\epsilon_1}{1-\epsilon_1}}) \cdot \max_{1 \leq l \leq N} |x_l^N|.$$

Since $G_{N,2}(x, y, k)$ is smooth in $|x-y|$ and is bounded, a straightforward calculation shows that $\widetilde{q}_2^N \in X$ as well and

$$\|\widetilde{q}_2^N\|_X \lesssim O(N^{\frac{\epsilon_0}{3} - \frac{\epsilon_1}{1-\epsilon_1}}) \cdot \max_{1 \leq l \leq N} |x_l^N|.$$

Thus, we have $\widetilde{q}^N \in X$ and

$$\|\widetilde{q}^N\|_X \lesssim \max_{1 \leq l \leq N} |p_l^N| \lesssim O(N^{\frac{\epsilon_0}{3} - \frac{\epsilon_1}{1-\epsilon_1}}) \cdot \max_{1 \leq l \leq N} |x_l^N|.$$

On the other hand, one can prove that there exists $\widetilde{e}^N \in X$ such that $\widetilde{e}^N(y_j^N) = e_j^N$ and $\|\widetilde{e}^N\|_X \lesssim N^{-\frac{1-\epsilon_0}{6}}\|u^i\|_X$; see [**96**]. Therefore,

$$\|\widetilde{e}^N\|_X + \|\widetilde{q}^N\|_X \lesssim N^{-\frac{1-\epsilon_0}{6}}\|u^i\|_X + O(N^{\frac{\epsilon_0}{3}-\frac{\epsilon_1}{1-\epsilon_1}}) \cdot \max_{1\leq l\leq N}|x_l^N|.$$

It then follows from Proposition 14.29 that

$$\max_{1\leq j\leq N}|r_j^N| \lesssim N^{-\frac{1-\epsilon_0}{6}}\|u^i\|_X + O(N^{\frac{\epsilon_0}{3}-\frac{\epsilon_1}{1-\epsilon_1}}) \cdot \max_{1\leq l\leq N}|x_l^N|. \tag{14.39}$$

Step 3. Note that $\max_{1\leq j\leq N}|\psi(y_j^N)|$ is bounded independently of N. We can derive from (14.39) that $\max_{1\leq j\leq N}|x_j^N|$ is also bounded independently of N, which further implies that

$$\max_{1\leq j\leq N}|r_j^N| \lesssim N^{-\min\{\frac{1-\epsilon_0}{6},\frac{\epsilon_0}{3}-\frac{\epsilon_1}{1-\epsilon_1}\}}.$$

This completes the proof of the theorem. □

As a consequence of the above result and (14.35), we have the following corollary.

COROLLARY 14.31. *The following estimate holds:*

$$\max_{1\leq j\leq N}\|\psi_j^N\|_{L^2(\partial D_j^N)} \lesssim s^{-\epsilon_1} \cdot \|u^i\|_X.$$

14.3.3.2. *Convergence of Micro-Field to the Effective One.* Let us consider the total field $u^N = u^i + \sum_{1\leq j\leq N}\mathcal{S}^k_{D_j^N}[\psi_j^N]$ outside the bubbles. Define

$$\widetilde{u}^N(x) = u^i(x) + \sum_{1\leq j\leq N} g\Gamma_k(x-y_j^N)x_j^N = u^i(x) + \frac{1}{N}\sum_{1\leq j\leq N}\beta_N\cdot\Lambda\cdot\Gamma_k(x-y_j^N)x_j^N, \tag{14.40}$$

and denote by

$$Y_{\epsilon_2}^N = \{x : |x-y_j^N| \geq \frac{1}{N^{1-\epsilon_2}} \text{ for all } 1\leq j\leq N\}$$

for some fixed constant $\epsilon_2 \in (0,\frac{1}{3})$. The reason for introducing the set $Y_{\epsilon_2}^N$ is that the convergence of the micro-field to the effective field does not hold near the bubbles because of the singularity of the Green function near the source point. However, it holds in the region away from the bubbles, which is characterized by $Y_{\epsilon_2}^N$.

LEMMA 14.32. *The following estimate holds uniformly for all $x \in Y_{\epsilon_2}^N$:*

$$|\widetilde{u}^N(x) - u^N(x)| \lesssim N^{\frac{\epsilon_0}{3}-\frac{\epsilon_1}{1-\epsilon_1}}.$$

PROOF. For each $x \in Y_{\epsilon_2}^N$, it is clear that

$$u^N(x) = u^i(x) + \sum_{1\leq j\leq N} u_j^{s,N}(x).$$

By Theorem 14.24, we have

$$
\begin{aligned}
u^N(x) &= u^i(x) \\
&\quad + \sum_{1\le j\le N} g\Gamma_k(x-y_j^N)\left(u_j^{i,N}(y_j^N) + O[N^{\frac{\epsilon_0}{3}-\frac{\epsilon_1}{1-\epsilon_1}} + N^{-\frac{\epsilon_1}{1-\epsilon_1}-\epsilon_2}]\cdot \max_{1\le l\le N}|u_j^{i,N}(y_j^N)|\right) \\
&= \widetilde{u}^N(x) + \sum_{1\le j\le N} g\Gamma_k(x-y_j^N)\cdot O[N^{\frac{\epsilon_0}{3}-\frac{\epsilon_1}{1-\epsilon_1}}]\cdot \max_{1\le l\le N}|u_j^{i,N}(y_j^N)| \\
&= \widetilde{u}^N(x) + \sum_{1\le j\le N} g\Gamma_k(x-y_j^N)\cdot O[N^{\frac{\epsilon_0}{3}-\frac{\epsilon_1}{1-\epsilon_1}}]\cdot \|u^i\|_X \\
&= \widetilde{u}^N(x) + \sum_{1\le j\le N} g\Gamma_k(x-y_j^N)\cdot O[N^{\frac{\epsilon_0}{3}-\frac{\epsilon_1}{1-\epsilon_1}}].
\end{aligned}
$$

On the other hand,

$$
\begin{aligned}
\sum_{1\le j\le N} |g\Gamma_k(x-y_j^N)| &= \frac{1}{N}\sum_{1\le j\le N} |\beta_N|\cdot\Lambda\cdot|\Gamma_k(x-y_j^N)| \\
&\lesssim \frac{1}{N}\cdot\sum_{1\le j\le N}\frac{1}{|x-y_j^N|} \\
&\lesssim \frac{1}{N}\max_{1\le j\le N}\frac{1}{|x-y_j^N|} + \frac{1}{N}\cdot\sum_{2r_N\le|x-y_j^N|}\frac{1}{|x-y_j^N|}\lesssim 1.
\end{aligned}
$$

Therefore,

$$
u^N(x) = \widetilde{u}^N(x) + N^{\frac{\epsilon_0}{3}-\frac{\epsilon_1}{1-\epsilon_1}}.
$$

This completes the proof of the Lemma. $\square$

Define

$$
w(x) = u^i(x) + \int_\Omega \Gamma_k(x-y)V(y)\psi(y)dy.
$$

LEMMA 14.33. *We have the following two results:*

(i) *For all $x\in Y_{\epsilon_2}^N$, the following estimate holds uniformly:*

$$
|\widetilde{u}^N(x) - w(x)| \lesssim N^{-\min\{\frac{1-\epsilon_0}{6},\frac{1-\epsilon_2}{3},\epsilon_2,\frac{\epsilon_0}{3}-\frac{\epsilon_1}{1-\epsilon_1}\}};
$$

(ii) *We have $w=\psi$.*

PROOF. For each $x\in Y_{\epsilon_2}^N$, choose $y_l^N\in\{y_j^N\}_{1\le j\le N}$ such that

$$
|x-y_l^N| = \min_{1\le j\le N}|x-y_j^N|.
$$

We have

$$\widetilde{u}^N(x) - w(x) = \frac{1}{N}\beta_N \cdot \Lambda \cdot \Gamma_k(x - y_l^N)x_l^N + \frac{1}{N}\sum_{j\neq l}\beta_N \cdot \Lambda \cdot \Gamma_k(x - y_j^N)x_j^N - \int_\Omega \Gamma_k(x-y)V(y)\psi(y)dy$$

$$= \left[\frac{1}{N}\sum_{j\neq l}\beta_N \cdot \Lambda \cdot \Gamma_k(y_l^N - y_j^N)\psi(y_j^N) - \int_\Omega \Gamma_k(y_l^N - y)V(y)\psi(y)dy\right]$$

$$+\frac{1}{N}\sum_{j\neq l}\beta_N \cdot \Lambda \cdot [\Gamma_k(x - y_j^N) - \Gamma_k(y_l^N - y_j^N)]\psi(y_j^N)$$

$$+\frac{1}{N}\sum_{j\neq l}\int_\Omega [\Gamma_k(x-y) - \Gamma_k(y_l^N - y)]V(y)\psi(y)dy$$

$$+\frac{1}{N}\sum_{j\neq l}\beta_N \cdot \Lambda \cdot \Gamma_k(x - y_j^N)(x_j^N - \psi(y_j^N)) + \frac{1}{N}\beta_N \cdot \Lambda \cdot \Gamma_k(x - y_l^N)x_l^N$$

$$=: e_1 + e_2 + e_3 + e_4 + e_5.$$

Let us now estimate e_j, $j = 1, \ldots, 5$ one by one.

First, by Assumption 14.17,

$$|e_1| \lesssim N^{-\frac{\eta}{3}} \cdot \|\psi\|_X \lesssim N^{-\frac{1-\epsilon_0}{6}}.$$

Second, we can show that

$$|e_2| \lesssim |x - y_l^N|^{1-\epsilon_2} \cdot \|\psi\|_X \lesssim N^{-\frac{1-\epsilon_2}{3}}\|\psi\|_X.$$

Third, by Lemma 14.27,

$$|e_3| \lesssim |x - y_l^N|^{1-\epsilon_2} \cdot \|\psi\|_X \lesssim N^{-\frac{1-\epsilon_2}{3}}\|\psi\|_X.$$

Fourth, note that

$$|e_4| \lesssim \frac{1}{N}\sum_{j\neq l}|\beta_N| \cdot \Lambda \cdot |\max_{1\leq j\leq N}|x_j^N - \psi(y_j^N)| \cdot \frac{1}{|x - y_j^N|}.$$

By Assumption 14.16 and Theorem 14.30, we have

$$|e_4| \lesssim \max_{1\leq j\leq N}|x_j^N - \psi(y_j^N)| \lesssim N^{-\min\{\frac{1-\epsilon_0}{6}, \frac{\epsilon_0}{3} - \frac{\epsilon_1}{1-\epsilon_1}\}}.$$

Finally, one can check that

$$|e_5| \lesssim \frac{1}{N} \cdot N^{1-\epsilon_2} \cdot \max_{1\leq j\leq N}\|x_j^N\| \lesssim N^{-\epsilon_2}.$$

Therefore,

$$\widetilde{u}^N(x) - w(x) = O(N^{-\min\{\frac{1-\epsilon_0}{6}, \frac{1-\epsilon_2}{3}, \epsilon_2, \frac{\epsilon_0}{3} - \frac{\epsilon_1}{1-\epsilon_1}\}}).$$

This complete the proof of statement (i).

We now turn to statement (ii). It is clear that w satisfies the equation

$$(\triangle + k^2)w = (\triangle + k^2)u^i + V\psi = V\psi.$$

Recall that

$$(\triangle + k^2)\psi - V\psi = (\triangle + k^2)u^i = 0.$$

Therefore, we have

$$(\triangle + k^2)(w - \psi) = 0.$$

On the other hand, it is easy to see that $w - \psi$ satisfies the radiation condition. The conclusion $w = \psi$ follows immediately. □

As a consequence of Lemma 14.33, we obtain the following theorem.

THEOREM 14.34. *Let $\omega < \omega_M$ and let V be defined by (14.37). Then under Assumptions 14.13–14.18, the solution to the scattering problem (14.19) converges to the solution to the wave equation*

$$(\triangle + k^2 - V)\psi = 0$$

together with the radiation condition imposed on $\psi - u^i$ at infinity, in the sense that for $x \in Y_{\epsilon_2}^N$, the following estimate holds uniformly:

$$|u^N(x) - \psi(x)| \lesssim N^{-\min\{\frac{1-\epsilon_0}{6}, \frac{1-\epsilon_2}{3}, \epsilon_2, \frac{\epsilon_0}{3} - \frac{\epsilon_1}{1-\epsilon_1}\}}.$$

The above theorem shows that under certain conditions, we can treat the bubbly fluid as an effective medium for acoustic wave propagation. Note that

$$\triangle + k^2 - V = \triangle + k^2(1 - \frac{1}{k^2}\beta \cdot \Lambda \cdot \widetilde{V}).$$

Thus, the effective medium can be characterized by the refractive index $1-\frac{1}{k^2}\beta\cdot\Lambda\cdot\widetilde{V}$. By our assumption, $k = O(1)$ and $\widetilde{V} = O(1)$. When $\beta \cdot \Lambda \gg 1$, we see that we have an effective high refractive index medium. As a consequence, this together with the main result in Chapter 9 gives a rigorous mathematical theory for the super-focusing experiment in [**350**].

Similarly, we have the following result for the case $\omega > \omega_M$.

THEOREM 14.35. *Let $\omega > \omega_M$ and assume that $V(x) > k^2$ almost everywhere in Ω. Then under Assumptions 14.13-14.18, the solution to the scattering problem (14.19) converges to the solution to the following dissipative equation*

$$(\triangle + k^2 - V)\psi = 0$$

together with the radiation condition imposed on $\psi - u^i$ at infinity, in the sense that for $x \in Y_{\epsilon_2}^N$, the following estimate holds uniformly:

$$|u^N(x) - \psi(x)| \lesssim N^{-\min\{\frac{1-\epsilon_0}{6}, \frac{1-\epsilon_2}{3}, \epsilon_2, \frac{\epsilon_0}{3} - \frac{\epsilon_1}{1-\epsilon_1}\}}.$$

Finally, we conclude this section with the following three important remarks.

REMARK 14.36. *At the resonant frequency $\omega = \omega_M$, the scattering coefficient g is of order one. Thus each bubble scatterer is a point source with magnitude one. As a consequence, the addition or removal of one bubble from the fluid affects the total field by a magnitude of the same order as the incident field. Therefore, we cannot expect any effective medium theory for the bubbly medium at this resonant frequency.*

REMARK 14.37. *The super-focusing (or equivalently super-resolution) theory, developed in this chapter for bubbly fluid seems to be different from the one developed for Helmholtz resonators and plasmonic nanoparticles. However, they are closely related. In Chapters* 9 *and* 13, *it is shown that super-focusing (or super-resolution) is due to sub-wavelength propagating resonant modes which are generated by the sub-wavelength resonators embedded in the background homogeneous medium. In those two cases, the region with sub-wavelength resonators has size smaller or much smaller than the incident wavelength, and the number of sub-wavelength resonators is not very large, and hence neither is the number of sub-wavelength resonant modes. As a result, an effective medium theory is not necessary or even true. However, in the case of bubbles in a fluid as considered in this chapter, the region with bubbles has size comparable to or greater than the incident wavelength. This together with the fact that the ratio between the size of the individual bubble and the incident wavelength near the Minnaert resonant frequency is extremely small, indicates that the number of bubbles can be very large, as in the experiment in* [**350**], *even though they are dilute. This large number of bubbles generates a large number of resonant modes which eventually yield a continuum limit in the form of an effective medium with high refractive index. In fact, these resonant modes can be obtained from the point interaction system (*14.36*). On the other hand, it is shown in Chapter* 9 *that super-focusing (or super-resolution) is possible in high refractive index media. In this regard, the effective medium theory developed in this chapter can be viewed as a bridge between the super-focusing (or super-resolution) theories in Chapters* 9 *and* 13.

REMARK 14.38. *In this section, we derived an effective medium theory for the case $\omega < \omega_M$ and a special case of $\omega > \omega_M$ with some additional assumptions. However, our results still hold for the case $\omega > \omega_M$ without any additional assumption, if we assume that the limiting system $I - \mathcal{T}$ has a trivial kernel. This assumption implies that the limiting system is well-posed.*

14.4. Sub-wavelength Phononic Bandgap Opening

In this section we investigate whether there is a possibility of sub-wavelength bandgap opening in bubble phononic crystals. We first formulate the spectral problem for a bubble phononic crystal. Then we derive an asymptotic formula for the quasi-periodic Minnaert resonances in terms of the contrast between the densities of the air inside the bubbles and the fluid outside the bubbles. We prove the existence of a sub-wavelength bandgap and estimate its width. We also consider the dilute regime where the volume fraction of the bubbles is small.

14.4.1. Problem Formulation. In this subsection, we first describe the bubble phononic crystal under consideration. Assume that the bubbles occupy $\cup_{n\in\mathbb{Z}^d}(D+n)$ for a bounded and simply connected domain D with $\partial D \in \mathcal{C}^{1,\eta}$ with $0 < \eta < 1$. As before, we denote by ρ_b and κ_b the density and the bulk modulus of the air inside the bubbles, respectively, and by ρ and κ the corresponding parameters for the background media and let v, v_b, k, and k_b be defined by (14.2). We also let the two dimensionless contrast parameters δ and τ be defined by (14.3).

To investigate the phononic gap of the bubble phononic crystal we consider the following $\alpha-$quasi-periodic equation in the unit cell $Y = [-1/2, 1/2]^d$ for $d = 2, 3$:

$$
(14.41) \qquad \begin{cases} \nabla \cdot \dfrac{1}{\rho} \nabla u + \dfrac{\omega^2}{\kappa} u = 0 \quad \text{in} \quad Y \backslash \overline{D}, \\ \nabla \cdot \dfrac{1}{\rho_b} \nabla u + \dfrac{\omega^2}{\kappa_b} u = 0 \quad \text{in} \quad D, \\ u_+ - u_- = 0 \quad \text{on} \quad \partial D, \\ \dfrac{1}{\rho} \dfrac{\partial u}{\partial \nu}\Big|_+ - \dfrac{1}{\rho_b} \dfrac{\partial u}{\partial \nu}\Big|_- = 0 \quad \text{on} \quad \partial D, \\ e^{-\sqrt{-1}\alpha \cdot x} u \text{ is periodic.} \end{cases}
$$

By choosing proper physical units, we may assume that the bubble size is of order one. We assume that the wave speeds outside and inside the bubbles are comparable to each other and that there is a large contrast in the bulk modulus, that is, condition (14.4) holds.

From Chapter 5, it is known that (14.41) has nontrivial solutions for discrete values of ω such as

$$0 \le \omega_1^\alpha \le \omega_2^\alpha \le \cdots$$

and we have the following band structure of propagating frequencies for the given periodic structure:

$$[0, \max_\alpha \omega_1^\alpha] \cup [\min_\alpha \omega_2^\alpha, \max_\alpha \omega_2^\alpha] \cup [\min_\alpha \omega_3^\alpha, \max_\alpha \omega_3^\alpha] \cup \cdots .$$

14.4.2. Sub-wavelength Bandgaps. We use the quasi-periodic single-layer potential introduced in Section 2.12 to represent the solution to the scattering problem (14.41) in $Y \setminus \overline{D}$. We look for a solution u of (14.41) of the form:

$$
(14.42) \qquad u = \begin{cases} \mathcal{S}_D^{\alpha,k}[\psi] & \text{in } Y \setminus \overline{D}, \\ \mathcal{S}_D^{k_b}[\psi_b] & \text{in } D, \end{cases}
$$

for some surface potentials $\psi, \psi_b \in L^2(\partial D)$. Using the jump relations for the single-layer potentials, one can show that (14.41) is equivalent to the boundary integral equation

$$
(14.43) \qquad \mathcal{A}(\omega, \delta)[\Psi] = 0,
$$

where

$$
\mathcal{A}(\omega, \delta) = \begin{pmatrix} \mathcal{S}_D^{k_b} & -\mathcal{S}_D^{\alpha,k} \\ -\frac{1}{2} + (\mathcal{K}_D^{k_b})^* & -\delta(\frac{1}{2} + (\mathcal{K}_D^{-\alpha,k})^*) \end{pmatrix}, \quad \Psi = \begin{pmatrix} \psi_b \\ \psi \end{pmatrix}.
$$

Again, we denote by $\mathcal{H} = L^2(\partial D) \times L^2(\partial D)$ and by $\mathcal{H}_1 = H^1(\partial D) \times L^2(\partial D)$. It is clear that $\mathcal{A}(\omega, \delta)$ is a bounded linear operator from $\mathcal{H}$ to $\mathcal{H}_1$, *i.e.* $\mathcal{A}(\omega, \delta) \in \mathcal{L}(\mathcal{H}, \mathcal{H}_1)$. Moreover, we can check that the characteristic values of $\mathcal{A}(\omega, \delta)$ can be written as

$$0 \le \omega_1^\alpha \le \omega_2^\alpha \le \cdots .$$

We first look at the limiting case when $\delta = 0$. The operator $\mathcal{A}(\omega, \delta)$ is a perturbation of

$$
(14.44) \qquad \mathcal{A}(\omega, 0) = \begin{pmatrix} \mathcal{S}_D^{k_b} & -\mathcal{S}_D^{\alpha,k} \\ -\frac{1}{2} + (\mathcal{K}_D^{k_b})^* & 0 \end{pmatrix}.
$$

We see that ω_0 is a characteristic value of $\mathcal{A}(\omega,0)$ if only if $(\omega_0 v_b^{-1})^2$ is a Neumann eigenvalue of D or $(\omega_0 v^{-1})^2$ is a Dirichlet eigenvalue of $Y\backslash D$ with α-quasi-periodicity on ∂Y. Since zero is a Neumann eigenvalue of D, $\omega = 0$ is a characteristic value for the holomorphic operator-valued function $\mathcal{A}(\omega,0)$. By noting that there is a positive lower bound for the other Neumann eigenvalues of D and all the Dirichlet eigenvalues of $Y\backslash D$ with α-quasi-periodicity on ∂Y, we can conclude the following result by the Gohberg-Sigal theory.

LEMMA 14.39. *For any δ sufficiently small, there exists one and only one characteristic value $\omega_0 = \omega_0(\delta)$ in a neighborhood of the origin in the complex plane to the holomorphic operator-valued function $\mathcal{A}(\omega,\delta)$. Moreover, $\omega_0(0) = 0$ and ω_0 depends on δ continuously.*

14.4.2.1. *The Asymptotic Behavior of ω_1^α.* In this section we assume $\alpha \neq 0$. We define

$$\mathcal{A}_0 := \mathcal{A}(0,0) = \begin{pmatrix} \mathcal{S}_D^0 & -\mathcal{S}_D^{\alpha,0} \\ -\frac{1}{2} + (\mathcal{K}_D^0)^* & 0 \end{pmatrix}, \tag{14.45}$$

Let $\mathcal{A}_0^* : \mathcal{H}_1 \to \mathcal{H}$ be the adjoint of $\mathcal{A}_0$. We choose an element $\psi_0 \in L^2(\partial D)$ such that

$$\Big(-\frac{1}{2}I + (\mathcal{K}_D^0)^*\Big)[\psi_0] = 0, \quad \int_{\partial D} \psi_0 = 1.$$

We recall the definition (14.10) of the capacity of the set D, $\mathrm{cap}(\partial D)$, which is equivalent to

$$\mathcal{S}_D^0[\psi_0] = -\mathrm{cap}(\partial D)^{-1} \quad \text{on } \partial D. \tag{14.46}$$

Then we can easily check that $\mathrm{Ker}(\mathcal{A}_0)$ and $\mathrm{Ker}(\mathcal{A}_0^*)$ are spanned respectively by

$$\Psi_0 = \begin{pmatrix} \psi_0 \\ \widetilde{\psi}_0 \end{pmatrix} \quad \text{and} \quad \Phi_0 = \begin{pmatrix} 0 \\ 1 \end{pmatrix},$$

where $\widetilde{\psi}_0 = (\mathcal{S}_D^{\alpha,0})^{-1}\mathcal{S}_D^0[\psi_0]$.

We now perturb $\mathcal{A}_0$ by a rank-one operator $\mathcal{P}_0$ from $\mathcal{H}$ to $\mathcal{H}_1$ given by $\mathcal{P}_0[\Psi] := (\Psi,\Psi_0)\Phi_0$, and denote it by $\widetilde{\mathcal{A}_0} = \mathcal{A}_0 + \mathcal{P}_0$.

LEMMA 14.40. *The followings hold:*

(i) $\widetilde{\mathcal{A}_0}[\Psi_0] = \|\Psi_0\|^2\Phi_0$, $\widetilde{\mathcal{A}_0}^*[\Phi_0] = \|\Phi_0\|^2\Psi_0$.

(ii) *The operator $\widetilde{\mathcal{A}_0}$ and its adjoint $\widetilde{\mathcal{A}_0}^*$ are invertible in $\mathcal{L}(\mathcal{H},\mathcal{H}_1)$ and $\mathcal{L}(\mathcal{H}_1,\mathcal{H})$, respectively.*

PROOF. By construction, and the fact that $\mathcal{S}_D^0$ is bijective from $L^2(\partial D)$ to $H^1(\partial D)$ (see Section 2.4), we can show that $\widetilde{\mathcal{A}_0}$ (hence $\widetilde{\mathcal{A}_0}^*$) is bijective. The fact that

$$\widetilde{\mathcal{A}_0}[\Psi_0] = \|\Psi_0\|^2\Phi_0$$

is direct. Finally, by noticing that $\mathcal{P}_0^*[\theta] = (\theta,\Phi_0)\Psi_0$, it follows that

$$\widetilde{\mathcal{A}_0}^*[\Phi_0] = \mathcal{P}_0^*[\Phi_0] = \|\Phi_0\|^2\Psi_0.$$

□

Using the results in Subsection 2.8.3, we can expand $\mathcal{A}(\omega,\delta)$ as

(14.47)
$$\mathcal{A}(\omega,\delta) := \mathcal{A}_0 + \mathcal{B}(\omega,\delta) = \mathcal{A}_0 + \omega\mathcal{A}_{1,0} + \omega^2\mathcal{A}_{2,0} + \omega^3\mathcal{A}_{3,0} + \delta\mathcal{A}_{0,1} + \delta\omega^2\mathcal{A}_{2,1} + O(|\omega|^4 + |\delta\omega^3|)$$

where

$$\mathcal{A}_{1,0} = \begin{pmatrix} v_b^{-1}\mathcal{S}_{D,1} & 0 \\ 0 & 0 \end{pmatrix}, \quad \mathcal{A}_{2,0} = \begin{pmatrix} v_b^{-2}\mathcal{S}_{D,2} & -v^{-2}\mathcal{S}_{D,1}^{\alpha} \\ v_b^{-2}\mathcal{K}_{D,2} & 0 \end{pmatrix}, \quad \mathcal{A}_{3,0} = \begin{pmatrix} v_b^{-3}\mathcal{S}_{D,3} & 0 \\ v_b^{-3}\mathcal{K}_{D,3} & 0 \end{pmatrix},$$

$$\mathcal{A}_{0,1} = \begin{pmatrix} 0 & 0 \\ 0 & -(\frac{1}{2} + (\mathcal{K}_D^{-\alpha,0})^*) \end{pmatrix}, \quad \mathcal{A}_{2,1} = \begin{pmatrix} 0 & 0 \\ 0 & -v^{-2}\mathcal{K}_{D,1}^{\alpha} \end{pmatrix}.$$

Since $\widetilde{\mathcal{A}_0} = \mathcal{A}_0 + \mathcal{P}_0$, the equation (14.43) is equivalent to

$$(\widetilde{\mathcal{A}_0} - \mathcal{P}_0 + \mathcal{B})[\Psi_0 + \Psi_1] = 0,$$

where

$$(\Psi_1, \Psi_0) = 0$$

Observe that the operator $\widetilde{\mathcal{A}_0} + \mathcal{B}$ is invertible for sufficiently small δ and ω. Applying $(\widetilde{\mathcal{A}_0} + \mathcal{B})^{-1}$ to both sides of the above equation leads to

$$\Psi_1 = (\widetilde{\mathcal{A}_0} + \mathcal{B})^{-1}\mathcal{P}_0[\Psi_0] - \Psi_0 = \|\Psi_0\|^2 (\widetilde{\mathcal{A}_0} + \mathcal{B})^{-1}[\Phi_0] - \Psi_0. \tag{14.48}$$

Using the condition $(\Psi_1, \Psi_0) = 0$, we deduce that (14.43) has a nontrivial solution if and only if

$$\widetilde{A}(\omega,\delta) := \|\Psi_0\|^2 \left(\langle(\widetilde{\mathcal{A}_0} + \mathcal{B})^{-1}[\Phi_0], \Psi_0\rangle - 1\right) = 0. \tag{14.49}$$

Let us calculate $A(\omega,\delta) := \widetilde{A}(\omega,\delta)\,\|\Phi_0\|^2$. Using the Neumann series

$$(\widetilde{\mathcal{A}_0} + \mathcal{B})^{-1} = \left(1 + \widetilde{\mathcal{A}_0}^{-1}\mathcal{B}\right)^{-1}\widetilde{\mathcal{A}_0}^{-1} = \left(1 - \widetilde{\mathcal{A}_0}^{-1}\mathcal{B} + \widetilde{\mathcal{A}_0}^{-1}\mathcal{B}\widetilde{\mathcal{A}_0}^{-1}\mathcal{B} - \ldots\right)\widetilde{\mathcal{A}_0}^{-1},$$

and the fact that $\widetilde{\mathcal{A}_0}^{-1}[\Phi_0] = \|\Psi_0\|^{-2}\Psi_0$ and $(\widetilde{\mathcal{A}_0}^{*})^{-1}[\Psi_0] = \|\Phi_0\|^{-2}\Phi_0$, we obtain that

$$\begin{aligned} A(\omega,\delta) = &-\omega\langle\mathcal{A}_{1,0}[\Psi_0],\Phi_0\rangle - \omega^2\langle\mathcal{A}_{2,0}[\Psi_0],\Phi_0\rangle - \omega^3\langle\mathcal{A}_{3,0}[\Psi_0],\Phi_0\rangle \\ &-\delta\langle\mathcal{A}_{0,1}[\Psi_0],\Phi_0\rangle + \omega^2\left\langle\mathcal{A}_{1,0}\widetilde{\mathcal{A}_0}^{-1}\mathcal{A}_{1,0}[\Psi_0],\Phi_0\right\rangle \\ &+\omega^3\left\langle\mathcal{A}_{1,0}\widetilde{\mathcal{A}_0}^{-1}\mathcal{A}_{2,0}[\Psi_0],\Phi_0\right\rangle + \omega^3\left\langle\mathcal{A}_{2,0}\widetilde{\mathcal{A}_0}^{-1}\mathcal{A}_{1,0}[\Psi_0],\Phi_0\right\rangle \\ &+\omega\delta\left\langle\mathcal{A}_{1,0}\widetilde{\mathcal{A}_0}^{-1}\mathcal{A}_{0,1}[\Psi_0],\Phi_0\right\rangle + \omega\delta\left\langle\mathcal{A}_{0,1}\widetilde{\mathcal{A}_0}^{-1}\mathcal{A}_{1,0}[\Psi_0],\Phi_0\right\rangle \\ &+\omega^3\left\langle\mathcal{A}_{1,0}\widetilde{\mathcal{A}_0}^{-1}\mathcal{A}_{1,0}\widetilde{\mathcal{A}_0}^{-1}\mathcal{A}_{1,0}[\Psi_0],\Phi_0\right\rangle + O(|\omega|^4 + |\delta|\,|\omega|^2 + |\delta|^2). \end{aligned}$$

It is clear that $\mathcal{A}_{1,0}^*[\Phi_0] = 0$. Consequently, the expression simplifies into

(14.50)
$$\begin{aligned} A(\omega,\delta) = &-\omega^2\langle\mathcal{A}_{2,0}[\Psi_0],\Phi_0\rangle - \omega^3\langle\mathcal{A}_{3,0}[\Psi_0],\Phi_0\rangle + \omega^3\left\langle\mathcal{A}_{2,0}\widetilde{\mathcal{A}_0}^{-1}\mathcal{A}_{1,0}[\Psi_0],\Phi_0\right\rangle \\ &-\delta\langle\mathcal{A}_{0,1}[\Psi_0],\Phi_0\rangle + \omega\delta\left\langle\mathcal{A}_{0,1}\widetilde{\mathcal{A}_0}^{-1}\mathcal{A}_{1,0}[\Psi_0],\Phi_0\right\rangle \\ &+ O(|\omega|^4 + |\delta|\,|\omega|^2 + |\delta|^2). \end{aligned}$$

We now calculate the five remaining terms.

• Calculation of $\langle \mathcal{A}_{2,0}[\Psi_0], \Phi_0\rangle$. Using Lemma 2.91, we get

$$\begin{aligned}\langle \mathcal{A}_{2,0}[\Psi_0], \Phi_0\rangle &= v_b^{-2}\langle \mathcal{K}_{D,2}[\psi_0], 1\rangle = v_b^{-2}\langle \psi_0, \mathcal{K}_{D,2}^*[1]\rangle \\ &= -v_b^{-2}\int_{\partial D}\psi_0(x)\int_D \Gamma_0(x-y)\mathrm{d}y\mathrm{d}\sigma(x) \\ &= -v_b^{-2}\int_D \mathcal{S}_D^0[\psi_0](x)\mathrm{d}x = \frac{v_b^{-2}|D|}{\mathrm{cap}(\partial D)},\end{aligned}$$

where we used the fact that $\mathcal{S}_D^0[\psi_0](x) = -\mathrm{cap}(\partial D)^{-1}$ for all $x \in D$.

• Calculation of $\langle \mathcal{A}_{3,0}[\Psi_0], \Phi_0\rangle$. Similarly, using Lemma 2.91, we get

$$\langle \mathcal{A}_{3,0}[\Psi_0], \Phi_0\rangle = v_b^{-3}\langle \psi_0, \mathcal{K}_{D,3}^*[1]\rangle = v_b^{-3}\left\langle \psi_0, \frac{\sqrt{-1}|D|}{4\pi}\right\rangle = \frac{\sqrt{-1}v_b^{-3}|D|}{4\pi}.$$

• Calculation of $\langle \mathcal{A}_{0,1}[\Psi_0], \Phi_0\rangle$. We directly have

$$\langle \mathcal{A}_{0,1}[\Psi_0], \Phi_0\rangle = -\left\langle \widetilde{\psi}_0, (1/2 + \mathcal{K}_D^{-\alpha,0})[1]\right\rangle.$$

• Calculation of $\left\langle \mathcal{A}_{0,1}\widetilde{\mathcal{A}_0}^{-1}\mathcal{A}_{1,0}[\Psi_0], \Phi_0\right\rangle$. We have

$$\begin{aligned}\mathcal{A}_{1,0}[\Psi_0] &= v_b^{-1}\begin{pmatrix}\mathcal{S}_{D,1}[\psi_0]\\ 0\end{pmatrix} = v_b^{-1}\frac{-\sqrt{-1}}{4\pi}\begin{pmatrix}1\\0\end{pmatrix}, \\ \mathcal{A}_{0,1}^*[\Phi_0] &= \begin{pmatrix}0\\ -(\frac{1}{2}+\mathcal{K}_D^{-\alpha,0})[1]\end{pmatrix}.\end{aligned}$$

Let us calculate $\widetilde{\mathcal{A}_0}^{-1}\begin{pmatrix}1\\0\end{pmatrix}$. We look for $\langle a\psi_0, b\widetilde{\psi}_0\rangle \in \mathcal{H}$ so that

$$\begin{pmatrix}1\\0\end{pmatrix} = (\mathcal{A}_0 + \mathcal{P}_0)\begin{pmatrix}a\psi_0\\ b\widetilde{\psi}_0\end{pmatrix} = \begin{pmatrix}(a-b)\mathcal{S}_D^0[\psi_0]\\ 0\end{pmatrix} + (a\|\psi_0\|^2 + b\|\widetilde{\psi}_0\|^2)\begin{pmatrix}0\\1\end{pmatrix}.$$

By solving the above equations directly, we obtain

$$\widetilde{\mathcal{A}_0}^{-1}\begin{pmatrix}1\\0\end{pmatrix} = \frac{\mathrm{cap}(\partial D)}{\|\psi_0\|^2 + \|\widetilde{\psi}_0\|^2}\begin{pmatrix}-\|\widetilde{\psi}_0\|^2\psi_0\\ \|\psi_0\|^2\widetilde{\psi}_0\end{pmatrix}. \tag{14.51}$$

It follows that

$$\left\langle \mathcal{A}_{0,1}\widetilde{\mathcal{A}_0}^{-1}\mathcal{A}_{1,0}[\Psi_0], \Phi_0\right\rangle = \frac{\sqrt{-1}v_b^{-1}\mathrm{cap}(\partial D)\|\psi_0\|^2\left\langle \widetilde{\psi}_0, (1/2 + \mathcal{K}_D^{-\alpha,0})[1]\right\rangle}{4\pi(\|\psi_0\|^2 + \|\widetilde{\psi}_0\|^2)}.$$

• Calculation of $\left\langle \mathcal{A}_{2,0}\widetilde{\mathcal{A}_0}^{-1}\mathcal{A}_{1,0}[\Psi_0], \Phi_0\right\rangle$. Using similar calculations, we obtain

$$\begin{aligned}\left\langle \mathcal{A}_{2,0}\widetilde{\mathcal{A}_0}^{-1}\mathcal{A}_{1,0}[\Psi_0], \Phi_0\right\rangle &= \left\langle \widetilde{\mathcal{A}_0}^{-1}\mathcal{A}_{1,0}[\Psi_0], \mathcal{A}_{2,0}^*[\Phi_0]\right\rangle \\ &= \frac{\sqrt{-1}v_b^{-3}\mathrm{cap}(\partial D)\|\widetilde{\psi}_0\|^2}{4\pi(\|\psi_0\|^2 + \|\widetilde{\psi}_0\|^2)}\langle \psi_0, \mathcal{K}_{D,2}^*[1]\rangle \\ &= \frac{\sqrt{-1}v_b^{-3}|D|\|\widetilde{\psi}_0\|^2}{4\pi(\|\psi_0\|^2 + \|\widetilde{\psi}_0\|^2)}.\end{aligned}$$

Considering the above results, we can derive from (14.50) that

(14.52)
$$\begin{aligned} A(\omega,\delta) &= -\omega^2\frac{v_b^{-2}|D|}{\mathrm{cap}(\partial D)} - \omega^3 v_b^{-3}\frac{\sqrt{-1}c_1|D|}{4\pi} + c_2\delta + \omega\delta\frac{\sqrt{-1}c_1c_2v_b^{-1}\mathrm{cap}(\partial D)}{4\pi} \\ &\quad +O(|\omega|^4 + |\delta|\,|\omega|^2 + |\delta|^2), \end{aligned}$$

where

(14.53)
$$c_1 := \frac{\|\psi_0\|^2}{\|\psi_0\|^2 + \|\widetilde{\psi}_0\|^2},$$

and

(14.54)
$$c_2 := \left\langle \widetilde{\psi}_0, \left(1/2 + \mathcal{K}_D^{-\alpha,0}\right)[1] \right\rangle.$$

We now solve $A(\omega,\delta) = 0$. It is clear that $\delta = O(\omega^2)$ and thus $\omega_0(\delta) = O(\sqrt{\delta})$. We write

$$\omega_0(\delta) = a_1\delta^{\frac{1}{2}} + a_2\delta + O(\delta^{\frac{3}{2}}),$$

and get

$$\begin{aligned} &-\frac{v_b^{-2}|D|}{\mathrm{cap}(\partial D)}\left(a_1\delta^{\frac{1}{2}} + a_2\delta + O(\delta^{\frac{3}{2}})\right)^2 - \frac{\sqrt{-1}c_1v_b^{-3}|D|}{4\pi}\left(a_1\delta^{\frac{1}{2}} + a_2\delta + O(\delta^{\frac{3}{2}})\right)^3 \\ &\quad + c_2\delta + \frac{\sqrt{-1}c_1c_2v_b^{-1}\mathrm{cap}(\partial D)}{4\pi}\left(a_1\delta^{\frac{3}{2}} + a_2\delta^2 + O(\delta^{\frac{5}{2}})\right) + O(\delta^2) = 0. \end{aligned}$$

From the coefficients of the δ and $\delta^{\frac{3}{2}}$ terms, we obtain

$$-a_1^2\frac{v_b^{-2}|D|}{\mathrm{cap}(\partial D)} + c_2 = 0$$

and

$$2a_1a_2\frac{-v_b^{-2}|D|}{\mathrm{cap}(\partial D)} - a_1^3\frac{\sqrt{-1}c_1v_b^{-3}|D|}{4\pi} + a_1\frac{\sqrt{-1}c_1c_2v_b^{-1}\mathrm{cap}(\partial D)}{4\pi} = 0,$$

which yields

$$a_1 = \pm\sqrt{\frac{c_2\mathrm{cap}(\partial D)}{|D|}}v_b \quad \text{and} \quad a_2 = 0.$$

Therefore, we obtain

THEOREM 14.41. *For $\alpha \neq 0$ and sufficiently small δ, we have*

(14.55)
$$\omega_1^\alpha = \omega_M\sqrt{c_2} + O(\delta^{3/2}),$$

where $\omega_M = \sqrt{\dfrac{\delta\mathrm{cap}(\partial D)}{|D|}}v_b$ is the (free space) Minnaert resonant frequency.

Let us define the α-quasi-periodic capacity by

(14.56)
$$\mathrm{cap}(\partial D,\alpha) := -\left\langle (\mathcal{S}_D^{\alpha,0})^{-1}[1], 1\right\rangle.$$

Then we have

$$\begin{aligned} c_2 &= -\frac{1}{\mathrm{cap}(\partial D)}\left\langle \left(1/2 + (\mathcal{K}_D^{-\alpha,0})^*\right)(\mathcal{S}_D^{\alpha,0})^{-1}[1], 1\right\rangle \\ &= -\frac{1}{\mathrm{cap}(\partial D)}\left\langle (\mathcal{S}_D^{\alpha,0})^{-1}[1], 1\right\rangle = \frac{\mathrm{cap}(\partial D,\alpha)}{\mathrm{cap}(\partial D)}, \end{aligned}$$

and (14.55) is written as

$$\omega_1^\alpha = \omega_{M,\alpha} + O(\delta^{3/2})$$

with

$$\omega_{M,\alpha} = \sqrt{\frac{\delta \mathrm{cap}(\partial D, \alpha)}{|D|}} v_b. \tag{14.57}$$

REMARK 14.42. *Note that formula (14.55) holds in the two-dimensional case, where ω_M is the (free space) Minnaert resonant frequency and c_2 is defined by (14.54). It is worth emphasizing that for $\alpha \neq 0$, the quasi-periodic single-layer operator $\mathcal{S}_D^{\alpha,0} : L^2(\partial D) \to H^1(\partial D)$ is invertible. Moreover, the definitions (14.46) and (14.56) of both the capacity and the α-quasi-periodic capacity remain valid; see* [**41**].

Now from (14.57), we can see that

$$\omega_{M,\alpha} \to 0$$

as $\alpha \to 0$ because $\big(1/2+(\mathcal{K}_D^{-\alpha,0})^*\big)(\mathcal{S}_D^{\alpha,0})^{-1}[1] \to 0$ and so $\mathrm{cap}(\partial D, \alpha) \to 0$ as $\alpha \to 0$. Moreover, due to our assumptions on the bubble size and the wave speeds inside and outside the bubbles, it is easy to see that $\omega_{M,\alpha}$ lies in a small neighborhood of zero.

We define $\omega_1^* := \max_\alpha \omega_{M,\alpha}$. Then we deduce the following regarding a sub-wavelength bandgap opening.

THEOREM 14.43. *For every $\epsilon > 0$, there exists $\delta_0 > 0$ and $\widetilde{\omega} > \omega_1^* + \epsilon$ such that*

$$[\omega_1^* + \epsilon, \widetilde{\omega}] \subset [\max_\alpha \omega_1^\alpha, \min_\alpha \omega_2^\alpha] \tag{14.58}$$

for $\delta < \delta_0$.

PROOF. Using $\omega_1^0 = 0$ and the continuity of ω_1^α in α and δ, we get α_0 and δ_1 such that $\omega_1^\alpha < \omega_1^*$ for every $|\alpha| < \alpha_0$ and $\delta < \delta_1$. Following the derivation of (14.55), we can check that it is valid uniformly in α as far as $|\alpha| \geq \alpha_0$. Thus there exists $\delta_0 < \delta_1$ such $\omega_1^\alpha \leq \omega_1^* + \epsilon$ for $|\alpha| \geq \alpha_0$. We have shown that $\max_\alpha \omega_1^\alpha \leq \omega_1^* + \epsilon$ for sufficiently small δ. To have $\min_\alpha \omega_2^\alpha > \omega_1^* + \epsilon$ for small δ, it is enough to check that $\mathcal{A}(\omega, \delta)$ has no small characteristic value other than ω_1^α. For α away from 0, we can see that it is true following the proof of Theorem 14.41. If $\alpha = 0$, we have

$$\mathcal{A}(\omega, \delta) = \mathcal{A}(\omega, 0) + O(\delta), \tag{14.59}$$

near ω_2^0 with $\delta = 0$. Since $\omega_2^0 \neq 0$, we have $\omega_2^0(\delta) > \omega_1^* + \epsilon$ for sufficiently small δ. Finally, using the continuity of ω_2^α in α, we obtain $\min_\alpha \omega_2^\alpha > \omega_1^* + \epsilon$ for small δ. This completes the proof. □

As shown in [**77**], the first Bloch eigenvalue ω_1^α attains its maximum ω_1^* at the corner M of the Brillouin zone.

THEOREM 14.44. *Assume that D is symmetric with respect to $\{x_j = 0\}$ for $j = 1, \ldots, d$. Then both* $\mathrm{cap}(\partial D, \alpha)$ *and ω_1^α attain their maxima at $\alpha^* = (\pi, \ldots, \pi)$.*

In [**77**], an asymptotic expansion of the Bloch eigenfunctions near the critical frequency ω_1^* has been also computed. It is proved that these eigenfunctions can be decomposed into two parts: one is slowly varying and satisfies a homogenized equation, while the other is periodic across each elementary cell and is varying at the microscopic scale.

14.4.2.2. *Dilute Case.* We emphasize that our calculations in the previous subsection hold even for the dilute case as long as δ/s^2 is small where s is the characteristic size of D.

We state an asymptotic behavior of $\mathrm{cap}(\partial D, \alpha)$ when $D = sB$ for a small $s > 0$. Note that $\mathrm{cap}(\partial D) = \mathrm{cap}(\partial(sB)) = s\, \mathrm{cap}(\partial B)$. Fix $c > 0$, the following holds.

LEMMA 14.45. *For $|\alpha| > c > 0$, we have*

$$\mathrm{cap}(\partial D, \alpha) = \mathrm{cap}(\partial D) - R_\alpha(0)\mathrm{cap}(\partial D)^2 + O(s^3), \tag{14.60}$$

where $R_\alpha(x) := G_\alpha(x) - \Gamma_0(x)$, where G_α is the quasi-periodic Green's function associated to the Laplacian defined by (2.134).

PROOF. Since $R_\alpha(x)$ is smooth and $R_\alpha(x) = R_\alpha(0) + O(|x|)$ as $|x| \to 0$, we have

$$\begin{aligned}
\mathcal{S}_D^{\alpha,0}[\phi](sx) &= s\int_{\partial B} \Gamma_0(x-y)\widetilde{\phi}(y)d\sigma(y) + s^2 R_\alpha(0)\int_{\partial B}\widetilde{\phi}(y)d\sigma(y) + O\left(s^3\|\widetilde{\phi}\|\right)\\
&= s\left(\mathcal{S}_B^0[\widetilde{\phi}] + sR_\alpha(0)\int_{\partial B}\widetilde{\phi} + O\left(s^2\|\widetilde{\phi}\|\right)\right),
\end{aligned}$$

with $\widetilde{\phi}(x) := \phi(sx)$. Then

$$\begin{aligned}
(\mathcal{S}_D^{\alpha,0})^{-1}[1](sx) &= s^{-1}\Bigg((\mathcal{S}_B^0)^{-1}[1] - sR_\alpha(0)(\mathcal{S}_B^0)^{-1}\left[\int_{\partial B}(\mathcal{S}_B^0)^{-1}[1]\right]\\
&\qquad + O\left(s^2\right)\Bigg),
\end{aligned}$$

and so

$$\begin{aligned}
\mathrm{cap}(\partial D, \alpha) &= -s^2\int_{\partial B}(\mathcal{S}_D^{\alpha,0})^{-1}[1](sx)\, d\sigma(x)\\
&= s\left(\mathrm{cap}(\partial B) - sR_\alpha(0)\mathrm{cap}(\partial B)^2 + O\left(s^2\right)\right)\\
&= \mathrm{cap}(\partial D) - R_\alpha(0)\mathrm{cap}(\partial D)^2 + O(s^3).
\end{aligned}$$

□

By this approximation we have $\omega_{M,\alpha} \approx \omega_M$ for α away from 0 and so $\omega_1^* \approx \omega_M$. Combined with Theorem 14.43 this means that there is a band gap opening slightly above the Minneart resonance frequency for a single bubble. It is coherent with results in Section 14.3 showing an effective medium theory for the bubbly fluid as the number of bubbles tends to infinity. It is shown there that just above the Minnaert resonant frequency, the obtained effective media can have a negative bulk modulus.

It is worth emphasizing that in the nondilute regime, the super-focusing property and the bandgap opening at sub-wavelength scales near the critical frequency ω_1^* hold true. Nevertheless, the mechanism is not due to effective (high-contrast below ω_1^* and negative above ω_1^*) properties of the medium. The classical homogenization theory is no longer valid in the nondilute case. A high-frequency homogenization approach for sub-wavelength resonators has been developed in [**77**]. As mentioned before, the Bloch eigenfunctions can be decomposed into two parts: one is slowly varying and satisfies a homogenized equation, while the other is periodic across each elementary cell and is varying at the microscopic scale. The microscopic oscillations of the second part justify the super-focusing phenomenon, while the exponential decay of the slowly varying part proves the bandgap opening above the critical frequency ω_1^*.

14.5. Double-Negative Refractive Index Phenomenon

In this section, we show that, using bubble dimers, the effective mass density and bulk modulus of the bubbly fluid can both be negative over a non empty range of frequencies. A bubble dimer is a system that consists of two identical bubbles, slightly separated from one another. It can be approximated as a point scatterer with monopole and dipole modes. It features two slightly different sub-wavelength resonances, called the hybridized Minnaert resonances. The hybridized Minnaert resonances are fundamentally different modes. The first mode is, as in the case of a single bubble, a monopole mode, while the second mode is a dipole mode. The resonance associated with the dipole mode is usually referred to as the anti-resonance.

For an appropriate volume fraction, when the excitation frequency is close to the anti-resonance, a double-negative effective mass density and bulk modulus for bubbly media consisting of a large number of bubble dimers with certain conditions on their distribution can be obtained. The dipole modes in the background medium contribute to the effective mass density while the monopole modes contribute to the effective bulk modulus.

The results of this section are from [**42**].

14.5.1. Resonance for a Dimer Consisting of two Identical Bubbles. In this subsection, we consider the quasi-static resonances of a bubble dimer in $\mathbb{R}^3$. Throughout, we denote by D the normalized bubble dimer which consists of two identical balls D_1 and D_2, both with volume one and material parameters ρ_b and κ_b. We assume that the two balls D_1 and D_2 are symmetric with respect to the origin. Moreover, D_1 and D_2 are aligned with respect to the x_1-axis.

Let $\psi_1,\,\psi_2 \in L^2(\partial D)$ be given by

$$\mathcal{S}_D^0[\psi_1] = \begin{cases} 1 & \text{on } \partial D_1, \\ 0 & \text{on } \partial D_2, \end{cases} \qquad \mathcal{S}_D^0[\psi_2] = \begin{cases} 0 & \text{on } \partial D_1, \\ 1 & \text{on } \partial D_2. \end{cases} \tag{14.61}$$

Then, it is easy to check that

$$\text{Ker}\left(-\frac{1}{2}I + \mathcal{K}_D^{0,*}\right) = \text{span}\;\{\psi_1, \psi_2\}. \tag{14.62}$$

We define the capacitance coefficients matrix $C = (C_{ij})$ by

$$C_{ij} := -\int_{\partial D_j} \psi_i, \quad i,j = 1,2.$$

We remark that the matrix C is positive definite and symmetric.

We now present the main result on the quasi-static resonances of the bubble dimer D, which can be proved analogously to Theorem 14.5.

THEOREM 14.46. *There are two quasi-static resonances with positive real part for the bubble dimer D. Moreover, they have the following asymptotic expansions as δ, defined by (4.53), goes to zero:*

$$\omega_1 = \sqrt{(C_{11}+C_{12})v_b}\,\delta^{\frac{1}{2}} - \sqrt{-1}\tau_1\delta + O(\delta^{3/2}), \tag{14.63}$$

$$\omega_2 = \sqrt{(C_{11}-C_{12})v_b}\,\delta^{\frac{1}{2}} + \delta^{3/2}\hat{\eta}_1 + \sqrt{-1}\delta^2\hat{\eta}_2 + O(\delta^{5/2}), \tag{14.64}$$

where

$$\tau_1 = \frac{v_b^2}{4\pi v}(C_{11}+C_{12})^2,$$

and $\hat{\eta}_1$ *and* $\hat{\eta}_2$ *are real numbers which are determined by* D, v, *and* v_b. *The two resonances* ω_1 *and* ω_2 *are called the hybridized resonances of the bubble dimmer* D. *The resonance* ω_2 *is referred to as the anti-resonance.*

14.5.2. The Point and Dipole Approximation of Bubble dimers. In this subsection we provide an approximate formula for the solution u to the scattering problem for the bubble dimer $D = D_1 \cup D_2$. We refer to [**42**] for its proof.

We need the following lemma which can be proved using a simple symmetry argument.

LEMMA 14.47. *Let* ψ_1 *and* ψ_2 *be defined by (*14.61*). We have*

$$\int_{\partial D} y_2(\psi_1 - \psi_2) d\sigma(y) = \int_{\partial D} y_3(\psi_1 - \psi_2) d\sigma(y) = 0,$$

while

$$\int_{\partial D} y_1(\psi_1 - \psi_2) d\sigma(y) = P,$$

for some nonzero real number P.

In the next theorem, we state that the bubble dimer can be approximated as a point scatterer with monopole and dipole modes.

THEOREM 14.48. *For* $\omega = O(\delta^{1/2})$ *and a given plane wave* u^i, *the solution* u *to (*14.1*) can be approximated as* $\delta \to 0$ *by*

$$u(x) - u^i(x) = g^0(\omega) u^i(0) \Gamma_k(x) + \nabla u^i(0) \cdot g^1(\omega) \nabla \Gamma_k(x) + O(\delta |x|^{-1}), \tag{14.65}$$

when $|x|$ *is sufficiently large, where*

$$g^0(\omega) := \frac{C(1,1)}{1 - \omega_1^2/\omega^2}(1 + O(\delta^{1/2})), \tag{14.66}$$

$$g^1(\omega) = (g^1_{ij}(\omega)), \; g^1_{ij}(\omega) := \int_{\partial D} (\mathcal{S}_D^0)^{-1}[x_i](y) y_j \, d\sigma(y) + \frac{\delta v_b^2}{|D|(\omega_2^2 - \omega^2)} P^2 \delta_{i1} \delta_{j1}. \tag{14.67}$$

14.5.3. Effective Medium Theory for a System of Bubble Dimers. We consider the scattering of an incident acoustic plane wave u^i by N identical bubble dimers with different orientations distributed in a homogeneous fluid in $\mathbb{R}^3$. The N identical bubble dimers are generated by scaling the normalized bubble dimer D by a factor s, and then rotating the orientation and translating the center. More precisely, the bubble dimers occupy the domain

$$D^N := \cup_{1 \le j \le N} D_j^N,$$

where $D_j^N = z_j^N + s R_{d_j^N} D$ for $1 \le j \le N$, with z_j^N being the center of the dimer D_j^N, s being the characteristic size, and $R_{d_j^N}$ being the rotation in $\mathbb{R}^3$ which aligns the dimer D_j^N in the direction d_j^N. Here, d_j^N is a vector of unit length in $\mathbb{R}^3$.

We assume that $0 < s \ll 1$, $N \gg 1$ and that $\{z_j^N : 1 \le j \le N\} \subset \Omega$ where Ω is a bounded domain. Let u^i be the incident wave which we assume to be a plane wave for simplicity. The scattering of acoustic waves by the bubble dimers can be modeled by the following system of equations:

$$
(14.68) \qquad \begin{cases}
\nabla \cdot \dfrac{1}{\rho} \nabla u^N + \dfrac{\omega^2}{\kappa} u^N = 0 \quad \text{in } \mathbb{R}^3 \backslash D^N, \\
\nabla \cdot \dfrac{1}{\rho_b} \nabla u^N + \dfrac{\omega^2}{\kappa_b} u^N = 0 \quad \text{in } D^N, \\
u_+^N - u_-^N = 0 \quad \text{on } \partial D^N, \\
\dfrac{1}{\rho} \dfrac{\partial u^N}{\partial \nu}\Big|_+ - \dfrac{1}{\rho_b} \dfrac{\partial u^N}{\partial \nu}\Big|_- = 0 \quad \text{on } \partial D^N, \\
u^N - u^i \text{ satisfies the Sommerfeld radiation condition,}
\end{cases}
$$

where u^N is the total field.

We make the following assumptions.

ASSUMPTION 14.49. (i) *$\delta = \mu^2 s^2$ for some positive number $\mu > 0$;*

(ii) *$\omega = \omega_2 + a s^2$ for some real number $a \neq \mu^3 \hat{\eta}_1$;*

(iii) *$sN = \Lambda$ for some positive number $\Lambda > 0$;*

(iv) *The bubble dimers are regularly distributed in the sense that*

$$
\min_{i \neq j} |z_i^N - z_j^N| \geq \eta N^{-\frac{1}{3}},
$$

for some constant η independent of N;

(v) *There exists a function $\tilde{V} \in \mathcal{C}^1(\bar{\Omega})$ such that for any $f \in \mathcal{C}^{0,\alpha}(\Omega)$ with $0 < \alpha \leq 1$,*

$$
\max_{1 \leq j \leq N} \Big| \frac{1}{N} \sum_{i \neq j} \Gamma_k(z_j^N - z_i^N) f(z_i^N) - \int_\Omega \Gamma_k(z_j^N - y) \tilde{V}(y) f(y) dy \Big| \leq C \frac{1}{N^{\frac{\alpha}{3}}} \|f\|_{\mathcal{C}^{0,\alpha}(\Omega)}
$$

for some constant C independent of N;

(vi) *There exists a matrix valued function $\tilde{B} \in \mathcal{C}^1(\bar{\Omega})$ such that for $f \in (\mathcal{C}^{0,\alpha}(\Omega))^3$ with $0 < \alpha \leq 1$,*

$$
\max_{1 \leq j \leq N} \Big| \frac{1}{N} \sum_{i \neq j} (f(z_i^N) \cdot d_i^N)(d_i^N \cdot \nabla \Gamma_k(z_i^N - z_j^N)) - \int_\Omega f(y) \tilde{B} \nabla_y \Gamma_k(y - z_j^N) dy \Big|
$$
$$
\leq C \frac{1}{N^{\frac{\alpha}{3}}} \|f\|_{\mathcal{C}^{0,\alpha}(\Omega)}
$$

for some constant C independent of N;

(vii) *There exists a constant $C > 0$ such that*

$$
\max_{1 \leq j \leq N} \frac{1}{N} \sum_{i \neq j} \frac{1}{|z_j^N - z_i^N|} \leq C, \qquad \max_{1 \leq j \leq N} \frac{1}{N} \sum_{i \neq j} \frac{1}{|z_j^N - z_i^N|^2} \leq C,
$$

for all $1 \leq j \leq N$.

We introduce the two constants

$$
\tilde{g}^0 = \frac{2(C_{11} + C_{12})}{1 - \omega_1^2/\omega_2^2}, \quad \tilde{g}^1 = \frac{\mu^2 v_b^2}{2|D|\omega_2(\mu^3 \hat{\eta}_1 - a)} P^2,
$$

and the two functions

$$
M_1 = \begin{cases} I & \text{in } \mathbb{R}^3 \setminus \Omega, \\ I - \Lambda \tilde{g}^1 \tilde{B} & \text{in } \Omega, \end{cases}
$$

and

$$M_2 = \begin{cases} k^2 & \text{in } \mathbb{R}^3 \setminus \Omega, \\ k^2 - \Lambda \tilde{g}^0 \tilde{V} & \text{in } \Omega. \end{cases}$$

The following result from [**42**] holds.

THEOREM 14.50. *Suppose that there exists a unique solution u to*

$$\nabla \cdot M_1(x) \nabla u(x) + M_2(x) u(x) = 0 \quad \text{in } \mathbb{R}^3, \tag{14.69}$$

such that $u - u^i$ satisfies the Sommerfeld radiation condition at infinity. Then, under Assumption 14.49, we have $u^N(x) \to u(x)$ uniformly for $x \in \Omega_N$.

REMARK 14.51. *If the bubble dimers are distributed such that $\tilde{B}$ is a positive matrix with $\tilde{B}(x) \geq C > 0$ for some constant C for all $x \in \Omega$, then we see that for ω in the form $\omega = \omega_2 + as^2$ with $a < \mu^3 \tilde{\eta}_1$, and sufficiently large Λ, both the matrix $I - \Lambda \tilde{g}^1 \tilde{B}$ and the scalar function $k^2 - \Lambda \tilde{g}^0 \tilde{V}$ are negative. Therefore, we obtain an effective double-negative medium with both negative mass density and negative bulk modulus. On the other hand, for ω in between ω_1 and ω_2 but away from the anti-resonance ω_2, $\tilde{g}^1$ may be small enough such that the matrix $I - \Lambda \tilde{g}^1 \tilde{B}$ is positive, while the matrix $k^2 - \Lambda \tilde{g}^0 \tilde{V}$ remains negative. Then the obtained effective medium has negative mass density and positive bulk modulus.*

14.6. Numerical Illustrations

In this section we present some numerical examples to illustrate our main findings in this chapter.

14.6.1. Minnaert Resonance Calculations. To validate the Minnaert resonance formula (14.17) in two dimensions we first determine the characteristic value ω_c of $\mathcal{A}(\omega, \delta)$ in (14.8) numerically. We then calculate the complex root ω_M of (14.17) that has a positive real part. Comparing ω_c and ω_M over a range of appropriate values of δ allows us to judge the accuracy of the formula.

In order to perform the analysis in the correct regime, which was described in Section 14.2, we take $\rho = \kappa = 10^3$ and $\rho_b = \kappa_b = c$, where c is chosen such that the wave speed in both air and water is of order 1 and $\delta \in \{10^{-i}\}$, $i \in \{1, \ldots, 5\}$. We use 2^9 points to discretize the unit circle used in the calculation of the layer potentials that form $\mathcal{A}$. Calculating ω_c is equivalent to determining the smallest ω such that $\mathcal{A}(\omega, \delta)$ has a zero eigenvalue. We have

$$\omega_c = \min_{\omega \in \mathbb{C}} \{\omega \mid \lambda(\omega) = 0\} \quad \lambda \in \sigma(\mathcal{A}(\omega, \delta)),$$

and we approach $\lambda(\omega) = 0$ as a complex root finding problem which can be calculated using Muller's method. Muller's method is applied again in order to find the root ω_M satisfying (14.17). The resonant frequencies ω_c and ω_M, along with the relative errors, for specific values of δ are given in Table 14.1. In Figure 14.1 it can be seen that the relative error becomes very small when $\delta \ll 1$, confirming the excellent accuracy of the formula. In particular, we note that when $\delta = 10^{-3}$, which is close to the usual contrast between water and air, the difference between ω_c and ω_M is negligible with a relative error of only 0.0652%.

The results are shown in in Table 14.1 and Figure 14.1. The corresponding Matlab code can be downloaded at `www.ams.org/bookpages/surv-235`.

δ	ω_c	ω_M	Relative error
10^{-2}	$0.075146 - 0.023976\sqrt{-1}$	$0.074681 - 0.023687\sqrt{-1}$	0.6727 %
10^{-3}	$0.021001 - 0.004513\sqrt{-1}$	$0.020987 - 0.004508\sqrt{-1}$	0.0652%
10^{-4}	$0.005950 - 0.000959\sqrt{-1}$	$0.005949 - 0.000959\sqrt{-1}$	0.0062%
10^{-5}	$0.001714 - 0.000221\sqrt{-1}$	$0.001714 - 0.000221\sqrt{-1}$	0.0030%

TABLE 14.1. A comparison between the characteristic value ω_c of $\mathcal{A}(\omega, \delta)$ and the root of the two dimensional resonance formula with positive real part ω_M, over several values of δ.

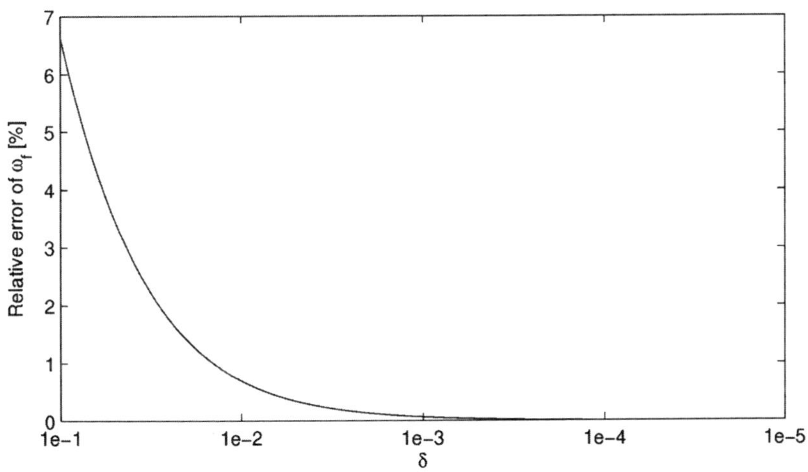

FIGURE 14.1. The relative error of the Minnaert resonance ω_c obtained by the two dimensional formula becomes negligible when we are in the appropriate high contrast regime.

14.6.2. Sub-wavelength Bandgap Calculations. In this subsection we perform numerical simulations in two dimensions to illustrate the main results of this chapter on sub-wavelength bandgap opening.

In view of the formula (14.55) for the α-quasi-periodic Minneart resonance, we want to compare $\omega^{\alpha}_{\text{approx}} := \omega_M \sqrt{\frac{\text{cap}(\partial D, \alpha)}{\text{cap}(\partial D)}}$ with the true α-quasi-periodic resonance $\omega^{\alpha}_{\text{exact}}$, which can be obtained through direct calculation of the minimum characteristic value of the operator $\mathcal{A}(\omega, \delta)$ in (14.6) using Muller's method described in Section 1.6.

We set the density and the bulk modulus of the bubbles to be $\rho_b = 1$ and $\kappa_b = 1$, respectively. In order to confirm that the formula becomes accurate in the appropriate regime, which features similar wavenumbers inside and outside the bubbles along with, in particular, a high contrast in the bulk modulii, we take the density and the bulk modulus of the background material to be $\rho = \kappa =$

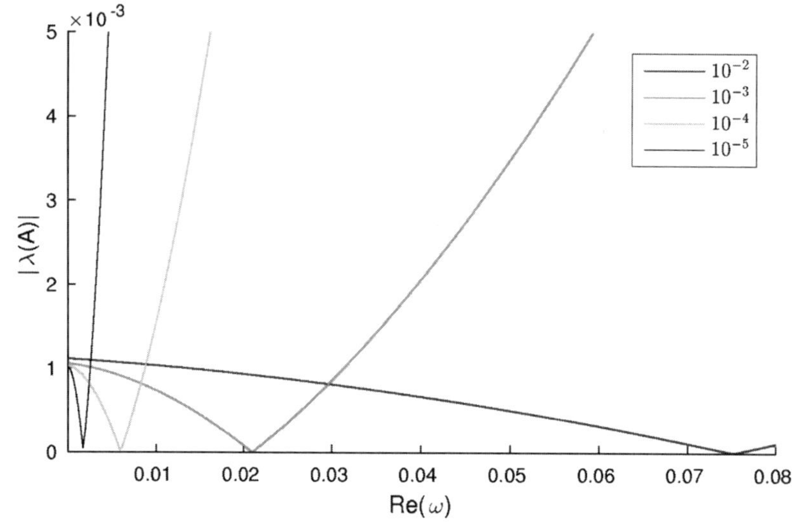

FIGURE 14.2. The Minnaert resonance for a single bubble of radius 1 in an infinite extent of liquid for contrast $\delta \in \{10^{-2}, 10^{-3}, 10^{-4}, 10^{-5}\}$.

$\delta^{-1} \in (10, 1000)$. We assume that the bubble represented by D is a disk of radius $R = 0.0125$. In Figure 14.3 we plot $\omega^{\alpha}_{\text{approx}}$ and $\omega^{\alpha}_{\text{exact}}$ against the contrast δ^{-1} and it is clear that the formula provides a highly accurate approximation when the contrast is sufficiently large.

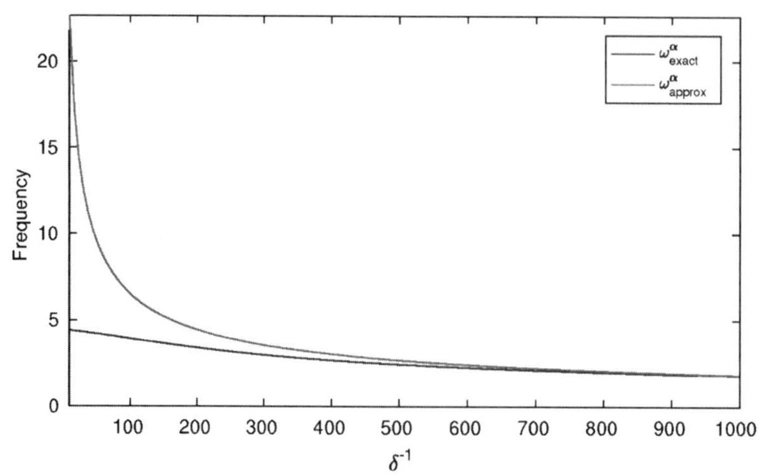

FIGURE 14.3. When the contrast δ^{-1} is sufficiently large, the α-quasi-periodic resonance ω_1^{α} given by Theorem 14.41 provides a highly accurate approximation of the true resonance $\omega^{\alpha}_{\text{exact}}$.

Next we present numerical examples to illustrate sub-wavelength bandgap openings. As D is a disk of radius R, we apply the multipole expansion method for computing the band structure (we refer to Subsection 5.4.2.3). As described in Subsection 5.4.2.1, the quasi-periodic Green's function is unsuitable for bandgap calculations due to the 'empty resonance' phenomenon. Therefore, we make use of the multiple expansion method which is efficient in the case of disk-shaped bubbles. Since $\mathcal{A}(\omega, \delta)$ is represented in terms of the single-layer potential only, it suffices to use the multipole expansion version of the single-layer potential derived in Subsection 5.4.2.3.

We first consider the dilute case. We set $R = 0.05$, $\rho = \kappa = 5000$ and $\rho_b = \kappa_b = 1$. In this case, we have $\delta = 0.0002$. Figure 14.4 shows the computed band structure. The points Γ, X and M represent $\alpha = (0, 0)$, $\alpha = (\pi, 0)$ and $\alpha = (\pi, \pi)$, respectively. We plot the first two characteristic values $\mathcal{A}(\omega, \delta)$ along the boundary of the triangle $\Gamma X M$. It can be seen that a sub-wavelength bandgap in the spectrum of $\mathcal{A}(\omega, \delta)$ does exist. Moreover, the bandgap between the first two bands is quite large. It is also worth mentioning that, by zooming in on the sub-wavelength bandgap (on the right in Figure 14.4), one can see that ω_1^* is attained at the point M (that is, $\alpha = (\pi, \pi)$). We used $N = 7$ for the truncation order of the cylindrical waves. Further numerical experiments indicate that this phenomenon is independent of the bubble radius or position.

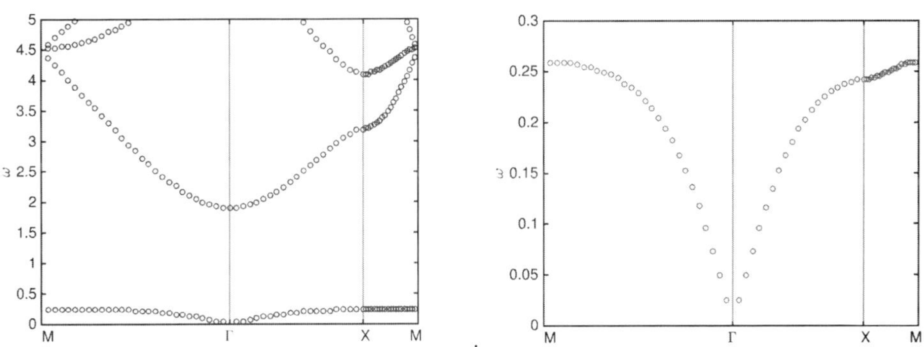

FIGURE 14.4. (Dilute case) The band structure of a square array of circular bubbles with radius $R = 0.05$ and contrast $\delta^{-1} = 5000$.

Next, in order to verify our conclusion from Lemma 14.45, namely that

$$\omega_1^* = \max_{\alpha} \omega_{M,\alpha} \approx \omega_M$$

when α is non-zero, we fix the contrast to be $\delta^{-1} = 1000$ and observe ω_1^* and ω_M over a range of bubble sizes in Figure 14.5.

Finally, we consider a non-dilute regime. We set $R = 0.25$ and $\rho = \kappa = 1000$ and $\rho_b = \kappa_b = 1$. In this case we have $\delta = 0.001$. Figure 14.6 shows the computed band structure. Again, a sub-wavelength bandgap can be observed. We used $N = 3$ for the truncation order of the cylindrical waves in the multipole expansion method.

14.6.3. Super-focusing Phenomenon. In this subsection, we perform a numerical investigation to explain the super-focusing phenomenon. We first calculate

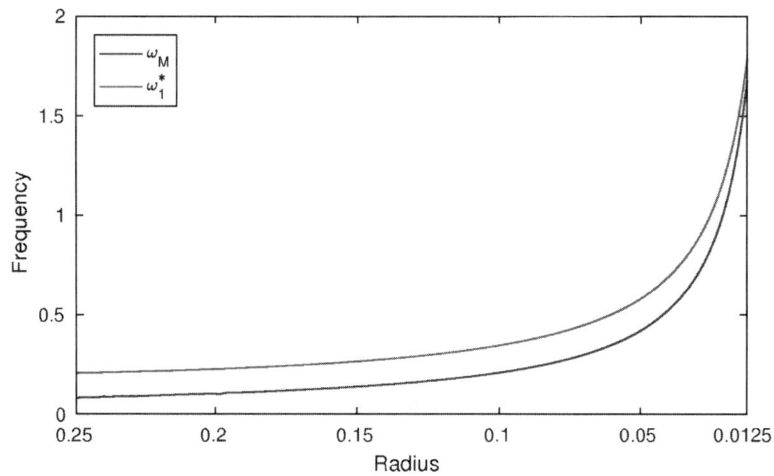

FIGURE 14.5. As the bubbles becomes smaller, the maximum frequency in the first band of the spectrum of the operator $\mathcal{A}(\omega, \delta)$, ω_1^*, approaches the Minnaert resonant frequency of a single bubble ω_M.

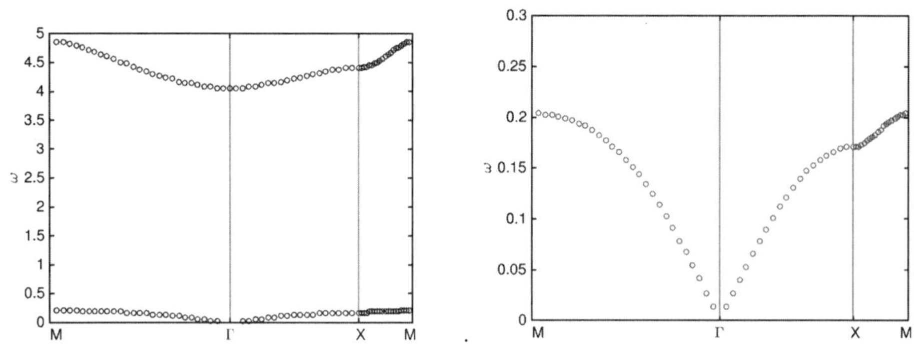

FIGURE 14.6. (Non-dilute case) The band structure of a square array of circular bubbles with radius $R = 0.25$ and contrast $\delta^{-1} = 1000$.

the imaginary part of the Green's function for a bubbly medium over a range of frequencies to ascertain the optimal frequency range for super-focusing. We then simulate wave propagation through the bubbly medium to evaluate the actual focusing performance.

We consider a source at the origin surrounded by a cube of uniformly distributed bubbles, for a range of gas volume fractions given by $\{0, 0.5, 1, 2\} \times 10^{-4}$. We denote by $\Phi_\omega(x)$ the Green's function in the presence of the bubbles at frequency ω.

In Figure 14.7 we plot $|\Im(\Phi_\omega(x))|$ when no bubbles are present, and in Figure 14.8 we plot $|\Im(\Phi_\omega(x))|$, for bubble clouds of increasing volume fraction.

When the volume fraction reaches 2×10^{-4} we obtain a very sharp peak over the origin just below the Minnaert resonance frequency, which indicates that focusing performance should be greatly enhanced in this frequency regime.

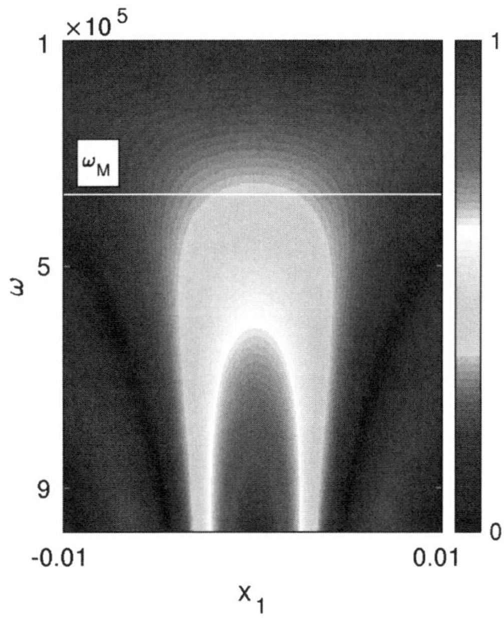

FIGURE 14.7. The normalized imaginary part of the Green's function when no bubbles are present in the liquid. The white line indicates the Minnaert resonance frequency $\omega_M \approx 3.7 \times 10^5$ kHz.

14.6.4. Double-Negative Refractive Index Phenomenon. In this subsection, we illustrate the double-negative refractive index phenomenon in bubbly media by numerical examples. We consider a cubic array of identical spherical bubble dimers. Suppose Ω is a cube with a side length of $L = 20$, *i.e.*, $\Omega = [0, 20]^3$. Let $a = 0.2$ and define a small cube $\Omega_a = [0, a]^3$. Then Ω can be considered as a union of small cubes as follows:

$$\Omega = \bigcup_{n_1,n_2,n_3=0,1,\ldots,99} \Omega_a + a(n_1, n_2, n_3).$$

We assume that a bubble dimer is centered in each of the small cubes. Then the total number of dimers is $N = 10^6$ and the periodicity of the dimer array is $a = 0.2$.

Recall that a bubble dimer is described by $z + sR_dD$, where z is the center of the dimer, s is its characteristic size, and R_d is a rotation in $\mathbb{R}^3$ which aligns the dimer in the direction d, where d is a unit vector. We set the characteristic size of the dimers to be $s = 0.1$. Since D has unit volume, the radius r_0 of the bubbles comprising the dimers is $r_0 = s(3/4\pi)^{1/3} \approx 0.005$.

We set $\rho_b = \kappa_b = 1$ and $\rho = \kappa = 5 \times 10^3$. Then $v = v_b = 1$, $k = k_b = \omega$, and $\delta = 2\times 10^{-4}$. We assume that the two bubbles comprising each dimer are separated by a distance of $l = 5r_0 \approx 0.0248$. Moreover, each dimer is randomly oriented so that the unit vector d is uniformly distributed on the unit sphere. Under these

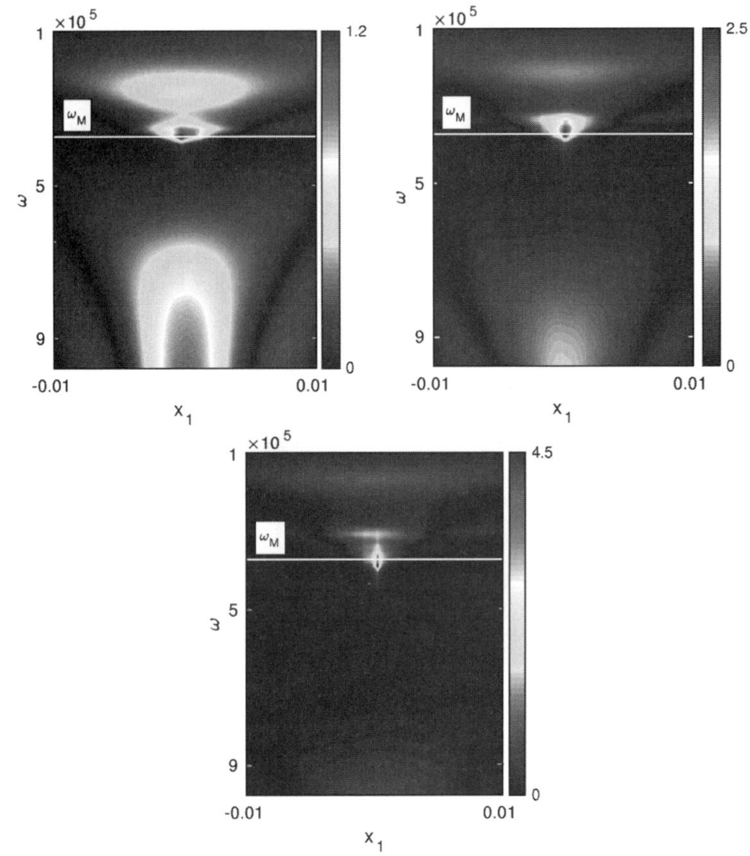

FIGURE 14.8. The imaginary part of the Green's function averaged over 25 bubble cloud configurations for increasing bubble concentrations. From left to right: the volume fraction is equal to 0.5×10^{-4}, 10^{-4}, and 2×10^{-4}. The results are normalized with respect to the background Green's function in Figure 14.7. The function peaks over the source at 0 when the excitation frequency is very near the Minnaert resonance frequency, indicating the potential for enhanced focusing.

assumptions, we can easily check that $\Lambda = 8 \times 10^3$, $\tilde{V} \approx |\Omega|^{-1} = 1.25 \times 10^{-4}$, and $\tilde{B} \approx (2|\Omega|)^{-1} I = 6.25 \times 10^{-5} I$.

Now we consider the effective properties of the homogenized media. Recall that the effective coefficients of the homogenized equation (14.69) are $k^2 - \Lambda \tilde{g}^0 \tilde{V}$ and $I - \Lambda \tilde{g}^1 \tilde{B}$. Using the above parameters, we have

$$
\begin{aligned}
& k^2 - \Lambda \tilde{g}^0 \tilde{V} \approx \omega^2 - \tilde{g}^0 = \omega^2 (1 - \tilde{g}^0/\omega^2), \\
(14.70) \qquad & I - \Lambda \tilde{g}^1 \tilde{B} \approx (1 - \tilde{g}^1/2) I.
\end{aligned}
$$

Note that the coefficient $I - \Lambda \tilde{g}^1 \tilde{B}$ can be roughly considered as a scalar quantity. The scattering functions $\tilde{g}_0$ and $\tilde{g}_1$ can be computed as follows. Since $\tilde{g}_0 \approx s g_0$ and

$\tilde{g}_1 \approx s g_1$, we have that

$$\tilde{g}^0(\omega, \delta, sR_dD) \approx \frac{2(C_{11}+C_{12})}{1-\omega_1^2/\omega^2},$$
$$\tilde{g}^1(\omega, \delta, sR_dD) \approx \frac{\delta \tilde{v}^2}{2(\omega_2^2-\omega^2)} P^2 d_i d_j.$$

The resonance frequencies ω_1 and ω_2 can be easily calculated using a standard multipole expansion together with a root finding method such as Muller's method [**59**]. We find that $\omega_1 \approx 4.6171 - 0.0926i$ and $\omega_2 \approx 5.3253 - 0.0005i$. Then it is simple to compute $\tilde{g}^0$ and $\tilde{g}^1$.

In the top figure of Figure 14.9, we plot the two effective coefficients as functions of frequency. Clearly, there is a narrow region contained in the interval $[5.2, 5.4]$ in which both of the coefficients are negative. So we expect that the negative refraction occurs in this frequency region.

We next consider the refractive index. In view of (14.69) and (14.70), the effective mass density ρ_{eff} and the effective bulk modulus κ_{eff} can be computed approximately by

$$\rho_{\text{eff}} \approx 1 - \tilde{g}^1/2, \quad \kappa_{\text{eff}} \approx (1-\tilde{g}^0/\omega^2)^{-1}.$$

As usual, we define the refractive index n_{eff} by

$$n_{\text{eff}} = \sqrt{\rho_{\text{eff}}}\sqrt{\kappa_{\text{eff}}^{-1}}.$$

In the bottom figure of Figure 14.9, we plot the refractive index as a function of frequency. It is clear that the refractive index becomes negative in a narrow region contained in the interval $[5.2, 5.4]$, as expected.

14.7. Concluding Remarks

In this chapter, we have investigated the acoustic wave propagation problem in bubbly media and rigorously derived the low-frequency Minnaert resonances. We have also derived an effective medium theory for acoustic wave propagation in bubbly fluid near Minnaert resonant frequency. We have shown that on the one hand, just below the Minnaert resonant frequency, the obtained effective media can have a high refractive index, while on the other hand, the obtained effective medium can be dissipative above the Minnaert resonant frequency. Furthermore, we have proved the existence of a sub-wavelength bandgap opening in bubble phononic crystals. Finally, we have derived a double-negative effective medium theory near the hybridized Minnaert resonances of bubble dimers. We have illustrated our findings in this chapter with a variety of numerical experiments.

Our results in this chapter shed light on the mechanism behind the extraordinary wave properties of bubbly fluids near sub-wavelength resonant frequencies. Following the various approaches described in this chapter, super-focusing properties, bandgap openings at the sub-wavelength scale, and double-negative refractive index phenomena can be obtained for general sub-wavelength resonators. We refer

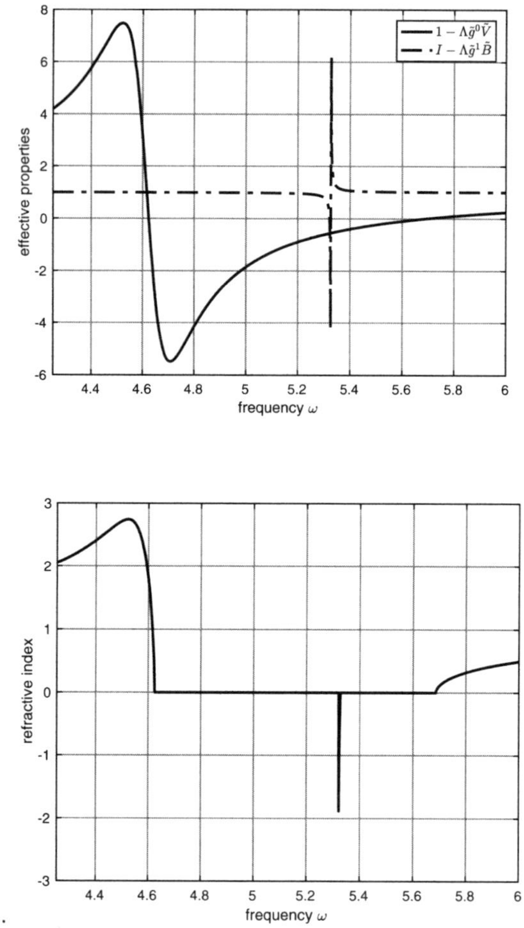

FIGURE 14.9. The effective properties of the homogenized media (top), and the refractive index (bottom).

the reader for instance to [**375**, **376**] for recent works on creating bandgaps using plasmonic resonances and to [**93**, **160**, **161**, **164**, **346**, **347**] for the derivation of effective properties of electromagnetic materials that consist of matrix of constant electromagnetic properties and a set of subwavelength resonators (plasmonic particles in a chiral medium, dielectric rodes with a strong permittivity, wires, split rings, ...). Similar results to those presented in this chapter can also be obtained in acoustics using Helmholtz resonators [**345**, **472**].

APPENDIX A

Spectrum of Self-Adjoint Operators

Given a linear self-adjoint operator L in the Hilbert space H with domain $D(L), \overline{D(L)} = H$, we define the resolvent set as

$$\rho(L) := \big\{z \in \mathbb{C} : (zI - L)^{-1}\text{exists as a bounded operator from } H \text{ to } D(L)\big\}.$$

Its complement $\sigma(L) = \mathbb{C} \setminus \rho(L)$ is the spectrum of L.

Since L is self-adjoint, $z \in \rho(L)$ if and only if there exists a constant $C(z)$ such

$$\|(zI - L)u\|_H \geq C(z)\|u\|_H$$

for all $u \in D(L)$. Moreover, $\sigma(L) \neq \emptyset$ and $\sigma(L) \subset \mathbb{R}$ and the following Weyl's criterion holds for characterizing $\sigma(l)$: $z \in \sigma(L)$ if and only if there exists $u_n \in D(L)$ such that

$$\lim_{n\to+\infty} \|(zI - L)u_n\|_H = 0.$$

We define the point spectrum $\sigma_p(L)$ of L as the set of eigenvalues of L:

$$\sigma_p(L) := \big\{z \in \sigma(L) : (zI - L)^{-1}\text{does not exist or equivalently } \operatorname{Ker}(zI - L) \neq \emptyset\big\}.$$

The complement $\sigma(L) \setminus \sigma_p(L)$ is the continuous spectrum $\sigma_c(L)$. If $z \in \sigma_c(L)$, then $(zI - L)^{-1}$ does exist but is not bounded. The discrete spectrum $\sigma_d(L)$ is defined by

$$\sigma_d(L) := \big\{z \in \sigma_p(L) : \dim \operatorname{Ker}(zI - L) < \infty \text{ and } z \text{ is isolated in } \sigma(L)\big\}.$$

The set $\sigma_{ess}(L) := \sigma(L) \setminus \sigma_d(L)$ is called the essential spectrum of L. Since L is self-adjoint, we have

$$\begin{aligned}\sigma_{ess}(L) &= \overline{\sigma_c(L)}\\ &\cup \big\{\text{ eigenvalues of infinite multiplicity and their accumulation points}\big\}\\ &\cup \big\{\text{ accumulation points of } \sigma_d(L)\big\}.\end{aligned}$$

A family of operators $\{\mathcal{E}(t)\}_{t=-\infty}^{+\infty}$ is called a spectral family (or a resolution of identity) if the following conditions are satisfied:

(i) $\mathcal{E}(t)$ is a projector for all $t \in \mathbb{R}$;

(ii) $\mathcal{E}(t) \leq \mathcal{E}(s)$ for all $t < s$;

(iii) $\{\mathcal{E}(t)\}$ is right continuous with respect to the strong topology, *i.e.*,

$$\lim_{s\to t+0} \|\mathcal{E}(s)u - \mathcal{E}(t)u\|_H = 0$$

for all $u \in H$.

(iv) $\{\mathcal{E}(t)\}$ is normalized as follows:

$$\lim_{t\to+\infty} \|\mathcal{E}(t)u - u\|_H = 0$$

for all $u \in H$.

We recall that for fixed $u, v \in H$, $\langle \mathcal{E}(t)u, v\rangle_H$ is a function of bounded variation with respect to t. Moreover, the self-adjoint operator L has a unique spectral representation, *i.e.*, there is a unique spectral family $\mathcal{E}(t)$ such that

$$Lu = \int_{-\infty}^{+\infty} t \, d\mathcal{E}(t)u$$

for all $u \in D(L)$. By the spectral resolution theorem, we have

$$(zI - L)^{-1} = \int_{-\infty}^{+\infty} \frac{1}{z-t} \, d\mathcal{E}(t) \tag{A.1}$$

for all $z \in \rho(L)$. Furthermore,

(i) $z \in \sigma_p(L)$ if and only if $\mathcal{E}(z) - \mathcal{E}(z-0) \neq 0$;
(ii) $z \in \sigma_c(L)$ if and only if $\mathcal{E}(z) - \mathcal{E}(z-0) = 0$.

Here $\mathcal{E}(z-0) := \lim_{\epsilon \to 0+} \mathcal{E}(z-\epsilon)$ in the sense of strong operator topology.

Let $\mathcal{C}(\sigma(L))$ be the set of continuous functions on $\sigma(L)$. We define

$$f(L) := \lim_{n \to +\infty} P_n(L)$$

with $\{P_n\}$ being a sequence of polynomials converging uniformly to f as $n \to +\infty$. Since for any $u \in H$ the function

$$f \mapsto \langle u, f(L)u\rangle_H$$

is a positive linear function on $\mathcal{C}(\sigma(L))$, there exists a unique Radon measure $\mu(u)$ on $\sigma(L)$ (called the spectral measure associated to u and L) such that

$$\int_{\sigma(L)} f \, d\mu(u) = \langle u, f(L)u\rangle_H$$

for all $f \in \mathcal{C}(\sigma(L))$. In particular, we have $\mu(u)(\sigma(L)) = \|u\|_H^2$, so $\mu(u)$ is a finite measure. Moreover, the measure $\mu(u)$ is invariant under linear transformations and can be decomposed into three parts:

$$\mu(u) = \mu_{ac} + \mu_{sc} + \mu_{pp},$$

where μ_{pp} is pure point measure, μ_{ac} is absolutely continuous, and μ_{sc} is singular both with respect to the Lebesgue measure. Let

$$\begin{aligned} H_{pp} &:= \big\{u \in H : \mu(u) \text{ is pure point}\big\}, \\ H_{ac} &:= \big\{u \in H : \mu(u) \text{ is absolutely continuous}\big\}, \\ H_{sc} &:= \big\{u \in H : \mu(u) \text{ is singulary continuous}\big\}. \end{aligned}$$

We have $H = H_{pp} \oplus H_{ac} \oplus H_{sc}$, where each subspace is invariant under L. Furthermore,

$$\sigma(L) = \overline{\sigma_{pp}}(L) \cup \sigma_{ac}(L) \cup \sigma_{sc}(L),$$

where

$$\overline{\sigma_{pp}}(L) = \sigma(L|_{H_{pp}}), \sigma_{ac}(L) = \sigma(L|_{H_{ac}}), \quad \text{and } \sigma_{sc}(L) = \sigma(L|_{H_{sc}}),$$

and the union may not be disjoint.

In terms of the spectral measure, (A.1) can be rewritten as

$$\langle u, (z-L)^{-1}u\rangle_H = \int_{\mathbb{R}} \frac{d\mu(u)(t)}{z-t},$$

which shows that $d\mu(u)(t) = \langle d\mathcal{E}(t)u, u\rangle_H$ for all $t \in \mathbb{R}$.

Now, suppose that the self-adjoint operator L is compact. Then

(i) L has a sequence of eigenvalues $\lambda_j \neq 0, j \in \mathbb{N}$, which can be enumerated in such a way that

$$|\lambda_0| \geq |\lambda_1| \geq \ldots \geq |\lambda_j| \geq \ldots;$$

(ii) If there are infinitely many eigenvalues then $\lim_{j\to+\infty} \lambda_j = 0$ and 0 is the only accumulation point of $\{\lambda_j\}_{j\in\mathbb{N}}$;

(iii) The multiplicity of λ_j is finite;

(iv) If φ_j is the normalized eigenvector for λ_j, then $\{\varphi_j\}_{j=0}^{+\infty}$ is an orthonormal basis on $R(L)$ and the spectral resolution theorem reduces to

$$Lu = \sum_{j=0}^{+\infty} \lambda_j \langle u, \varphi_j \rangle_H \varphi_j, \quad u \in H.$$

(v) $\sigma(L) = \{0, \lambda_0, \lambda_1, \ldots, \lambda_j, \ldots\}$ while 0 is not necessarily an eigenvalue of L.

APPENDIX B

Optimal Control and Level Set Representation

In this appendix we describe the optimal control approach and the level set representation used for solving optimal design problems in photonics and phononics.

B.1. Optimal Control Scheme

Let H be a Banach space. In photonics and phononics, H stands either for a set of admissible material properties or for a set of geometric shapes. Consider a discrepancy functional $J(u(h))$ depending on $h \in H$ via the solution $u(h)$ to a system where h acts as a parameter, say: $A(h)u(h) = g$. Here, g represents the data. In order to minimize J we need to compute its Fréchet derivative

$$\frac{\partial J}{\partial u}(u(h))\frac{\partial u}{\partial h},$$

which is not explicit in h. The introduction of the adjoint system

$$A(h)^* p(h) = \frac{\partial J}{\partial u}(u(h)), \tag{B.1}$$

where $A(h)^*$ denotes the adjoint of $A(h)$ makes this explicit. Multiplying (B.1) by $\frac{\partial u}{\partial h}\delta h$ we obtain

$$\frac{\partial J}{\partial u}(u(h))\frac{\partial u}{\partial h}\delta h = -p(h)\frac{\partial A}{\partial h}\delta h u(h),$$

and therefore, the Fréchet derivative of J is given by

$$-p(h)\frac{\partial A^*}{\partial h}u(h).$$

B.2. Level Set Method

Let H be a set of geometric shapes and consider the minimization over H of a discrepancy functional J. The main idea of the level set approach is to represent the domain D as the zero level set of a continuous function ϕ, *i.e.*,

$$D = \Big\{x : \phi(x) < 0\Big\},$$

to work with function ϕ instead of D, and to derive an evolution equation for ϕ to solve the minimization problem. In fact, by allowing additional time-dependence of ϕ, we can compute the geometric motion of D in time by evolving the level set function ϕ. A geometric motion with normal velocity $V = V(x,t)$ can be realized by solving the Hamilton-Jacobi equation

$$\frac{\partial \phi}{\partial t} + V|\nabla \phi| = 0. \tag{B.2}$$

Minimization within the level set framework consists of choosing a velocity V driving the evolution towards a minimum (or at least increasing the discrepancy functional we want to minimize).

Consider the geometry of the zero level set

$$\partial D = \left\{ x : \phi(x) = 0 \right\},$$

under a variation of ϕ. Suppose that $\phi(x)$ is perturbed by a small variation $\delta\phi(x)$. Let δx be the resulting variation of the point x. By taking the variation of the equation $\phi(x) = 0$, we find

$$\delta\phi = -\nabla\phi \cdot \delta x \,. \tag{B.3}$$

Observe that the unit outward normal at x is given by

$$\nu(x) = \frac{\nabla\phi(x)}{|\nabla\phi(x)|} \,.$$

Now, if t represents time, then the function ϕ depends on both x and t. We use the notation

$$\partial D(t) = \left\{ x : \phi(x,t) = 0 \right\}.$$

Assume that each point $x \in \partial D(t)$ moves perpendicular to the curve. That is, the variation δx satisfies

$$\delta x = V(x,t) \frac{\nabla\phi(x,t)}{|\nabla\phi(x,t)|} \,.$$

Suppose that $J := ||A(f) - g||^2$ and the minimization is performed over piecewise functions $f = f_+ \chi(\mathbb{R}^d \setminus \overline{D}) + f_- \chi(D)$ with $f_\pm$ being given constants. The minimal requirement for the variations of $\phi(x,t)$ is that J be a decreasing function of t. The directional derivative of the function J in the direction δf is given by

$$\delta J(f) = J'(f)\delta f = 2R_f^* \Big(g - A(f) \Big) \delta f \,,$$

where J' is the Fréchet derivative of J and R_f^* is the Fréchet derivative of $A(f)$. Since δf is a measure on ∂D given by

$$\delta f = (f_+ - f_-) \delta x \cdot \nu(x) \,,$$

we have

$$\delta f = (f_+ - f_-) \frac{\nabla\phi(x)}{|\nabla\phi(x)|} \cdot \delta x \Big|_{x \in \partial D} \,. \tag{B.4}$$

Hence,

$$\delta J(f) = (f_+ - f_-) J'(f) V \,,$$

and therefore, in order to make $\delta J(f)$ negative, we can choose

$$V(x,t) = (f_+ - f_-) R_f^* \Big(g - A(f) \Big) \,. \tag{B.5}$$

As (B.5) is only valid for $x \in \partial D$, a velocity extension to the entire domain should be performed. This leads to the Hamilton-Jacobi equation (B.2) for $\phi(x,t)$ with the initial condition $\phi(x,0) = \phi_0(x)$, and thus the problem of maximizing $J(f)$ is converted into a level set form.

B.3. Shape Derivatives

The shape derivative measures the sensitivity of boundary perturbations. It is defined as follows. Let $\theta \in W^{1,\infty}(\mathbb{R}^2)^2$ be such that $||\theta||_{W^{1,\infty}} < 1$, where $W^{1,\infty}$ is defined by (2.1). Consider the perturbation under the map θ:

$$D_\theta = (I + \theta)D,$$

where I is the identity map. In other words, the set D_θ is defined as

$$D_\theta = \Big\{ x + \theta(x) : x \in D \Big\}. \tag{B.6}$$

The shape derivative of an objective shape functional $\mathcal{J} : \mathbb{R}^2 \to \mathbb{R}$ at D is defined as the Fréchet differential of $\theta \mapsto \mathcal{J}(D_\theta)$ at 0. The vector θ can be viewed as a vector field advecting the reference domain D. The shape derivative $d_{\mathcal{S}}\mathcal{J}$ depends only on $\theta \cdot \nu$ on the boundary ∂D because the shape of D does not change at all if θ is lying on the tangential direction of the boundary ∂D.

Bibliography

[1] T. Abbas, H. Ammari, G. Hu, A. Wahab, and J. C. Ye, *Two-dimensional elastic scattering coefficients and enhancement of nearly elastic cloaking*, J. Elasticity **128** (2017), no. 2, 203–243, DOI 10.1007/s10659-017-9624-7. MR3669191

[2] T. Abboud and H. Ammari, *Diffraction at a curved grating: TM and TE cases, homogenization*, J. Math. Anal. Appl. **202** (1996), no. 3, 995–1026, DOI 10.1006/jmaa.1996.0357. MR1408364

[3] M. Abramowitz and I. A. Stegun (eds.), *Handbook of mathematical functions, with formulas, graphs, and mathematical tables*, Dover Publications, Inc., New York, 1966. MR0208797

[4] E. Akhmetgaliyev and O. P. Bruno, *Regularized integral formulation of mixed Dirichlet-Neumann problems*, J. Integral Equations Appl. **29** (2017), no. 4, 493–529, DOI 10.1216/JIE-2017-29-4-493. MR3722840

[5] E. Akhmetgaliyev, O. P. Bruno, and N. Nigam, *A boundary integral algorithm for the Laplace Dirichlet-Neumann mixed eigenvalue problem*, J. Comput. Phys. **298** (2015), 1–28, DOI 10.1016/j.jcp.2015.05.016. MR3374535

[6] G. Allaire and M. Amar, *Boundary layer tails in periodic homogenization* (English, with English and French summaries), ESAIM Control Optim. Calc. Var. **4** (1999), 209–243, DOI 10.1051/cocv:1999110. MR1696289

[7] M. S. Agranovich, B. A. Amosov, and M. Levitin, *Spectral problems for the Lamé system with spectral parameter in boundary conditions on smooth or nonsmooth boundary*, Russ. J. Math. Phys. **6** (1999), no. 3, 247–281. MR1816945

[8] J. F. Ahner and G. C. Hsiao, *On the two-dimensional exterior boundary-value problems of elasticity*, SIAM J. Appl. Math. **31** (1976), no. 4, 677–685, DOI 10.1137/0131060. MR0426573

[9] E. Akhmetgaliyev, O. P. Bruno, and N. Nigam, *A boundary integral algorithm for the Laplace Dirichlet-Neumann mixed eigenvalue problem*, J. Comput. Phys. **298** (2015), 1–28, DOI 10.1016/j.jcp.2015.05.016. MR3374535

[10] K. Aki and P. G. Richards, *Quantitative Seismology*, Vol. 1, W. H. Freeman & Co., San Francisco, 1980.

[11] H.-D. Alber, *A quasi-periodic boundary value problem for the Laplacian and the continuation of its resolvent*, Proc. Roy. Soc. Edinburgh Sect. A **82** (1978/79), no. 3-4, 251–272, DOI 10.1017/S0308210500011239. MR532908

[12] J. H. Albert, *Genericity of simple eigenvalues for elliptic PDE's*, Proc. Amer. Math. Soc. **48** (1975), 413–418, DOI 10.2307/2040275. MR0385934

[13] J. H. Albert, *Generic properties of eigenfunctions of elliptic partial differential operators*, Trans. Amer. Math. Soc. **238** (1978), 341–354, DOI 10.2307/1997812. MR0471000

[14] O. Alexandrov and F. Santosa, *A topology-preserving level set method for shape optimization*, J. Comput. Phys. **204** (2005), no. 1, 121–130, DOI 10.1016/j.jcp.2004.10.005. MR2121907

[15] G. Allaire, S. Aubry, and F. Jouve, *Eigenfrequency optimization in optimal design*, Comput. Methods Appl. Mech. Engrg. **190** (2001), no. 28, 3565–3579, DOI 10.1016/S0045-7825(00)00284-X. MR1819157

[16] G. Allaire, F. de Gournay, F. Jouve, and A.-M. Toader, *Structural optimization using topological and shape sensitivity via a level set method*, Control Cybernet. **34** (2005), no. 1, 59–80. MR2211063

[17] G. Allaire, F. Jouve, and A.-M. Toader, *Structural optimization using sensitivity analysis and a level-set method*, J. Comput. Phys. **194** (2004), no. 1, 363–393, DOI 10.1016/j.jcp.2003.09.032. MR2033390

[18] A. Alù, *Mantle cloak: Invisibility induced by a surface*, Phys. Rev. B **80** (2009), 245115.

[19] H. Ammari, *Uniqueness theorems for an inverse problem in a doubly periodic structure*, Inverse Problems **11** (1995), no. 4, 823–833. MR1346000

[20] H. Ammari, *An introduction to mathematics of emerging biomedical imaging*, Mathématiques & Applications (Berlin) [Mathematics & Applications], vol. 62, Springer, Berlin, 2008. MR2440857

[21] H. Ammari, E. Beretta, E. Francini, H. Kang, and M. Lim, *Optimization algorithm for reconstructing interface changes of a conductivity inclusion from modal measurements*, Math. Comp. **79** (2010), no. 271, 1757–1777, DOI 10.1090/S0025-5718-10-02344-6. MR2630011

[22] H. Ammari, E. Beretta, E. Francini, H. Kang, and M. Lim, *Reconstruction of small interface changes of an inclusion from modal measurements II: The elastic case* (English, with English and French summaries), J. Math. Pures Appl. (9) **94** (2010), no. 3, 322–339, DOI 10.1016/j.matpur.2010.02.001. MR2679030

[23] H. Ammari, N. Béreux, and J.-C. Nédélec, *Resonances for Maxwell's equations in a periodic structure*, Japan J. Indust. Appl. Math. **17** (2000), no. 2, 149–198, DOI 10.1007/BF03167342. MR1769190

[24] H. Ammari, E. Bonnetier, F. Triki, and M. Vogelius, *Elliptic estimates in composite media with smooth inclusions: an integral equation approach* (English, with English and French summaries), Ann. Sci. Éc. Norm. Supér. (4) **48** (2015), no. 2, 453–495, DOI 10.24033/asens.2249. MR3346176

[25] H. Ammari, E. Bretin, J. Garnier, H. Kang, H. Lee, and A. Wahab, *Mathematical methods in elasticity imaging*, Princeton Series in Applied Mathematics, Princeton University Press, Princeton, NJ, 2015. MR3309240

[26] H. Ammari, E. Bretin, J. Garnier, and A. Wahab, *Time-reversal algorithms in viscoelastic media*, European J. Appl. Math. **24** (2013), no. 4, 565–600, DOI 10.1017/S0956792513000107. MR3082867

[27] H. Ammari, P. Calmon, and E. Iakovleva, *Direct elastic imaging of a small inclusion*, SIAM J. Imaging Sci. **1** (2008), no. 2, 169–187, DOI 10.1137/070696076. MR2486036

[28] H. Ammari, Y. Capdeboscq, H. Kang, E. Kim, and M. Lim, *Attainability by simply connected domains of optimal bounds for the polarization tensor*, European J. Appl. Math. **17** (2006), no. 2, 201–219, DOI 10.1017/S0956792506006541. MR2266483

[29] H. Ammari, Y. T. Chow, K. Liu, and J. Zou, *Optimal shape design by partial spectral data*, SIAM J. Sci. Comput. **37** (2015), no. 6, B855–B883, DOI 10.1137/130942498. MR3425377

[30] H. Ammari, Y. T. Chow, and J. Zou, *The concept of heterogeneous scattering coefficients and its application in inverse medium scattering*, SIAM J. Math. Anal. **46** (2014), no. 4, 2905–2935, DOI 10.1137/130941468. MR3249363

[31] H. Ammari, Y. T. Chow, and J. Zou, *Phased and phaseless domain reconstructions in the inverse scattering problem via scattering coefficients*, SIAM J. Appl. Math. **76** (2016), no. 3, 1000–1030, DOI 10.1137/15M1043959. MR3505314

[32] H. Ammari, Y. T. Chow, and J. Zou, *Super-resolution in imaging high contrast targets from the perspective of scattering coefficients*, J. Math. Pures Appl. (9) **111** (2018), 191–226, DOI 10.1016/j.matpur.2017.09.008. MR3760753

[33] H. Ammari, G. Ciraolo, H. Kang, H. Lee, and K. Yun, *Spectral analysis of the Neumann-Poincaré operator and characterization of the stress concentration in anti-plane elasticity*, Arch. Ration. Mech. Anal. **208** (2013), no. 1, 275–304, DOI 10.1007/s00205-012-0590-8. MR3021549

[34] H. Ammari, G. Ciraolo, H. Kang, H. Lee, and G. W. Milton, *Spectral theory of a Neumann-Poincaré-type operator and analysis of cloaking due to anomalous localized resonance*, Arch. Ration. Mech. Anal. **208** (2013), no. 2, 667–692, DOI 10.1007/s00205-012-0605-5. MR3035988

[35] H. Ammari, G. Ciraolo, H. Kang, H. Lee, and G. W. Milton, *Spectral theory of a Neumann-Poincaré-type operator and analysis of cloaking by anomalous localized resonance II*, Inverse problems and applications, Contemp. Math., vol. 615, Amer. Math. Soc., Providence, RI, 2014, pp. 1–14, DOI 10.1090/conm/615/12244. MR3221596

[36] H. Ammari, G. Ciraolo, H. Kang, H. Lee, and G.W. Milton, *Anomalous localized resonance using a folded geometry in three dimensions*, Proc. Royal Soc. A **469** (2013), 20130048.

[37] H. Ammari, Y. Deng, and P. Millien, *Surface plasmon resonance of nanoparticles and applications in imaging*, Arch. Ration. Mech. Anal. **220** (2016), no. 1, 109–153, DOI 10.1007/s00205-015-0928-0. MR3458160

[38] H. Ammari, B. Fitzpatrick, D. Gontier, H. Lee, and H. Zhang, *Minnaert resonances for acoustic waves in bubbly media*, Ann. Inst. H. Poincaré Anal. Non Linéaire, `https://doi.org/10.1016/j.anihpc.2018.03.007`.

[39] H. Ammari, B. Fitzpatrick, D. Gontier, H. Lee, and H. Zhang, *A mathematical and numerical framework for bubble meta-screens*, SIAM J. Appl. Math. **77** (2017), no. 5, 1827–1850, DOI 10.1137/16M1090235. MR3717823

[40] H. Ammari, B. Fitzpatrick, D. Gontier, H. Lee, and H. Zhang, *Sub-wavelength focusing of acoustic waves in bubbly media*, Proc. A. **473** (2017), no. 2208, 20170469, 17. MR3749985

[41] H. Ammari, B. Fitzpatrick, H. Lee, S. Yu, and H. Zhang, *Subwavelength phononic bandgap opening in bubbly media*, J. Differential Equations **263** (2017), no. 9, 5610–5629, DOI 10.1016/j.jde.2017.06.025. MR3688425

[42] H. Ammari, B. Fitzpatrick, H. Lee, S. Yu, and H. Zhang, *Double-negative acoustic metamaterials*, to appear (arXiv:1709.08177).

[43] H. Ammari, J. Garnier, W. Jing, H. Kang, M. Lim, K. Sølna, and H. Wang, *Mathematical and statistical methods for multistatic imaging*, Lecture Notes in Mathematics, vol. 2098, Springer, Cham, 2013. MR3135281

[44] H. Ammari, J. Garnier, V. Jugnon, H. Kang, H. Lee, and M. Lim, *Enhancement of near-cloaking. Part III: Numerical simulations, statistical stability, and related questions*, Multi-scale and high-contrast PDE: from modelling, to mathematical analysis, to inversion, Contemp. Math., vol. 577, Amer. Math. Soc., Providence, RI, 2012, pp. 1–24, DOI 10.1090/conm/577/11460. MR2985063

[45] H. Ammari, J. Garnier, H. Kang, M. Lim, and S. Yu, *Generalized polarization tensors for shape description*, Numer. Math. **126** (2014), no. 2, 199–224, DOI 10.1007/s00211-013-0561-5. MR3150221

[46] H. Ammari, J. Garnier, H. Kang, W.-K. Park, and K. Sølna, *Imaging schemes for perfectly conducting cracks*, SIAM J. Appl. Math. **71** (2011), no. 1, 68–91, DOI 10.1137/100800130. MR2765649

[47] H. Ammari, E. Iakovleva, and D. Lesselier, *Two numerical methods for recovering small inclusions from the scattering amplitude at a fixed frequency*, SIAM J. Sci. Comput. **27** (2005), no. 1, 130–158, DOI 10.1137/040612518. MR2201178

[48] H. Ammari, E. Iakovleva, and D. Lesselier, *A MUSIC algorithm for locating small inclusions buried in a half-space from the scattering amplitude at a fixed frequency*, Multiscale Model. Simul. **3** (2005), no. 3, 597–628, DOI 10.1137/040610854. MR2136165

[49] H. Ammari, E. Iakovleva, D. Lesselier, and G. Perrusson, *MUSIC-type electromagnetic imaging of a collection of small three-dimensional inclusions*, SIAM J. Sci. Comput. **29** (2007), no. 2, 674–709, DOI 10.1137/050640655. MR2306264

[50] H. Ammari, K. Kalimeris, H. Kang, and H. Lee, *Layer potential techniques for the narrow escape problem* (English, with English and French summaries), J. Math. Pures Appl. (9) **97** (2012), no. 1, 66–84, DOI 10.1016/j.matpur.2011.09.011. MR2863765

[51] H. Ammari and H. Kang, *Reconstruction of small inhomogeneities from boundary measurements*, Lecture Notes in Mathematics, vol. 1846, Springer-Verlag, Berlin, 2004. MR2168949

[52] H. Ammari and H. Kang, *Polarization and moment tensors with applications to inverse problems and effective medium theory*, Applied Mathematical Sciences, vol. 162, Springer, New York, 2007. MR2327884

[53] H. Ammari and H. Kang, *Boundary layer techniques for solving the Helmholtz equation in the presence of small inhomogeneities*, J. Math. Anal. Appl. **296** (2004), no. 1, 190–208, DOI 10.1016/j.jmaa.2004.04.003. MR2070502

[54] H. Ammari, H. Kang, and K. Kim, *Polarization tensors and effective properties of anisotropic composite materials*, J. Differential Equations **215** (2005), no. 2, 401–428, DOI 10.1016/j.jde.2004.09.010. MR2147467

[55] H. Ammari, H. Kang, E. Kim, and H. Lee, *Vibration testing for anomaly detection*, Math. Methods Appl. Sci. **32** (2009), no. 7, 863–874, DOI 10.1002/mma.1070. MR2507936

[56] H. Ammari, H. Kang, E. Kim, and J.-Y. Lee, *The generalized polarization tensors for resolved imaging. Part II: Shape and electromagnetic parameters reconstruction of an electromagnetic inclusion from multistatic measurements*, Math. Comp. **81** (2012), no. 278, 839–860, DOI 10.1090/S0025-5718-2011-02534-2. MR2869039

[57] H. Ammari, H. Kang, E. Kim, H. Lee, and K. Louati, *Vibration analysis for detecting internal corrosion*, Stud. Appl. Math. **122** (2009), no. 1, 85–104, DOI 10.1111/j.1467-9590.2008.00424.x. MR2488327

[58] H. Ammari, H. Kang, E. Kim, K. Louati, and M. S. Vogelius, *A MUSIC-type algorithm for detecting internal corrosion from electrostatic boundary measurements*, Numer. Math. **108** (2008), no. 4, 501–528, DOI 10.1007/s00211-007-0130-x. MR2369202

[59] H. Ammari, H. Kang, and H. Lee, *Layer potential techniques in spectral analysis*, Mathematical Surveys and Monographs, vol. 153, American Mathematical Society, Providence, RI, 2009. MR2488135

[60] H. Ammari, H. Kang, H. Lee, J. Lee, and M. Lim, *Optimal estimates for the electric field in two dimensions* (English, with English and French summaries), J. Math. Pures Appl. (9) **88** (2007), no. 4, 307–324, DOI 10.1016/j.matpur.2007.07.005. MR2384571

[61] H. Ammari, H. Kang, and H. Lee, *Asymptotic expansions for eigenvalues of the Lamé system in the presence of small inclusions*, Comm. Partial Differential Equations **32** (2007), no. 10-12, 1715–1736, DOI 10.1080/03605300600910266. MR2372485

[62] H. Ammari, H. Kang, and H. Lee, *Asymptotic analysis of high-contrast phononic crystals and a criterion for the band-gap opening*, Arch. Ration. Mech. Anal. **193** (2009), no. 3, 679–714, DOI 10.1007/s00205-008-0179-4. MR2525115

[63] H. Ammari, H. Kang, H. Lee, and M. Lim, *Enhancement of near cloaking using generalized polarization tensors vanishing structures. Part I: The conductivity problem*, Comm. Math. Phys. **317** (2013), no. 1, 253–266, DOI 10.1007/s00220-012-1615-8. MR3010374

[64] H. Ammari, H. Kang, H. Lee, and M. Lim, *Enhancement of near-cloaking. Part II: The Helmholtz equation*, Comm. Math. Phys. **317** (2013), no. 2, 485–502, DOI 10.1007/s00220-012-1620-y. MR3010192

[65] H. Ammari, H. Kang, H. Lee, M. Lim, and S. Yu, *Enhancement of near cloaking for the full Maxwell equations*, SIAM J. Appl. Math. **73** (2013), no. 6, 2055–2076, DOI 10.1137/120903610. MR3127003

[66] H. Ammari, H. Kang, and M. Lim, *Effective parameters of elastic composites*, Indiana Univ. Math. J. **55** (2006), no. 3, 903–922, DOI 10.1512/iumj.2006.55.2681. MR2244590

[67] H. Ammari, H. Kang, and M. Lim, *Gradient estimates for solutions to the conductivity problem*, Math. Ann. **332** (2005), no. 2, 277–286, DOI 10.1007/s00208-004-0626-y. MR2178063

[68] H. Ammari, H. Kang, M. Lim, and H. Zribi, *Layer potential techniques in spectral analysis. Part I: Complete asymptotic expansions for eigenvalues of the Laplacian in domains with small inclusions*, Trans. Amer. Math. Soc. **362** (2010), no. 6, 2901–2922, DOI 10.1090/S0002-9947-10-04695-7. MR2592941

[69] H. Ammari, H. Kang, M. Lim, and H. Zribi, *Conductivity interface problems. I. Small perturbations of an interface*, Trans. Amer. Math. Soc. **362** (2010), no. 5, 2435–2449, DOI 10.1090/S0002-9947-09-04842-9. MR2584606

[70] H. Ammari, H. Kang, M. Lim, and H. Zribi, *The generalized polarization tensors for resolved imaging. Part I: Shape reconstruction of a conductivity inclusion*, Math. Comp. **81** (2012), no. 277, 367–386, DOI 10.1090/S0025-5718-2011-02533-0. MR2833499

[71] H. Ammari, H. Kang, G. Nakamura, and K. Tanuma, *Complete asymptotic expansions of solutions of the system of elastostatics in the presence of an inclusion of small diameter and detection of an inclusion*, J. Elasticity **67** (2002), no. 2, 97–129 (2003), DOI 10.1023/A:1023940025757. MR1985444

[72] H. Ammari, H. Kang, and F. Santosa, *Scattering of electromagnetic waves by thin dielectric planar structures*, SIAM J. Math. Anal. **38** (2006), no. 4, 1329–1342, DOI 10.1137/040618382. MR2274486

[73] H. Ammari, H. Kang, S. Soussi, and H. Zribi, *Layer potential techniques in spectral analysis. II. Sensitivity analysis of spectral properties of high contrast band-gap materials*, Multiscale Model. Simul. **5** (2006), no. 2, 646–663, DOI 10.1137/050646287. MR2247766

[74] H. Ammari, H. Kang, and K. Touibi, *Boundary layer techniques for deriving the effective properties of composite materials*, Asymptot. Anal. **41** (2005), no. 2, 119–140. MR2129229

[75] H. Ammari, H. Kang, and H. Zribi, *Sensitivity analysis with respect to the electrical conductivity*, J. Comput. Math. **25** (2007), no. 3, 244–251. MR2320232

[76] H. Ammari and A. Khelifi, *Electromagnetic scattering by small dielectric inhomogeneities* (English, with English and French summaries), J. Math. Pures Appl. (9) **82** (2003), no. 7, 749–842, DOI 10.1016/S0021-7824(03)00033-3. MR2005296

[77] H. Ammari, H. Lee, and H. Zhang, *High frequency homogenization of bubbly crystals*, to appear (arXiv:1708.07955).

[78] H. Ammari, P. Millien, M. Ruiz, and H. Zhang, *Mathematical analysis of plasmonic nanoparticles: the scalar case*, Arch. Ration. Mech. Anal. **224** (2017), no. 2, 597–658, DOI 10.1007/s00205-017-1084-5. MR3614756

[79] H. Ammari and S. Moskow, *Asymptotic expansions for eigenvalues in the presence of small inhomogeneities*, Math. Methods Appl. Sci. **26** (2003), no. 1, 67–75, DOI 10.1002/mma.343. MR1943110

[80] H. Ammari, M. Putinar, M. Ruiz, S. Yu, and H. Zhang, Shape reconstruction of nanoparticles from their associated plasmonic resonances, J. Math. Pures Appl., `https://doi.org/10.1016/j.matpur.2017.09.003`.

[81] H. Ammari, F. Romero, and M. Ruiz, *Heat generation with plasmonic nanoparticles*, Multiscale Model. Simul. **16** (2018), no. 1, 356–384, DOI 10.1137/17M1125893. MR3766961

[82] H. Ammari, M. Ruiz, W. Wu, S. Yu, and H. Zhang, *Mathematical and numerical framework for metasurfaces using thin layers of periodically distributed plasmonic nanoparticles*, Proc. A. **472** (2016), no. 2193, 20160445, 14, DOI 10.1098/rspa.2016.0445. MR3562217

[83] H. Ammari, M. Ruiz, S. Yu, and H. Zhang, *Reconstructing fine details of small objects by using plasmonic spectroscopic data*, SIAM J. Imaging Sci. **11** (2018), no. 1, 1–23, DOI 10.1137/17M1126540. MR3742683

[84] H. Ammari, M. Ruiz, S. Yu, and H. Zhang, *Reconstructing fine details of small objects by using plasmonic spectroscopic data*, SIAM J. Imaging Sci. **11** (2018), no. 1, 1–23, DOI 10.1137/17M1126540. MR3742683

[85] H. Ammari, M. Ruiz, S. Yu, and H. Zhang, *Mathematical analysis of plasmonic resonances for nanoparticles: the full Maxwell equations*, J. Differential Equations **261** (2016), no. 6, 3615–3669, DOI 10.1016/j.jde.2016.05.036. MR3527640

[86] H. Ammari and F. Santosa, *Guided waves in a photonic bandgap structure with a line defect*, SIAM J. Appl. Math. **64** (2004), no. 6, 2018–2033, DOI 10.1137/S0036139902404025. MR2110347

[87] H. Ammari, M. P. Tran, and H. Wang, *Shape identification and classification in echolocation*, SIAM J. Imaging Sci. **7** (2014), no. 3, 1883–1905, DOI 10.1137/14096164X. MR3264567

[88] H. Ammari and F. Triki, *Resonances for microstrip transmission lines*, SIAM J. Appl. Math. **64** (2003/04), no. 2, 601–636, DOI 10.1137/S0036139902418390. MR2049666

[89] H. Ammari and F. Triki, *Splitting of resonant and scattering frequencies under shape deformation*, J. Differential Equations **202** (2004), no. 2, 231–255, DOI 10.1016/j.jde.2004.02.017. MR2068440

[90] H. Ammari, M. S. Vogelius, and D. Volkov, *Asymptotic formulas for perturbations in the electromagnetic fields due to the presence of inhomogeneities of small diameter. II. The full Maxwell equations*, J. Math. Pures Appl. (9) **80** (2001), no. 8, 769–814, DOI 10.1016/S0021-7824(01)01217-X. MR1860816

[91] H. Ammari and D. Volkov, *Asymptotic formulas for perturbations in the eigenfrequencies of the full Maxwell equations due to the presence of imperfections of small diameter*, Asymptot. Anal. **30** (2002), no. 3-4, 331–350. MR1932037

[92] H. Ammari and D. Volkov, *The leading-order term in the asymptotic expansion of the scattering amplitude of a collection of finite number of dielectric inhomogeneities of small diameter*, Intern. J. Multiscale Comput. Eng. **3** (2005), 285–296.

[93] H. Ammari, W. Wu, and S. Yu, Double-negative electromagnetic metamaterials due to chirality, to appear (arXiv:1712.02863).

[94] H. Ammari and H. Zhang, *A mathematical theory of super-resolution by using a system of sub-wavelength Helmholtz resonators*, Comm. Math. Phys. **337** (2015), no. 1, 379–428, DOI 10.1007/s00220-015-2301-4. MR3324165

[95] H. Ammari and H. Zhang, *Super-resolution in high-contrast media*, Proc. A. **471** (2015), no. 2178, 20140946, 11, DOI 10.1098/rspa.2014.0946. MR3367310

[96] H. Ammari and H. Zhang, *Effective medium theory for acoustic waves in bubbly fluids near Minnaert resonant frequency*, SIAM J. Math. Anal. **49** (2017), no. 4, 3252–3276, DOI 10.1137/16M1078574. MR3690650

[97] K. Ando, Y.-G. Ji, H. Kang, K. Kim, and S. Yu, *Spectral properties of the Neumann-Poincaré operator and cloaking by anomalous localized resonance for the elasto-static system*, European J. Appl. Math. **29** (2018), no. 2, 189–225, DOI 10.1017/S0956792517000080. MR3769919

[98] K. Ando and H. Kang, *Analysis of plasmon resonance on smooth domains using spectral properties of the Neumann-Poincaré operator*, J. Math. Anal. Appl. **435** (2016), no. 1, 162–178, DOI 10.1016/j.jmaa.2015.10.033. MR3423389

[99] K. Ando, H. Kang, K. Kim, and S. Yu, *Spectrum of Neumann-Poincaré operator on annuli and cloaking by anomalous localized resonance for linear elasticity*, SIAM J. Math. Anal. **49** (2017), no. 5, 4232–4250, DOI 10.1137/17M1114089. MR3717815

[100] K. Ando, H. Kang, and Y. Miyanishi, *Elastic Neumann-Poincaré operators on three dimensional smooth domains: Polynomial compactness and spectral structure*, Int. Math. Res. Not. IMRN, `https://doi.org/10.1093/imrn/rnx258`.

[101] K. Ando, H. Kang, and H. Liu, *Plasmon resonance with finite frequencies: a validation of the quasi-static approximation for diametrically small inclusions*, SIAM J. Appl. Math. **76** (2016), no. 2, 731–749, DOI 10.1137/15M1025943. MR3479707

[102] T. Arens, *The scattering of plane elastic waves by a one-dimensional periodic surface*, Math. Methods Appl. Sci. **22** (1999), no. 1, 55–72, DOI 10.1002/(SICI)1099-1476(19990110)22:1⟨55::AID-MMA20⟩3.3.CO;2-K. MR1669572

[103] T. Arens, Scattering by biperiodic layered media: The integral equation approach, Habilitation treatise, KIT Karlsruhe, 2010.

[104] T. Arens, K. Sandfort, S. Schmitt, and A. Lechleiter, *Analysing Ewald's method for the evaluation of Green's functions for periodic media*, IMA J. Appl. Math. **78** (2013), no. 3, 405–431, DOI 10.1093/imamat/hxr057. MR3064253

[105] S. Arhab, G. Soriano, Y. Ruan, G. Maire, A. Talneau, D. Sentenac, P.C. Chaumet, K. Belkebir, and H. Giovannini, *Nanometric resolution with far-field optical profilometry*, Phys. Rev. Lett. **111** (2013), 053902.

[106] D. H. Armitage and S. J. Gardiner, *Classical potential theory*, Springer Monographs in Mathematics, Springer-Verlag London, Ltd., London, 2001. MR1801253

[107] K. Astala, L. Päivärinta, and E. Saksman, *The finite Hilbert transform in weighted spaces*, Proc. Roy. Soc. Edinburgh Sect. A **126** (1996), no. 6, 1157–1167, DOI 10.1017/S0308210500023337. MR1424219

[108] C. E. Athanasiadis, G. Pelekanos, V. Sevroglou, and I. G. Stratis, *On the scattering of two-dimensional elastic point sources and related near-field inverse problems for small discs*, Proc. Roy. Soc. Edinburgh Sect. A **139** (2009), no. 4, 719–741, DOI 10.1017/S0308210507001059. MR2520552

[109] C. Athanasiadis, V. Sevroglou, and I. G. Stratis, *Scattering relations for point-generated dyadic fields in two-dimensional linear elasticity*, Quart. Appl. Math. **64** (2006), no. 4, 695–710, DOI 10.1090/S0033-569X-06-01041-0. MR2284466

[110] C. E. Athanasiadis, V. Sevroglou, and I. G. Stratis, *3D elastic scattering theorems for point-generated dyadic fields*, Math. Methods Appl. Sci. **31** (2008), no. 8, 987–1003, DOI 10.1002/mma.964. MR2418675

[111] H.A. Atwater and A. Polman, *Plasmonics for improved photovoltaic devices*, Nature Materials **9** (2010), 205–213.

[112] A. Ávila, G. Griso, and B. Miara, *Bandes phononiques interdites en élasticité linéarisée* (French, with English and French summaries), C. R. Math. Acad. Sci. Paris **340** (2005), no. 12, 933–938, DOI 10.1016/j.crma.2005.04.026. MR2152282

[113] W. Axmann and P. Kuchment, *An efficient finite element method for computing spectra of photonic and acoustic band-gap materials. I. Scalar case*, J. Comput. Phys. **150** (1999), no. 2, 468–481, DOI 10.1006/jcph.1999.6188. MR1684898

[114] G. Baffou, C. Girard, and R. Quidant, *Mapping heat origin in plasmonic structures*, Phys. Rev. Lett. **104** (2010), 136805.

[115] H. T. Banks, M. L. Joyner, B. Wincheski, and W. P. Winfree, *Real time computational algorithms for eddy-current-based damage detection*, Inverse Problems **18** (2002), no. 3, 795–823, DOI 10.1088/0266-5611/18/3/318. MR1910203

[116] E. S. Bao, Y. Y. Li, and B. Yin, *Gradient estimates for the perfect conductivity problem*, Arch. Ration. Mech. Anal. **193** (2009), no. 1, 195–226, DOI 10.1007/s00205-008-0159-8. MR2506075

[117] E. S. Bao, Y. Y. Li, and B. Yin, *Gradient estimates for the perfect and insulated conductivity problems with multiple inclusions*, Comm. Partial Differential Equations **35** (2010), no. 11, 1982–2006, DOI 10.1080/03605300903564000. MR2754076

[118] G. Bao, *A uniqueness theorem for an inverse problem in periodic diffractive optics*, Inverse Problems **10** (1994), no. 2, 335–340. MR1269011

[119] G. Bao, *Finite element approximation of time harmonic waves in periodic structures*, SIAM J. Numer. Anal. **32** (1995), no. 4, 1155–1169, DOI 10.1137/0732053. MR1342287

[120] G. Bao, *An inverse diffraction problem in periodic structures*, Mathematical and numerical aspects of wave propagation (Mandelieu-La Napoule, 1995), SIAM, Philadelphia, PA, 1995, pp. 694–703. MR1328231

[121] G. Bao, *Numerical analysis of diffraction by periodic structures: TM polarization*, Numer. Math. **75** (1996), no. 1, 1–16, DOI 10.1007/s002110050227. MR1417860

[122] G. Bao, *Variational approximation of Maxwell's equations in biperiodic structures*, SIAM J. Appl. Math. **57** (1997), no. 2, 364–381, DOI 10.1137/S0036139995279408. MR1438758

[123] G. Bao and E. Bonnetier, *Optimal design of periodic diffractive structures*, Appl. Math. Optim. **43** (2001), no. 2, 103–116, DOI 10.1007/s00245-001-0004-z. MR1814589

[124] G. Bao, Y. Cao, and H. Yang, *Numerical solution of diffraction problems by a least-squares finite element method*, Math. Methods Appl. Sci. **23** (2000), no. 12, 1073–1092, DOI 10.1002/1099-1476(200008)23:12⟨1073::AID-MMA152⟩3.0.CO;2-D. MR1773923

[125] G. Bao and Y. Chen, *A nonlinear grating problem in diffractive optics*, SIAM J. Math. Anal. **28** (1997), no. 2, 322–337, DOI 10.1137/S0036141095284461. MR1434038

[126] G. Bao and D. C. Dobson, *Second harmonic generation in nonlinear optical films*, J. Math. Phys. **35** (1994), no. 4, 1622–1633, DOI 10.1063/1.530559. MR1267910

[127] G. Bao and D. C. Dobson, *Diffractive optics in nonlinear media with periodic structure*, European J. Appl. Math. **6** (1995), no. 6, 573–590, DOI 10.1017/S0956792500002072. MR1364718

[128] G. Bao and D. C. Dobson, *Modeling and optimal design of diffractive optical structures*, Surveys Math. Indust. **8** (1998), no. 1, 37–62. MR1658983

[129] G. Bao and D. C. Dobson, *Variational methods for diffractive optics modeling*, Mathematical modeling in optical science, Frontiers Appl. Math., vol. 22, SIAM, Philadelphia, PA, 2001, pp. 37–69. MR1831329

[130] G. Bao and D. C. Dobson, *On the scattering by a biperiodic structure*, Proc. Amer. Math. Soc. **128** (2000), no. 9, 2715–2723, DOI 10.1090/S0002-9939-00-05509-X. MR1694448

[131] G. Bao, D. C. Dobson, and J. A. Cox, *Mathematical studies in rigorous grating theory*, J. Opt. Soc. Amer. A **12** (1995), no. 5, 1029–1042, DOI 10.1364/JOSAA.12.001029. MR1335399

[132] G. Bao and A. Friedman, *Inverse problems for scattering by periodic structures*, Arch. Rational Mech. Anal. **132** (1995), no. 1, 49–72, DOI 10.1007/BF00390349. MR1360080

[133] G. Bao, J. Lin, and F. Triki, *A multi-frequency inverse source problem*, J. Differential Equations **249** (2010), no. 12, 3443–3465, DOI 10.1016/j.jde.2010.08.013. MR2737437

[134] G. Bao and Z. Zhou, *An inverse problem for scattering by a doubly periodic structure*, Trans. Amer. Math. Soc. **350** (1998), no. 10, 4089–4103, DOI 10.1090/S0002-9947-98-02227-2. MR1487607

[135] G. Bao and H. Yang, *A least-squares finite element analysis for diffraction problems*, SIAM J. Numer. Anal. **37** (2000), no. 2, 665–682, DOI 10.1137/S0036142998342380. MR1740763

[136] G. Bao and H. Liu, *Nearly cloaking the electromagnetic fields*, SIAM J. Appl. Math. **74** (2014), no. 3, 724–742, DOI 10.1137/130939298. MR3213806

[137] G. Bao, H. Liu, and J. Zou, *Nearly cloaking the full Maxwell equations: cloaking active contents with general conducting layers* (English, with English and French summaries), J. Math. Pures Appl. (9) **101** (2014), no. 5, 716–733, DOI 10.1016/j.matpur.2013.10.010. MR3192429

[138] G. Bao, H. Zhang, and J. Zou, *Unique determination of periodic polyhedral structures by scattered electromagnetic fields*, Trans. Amer. Math. Soc. **363** (2011), no. 9, 4527–4551, DOI 10.1090/S0002-9947-2011-05334-1. MR2806682

[139] G. Bao, H. Zhang, and J. Zou, *Unique determination of periodic polyhedral structures by scattered electromagnetic fields II: The resonance case*, Trans. Amer. Math. Soc. **366** (2014), no. 3, 1333–1361, DOI 10.1090/S0002-9947-2013-05761-3. MR3145733

[140] A. Barnett and L. Greengard, *A new integral representation for quasi-periodic fields and its application to two-dimensional band structure calculations*, J. Comput. Phys. **229** (2010), no. 19, 6898–6914, DOI 10.1016/j.jcp.2010.05.029. MR2677757

[141] A. Barnett and L. Greengard, *A new integral representation for quasi-periodic scattering problems in two dimensions*, BIT **51** (2011), no. 1, 67–90, DOI 10.1007/s10543-010-0297-x. MR2784653

[142] H. Bart, M. A. Kaashoek, and D. C. Lay, *The integral formula for the reduced algebraic multiplicity of meromorphic operator functions*, Proc. Edinburgh Math. Soc. (2) **21** (1978/79), no. 1, 65–71, DOI 10.1017/S0013091500015893. MR0500205

[143] J. T. Beale, *Scattering frequencies of reasonators*, Comm. Pure Appl. Math. **26** (1973), 549–563, DOI 10.1002/cpa.3160260408. MR0352730

[144] M. P. Bendsøe and O. Sigmund, *Topology optimization. Theory, methods and applications*, Springer-Verlag, Berlin, 2003. MR2008524

[145] E. Beretta and E. Francini, *Asymptotic formulas for perturbations in the electromagnetic fields due to the presence of thin inhomogeneities*, Inverse problems: Theory and applications (Cortona/Pisa, 2002), Contemp. Math., vol. 333, Amer. Math. Soc., Providence, RI, 2003, pp. 49–62, DOI 10.1090/conm/333/05953. MR2032006

[146] E. Beretta and E. Francini, *An asymptotic formula for the displacement field in the presence of thin elastic inhomogeneities*, SIAM J. Math. Anal. **38** (2006), no. 4, 1249–1261, DOI 10.1137/050648596. MR2274482

[147] E. Beretta, E. Francini, and M. S. Vogelius, *Asymptotic formulas for steady state voltage potentials in the presence of thin inhomogeneities. A rigorous error analysis* (English, with English and French summaries), J. Math. Pures Appl. (9) **82** (2003), no. 10, 1277–1301, DOI 10.1016/S0021-7824(03)00081-3. MR2020923

[148] E. Beretta, A. Mukherjee, and M. Vogelius, *Asymptotic formulas for steady state voltage potentials in the presence of conductivity imperfections of small area*, Z. Angew. Math. Phys. **52** (2001), no. 4, 543–572, DOI 10.1007/PL00001561. MR1856987

[149] J. Bergh and J. Löfström, *Interpolation spaces. An introduction*, Springer-Verlag, Berlin-New York, 1976. Grundlehren der Mathematischen Wissenschaften, No. 223. MR0482275

[150] G. Besson, *Comportement asymptotique des valeurs propres du laplacien dans un domaine avec un trou* (French, with English summary), Bull. Soc. Math. France **113** (1985), no. 2, 211–230. MR820320

[151] D. Boffi, M. Conforti, and L. Gastaldi, *Modified edge finite elements for photonic crystals*, Numer. Math. **105** (2006), no. 2, 249–266, DOI 10.1007/s00211-006-0037-y. MR2262758

[152] E. Bonnetier and F. Triki, *On the spectrum of the Poincaré variational problem for two close-to-touching inclusions in 2D*, Arch. Ration. Mech. Anal. **209** (2013), no. 2, 541–567, DOI 10.1007/s00205-013-0636-6. MR3056617

[153] E. Bonnetier and H. Zhang, *Characterization of the essential spectrum of the Neumann-Poincaré operator in 2D domains with corner via Weyl sequences*, Revista Matematica Iberoamericana, to appear (arxiv: 1702.08127).

[154] L. Borcea and J. Garnier, *Robust imaging with electromagnetic waves in noisy environments*, Inverse Problems **32** (2016), no. 10, 105010, 30, DOI 10.1088/0266-5611/32/10/105010. MR3627034

[155] P.I. Borel, A. Harpøth, L.H. Frandsen, M. Kristensen, P. Shi, J.S. Jensen, and O. Sigmund, *Topology optimization and fabrication of photonic crystal structures*, Optics Express **12** (2004), 1996–2001.

[156] M. Born and E. Wolf, *Principles of optics: Electromagnetic theory of propagation, interference and diffraction of light*, With contributions by A. B. Bhatia, P. C. Clemmow, D. Gabor, A. R. Stokes, A. M. Taylor, P. A. Wayman and W. L. Wilcock. Third revised edition, Pergamon Press, Oxford-New York-Paris, 1965. MR0198807

[157] D. Borthwick and P. Perry, *Scattering poles for asymptotically hyperbolic manifolds*, Trans. Amer. Math. Soc. **354** (2002), no. 3, 1215–1231, DOI 10.1090/S0002-9947-01-02906-3. MR1867379

[158] J. M. Borwein, M. L. Glasser, R. C. McPhedran, J. G. Wan, and I. J. Zucker, *Lattice sums then and now*, Encyclopedia of Mathematics and its Applications, vol. 150, Cambridge University Press, Cambridge, 2013. With a foreword by Helaman Ferguson and Claire Ferguson. MR3135109

[159] L.C. Botten, R.C. McPhedran, N.A. Nicorovici, A.A. Asatryan, C.M. de Sterke, P.A. Robinson, K. Busch, G.H. Smith, and T.N. Langtry, *Rayleigh multipole methods for photonic crystal calculations*, Prog. Electromag. Res. **41** (2003), 21–60.

[160] G. Bouchitté, C. Bourel, and D. Felbacq, *Homogenization near resonances and artificial magnetism in three dimensional dielectric metamaterials*, Arch. Ration. Mech. Anal. **225** (2017), no. 3, 1233–1277, DOI 10.1007/s00205-017-1132-1. MR3667288

[161] G. Bouchitté, C. Bourel, and L. Manca, *Resonant effects in random dielectric structures*, ESAIM Control Optim. Calc. Var. **21** (2015), no. 1, 217–246, DOI 10.1051/cocv/2014026. MR3348421

[162] G. Bouchitté and D. Felbacq, *Homogenization near resonances and artificial magnetism from dielectrics* (English, with English and French summaries), C. R. Math. Acad. Sci. Paris **339** (2004), no. 5, 377–382, DOI 10.1016/j.crma.2004.06.018. MR2092467

[163] G. Bouchitté and B. Schweizer, *Homogenization of Maxwell's equations in a split ring geometry*, Multiscale Model. Simul. **8** (2010), no. 3, 717–750, DOI 10.1137/09074557X. MR2609637

[164] G. Bouchitté and B. Schweizer, *Plasmonic waves allow perfect transmission through sub-wavelength metallic gratings*, Netw. Heterog. Media **8** (2013), no. 4, 857–878, DOI 10.3934/nhm.2013.8.857. MR3180689

[165] R. Brown, P. D. Hislop, and A. Martinez, *Eigenvalues and resonances for domains with tubes: Neumann boundary conditions*, J. Differential Equations **115** (1995), no. 2, 458–476, DOI 10.1006/jdeq.1995.1023. MR1310941

[166] M. Brühl, M. Hanke, and M. S. Vogelius, *A direct impedance tomography algorithm for locating small inhomogeneities*, Numer. Math. **93** (2003), no. 4, 635–654, DOI 10.1007/s002110200409. MR1961882

[167] O.P. Bruno and B. Delourme, *Rapidly convergent two-dimensional quasi-periodic Green function throughout the spectrum including Wood anomalies*, J. Comput. Phys. **262** (2014), 262–290.

[168] O. P. Bruno and A. G. Fernandez-Lado, *Rapidly convergent quasi-periodic Green functions for scattering by arrays of cylinders—including Wood anomalies*, Proc. A. **473** (2017), no. 2199, 20160802, 23, DOI 10.1098/rspa.2016.0802. MR3641599

[169] O. Bruno and S. Lintner, *Superlens-cloaking of small dielectric bodies in the quasistatic regime*, J. Appl. Phys. **102** (2007), 124502.

[170] O. P. Bruno and F. Reitich, *Solution of a boundary value problem for the Helmholtz equation via variation of the boundary into the complex domain*, Proc. Roy. Soc. Edinburgh Sect. A **122** (1992), no. 3-4, 317–340, DOI 10.1017/S0308210500021132. MR1200203

[171] O. Bruno and F. Reitich, *Numerical solution of diffraction problems: A method of variation of boundaries*, J. Opt. Soc. Amer. A. **10** (1993), 1168–1175.

[172] O. Bruno and F. Reitich, *Numerical solution of diffractive problems: A method of variation of boundaries II*, Dielectric gratings, Pade approximants and singularities, J. Opt. Soc. Amer. A. **10** (1993), 2307–2316.

[173] O. Bruno and F. Reitich, *Accurate calculation of diffractive grating efficiencies*, SPIE, 1919, Mathematics in Smart Structures (1993), 236–247.

[174] O. Bruno and F. Reitich, *Numerical solution of diffraction problems: A method of variation of boundaries III. Doubly-periodic gratings*, J. Opt. Soc. Amer. **10** (1993), 2551–2562.

[175] O. P. Bruno and F. Reitich, *Approximation of analytic functions: a method of enhanced convergence*, Math. Comp. **63** (1994), no. 207, 195–213, DOI 10.2307/2153569. MR1240654

[176] O. P. Bruno and F. Reitich, *Boundary-variation solution of eigenvalue problems for elliptic operators*, J. Fourier Anal. Appl. **7** (2001), no. 2, 169–187, DOI 10.1007/BF02510422. MR1817674

[177] O. P. Bruno, S. P. Shipman, C. Turc, and S. Venakides, *Three-dimensional quasi-periodic shifted Green function throughout the spectrum, including Wood anomalies*, Proc. A. **473** (2017), no. 2207, 20170242, 18. MR3741839

[178] O. P. Bruno, S. P. Shipman, C. Turc, and S. Venakides, *Superalgebraically convergent smoothly windowed lattice sums for doubly periodic Green functions in three-dimensional*

space, Proc. A. **472** (2016), no. 2191, 20160255, 19, DOI 10.1098/rspa.2016.0255. MR3540739

[179] K. Bryan and T. Leise, *Impedance imaging, inverse problems, and Harry Potter's cloak*, SIAM Rev. **52** (2010), no. 2, 359–377, DOI 10.1137/090757873. MR2646807

[180] A. Buffa, M. Costabel, and D. Sheen, *On traces for* $\mathbf{H}(\mathbf{curl}, \Omega)$ *in Lipschitz domains*, J. Math. Anal. Appl. **276** (2002), no. 2, 845–867, DOI 10.1016/S0022-247X(02)00455-9. MR1944792

[181] V. I. Burenkov and P. D. Lamberti, *Spectral stability of Dirichlet second order uniformly elliptic operators*, J. Differential Equations **244** (2008), no. 7, 1712–1740, DOI 10.1016/j.jde.2007.12.009. MR2404437

[182] M. Burger, B. Hackl, and W. Ring, *Incorporating topological derivatives into level set methods*, J. Comput. Phys. **194** (2004), no. 1, 344–362, DOI 10.1016/j.jcp.2003.09.033. MR2033389

[183] M. Burger, S.J. Osher, and E. Yablonovitch, *Inverse problem techniques for the design of photonic crystals*, IEICE Trans. Electron., E87-C (2004), 258–265.

[184] M. Burger and S. J. Osher, *A survey on level set methods for inverse problems and optimal design*, European J. Appl. Math. **16** (2005), no. 2, 263–301, DOI 10.1017/S0956792505006182. MR2155293

[185] G. Buttazzo and R. V. Kohn, *Reinforcement by a thin layer with oscillating thickness*, Appl. Math. Optim. **16** (1987), no. 3, 247–261, DOI 10.1007/BF01442194. MR901816

[186] E. J. Candès and C. Fernandez-Granda, *Towards a mathematical theory of super-resolution*, Comm. Pure Appl. Math. **67** (2014), no. 6, 906–956, DOI 10.1002/cpa.21455. MR3193963

[187] E. J. Candès and C. Fernandez-Granda, *Super-resolution from noisy data*, J. Fourier Anal. Appl. **19** (2013), no. 6, 1229–1254, DOI 10.1007/s00041-013-9292-3. MR3132912

[188] Y. Capdeboscq and H. Kang, *Improved Hashin-Shtrikman bounds for elastic moment tensors and an application*, Appl. Math. Optim. **57** (2008), no. 2, 263–288, DOI 10.1007/s00245-007-9022-9. MR2386106

[189] Y. Capdeboscq and M. S. Vogelius, *A review of some recent work on impedance imaging for inhomogeneities of low volume fraction*, Partial differential equations and inverse problems, Contemp. Math., vol. 362, Amer. Math. Soc., Providence, RI, 2004, pp. 69–87, DOI 10.1090/conm/362/06606. MR2091492

[190] F. Capolino, D. R. Wilton, and W. A. Johnson, *Efficient computation of the 2-D Green's function for 1-D periodic structures using the Ewald method*, IEEE Trans. Antennas and Propagation **53** (2005), no. 9, 2977–2984, DOI 10.1109/TAP.2005.854556. MR2228979

[191] T. Carleman, *Sur le résolution des certaines équations intégrales*, Arkiv. Mat. Astron. Fys. **16** (1922), 1–19.

[192] D. J. Cedio-Fengya, S. Moskow, and M. S. Vogelius, *Identification of conductivity imperfections of small diameter by boundary measurements. Continuous dependence and computational reconstruction*, Inverse Problems **14** (1998), no. 3, 553–595, DOI 10.1088/0266-5611/14/3/011. MR1629995

[193] A. Chai, M. Moscoso, and G. Papanicolaou, *Robust imaging of localized scatterers using the singular value decomposition and* ℓ_1 *minimization*, Inverse Problems **29** (2013), no. 2, 025016, 28, DOI 10.1088/0266-5611/29/2/025016. MR3020437

[194] T. K. Chang and H. J. Choe, *Spectral properties of the layer potentials associated with elasticity equations and transmission problems on Lipschitz domains*, J. Math. Anal. Appl. **326** (2007), no. 1, 179–191, DOI 10.1016/j.jmaa.2006.03.001. MR2277775

[195] X. Chen and A. Friedman, *Maxwell's equations in a periodic structure*, Trans. Amer. Math. Soc. **323** (1991), no. 2, 465–507, DOI 10.2307/2001542. MR1010883

[196] H. Cheng, *On the method of images for systems of closely spaced conducting spheres*, SIAM J. Appl. Math. **61** (2000/01), no. 4, 1324–1337, DOI 10.1137/S0036139999364992. MR1813682

[197] H. Cheng, W.Y. Crutchfield, M. Doery, and L. Greengard, *Fast, accurate integral equation methods for the analysis of photonic crystal fibers I: Theory*, Optics Express **12** (2004), 3791–3805.

[198] H. Cheng and L. Greengard, *A method of images for the evaluation of electrostatic fields in systems of closely spaced conducting cylinders*, SIAM J. Appl. Math. **58** (1998), no. 1, 122–141, DOI 10.1137/S0036139996297614. MR1610021

[199] M. H. Cho, W. Cai, and T.-H. Her, *A boundary integral equation method for photonic crystal fibers*, J. Sci. Comput. **28** (2006), no. 2-3, 263–278, DOI 10.1007/s10915-006-9080-1. MR2272632

[200] D. Chung, H. Kang, K. Kim, and H. Lee, *Cloaking due to anomalous localized resonance in plasmonic structures of confocal ellipses*, SIAM J. Appl. Math. **74** (2014), no. 5, 1691–1707, DOI 10.1137/140956762. MR3270192

[201] F. H. Clarke, Yu. S. Ledyaev, R. J. Stern, and P. R. Wolenski, *Nonsmooth analysis and control theory*, Graduate Texts in Mathematics, vol. 178, Springer-Verlag, New York, 1998. MR1488695

[202] R. R. Coifman, A. McIntosh, and Y. Meyer, *L'intégrale de Cauchy définit un opérateur borné sur L^2 pour les courbes lipschitziennes* (French), Ann. of Math. (2) **116** (1982), no. 2, 361–387, DOI 10.2307/2007065. MR672839

[203] R.E. Collin, *Field Theory of Guided Waves*, Second Edition, IEEE Press, New York, 1991.

[204] D. Colton and R. Kress, *Inverse acoustic and electromagnetic scattering theory*, 2nd ed., Applied Mathematical Sciences, vol. 93, Springer-Verlag, Berlin, 1998. MR1635980

[205] J. M. Combes and L. Thomas, *Asymptotic behaviour of eigenfunctions for multiparticle Schrödinger operators*, Comm. Math. Phys. **34** (1973), 251–270. MR0391792

[206] J. A. Cox, *Inverse and optimal design problems for imaging and diffractive optical systems*, in Inverse Problems and Optimal Design in Industry, ed. H. Engl and J. McLaughlin, 27–36, B.G. Teubner, Stuttgart, 1994.

[207] J. A. Cox and D. Dobson, *An integral equation method for biperiodic diffraction structures*, in J. Lerner and W. McKinney, ed., "International Conference on the Applications and Theory of Periodic Structures", Proc. SPIE 1545 (1991), 106–113

[208] J. A. Cox and D. Dobson, *Mathematical modeling for diffractive optics*, in "Diffractive and Miniaturized Optics (Critical Reviews)" Sing Lee, ed., SPIE vol. CR-49, 32–53 (1994).

[209] G. Courtois, *Spectrum of manifolds with holes*, J. Funct. Anal. **134** (1995), no. 1, 194–221, DOI 10.1006/jfan.1995.1142. MR1359926

[210] S.J. Cox and D.C. Dobson, *Band structure optimization of two-dimensional photonic crystals in H-polarization*, J. Comput. Phys. **158** (2000), 214–224.

[211] W.Y. Crutchfield, H. Cheng, and L. Greengard, *Sensitivity analysis of photonic crystal fiber*, Optics Express **12** (2004), 4220–4226.

[212] B. E. J. Dahlberg, C. E. Kenig, and G. C. Verchota, *Boundary value problems for the systems of elastostatics in Lipschitz domains*, Duke Math. J. **57** (1988), no. 3, 795–818, DOI 10.1215/S0012-7094-88-05735-3. MR975122

[213] D. Daners, *Dirichlet problems on varying domains*, J. Differential Equations **188** (2003), no. 2, 591–624, DOI 10.1016/S0022-0396(02)00105-5. MR1955096

[214] G. Dassios, *Ellipsoidal harmonics: Theory and applications*, Encyclopedia of Mathematics and its Applications, vol. 146, Cambridge University Press, Cambridge, 2012. MR2977792

[215] A. Dienstfrey, F. Hang, and J. Huang, *Lattice sums and the two-dimensional, periodic Green's function for the Helmholtz equation*, R. Soc. Lond. Proc. Ser. A Math. Phys. Eng. Sci. **457** (2001), no. 2005, 67–85, DOI 10.1098/rspa.2000.0656. MR1843935

[216] D.C. Dobson, An efficient method for band structure calculations in 2D photonic crystals, J. Comput. Phys., 149 (1999), 363–376.

[217] D. C. Dobson, *Optimal design of periodic antireflective structures for the Helmholtz equation*, European J. Appl. Math. **4** (1993), no. 4, 321–339, DOI 10.1017/S0956792500001169. MR1251818

[218] D. C. Dobson, *A variational method for electromagnetic diffraction in biperiodic structures*, RAIRO Modél. Math. Anal. Numér. **28** (1994), no. 4, 419–439, DOI 10.1051/m2an/1994280404191. MR1288506

[219] D. C. Dobson, *Optimal shape design of blazed diffraction gratings*, Appl. Math. Optim. **40** (1999), no. 1, 61–78, DOI 10.1007/s002459900116. MR1686606

[220] D. Dobson and A. Friedman, *The time-harmonic Maxwell equations in a doubly periodic structure*, J. Math. Anal. Appl. **166** (1992), no. 2, 507–528, DOI 10.1016/0022-247X(92)90312-2. MR1160941

[221] D. C. Dobson and L. B. Simeonova, *Optimization of periodic composite structures for subwavelength focusing*, Appl. Math. Optim. **60** (2009), no. 1, 133–150, DOI 10.1007/s00245-008-9063-8. MR2511789

[222] D. C. Dobson, J. Gopalakrishnan, and J. E. Pasciak, *An efficient method for band structure calculations in 3D photonic crystals*, J. Comput. Phys. **161** (2000), no. 2, 668–679, DOI 10.1006/jcph.2000.6521. MR1764252

[223] D. C. Dobson and J. E. Pasciak, *Analysis of an algorithm for computing electromagnetic Bloch modes using Nedelec spaces*, Comput. Methods Appl. Math. **1** (2001), no. 2, 138–153. MR1854308

[224] O. Dorn and D. Lesselier, *Level set methods for inverse scattering*, Inverse Problems **22** (2006), no. 4, R67–R131, DOI 10.1088/0266-5611/22/4/R01. MR2249455

[225] O. Dorn, E. L. Miller, and C. M. Rappaport, *A shape reconstruction method for electromagnetic tomography using adjoint fields and level sets. Electromagnetic imaging and inversion of the Earth's subsurface*, Inverse Problems **16** (2000), no. 5, 1119–1156, DOI 10.1088/0266-5611/16/5/303. MR1798351

[226] K. Dossou, M. A. Byrne, and L. C. Botten, *Finite element computation of grating scattering matrices and application to photonic crystal band calculations*, J. Comput. Phys. **219** (2006), no. 1, 120–143, DOI 10.1016/j.jcp.2006.03.029. MR2273370

[227] E.N. Economou and M. Sigalas, *Stop bands for elastic waves in periodic composite materials*, J. Acoust. Soc. Am. **95** (4) (1994), 1734–1740.

[228] D. Elliott and S. Okada, *The finite Hilbert transform and weighted Sobolev spaces*, Math. Nachr. **266** (2004), 34–47, DOI 10.1002/mana.200310142. MR2040334

[229] J. Elschner and G. Hu, *Variational approach to scattering of plane elastic waves by diffraction gratings*, Math. Methods Appl. Sci. **33** (2010), no. 16, 1924–1941, DOI 10.1002/mma.1305. MR2743411

[230] J. Elschner and G. Hu, *Scattering of plane elastic waves by three-dimensional diffraction gratings*, Math. Models Methods Appl. Sci. **22** (2012), no. 4, 1150019, 34 pp., DOI 10.1142/S0218202511500199. MR2902151

[231] J. Elschner and G. Hu, *An optimization method in inverse elastic scattering for one-dimensional grating profiles*, Commun. Comput. Phys. **12** (2012), no. 5, 1434–1460, DOI 10.4208/cicp.220611.130112a. MR2927998

[232] C. Engström, H. Langer, and C. Tretter, *Rational eigenvalue problems and applications to photonic crystals*, J. Math. Anal. Appl. **445** (2017), no. 1, 240–279, DOI 10.1016/j.jmaa.2016.07.048. MR3543766

[233] L. Escauriaza and J. K. Seo, *Regularity properties of solutions to transmission problems*, Trans. Amer. Math. Soc. **338** (1993), no. 1, 405–430, DOI 10.2307/2154462. MR1149120

[234] P.P. Ewald, *The calculation of optical and electrostatic grid potential*, Ann. Phys **64** (1921), 253–287.

[235] S. Fan, P.R. Villeneuve, and J.D. Joannopoulos, Channel-drop filters in photonic crystals, Optics Express, 3 (1998), 4–11.

[236] T. Feng and H. Kang, *Spectrum of the Neumann-Poincaré operator for ellipsoids and tunability*, Integral Equations Operator Theory **84** (2016), no. 4, 591–599, DOI 10.1007/s00020-016-2280-7. MR3483876

[237] C. A. Fernández, *Resonances in scattering by a resonator*, Indiana Univ. Math. J. **34** (1985), no. 1, 115–125, DOI 10.1512/iumj.1985.34.34007. MR773397

[238] C. Fernández and G. P. Menzala, *Resonances of an elastic resonator*, Appl. Anal. **76** (2000), no. 1-2, 41–49, DOI 10.1080/00036810008840864. MR1802818

[239] A. Figotin and P. Kuchment, *Band-gap structure of spectra of periodic dielectric and acoustic media. I. Scalar model*, SIAM J. Appl. Math. **56** (1996), no. 1, 68–88, DOI 10.1137/S0036139994263859. MR1372891

[240] A. Figotin and P. Kuchment, *Band-gap structure of spectra of periodic dielectric and acoustic media. II. Two-dimensional photonic crystals*, SIAM J. Appl. Math. **56** (1996), no. 6, 1561–1620, DOI 10.1137/S0036139995285236. MR1417473

[241] A. Figotin and P. Kuchment, *Spectral properties of classical waves in high-contrast periodic media*, SIAM J. Appl. Math. **58** (1998), no. 2, 683–702, DOI 10.1137/S0036139996297249. MR1617610

[242] N. Filonov, *Gaps in the spectrum of the Maxwell operator with periodic coefficients*, Comm. Math. Phys. **240** (2003), no. 1-2, 161–170, DOI 10.1007/s00220-003-0904-7. MR2004984

[243] G. B. Folland, *Introduction to partial differential equations*, 2nd ed., Princeton University Press, Princeton, NJ, 1995. MR1357411

[244] S. P. Fortes, R. P. Lipton, and S. P. Shipman, *Convergent power series for fields in positive or negative high-contrast periodic media*, Comm. Partial Differential Equations **36** (2011), no. 6, 1016–1043, DOI 10.1080/03605302.2010.531860. MR2765428

[245] S. P. Fortes, R. P. Lipton, and S. P. Shipman, *Sub-wavelength plasmonic crystals: dispersion relations and effective properties*, Proc. R. Soc. Lond. Ser. A Math. Phys. Eng. Sci. **466** (2010), no. 2119, 1993–2020, DOI 10.1098/rspa.2009.0542. MR2652730

[246] L. Friedlander, *On the density of states of periodic media in the large coupling limit*, Comm. Partial Differential Equations **27** (2002), no. 1-2, 355–380, DOI 10.1081/PDE-120002790. MR1886963

[247] A. Friedman, *Detection of mines by electric measurements*, SIAM J. Appl. Math. **47** (1987), no. 1, 201–212, DOI 10.1137/0147012. MR873244

[248] A. Friedman and B. Gustafsson, *Identification of the conductivity coefficient in an elliptic equation*, SIAM J. Math. Anal. **18** (1987), no. 3, 777–787, DOI 10.1137/0518059. MR883568

[249] A. Friedman and M. Vogelius, *Identification of small inhomogeneities of extreme conductivity by boundary measurements: a theorem on continuous dependence*, Arch. Rational Mech. Anal. **105** (1989), no. 4, 299–326, DOI 10.1007/BF00281494. MR973245

[250] D. Fujiwara and S. Ozawa, *The Hadamard variational formula for the Green functions of some normal elliptic boundary value problems*, Proc. Japan Acad. Ser. A Math. Sci. **54** (1978), no. 8, 215–220. MR517324

[251] R. R. Gadyl′shin, *On the influence of the location and shape of a hole on the properties of the acoustic Helmholtz resonator* (Russian, with English and Russian summaries), Teoret. Mat. Fiz. **93** (1992), no. 1, 107–118, DOI 10.1007/BF01016473; English transl., Theoret. and Math. Phys. **93** (1992), no. 1, 1151–1159 (1993). MR1226213

[252] R. R. Gadyl′shin, *Surface potentials and the method of matching asymptotic expansions in the Helmholtz resonator problem* (Russian, with Russian summary), Algebra i Analiz **4** (1992), no. 2, 88–115; English transl., St. Petersburg Math. J. **4** (1993), no. 2, 273–296. MR1182395

[253] R. R. Gadyl′shin, *On scattering frequencies of acoustic resonator* (English, with English and French summaries), C. R. Acad. Sci. Paris Sér. I Math. **316** (1993), no. 9, 959–963. MR1218297

[254] R. R. Gadyl′shin, *On acoustic Helmholtz resonator and on its electromagnetic analog*, J. Math. Phys. **35** (1994), no. 7, 3464–3481, DOI 10.1063/1.530478. MR1279315

[255] R. R. Gadyl′shin, *Asymptotics of scattering frequencies with small imaginary parts for an acoustic resonator* (English, with English and French summaries), RAIRO Modél. Math. Anal. Numér. **28** (1994), no. 6, 761–780, DOI 10.1051/m2an/1994280607611. MR1302423

[256] R.P. Gadyl'shin, *Existence and asymptotics of poles with small imaginary part for the Helmholtz resonator*, Russ. Math. Surv. **52** (1997), 1–72.

[257] R. R. Gadyl′shin, *Asymptotics of poles with small imaginary part for a complete electromagnetic analog of the Helmholtz resonator*, Russian J. Math. Phys. **5** (1997), no. 3, 307–320 (1998). MR1605633

[258] R. R. Gadyl′shin, *Asymptotics of the eigenvalues of a boundary value problem with rapidly oscillating boundary conditions* (Russian, with Russian summary), Differ. Uravn. **35** (1999), no. 4, 540–551, 575; English transl., Differential Equations **35** (1999), no. 4, 540–551. MR1719780

[259] R. R. Gadyl′shin, *On the asymptotics of eigenvalues for a periodically fixed membrane* (Russian, with Russian summary), Algebra i Analiz **10** (1998), no. 1, 3–19; English transl., St. Petersburg Math. J. **10** (1999), no. 1, 1–14. MR1618345

[260] R.R. Gadyl'shin and A.M. Il'in, *Asymptotic behavior of the eigenvalues of the Dirichlet problem in a domain with a narrow slit*, Sbornik Math. **189** (1998), 503–526.

[261] P. R. Garabedian and M. Schiffer, *Convexity of domain functionals*, J. Analyse Math. **2** (1953), 281–368, DOI 10.1007/BF02825640. MR0060117

[262] F.J. Garcia de Abajo, *Colloquium: Light scattering by particle and hole arrays*, Rev. Modern Phys. **79** (2007), 1267–1290.

[263] R. Giannini, C.V. Hafner, and J.F. Löffler, *Scaling behavior of individual nanoparticle plasmon resonances*, J. Phys. Chem. C, **119** (2015), 6138–6147.

[264] D. Gilbarg and N. S. Trudinger, *Elliptic partial differential equations of second order*, 2nd ed., Grundlehren der Mathematischen Wissenschaften [Fundamental Principles of Mathematical Sciences], vol. 224, Springer-Verlag, Berlin, 1983. MR737190

[265] A. Gillman and A. Barnett, *A fast direct solver for quasi-periodic scattering problems*, J. Comput. Phys. **248** (2013), 309–322, DOI 10.1016/j.jcp.2013.04.015. MR3066154

[266] I.T.S. Gohberg and E.I. Sigal, *Operator extension of the logarithmic residue theorem and Rouché's theorem*, (Russian) Mat. Sb. (N.S.) **84** (1971), 607–642; translation in Math. USSR-Sb., 13 (1971), 603–625.

[267] J. Gopalakrishnan, S. Moskow, and F. Santosa, *Asymptotic and numerical techniques for resonances of thin photonic structures*, SIAM J. Appl. Math. **69** (2008), no. 1, 37–63, DOI 10.1137/070701388. MR2447938

[268] F. de Gournay, *Velocity extension for the level-set method and multiple eigenvalues in shape optimization*, SIAM J. Control Optim. **45** (2006), no. 1, 343–367, DOI 10.1137/050624108. MR2225309

[269] I. S. Gradshteyn and I. M. Ryzhik, *Table of integrals, series, and products*, 8th ed., Elsevier/Academic Press, Amsterdam, 2015. Translated from the Russian; Translation edited and with a preface by Daniel Zwillinger and Victor Moll; Revised from the seventh edition [MR2360010]. MR3307944

[270] M. Grasmair, M. Haltmeier, and O. Scherzer, *Necessary and sufficient conditions for linear convergence of ℓ^1-regularization*, Comm. Pure Appl. Math. **64** (2011), no. 2, 161–182, DOI 10.1002/cpa.20350. MR2766525

[271] D. S. Grebenkov and B.-T. Nguyen, *Geometrical structure of Laplacian eigenfunctions*, SIAM Rev. **55** (2013), no. 4, 601–667, DOI 10.1137/120880173. MR3124880

[272] L. Greengard and M. C. Kropinski, *Integral equation methods for Stokes flow in doubly-periodic domains*, J. Engrg. Math. **48** (2004), no. 2, 157–170, DOI 10.1023/B:ENGI.0000011923.59797.92. MR2033414

[273] A. Greenleaf, Y. Kurylev, M. Lassas, and G. Uhlmann, *Cloaking devices, electromagnetic wormholes, and transformation optics*, SIAM Rev. **51** (2009), no. 1, 3–33, DOI 10.1137/080716827. MR2481110

[274] D. Grieser, *The plasmonic eigenvalue problem*, Rev. Math. Phys. **26** (2014), no. 3, 1450005, 26 pp., DOI 10.1142/S0129055X14500056. MR3195185

[275] R. Griesmaier, *An asymptotic factorization method for inverse electromagnetic scattering in layered media*, SIAM J. Appl. Math. **68** (2008), no. 5, 1378–1403, DOI 10.1137/060677021. MR2407129

[276] R. Griesmaier and M. Hanke, *Polarization tensors of planar domains as functions of the admittivity contrast*, Appl. Anal. **96** (2017), no. 6, 970–987, DOI 10.1080/00036811.2016.1169528. MR3621942

[277] L. Guillopé and M. Zworski, *Scattering asymptotics for Riemann surfaces*, Ann. of Math. (2) **145** (1997), no. 3, 597–660, DOI 10.2307/2951846. MR1454705

[278] J. Hadamard, *Mémoire sur le problème d'analyse relatif à l'équilibre des plaques élastiques encastrées*, *OEuvres de Jacques Hadamard*, 515–631, Vol. 2, Ed. CNRS, Paris, 1968.

[279] M. A. Haider, S. P. Shipman, and S. Venakides, *Boundary-integral calculations of two-dimensional electromagnetic scattering in infinite photonic crystal slabs: channel defects and resonances*, SIAM J. Appl. Math. **62** (2002), no. 6, 2129–2148, DOI 10.1137/S003613990138531X. MR1918310

[280] J. K. Hale and S. M. Verduyn Lunel, *Introduction to functional-differential equations*, Applied Mathematical Sciences, vol. 99, Springer-Verlag, New York, 1993. MR1243878

[281] S. Halkjær, O. Sigmund, and J.S. Jensen, *Maximizing band gaps in plate structures*, Struct. Multidisc. Optim. **32** (2006), 263–275.

[282] S. Halkjær, O. Sigmund, and J.S. Jensen, *Inverse design of phononic crystals by topology optimization*, Zeitschrift fur Kristallographie **220** (2005), 895–905.

[283] J. G. Harris, *Linear elastic waves*, Cambridge Texts in Applied Mathematics, Cambridge University Press, Cambridge, 2001. MR1849581

[284] H. Hashemi, A. Oskooi, J.D. Joannopoulos, and S.G. Johnson, *General scaling limitations of ground-plane and isolated-object cloaks*, Phys. Rev. A **84** (2011), 023815.

[285] Z. Hashin and S. Shtrikman, *A variational approach to the theory of the elastic behaviour of multiphase materials*, J. Mech. Phys. Solids **11** (1963), 127–140, DOI 10.1016/0022-5096(63)90060-7. MR0159459

[286] H. Hasimoto, *On the periodic fundamental solutions of the Stokes' equations and their application to viscous flow past a cubic array of spheres*, J. Fluid Mech. **5** (1959), 317–328, DOI 10.1017/S0022112059000222. MR0102285

[287] L. He, C.-Y. Kao, and S. Osher, *Incorporating topological derivatives into shape derivatives based level set methods*, J. Comput. Phys. **225** (2007), no. 1, 891–909, DOI 10.1016/j.jcp.2007.01.003. MR2346703

[288] P. Heider, *Computation of scattering resonances for dielectric resonators*, Comput. Math. Appl. **60** (2010), no. 6, 1620–1632, DOI 10.1016/j.camwa.2010.06.044. MR2679129

[289] J. Helsing, *The effective conductivity of arrays of squares: Large random unit cells and extreme contrast ratios*, J. Comput. Phys. **230** (2011), 7533–7547.

[290] J. Helsing, *Solving integral equations on piecewise smooth boundaries using the RCIP method: a tutorial*, Abstr. Appl. Anal. (2013), Art. ID 938167, 20 pp. MR3045054

[291] J. Helsing, H. Kang, and M. Lim, *Classification of spectra of the Neumann-Poincaré operator on planar domains with corners by resonance* (English, with English and French summaries), Ann. Inst. H. Poincaré Anal. Non Linéaire **34** (2017), no. 4, 991–1011, DOI 10.1016/j.anihpc.2016.07.004. MR3661868

[292] J. Helsing and K.-M. Perfekt, *On the polarizability and capacitance of the cube*, Appl. Comput. Harmon. Anal. **34** (2013), no. 3, 445–468, DOI 10.1016/j.acha.2012.07.006. MR3027912

[293] J. Helsing and R. Ojala, *Corner singularities for elliptic problems: integral equations, graded meshes, quadrature, and compressed inverse preconditioning*, J. Comput. Phys. **227** (2008), no. 20, 8820–8840, DOI 10.1016/j.jcp.2008.06.022. MR2459537

[294] R. Hempel and I. Herbst, *Strong magnetic fields, Dirichlet boundaries, and spectral gaps*, Comm. Math. Phys. **169** (1995), no. 2, 237–259. MR1329195

[295] R. Hempel and K. Lienau, *Spectral properties of periodic media in the large coupling limit*, Comm. Partial Differential Equations **25** (2000), no. 7-8, 1445–1470, DOI 10.1080/03605300008821555. MR1765136

[296] E. Hille, *Analytic Function Theory, Volume II*, Blaisdell, 1962.

[297] P. D. Hislop and A. Martinez, *Scattering resonances of a Helmholtz resonator*, Indiana Univ. Math. J. **40** (1991), no. 2, 767–788, DOI 10.1512/iumj.1991.40.40034. MR1119196

[298] P. D. Hislop and I. M. Sigal, *Introduction to spectral theory. With applications to Schrödinger operators*, Applied Mathematical Sciences, vol. 113, Springer-Verlag, New York, 1996. MR1361167

[299] K.M. Ho, C.T. Chan, and C.M. Soukoulis, *Existence of a photonic gap in periodic dielectric structures*, Phys. Rev. Lett. **65** (1990), 3152–3155.

[300] A. T. de Hoop, *On the plane-wave extinction cross-section of an obstacle*, Appl. Sci. Res. B **7** (1959), 463–469. MR0106017

[301] G. Hu and H. Liu, *Nearly cloaking the elastic wave fields* (English, with English and French summaries), J. Math. Pures Appl. (9) **104** (2015), no. 6, 1045–1074, DOI 10.1016/j.matpur.2015.07.004. MR3420410

[302] A. M. Il′in, *A boundary value problem for an elliptic equation of second order in a domain with a narrow slit. I. The two-dimensional case* (Russian), Mat. Sb. (N.S.) **99(141)** (1976), no. 4, 514–537. MR0407439

[303] A. M. Il′in, *Matching of asymptotic expansions of solutions of boundary value problems*, Translations of Mathematical Monographs, vol. 102, American Mathematical Society, Providence, RI, 1992. Translated from the Russian by V. Minachin [V. V. Minakhin]. MR1182791

[304] G. Inglese, *An inverse problem in corrosion detection*, Inverse Problems **13** (1997), no. 4, 977–994, DOI 10.1088/0266-5611/13/4/006. MR1463588

[305] V. Isakov, *Inverse source problems*, Mathematical Surveys and Monographs, vol. 34, American Mathematical Society, Providence, RI, 1990. MR1071181

[306] V. Isakov, *Inverse problems for partial differential equations*, Applied Mathematical Sciences, vol. 127, Springer-Verlag, New York, 1998. MR1482521

[307] P. K. Jain, K. S. Lee, I. H. El-Sayed, and M. A. El-Sayed, *Calculated absorption and scattering properties of gold nanoparticles of different size, shape, and composition: Applications in biomedical imaging and biomedicine*, J. Phys. Chem. B **110** (2006), 7238–7248.

[308] J. S. Jensen and N. L. Pedersen, *On maximal eigenfrequency separation in two-material structures: the 1D and 2D scalar cases*, J. Sound Vibrat. **289** (2006), 967–986.

[309] J. S. Jensen and O. Sigmund, *Systematic design of photonic crystal structures using topology optimization: low-loss waveguide bends*, Appl. Phys. Lett. **84** (2004), 2022–2024.

[310] D. Jerison and N. Nadirashvili, *The "hot spots" conjecture for domains with two axes of symmetry*, J. Amer. Math. Soc. **13** (2000), no. 4, 741–772, DOI 10.1090/S0894-0347-00-00346-5. MR1775736

[311] S. Jiang and V. Rokhlin, *Second kind integral equations for the classical potential theory on open surfaces. I. Analytic apparatus*, J. Comput. Phys. **191** (2003), no. 1, 40–74, DOI 10.1016/S0021-9991(03)00304-8. MR2008486

[312] S. Jiang and V. Rokhlin, *Second kind integral equations for the classical potential theory on open surfaces. II*, J. Comput. Phys. **195** (2004), no. 1, 1–16, DOI 10.1016/j.jcp.2003.10.001. MR2043129

[313] S. G. Johnson and J. D. Joannopoulos, *Photonic Crystals. The Road from Theory to Practice*, Kluwer Acad. Publ., Boston, 2002.

[314] R. E. Jorgenson and R. Mittra, *Efficient calculation of the free-space periodic Green's function*, IEEE Trans. Antennas and Propagation **38** (1990), no. 5, 633–642, DOI 10.1109/8.53491. MR1050176

[315] I.V. Kamotski and V.P. Smyshlyaev, *Localised modes due to defects in high contrast periodic media via homogenization*, BICS preprint 3/06.

[316] H. Kang, K. Kim, H. Lee, J. Shin, and S. Yu, *Spectral properties of the Neumann-Poincaré operator and uniformity of estimates for the conductivity equation with complex coefficients*, J. Lond. Math. Soc. (2) **93** (2016), no. 2, 519–545, DOI 10.1112/jlms/jdw003. MR3483126

[317] H. Kang, H. Lee, and K. Yun, *Optimal estimates and asymptotics for the stress concentration between closely located stiff inclusions*, Math. Ann. **363** (2015), no. 3-4, 1281–1306, DOI 10.1007/s00208-015-1203-2. MR3412359

[318] H. Kang, M. Lim, and S. Yu, *Spectral resolution of the Neumann-Poincaré operator on intersecting disks and analysis of plasmon resonance*, Arch. Ration. Mech. Anal. **226** (2017), no. 1, 83–115, DOI 10.1007/s00205-017-1129-9. MR3686000

[319] H. Kang and G. W. Milton, *On conjectures of Polya-Szegő and Eshelby*, Inverse problems, multi-scale analysis and effective medium theory, Contemp. Math., vol. 408, Amer. Math. Soc., Providence, RI, 2006, pp. 75–80, DOI 10.1090/conm/408/07687. MR2262351

[320] H. Kang and G. W. Milton, *Solutions to the Pólya-Szegö conjecture and the weak Eshelby conjecture*, Arch. Ration. Mech. Anal. **188** (2008), no. 1, 93–116, DOI 10.1007/s00205-007-0087-z. MR2379654

[321] H. Kang and J. K. Seo, *Inverse conductivity problem with one measurement: uniqueness of balls in* $\mathbb{R}^3$, SIAM J. Appl. Math. **59** (1999), no. 5, 1533–1539, DOI 10.1137/S0036139997324595. MR1699026

[322] C.Y. Kao, S. Osher, and E. Yablonovich, *Maximizing band gaps in two-dimensional photonic crystals by using level set methods*, Appl. Phys. B **81** (2005), 235–244.

[323] C.-Y. Kao and F. Santosa, *Maximization of the quality factor of an optical resonator*, Wave Motion **45** (2008), no. 4, 412–427, DOI 10.1016/j.wavemoti.2007.07.012. MR2406839

[324] Y. E. Karpeshina, *Perturbation theory for the Schrödinger operator with a periodic potential*, Lecture Notes in Mathematics, vol. 1663, Springer-Verlag, Berlin, 1997. MR1472485

[325] Y. Karpeshina, *Spectral properties of the periodic magnetic Schrödinger operator in the high-energy region. Two-dimensional case*, Comm. Math. Phys. **251** (2004), no. 3, 473–514, DOI 10.1007/s00220-004-1129-0. MR2102328

[326] T. Kato, *Perturbation theory for linear operators*, 2nd ed., Springer-Verlag, Berlin-New York, 1976. Grundlehren der Mathematischen Wissenschaften, Band 132. MR0407617

[327] P. Kaup and F. Santosa, *Nondestructive evaluation of corrosion damage using electrostatic measurements*, J. Nondestructive Eval. **14** (1995), 127–136.

[328] P. Kaup, F. Santosa, and M. Vogelius, *A method for imaging corrosion damage in thin plates from electrostatic data*, Inverse Problems **12** (1996), 279–293.

[329] M. V. Keldyš, *On the characteristic values and characteristic functions of certain classes of non-self-adjoint equations* (Russian), Doklady Akad. Nauk SSSR (N.S.) **77** (1951), 11–14. MR0041353

[330] H. Kettunen, M. Lassas, and P. Ola, *On absence and existence of the anomalous localized resonance without the quasi-static approximation*, SIAM J. Appl. Math. **78** (2018), no. 1, 609–628, DOI 10.1137/16M1097055. MR3766974

[331] D. Khavinson, M. Putinar, and H. S. Shapiro, *Poincaré's variational problem in potential theory*, Arch. Ration. Mech. Anal. **185** (2007), no. 1, 143–184, DOI 10.1007/s00205-006-0045-1. MR2308861

[332] L. Klinkenbusch, *Scattering of an arbitrary plane electromagnetic wave by a finite elliptic cone*, Archiv für Elektrotechnik **76** (1993), 181–193.

[333] R. V. Kohn, J. Lu, B. Schweizer, and M. I. Weinstein, *A variational perspective on cloaking by anomalous localized resonance*, Comm. Math. Phys. **328** (2014), no. 1, 1–27, DOI 10.1007/s00220-014-1943-y. MR3196978

[334] R. V. Kohn, D. Onofrei, M. S. Vogelius, and M. I. Weinstein, *Cloaking via change of variables for the Helmholtz equation*, Comm. Pure Appl. Math. **63** (2010), no. 8, 973–1016, DOI 10.1002/cpa.20326. MR2642383

[335] R. V. Kohn, H. Shen, M. S. Vogelius, and M. I. Weinstein, *Cloaking via change of variables in electric impedance tomography*, Inverse Problems **24** (2008), no. 1, 015016, 21, DOI 10.1088/0266-5611/24/1/015016. MR2384775

[336] V. Kozlov, *On the Hadamard formula for nonsmooth domains*, J. Differential Equations **230** (2006), no. 2, 532–555, DOI 10.1016/j.jde.2006.08.004. MR2269932

[337] V. A. Kozlov and J. Rossmann, *Singularities of solutions of elliptic boundary value problems near conical points*, Math. Nachr. **170** (1994), 161–181, DOI 10.1002/mana.19941700113. MR1302373

[338] R. Kress, *Numerical analysis*, Graduate Texts in Mathematics, vol. 181, Springer-Verlag, New York, 1998. MR1621952

[339] A. Krishnan, T. Thio, T.J. Kim, H.J. Lezec, T.W. Ebbessen, P.A. Wolff, J. Pendry, L. Martin-Moreno, and F.J. Garcia-Vidal, *Evanescently coupled resonance in surface plasmon enhanced transmission*, Optics Comm. **200** (2001), 1–7.

[340] P. Kuchment, *Floquet theory for partial differential equations*, Operator Theory: Advances and Applications, vol. 60, Birkhäuser Verlag, Basel, 1993. MR1232660

[341] P. Kuchment, *The mathematics of photonic crystals*, Mathematical modeling in optical science, Frontiers Appl. Math., vol. 22, SIAM, Philadelphia, PA, 2001, pp. 207–272. MR1831334

[342] V. D. Kupradze, T. G. Gegelia, M. O. Basheleĭshvili, and T. V. Burchuladze, *Three-dimensional problems of the mathematical theory of elasticity and thermoelasticity*, Translated from the second Russian edition, North-Holland Series in Applied Mathematics and Mechanics, vol. 25, North-Holland Publishing Co., Amsterdam-New York, 1979. Edited by V. D. Kupradze. MR530377

[343] M.S. Kushwaha, P. Halevi, G. Martinez, L. Dobryzynski, and B. Djafari-Rouhani, *Theory of acoustic band structures of periodic elastic composites*, Phys. Rev. B **49** (1994), 2313–2322.

[344] O. Kwon, J. K. Seo, and J.-R. Yoon, *A real-time algorithm for the location search of discontinuous conductivities with one measurement*, Comm. Pure Appl. Math. **55** (2002), no. 1, 1–29, DOI 10.1002/cpa.3009. MR1857878

[345] A. Lamacz and B. Schweizer, *Effective acoustic properties of a meta-material consisting of small Helmholtz resonators*, Discrete Contin. Dyn. Syst. Ser. S **10** (2017), no. 4, 815–835, DOI 10.3934/dcdss.2017041. MR3640539

[346] A. Lamacz and B. Schweizer, *A negative index meta-material for Maxwell's equations*, SIAM J. Math. Anal. **48** (2016), no. 6, 4155–4174, DOI 10.1137/16M1064246. MR3580817

[347] A. Lamacz and B. Schweizer, *Effective Maxwell equations in a geometry with flat rings of arbitrary shape*, SIAM J. Math. Anal. **45** (2013), no. 3, 1460–1494, DOI 10.1137/120874321. MR3056753

[348] P. D. Lamberti and M. Lanza de Cristoforis, *A real analyticity result for symmetric functions of the eigenvalues of a domain dependent Dirichlet problem for the Laplace operator*, J. Nonlinear Convex Anal. **5** (2004), no. 1, 19–42. MR2049054

[349] N. S. Landkof, *Foundations of modern potential theory*, Springer-Verlag, New York-Heidelberg, 1972. Translated from the Russian by A. P. Doohovskoy; Die Grundlehren der mathematischen Wissenschaften, Band 180. MR0350027

[350] M. Lanoy, R. Pierrat, F. Lemoult, M. Fink, V. Leroy, A Tourin, *Subwavelength focusing in bubbly media using broadband time reversal*, Phys. Rev. B **91.22** (2015), 224202.

[351] A. Laptev, A. Peicheva, and A. Shlapunov, *Finding eigenvalues and eigenfunctions of the Zaremba problem for the circle*, Complex Anal. Oper. Theory **11** (2017), no. 4, 895–926, DOI 10.1007/s11785-016-0603-y. MR3626680

[352] P. D. Lax, *Functional analysis*, Pure and Applied Mathematics (New York), Wiley-Interscience [John Wiley & Sons], New York, 2002. MR1892228

[353] N. N. Lebedev, *Special functions and their applications*, Revised English edition. Translated and edited by Richard A. Silverman, Prentice-Hall, Inc., Englewood Cliffs, N.J., 1965. MR0174795

[354] O. K. Lee, H. Kang, J. C. Ye, and M. Lim, *A non-iterative method for the electrical impedance tomography based on joint sparse recovery*, Inverse Problems **31** (2015), no. 7, 075002, 23, DOI 10.1088/0266-5611/31/7/075002. MR3366751

[355] J. Lekner, *Capacitance coefficients of two spheres*, Jour. Electrostatics **69** (2011), 11–14.

[356] F. Lemoult, M. Fink and G. Lerosey, *Acoustic resonators for far-field control of sound on a subwavelength scale*, Phys. Rev. Lett. **107** (2011), 064301.

[357] F. Lemoult, G. Lerosey, J. de Rosny, and M. Fink, *Time reversal in subwavelength-scaled resonant media: beating the diffraction limit*, Int. J. Microwave Sci. Tech. **2011** (2011), 425710.

[358] F. Lemoult, A. Ourir, J. de Rosny, A. Tourin, M. Fink, and G. Lerosey, *Resonant metalenses for breaking the diffraction barrier*, Phys. Rev. Lett. **104** (2010), 203901.

[359] G. Lerosey, J. de Rosny, A. Tourin, and M. Fink, *Focusing beyond the diffraction limit with far-field time reversal*, Science **315** (2007), 1120–1122.

[360] V. Leroy, A. Bretagne, M. Fink, A. Tourin, H. Willaime and P. Tabeling, *Design and characterization of bubble phononic crystals*, Appl. Phys. Lett. **95.17** (2009), 171904.

[361] V. Leroy, M. Devaud, and J. C. Bacri, *The air bubble: experiments on an usual harmonic oscillator*, Am. J. Phys. **10** (2002), 1012–1019.

[362] V. Leroy, A. Strybulevych, M. Lanoy, F. Lemoult, A. Tourin, and J. H. Page, *Superabsorption of acoustic waves with bubble metascreens*, Phys. Rev. B **91.2** (2015), 020301.

[363] V. Leroy, A. Strybulevych , M. G. Scanlon, and J.H. Page, *Transmission of ultrasound through a single layer of bubbles*, Europ. Phys. Jour. E **29.1** (2009), 123–130.

[364] P.-D. Létourneau, Y. Wu, G. Papanicolaou, J. Garnier, and E. Darve, *A numerical study of super-resolution through fast 3D wideband algorithm for scattering in highly-heterogeneous media*, Wave Motion **70** (2017), 113–134, DOI 10.1016/j.wavemoti.2016.08.012. MR3621194

[365] J. Li, *A literature survey of mathematical study of metamaterials*, Int. J. Numer. Anal. Model. **13** (2016), no. 2, 230–243. MR3421776

[366] H. Li and H. Liu, *On anomalous localized resonance for the elastostatic system*, SIAM J. Math. Anal. **48** (2016), no. 5, 3322–3344, DOI 10.1137/16M1059023. MR3549022

[367] Y. Y. Li and M. Vogelius, *Gradient estimates for solutions to divergence form elliptic equations with discontinuous coefficients*, Arch. Ration. Mech. Anal. **153** (2000), no. 2, 91–151, DOI 10.1007/s002050000082. MR1770682

[368] Y. Li and L. Nirenberg, *Estimates for elliptic systems from composite material*, Comm. Pure Appl. Math. **56** (2003), no. 7, 892–925, DOI 10.1002/cpa.10079. Dedicated to the memory of Jürgen K. Moser. MR1990481

[369] J. Lin and H. Zhang, *Scattering by a periodic array of subwavelength slits II: surface bound state, total transmission and field enhancement in homogenization regimes*, Multiscale Model. Simul., to appear (arXiv:1706.02126).

[370] J. Lin and H. Zhang, *Scattering by a periodic array of subwavelength slits I: field enhancement in the diffraction regime*, Multiscale Model. Simul., to appear (arXiv:1706.02122).

[371] J. Lin and H. Zhang, *Scattering and field enhancement of a perfect conducting narrow slit*, SIAM J. Appl. Math. **77** (2017), no. 3, 951–976, DOI 10.1137/16M1094464. MR3662042

[372] S. Link and M.A. El-Sayed, *Shape and size dependence of radiative, non-radiative and photothermal properties of gold nanocrystals*, Int. Rev. Phys. Chem. **19** (2000), 409–453.

[373] C. M. Linton, *The Green's function for the two-dimensional Helmholtz equation in periodic domains*, J. Engrg. Math. **33** (1998), no. 4, 377–402, DOI 10.1023/A:1004377501747. MR1638860

[374] R. Lipton, *Inequalities for electric and elastic polarization tensors with applications to random composites*, J. Mech. Phys. Solids **41** (1993), no. 5, 809–833, DOI 10.1016/0022-5096(93)90001-V. MR1214019

[375] R. Lipton and R. Viator Jr., *Creating band gaps in periodic media*, Multiscale Model. Simul. **15** (2017), no. 4, 1612–1650, DOI 10.1137/16M1083396. MR3723313

[376] R. Lipton and R. Viator Jr., *Bloch waves in crystals and periodic high contrast media*, ESAIM Math. Model. Numer. Anal. **51** (2017), no. 3, 889–918, DOI 10.1051/m2an/2016046. MR3666650

[377] L. P. Liu, *Solutions to the Eshelby conjectures*, Proc. R. Soc. Lond. Ser. A Math. Phys. Eng. Sci. **464** (2008), no. 2091, 573–594, DOI 10.1098/rspa.2007.0219. MR2426985

[378] J. López-Gómez and C. Mora-Corral, *Algebraic multiplicity of eigenvalues of linear operators*, Operator Theory: Advances and Applications, vol. 177, Birkhäuser Verlag, Basel, 2007. MR2337426

[379] E. Lund and N. Olhoff, *Shape design sensitivity analysis of eigenvalues using "exact" numerical differentiation of finite element matrices*, Struct. Optim. **8** (1994), 52–59.

[380] B. Luong and F. Santosa, *Quantitative imaging of corrosion in plates by eddy current methods*, SIAM J. Appl. Math. **58** (1998), no. 5, 1509–1531, DOI 10.1137/S0036139996313356. MR1631097

[381] Z. D. Ma, N. Kikuchi, and I. Higiwara, *Structural topology and shape optimization for a frequency response problem*, Comput. Mech. **13** (1993), no. 3, 157–174, DOI 10.1007/BF00370133. MR1256535

[382] R. Magnus, *The spectrum and eigenspaces of a meromorphic operator-valued function*, Proc. Roy. Soc. Edinburgh Sect. A **127** (1997), no. 5, 1027–1051, DOI 10.1017/S0308210500026871. MR1475644

[383] A. S. Markus, *Introduction to the spectral theory of polynomial operator pencils*, Translations of Mathematical Monographs, vol. 71, American Mathematical Society, Providence, RI, 1988. Translated from the Russian by H. H. McFaden; Translation edited by Ben Silver; With an appendix by M. V. Keldysh. MR971506

[384] P. G. Martinsson and A. B. Movchan, *Vibrations of lattice structures and phononic band gaps*, Quart. J. Mech. Appl. Math. **56** (2003), no. 1, 45–64, DOI 10.1093/qjmam/56.1.45. MR1993981

[385] K.H. Matlack, A. Bauhofer, S. Krödel, A. Palermo, and C. Daraio, *Composite 3D-printed metastructures for low-frequency and broadband vibration absorption*, Proc. Natl. Acad. Sci. USA **113** (2016), 8386–8390.

[386] I.D. Mayergoyz, D.R. Fredkin, and Z. Zhang, *Electrostatic (plasmon) resonances in nanoparticles*, Phys. Rev. B **72** (2005), 155412.

[387] I.D. Mayergoyz and Z. Zhang, *Numerical analysis of plasmon resonances in nanoparticules*, IEEE Trans. Mag. **42** (2006), 759–762.

[388] V.G. Maz'ya, S.A. Nazarov, and B.A. Plamenevskii, *Asymptotic expansions of eigenvalues of boundary value problems for the Laplace operator in domains with small holes*, Math. USSR **24** (1985), 321–345.

[389] V. Maz′ya, S. Nazarov, and B. Plamenevskij, *Asymptotic theory of elliptic boundary value problems in singularly perturbed domains. Vol. I*, Operator Theory: Advances and Applications, vol. 111, Birkhäuser Verlag, Basel, 2000. Translated from the German by Georg Heinig and Christian Posthoff. MR1779977

[390] W. McLean, *Strongly elliptic systems and boundary integral equations*, Cambridge University Press, Cambridge, 2000. MR1742312

[391] R. C. McPhedran, N. A. Nicorovici, L. C. Botten, and K.-D. Bao, *Green's function, lattice sums and Rayleigh's identity for a dynamic scattering problem*, Wave propagation in complex media (Minneapolis, MN, 1994), IMA Vol. Math. Appl., vol. 96, Springer, New York, 1998, pp. 155–186, DOI 10.1007/978-1-4612-1678-0_8. MR1489749

[392] I. McGillivray, *Capacitary asymptotic expansion of the groundstate to second order*, Comm. Partial Differential Equations **23** (1998), no. 11-12, 2219–2252, DOI 10.1080/03605309808821414. MR1662184

[393] T. Meklachi, G. W. Milton, D. Onofrei, A. E. Thaler, and G. Funchess, *Sensitivity of anomalous localized resonance phenomena with respect to dissipation*, Quart. Appl. Math. **74** (2016), no. 2, 201–234, DOI 10.1090/qam/1408. MR3505601

[394] R. B. Melrose, *Geometric scattering theory*, Stanford Lectures, Cambridge University Press, Cambridge, 1995. MR1350074

[395] R. Mennicken and M. Möller, *Root functions, eigenvectors, associated vectors and the inverse of a holomorphic operator function*, Arch. Math. (Basel) **42** (1984), no. 5, 455–463, DOI 10.1007/BF01190696. MR756699

[396] O.D. Miller, C.W. Hsu, M.T.H. Reid, W. Qiu, B.G. DeLacy, J.D. Joannopoulos, M. Soljacić, and S. G. Johnson, *Fundamental limits to extinction by metallic nanoparticles*, Phys. Rev. Lett. **112** (2014), 123903.

[397] G. W. Milton, *The theory of composites*, Cambridge Monographs on Applied and Computational Mathematics, vol. 6, Cambridge University Press, Cambridge, 2002. MR1899805

[398] G.W. Milton, M. Briane, and J.R. Willis, *On cloaking for elasticity and physical equations with a transformation invariant form*, New J. Phys. **8** (2006), 248.

[399] G. W. Milton and N.-A. P. Nicorovici, *On the cloaking effects associated with anomalous localized resonance*, Proc. R. Soc. Lond. Ser. A Math. Phys. Eng. Sci. **462** (2006), no. 2074, 3027–3059, DOI 10.1098/rspa.2006.1715. MR2263683

[400] G. W. Milton, N.-A. P. Nicorovici, R. C. McPhedran, and V. A. Podolskiy, *A proof of superlensing in the quasistatic regime, and limitations of superlenses in this regime due to anomalous localized resonance*, Proc. R. Soc. Lond. Ser. A Math. Phys. Eng. Sci. **461** (2005), no. 2064, 3999–4034, DOI 10.1098/rspa.2005.1570. MR2186014

[401] M. Minnaert, *On musical air-bubbles and the sounds of running water*, London Edinburgh, Dublin Philos. Mag. and J. Sci. **16** (1933), 235–248.

[402] I. Mitrea, *Spectral radius properties for layer potentials associated with the elastostatics and hydrostatics equations in nonsmooth domains*, J. Fourier Anal. Appl. **5** (1999), no. 4, 385–408, DOI 10.1007/BF01259379. MR1700092

[403] P.E. Mix, *Introduction to Nondestructive Testing: A Training Guide*, 2nd Ed., Wiley-Interscience, Hoboken, N.J., 2005.

[404] P. Monk, *Finite element methods for Maxwell's equations*, Numerical Mathematics and Scientific Computation, Oxford University Press, New York, 2003. MR2059447

[405] Y. Montelongo, J.O. Tenorio-Pearl, C. Williams, S. Zhang, W.I. Milne, and T.D. Wilkinson, *Plasmonic nanoparticle scattering for color holograms*, Proc. Natl. Acad. Sci. USA **111** (2014), 12679–12683.

[406] P. M. Morse and H. Feshbach, *Methods of theoretical physics. 2 volumes*, McGraw-Hill Book Co., Inc., New York-Toronto-London, 1953. MR0059774

[407] S. Moskow, F. Santosa, and J. Zhang, *An approximate method for scattering by thin structures*, SIAM J. Appl. Math. **66** (2005), no. 1, 187–205, DOI 10.1137/040617388. MR2179748

[408] A. B. Movchan, R. C. McPhedran, and N. A. Nicorovici, *Green's tensors and lattice sums for elastostatics and elastodynamics*, Proc. Roy. Soc. London Ser. A **453** (1997), no. 1958, 643–662, DOI 10.1098/rspa.1997.0036. MR1920883

[409] A. B. Movchan and S. K. Serkov, *The Pólya-Szegő matrices in asymptotic models of dilute composites*, European J. Appl. Math. **8** (1997), no. 6, 595–621, DOI 10.1017/S095679259700315X. MR1608611

[410] N. I. Muskhelishvili, *Singular integral equations. Boundary problems of function theory and their application to mathematical physics*, Noordhoff International Publishing, Leyden, 1977. Revised translation from the Russian, edited by J. R. M. Radok; Reprinted of the 1958 edition. MR0438058

[411] J.-C. Nédélec, *Acoustic and electromagnetic equations. Integral representations for harmonic problems*, Applied Mathematical Sciences, vol. 144, Springer-Verlag, New York, 2001. MR1822275

[412] J.-C. Nédélec and F. Starling, *Integral equation methods in a quasi-periodic diffraction problem for the time-harmonic Maxwell's equations*, SIAM J. Math. Anal. **22** (1991), no. 6, 1679–1701, DOI 10.1137/0522104. MR1129404

[413] N.A. Nicorovici, R.C. McPhedran, and L.C. Botten, *Photonic band gaps for arrays of perfectly conducting cylinders*, Phys. Rev. E **52** (1995), 1135–1145.

[414] N.-A.P. Nicorovici, R.C. McPhedran, S. Enoch, and G. Tayeb, *Finite wavelength cloaking by plasmonic resonance*, New. J. Phys. **10** (2008), 115020.

[415] N.-A.P. Nicorovici, R.C. McPhedran, and G.W. Milton, *Optical and dielectric properties of partially resonant composites*, Phys. Rev. B **49** (1994), 8479–8482.

[416] N.-A.P. Nicorovici, G.W. Milton, R.C. McPhedran, and L.C. Botten, *Quasistatic cloaking of two-dimensional polarizable discrete systems by anomalous resonance*, Optics Express **15** (2007), 6314–6323.

[417] A. Noll, *Domain perturbations, shift of eigenvalues and capacity*, J. Funct. Anal. **170** (2000), no. 1, 246–263, DOI 10.1006/jfan.1999.3496. MR1736203

[418] A. N. Norris and A. L. Shuvalov, *Elastic cloaking theory*, Wave Motion **48** (2011), no. 6, 525–538, DOI 10.1016/j.wavemoti.2011.03.002. MR2811881

[419] A.N. Norris and G. Wickham, *Elastic Helmholtz resonators*, J. Acoust. Soc. Amer. **93** (1993), 617–630.

[420] A. A. Novotny, R. A. Feijóo, E. Taroco, and C. Padra, *Topological sensitivity analysis*, Comput. Methods Appl. Mech. Engrg. **192** (2003), no. 7-8, 803–829, DOI 10.1016/S0045-7825(02)00599-6. MR1954213

[421] A. Novotny, R. Feijóo, E. Taroco, and C. Padra, *Topological-shape sensitivity method: theory and application*, IUTAM Symposium on Topological Design Optimization of Structures, Machines and Materials: Status and Perspectives, M.P. Bendsøe et al. (eds), 2006, 469–478.

[422] S. Okada and D. Elliott, *Hölder continuous functions and the finite Hilbert transform*, Math. Nachr. **169** (1994), 219–233, DOI 10.1002/mana.19941690116. MR1292808

[423] F. W. J. Olver, D. W. Lozier, R. F. Boisvert, and C. W. Clark (eds.), *NIST handbook of mathematical functions*, U.S. Department of Commerce, National Institute of Standards and Technology, Washington, DC; Cambridge University Press, Cambridge, 2010. With 1 CD-ROM (Windows, Macintosh and UNIX). MR2723248

[424] S. Oroskar, D.R. Jackson, D.R. Wilton, *Efficient computation of the 2D periodic Green's function using the Ewald method*, J. Comput. Phys. **219** (2006), 899–911.

[425] J. E. Osborn, *Spectral approximation for compact operators*, Math. Comput. **29** (1975), 712–725. MR0383117

[426] S. Osher and R. Fedkiw, *Level set methods and dynamic implicit surfaces*, Applied Mathematical Sciences, vol. 153, Springer-Verlag, New York, 2003. MR1939127

[427] S. J. Osher and F. Santosa, *Level set methods for optimization problems involving geometry and constraints. I. Frequencies of a two-density inhomogeneous drum*, J. Comput. Phys. **171** (2001), no. 1, 272–288, DOI 10.1006/jcph.2001.6789. MR1843648

[428] S. Osher and J. A. Sethian, *Fronts propagating with curvature-dependent speed: algorithms based on Hamilton-Jacobi formulations*, J. Comput. Phys. **79** (1988), no. 1, 12–49, DOI 10.1016/0021-9991(88)90002-2. MR965860

[429] S. Ozawa, *Singular variation of domains and eigenvalues of the Laplacian*, Duke Math. J. **48** (1981), no. 4, 767–778, DOI 10.1215/S0012-7094-81-04842-0. MR782576

[430] S. Ozawa, *Point interaction potential approximation for* $(-\Delta+U)^{-1}$ *and eigenvalues of the Laplacian on wildly perturbed domain*, Osaka J. Math. **20** (1983), no. 4, 923–937. MR727440

[431] S. Ozawa, *Electrostatic capacity and eigenvalues of the Laplacian*, J. Fac. Sci. Univ. Tokyo Sect. IA Math. **30** (1983), no. 1, 53–62. MR700595

[432] S. Ozawa, *An asymptotic formula for the eigenvalues of the Laplacian in a three-dimensional domain with a small hole*, J. Fac. Sci. Univ. Tokyo Sect. IA Math. **30** (1983), no. 2, 243–257. MR722496

[433] S. Ozawa, *Spectra of domains with small spherical Neumann boundary*, J. Fac. Sci. Univ. Tokyo Sect. IA Math. **30** (1983), no. 2, 259–277. MR722497

[434] S. Ozawa, *Asymptotic property of an eigenfunction of the Laplacian under singular variation of domains—the Neumann condition*, Osaka J. Math. **22** (1985), no. 4, 639–655. MR815438

[435] S. Ozawa, *Eigenvalues of the Laplacian under singular variation of domains—the Robin problem with obstacle of general shape*, Proc. Japan Acad. Ser. A Math. Sci. **72** (1996), no. 6, 124–125. MR1404488

[436] E.D. Palik, Handbook of Optical Constants of Solids (Academic, 1985).

[437] S. Palomba, L. Novotny, and R.E. Palmer, *Blue-shifted plasmon resonance of individual size-selected gold nanoparticles*, Optics Commun. **281** (2008), 480–483.

[438] G. C. Papanicolaou, *Diffusion in random media*, Surveys in applied mathematics, Vol. 1, Surveys Appl. Math., vol. 1, Plenum, New York, 1995, pp. 205–253. MR1366209

[439] V. G. Papanicolaou, *Ewald's method revisited: rapidly convergent series representations of certain Green's functions*, J. Comput. Anal. Appl. **1** (1999), no. 1, 105–114, DOI 10.1023/A:1022622721152. MR1758787

[440] M.-J. Park and B. Lee, *Efficient calculation of the Green's function for 2-D periodic line sources using Ewald method*, IEEE Antenn. Wire. Propag. Lett. **8** (2009), 565–567.

[441] W. J. Parnell, *Nonlinear pre-stress for cloaking from antiplane elastic waves*, Proc. R. Soc. Lond. Ser. A Math. Phys. Eng. Sci. **468** (2012), no. 2138, 563–580, DOI 10.1098/rspa.2011.0477. MR2874051

[442] L. Parnovski and A. V. Sobolev, *Lattice points, perturbation theory and the periodic polyharmonic operator*, Ann. Henri Poincaré **2** (2001), no. 3, 573–581, DOI 10.1007/PL00001046. MR1846857

[443] L. Parnovski and A. V. Sobolev, *On the Bethe-Sommerfeld conjecture for the polyharmonic operator*, Duke Math. J. **107** (2001), no. 2, 209–238, DOI 10.1215/S0012-7094-01-10721-7. MR1823047

[444] L. Parnovski, *Bethe-Sommerfeld conjecture*, Ann. Henri Poincaré **9** (2008), no. 3, 457–508, DOI 10.1007/s00023-008-0364-x. MR2419769

[445] J.B. Pendry, A.I. Fernandez-Dominguez, Y. Luo, and R. Zhao, *Capturing photons with transformation optics*, Nature Phys. **9** (2013), 518–522.

[446] J.B. Pendry, Y. Luo, and R. Zhao, *Transforming the optical landscape*, Science **348** (2015), 521–524.

[447] J. B. Pendry, A. Aubry, D. R. Smith, and S. A. Maier, *Transformation optics and subwavelength control of light*, Science **337** (2012), no. 6094, 549–552, DOI 10.1126/science.1220600. MR2986782

[448] K.-M. Perfekt and M. Putinar, *Spectral bounds for the Neumann-Poincaré operator on planar domains with corners*, J. Anal. Math. **124** (2014), 39–57, DOI 10.1007/s11854-014-0026-5. MR3286048

[449] K.-M. Perfekt and M. Putinar, *The essential spectrum of the Neumann-Poincaré operator on a domain with corners*, Arch. Ration. Mech. Anal. **223** (2017), no. 2, 1019–1033, DOI 10.1007/s00205-016-1051-6. MR3590671

[450] R. Petit (ed.), *Electromagnetic theory of gratings*, Topics in Current Physics, vol. 22, Springer-Verlag, Berlin-New York, 1980. MR609533

[451] L. Poladian, *Effective Transport and Optical Properties of Composite Materials*, Ph.D. thesis, University of Sydney, 1990.

[452] L. Poladian, *General theory of electrical images in sphere pairs*, Quart. J. Mech. Appl. Math. **41** (1988), no. 3, 395–417, DOI 10.1093/qjmam/41.3.395. MR967057

[453] L. Poladian, *Asymptotic behaviour of the effective dielectric constants of composite materials*, Proc. Roy. Soc. A, **426** (1988), 343–359.

[454] M. Yu. Planida, *Asymptotics of the eigenvalues of a Laplacian with singular perturbed boundary conditions on narrow and thin sets* (Russian, with Russian summary), Mat. Sb. **196** (2005), no. 5, 83–120, DOI 10.1070/SM2005v196n05ABEH000897; English transl., Sb. Math. **196** (2005), no. 5-6, 703–741. MR2154783

[455] S. B. Platts, N. V. Movchan, R. C. McPhedran, and A. B. Movchan, *Two-dimensional phononic crystals and scattering of elastic waves by an array of voids*, R. Soc. Lond. Proc. Ser. A Math. Phys. Eng. Sci. **458** (2002), no. 2026, 2327–2347, DOI 10.1098/rspa.2002.0960. MR1932361

[456] J. Plemelj, *Potentialtheoretische Untersuchungen*, Leipzig: B.G. Teubner, 1911.

[457] I. Ponomarev, *Separation of variables in the computation of spectra in 2-D photonic crystals*, SIAM J. Appl. Math. **61** (2000/01), no. 4, 1202–1218, DOI 10.1137/S0036139997320895. MR1813676

[458] C. G. Poulton, A. B. Movchan, R. C. McPhedran, N. A. Nicorovici, and Y. A. Antipov, *Eigenvalue problems for doubly periodic elastic structures and phononic band gaps*, R. Soc. Lond. Proc. Ser. A Math. Phys. Eng. Sci. **456** (2000), no. 2002, 2543–2559, DOI 10.1098/rspa.2000.0624. MR1796495

[459] C.W. Qiu, L. Hu, B. Zhang, B.I. Wu, S.G. Johnson, and J.D. Joannopoulos, *Spherical cloaking using nonlinear transformations for improved segmentation into concentric isotropic coatings*, Optics Express **17** (2009), 13467–13478.

[460] J. Rauch and M. Taylor, *Potential and scattering theory on wildly perturbed domains*, J. Funct. Anal. **18** (1975), 27–59, DOI 10.1016/0022-1236(75)90028-2. MR0377303

[461] M. Reed and B. Simon, *Methods of modern mathematical physics. IV. Analysis of operators*, Academic Press [Harcourt Brace Jovanovich, Publishers], New York-London, 1978. MR0493421

[462] F. Rellich, *Perturbation theory of eigenvalue problems*, Gordon and Breach Science Publishers, New York-London-Paris, 1969. MR0240668

[463] G. F. Roach, *Green's functions. Introductory theory with applications*, Van Nostrand Reinhold Co., London-New York-Melbourne, 1970. MR0277804

[464] J. de Rosny, G. Lerosey, A. Tourin, and M. Fink, *Time reversal of electromagnetic waves*, Modeling and computations in electromagnetics, Lect. Notes Comput. Sci. Eng., vol. 59, Springer, Berlin, 2008, pp. 187–202, DOI 10.1007/978-3-540-73778-0_7. MR2766830

[465] K. Sakoda, *Optical Properties of Photonic Crystals*, Springer-Verlag, Berlin, 2001.

[466] J. Sanchez Hubert and E. Sánchez-Palencia, *Vibration and coupling of continuous systems. Asymptotic methods*, Springer-Verlag, Berlin, 1989. MR996423

[467] F. Santosa, *A level-set approach for inverse problems involving obstacles*, ESAIM Contrôle Optim. Calc. Var. **1** (1995/96), 17–33. MR1382514

[468] J. Saranen and G. Vainikko, *Periodic integral and pseudodifferential equations with numerical approximation*, Springer Monographs in Mathematics, Springer-Verlag, Berlin, 2002. MR1870713

[469] D. Sarid and W.A. Challener, *Modern Introduction to Surface Plasmons: Theory, Mathematical Modeling, and Applications*, Cambridge University Press, New York, 2010.

[470] L.B. Scaffardi and J.O. Tocho, *Size dependence of refractive index of gold nanoparticles*, Nanotech. **17** (2006), 1309–1315.

[471] G. Schwarz, *Hodge decomposition—a method for solving boundary value problems*, Lecture Notes in Mathematics, vol. 1607, Springer-Verlag, Berlin, 1995. MR1367287

[472] B. Schweizer, *The low-frequency spectrum of small Helmholtz resonators*, Proc. A. **471** (2015), no. 2174, 20140339, 18, DOI 10.1098/rspa.2014.0339. MR3303396

[473] J. A. Sethian, *Level set methods and fast marching methods. Evolving interfaces in computational geometry, fluid mechanics, computer vision, and materials science*, 2nd ed., Cambridge Monographs on Applied and Computational Mathematics, vol. 3, Cambridge University Press, Cambridge, 1999. MR1700751

[474] J. A. Sethian and A. Wiegmann, *Structural boundary design via level set and immersed interface methods*, J. Comput. Phys. **163** (2000), no. 2, 489–528, DOI 10.1006/jcph.2000.6581. MR1783559

[475] J. Selden, *Periodic operators in high-contrast media and the integrated density of states function*, Comm. Partial Differential Equations **30** (2005), no. 7-9, 1021–1037, DOI 10.1081/PDE-200064440. MR2180293

[476] G.P. Sendeckyj, *Elastic inclusion problems in plane elastostatics*, Int. J. Solids Struct. **6** (1970), 1535–1543.

[477] V. Sevroglou and G. Pelekanos, *Two-dimensional elastic Herglotz functions and their applications in inverse scattering*, J. Elasticity **68** (2002), no. 1-3, 123–144, DOI 10.1023/A:1026059224433. MR2024308

[478] V.P. Shestopalov and Y.V. Shestopalov, *Spectral Theory and Excitation of Open Structures*, IEEE Press, London, 1996.

[479] S. P. Shipman and S. Venakides, *Resonance and bound states in photonic crystal slabs*, SIAM J. Appl. Math. **64** (2003), no. 1, 322–342, DOI 10.1137/S0036139902411120. MR2029138

[480] M.M. Sigalas and E.N. Economou, *Elastic and acoustic wave band structure*, J. Sound. Vib. **158** (1992), 377–382.

[481] O. Sigmund and J. S. Jensen, *Systematic design of phononic band-gap materials and structures by topology optimization*, R. Soc. Lond. Philos. Trans. Ser. A Math. Phys. Eng. Sci. **361** (2003), no. 1806, 1001–1019, DOI 10.1098/rsta.2003.1177. MR1995446

[482] B. Simon, *Fifty years of eigenvalue perturbation theory*, Bull. Amer. Math. Soc. (N.S.) **24** (1991), no. 2, 303–319, DOI 10.1090/S0273-0979-1991-16020-9. MR1066108

[483] M. M. Skriganov, *The spectrum band structure of the three-dimensional Schrödinger operator with periodic potential*, Invent. Math. **80** (1985), no. 1, 107–121, DOI 10.1007/BF01388550. MR784531

[484] M.M. Skriganov, *Geometric and arithmetic methods in the spectral theory of multidimensional periodic operators*, Trudy Mat. Inst. Steklov **171** (1985), 1–117. MR798454; translation in Proc. Steklov Inst. Math., 1987, no. 2 (171).

[485] D. Slepian, *Some comments on Fourier analysis, uncertainty and modeling*, SIAM Rev. **25** (1983), no. 3, 379–393, DOI 10.1137/1025078. MR710468

[486] D.R. Smith, J.B. Pendry, and M.C.K. Wiltshire, *Metamaterials and negative refractive index*, Science **305** (2004), 788–792.

[487] S. Soussi, *Modeling photonic crystal fibers*, Adv. in Appl. Math. **36** (2006), no. 3, 288–317, DOI 10.1016/j.aam.2005.06.002. MR2219948

[488] H.S. Sözüer, J.W. Haus, and R. Inguva, *Photonic bands: Convergence problems with the plane-wave method*, Phys. Rev. B **45** (1992), 13962–13972.

[489] A. Spence and C. Poulton, *Photonic band structure calculations using nonlinear eigenvalue techniques*, J. Comput. Phys. **204** (2005), no. 1, 65–81, DOI 10.1016/j.jcp.2004.09.016. MR2121904

[490] E. M. Stein and R. Shakarchi, *Complex analysis*, Princeton Lectures in Analysis, vol. 2, Princeton University Press, Princeton, NJ, 2003. MR1976398

[491] S. Steinberg, *Meromorphic families of compact operators*, Arch. Rational Mech. Anal. **31** (1968/1969), 372–379, DOI 10.1007/BF00251419. MR0233240

[492] J. Stoer and R. Bulirsch, *Introduction to numerical analysis*, 2nd ed., Texts in Applied Mathematics, vol. 12, Springer-Verlag, New York, 1993. Translated from the German by R. Bartels, W. Gautschi and C. Witzgall. MR1295246

[493] Y. Tanaka, T. Yano, and S.-i. Tamura, *Surface guided waves in two-dimensional phononic crystals*, Wave Motion **44** (2007), no. 6, 501–512, DOI 10.1016/j.wavemoti.2007.02.009. MR2331376

[494] M. E. Taylor, *Partial differential equations. II. Qualitative studies of linear equations*, Applied Mathematical Sciences, vol. 116, Springer-Verlag, New York, 1996. MR1395149

[495] M. E. Taylor, *Tools for PDE. Pseudodifferential operators, paradifferential operators, and layer potentials*, Mathematical Surveys and Monographs, vol. 81, American Mathematical Society, Providence, RI, 2000. MR1766415

[496] E.L. Thomas, *Bubbly but quiet*, Nature **462** (2009), 990–991.

[497] R. H. Torres, *Maxwell's equations and dielectric obstacles with Lipschitz boundaries*, J. London Math. Soc. (2) **57** (1998), no. 1, 157–169, DOI 10.1112/S0024610798005900. MR1624738

[498] S.A. Tretyakov, *Metasurfaces for general transformations of electromagnetic fields*, Proc. Royal Soc. A, **373** (2015), 20140362.

[499] K. Uhlenbeck, *Generic properties of eigenfunctions*, Amer. J. Math. **98** (1976), no. 4, 1059–1078, DOI 10.2307/2374041. MR0464332

[500] Y.A. Urzhumov, N.B. Kundtz, D.R. Smith, and J.B. Pendry, *Cross-section comparisons of cloaks designed by transformation optical and optical conformal mapping approaches*, J. Opt. **13** (2011), 024002.

[501] V. Varatharajulu and Y. H. Pao, *Scattering matrix for elastic waves. I. Theory*, J. Acoust. Soc. Amer. **60** (1976), no. 3, 556–566, DOI 10.1121/1.381129. MR0431873

[502] J. Vasseur, P. Deymier, G. Frantziskonis, G. Hong, B. Djafari-Rouhani, and L. Dobrzynski, *Experimental evidence for the existence of absolute acoustic band gaps in two-dimensional periodic composite media*, J. Phys. Condens. Matter **10** (1998), 6051–6064.

[503] S. Venakides, M. A. Haider, and V. Papanicolaou, *Boundary integral calculations of two-dimensional electromagnetic scattering by photonic crystal Fabry-Perot structures*, SIAM J. Appl. Math. **60** (2000), no. 5, 1686–1706, DOI 10.1137/S0036139999350779. MR1761767

[504] G. Verchota, *Layer potentials and regularity for the Dirichlet problem for Laplace's equation in Lipschitz domains*, J. Funct. Anal. **59** (1984), no. 3, 572–611, DOI 10.1016/0022-1236(84)90066-1. MR769382

[505] M. Vogelius and J.-M. Xu, *A nonlinear elliptic boundary value problem related to corrosion modeling*, Quart. Appl. Math. **56** (1998), no. 3, 479–505, DOI 10.1090/qam/1637048. MR1637048

[506] K. Wapenaar and H. Douma, *A unified optical theorem for scalar and vectorial wave fields*, J. Acoust. Soc. Am. **131** (2012), 3611–3626.

[507] M. J. Ward and J. B. Keller, *Strong localized perturbations of eigenvalue problems*, SIAM J. Appl. Math. **53** (1993), no. 3, 770–798, DOI 10.1137/0153038. MR1218383

[508] Ch. Weber, *A local compactness theorem for Maxwell's equations*, Math. Methods Appl. Sci. **2** (1980), no. 1, 12–25, DOI 10.1002/mma.1670020103. MR561375

[509] C. H. Wilcox, *Scattering theory for diffraction gratings*, Applied Mathematical Sciences, vol. 46, Springer-Verlag, New York, 1984. MR725334

[510] E. Yablonovitch, *Inhibited spontaneous emission in solid-state physics and electronics*, Phys. Rev. Lett. **58** (1987), 2059–2062.

[511] J. Yoo, Y. Jung, M. Lim, J. C. Ye, and A. Wahab, *A joint sparse recovery framework for accurate reconstruction of inclusions in elastic media*, SIAM J. Imaging Sci. **10** (2017), no. 3, 1104–1138, DOI 10.1137/16M110318X. MR3681364

[512] K. Yosida, *Functional analysis*, 4th ed., Springer-Verlag, New York-Heidelberg, 1974. Die Grundlehren der mathematischen Wissenschaften, Band 123. MR0350358

[513] S. Yu and H. Ammari, *Plasmonic interaction between nanospheres*, SIAM Rev. **60** (2018), no. 2, 356–385, DOI 10.1137/17M1115319. MR3797722

[514] S. Yu and M. Lim, *Shielding at a distance due to anomalous resonance*, New J. Phys. **19** (2017), 033018.

[515] K. Yun, *An optimal estimate for electric fields on the shortest line segment between two spherical insulators in three dimensions*, J. Differential Equations **261** (2016), no. 1, 148–188, DOI 10.1016/j.jde.2016.03.005. MR3487255

[516] M. Zhao, Y. Capdeville, and H. Zhang, *Direct numerical modeling of time-reversal acoustic subwavelength focusing*, Wave Motion **67** (2016), 102–115, DOI 10.1016/j.wavemoti.2016.07.010. MR3546261

[517] Q. Zhao, J. Zhou, F. Zhang, and D. Lippens, *Mie resonance-based dielectric metamaterials*, Materials Today **12** (2009), 60–69.

[518] V. V. Zhikov, *On an extension and an application of the two-scale convergence method* (Russian, with Russian summary), Mat. Sb. **191** (2000), no. 7, 31–72, DOI 10.1070/SM2000v191n07ABEH000491; English transl., Sb. Math. **191** (2000), no. 7-8, 973–1014. MR1809928

[519] V. V. Zhikov, *Gaps in the spectrum of some elliptic operators in divergent form with periodic coefficients* (Russian), Algebra i Analiz **16** (2004), no. 5, 34–58, DOI 10.1090/S1061-0022-05-00878-2; English transl., St. Petersburg Math. J. **16** (2005), no. 5, 773–790. MR2106666

[520] L. Zhou, X. Huang, Y. Zhang, S.-T. Chui, *Resonance properties of metallic ring systems*, Materials Today **12** (2009), 52–59.

[521] J. M. Ziman, *Principles of the theory of solids*, 2nd ed., Cambridge University Press, London, 1972. MR0345569

[522] *Photonic Band Gap Links*, `http://www.pbglink.com`.

Index

Selected Published Titles in This Series

235 **Habib Ammari, Brian Fitzpatrick, Hyeonbae Kang, Matias Ruiz, Sanghyeon Yu, and Hai Zhang,** Mathematical and Computational Methods in Photonics and Phononics, 2018

232 **Dmitry Khavinson and Erik Lundberg,** Linear Holomorphic Partial Differential Equations and Classical Potential Theory, 2018

231 **Eberhard Kaniuth and Anthony To-Ming Lau,** Fourier and Fourier-Stieltjes Algebras on Locally Compact Groups, 2018

230 **Stephen D. Smith,** Applying the Classification of Finite Simple Groups, 2018

229 **Alexander Molev,** Sugawara Operators for Classical Lie Algebras, 2018

228 **Zhenbo Qin,** Hilbert Schemes of Points and Infinite Dimensional Lie Algebras, 2018

227 **Roberto Frigerio,** Bounded Cohomology of Discrete Groups, 2017

226 **Marcelo Aguiar and Swapneel Mahajan,** Topics in Hyperplane Arrangements, 2017

225 **Mario Bonk and Daniel Meyer,** Expanding Thurston Maps, 2017

224 **Ruy Exel,** Partial Dynamical Systems, Fell Bundles and Applications, 2017

223 **Guillaume Aubrun and Stanisław J. Szarek,** Alice and Bob Meet Banach, 2017

222 **Alexandru Buium,** Foundations of Arithmetic Differential Geometry, 2017

221 **Dennis Gaitsgory and Nick Rozenblyum,** A Study in Derived Algebraic Geometry, 2017

220 **A. Shen, V. A. Uspensky, and N. Vereshchagin,** Kolmogorov Complexity and Algorithmic Randomness, 2017

219 **Richard Evan Schwartz,** The Projective Heat Map, 2017

218 **Tushar Das, David Simmons, and Mariusz Urbański,** Geometry and Dynamics in Gromov Hyperbolic Metric Spaces, 2017

217 **Benoit Fresse,** Homotopy of Operads and Grothendieck–Teichmüller Groups, 2017

216 **Frederick W. Gehring, Gaven J. Martin, and Bruce P. Palka,** An Introduction to the Theory of Higher-Dimensional Quasiconformal Mappings, 2017

215 **Robert Bieri and Ralph Strebel,** On Groups of PL-homeomorphisms of the Real Line, 2016

214 **Jared Speck,** Shock Formation in Small-Data Solutions to 3D Quasilinear Wave Equations, 2016

213 **Harold G. Diamond and Wen-Bin Zhang (Cheung Man Ping),** Beurling Generalized Numbers, 2016

212 **Pandelis Dodos and Vassilis Kanellopoulos,** Ramsey Theory for Product Spaces, 2016

211 **Charlotte Hardouin, Jacques Sauloy, and Michael F. Singer,** Galois Theories of Linear Difference Equations: An Introduction, 2016

210 **Jason P. Bell, Dragos Ghioca, and Thomas J. Tucker,** The Dynamical Mordell–Lang Conjecture, 2016

209 **Steve Y. Oudot,** Persistence Theory: From Quiver Representations to Data Analysis, 2015

208 **Peter S. Ozsváth, András I. Stipsicz, and Zoltán Szabó,** Grid Homology for Knots and Links, 2015

207 **Vladimir I. Bogachev, Nicolai V. Krylov, Michael Röckner, and Stanislav V. Shaposhnikov,** Fokker–Planck–Kolmogorov Equations, 2015

206 **Bennett Chow, Sun-Chin Chu, David Glickenstein, Christine Guenther, James Isenberg, Tom Ivey, Dan Knopf, Peng Lu, Feng Luo, and Lei Ni,** The Ricci Flow: Techniques and Applications: Part IV: Long-Time Solutions and Related Topics, 2015

For a complete list of titles in this series, visit the
AMS Bookstore at **www.ams.org/bookstore/survseries/**.